Lecture Notes in Computational Science and Engineering

40

Ralf Kornhuber
Ronald Hoppe
Jacques Périaux
Olivier Pironneau
Olof Widlund
Jinchao Xu
Editors

Domain Decomposition Methods in Science and Engineering

With 184 Figures

Springer

Editors

Ralf Kornhuber
Fachbereich Mathematik und Informatik
Freie Universität Berlin
Arnimallee 2-6
14195 Berlin, Germany
e-mail: kornhuber@math.fu-berlin.de

Ronald Hoppe
Department of Mathematics
University of Houston
Philip G. Hoffman Hall 651
77204-3008 Houston, TX, USA
e-mail: rohop@math.uh.edu

Jacques Périaux
Dassault Aviation
78 Quai Marcel Dassault
Cedex 300
92552 St. Cloud, France
e-mail:
jacques.periaux@dassault-aviation.fr

Olivier Pironneau
Laboratoire Jacques-Louis Lions
Université Paris VI
175 Rue de Chevaleret
75013 Paris, France
e-mail:
olivier.pironneau@ann.jussieu.fr

Olof Widlund
Courant Institute of Mathematical Science
New York University
Mercer Street 251
10012 New York, USA
e-mail: widlund@cims.nyu.edu

Jinchao Xu
Department of Mathematics
Eberly College of Science
Pennsylvania Sate University
McAllister Building 218
16802 University Park, PA, USA
e-mail: xu@math.psu.edu

Library of Congress Control Number: 2004111962

Mathematics Subject Classification (2000): 65M55, 65Y05

ISSN 1439-7358
ISBN 3-540-22523-4 Springer Berlin Heidelberg New York

Springer is a part of Springer Science+Business Media
springeronline.com

Printed in Germany

Cover Design: Friedhelm Steinen-Broo, Estudio Calamar, Spain
Cover production: *design & production*, Heidelberg
Typeset by the authors using a Springer TeX macro package
Production: LE-TeX Jeloneck, Schmidt & Vöckler GbR, Leipzig

Printed on acid-free paper 46/3142/YL - 5 4 3 2 1 0

Preface

This volume contains a selection of 72 papers presented at the 15th International Conference on Domain Decomposition which was hosted by Freie Universität Berlin (FU) in cooperation with Zuse Institute Berlin (ZIB), Weierstrass Institute Berlin (WIAS) and the DFG Research Center 'Mathematics for Key Technologies' in Berlin, Germany, July 21 - 25, 2003. The attendance of 167 scientists from 24 countries accentuates the relevance of this series of almost annual meetings. In addition, an introductory tutorial by William D. Gropp and David E. Keyes arranged in the run up to the conference attracted 31 participants from all parts of the world, most of which were students. The conference itself included 15 plenary lectures delivered by leading experts in the field, 12 Minisymposia, 37 contributed talks and a poster session. A total of 144 presentations made this meeting one of the largest in the series of domain decomposition conferences. Since three parallel sessions were employed in order to accommodate as many presenters as possible, attendees and non-attendees alike may turn to this volume to keep up with future trends that might be guessed from the diversity of subjects.

Domain decomposition conferences have become the most important market place world wide for exchanging and discussing new ideas about the old algorithmic paradigm of 'divide and conquer'. Much of this reputation stems from the close interaction of experts in numerical analysis and practitioners from various fields of application concerning *fast and reliable iterative methods* for discretized partial differential equations: Schwarz methods and substructuring techniques form today's basis for large scale parallel computing. The unified view on the decomposition into subdomains and the decomposition into frequencies in terms of abstract Schwarz methods or subspace correction bridged the gap between domain decomposition and multigrid. Sophisticated finite element tearing and interconnecting techniques opened new perspectives (not only) in linear elasticity.

While classical domain decomposition concentrates on a given discretized PDE, coupling/decoupling techniques have meanwhile been applied successfully to derive efficient solution procedures including the *discretization* itself:

Mortar finite elements are most famous for their flexibility, e.g., with respect to non-matching grids, a property which is particularly attractive in multi-body contact. Other promising results concern the fast solution of time-dependent problems by waveform relaxations with optimized coupling conditions or by parareal algorithms.

The two latter approaches are motivated by parallel computation. On the other hand, it is the underlying physical background that motivates, e.g., the splitting of problems on an unbounded domain into a bounded and an unbounded part and gives rise to different discretizations in these subdomains together with suitable coupling conditions. Many other physical problems involve the localisation of the physics and their transient variability across the geometric domain. For the mathematical description of such heterogeneous processes it is important to understand various options of coupling subdomains in relation to the overall multi-physics problem. In this way, heterogeneous domain decomposition can be regarded as a new and promising approach to the *mathematical modeling* of complex phenomena on multiple scales.

This volume reviews recent developments in mathematical modeling, discretization, and fast and reliable solution by domain decomposition or related techniques, including implementation issues. Applications comprise biocomputing, computational mechanics, combustion, electromagnetics, electronic packaging, electrodynamics, fluid dynamics, medicine, metallurgy, microwave technology, optimal control, porous media flow, and voice generation. For the convenience of readers coming recently into the subject, a bibliography of previous proceedings is provided below, along with some major recent review articles and related special interest volumes. This list will inevitably be found embarrassingly incomplete. (No attempt has been made to supplement this list with the larger and closely related literature of multigrid and general iterative methods, except for the books by Hackbusch and Saad, which have significant domain decomposition components.)

P. Bjørstad, M. Espedal, and D. Keyes, editors. *Proc. Ninth Int. Conf. on Domain Decomposition Methods for Partial Differential Equations*, Ullensvang, 1997. Wiley, New York, 1999.

T. Chan, R. Glowinski, J. Périaux, and O. Widlund, editors. *Proc. Second Int. Symp. on Domain Decomposition Methods for Partial Differential Equations*, Los Angeles, 1988. SIAM, Philadelphia, 1989.

T. Chan, R. Glowinski, J. Périaux, and O. Widlund, editors. *Proc. Third Int. Symp. on Domain Decomposition Methods for Partial Differential Equations*, Houston, 1989. SIAM, Philadelphia, 1990.

T. Chan, T. Kako, H. Kawarada, and O. Pironneau, editors. *Proc. Twelfth Int. Conf. on Domain Decomposition Methods for Partial Differential Equations*, Chiba, 1999. DDM.org, Bergen, 2001.

T. Chan and T. Mathew. Domain decomposition algorithms. *Acta Numerica*, pages 61–143, 1994.

M. Débit, M. Garbey, R. Hoppe, D. Keyes, Y. Kuznetsov, and J. Périaux, editors. *Proc. Thirteenth Int. Conf. on Domain Decomposition Methods for Partial Differential Equations*, Lyon, 2000. CINME, Barcelona, 2002.

C. Farhat and F.-X. Roux. Implicit parallel processing in structural mechanics. *Computational Mechanics Advances*, 2:1–124, 1994.

R. Glowinski, G. Golub, G. Meurant, and J. Périaux, editors. *Proc. First Int. Symp. on Domain Decomposition Methods for Partial Differential Equations*, Paris, 1987. SIAM, Philadelphia, 1988.

R. Glowinski, Y. Kuznetsov, G. Meurant, J. Périaux, and O. Widlund, editors. *Proc. Fourth Int. Symp. on Domain Decomposition Methods for Partial Differential Equations*, Moscow, 1990. SIAM, Philadelphia, 1991.

R. Glowinski, J. Périaux, Z.-C. Shi, and O. Widlund, editors. *Proc. Eighth Int. Conf. on Domain Decomposition Methods for Partial Differential Equations*, Beijing, 1995. Wiley, Strasbourg, 1997.

W. Hackbusch. *Iterative Methods for Large Sparse Linear Systems*. Springer, Heidelberg, 1993.

I. Herrera, D. Keyes, O. Widlund, and R. Yates, editors. *Proc. Fourteenth Int. Conf. on Domain Decomposition Methods in Science and Engineering*, Cocoyoc, 2002. UNAM, Mexico City, 2003.

D. Keyes, T. Chan, G. Meurant, J. Scroggs, and R. Voigt, editors. *Proc. Fifth Int. Conf. on Domain Decomposition Methods for Partial Differential Equations*, Norfolk, 1991. SIAM, Philadelphia, 1992.

D. Keyes, Y. Saad, and D. Truhlar, editors. *Domain-based Parallelism and Problem Decomposition Methods in Science and Engineering*, 1995. SIAM, Philadelphia.

D. Keyes and J. Xu, editors. *Proc. Seventh Int. Conf. on Domain Decomposition Methods for Partial Differential Equations*, PennState, 1993. AMS, Providence, 1995.

B. Khoromskij and G. Wittum. *Numerical Solution of Elliptic Differential Equations by Reduction to the Interface*. Springer, 2004.

C.-H. Lai, P. Bjørstad, M. Cross, and O. Widlund, editors. *Proc. Eleventh Int. Conf. on Domain Decomposition Methods for Partial Differential Equations*, Greenwich, 1999. DDM.org, Bergen, 2000.

P. Le Tallec. Domain decomposition methods in computational mechanics. *Computational Mechanics Advances*, 2:121–220, 1994.

J. Mandel, C. Farhat, and X.-C. Cai, editors. *Proc. Tenth Int. Conf. on Domain Decomposition Methods for Partial Differential Equations*, Boulder, 1998. AMS, Providence, 1999.

L. Pavarino and A. Toselli. *Recent Developments in Domain Decomposition Methods*, volume 23 of *Lecture Notes in Computational Science & Engineering*. Springer, 2002.

A. Quarteroni, J. Périaux, Y. Kuznetsov, and O. Widlund, editors. *Proc. Sixth Int. Conf. on Domain Decomposition Methods for Partial Differential Equations*, Como, 1992. AMS, Providence, 1994.
A. Quarteroni and A. Valli. *Domain Decomposition Methods for Partial Differential Equations.* Oxford, 1999.
Y. Saad. *Iterative Methods for Sparse Linear Systems.* PWS, Boston, 1996.
B. Smith, P. Bjørstad, and W. Gropp. *Domain Decomposition: Parallel Multilevel Algorithms for Elliptic Partial Differential Equations.* Cambridge Univ. Press, Cambridge, 1996.
A. Toselli and O. Widlund. *Domain Decomposition Methods.* Springer, 2004.
B. Wohlmuth. *Discretization Methods and Iterative Solvers on Domain Decomposition.* Springer, 2001.
J. Xu. Iterative methods by space decomposition and subspace correction. *SIAM Review*, 34:581–613, 1991.

We also recommend the homepage for domain decomposition on the World Wide Web `www.ddm.org` maintained by Martin Gander. This site features links to past and future conferences, a growing number of conference proceedings together with updated bibliographic and personal information pertaining to domain decomposition.

We wish to thank all members of the Scientific Committee for Domain Decomposition Conferences, and in particular the chair Ronald H.W. Hoppe, for their help in setting the scientific direction of this conference. We are also grateful to the organizers of the minisymposia for shaping the profile of the scientific program and attracting high-quality presentations. The conference offered a fruitful integration of scientific excellence of speakers with a great level of interaction not only during the sessions but also along the friendly conference dinner under the 'communication tent', bringing a pleasant and relaxed atmosphere for exchanging information among attendees and lecturers. The local organization was carried out by a wonderful team of almost 50 members of FU Berlin, Zuse Institute Berlin (ZIB), and Weierstrass Institute Berlin (WIAS). We thank all members of the local organizing committee chaired by Ralf Kornhuber and, most notably, the conference manager Sabrina Nordt for perfectly taking care of all aspects of preparing and running DD15.

We gratefully acknowledge the financial and logistic support of this conference by FU Berlin, WIAS, ZIB, the German Research Foundation (DFG), and, last but not least, by the DFG Research Center 'Mathematics for Key Technologies'.

The timely production of these proceedings would not have been possible without the excellent cooperation of the authors and the referees. We would like to thank all of them for their graceful and timely response to our various demands. Special thanks are due to the technical editors Rainer Roitzsch and Uwe Pöhle for patiently eliminating all kinds of bugs from the final LaTeX source and for presenting these proceedings on the web. Finally we wish to thank Martin Peters and Thanh-Ha Le Thi from Springer for a friendly, reliable, and efficient collaboration.

Ralf Kornhuber
Berlin, Germany

Ronald H.W. Hoppe
Augsburg, Germany and Houston, USA

Jacques Périaux
Paris, France

Olivier Pironneau
Paris, France

Olof B. Widlund
New York, USA

Jinchao Xu
PennState, USA

Contents

Part I Invited Talks

Part I

Invited Talks

Non-matching Grids and Lagrange Multipliers

S. Bertoluzza[1], F. Brezzi[1,2], L.D. Marini[1,2], and G. Sangalli[1]

[1] Istituto di Matematica Applicata e Tecnologie Informatiche del C.N.R., Pavia (http://www.imati.cnr.it/{~aivlis,~brezzi,~marini,~sangalli})
[2] Università di Pavia, Dipartimento di Matematica

Summary. In this paper we introduce a variant of the three-field formulation where we use only two sets of variables. Considering, to fix the ideas, the homogeneous Dirichlet problem for $-\Delta u = g$ in Ω, our variables are *i)* an approximation ψ_h of u on the *skeleton* (the union of the interfaces of the sub-domains) on an independent grid (that could often be uniform), and *ii)* the approximations u_h^s of u in each sub-domain Ω^s (each on its own grid). The novelty is in the way to derive, from ψ_h, the values of each trace of u_h^s on the boundary of each Ω^s. We do it by solving an auxiliary problem on each $\partial\Omega^s$ that resembles the mortar method but is more flexible. Optimal error estimates are proved under suitable assumptions.

1 Introduction

Assume, for simplicity, that we have to solve the model problem

$$\text{find } u \in H_0^1(\Omega) \text{ such that} \qquad -\Delta u = g \text{ in } \Omega \quad \text{with } u = 0 \text{ on } \partial\Omega \tag{1}$$

on a polygonal or polyhedral domain $\Omega \subset \mathbb{R}^n$, $n = 2, 3$, where g is a given function sufficiently regular in Ω. In order to apply a Domain Decomposition technique we split Ω into sub-domains Ω^s $(s = 1, 2, ..., S)$ and we consider the *skeleton*

$$\Sigma := \cup_s \Gamma^s, \qquad \text{with} \qquad \Gamma^s \equiv \partial\Omega^s. \tag{2}$$

For the sake of simplicity we will use a three-dimensional notation, and speak therefore of *faces, edges* and *vertices*. The change of terminology in the polygonal case is obvious and left to the reader. On Σ we consider

$$\Phi := \{\varphi \in L^2(\Sigma) : \exists v \in H_0^1(\Omega) \text{ with } \varphi = v_{|\Sigma}\} \equiv H_0^1(\Omega)_{|\Sigma} \equiv H^{1/2}(\Sigma). \tag{3}$$

In each Ω^s we consider instead

$$V^s := \{v^s \in H^1(\Omega^s) \text{ such that } \exists v \in H_0^1(\Omega) \text{ with } v^s = v_{|\Omega^s}\}, \tag{4}$$

that can also be seen as the set of functions in $H^1(\Omega^s)$ that vanish at the intersection (if any) of Γ^s with $\partial\Omega$. In its turn, $H^1_0(\Omega)$ could be identified with a subspace of

$$V := \{u \in L^2(\Omega),\ u|_{\Omega^s} \in V^s\}, \tag{5}$$

and in particular, setting $v^s := v|_{\Omega^s}$ we can write

$$H^1_0(\Omega) \simeq \{v \in V \text{ such that } \exists \varphi \in \Phi \text{ with } v^s = \varphi \text{ on } \Gamma^s, \quad s = 1, ..., S\}. \tag{6}$$

For each s we will also introduce the trace space $\Theta^s = H^{1/2}(\Gamma^s)$, and we set $\Theta = \prod_s \Theta^s$. For $v \in V$, $\theta = (\theta^1, \dots, \theta^s) \in \Theta$ we will write

$$v|_\Sigma = \theta \text{ to indicate that } \theta^s = v^s|_{\Gamma^s} \text{ (with } v^s = v|_{\Omega^s}), \quad s = 1, \dots, S.$$

When discretizing the problem, we assume to be given a decomposition $\mathcal{T}_\delta^\Sigma$ of Σ and a corresponding space $\Phi_\delta \subset \Phi$ of piecewise polynomials. We also assume that in each Ω^s we are given a decomposition $\mathcal{T}_h^s \equiv \mathcal{T}_h^{\Omega^s}$ with a corresponding space $V_h^s \subset V^s$ of piecewise polynomials, and we set

$$V_h := \{v \in V \text{ such that } v|_{\Omega^s} \in V_h^s\}. \tag{7}$$

It is clear that each decomposition $\mathcal{T}_h^s$ will induce a decomposition $\mathcal{T}_h^{\Gamma^s}$ on Γ^s and a corresponding space of traces $\Theta_h^s \subset \Theta^s$. On the other hand the restriction of $\mathcal{T}_\delta^\Sigma$ to Γ^s also induces a decomposition $\mathcal{T}_\delta^{\Gamma^s}$ of Γ^s and another space of piecewise polynomials Φ_δ^s made by the restrictions of the functions in Φ_δ to Γ^s. Hence, on each Γ^s we have *two* decompositions (one coming from $\mathcal{T}_\delta^\Sigma$ and one from $\mathcal{T}_h^s$) and two spaces of piecewise polynomial functions (one from Φ_δ and one from V_h^s). Note, incidentally, that on each face f belonging to two different sub-domains we will have *three* decompositions and three spaces: one from Σ and the other two from the two sub-domains.

The first basic idea of our method is to design for every sub-domain Ω^s a linear operator $\mathcal{G}^s$ (the *generation operator*) that maps every *mother* $\varphi_\delta \in \Phi_\delta$ into an element (*daughter*) $\theta_h^s = \mathcal{G}^s(\varphi_\delta) \in \Theta_h^s$. Together with the individual $\mathcal{G}^s$ we consider a *global* operator $\mathcal{G}$ defined as

$$\mathcal{G}(\varphi_\delta) = (\theta_h^1, \dots, \theta_h^S) \in \Theta_h \quad \text{with } \theta^s = \mathcal{G}^s(\varphi_\delta). \tag{8}$$

The way to construct the operators $\mathcal{G}^s$ constitutes the second basic idea of this paper, and will be described in a while.

Once we have the operators $\mathcal{G}^s$ we can consider the subspace $\mathcal{S}_h$ of V_h made of *sisters* (that is, daughters of the same mother):

$$\mathcal{S}_h := \{v_h \in V_h \text{ such that } \exists \varphi_\delta \in \Phi_\delta \text{ with } v_{h|\Sigma} = \mathcal{G}(\varphi_\delta)\} \subseteq V. \tag{9}$$

We point out that in our previous definitions we consider as *daughter*, at the same time, an element θ_h^s $(= v_h^s|_{\Gamma^s})$ of Θ_h^s, and any function $v_h^s \in V_h^s$ having that same trace. It is clear, comparing (9) with (6), that $\mathcal{S}_h$ can be seen as a nonconforming approximation of $H_0^1(\Omega)$. This allows us to consider the following discrete formulation. We set

$$a_s(u,v) := \int_{\Omega^s} \nabla u \cdot \nabla v \mathrm{d}x \qquad \text{and} \qquad a(u,v) := \sum_{s=1}^{S} a_s(u^s, v^s) \tag{10}$$

and we look for $u_h \in \mathcal{S}_h$ such that

$$a(u_h, v_h) = \int_\Omega g\, v_h \mathrm{d}x \quad \forall v_h \in \mathcal{S}_h. \tag{11}$$

It is clear that, under reasonable assumptions on the subspaces Φ_δ and V_h^s and on the generation operators $\mathcal{G}^s$, problem (11) will have good stability and accuracy properties.

The idea of imposing weak continuity by introducing the space Φ_δ and define a nonconforming approximation of $H_0^1(\Omega)$ by taking the subset of V_h whose elements take (in some weak sense) value $\varphi_h \in \Phi_\delta$ is one of the main ideas of the three field formulation (Brezzi and Marini [1994]). Following that approach, for each sub-domain Ω^s we could take a space M_h^s of Lagrange multipliers, and, for every $\varphi_\delta \in \Phi_\delta$, we could define $\mathcal{G}^s(\varphi_\delta) \in \Theta_h^s$ by

$$\int_{\Gamma^s} (\varphi_\delta - \mathcal{G}^s(\varphi_\delta))\mu_h^s \,\mathrm{d}x = 0 \qquad \forall \mu_h^s \in M_h^s. \tag{12}$$

In general, however, equation (12) does not define $\mathcal{G}^s(\varphi_delta)$ uniquely, even when the spaces M_h^s and Θ_h^s satisfy the required *inf-sup* condition (see (24)). Though this is not a problem in the definition and in the analysis of the three field formulation, we would like to point out that having the trace of the elements v_h^s on Γ^s somehow uniquely determined by an element of Φ_δ has some clear advantage from the point of view of implementation. In particular it allows to use standard Dirichlet solvers (which can easily be found already implemented and whose optimization is well understood) as a brick for treating the equation in the subdomain. In order for $\mathcal{G}^s(\varphi_\delta)$ to be uniquely determined by (12) the spaces M_h^s and Θ_h^s must have the same dimension. A simple minded choice is $M_h^s \equiv \Theta_h^s$, that guarantees existence and uniqueness of the solution of (12) together with optimal stability and accuracy properties of the projector $\mathcal{G}^s$. This choice however is not the optimal one: in fact, during the estimate of the error for problem (11), there seems to be no way to get rid of a term like

$$\sum_s \int_{\Gamma^s} \frac{\partial u}{\partial \mathbf{n}_s} (\varphi_\delta - \mathcal{G}^s(\varphi_\delta))\,\mathrm{d}x. \tag{13}$$

An obvious way to treat the term in (13) is to use the fact that $\varphi_\delta - \mathcal{G}^s(\varphi_\delta)$ is orthogonal to all elements in M_h^s, so that we can subtract from $\partial u/\partial \mathbf{n}_s$ any element of M_h^s. In particular we are interested in subtracting a suitable approximation $\mu_I^s \simeq \partial u/\partial \mathbf{n}_s$. It is then crucial to be able to find in M_h^s a μ_I^s that approximates $\partial u/\partial \mathbf{n}_s$ with the needed order. However, $\partial u/\partial \mathbf{n}_s$ is *discontinuous* passing from one face to another of the same Ω^s. And if the space M_h^s is made of *continuous* functions (as it would be with the choice $M_h^s \equiv \Theta_h^s$), then the order of approximation (say, in $H^{-1/2}(\partial\Omega^s)$) cannot be better than $O(h)$ (and actually with some additional logarithmic loss, as $O(h|\lg h|)$. Hence, we do need an M_h^s made of functions that can be discontinuous when passing from one face to another of the same Ω^s. The requirement to contain a suitable amount of discontinuities and the one to have the same dimension of Θ_h^s seem very difficult to conciliate. Actually, a quite similar difficulty is met in the *mortar method*, (see e.g. Bernardi et al. [1993], Belgacem and Maday [1997], Hoppe et al. [1998], Wohlmuth [2001]), in particular in three dimensions. There, the requirement that M_h^s have the same dimension as Θ_h^s is relaxed as little as possible. The values of a "weakly continuous" function v_h^s at nodes which are interior to the faces of Γ^s on the slave sides are uniquely determined by the weak continuity equation, while the degrees of freedom corresponding to nodes on the edges of Γ^s (whose union forms the so called wirebasket) are free. We point out that the mortar method can be described in the framework given here provided we relax the assumption $\Phi_\delta \subset H^{1/2}(\Sigma)$ by allowing the functions ϕ_δ to be discontinuous across the "wirebasket": Φ_δ would correspond to the traces of v_h on the "master sides" (or "mortars") and $\mathcal{G}^s$ being defined as the identity on master sides and to one of the available mortar projections on "slave sides".

The idea, here, is to give up the equality of the dimensions but still obtain a well defined operator $\mathcal{G}^s$, by changing (12) in a slightly more complicated formulation, involving an additional Lagrange multiplier. Let us see the main features of this path.

We choose first a space M_h^s having in mind the fact that we must be able to use it for approximating $\partial u/\partial \mathbf{n}_s$ with the right order. We also need its dimension to be smaller than (or equal to) that of Θ_h^s. Then we change (12) in the following way. For every $\varphi_\delta \in \Phi_\delta$ we look for *a pair* $(\widetilde{\theta_h^s}, \widetilde{\mu_h^s})$ in $\Theta_h^s \times M_h^s$ such that

$$\int_{\Gamma^s} (\varphi_\delta - \widetilde{\theta_h^s})\, \mu_h^s \,\mathrm{d}x = 0 \qquad \forall \mu_h^s \in M_h^s \tag{14}$$

and

$$\sum_{T \in \mathcal{T}_h^{\Gamma^s}} \int_T h_T^{-1} (\varphi_\delta - \widetilde{\theta_h^s})\, \theta_h^s \,\mathrm{d}x + \int_{\Gamma^s} \widetilde{\mu_h^s}\, \theta_h^s \,\mathrm{d}x = 0 \qquad \forall \theta_h^s \in \Theta_h^s. \tag{15}$$

Then we set

$$\mathcal{G}^s(\varphi_\delta) := \widetilde{\theta_h^s}. \tag{16}$$

It is clear that in (14)-(15) the number of equations will always be equal to the number of unknowns. It is also clear that if (by shear luck) we have $\varphi_\delta|_{\Gamma^s} \in \Theta_h^s$ then $\mathcal{G}^s(\varphi_\delta) = \varphi_\delta|_{\Gamma^s}$ (and $\widetilde{\mu_h^s} = 0$). This will, in the end, provide for the new approach (14)-(16) an optimal order of accuracy (as we had for the previous simple-minded (12)). It is, finally, also obvious that some sort of *inf-sup* condition will be needed in order to ensure existence and uniqueness of the solution of (14)-(15), unless some suitable additional stabilization is introduced. However, as we shall see, the possibility of escaping the cage of the equal dimensionality of M_h^s and Θ_h^s opens a whole lot of interesting possibilities.

In this paper we will follow the path indicated above. In the next section we will make precise all the necessary assumptions, and in Section 3 we will derive abstract error bounds for problem (11) when the operators $\mathcal{G}^s$ are constructed as in (14)-(16). In Section 4 we will present some possible choices for the finite element spaces and discuss their stability and accuracy properties. In particular we will show that the simple choice of using totally discontinuous functions for M_h^s, stabilizing the problem with suitable *boundary bubbles*, leads to a problem with optimal convergence properties and, at the same time, a very simple implementation. This is reminiscent of what has been done for instance in Baiocchi et al. [1992], Brezzi et al. [1997], Buffa [2002], and Brezzi and Marini [2000], but simpler and more effective. Finally, in the last section we briefly discuss some possible variants/extensions, in particular regarding the possibility of using discontinuous mothers.

2 Assumptions on the decomposition and on the discretizations

We consider now the assumptions to be made on the decomposition and on the discretizations.

Assumptions on Ω and on the domain decomposition

We assume that Ω is an open polyhedron, that each Ω^s, for $s = 1, ..., S$, is also an open polyhedron, that the intersection of two different Ω^s is empty, and that the union of the closures of all Ω^s is the closure of Ω. As in (2) the *skeleton* Σ will be the union of the boundaries $\partial\Omega^s$. We do not assume that this decomposition is *compatible*. This means that we do not assume that the intersection of the closure of two different Ω^s is either a common face, or a common edge, or a common vertex. For simplicity we assume however that the number S of subdomains is fixed once and for all, and we do not keep track of the dependency of the various constants on S.

Assumptions on the decomposition $\mathcal{T}_\delta^\Sigma$

We assume that we are given a sequence $\{\mathcal{T}_\delta^\Sigma\}_\delta$ of decompositions of Σ. Each decomposition $\mathcal{T}_\delta^\Sigma$ is made of open triangles, in such a way that the intersection of two different triangles is empty, and the union of the closures of all triangles is Σ. We assume *compatibility*, that is we assume that the intersection of the closures of two different triangles is either empty, a common edge or a common vertex. We also assume, as usual, *shape regularity*, for instance by assuming that the ratio between the diameter of each triangle and the radius of its biggest inscribed circle is $\leq \kappa_0$, with κ_0 independent of δ. Finally we assume *quasi-uniformity*: there exists a constant q, independent of δ such that, if δ_T^{min} and δ_T^{max} are the minimum and the maximum diameters (respectively) of the triangles in $\mathcal{T}_\delta^\Sigma$, then $\delta_T^{min} \geq q\, \delta_T^{max}$.

Assumptions on the decompositions $\mathcal{T}_h^s$ (and $\mathcal{T}_h^{\Gamma^s}$)

We assume that we are given, for each $s = 1, ..., S$, a sequence $\{\mathcal{T}_h^s\}_h$ of decompositions of Ω^s. Each decomposition is made of open tetrahedra in such a way that the intersection of two different tetrahedra is empty, and the union of the closures of all tetrahedra is Ω^s. We also assume *compatibility*: the intersection of the closures of two different tetrahedra is either empty, a common face, a common edge, or a common vertex. Finally we assume *shape regularity*, for instance by assuming that the ratio between the diameter of each tetrahedron and the radius of its biggest inscribed sphere is $\leq \kappa_1$, with κ_1 independent of h. We point out that we do not assume quasi-uniformity for the meshes $\mathcal{T}_h^s$. We recall that the triangulation $\mathcal{T}_h^{\Gamma^s}$ is the restriction on Γ^s of $\mathcal{T}_h^s$.

Assumptions on the discretizations Φ_δ, V_h^s, and M_h^s

We assume that for each δ and for each $T \in \mathcal{T}_\delta^\Sigma$ we are given a space of polynomials $\mathcal{P}_T$. The space Φ_δ will then be defined as

$$\Phi_\delta := \{\varphi \in \Phi \text{ such that } \varphi_{|T} \in \mathcal{P}_T, \quad T \in \mathcal{T}_\delta^\Sigma\}, \tag{17}$$

where Φ is always given by (3). Similarly we assume that for each s, for each h, and for each $K \in \mathcal{T}_h^s$ we are given a space of polynomials $\mathcal{P}_K$. The space V_h^s will then be defined as

$$V_h^s := \{v^s \in V^s \text{ such that } v_{|K}^s \in \mathcal{P}_K, \quad K \in \mathcal{T}_h^s\}, \tag{18}$$

where V^s is still given by (4).

The corresponding restrictions of the above spaces to each Γ^s are defined as in the previous section, namely

$$\Phi_\delta^s := (\Phi_\delta)_{|\Gamma^s} \qquad \text{and} \qquad \Theta_h^s := (V_h^s)_{|\Gamma^s}, \quad s = 1, ..., S. \tag{19}$$

We assume that there exist bounded lifting operators from Θ_h^s to V_h^s. More precisely, for all $s = 1, \ldots, S$, for all $\theta_h^s \in \Theta_h^s$ there exists $w_h^s \in V_h^s$ such that

$$w_h^s|_{\Gamma^s} = \theta_h^s \qquad \text{and} \qquad \|w_h^s\|_{1,\Omega^s} \leq C\|\theta_h^s\|_{H^{1/2}(\Gamma^s)}. \tag{20}$$

Finally we assume that for each s, for each h, and for each $T \in \mathcal{T}_h^{\Gamma^s}$ we are given a space of polynomials $\mathcal{Q}_T$. The space M_h^s will then be defined as

$$M_h^s := \{\mu \in L^2(\Gamma^s) \text{ such that } \mu_{|T} \in \mathcal{Q}_T, \quad T \in \mathcal{T}_h^{\Gamma^s}\}. \tag{21}$$

If we like, we can also add some continuity requirements to (21). In view of the discussion of the previous section, however, it would be unwise to force continuity in the passage from one face to another. In order for the bilinear form a to ve coercive in a suitable space, we make the following minimal assumption on M_h^s:

$$\text{for every } \Omega^s \text{ the space } M_h^s \text{ contains the constants on } \Gamma^s. \tag{22}$$

Moreover, for simplicity, we assume that there exists an integer number κ such that all the spaces $\mathcal{P}_T$, $\mathcal{P}_K$, and $\mathcal{Q}_T$ verify

$$\mathcal{P}_T \subseteq \mathbb{P}_\kappa(T), \qquad \mathcal{P}_K \subseteq \mathbb{P}_\kappa(K), \qquad \mathcal{Q}_T \subseteq \mathbb{P}_\kappa(T),$$

where $\mathbb{P}_\kappa(\omega)$ is the space of polynomials of degree $\leq \kappa$ on ω. Using the notation of Brezzi and Fortin [1991a] for the usual Lagrange finite element spaces we can then write

$$V_h^s \subseteq \mathcal{L}_\kappa^1(\mathcal{T}_h^s), \quad \Theta_h^s \subseteq \mathcal{L}_\kappa^1(\mathcal{T}_h^{\Gamma^s}), \quad M_h^s \subseteq \mathcal{L}_\kappa^0(\mathcal{T}_h^{\Gamma^s}), \quad \Phi_\delta \subseteq \mathcal{L}_\kappa^1(\mathcal{T}_\delta^{\Sigma}).$$

The operators $\mathcal{G}^s$ and the compatibility assumptions among the discretizations

Having defined the spaces Θ_h^s and M_h^s we can now consider the operators $\mathcal{G}^s$ (that will always be given by (14)-(16)) together with the global operator $\mathcal{G}$ (still given by (8)). Once we have the operators $\mathcal{G}^s$ and $\mathcal{G}$, we can define the space of sisters $\mathcal{S}_h$, always as in (9). In $\mathcal{S}_h$ we define:

$$\|v_h\|^2 := \sum_{s=1}^{S} \|\nabla v_h^s\|_{0,\Omega^s}^2 \tag{23}$$

We can now turn to the more important assumptions, that will require some compatibility conditions among the spaces Φ_δ^s, Θ_h^s and M_h^s.

Our first assumption will deal with the well-posedness of the problem (14)-(16). As this is a problem in classical mixed form, we have no real escape but assuming an *inf-sup* condition on the spaces Θ_h^s and M_h^s:

$\exists \beta > 0$ such that $\forall s = 1, ..., S$ and $\forall h > 0$

$$\inf_{\mu_h^s \in M_h^s \setminus \{0\}} \sup_{\theta_h^s \in \Theta_h^s \setminus \{0\}} \frac{\int_{\Gamma^s} \theta_h^s \, \mu_h^s \, dx}{\|\theta_h^s\|_{h,\frac{1}{2},\Gamma^s} \, \|\mu_h^s\|_{h,-\frac{1}{2},\Gamma^s}} > \beta, \tag{24}$$

were the norms in the denominator of (24) are defined, for any real r, as

$$\|\theta_h^s\|_{h,r,\Gamma^s}^2 := \sum_{T\in\mathcal{T}_h^{\Gamma^s}} h_T^{-2r}\|\theta_h^s\|_{0,T}^2 \tag{25}$$

and h_T is the diameter of T. Condition (24) will be, in a sense, the only nontrivial assumption that we have to take into account in the definition of our spaces V_h^s and M_h^s. However, in the next section, we are going to see some families of elements where (24) can be checked rather easily.

Our last assumption will deal with the *bound on the mother*. We point out that, so far, we did not assume that an element of the space of sisters $\mathcal{S}_h$ had a unique mother. Indeed, we do not need it. Strictly speaking, we only need that

$$\exists\gamma > 0 \text{ such that: } \forall v_h \in \mathcal{S}_h, \exists\varphi_\delta \in \Phi_\delta \text{ with } \mathcal{G}(\varphi_\delta) = v_{h|\Sigma} \text{ and}$$

$$\|\varphi\|_\Phi^2 := \sum_{s=1}^{S} |\varphi_\delta|_{H^{1/2}(\Gamma^s)}^2 \leq \gamma^2 |\!|\!| v_h |\!|\!|^2. \tag{26}$$

We point out that $\|\cdot\|_\Phi$ is indeed a norm on Φ, since the elements of Φ vanish on $\partial\Omega$ (see Bertoluzza [2003]). One of the consequences of (26) is that the seminorm $|\!|\!|\cdot|\!|\!|$ is indeed a norm. In fact, given $v_h \in \mathcal{S}_h$ and letting $\varphi \in \Phi_\delta$ given by (26), provided (22) holds, it can be shown (see Bertoluzza [2003]) that

$$\|v_h\|_{0,\Omega} \leq C(|\!|\!| v_h |\!|\!| + \|\varphi\|_\Phi) \leq C|\!|\!| v_h |\!|\!|. \tag{27}$$

We shall discuss in the following sections whether and when this assumption is satisfied. We anticipate however that this will be another easy condition, that could be roughly summarized by: on each face f of each $\partial\Omega^s$ the mesh $\mathcal{T}_\delta^{\Gamma^s}$ (induced by $\mathcal{T}_\delta^{\Sigma}$) is coarser than the two meshes $\mathcal{T}_h^{\Gamma^s}$ (induced by the two $\mathcal{T}_h^s$ relative to the sub-domains having f in common).

3 Basic Error Estimates

We are now ready to analyze the problem (11) and derive abstract error estimates for it.

We start by looking in more detail to the operator $\mathcal{G}^s$. Thanks to the classical theory of mixed finite element (see Brezzi and Fortin [1991b]) we can prove the following Lemma.

Lemma 1. *Assume that the inf-sup condition* (24) *is satisfied, and let* $\varphi \in L^2(\Sigma)$; *then for every* $s = 1, ..., S$

$$\|\mathcal{G}^s(\varphi)\|_{h,\frac{1}{2},\Gamma^s} \leq C\|\varphi\|_{h,\frac{1}{2},\Gamma^s}. \tag{28}$$

We point out that the norm $\|\cdot\|_{h,\frac{1}{2},\Gamma^s}$, induced by the bilinear form $(u,v) \to \sum_{T\in\mathcal{T}_h^{\Gamma^s}} \int_T h_T^{-1} u\, v\, dx$ plays the role of a discrete $H^{1/2}(\Gamma^s)$ norm. Indeed we have the following lemma.

Lemma 2. *The following inverse inequality holds: for all $\theta_h^s \in \Theta_h^s$*

$$\|\theta_h^s\|_{H^{1/2}(\Gamma^s)} \leq C\|\theta_h^s\|_{h,\frac{1}{2},\Gamma^s} \tag{29}$$

Proof. We shall actually prove that (29) holds for all $\theta_h^s \in \mathcal{L}_\kappa^1(\mathcal{T}_h^{\Gamma^s})$. It is well known that a function in $\mathcal{L}_\kappa^1(\mathcal{T}_h^s)$ is uniquely identified by its values at a set $\{x_i\}_i$ of nodes corresponding to the canonical Lagrange basis. Let $\theta_h^s \in \mathcal{L}_\kappa^1(\mathcal{T}_h^{\Gamma^s})$ and let $w_h \in \mathcal{L}_\kappa^1(\mathcal{T}_h^s)$ be its finite element lifting, i.e., the function verifying $w_h^s(x_i) = \theta_h^s(x_i)$ at all nodes on Γ^s and $w_h^s(x_i) = 0$ at all other nodes. Clearly, $\|\theta_h^s\|_{H^{1/2}(\Gamma^s)} \leq C\|w_h^s\|_{H^1(\Omega^s)}$. Let us then bound the $H^1(\Omega^s)$ norm of w_h^s. By definition w_h^s is different from 0 only on those tetrahedra $T \in \mathcal{T}_h^s$ which are adjacent to the boundary. Let K be one of such tetrahedra and let $T_i \in \mathcal{T}_h^{\Gamma^s}$, $i = 1, \ldots, m$ be the triangles that share one or more nodes with K. Thanks to usual arguments, we can write:

$$\|w_h^s\|_{H^1(K)}^2 \leq Ch_K^{-1}\|w_h^s\|_{L^2(\partial K)}^2 \leq C\sum_{i=1}^m h_{T_i}^{-1}\|w_h^s\|_{L^2(T_i)}^2.$$

Adding with respect to all elements K adjacent to Γ^s, we obtain that

$$\|w_h^s\|_{H^1(\Omega^s)} \leq C\|\theta_h^s\|_{h,\frac{1}{2},\Gamma^s},$$

which implies (29).

Remark 1. Note that if we had assumed the quasi-uniformity of the triangulation $\mathcal{T}_h^{\Gamma^s}$, then (29) could easily be obtained by space interpolation, using the standard inverse inequality between the H^1 and the L^2 norms. This is however not the case, and in the above proof we only made use of the regularity of the mesh.

Lemma 2 trivially implies the continuity of $\mathcal{G}^s$ from $L^2(\Gamma^s)$ (endowed with the norm $\|\cdot\|_{h,\frac{1}{2},\Gamma^s}$), to $H^{1/2}(\Gamma^s)$. However a stronger result holds, stated in the following theorem

Theorem 1. *$\mathcal{G}^s(\cdot)$ is continuous from $H^{1/2}(\Gamma^s)$ to $H^{1/2}(\Gamma^s)$:*

$$\|\mathcal{G}^s(\varphi)\|_{H^{1/2}(\Gamma^s)} \leq C\|\varphi\|_{H^{1/2}(\Gamma^s)}. \tag{30}$$

Proof. First, we introduce the Clément interpolant $\theta_I^s \in \Theta_h^s$ of $\theta^s = \varphi|_{\Gamma^s}$, which gives (see Clément [1975])

$$\begin{aligned} \|\theta_I^s\|_{H^{1/2}(\Gamma^s)} &\leq C\|\theta^s\|_{H^{1/2}(\Gamma^s)} \\ \|\theta^s - \theta_I^s\|_{h,\frac{1}{2},\Gamma^s} &\leq C\|\theta^s\|_{H^{1/2}(\Gamma^s)}. \end{aligned} \tag{31}$$

Since $\mathcal{G}^s(\cdot)$ is linear and using the triangle inequality, we have

$$\|\mathcal{G}^s(\theta^s)\|_{H^{1/2}(\Gamma^s)} \leq \|\mathcal{G}^s(\theta^s - \theta_I^s)\|_{H^{1/2}(\Gamma^s)} + \|\mathcal{G}^s(\theta_I^s)\|_{H^{1/2}(\Gamma^s)} = I + II.$$

Making use of Lemma 2, Lemma 1 and (31), we get

$$
\begin{aligned}
I = \|\mathcal{G}^s(\theta - \theta_I^s)\|_{H^{1/2}(\Gamma^s)} &\le C\|\mathcal{G}^s(\theta - \theta_I^s)\|_{h,\frac{1}{2},\Gamma^s} \\
&\le C\|\theta - \theta_I^s\|_{h,\frac{1}{2},\Gamma^s} \\
&\le C\|\theta\|_{H^{1/2}(\Gamma^s)}.
\end{aligned}
$$

Moreover, since $\mathcal{G}^s(\theta_I^s) = \theta_I^s$ and using (31), we have

$$II = \|\mathcal{G}^s(\theta_I^s)\|_{H^{1/2}(\Gamma^s)} \le C\|\theta\|_{H^{1/2}(\Gamma^s)},$$

giving (30).

We can now prove our error estimate. From the definition (10) and assumption (27) we easily get that problem (11) has a unique solution. Let now ψ_I be an interpolant of the exact solution u in Φ_δ. For every Ω^s $(s = 1, ..., S)$ let $u_I^s \in V_h^s$ be defined as the unique solution of

$$
\begin{cases}
u_I^s = \mathcal{G}^s(\psi_I) \text{ on } \Gamma^s \\
a_s(u_I^s, v_h^s) = \int_{\Omega^s} g\, v_h^s \,\mathrm{d}x \qquad \forall v_h^s \in V_h^s \cap H_0^1(\Omega^s).
\end{cases}
\tag{32}
$$

It is obvious that (32) has a unique solution. Let u_I be equal to u_I^s in each Ω^s $(s = 1, ..., S)$. It is clear that $u_I \in \mathcal{S}_h$. We now set $e_h := u_I - u_h \in \mathcal{S}_h$. Using the the definition (10) and adding and subtracting u we have:

$$|\!|\!|e_h|\!|\!|^2 = a(e_h, e_h) = a(u_I - u, e_h) + a(u - u_h, e_h) =: I + II. \tag{33}$$

Using (11) and integrating $a(u, e_h)$ by parts in each Ω^s we obtain

$$
\begin{aligned}
II = a(u - u_h, e_h) &= -\sum_{s=1}^{S}\int_{\Omega^s} g\, e_h^s \,\mathrm{d}x + \sum_{s=1}^{S}\int_{\Gamma^s} \frac{\partial u}{\partial \mathbf{n}^s} e_h^s \,\mathrm{d}x - \sum_{s=1}^{S}\int_{\Omega^s} g\, e_h^s \,\mathrm{d}x \\
&= \sum_{s=1}^{S}\int_{\Gamma^s} \frac{\partial u}{\partial \mathbf{n}^s} e_h^s \,\mathrm{d}x.
\end{aligned}
\tag{34}
$$

As $e_h \in \mathcal{S}_h$, and using assumption (26) there will be a mother $\eta_\delta \in \Phi_\delta$ with $\|\eta_\delta\|_\Phi \le C|\!|\!|e_h|\!|\!|$, such that $\mathcal{G}(\eta_\delta) = e_h|_\Sigma$. Hence the continuity of $\partial u/\partial \mathbf{n}$, and the fact that η_δ is single-valued on the skeleton Σ yield

$$II = \sum_{s=1}^{S}\int_{\Gamma^s} \frac{\partial u}{\partial \mathbf{n}^s}(e_h^s - \eta_\delta)\,\mathrm{d}x = \sum_{s=1}^{S}\int_{\Gamma^s} \frac{\partial u}{\partial \mathbf{n}^s}(\mathcal{G}^s(\eta_\delta) - \eta_\delta)\,\mathrm{d}x. \tag{35}$$

We can now use the definition of $\mathcal{G}^s$ (see (15)) and subtract from $\partial u/\partial \mathbf{n}$ its best approximation μ_I^s, thus obtaining

$$II = \sum_{s=1}^{S}\int_{\Gamma^s} \left(\frac{\partial u}{\partial \mathbf{n}^s} - \mu_I^s\right)(\mathcal{G}^s(\eta_\delta) - \eta_\delta)\,\mathrm{d}x. \tag{36}$$

We remember now that $\mathcal{G}^s(\eta_\delta) = e_h^s$ on Γ^s for all s. We also point out that (thanks to (22)) we can assume that the mean value of $\partial u/\partial \mathbf{n}^s - \mu_I^s$ on each Γ^s is zero, so that we can use the $H^{1/2}$-seminorm of e^s and η_δ instead of the norm in the estimate. Then we use Cauchy-Schwarz inequality, we use (26) for η_δ, and (27) standard trace inequality in each Ω^s for e^s to obtain

$$II \leq \sum_{s=1}^{S} \|\frac{\partial u}{\partial \mathbf{n}^s} - \mu_I^s\|_{H^{-1/2}(\Gamma^s)} \left(|e^s|_{H^{1/2}(\Gamma^s)} + |\eta_\delta|_{H^{1/2}(\Gamma^s)}\right)$$
$$\leq \Big(\sum_{s=1}^{S} \|\frac{\partial u}{\partial \mathbf{n}^s} - \mu_I^s\|^2_{H^{-1/2}(\Gamma^s)}\Big)^{1/2} \|e_h\|. \quad (37)$$

It remains to estimate I. After the obvious

$$I = a\,(u_I - u, e_h) \leq \|u_I - u\|\,\|e_h\| \quad (38)$$

we have to estimate $\|u - u_I\|$. Using the definition (32) of u_I^s we can apply the usual theory for estimating the error for each Dirichlet problem in Ω^s. Thanks to (20) we have first

$$\|u - u_I^s\|_{1,\Omega^s} \leq C\left(\inf_{v_h^s \in V_h^s} \|u - v_h^s\|_{1,\Omega^s} + \|u - u_I^s\|_{H^{1/2}(\Gamma^s)}\right). \quad (39)$$

It is then clear that the crucial step is to estimate $\|u - u_I^s\|_{H^{1/2}(\Gamma^s)}$, for each s.

To this aim let us introduce an interpolant $\chi_I^s \in \Theta_h^s$ of $u|_{\Gamma^s}$. We can write

$$\|u - u_I^s\|_{H^{1/2}(\Gamma^s)} \equiv \|u - \mathcal{G}^s(\psi_I)\|_{H^{1/2}(\Gamma^s)}$$
$$\leq \|u - \chi_I^s\|_{H^{1/2}(\Gamma^s)} + \|\chi_I^s - \mathcal{G}^s(u)\|_{H^{1/2}(\Gamma^s)} \quad (40)$$
$$+\|\mathcal{G}^s(u) - \mathcal{G}^s(\psi_I)\|_{H^{1/2}(\Gamma^s)} \quad (41)$$

Since $\chi_I^s = \mathcal{G}^s(\chi_I^s)$ and using Theorem 1, we easily get $\|\chi_I^s - \mathcal{G}^s(u)\|_{H^{1/2}(\Gamma^s)} = \|\mathcal{G}^s(\chi_I^s - u)\|_{H^{1/2}(\Gamma^s)} \leq C\|u - \chi_I^s\|_{H^{1/2}(\Gamma^s)}$. By a similar argument we obtain $\|\mathcal{G}^s(u - \psi_I)\|_{H^{1/2}(\Gamma^s)} \leq \|u - \psi_I\|_{H^{1/2}(\Gamma^s)}$.

We can then collect (33)-(38) and (39)–(41) in the following theorem.

Theorem 2. *Assume that the assumptions of Section 2 on the decomposition and on the discretizations are satisfied. Assume that the operators $\mathcal{G}^s$ are constructed as in (14)-(16). Let u be the exact solution of (1) and u_h be the solution of (11). Then we have*

$$\|u - u_h\|^2 \leq C \sum_{s=1}^{S} \Big(\inf_{v_h^s \in V_h^s} \|u - v_h^s\|^2_{1,\Omega^s} + \inf_{\mu_h^s \in M_h^s} \|\frac{\partial u}{\partial \mathbf{n}^s} - \mu_h^s\|^2_{H^{-1/2}(\Gamma^s)}\Big)$$
$$+ \inf_{\varphi_\delta \in \Phi_\delta} \|u - \varphi_\delta\|^2_{H^{1/2}(\Sigma)}. \quad (42)$$

4 Examples and Remarks

In this section we want to show an example of finite element discretizations that satisfy the abstract assumptions of Section 2, and derive the corresponding error bounds in terms of suitable powers of h.

We do not discuss the assumptions on the decomposition of Ω into the Ω^s. We just remark once more that it does not need to be *compatible*: for instance, the intersection of the closures of two different Ω^s can be a face of one of them and only a piece of a face of the other.

We discuss instead the choice of the finite element spaces Φ_δ, V_h^s, and M_h^s.

Assume that we are given an integer number $k \geq 1$.

For every T in the triangulation $\mathcal{T}_\delta^\Sigma$ of the skeleton Σ we choose $\mathcal{P}_T := \mathbb{P}_k(T)$, the space of polynomials of degree $\leq k$ on T. The space Φ_δ, according to (17), becomes then

$$\Phi_\delta := \{\varphi \in \Phi \text{ such that } \varphi_{|T} \in \mathbb{P}_k(T), \quad T \in \mathcal{T}_\delta^\Sigma\} = \mathcal{L}_k^1(\mathcal{T}_\delta^\Sigma) \cap \Phi \tag{43}$$

(we recall that the elements of Φ have to vanish on $\partial\Omega$ so we need to take the intersection of $\mathcal{L}_k^1(\mathcal{T}_\delta^\Sigma)$ with Φ in order to properly define Φ_δ). For each s and for every T in the triangulation $\mathcal{T}_h^{\Gamma^s}$ of Γ^s we take instead as $\mathcal{Q}_T$ the space $\mathcal{Q}_T := \mathbb{P}_{k-1}(T)$. According to (21) the space M_h^s becomes then

$$M_h^s := \{\mu \in L^2(\Gamma^s) : \ \mu_{|T} \in \mathbb{P}_{k-1}(T), \quad T \in \mathcal{T}_h^{\Gamma^s}\} = \mathcal{L}_k^0(\mathcal{T}_h^{\Gamma^s}). \tag{44}$$

We point out that Φ_δ is made of continuous functions, while M_h^s is made of functions that are, a priori, totally discontinuous from one element to another.

The choice of each V_h^s will be slightly more elaborate. For each tetrahedron $K \in \mathcal{T}_h^s$ with no faces belonging to Γ^s we take $\mathcal{P}_K := \mathbb{P}_k$. If instead K has a face f on Γ^s we consider the cubic function b_f on K that vanishes on the three remaining internal faces of K, and we augment the space $\mathbb{P}_k$ with the space B_{k+2}^f obtained multiplying b_f times the functions in $\mathcal{Q}_f \equiv \mathbb{P}_{k-1}(f)$ (that is the space of polynomials of degree $\leq k-1$ on f: remember that the face f will be one of the triangles $T \in \mathcal{T}_h^{\Gamma^s}$). If K has another face on Γ^s we repeat the operation, augmenting further the space $\mathbb{P}_k$. In summary

$$\mathcal{P}_K := \mathbb{P}_k(K) + \{\bigoplus_{f\subset\Gamma^s} B_{k+2}^f\} \equiv \mathbb{P}_k + \{\bigoplus_{f\subset\Gamma^s} b_f\mathbb{P}_{k-1}(f)\}. \tag{45}$$

We note that $\bigoplus b_f\mathbb{P}_{k-1}(f)$ is a direct sum, but its sum with $\mathbb{P}_k(K)$ is not direct whenever $k \geq 3$. This however will not be a problem for the following developments.

We can now discuss the various abstract assumptions that have been made in Section 2. To start with, condition (22) is obviously satisfied. Similarly, (20) holds as shown for instance in Bernardi et al. [to appear]. We consider then the *inf-sup* condition (24).

Lemma 3. *Let* M_h^s *and* Θ_h^s *be constructed as in* (44) *and in* (19) *with* (45), *respectively. Then the inf-sup condition* (24) *holds true.*

Proof. For every $\mu_h^s \in M_h^s$ we construct $v_h^s \in V_h^s$ as

$$v_h^s = \sum_{T \in \mathcal{T}_h^{\Gamma^s}} h_T\, b_T\, \mu_h^s \tag{46}$$

where as before b_T is the cubic function on K (the tetrahedron having T as one of its faces) vanishing on the other three faces of K and having mean value 1 on T. It is not too difficult to check that

$$\|\mu_h^s\|_{h,-\frac{1}{2},\Gamma^s}\, \|v_h^s\|_{h,\frac{1}{2},\Gamma^s} \le C \int_{\mathcal{T}_h^{\Gamma^s}} v_h^s \mu_h^s \tag{47}$$

that is precisely the *inf-sup* condition (24) that we need.

We consider now the other *inf-sup* that is involved in the present scheme (although we did not write it as an *inf-sup*), that is the bound on the mother (26). By applying the technique of Babuska [1973] it is not difficult to realize that if $\mathcal{T}_\delta^\Sigma$ is "coarse enough" on each face, compared with the meshes of the two sub-domains having that face in common, then

$$\inf_{\varphi_\delta \in \mathcal{L}_k^1(\mathcal{T}_\delta^{\Gamma^s}) \setminus \{0\}} \;\sup_{\mu_h^s \in M_h^s \setminus \{0\}} \frac{\int_{\Gamma^s} \varphi_\delta\, \mu_h^s \,\mathrm{d}x}{\|\mu_h^s\|_{H^{-1/2}(\Gamma^s)}\, \|\varphi_\delta\|_{H^{1/2}(\Gamma^s)}} > \gamma_0. \tag{48}$$

It is now easy to see that (48) implies (26): let $v_h^s \in \mathcal{S}_h$, then, by definition, there exists $\varphi_\delta \in \Phi_\delta$ such that $v_h^s|_\Sigma = \varphi_\delta$. Letting $\breve{\varphi}^s = (1/|\Gamma^s|) \int_{\Gamma^s} \varphi_\delta \mathrm{d}x$ we have that $\varphi_\delta - \breve{\varphi}^s \in \mathcal{L}_k^1(\mathcal{T}_\delta^{\Gamma^s})$. Let now $\mu_h^* \in M_h^s$ be the element that realizes the supremum in (48) for such an element of $\mathcal{L}_k^1(\mathcal{T}_\delta^{\Gamma^s})$. Using (48), and then (14), we obtain

$$\gamma_0 \|\mu_h^*\|_{H^{-1/2}(\Gamma^s)}\, \|\varphi_\delta - \breve{\varphi}^s\|_{H^{1/2}(\Gamma^s)} \le \int_{\Gamma^s} \mu_h^* \,(\varphi_\delta - \breve{\varphi}^s)\,\mathrm{d}x \tag{49}$$

$$= \int_{\Gamma^s} \mu_h^*\, \mathcal{G}^s(\varphi_\delta - \breve{\varphi}^s)\,\mathrm{d}x. \tag{50}$$

Now, since $\varphi_\delta - \breve{\varphi}^s$ has zero mean value on Γ^s, the same is true for $\mathcal{G}^s(\varphi_\delta - \breve{\varphi}^s)$ (see (14) and (22)). Then, denoting by $\breve{v}^s = (1/|\Gamma^s|) \int_{\Gamma^s} v_h^s \mathrm{d}x$ the average of v_h^s on Γ^s, we have

$$\begin{aligned}\int_{\Gamma^s} \mu_h^*\, \mathcal{G}^s(\varphi_\delta - \breve{\varphi}^s)\,\mathrm{d}x &= \int_{\Gamma^s} \mu_h^*\,(v_h^s - \breve{v}^s)\,\mathrm{d}x \\ &\le \|\mu_h^*\|_{H^{-1/2}(\Gamma^s)}\, |v_h^s|_{H^{1/2}(\Gamma^s)} \\ &\le \|\mu_h^*\|_{H^{-1/2}(\Gamma^s)}\, |v_h^s|_{1,\Omega^s}\end{aligned}$$

that, since $|\varphi_\delta|_{H^{1/2}(\Gamma^s)} = |\varphi_\delta - \breve{\varphi}^s|_{H^{1/2}(\Gamma^s)} \simeq \|\varphi_\delta - \breve{\varphi}^s\|_{H^{1/2}(\Gamma^s)}$, joined with (49) immediately implies (26).

We can collect the previous results, together with the abstract error estimates of the previous section, in the following theorem.

Theorem 3. *Assume that the assumptions on the decompositions $\mathcal{T}_\delta^\Sigma$ and $\mathcal{T}_h^s$ of Section 2 are satisfied, and assume that the spaces Φ_δ, M_h^s and V_h^s are defined as in* (43), (44) *and* (18) *with* (45), *respectively. Assume finally that* (48) *holds. Then we have*

$$|||u_h - u||| \leq C\,(|h|^k + |\delta|^k)\|u\|_{k+1,\Omega} \tag{51}$$

The proof follows immediately from Theorem 1, the results of this section, and usual approximation estimates.

We end this section with some observations on the actual implementation of the method when the bubble stabilization (45) is used.

Indeed, let us see how the computation of the generation operators $\mathcal{G}^s$ can be performed in practice. Assume that we are given a function φ in, say, $L^2(\Gamma^s)$. We recall that, to compute $\widetilde{\theta_h^s} = \mathcal{G}^s(\varphi)$, we have to find the pair $(\widetilde{\theta_h^s}, \widetilde{\mu_h^s}) \in \Theta_h^s \times M_h^s$ such that

$$\int_{\Gamma^s} (\varphi - \widetilde{\theta_h^s})\,\mu_h^s \,\mathrm{d}x = 0 \qquad \forall \mu_h^s \in M_h^s, \tag{52}$$

$$\sum_{T \in \mathcal{T}_h^{\Gamma^s}} \int_T h_T^{-1}(\varphi - \widetilde{\theta_h^s})\,\theta_h^s \,\mathrm{d}x + \int_{\Gamma^s} \widetilde{\mu_h^s}\,\theta_h^s \,\mathrm{d}x = 0 \qquad \forall \theta_h^s \in \Theta_h^s. \tag{53}$$

We also recall that, with the choice (45), the space Θ_h^s can be written as $\Theta_h^s = \mathcal{L}_k^1(\mathcal{T}_h^{\Gamma^s}) + B_{k+2}(\mathcal{T}_h^{\Gamma^s})$ where $\mathcal{L}_k^1(\mathcal{T}_h^{\Gamma^s})$ is, as before, the space of continuous piecewise polynomials of degree k on the mesh $\mathcal{T}_h^{\Gamma^s}$, and $B_{k+2}(\mathcal{T}_h^{\Gamma^s})$ is the space of bubbles of degree $k+2$, always on $\mathcal{T}_h^{\Gamma^s}$. In order to write is as a *direct sum* we introduce the space

$$W^s = \{\theta_h^s \in \Theta_h^s \text{ such that } \int_{\Gamma^s} \theta_h^s\,\mu_h^s \mathrm{d}x = 0\ \forall \mu_h^s \in M_h^s\} \tag{54}$$

We can then split *in a unique way* $\widetilde{\theta_h^s} = \widetilde{w} + \widetilde{b}$ with $\widetilde{w} \in W^s$ and $\widetilde{b}$ in $B_{k+2}(\mathcal{T}_h^{\Gamma^s})$. It is now clear that $\widetilde{b}$ can be computed immediately from (52) that becomes:

$$\int_{\Gamma^s} (\varphi - \widetilde{b})\,\mu_h^s \,\mathrm{d}x = 0 \qquad \forall \mu_h^s \in M_h^s. \tag{55}$$

Once $\widetilde{b}$ is known, one can compute $\widetilde{w}$ from (53) that easily implies

$$\sum_{T \in \mathcal{T}_h^{\Gamma^s}} \int_T h_T^{-1}(\varphi - \widetilde{w})\,w, \mathrm{d}x = \sum_{T \in \mathcal{T}_h^{\Gamma^s}} \int_T h_T^{-1}\widetilde{b}\,w, \mathrm{d}x \qquad \forall w \in W^s. \tag{56}$$

In this way the saddle point problem (52)-(53) splits into two smaller subproblems, each with a symmetric and positive definite matrix. In particular (55) can be solved element by element, so that (56) is the only true system to be solved.

5 Relaxing the continuity of the *Mothers*

One of the main advantages of the present method (and in general of all non conforming domain decomposition methods) is the freedom given by the possibility of meshing and treating each sub-domain independently of the others. In our approach however, the discretization Φ_δ of $H^{1/2}(\Sigma)$ is required to be continuous. Such request can be relaxed by defining Φ_δ face by face and asking for continuity within each face but allowing the elements of Φ_δ to jump across the boundary between two adjacent faces. More precisely, considering a splitting of the skeleton Σ in disjoint faces $\Sigma = \cup f$ (with $f = \Gamma^s \cap \Gamma^\ell$ for some $s, \ell = 1, \dots, S$) we can introduce for each face a family of triangulations $\mathcal{T}_\delta^f$ and consider a corresponding space $\Phi_\delta^f \subset H^{1/2}(f)$ of piecewise polynomials. The global space Φ_δ could then be defined by

$$\Phi_\delta = \{\varphi_\delta \in L^2(\Sigma) \text{ with } \varphi_\delta|_f \in \Phi_\delta^f \text{ for all faces } f \text{ of } \Sigma\}.$$

Such a choice has several advantages, in particular from the point of view of implementation. Each face can be meshed independently of the other faces. Moreover, each node on Σ belongs to only one face f and therefore it only "sees" two sub-domains. This greatly simplifies the data structure needed for describing the elements of Φ_δ and the manipulations of such elements and of their interaction with other elements.

The analysis presented in the previous section needs then to be modified in order to take the discontinuity of the mothers into account. In particular, if the elements of Φ_δ are discontinuous, the space Φ_δ is not a subspace of $H^{1/2}(\Sigma)$, and therefore bounds like $\|\mathcal{G}^s(u - \psi_I)\|_{H^{1/2}(\Gamma^s)} \le \|u - \psi_I\|_{H^{1/2}(\Gamma^s)}$ would not make sense. A completely revised analysis is carried out in a further work in preparation, and results in an almost optimal estimate (with the loss of a logarithmic factor). We just point out that the analysis of Section 3 could still be applied if the space $\Phi_\delta^c = \Phi_\delta \cap H^{1/2}(\Sigma)$ has good approximation properties. Such space is the one where one should choose the best approximation ψ_I. This is indeed a very special case: in general such space does not provide a good approximation. It may very well happen that it contains only the function $\varphi_\delta = 0$. A case in which the space Φ_δ^c does provide good approximation is the case in which the meshes on two adjacent faces share a sufficiently fine set of common nodes (in particular the case when, restricted to the common edge, the nodes of the two (or more) meshes are one a subset of the other). Though this is quite an heavy restriction to the freedom given by the possibility of using discontinuous mothers, such a case would still have many advantages from the implementation point of view, while retaining the optimal error estimate. Remark that the subspace Φ_δ^c would only be used for analyzing the method, while its implementation fully relies on the discontinuous space Φ_δ.

References

I. Babuska. The finite element method with lagrangian multipliers. *Numer. Math.*, 20:179–192, 1973.

C. Baiocchi, F. Brezzi, and L. D. Marini. Stabilization of galerkin methods and applications to domain decomposition. In A. Bensoussan et al., editor, *Future Tendencies in Computer Science, Control and Applied Mathematics*, volume 653 of *Lecture Notes in Computer Science*, pages 345–355. Springer-Verlag, 1992.

F. B. Belgacem and Y. Maday. The mortar element method for three dimensional finite elements. *RAIRO Mathematical Modelling and Numerical Analysis*, 31(2):289–302, 1997.

C. Bernardi, Y. Maday, and A. T. Patera. Domain decomposition by the mortar element method. In H. K. ans M. Garbey, editor, *Asymptotic and Numerical Methods for Partial Differential Equations with Critical Parameters*, pages 269–286. N.A.T.O. ASI, Kluwer Academic Publishers, 1993.

C. Bernardi, Y. Maday, and F. Rapetti. *Discrétisations variationnelles de problèmes aux limites elliptiques.* Mathématiques et Applications. SMAI, to appear.

S. Bertoluzza. Analysis of a stabilized three-fields domain decomposition method. *Numer. Math.*, 93(4):611–634, 2003. ISSN 0029-599X.

F. Brezzi and M. Fortin. *Mixed and Hybrid Finite Element Methods.* Springer-Verlag, New-York, 1991a.

F. Brezzi and M. Fortin. *Mixed and hybrid finite element methods.* Springer-Verlag, New York, 1991b. ISBN 0-387-97582-9.

F. Brezzi, L. P. Franca, L. D. Marini, and A. Russo. Stabilization techniques for domain decomposition methods with nonmatching grids. In *Proc. from the IX International Conference on Domain Decomposition Methods, June 1996, Bergen, Norway*, 1997.

F. Brezzi and L. D. Marini. A three field domain decomposition method. *Contemp. Math.*, 157:27–34, 1994.

F. Brezzi and L. D. Marini. Error estimates for the three-field formulation with bubble stabilization. *Math. of Comp.*, 70:911–934, 2000.

A. Buffa. Error estimates for a stabilized domain decomposition method with nonmatching grids. *Numer. Math.*, 90(4):617–640, 2002.

P. Clément. Approximation by finite element functions using local regularization. *RAIRO Anal. Numér.*, 9:77–84, 1975.

R. Hoppe, Y. Iliash, Y. Kuznetsov, Y. Vassilevski, and B. Wohlmuth. Analysis and parallel implementation of adaptive mortar element methods. *East West J. Num. An.*, 6(3):223–248, 1998.

B. Wohlmuth. *Discretization Methods and Iterative Solvers Based on Domain Decomposition*, volume 17 of *Lecture Notes in Computational Science and Engineering.* Springer, 2001.

A FETI Method for a Class of Indefinite or Complex Second- or Fourth-Order Problems

Charbel Farhat[1], Jing Li[2], Michel Lesoinne[1], and Philippe Avery[1]

[1] University of Colorado at Boulder, Department of Aerospace Engineering Sciences (`http://caswww.colorado.edu/~charbel/`)
[2] Kent State University, Department of Mathematical Sciences

Summary. The FETI-DP domain decomposition method is extended to address the iterative solution of a class of indefinite problems of the form $(\mathbf{K}-\sigma^2\mathbf{M})\mathbf{x}=\mathbf{b}$, and a class of complex problems of the form $(\mathbf{K}-\sigma^2\mathbf{M}+i\sigma\mathbf{D})\mathbf{x}=\mathbf{b}$, where $\mathbf{K}$, $\mathbf{M}$, and $\mathbf{D}$ are three real symmetric positive semi-definite matrices arising from the finite element discretization of either second-order elastodynamic problems or fourth-order plate and shell dynamic problems, i is the imaginary complex number, and σ is a positive real number.

1 Introduction

Real linear or linearized systems of equations of the form

$$(\mathbf{K}-\sigma^2\mathbf{M})\mathbf{x}=\mathbf{b} \tag{1}$$

and complex linear or linearized systems of equations of the form

$$(\mathbf{K}-\sigma^2\mathbf{M}+i\sigma\mathbf{D})\mathbf{x}=\mathbf{b} \tag{2}$$

are frequent in computational structural dynamics. Eq. (1) is encountered, for example, in the finite element (FE) simulation of the forced response of an undamped mechanical system to a periodic excitation . In that case, $\mathbf{K}$ and $\mathbf{M}$ are the FE stiffness and mass matrices of the considered mechanical system, respectively, σ is the circular frequency of the external periodic excitation, $\mathbf{b}$ is its amplitude, $(\mathbf{K}-\sigma^2\mathbf{M})$ is the impedance of the mechanical system, and $\mathbf{x}$ is the amplitude of its forced response. Such problems also arise during the solution by an inverse shifted method of the generalized symmetric eigenvalue problem $\mathbf{Kx}=\omega^2\mathbf{Mx}$ associated with an undamped mechanical system. In that example, $\mathbf{K}$ and $\mathbf{M}$ have the same meaning as in the previous case, $(\omega^2, \mathbf{x})$ is a desired pair of eigenvalue and eigenvector representing the square of a natural circular frequency and the corresponding natural vibration mode of the undamped mechanical system, respectively, and the shift σ^2 is introduced to obtain quickly the closest eigenvalues to σ^2. In both examples mentioned here, the matrices $\mathbf{K}$ and $\mathbf{M}$ are symmetric positive semi-definite, and therefore $(\mathbf{K} - \sigma^2\mathbf{M})$ rapidly becomes indefinite when σ is increased. Eq. (2) is encountered in similar problems when the mechanical system is damped, in which case i denotes the pure imaginary number satisfying $i^2 = -1$ and $\mathbf{D}$ denotes the FE damping matrix and is also symmetric positive semi-definite.

Domain decomposition based preconditioned conjugate gradient (PCG) methods have emerged as powerful equation solvers in this field on both sequential and parallel computing platforms. While most successful domain decomposition methods (DDMs) have been designed for the solution of symmetric positive (semi-) definite systems, some have targeted indefinite problems of the form given in (1) (Cai and Widlund [1992]) . The objective of this paper is to present an alternative DDM that addresses both classes of indefinite (1) and complex (2) problems, that is based on the FETI-DP (Farhat, Lesoinne and Pierson [2000], Farhat et al. [2001]) DDM, and that is scalable when $\mathbf{K}$, $\mathbf{M}$, and $\mathbf{D}$ result from the FE discretization of second-order elastodynamic problems and fourth-order plate and shell dynamic problems.

2 The FETI-DP method

The dual-primal finite element tearing and interconnecting method (FETI-DP) (Farhat, Lesoinne and Pierson [2000], Farhat et al. [2001]) is a third-generation FETI method (for example, see Farhat [1991], Farhat and Roux [1991]) developed for the scalable and fast iterative solution of systems of equations arising from the FE discretization of static, dynamic, second-order, and fourth-order elliptic partial differential equations (PDEs). When equipped with the Dirichlet preconditioner (Farhat, Mandel and Roux [1994]) and applied to fourth-order or two-dimensional second-order problems, the condition number κ of its interface problem grows asymptotically as (Mandel and Tezaur [2001])

$$\kappa = \mathcal{O}\ (1 + \log^m \frac{H}{h}\ ), \quad m \leq 2, \tag{3}$$

where H and h denote the subdomain and mesh sizes, respectively. When equipped with the same Dirichlet preconditioner and an auxiliary coarse problem constructed by enforcing some set of optional constraints at the subdomain interfaces (Farhat et al. [2001]), the condition number estimate (3) also holds for second-order scalar elliptic problems (Klawonn, Widlund and Dryja [2002]). The result (3) proves the numerical scalability of the FETI methodology with respect to all of the problem size, the subdomain size, and the number of subdomains. More specifically, it suggests that one can expect the FETI-DP method to solve small-scale and large-scale problems in similar iteration counts. This in turn suggests that when the FETI-DP method is well-implemented on a parallel processor, it should be capable of solving an n-times larger problem using an n-times larger number of processors in almost a constant CPU time. This was demonstrated in practice for many complex structural mechanics problems (for example, see Farhat, Lesoinne and Pierson [2000] and Farhat et al. [2001] and the references cited therein).

Next, the FETI-DP method is overviewed in the context of the generic symmetric positive semi-definite (static) problem

$$\mathbf{K}\mathbf{x} = \mathbf{b}, \tag{4}$$

where $\mathbf{K}$ has the same meaning as in problems (1,2) and $\mathbf{b}$ is an arbitrary vector, in order to keep this paper as self-contained as possible.

2.1 Non-overlapping domain decomposition and notation

Let Ω denote the computational support of a second- or fourth-order problem whose discretization leads to problem (4), $\{\Omega^{(s)}\}_{s=1}^{N_s}$ denote its decomposition into N_s subdomains with matching interfaces $\Gamma^{(s,q)} = \partial\Omega^{(s)} \bigcap \partial\Omega^{(q)}$, and let $\Gamma = \bigcup\limits_{s=1,q>s}^{s=N_s} \Gamma^{(s,q)}$ denote the global interface of this decomposition. *In the remainder of this paper, each interface $\Gamma^{(s,q)}$ is referred to as an "edge", whether Ω is a two- or three-dimensional domain.* Let also $\mathbf{K}^{(s)}$ and $\mathbf{b}^{(s)}$ denote the contributions of subdomain $\Omega^{(s)}$ to $\mathbf{K}$ and $\mathbf{b}$, respectively, and let $\mathbf{x}^{(s)}$ denote the vector of dof associated with it.

Let N_c of the N_I nodes lying on the global interface Γ be labeled "corner" nodes (see Fig. 1), Γ_c denote the set of these corner nodes, and let $\Gamma' = \Gamma \backslash \Gamma_c$. If in each subdomain $\Omega^{(s)}$ the unknowns are partitioned into global corner dof designated by the subscript c, and "remaining" dof designated by the subscript r, $\mathbf{K}^{(s)}$, $\mathbf{x}^{(s)}$ and $\mathbf{b}^{(s)}$ can be partitioned as follows

$$\mathbf{K}^{(s)} = \begin{bmatrix} \mathbf{K}_{rr}^{(s)} & \mathbf{K}_{rc}^{(s)} \\ \mathbf{K}_{rc}^{(s)T} & \mathbf{K}_{cc}^{(s)} \end{bmatrix}, \qquad \mathbf{x}^{(s)} = \begin{bmatrix} \mathbf{x}_r^{(s)} \\ \mathbf{x}_c^{(s)} \end{bmatrix} \qquad \text{and} \qquad \mathbf{b}^{(s)} = \begin{bmatrix} \mathbf{b}_r^{(s)} \\ \mathbf{b}_c^{(s)} \end{bmatrix}. \tag{5}$$

The r-type dof can be further partitioned into "interior" dof designated by the subscript i, and subdomain interface "boundary" dof designated by the subscript b. Hence, $\mathbf{x}_r^{(s)}$ and $\mathbf{b}_r^{(s)}$ can be further partitioned as follows

$$\mathbf{x}_r^{(s)} = \begin{bmatrix} \mathbf{x}_i^{(s)} & \mathbf{x}_b^{(s)} \end{bmatrix}^T \qquad \text{and} \qquad \mathbf{b}_r^{(s)} = \begin{bmatrix} \mathbf{b}_i^{(s)} & \mathbf{b}_b^{(s)} \end{bmatrix}^T, \tag{6}$$

where the superscript T designates the transpose.

Let $\mathbf{x}_c$ denote the global vector of corner dof, and $\mathbf{x}_c^{(s)}$ denote its restriction to $\Omega^{(s)}$. Let also $\mathbf{B}_r^{(s)}$ and $\mathbf{B}_c^{(s)}$ be the two subdomain Boolean matrices defined by

$$\mathbf{B}_r^{(s)}\mathbf{x}_r^{(s)} = \pm\mathbf{x}_b^{(s)} \qquad \text{and} \qquad \mathbf{B}_c^{(s)}\mathbf{x}_c = \mathbf{x}_c^{(s)}, \tag{7}$$

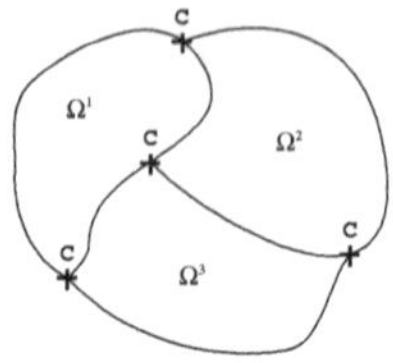

Fig. 1. Example of a definition of corner points

where the $\pm$ sign is set by any convention that implies that $\sum_{s=1}^{N_s} \mathbf{B}_r^{(s)} \mathbf{x}_r^{(s)}$ represents the *jump* of the solution $\mathbf{x}$ across the subdomain interfaces. Finally, let

$$\mathbf{b}_c = \sum_{s=1}^{N_s} \mathbf{B}_c^{(s)^T} \mathbf{b}_c^{(s)}. \tag{8}$$

In Farhat, Lesoinne and Pierson [2000] and Farhat et al. [2001], it was shown that solving problem (4) is equivalent to solving the following domain-decomposed problem

$$\mathbf{K}_{rr}^{(s)}\,\mathbf{x}_r^{(s)} + \mathbf{K}_{rc}^{(s)}\mathbf{B}_c^{(s)}\,\mathbf{x}_c + \mathbf{B}_r^{(s)^T}\lambda + \mathbf{B}_r^{(s)^T}\mathbf{Q}_b\,\mu = \mathbf{b}_r^{(s)}, \quad s = 1, ..., N_s \tag{9}$$

$$\sum_{s=1}^{N_s} \mathbf{B}_c^{(s)^T} \mathbf{K}_{rc}^{(s)^T}\,\mathbf{x}_r^{(s)} + \sum_{s=1}^{N_s} \mathbf{B}_c^{(s)^T} \mathbf{K}_{cc}^{(s)} \mathbf{B}_c^{(s)}\,\mathbf{x}_c = \mathbf{b}_c, \tag{10}$$

$$\sum_{s=1}^{N_s} \mathbf{B}_r^{(s)} \mathbf{x}_r^{(s)} = 0, \tag{11}$$

$$\mathbf{Q}_b^T \sum_{s=1}^{N_s} \mathbf{B}_r^{(s)} \mathbf{x}_r^{(s)} = 0, \tag{12}$$

where λ is an N_λ-long vector of Lagrange multipliers introduced on Γ' to enforce the continuity (11) of the solution $\mathbf{x}$, and μ is another vector of Lagrange multipliers introduced to enforce the optional linear constraints (12). These optional constraints, a concept first developed in Farhat, Chen, Risler and Roux [1998], are associated with a matrix $\mathbf{Q}_b$ with $N_Q < N_\lambda$ columns defined on Γ'. The word "optional" refers to the fact that Eq. (12) and the vector of Lagrange multipliers μ are not necessarily needed for formulating the above domain-decomposed problem. Indeed, since the solution of problem (4) is continuous across the subdomain interfaces, it satisfies Eq. (11) and therefore satisfies Eq. (12) for any matrix $\mathbf{Q}_b$.

The domain-decomposed problem (9–12) was labeled "dual-primal" in Farhat, Lesoinne and Pierson [2000] and Farhat et al. [2001] because it is formulated in terms of two different types of global unknowns: the dual Lagrange multipliers represented by the vector λ, and the primal corner dof represented by the vector $\mathbf{x}_c$.

In the remainder of this paper, the j-th column of $\mathbf{Q}_b$ is denoted by $\mathbf{q}_j$.

2.2 Interface and coarse problems

Let

$$\widetilde{\mathbf{K}}_{cc} = \begin{bmatrix} \mathbf{K}_{cc} & 0 \\ 0 & 0 \end{bmatrix}, \qquad \mathbf{d}_r = \sum_{s=1}^{N_s} \mathbf{B}_r^{(s)} \mathbf{K}_{rr}^{(s)^{-1}} \mathbf{b}_r^{(s)},$$

$$\text{and} \qquad \mathbf{b}_c^* = \mathbf{b}_c - \sum_{s=1}^{N_s} (\mathbf{K}_{rc}^{(s)} \mathbf{B}_c^{(s)})^T \mathbf{K}_{rr}^{(s)^{-1}} \mathbf{b}_r^{(s)}. \tag{13}$$

After some algebraic manipulations aimed at eliminating symbolically $\mathbf{x}_r^{(s)}$, $s = 1, ..., N_s$, $\mathbf{x}_c$, and μ, the domain-decomposed problem (9–12) can be transformed into the following symmetric positive semi-definite interface problem

$$(\mathbf{F}_{I_{rr}} + \widetilde{\mathbf{F}}_{I_{rc}} \widetilde{\mathbf{K}}_{cc}^{*^{-1}} \widetilde{\mathbf{F}}_{I_{rc}}^T)\lambda = \mathbf{d}_r - \widetilde{\mathbf{F}}_{I_{rc}} \widetilde{\mathbf{K}}_{cc}^{*^{-1}} \tilde{\mathbf{b}}_c^*, \tag{14}$$

where

$$\mathbf{F}_{I_{rr}} = \sum_{s=1}^{N_s} \mathbf{B}_r^{(s)} \mathbf{K}_{rr}^{(s)^{-1}} \mathbf{B}_r^{(s)^T}, \qquad \widetilde{\mathbf{F}}_{I_{rc}} = \sum_{s=1}^{N_s} \mathbf{B}_r^{(s)} \mathbf{K}_{rr}^{(s)^{-1}} \widetilde{\mathbf{K}}_{rc}^{(s)} \mathbf{B}_c^{(s)},$$

$$\widetilde{\mathbf{K}}_{rc}^{(s)} = \begin{bmatrix} \mathbf{K}_{rc}^{(s)} \mathbf{B}_c^{(s)} & \mathbf{B}_r^{(s)^T} \mathbf{Q}_b \end{bmatrix}, \qquad \tilde{\mathbf{b}}_c^* = \begin{bmatrix} \mathbf{b}_c^* \\ -\mathbf{Q}_b^T \mathbf{d}_r \end{bmatrix},$$

$$\widetilde{\mathbf{K}}_{cc}^* = \widetilde{\mathbf{K}}_{cc} -$$

$$\begin{bmatrix} \sum_{s=1}^{N_s} (\mathbf{K}_{rc}^{(s)} \mathbf{B}_c^{(s)})^T \mathbf{K}_{rr}^{(s)^{-1}} (\mathbf{K}_{rc}^{(s)} \mathbf{B}_c^{(s)}) & \sum_{s=1}^{N_s} (\mathbf{K}_{rc}^{(s)} \mathbf{B}_c^{(s)})^T \mathbf{K}_{rr}^{(s)^{-1}} (\mathbf{B}_r^{(s)^T} \mathbf{Q}_b) \\ \sum_{s=1}^{N_s} (\mathbf{B}_r^{(s)^T} \mathbf{Q}_b)^T \mathbf{K}_{rr}^{(s)^{-1}} (\mathbf{K}_{rc}^{(s)} \mathbf{B}_c^{(s)}) & \sum_{s=1}^{N_s} (\mathbf{B}_r^{(s)^T} \mathbf{Q}_b)^T \mathbf{K}_{rr}^{(s)^{-1}} (\mathbf{B}_r^{(s)^T} \mathbf{Q}_b) \end{bmatrix}. \tag{15}$$

The FETI-DP method is a DDM which solves the original problem (4) by applying a PCG algorithm to the solution of the corresponding dual interface problem (14). At the n-th PCG iteration, the matrix-vector product $(\mathbf{F}_{I_{rr}} + \widetilde{\mathbf{F}}_{I_{rc}} \widetilde{\mathbf{K}}_{cc}^{*^{-1}} \widetilde{\mathbf{F}}_{I_{rc}}^T)\lambda^n$ incurs the solution of an auxiliary problem of the form

$$\widetilde{\mathbf{K}}_{cc}^* \mathbf{z} = \widetilde{\mathbf{F}}_{I_{rc}}^T \lambda^n. \tag{16}$$

From the fifth of Eqs. (15), it follows that the size of this auxiliary problem is equal to the sum of the number of corner dof, N_c^{dof}, and the number of columns of the matrix $\mathbf{Q}_b$, N_Q.

For $N_Q = 0$ — that is, for $\mathbf{Q}_b = 0$, the auxiliary problem (16) is a coarse problem, and $\widetilde{\mathbf{K}}_{cc}^*$ is a sparse matrix whose pattern is that of the stiffness matrix obtained when each subdomain is treated as a "superelement" whose nodes are its corner nodes. This coarse problem ensures that the FETI-DP method equipped with the Dirichlet preconditioner (see Section 2.3) is numerically scalable for fourth-order plate and shell problems, and two-dimensional second-order elasticity problems (Farhat et al. [2001], Mandel and Tezaur [2001]). However, for $\mathbf{Q}_b = 0$, the FETI-DP method equipped with the Dirichlet preconditioner is not numerically scalable for three-dimensional second-order problems.

For any choice of $\mathbf{Q}_b \neq 0$, $\widetilde{\mathbf{K}}^*_{cc}$ remains a sparse matrix. If $\mathbf{Q}_b$ is constructed edge-wise — that is, if each column of $\mathbf{Q}_b$ is constructed as the restriction of some operator to a specific edge of Γ' — the sparsity pattern of $\widetilde{\mathbf{K}}^*_{cc}$ becomes that of a stiffness matrix obtained by treating each subdomain as a superelement whose nodes are its corner nodes augmented by virtual mid-side nodes. The number of dof attached to each virtual mid-side node is equal to the number of columns of $\mathbf{Q}_b$ associated with the edge on which lies this mid-side node. If N_Q is kept relatively small, the auxiliary problem (16) remains a relatively small coarse problem. This coarse problem was labeled "augmented" coarse problem in Farhat, Lesoinne and Pierson [2000] in order to distinguish it from the smaller coarse problem obtained with $\mathbf{Q}_b = 0$. Furthermore, each column of $\mathbf{Q}_b$ is referred to as an "augmentation coarse mode". When these augmentation coarse modes are chosen as the translational rigid body modes of each edge of Γ', the FETI-DP method equipped with the Dirichlet preconditioner becomes numerically scalable for three-dimensional second-order problems (Klawonn, Widlund and Dryja [2002]).

2.3 Local preconditioning

Two local preconditioners have been developed so far for the FETI-DP method:

1. The Dirichlet preconditioner which can be written as

$$\overline{\mathbf{F}}_{I_{rr}}^{D^{-1}} = \sum_{s=1}^{N_s} \mathbf{W}^{(s)}\mathbf{B}_r^{(s)} \begin{bmatrix} 0 & 0 \\ 0 & \mathbf{S}_{bb}^{(s)} \end{bmatrix} \mathbf{B}_r^{(s)^T}\mathbf{W}^{(s)},$$

$$\text{where} \qquad \mathbf{S}_{bb}^{(s)} = \mathbf{K}_{bb}^{(s)} - \mathbf{K}_{ib}^{(s)^T}\mathbf{K}_{ii}^{(s)^{-1}}\mathbf{K}_{ib}^{(s)}, \qquad (17)$$

   the subscripts i and b have the same meaning as in Section 2.1, and $\mathbf{W}^{(s)}$ is a subdomain diagonal scaling matrix that accounts for possible subdomain heterogeneities (Rixen and Farhat [1999]). This preconditioner is mathematically optimal in the sense that it leads to the condition number estimate (3).
2. The lumped preconditioner which can be written as

$$\overline{\mathbf{F}}_{I_{rr}}^{L^{-1}} = \sum_{s=1}^{N_s} \mathbf{W}^{(s)}\mathbf{B}_r^{(s)} \begin{bmatrix} 0 & 0 \\ 0 & \mathbf{K}_{bb}^{(s)} \end{bmatrix} \mathbf{B}_r^{(s)^T}\mathbf{W}^{(s)}. \qquad (18)$$

   This preconditioner is not mathematically optimal in the sense defined above; however, it decreases the cost of each iteration in comparison with the Dirichlet preconditioner often with a modest increase in the iteration count.

3 The FETI-DPH method

In the context of Eq. (1), $\mathbf{K}_{rr}^{(s)}$ becomes $\mathbf{K}_{rr}^{(s)} - \sigma^2 \mathbf{M}_{rr}^{(s)}$. Hence, the extension of the FETI-DP method to problems of the form given in (1) or (2) requires addressing the following issues:

1. $\mathbf{K}_{rr}^{(s)} - \sigma^2 \mathbf{M}_{rr}^{(s)}$ is indefinite and therefore the dual interface problem (14) is indefinite.
2. Independently of which interface points are chosen as corner points, $\mathbf{K}_{rr}^{(s)} - \sigma^2 \mathbf{M}_{rr}^{(s)}$ is in theory singular when σ^2 coincides with an eigenvalue of the pencil $(\mathbf{K}_{rr}^{(s)}, \mathbf{M}_{rr}^{(s)})$.
3. How to construct augmentation coarse modes and extended Dirichlet and lumped preconditioners that address the specifics of problems (1,2).

For problems of the form given in (2), only the third issue is relevant. The first issue can be addressed by solving the dual interface problem (14) by a preconditioned generalized minimum residual (PGMRES) algorithm rather than a PCG algorithm. The second and third issues were addressed in Farhat, Macedo and Lesoinne [2000] in the context of the basic FETI method and acoustic scattering applications — that is, for the exterior Helmholtz *scalar* problem where $\sigma^2 = k^2$ and k denotes the wave number. More specifically, a regularization procedure was developed in that reference to prevent all subdomain problems from being singular for any value of the wave number k, without destroying the sparsity of the local matrices $\mathbf{K}_{rr}^{(s)} - k^2 \mathbf{M}_{rr}^{(s)}$ and without affecting the solution of the original problem (1). Furthermore, for the scalar Helmholtz equation, the coarse modes were chosen in Farhat, Macedo and Lesoinne [2000] as plane waves of the form $e^{ik\theta_j^T X_b}$, $j = 1, 2, \cdots$, where θ_j denotes a direction of wave propagation and X_b the coordinates of a node on Γ. The resulting DDM was named the FETI-H method (H for Helmholtz).

Unfortunately, the regularization procedure characterizing the FETI-H method transforms each real subdomain problem associated with Eq. (1) into a complex subdomain problem. For acoustic scattering applications, this is not an issue because the Sommerfeld radiation condition causes the original problem to be in the complex domain. However, for real-valued problems such as those represented by Eq. (1), the regularization procedure of the FETI-H method is unjustifiable from computational resource and performance viewpoints.

In practice, experience reveals that $\mathbf{K}_{rr}^{(s)} - \sigma^2 \mathbf{M}_{rr}^{(s)}$ is non-singular as long as $\mathbf{K}_{rr}^{(s)}$ is non-singular. This observation is exploited here to design an extension of the FETI-DP method for indefinite problems of the form given in (1) and complex problems of the form given in (2) by:

1. Replacing the PCG solver by the PGMRES solver.
2. Adapting the Dirichlet and lumped preconditioners to exploit an interesting characteristic of problems (1,2).

3. Constructing a new augmentation coarse space that is effective for second-order elastodynamic problems as well as fourth-order plate and shell dynamic problems.

The extension of FETI-DP outlined above is named here the FETI-DPH method.

3.1 Adapted Dirichlet and lumped preconditioners

Consider the subdomain (impedance) matrix

$$\mathbf{Z}_{rr}^{(s)} = \mathbf{K}_{rr}^{(s)} - \sigma^2 \mathbf{M}_{rr}^{(s)}. \tag{19}$$

For the applications outlined in the introduction, $\mathbf{M}_{rr}^{(s)}$ is a mass matrix; hence, in three dimensions and at the element level, this matrix is proportional to h^3. On the other hand, for the same applications, $\mathbf{K}_{rr}^{(s)}$ is a stiffness matrix; in three dimensions and at the element level, it is proportional to h for second-order elasticity problems, and to $1/h^2$ for fourth-order plate and shell problems. It follows that for a sufficiently fine mesh, $\mathbf{Z}_{rr}^{(s)}$ is dominated by $\mathbf{K}_{rr}^{(s)}$. These observations, the optimality of the Dirichlet preconditioner and the computational efficiency of the lumped preconditioner established for the solution of problem (4) suggest preconditioning the local matrices $\mathbf{Z}_{rr}^{(s)}$ by Dirichlet and lumped constructs that are based on $\mathbf{K}_{rr}^{(s)}$ (see Section 2.3) and not $\mathbf{Z}_{rr}^{(s)}$. When Rayleigh damping is used,

$$\mathbf{D}_{rr}^{(s)} = c_K \mathbf{K}_{rr}^{(s)} + c_M \mathbf{M}_{rr}^{(s)}, \tag{20}$$

where c_K and c_M are two real constants, and the same reasoning can be invoked to advocate preconditioning the local matrices

$$\mathbf{Z}_{rr}^{(s)} = \mathbf{K}_{rr}^{(s)} - \sigma^2 \mathbf{M}_{rr}^{(s)} + i\sigma \mathbf{D}_{rr}^{(s)} \tag{21}$$

by Dirichlet and lumped constructs that are based on $(1 + i\sigma c_K)\mathbf{K}_{rr}^{(s)}$ and not $\mathbf{Z}_{rr}^{(s)}$.

Finally, it is pointed out that the ad-hoc reasoning outlined above can be mathematically justified, at least in the context of the scalar Helmholtz equation (for example, see Klawonn [1995] and the references cited therein).

3.2 Wavy augmented coarse problem

Let $\mathbf{r}$ denote the residual associated with the iterative solution of the dual interface problem (14). From Eqs. (9–12) and Eq. (14), it follows that

$$\mathbf{r} = \mathbf{d}_r - \widetilde{\mathbf{F}}_{I_{rc}} \widetilde{\mathbf{K}}_{cc}^{*^{-1}} \widetilde{\mathbf{b}}_c^* - (\mathbf{F}_{I_{rr}} + \widetilde{\mathbf{F}}_{I_{rc}} \widetilde{\mathbf{K}}_{cc}^{*^{-1}} \widetilde{\mathbf{F}}_{I_{rc}}^T)\lambda = \sum_{s=1}^{N_s} \mathbf{B}_r^{(s)} \mathbf{x}_r^{(s)}, \tag{22}$$

which shows that the residual $\mathbf{r}$ represents the jump of the iterate solution across the subdomain interfaces.

From Eq. (12), Eq. (15), Eq. (14) and Eq. (11), it follows that at each iteration of the PGMRES algorithm applied to the solution of problem (14), FETI-DPH forces the jump of the solution across the subdomain interfaces to be orthogonal to the subspace represented by the matrix $\mathbf{Q}_b$. This feature is a strategy for designing an auxiliary coarse problem which, when $\mathbf{Q}_b$ is well chosen, accelerates the convergence of a DDM (Farhat, Chen, Risler and Roux [1998]). In this work, the search for a suitable matrix $\mathbf{Q}_b$ is driven by the following reasoning. Suppose that the space of traces on Γ' of the desired solution of problem (1) is spanned by a set of orthogonal vectors $\{\mathbf{v}_{j_E}\}_{j=1}^{N_\lambda}$, where the subscript E indicates that $\mathbf{v}_{j_E}$ is non-zero only on edge $E \in \Gamma'$. Then, the residual $\mathbf{r}$ defined in Eq. (22) can be written as

$$\mathbf{r} = \sum_{j=1}^{N_\lambda} \alpha_j \mathbf{v}_{j_E}, \tag{23}$$

where $\{\alpha_j\}_{j=1}^{N_\lambda}$ is a set of real coefficients. If each augmentation coarse mode is chosen as

$$\mathbf{q}_j = \mathbf{v}_{j_E}, \qquad j = 1, \cdots, N_Q, \tag{24}$$

Eq. (12) simplifies to

$$\alpha_j = 0, \qquad j = 1, \cdots, N_Q. \tag{25}$$

In that case, Eq. (25) implies that at each iteration of the PGMRES algorithm, the first N_Q components of the residual $\mathbf{r}$ in the basis $\{\mathbf{v}_{j_E}\}_{j=1}^{N_\lambda}$ are zero. If a few vectors $\{\mathbf{v}_{j_E}\}_{j=1}^{N_Q}$, $N_Q << N_\lambda$, that dominate the expansion (23) can be found, then choosing these vectors as coarse augmentation modes can be expected to accelerate the convergence of the iterative solution of the dual interface problem (14). Hence, it remains to exhibit such a set of orthogonal vectors $\mathbf{v}_{j_E}$ and construct a computationally efficient matrix $\mathbf{Q}_b$.

A second-order elastodynamic problem is governed by Navier's displacement equations of motion

$$\mu \Delta u + (\Lambda + \mu)\nabla(\nabla \cdot u) + b = \rho \frac{\partial^2 u}{\partial t^2}, \tag{26}$$

where $u \in \mathbb{R}^3$ denotes the displacement (vector) field of the elastodynamic system, Λ and μ its Lamé moduli, $b \in \mathbb{R}^3$ its body forces, ρ its density, and t denotes time. If a harmonic motion is assumed, — that is, if

$$u(X,t) = v(X)e^{-i\omega t}, \tag{27}$$

where $X \in \mathbb{R}^3$ denotes the spatial variables, and ω denotes a circular frequency, the homogeneous form of Eq. (26) becomes

$$\mu\Delta v + (\Lambda + \mu)\nabla(\nabla \cdot v) + \rho\omega^2 v = 0. \tag{28}$$

The free-space solutions of the above vector equation are

$$v = a_p \sin(k_p\theta \cdot X), \quad v = a_p \cos(k_p\theta \cdot X), \tag{29}$$
$$v = a_{s_1} \sin(k_s\theta \cdot X), \quad v = a_{s_1} \cos(k_s\theta \cdot X), \tag{30}$$
$$v = a_{s_2} \sin(k_s\theta \cdot X), \quad v = a_{s_2} \cos(k_s\theta \cdot X), \tag{31}$$

where $\theta \in \mathbb{R}^3$ is an arbitrary vector of unit length ($\|\theta\|_2 = 1$), $a_p \in \mathbb{R}^3$ is a vector that is parallel to θ, $(a_{s_1}, a_{s_2}) \in \mathbb{R}^3 \times \mathbb{R}^3$ are two independent vectors in the plane orthogonal to θ,

$$k_p = \sqrt{\frac{\rho\omega^2}{\Lambda + 2\mu}}, \quad \text{and} \quad k_s = \sqrt{\frac{\rho\omega^2}{\mu}}. \tag{32}$$

The free-space solutions (29) are known as the elastic pressure or longitudinal waves, and the free-space solutions (30) and (31) are known as the elastic shear or transverse waves.

Consider next the following fourth-order PDE associated with a given elastic body

$$\Delta^2 u - \frac{m}{D}\omega^2 u = 0, \qquad \text{where} \quad m = \rho\tau, \quad D = \frac{E\tau^3}{12(1-\nu^2)}, \tag{33}$$

E denotes the Young modulus of the elastic body, ν its Poisson ratio, τ its thickness, and all other variables have the same meaning as before. The reader can check that the free-space solutions (29,30,31) with

$$k_p = k_s =^4 \sqrt{\frac{m}{D}\omega^2} \tag{34}$$

are also free-space solutions of Eq. (33). The PDE (33) can model the harmonic transverse motion of a plate. In that case, u is a scalar representing the transverse displacement field. However, for the purpose of constructing an augmented coarse problem for the FETI-DPH method, and only for this purpose, it is assumed here that when $u \in \mathbb{R}^3$, Eq. (33) models the harmonic motion of a shell in all three dimensions.

Hence, a general solution of either Eq. (28) or Eq. (33) can be written as

$$\begin{aligned} v = & \sum_{j=1}^{\infty} \left\{ a_{p_j} \left(c_{1_j} \sin(k_p\theta_j \cdot X) + c_{2_j} \cos(k_p\theta_j \cdot X) \right) \right\} \\ & + \sum_{j=1}^{\infty} \left\{ a_{s_{1_j}} \left(c_{3_j} \sin(k_s\theta_j \cdot X) + c_{4_j} \cos(k_s\theta_j \cdot X) \right) \right\} \\ & + \sum_{j=1}^{\infty} \left\{ a_{s_{2_j}} \left(c_{5_j} \sin(k_s\theta_j \cdot X) + c_{6_j} \cos(k_s\theta_j \cdot X) \right) \right\}, \end{aligned} \tag{35}$$

where $\theta_j \in \mathbb{R}^3$ is an arbitrary vector of unit length defining the direction of propagation of an elastic pressure or shear wave, c_{1_j}, c_{2_j}, c_{3_j}, c_{4_j}, c_{5_j}, and c_{6_j} are real coefficients, and k_p and k_s are given by Eq. (32) for a second-order elastodynamic problem and by Eq. (34) for a fourth-order plate or shell dynamic problem. From Eq. (35) and Eq. (24), it follows that the desired matrix $\mathbf{Q}_b$ is composed of blocks of six columns. The columns of each block are associated with one direction of propagation θ_j and one edge E of the mesh partition, and can be written as

$$\begin{aligned}
\mathbf{q}_{b_l}\begin{bmatrix}3(m-1)+1\\3(m-1)+2\\3(m-1)+3\end{bmatrix} &= a_{p_j}\ \sin(k_p\theta_j\cdot X_m),\\
\mathbf{q}_{b_{l+1}}\begin{bmatrix}3(m-1)+1\\3(m-1)+2\\3(m-1)+3\end{bmatrix} &= a_{p_j}\ \cos(k_p\theta_j\cdot X_m),\\
\cdots \quad & \cdots\\
\mathbf{q}_{b_{l+4}}\begin{bmatrix}3(m-1)+1\\3(m-1)+2\\3(m-1)+3\end{bmatrix} &= a_{s2_j}\ \sin(k_s\theta_j\cdot X_m),\\
\mathbf{q}_{b_{l+5}}\begin{bmatrix}3(m-1)+1\\3(m-1)+2\\3(m-1)+3\end{bmatrix} &= a_{s2_j}\ \cos(k_s\theta_j\cdot X_m),\\
l = 6(j-1)+1, \qquad & m = 1,\cdots,N_I-N_c,
\end{aligned} \tag{36}$$

where $\mathbf{q}_b[3(m-1)+1]$ designates the entry of $\mathbf{q}_b$ associated with the dof in the x-direction attached to the m-th node on an edge $E \in \Gamma'$, $\mathbf{q}_b[3(m-1)+2]$ designates the entry associated with the dof along the y-direction, $\mathbf{q}_b[3(m-1)+3]$ designates the entry associated with the dof along the z-direction, and $X_m \in \mathbb{R}^3$ denotes the coordinates of this m-th node. Hence, if N_E denotes the number of edges of the mesh partition, and N_θ the number of considered directions of wave propagation, the total number of augmentation coarse modes is given in general by $N_Q = 6N_EN_\theta$. To these modes can be added the edge-based translational rigid body modes as these are free-space solutions of Eq. (28) when $\omega = 0$.

In this paper, the number of directions is limited by $N_\theta^{max} = 13$, and the directions θ_j are generated as follows. A generic cube is discretized into $3\times3\times3$ points. A direction θ_j is defined by connecting the center point to any of the other 26 points lying on a face of the cube. Since each direction θ_j is used to define both a cosine and a sine mode, only one direction θ_j is retained for each pair of opposite directions, which results in a maximum of 13 directions.

4 Performance studies and preliminary conclusions

Here, the FETI-DPH method is applied to the solution of various problems of the form given in (1) or (2) and associated with: (a) the discretization by quadratic tetrahedral elements (10 nodes per element) of a wheel carrier fixed at a few of its nodes, and (b) the discretization by linear triangular shell elements of an alloy wheel clamped at a few center points (Fig. 2). When the structure is assumed to be damped, Eq. (20) is used to construct D and c_K and c_M are determined by requiring that the critical damping ratio of the first 10 modes of the structure be equal in a least-squares sense to a specified value, ξ. In all problems, the shift is set to $\sigma^2 = \omega^2 = 4\pi^2 f^2$, where ω^2 is the square of a (possibly natural) circular frequency of the structure and f is the corresponding frequency in Hz. To help the reader appreciate the magnitude of a chosen shift value, the natural frequencies of both structures are characterized in Table 1. In order to investigate the performance, potential, and various scalability properties of the FETI-DPH method, various values of σ^2 are considered, three meshes with different resolutions are employed for the wheel carrier second-order problem (504,375 dof, 1,317,123 dof, and 2,091,495 dof), and one mesh with 936,102 dof is employed for the alloy wheel fourth-order shell problem. In all cases, the right-sides of problems (1,2) are generated by a distributed load, the computations are performed on a Silicon Graphics Origin 3800 system with 40 R12000 400 MHz processors, and convergence is declared when the relative residual satisfies

$$RE^n = \frac{\|(\mathbf{K} - \sigma^2\mathbf{M} + i\sigma\mathbf{D})\ \mathbf{x}^n - \mathbf{b}\|_2}{\|\mathbf{b}\|_2} \leq 10^{-6}. \qquad (37)$$

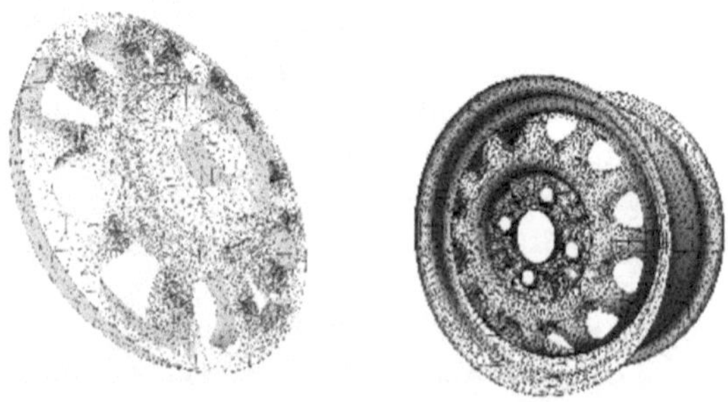

Fig. 2. FE discretizations of a wheel carrier (left) and an alloy wheel (right)

First, attention is directed to the wheel carrier undamped problem, and for each generated mesh, N_s is chosen to keep the subdomain problem size constant. Two frequencies, 500 KHz and 2 MHz, are considered: the latter value of the shift σ^2 arises, for example, when exciting the structure by its $200-th$ natural frequency, or shifting around it during the solution of an

Table 1. Eigenvalue/Frequency partial spectrum of the pencil $(\mathbf{K}, \mathbf{M})$

	Wheel Carrier	$(2^{nd}$-order)	Alloy Wheel	$(4^{th}$-order)
Mode Number	Eigenvalue (ω^2)	Frequency	Eigenvalue (ω^2)	Frequency
1	2.6e+11	8.2e+04 Hz	7.6e+05	1.4e+02 Hz
100	5.2e+13	1.1e+06 Hz	1.0e+09	5.1e+03 Hz
200	1.6e+14	2.0e+06 Hz	3.0e+09	8.7e+03 Hz
300	2.8e+14	2.6e+06 Hz	5.7e+09	1.2e+04 Hz
400	4.0e+14	3.2e+06 Hz	9.5e+09	1.5e+04 Hz
500	5.1e+14	3.5e+06 Hz		
600	6.0e+14	3.9e+06 Hz		

eigenvalue problem. The number of wave directions is set to $N_\theta = 2$, and the three translational rigid body modes are included in the construction of the augmentation matrix $\mathbf{Q}_b$. The performance results of the FETI-DPH solver obtained on $N_p = 12$ processors are reported in Table 2 where N_{itr} records the iteration count. For each considered frequency, the iteration count associated with the chosen number of subdomains and chosen preconditioner is almost independent of the mesh size, which highlights the numerical scalability of the FETI-DPH method with respect to both the subdomain problem size and the total problem size. For this second-order problem, the lumped and Dirichlet preconditioners deliver similar CPU performances; hence, the lumped preconditioner is preferable since it requires less memory.

Table 2. Performance of the FETI-DPH solver: wheel carrier, undamped, 2nd-order problem; fixed subdomain problem size; $N_\theta = 2$ (+ the three translational rigid body modes); $N_p = 12$

Frequency	Shift (σ^2)	Mesh size	N_s	N_{itr} Lumped	CPU Lumped	N_{itr} Dirichlet	CPU Dirichlet
		504,375 dof	250	63	64 s.	45	60 s.
5×10^5 Hz	9.8e+12	1,317,123 dof	600	70	207 s.	53	206 s.
		2,091,495 dof	950	60	364 s.	45	358 s.
		504,375 dof	250	137	123 s.	105	119 s.
2×10^6 Hz	1.6e+14	1,317,123 dof	600	174	483 s.	140	491 s.
		2,091,495 dof	950	151	901 s.	118	887 s.

To illustrate the performance of the FETI-DPH solver for problems of the form given in (2), the wheel carrier is next assumed to have a Rayleigh damping. The mesh with $N_{dof} = 1,317,123$ is considered, the number of subdomains is set to $N_s = 600$, the shift is set to $\sigma^2 = 10^5$ Hz, the number of wave directions is set to $N_\theta = 2$, the three translational rigid body modes are included in the construction of the augmentation matrix $\mathbf{Q}_b$, and the number of processors is set to $N_p = 16$. For these parameters, the performance results of FETI-DPH equipped with the lumped preconditioner are reported in Table 3 for the undamped case ($\xi = 0$), and for realistic damping scenarios ($\xi = 1\%$, $\xi = 2\%$, and $\xi = 5\%$). These results suggest that the intrinsic performance of FETI-DPH improves with the amount of damping. For the undamped case, FETI-DPH operates in the real domain. This explains why in that case, each iteration is 2.7 times faster than in the damped case where FETI-DPH operates in the complex plane.

Table 3. Performance of the FETI-DPH solver: wheel carrier, damped, 2nd-order problem; $N_{dof} = 1,317,123$; $N_s = 600$; $\sigma^2 = 10^5$ Hz; lumped preconditioner; $N_\theta = 2$ (+ the three translational rigid body modes); $N_p = 16$

ξ	c_K	c_M	N_{itr}	CPU
0%	0	0	62	182 s.
1%	3.42e-6	17.9	51	403 s.
2%	6.85e-6	35.8	49	394 s.
5%	1.71e-5	89.5	48	384 s.

Next, attention is directed to the undamped alloy wheel problem to investigate the performance for a fourth-order shell problem of the FETI-DPH solver equipped with the Dirichlet preconditioner. Two different frequencies, 5 KHz and 20 KHz, are considered: the upper value of the shift σ^2 arises, for example, when exciting the considered alloy wheel by a frequency that is higher than its $400-th$ natural frequency, or shifting around that frequency during the solution of an eigenvalue problem. The number of subdomains is varied between $N_s = 100$ and $N_s = 400$ and the number of processors is fixed to $N_p = 8$. Table 4, where N_{coarse} denotes the total size of the augmented coarse problem, contrasts for each value of N_s the performance of FETI-DP (with PGMRES as a solver) and the best performance of FETI-DPH obtained by varying N_θ. The reported performance results suggest that the FETI-DPH solver is numerically scalable for dynamic shell problems of the form given in (1). They also highlight the superiority of FETI-DPH over FETI-DP which fails to converge in a reasonable iteration count for large values of the shift σ^2.

Acknowledgement. This research was supported by the Sandia National Laboratories under Contract No. 31095. The first author also acknowledges the support by the Sandia National Laboratories under Contract No. 29341, and the second author also acknowledges partial support by the National Science Foundation under Grant No. DMS-0209297. Any opinions, findings, and conclusions or recommendations expressed in this material are those of the authors and do not necessarily reflect the views of the Sandia National Laboratories or the National Science Foundation.

Table 4. FETI-DPH vs. FETI-DP: alloy wheel, undamped, 4th-order problem; $N_{dof} = 936,102$; Dirichlet preconditioner; $N_p = 8$

Frequency	Shift (σ^2)	N_s	N_θ	N_{coarse}	N_{itr}	CPU
5×10^3 Hz	9.8e+8	100	0	3,258	347	534 s.
		100	3	7,275	122	265 s.
		200	0	6,372	236	301 s.
		200	2	11,853	116	200 s.
		400	0	12,129	226	317 s.
		400	2	21,924	123	271 s.
2×10^4 Hz	1.6e+10	100	0	3,258	>400	–
		100	5	9,512	330	680 s.
		200	0	6,372	>400	–
		200	5	17,581	261	564 s.
		400	0	12,129	>400	–
		400	3	27,270	265	706 s.

References

X. C. Cai and O. Widlund. Domain decomposition algorithms for indefinite elliptic problems. *SIAM J. Sci. Statist. Comput.*, 13:243–258, 1992.

C. Farhat, M. Lesoinne, and K. Pierson. A scalable dual-primal domain decomposition method. *Numer. Lin. Alg. Appl.*, 7:687–714, 2000.

C. Farhat, M. Lesoinne, P. LeTallec, K. Pierson, and D. Rixen. FETI-DP: a dual-primal unified FETI method - part I: a faster alternative to the two-level FETI method. *Internat. J. Numer. Meths. Engrg.*, 50:1523–1544, 2001.

C. Farhat. A Lagrange multiplier based divide and conquer finite element algorithm. *J. Comput. Sys. Engrg.*, 2:149–156, 1991.

C. Farhat and F. X. Roux. A method of finite element tearing and interconnecting and its parallel solution algorithm, *Internat. J. Numer. Meths. Engrg.*, 32:1205–1227, 1991.

C. Farhat, J. Mandel and F. X. Roux. Optimal convergence properties of the FETI domain decomposition method. *Comput. Meths. Appl. Mech. Engrg.*, 115:367–388, 1994.

J. Mandel and R. Tezaur, On the convergence of a dual-primal substructuring method. *Numer. Math.*, 88:543–558, 2001.

A. Klawonn, O. B. Widlund and M. Dryja. Dual-primal FETI methods for three-dimensional elliptic problems with heterogeneous coefficients. *SIAM J. Numer. Anal.*, 40:159–179, 2002.

C. Farhat, P. S. Chen, F. Risler and F. X. Roux. A unified framework for accelerating the convergence of iterative substructuring methods with Lagrange multipliers. *Internat. J. Numer. Meths. Engrg.*, 42:257–288, 1998.

D. Rixen and C. Farhat. A simple and efficient extension of a class of substructure based preconditioners to heterogeneous structural mechanics problems. *Internat. J. Numer. Meths. Engrg.*, 44:489–516, 1999.

C. Farhat, A. Macedo and M. Lesoinne. A two-level domain decomposition method for the iterative solution of high frequency exterior Helmholtz problems. *Numer.Math.*, 85:283–308, 2000.

A. Klawonn. *Preconditioners for Indefinite Problems.* Ph. D. Thesis, Westfalische, Wilhelms-Universitat, Munster, 1995.

Hybrid Schwarz-Multigrid Methods for the Spectral Element Method: Extensions to Navier-Stokes

Paul F. Fischer[1] and James W. Lottes[2]

[1] Argonne National Laboratory, Mathematics and Computer Science Division (`http://www.mcs.anl.gov/~fischer/`)
[2] University of Illinois, Dept. of Theoretical and Applied Mechanics

Summary. The performance of multigrid methods for the standard Poisson problem and for the consistent Poisson problem arising in spectral element discretizations of the Navier-Stokes equations is investigated. It is demonstrated that overlapping additive Schwarz methods are effective smoothers, provided that the solution in the overlap region is weighted by the inverse counting matrix. It is also shown that spectral element based smoothers are superior to those based upon finite element discretizations. Results for several large 3D Navier-Stokes applications are presented.

1 Introduction

The spectral element method (SEM) is a high-order weighted residual technique that combines the geometric flexibility of finite elements with the rapid convergence properties and tensor-product efficiencies of global spectral methods. Globally, elements are coupled in an unstructured framework with interelement coupling enforced through standard matching of nodal interface values. Locally, functions are represented as tensor products of stable Nth-order Lagrangian interpolants based on Gauss-Lobatto (GL) or Gauss (G) quadrature points. For problems having smooth solutions, such as the incompressible Navier-Stokes equations, the SEM converges exponentially fast with the local approximation order N. Because of its minimal numerical dissipation and dispersion, the SEM is particularly well suited for the simulation of flows at transitional Reynolds numbers, where physical dissipation is small and turbulence-model dissipation is absent.

The two-level hierarchy of the spectral element discretization provides a natural route to domain decomposition with several benefits. The loose C^0 interelement coupling implies that the stencil depth does not increase with approximation order, so that interprocessor communication is minimal. The local tensor-product structure allows matrix-vector products to be recast as cache-efficient *matrix-matrix* products and also allows local subdomain problems to be solved efficiently with fast tensor-product solvers. Finally, the high-order polynomial expansions provide a readily available sequence of nested grids (obtained through successive reductions in polynomial degree) for use in multilevel solvers.

This paper presents recent developments in spectral element multigrid (SEMG) methods. Our point of departure is the original work of Rønquist and Patera [1987] and Maday and Muñoz [1988], who developed variational SEMG for the two-dimensional Poisson problem using intra-element prolongation/restriction operators coupled with Jacobi smoothing. The high-aspect-ratio cells present in the tensor-product GL grid are a well-known source of difficulty in spectral multigrid methods and have drawn much attention over the past decade. We have developed multigrid smoothers in Lottes and Fischer [2004] based on the overlapping additive Schwarz method of Dryja and Widlund [1987] and Fischer et al. [2000]. We bypass the high-aspect-ratio cell difficulty by solving the local problems directly using fast tensor-product solvers; this approach ensures that the smoother cost does not exceed the cost of residual evaluation. Here, we extend our SEMG approach from the two-dimensional Laplacian to the more difficult consistent Poisson operator that governs the pressure in the mixed $\mathbb{P}_N - \mathbb{P}_{N-2}$ spectral element formulation of Maday and Patera [1989].

In the next section, we introduce the SE discretization for a model Poisson problem. The basic elements of our multilevel iterative procedures are presented in Section 3, along with results for the Poisson problem. Extensions to unsteady Navier-Stokes applications are described in Section 4.

2 Discretization of the Poisson Problem

The spectral element discretization of the Poisson problem in $\mathbb{R}^d$ is based on the weighted residual formulation: *Find* $u \in X_N$ *such that*

$$(\nabla v, \nabla u)_{GL} = (v, f)_{GL} \qquad \forall v \in X_N. \tag{1}$$

The inner product $(.,.)_{GL}$ refers to the Gauss-Lobatto-Legendre (GL) quadrature associated with the space $X_N := [Z_N \cap H_0^1(\Omega)]$, where $Z_N := \{v \in L^2(\Omega) | v_{|\Omega^e} \in \mathbb{P}_N(\Omega^e)\}$. Here, L^2 is the space of square integrable functions on Ω; H_0^1 is the space of functions in L^2 that vanish on the boundary and whose first derivative is also in L^2; and $\mathbb{P}_N(\Omega^e)$ is the space of functions on Ω^e whose image is a tensor-product polynomial of degree $\leq N$ in the reference domain, $\hat{\Omega} := [-1,1]^d$. For $d = 2$, a typical element in X_N is written

$$u(\mathbf{x}^e(r,s))|_{\Omega^e} = \sum_{i=0}^{N} \sum_{j=0}^{N} u_{ij}^e h_i^N(r) h_j^N(s), \tag{2}$$

where u_{ij}^e is the nodal basis coefficient; $h_i^N \in \mathbb{P}_N$ is the Lagrange polynomial satisfying $h_i^N(\xi_j) = \delta_{ij}$, where ξ_j, $j = 0, \ldots, N$ are the the GL quadrature

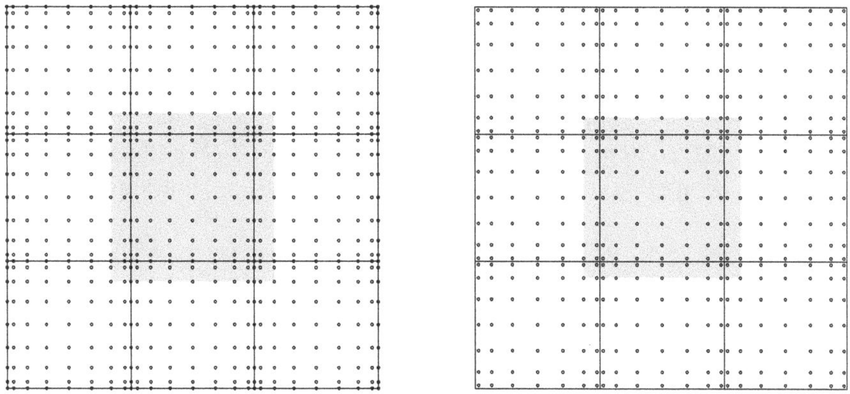

Fig. 1. Spectral element configuration ($E = 9$, $N = 8$) showing Lagrange interpolation points for functions in X^N (left) and Y^N (right). The shaded regions illustrate the "minimal overlap" domain extension for the overlapping Schwarz smoothers

points (the zeros of $(1-\xi^2)L'_N(\xi)$, where L_N is the Legendre polynomial of degree N) and δ_{ij} is the Kronecker delta function; and $\mathbf{x}^e(r,s)$ is the coordinate mapping from $\widehat{\Omega}$ to Ω^e. We assume $\Omega = \cup_{e=1}^{E}\Omega^e$ and that the intersection of two subdomains (spectral elements) is an entire edge, a single vertex, or void. Function continuity ($u \in H^1$) is enforced by ensuring that nodal values on element boundaries coincide with those on adjacent elements. Figure 1 illustrates a spectral element decomposition of the square using $E = 9$ elements. The Gauss-Lobatto-based mesh on the left shows the nodal distribution for X_N. The Gauss-based mesh on the right is used for functions in Y_N, which will be introduced in the context of the Stokes discretization in Section 4.

Computational Preliminaries. Because we employ iterative solvers, we need an efficient procedure for evaluating matrix-vector products associated with the bilinear forms in (1). As noted by Orszag [1980], tensor-product bases play a key role in this respect, particularly for large N (i.e., $N \geq 8$). Here, we introduce several points that are central to our element-based solution strategy.

As with standard finite element methods, we assume availability of both local element-based and global mesh-based node numberings, with the local-to-global map given by $q(i_1,\ldots,i_d,e) \in \{1,\ldots,\bar{n}\}$, for $i_k \in \{0,\ldots,N\}$, $k \in \{1,\ldots,d\}$, and $e \in \{1,\ldots,E\}$, where $\bar{n}$ is the number of distinct global nodes. Let Q^T be the $\bar{n}\times E(N+1)^d$ matrix with columns $\underline{\hat{e}}_{q(i_1,\ldots,i_d,e)}$, where $\underline{\hat{e}}_q$ denotes the qth column of the $\bar{n}\times\bar{n}$ identity matrix. Then the matrix-vector product $\underline{u}_L = Q\underline{u}$ represents a global-to-local mapping for any function $u(\mathbf{x}) \in X^N$, and the bilinear form on the left of (1) can be written

$$(\nabla v, \nabla u) = \underline{v}^T Q^T A_L Q \underline{u}, \tag{3}$$

where A_L=block-diag$(A^e)_{e=1}^E$ is the unassembled stiffness matrix comprising the local stiffness matrices, A^e, and $Q^T$and Q correspond to respective gather and scatter operations. In practice the global stiffness matrix, $A := Q^T A_L Q$, is never formed. One simply effects the *action* of A by applying each matrix to a vector through appropriate subroutine calls.

In the SEM, computational efficiency dictates that *local* stiffness matrices should also be applied in matrix-free form. The local contributions to (3) are

$$(\nabla v, \nabla u)_{GL}^e = (\underline{v}^e)^T A^e \underline{u}^e = (\underline{v}^e)^T \begin{pmatrix} D_1 \\ D_2 \end{pmatrix}^T \begin{pmatrix} G_{11}^e & G_{12}^e \\ G_{12}^e & G_{22}^e \end{pmatrix} \begin{pmatrix} D_1 \\ D_2 \end{pmatrix} \underline{u}^e, \quad (4)$$

with respective geometric factors and derivative operators,

$$G_{ij}^e := \left(\widehat{B} \otimes \widehat{B}\right) \left[\sum_{k=1}^{d} \frac{\partial r_i}{\partial x_k} \frac{\partial r_j}{\partial x_k} \right]^e J^e, \quad D_1 := (I \otimes \widehat{D}), \quad D_2 := (\widehat{D} \otimes I). \quad (5)$$

Here, $\underline{v}^e$ and $\underline{u}^e$ are vectors containing the lexicographically ordered nodal basis coefficients $\{v_{ij}^e\}$ and $\{u_{ij}^e\}$, respectively; $\widehat{B}$=diag$(\rho_k)_{k=0}^N$ is the one-dimensional mass matrix composed of the GL quadrature weights; and $\widehat{D}$ is the one-dimensional derivative matrix with entries

$$\widehat{D}_{ij} = \left. \frac{dh_j}{dr} \right|_{\xi_i}, \qquad i, j \in \{0, \ldots, N\}^2.$$

The Jacobian, J^e, and metric terms (in brackets in (5)) are evaluated pointwise at each GL quadrature point, (ξ_p, ξ_q), so that each of the composite geometric matrices, G_{ij}^e, is diagonal.

The presence of the cross terms, G_{12}^e, implies that A^e is full and requires storage of $(N+1)^4$ nonzeros for *each* spectral element if explicitly formed. In the spectral element method, this excessive storage and work overhead is avoided by retaining the factored form (5), which requires (to leading order) storage of only $3E(N+1)^2$ nonzeros and work of $\approx 8E(N+1)^3$ per matrix-vector product. The savings is more significant in 3D, where the respective storage and work complexities are $6E(N+1)^3$ and $\approx 12E(N+1)^4$ for the factored form, versus $O(EN^6)$ if A is explicitly formed. Moreover, the leading order work terms for the factored form can be cast as efficient matrix-matrix products, as discussed in detail by Deville et al. [2002]. These complexity savings can be extended to all system matrices and are the basis for efficient realizations of high-order weighted residual techniques.

If Ω^e is a regular parallelepiped, the local stiffness matrix simplifies to a separable form. For example, for an $L_x^e \times L_y^e$ rectangle, one would have

$$A^e = \frac{L_y^e}{L_x^e} \widehat{B} \otimes \widehat{A} + \frac{L_y^e}{L_x^e} \widehat{A} \otimes \widehat{B}, \qquad \widehat{A} := \widehat{D}^T \widehat{B} \widehat{D}. \quad (6)$$

This form has a readily computable (pseudo-) inverse given by the fast diagonalization method (FDM) of Lynch et al. [1964],

$$A_e^{-1} = (S \otimes S) \left[\frac{L_y^e}{L_x^e} I \otimes \Lambda + \frac{L_y^e}{L_x^e} \Lambda \otimes I \right]^{-1} (S^T \otimes S^T), \tag{7}$$

where S is the matrix of eigenvectors and Λ the matrix of eigenvalues satisfying $\widehat{A}S = \widehat{B}S\Lambda$ and $S^T\widehat{B}S = I$. The bracketed term in (7) is diagonal, and its pseudo-inverse is computed by inverting nonzero elements and retaining zeros elsewhere. For arbitrarily deformed elements, the discrete Laplacian cannot be expressed in the tensor-product form (6), and the FDM cannot be used. For the purposes of a preconditioner, however, it suffices to apply the FDM to a regular parallelepiped of equivalent size, as demonstrated in Couzy [1995] and Fischer et al. [2000]. Similar strategies for the case of nonconstant coefficients are discussed by Shen [1996].

3 Multilevel Solvers

We are interested in methods for solving the global system $A\underline{u} = \underline{g}$. To introduce notation, we consider the two-level multigrid sweep.

Procedure Two-Level: (8)

$$\begin{aligned}
&i) \quad \underline{u}^{k+1} = \underline{u}^k + \sigma M(\underline{g} - A\underline{u}^k), \qquad k = 0, \ldots, m_d - 1 \\
&ii) \quad \underline{r} = \underline{g} - A\underline{u}^{m_d} \\
&iii) \quad \underline{\tilde{e}} = \sigma_C P A_C^{-1} P^T \underline{r} \\
&iv) \quad \underline{\tilde{u}}^0 = \underline{u}^{m_d} + \underline{\tilde{e}} \\
&v) \quad \underline{\tilde{u}}^{k+1} = \underline{\tilde{u}}^k + \sigma M(\underline{g} - A\underline{\tilde{u}}^k), \qquad k = 0, \ldots, m_u - 1 \\
&vi) \quad \text{If } ||A\underline{\tilde{u}}^{m_u} - \underline{g}|| < \text{tol, set } \underline{u} := \underline{\tilde{u}}^{m_u}, \textit{quit}. \\
&\qquad \text{Else, } \underline{u}^0 := \underline{\tilde{u}}^{m_u}, \text{ go to } (i).
\end{aligned}$$

Here M is the smoother, σ and σ_C are relaxation parameters, and m_d and m_u are the number of smoothing steps on the downward and upward legs of the cycle, respectively. Steps (i) and (v) are designed to eliminate high-frequency error components that cannot be represented on the coarse grid. The idea is that the error after (ii), $\underline{e} := A^{-1}\underline{r}$, should be well approximated by $\underline{\tilde{e}}$, which lies in the coarse-grid space represented by the columns of P. The coarse-grid problem, A_C^{-1}, is solved directly, if A_C is sufficiently sparse, or approximated by recursively applying the two-level procedure to the A_C system, giving rise to the multigrid "V" cycle. The prolongation matrix P interpolates from the coarse space to the fine nodes using the local tensor-product basis functions for the coarse space.

If the two-level procedure is used as a preconditioner, we take $\underline{u}^0 = 0$, $m_d = 1$, and $m_u = 0$, and the procedure simplifies to the following.

Procedure Preconditioner: (9)

$$
\begin{aligned}
&i)\ \underline{u}^1 = \sigma M \underline{g},\\
&ii)\ \underline{r} = \underline{g} - A\underline{u}^1\\
&iii)\ \underline{\tilde{e}} = \sigma_C P A_C^{-1} P^T \underline{r}\\
&iv)\ \underline{u} = \underline{u}^1 + \underline{\tilde{e}},\ \text{return.}
\end{aligned}
$$

The preconditioner can be viewed either as an application of the multigrid V-cycle or as a two-level multiplicative Schwarz method (Smith et al. [1996]). By simply replacing (*ii*) with $\underline{r} = \underline{g}$, we obtain a two-level additive Schwarz method, which has the advantage of avoiding an additional multiplication by A. This savings is important in the Navier-Stokes applications that we consider in Section 4.

Smoothers for the Poisson Problem. Here, we review the SEMG smoothing strategies considered for the Poisson problem in Lottes and Fischer [2004]. Our original intent was to base the smoother, M, on the additive overlapping Schwarz method of Dryja and Widlund [1987], with local subdomain problems discretized by finite elements (FEs) having nodes coincident with the GL nodes, as considered by Casarin [1997], Fischer [1997], and Pahl [1993]. By using the fast diagonalization method to solve the local problems, however, we are freed from the constraint of using FE-based preconditioners because the cost depends only on the use of tensor-product forms and not on the sparsity of the originating operator. Hence, we are able to consider subdomain problems derived as restrictions of the originating spectral element matrix, A, as first suggested by Casarin [1997].

The use of Schwarz-based smoothing, which is arguably more expensive than traditional smoothers, is motivated by several factors. First, it is not practical to apply Gauss-Seidel smoothing in the SEM because the matrix entries are not available (see (4)). The alternative of pointwise-Jacobi smoothing was shown by Rønquist [1988] and Maday et al. [1992] not to scale for $d > 1$. Specifically, the authors demonstrated a convergence factor of $\rho = 0.75$ for $d = 1$, but only $\rho = 1 - c/N \log N$ for $d = 2$. Second, while one can exploit the SE-FE spectral equivalence established by Orszag [1980] to ostensibly convert the SE problem into a FE problem and then apply traditional multigrid, the FE problem inherits the difficulties of its SE counterpart, namely, the high-aspect ratio cells that arise from the tensor-product of the one-dimensional Gauss-Lobatto grids. Moreover, even if the GL-based FE problem could be solved with low work, the iteration count would still be higher than what is observed for the Schwarz-based approach. Third, to minimize cost, it is reasonable to have a smoother whose cost is on par with that of residual evaluation if it can substantially reduce the iteration count.

We illustrate the problem of high-aspect ratio cells by considering application of the two-level procedure (8) to the model Poisson problem (1) discretized on the unit square with an $8N \times 8N$ array of bilinear finite elements.

Table 1. MG method on FE problem

FE Spacing	No.	Smoother/ Preconditioner	Coarse Space	Iterations, 10^{-11} Reduction $N=4$	$N=8$	$N=12$	$N=16$
Uniform	a	Jacobi	$N/2$	39	38	38	38
	b	GSRB	$N/2$	9	9	9	9
	c	H Schwarz	$N/2$	40	41	42	42
	d	H Schwarz (W)	$N/2$	7	7	6	6
SE	e	Jacobi	$N/2$	41	84	148	219
	f	GSRB	$N/2$	11	28	46	65
	g	H Schwarz	$N/2$	40	43	47	52
	h	H Schwarz (W)	$N/2$	6	7	7	9

Iteration counts for four different smoothing strategies are shown in Table 1. Jacobi implies M^{-1}:=diag(A); GSRB is a Gauss-Seidel sweep with the nodes ordered into two maximally independent ("red-back") subsets; and H Schwarz and H Schwarz(W) correspond to the hybrid Schwarz-based smoothers introduced below. In all cases, σ is chosen such that the maximum eigenvalue of σMA is unity, and $\sigma_C = 1$. The coarse system is solved directly and is based on the same FE discretization, save that, in each direction, every other nodal point is discarded. The first set of results is for uniformly sized elements of length $1/8N$ on each side. Resolution-independent convergence is obtained for each of the smoothing strategies, with GSRB and H Schwarz(W) being the most competitive. Although H Schwarz(W) has a lower iteration count, GSRB requires less work per iteration, and the two are roughly equal in computational cost. The second set of results is for an $8N \times 8N$ array of bilinear elements whose vertices coincide with the GL node spacing associated with an 8×8 array of spectral elements of order N. In this case, the pointwise Jacobi and GSRB smoothers break down as N is increased. Only H Schwarz(W) retains performance comparable to the uniform grid case. We note that line-based relaxation strategies proposed by Shen et al. [2000] and Beuchler [2002] also compensate for the high-aspect-ratio cell difficulty. For the values of N considered here, however, the hybrid Schwarz approach is likely to be faster, at least on cache-based architectures, where the matrix-matrix product-oriented fast-diagonalization method is very effective.

Our hybrid Schwarz strategy is based on a multiplicative combination of an additive Schwarz smoother at the fine scale and a coarse-grid correction. The smoother, originally due to Dryja and Widlund [1987], is written as

$$M := \sum_{e=1}^{E} R_e^T A_e^{-1} R_e. \tag{10}$$

Here, R_e is the standard Boolean restriction matrix that extracts from a global nodal vector those values associated with the *interior* of the extended subdomain $\bar{\Omega}^e$. In all cases, $\bar{\Omega}^e$ is an extension of Ω^e that includes a single row

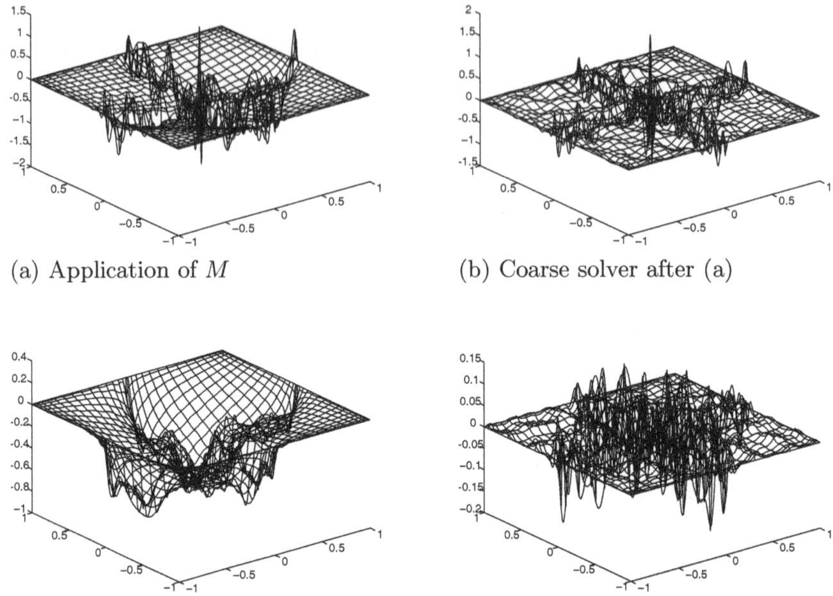

(a) Application of M

(b) Coarse solver after (a)

(c) Application of WM

(d) Coarse solver after (c)

Fig. 2. Error plots for the hybrid Schwarz preconditioner and coarse solve, with $N_C = N/2$ and $(E, N) = (4, 16)$, applied to a random initial guess

(or plane, in 3D) of nodal values in each of $2d$ directions. as illustrated in Fig. 1 (left). R_e^T extends by zero the vector of nodal values interior to $\bar{\Omega}^e$ to a full length vector. Multiplication by A_e^{-1} is effected by using the fast diagonalization method similar to (7). In a preprocessing step, one assembles *one-dimensional* stiffness and mass matrices, A_* and B_* ($* = x$, y or z), for each space dimension, $1, \ldots, d$; restricts these to the relevant ranges using a one-dimensional restriction matrix R_*^e; and solves an eigenvalue problem of the form $((R_*^e)^T A_* R_*^e) S_*^e = ((R_*^e)^T B_* R_*^e) S_*^e \Lambda_*^e$ to obtain the requisite eigenpairs (S_*^e, Λ_*^e). Because the spectral elements are compactly supported, the preprocessing step requires knowledge only of the size of the elements on either side of Ω^e, in each of the d directions. For subdomains that are not rectilinear, A_e is based on average lengths in each direction.

We have found it important to weight the solutions in the overlap region by the inverse of the diagonal counting matrix

$$C := \sum_{e=1}^{E} R_e^T R_e. \tag{11}$$

The entries of C enumerate the number of subdomains that share a particular vertex. Setting $W = C^{-1}$ gives rise to the weighted overlapping Schwarz smoother $M_W := WM$. Although convergence theory for the weighted

Schwarz method is yet to be developed, the methodology of Frommer and Szyld [2001] should be applicable to this setting as well. In addition to reducing the maximum eigenvalue of MA (which, by simple counting arguments, is $\max C_{ii}$; see Smith et al. [1996]), multiplication by W significantly improves the smoothing performance of the additive Schwarz step. This latter point is illustrated in Fig. 2, which shows the error when the two-level preconditioner (9) is applied to random right-hand-side vector for a 2×2 array of spectral elements with $N=16$. Figure 2(a) shows the error after a single application of the additive Schwarz smoother (10), with $\sigma=1$. While the solution is smooth in the interior, there is significant undamped error along the interface, particularly at the cross point. As noted by Lottes and Fischer [2004], the error along the interface can be reduced by choosing $\sigma = 1/4$, but the overall error is no longer smooth. In either case, the subsequent coarse-grid correction does not yield a significant error reduction. By contrast, the error after application of M_W, seen in Fig. 2(c), is relatively smooth, and the coarse-grid correction is very effective. Comparing the magnitudes in Figs. 2(b) and 2(d), one sees a tenfold reduction in the error through the introduction of W.

Table 2 presents convergence results for the Poisson problem on the square discretized with an 8×8 array of spectral elements. Case 2(a) shows results for the unweighted additive Schwarz preconditioner using an FE-based smoother. This scheme is the Poisson equivalent to the method developed by Fischer et al. [2000] for the pressure subproblem considered in the next section. For all the other cases, A_e is based on a restriction of A rather than on an FE discretization. Case 2(b) shows that this simple change considerably reduces the iteration count. Enriching the coarse space from $N_C = 1$ to $N/2$ and incorporating the weight matrix W yields further reductions in iteration count and work. (Because of symmetry requirements, W is applied as a pre- and postmultiplication by $W^{1/2}$ for the preconditioned conjugate gradient, PCG, cases). The work shown in the last column of Table 2 is an estimate of the number of equivalent matrix-vector products required to reduce the error by 10^{-11}. Rather than attempting to symmetrize the hybrid Schwarz method (9), we simply switched to GMRES, which allowed W to be applied directly during the summation of the overlapping solutions. Comparison of cases 2(f) and 2(h) underscores the importance of weighting.

4 Extension to Navier-Stokes

Efficient solution of the incompressible Navier-Stokes equations in complex domains depends on the availability of fast solvers for sparse linear systems. For unsteady flows, the pressure operator is the leading contributor to stiffness, as the characteristic propagation speed is infinite. Our pressure solution procedure involves two stages. First, we exploit the fact that we solve similar problems from one step to the next by projecting the current solution onto a subspace of previous solutions to generate a high-quality initial approxima-

Table 2. Iteration count for E=8×8 SE problem

Method	No.	Smoother/ Preconditioner	Coarse Space	Iterations, 10^{-11} Reduction $N=4$	$N=8$	$N=12$	$N=16$	Work $N=16$
PCG	a	A Schwarz (FE)	1	28	35	46	58	116
	b	A Schwarz	1	25	27	35	43	86
	c	A Schwarz	$N/2$	26	26	26	27	81
	d	A Schwarz (W)	1	17	24	33	43	86
	e	A Schwarz (W)	$N/2$	16	21	22	24	72
MG/ GMRES	f	H Schwarz	$N/2$	21	23	24	25	100
	g	H Schwarz (W)	1	14	20	29	36	108
	h	H Schwarz (W)	$N/2$	13	12	12	13	52

tion, as outlined in Fischer [1998]. We then compute the correction to this approximation using a scalable iterative solver. Here, we extend the multigrid methods presented in the preceding sections to computation of the pressure in SE-based simulations of incompressible flows.

To introduce notation, we review the Navier-Stokes discretization presented in detail in Fischer [1997]. The temporal discretization is based on a semi-implicit scheme in which the nonlinear term is treated explicitly and the remaining unsteady Stokes problem is solved implicitly. Our spatial discretization is based on the $\mathbb{P}_N - \mathbb{P}_{N-2}$ spectral element method of Maday and Patera [1989]. Assuming $\mathbf{f}^n$ incorporates all terms explicitly known at time t^n, the $\mathbb{P}_N - \mathbb{P}_{N-2}$ formulation of the Navier-Stokes problem reads: *Find* $(\mathbf{u}^n, p^n) \in X_N \times Y_N$ *such that*

$$\begin{aligned} \frac{1}{Re}(\nabla \mathbf{v}, \nabla \mathbf{u}^n)_{GL} + \frac{1}{\Delta t}(\mathbf{v}, \mathbf{u}^n)_{GL} - (\nabla \cdot \mathbf{v}, p^n)_G &= (\mathbf{v}, \mathbf{f}^n)_{GL}, \\ (q, \nabla \cdot \mathbf{u}^n)_G &= 0, \end{aligned} \qquad (12)$$

$\forall\, (\mathbf{v}, q) \in X_N \times Y_N$. The inner products $(.,.)_{GL}$ and $(.,.)_G$ refer to the Gauss-Lobatto-Legendre (GL) and Gauss-Legendre (G) quadratures associated with the spaces $X_N := [Z_N \cap H_0^1(\Omega)]^d$ and $Y_N := Z_{N-2}$, respectively, and Z_N is the space introduced in conjunction with (1). The local velocity basis is given, componentwise, by the form (2). The pressure is similar, save that the nodal interpolants are based on the N-1 Gauss points, $\eta_i \in (-1, 1)$, as illustrated in Fig. 1 (right).

Insertion of the SEM bases into (12) yields a discrete Stokes system to be solved at each step:

$$\mathbf{H}\,\underline{\mathbf{u}}^n - \mathbf{D}^T \underline{p}^n = \mathbf{B}\,\underline{\mathbf{f}}^n, \;\; \mathbf{D}\,\underline{\mathbf{u}}^n = 0. \qquad (13)$$

$\mathbf{H} = \frac{1}{Re}\mathbf{A} + \frac{1}{\Delta t}\mathbf{B}$ is the discrete equivalent of the Helmholtz operator, $(-\frac{1}{Re}\nabla^2 + \frac{1}{\Delta t})$; $-\mathbf{A}$ is the discrete Laplacian; $\mathbf{B}$ is the (diagonal) mass matrix associated with the velocity mesh; $\mathbf{D}$ is the discrete divergence operator, and $\mathbf{f}^n$ accounts for the explicit treatment of the nonlinear terms. Note that the Galerkin approach implies that the governing system in (13) is symmetric and that the matrices $\mathbf{H}$, $\mathbf{A}$, and $\mathbf{B}$ are all symmetric positive definite. We have used bold capital letters to indicate matrices that interact with vector fields.

The Stokes system (13) is advanced by using the operator splitting approach presented by Maday et al. [1990] and Perot [1993]. One first solves

$$\mathbf{H}\,\underline{\hat{\mathbf{u}}} = \mathbf{B}\,\underline{\mathbf{f}}^n + \mathbf{D}^T\,\underline{p}^{n-1}, \tag{14}$$

which is followed by a pressure correction step

$$E\,\delta\underline{p}^n = -\frac{1}{\Delta t}\mathbf{D}\underline{\hat{\mathbf{u}}}, \qquad \underline{\mathbf{u}}^n = \underline{\hat{\mathbf{u}}} + \Delta t\mathbf{B}^{-1}\mathbf{D}^T\delta\underline{p}^n, \qquad \underline{p}^n = \underline{p}^{n-1} + \delta\underline{p}^n, \tag{15}$$

where $E := \mathbf{D}\mathbf{B}^{-1}\mathbf{D}^T$ is the Stokes Schur complement governing the pressure in the absence of the viscous term.

E is the consistent Poisson operator for the pressure and is spectrally equivalent to A. Through a series of tests that will be reported elsewhere, we have found the following to be an effective multilevel strategy for solving E. We employ (9) to precondition GMRES with a weighted additive Schwarz smoother. The local problems are based on $E_e := \tilde{R}_e E \tilde{R}_e^T$, where the subdomains defined by the restriction matrices $\tilde{R}_e$ correspond to the minimal-overlap extension illustrated in Fig. 1 (right). The coarse-grid problem, A_C, is based on A with $N_C = N/2$ (typically), which was found not only to be cheaper but also better at removing errors along the element interfaces. At all intermediate levels, A_C^{-1} is approximated with a single V-cycle (8).

The local problems are solved using the fast diagonalization method, which requires that E_e (and therefore E) be separable. In two dimensions, we need to cast E in the form

$$E \;=\; J_y \otimes E_x \;+\; E_y \otimes J_x. \tag{16}$$

For simplicity, we assume that we have a single element with $\Omega = \widehat{\Omega}$ and ignore the details of boundary conditions. From the definition of E, we have

$$E \;=\; D_x B^{-1} D_x^T \;+\; D_y B^{-1} D_y^T. \tag{17}$$

The divergence and inverse mass matrices have the tensor-product forms

$$D_x = (\widetilde{B} \otimes \widetilde{B})(\widetilde{J} \otimes \widetilde{D}), \quad D_y = (\widetilde{B} \otimes \widetilde{B})(\widetilde{D} \otimes \widetilde{J}), \quad B^{-1} = (\widehat{B}^{-1} \otimes \widehat{B}^{-1}) \tag{18}$$

Here, $\widetilde{B}$=diag$(\tilde{\rho}_i)_{i=1}^{N-1}$ consists of the Gauss-Legendre quadrature weights, and $\widetilde{J}$ and $\widetilde{D}$ are respective interpolation and derivative matrices mapping from the GL points to the G points,

$$\widetilde{J}_{ij} = h_j^N(\eta_i), \qquad\qquad \widetilde{D}_{ij} = \left.\frac{dh_j^N}{dr}\right|_{r=\eta_i}. \tag{19}$$

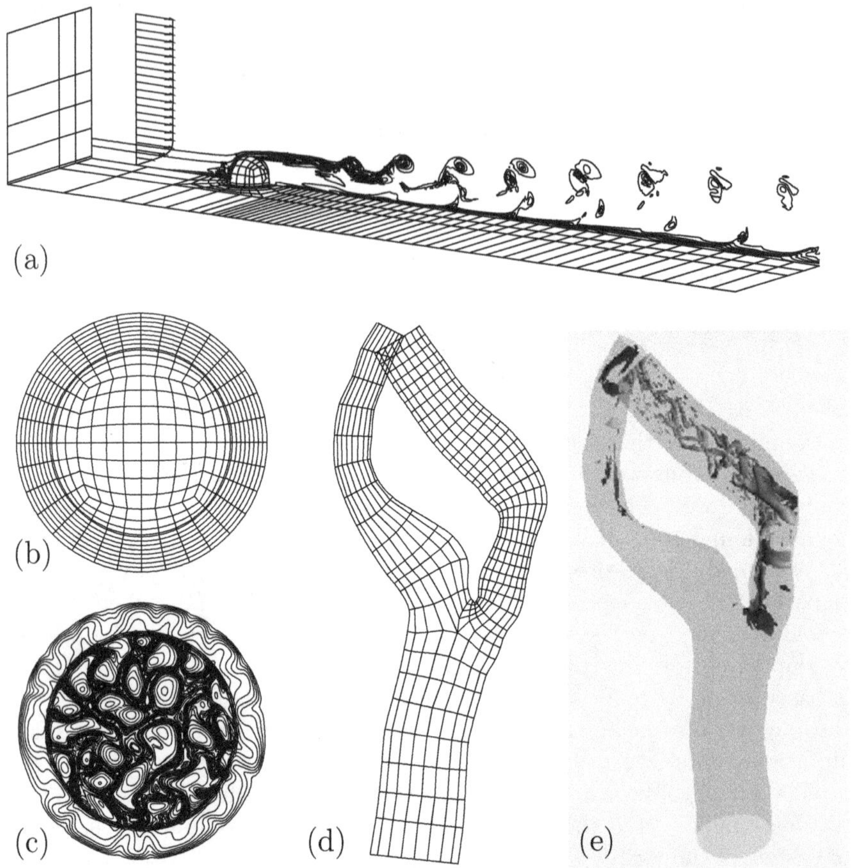

Fig. 3. SE Navier-Stokes examples: (a) $E = 1021$ mesh, inlet profile, and vorticity contours for roughness element; (b) $E = 1536$ mesh and (c) temperature contours for buoyancy driven convection; (d) $E = 2544$ mesh and (e) coherent structures for flow in a diseased carotid artery

Inserting (18) into (17) yields the desired form (16) with

$$J_x = J_y = \widetilde{B}\widetilde{J}\widehat{B}^{-1}\widetilde{J}^T\widetilde{B}^T, \qquad E_x = E_y = \widetilde{B}\widetilde{D}\widehat{B}^{-1}\widetilde{D}^T\widetilde{B}^T. \tag{20}$$

The extension to multiple elements follows by recognizing that the gather-scatter operator used to assemble the local matrices can be written as $Q = Q_y \otimes Q_x$ for a tensor-product array of elements. Following our element-centric solution strategy, we thus generate E_e by viewing Ω^e as being embedded in a 3^d array of rectilinear elements of known dimensions. Unlike A_e, the entries of E_e are also influenced by the "neighbors of neighbors." This influence, however, is small and is neglected.

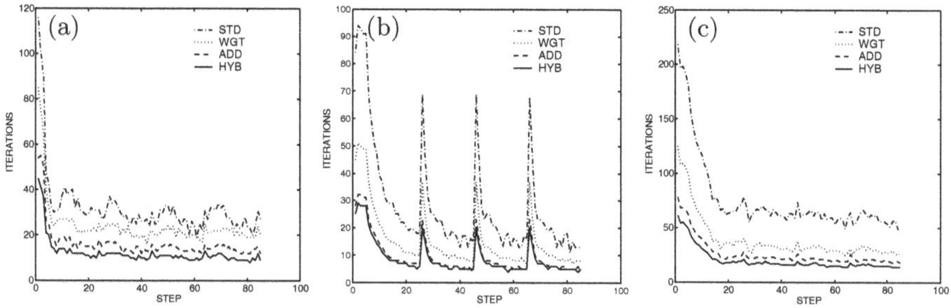

Fig. 4. Iteration histories for FE-based two-level (STD), weighted SE-based two-level (WGT), additive multilevel (ADD), and multiplicative multilevel (HYB) schemes for spectral elements simulations of order N=9: (a) hairpin vortex, E=1021; (b) hemispherical convection, E=1536; and (c) carotid artery simulation, E=2544

Navier-Stokes Results. We turn now to application of spectral element multigrid (SEMG) to the simulation of unsteady incompressible flows. In numerous 2D and 3D Navier-Stokes test problems, we have found the additive variant of the procedure outlined in the preceding section to be roughly two to three times faster than the two-level additive Schwarz method developed in Fischer et al. [2000]. A sample of these results is presented below.

We consider the three test cases shown in Fig. 3. The first case, Fig. 3(a), is boundary-layer flow past a hemispherical roughness element at Reynolds number Re=1000 (based on roughness height). The flow generates a pre-transitional chain of hairpin vortices evidenced by the spanwise vorticity contours shown in the symmetry plane. The second example, Fig. 3(b)-(c), is buoyancy-driven convection in a rotating hemispherical shell having inner radius 2.402 and outer radius 3.3. The Rayleigh number (based on shell thickness) is Ra=20,000 and the Taylor number is Ta=160,000. The third case, Fig. 3(d)-(e), simulates transitional flow in a diseased carotid artery. The severe stenosis in the internal (right) branch results in high flow velocities and, ultimately, transition to turbulence. Figure 3(e) shows the coherent structures that arise just before peak systole.

Figure 4 shows the pressure iteration history for the first 85 timesteps of the three examples, using the initial conditions of Fig. 3. For all cases, N=9 and the coarse problem is based on linear elements whose vertices are derived from an oct-refinement of the SE mesh. Four methods are considered: STD refers to the two-level additive Schwarz method of Fischer et al. [2000]; WGT is the same as STD, save that E_e is based on a restriction of E, rather than an FE-based discretization of the Poisson problem, and that the weight matrix W is included; ADD is the same as WGT, save that three levels are employed, with N_{mid}=5; HYB is the same as ADD, save that the multiplicative variant of (9) is used. PCG is used for STD and WGT, whereas GMRES is used for ADD and HYB. Although HYB requires fewer iterations, ADD is the fastest method

Table 3. ADD Avg. Iteration Count for Navier-Stokes Examples

Problem	N=5	N=7	N=9	N=11	N=13	N=15	N=17
Hairpin Vortex	9.8	11.1	15.1	17.5	20.4	23.5	26.1
Spherical Conv.	8.2	7.8	8.3	8.9	9.9	10.9	11.6
Carotid Artery	18.5	20.6	23.7	26.0	29.3	32.5	36.0
Carotid (WGT)	16.5	22.2	30.0	39.5	48.4	59.4	65.8

because it requires only one product in E per iteration. The prominent spikes in Fig. 4(b) result from resetting the projection basis set (Fischer [1998]).

The scalability of the three-level ADD method is illustrated in Table 3, which shows the average iteration count over the last 20 steps for varying N with $N_{\text{mid}}=N/2$. Order-independence is not assured in complex domains, particularly if the mesh contains high aspect-ratio elements (Fischer [1997]). The performance of ADD is nonetheless quite reasonable when one considers that the number of pressure nodes varies by a factor of 64 in moving from $N =$ 5 to 17. For purposes of comparison, results for the WGT method are shown for the carotid. The additional level of ADD clearly reduces order dependence.

Acknowledgement. This work was supported by the Mathematical, Information, and Computational Sciences Division subprogram of the Office of Advanced Scientific Computing Research, Office of Science, U.S. Department of Energy, under Contract W-31-109-Eng-38.

References

S. Beuchler. Multigrid solver for the inner problem in domain decomposition methods for p-fem. *SIAM J. Numer. Anal.*, 40:928–944, 2002.

M. A. Casarin. Quasi-optimal Schwarz methods for the conforming spectral element discretization. *SIAM J. Numer. Anal.*, 34:2482–2502, 1997.

W. Couzy. *Spectral Element Discretization of the Unsteady Navier-Stokes Equations and its Iterative Solution on Parallel Computers.* PhD thesis, Swiss Federal Institute of Technology-Lausanne, 1995. Thesis nr. 1380.

M. O. Deville, P. F. Fischer, and E. H. Mund. *High-order methods for incompressible fluid flow.* Cambridge University Press, Cambridge, 2002.

M. Dryja and O. B. Widlund. An additive variant of the Schwarz alternating method for the case of many subregions. Technical Report TR 339, Courant Inst., NYU, 1987. Dept. Comp. Sci.

P. F. Fischer. An overlapping Schwarz method for spectral element solution of the incompressible Navier-Stokes equations. *J. Comput. Phys.*, 133:84–101, 1997.

P. F. Fischer. Projection techniques for iterative solution of $Ax = b$ with successive right-hand sides. *Comput. Methods Appl. Mech. Engrg.*, 163: 193–204, 1998.

P. F. Fischer, N. I. Miller, and H. M. Tufo. An overlapping Schwarz method for spectral element simulation of three-dimensional incompressible flows. In P. Bjørstad and M. Luskin, editors, *Parallel Solution of Partial Differential Equations*, pages 158–180, Berlin, 2000. Springer.

A. Frommer and D. B. Szyld. An algebraic convergence theory for restricted additive Schwarz methods using weighted max norms. *SIAM Journal on Numerical Analysis*, 39:463–479, 2001.

J. W. Lottes and P. F. Fischer. Hybrid multigrid/Schwarz algorithms for the spectral element method. *J. Sci. Comput.*, (to appear), 2004.

R. E. Lynch, J. R. Rice, and D. H. Thomas. Direct solution of partial difference equations by tensor product methods. *Numer. Math.*, 6:185–199, 1964.

Y. Maday and R. Muñoz. Spectral element multigrid: Numerical analysis. *J. Sci. Comput.*, 3:323–354, 1988.

Y. Maday, R. Muñoz, A. T. Patera, and E. M. Rønquist. Spectral element multigrid methods. In P. de Groen and R. Beauwens, editors, *Proc. of the IMACS Int. Symposium on Iterative Methods in Linear Algebra, Brussels, 1991*, pages 191–201, Amsterdam, 1992. Elsevier.

Y. Maday and A. T. Patera. Spectral element methods for the Navier-Stokes equations. In A. K. Noor and J. T. Oden, editors, *State-of-the-Art Surveys in Computational Mechanics*, pages 71–143. ASME, New York, 1989.

Y. Maday, A. T. Patera, and E. M. Rønquist. An operator-integration-factor splitting method for time-dependent problems: Application to incompressible fluid flow. *J. Sci. Comput.*, 5:263–292, 1990.

S. A. Orszag. Spectral methods for problems in complex geometry. *J. Comput. Phys.*, 37:70–92, 1980.

S. S. Pahl. Schwarz type domain decomposition methods for spectral element discretizations. Master's thesis, Univ. of Witwatersrand, Johannesburg, South Africa, 1993. Dept. of Computational and Applied Math.

J. B. Perot. An analysis of the fractional step method. *J. Comp. Phys.*, 108: 51–58, 1993.

E. Rønquist. *Optimal Spectral Element Methods for the Unsteady Three-Dimensional Incompressible Navier-Stokes Equations.* PhD thesis, Massachusetts Institute of Technology, 1988. Cambridge, MA.

E. M. Rønquist and A. T. Patera. Spectral element multigrid, I:. Formulation and numerical results. *J. Sci. Comput.*, 2:389–406, 1987.

J. E. Shen. Efficient Chebyshev-Legendre Galerkin methods for elliptic problems. In A. V. Ilin and L. R. Scott, editors, *Third Int. Conference on Spectral and High Order Methods*, pages 233–239. Houston Journal of Mathematics, 1996.

J. E. Shen, F. Wang, and J. Xu. A finite element multigrid preconditioner for Chebyshev-collocation methods. *Appl. Numer. Math.*, 33:471–477, 2000.

B. Smith, P. Bjørstad, and W. Gropp. *Domain Decomposition: Parallel Multilevel Methods for Elliptic PDEs.* Cambridge University Press, Cambridge, 1996.

Numerical Approximation of Dirichlet-to-Neumann Mapping and its Application to Voice Generation Problem

Takashi Kako and Kentarou Touda

The University of Electro-Communications, Department of Computer Science
(`http://www.im.uec.ac.jp/~kako/EngIntro.html`)

Summary. In this paper, we treat the numerical method for the Helmholtz equation in unbounded region with simple cylindrical or spherical shape outside some bounded region and apply the method to voice generation problem. The numerical method for the Helmholtz equation in unbounded region is based on the domain decomposition technique to divide the region into a bounded region and the rest unbounded one. We then treat the approximation of the artificial boundary condition given through the DtN mapping on the artificial boundary. We apply the finite element approximation to discretize the problem. In applying the method to the voice generation problem, it is essential to compute the frequency response function or the formant curve. We give variational formulas for the resolvent poles with respect to the variation of vocal tract boundary which determine the peaks of frequency response function known as formants, and we propose the use of variational formulas to design the location of formants.

1 Numerical Method for Exterior Helmholtz Problem

1.1 Formulation of Exterior Helmholtz Problem

We consider the following 2-dimensional exterior Helmholtz equation with unknown function u as the mathematical model for the time stationary problem of outgoing or radiation sound wave propagation in unbounded region outside an obstacle:

$$-\Delta u - k^2 u = 0 \qquad \text{in } \Omega = R^2 \backslash \mathcal{O}, \tag{1}$$

$$\frac{\partial u}{\partial n} = g \qquad \text{on } \partial\Omega, \tag{2}$$

$$\lim_{r\to\infty} \sqrt{r}\left(\frac{\partial u}{\partial r} - \mathbf{i}ku\right) = 0, \qquad \mathbf{i} = \sqrt{-1}, \tag{3}$$

where Ω is the interior of the complement of a bounded obstacle $\mathcal{O}$ in R^2 with smooth boundary $\partial\Omega$ on which the Neumann boundary condition (2) is imposed with an inhomogeneous data g. A real constant k is called a wave number and the condition (3) is the Sommerfeld radiation condition at infinity which excludes any unphysical incoming wave. In the case with tubular cylindrical outside region, a similar formulation is possible with necessary modification of boundary condition and radiation condition (see Section 2.2).

Related to this problem in unbounded region, we introduce a circular artificial boundary Γ_R with radius R and decompose the original domain into two sub-domains. We then consider the boundary value problem in the part of bounded sub-domain given as

$$-\Delta u - k^2 u = 0 \qquad \text{in } \Omega_R \equiv \Omega \cap B_R, \tag{4}$$

$$\frac{\partial u}{\partial n} = g \qquad \text{on } \partial\Omega, \tag{5}$$

$$\frac{\partial u}{\partial r} = Mu \qquad \text{on } \Gamma_R, \tag{6}$$

where $B_R \supset\supset \mathcal{O}$ is a circular domain with radius R bounded by an artificial circular boundary Γ_R: $B_R = \{(x,y) \mid r \equiv \sqrt{x^2+y^2} < R\}$, and M is a differential or pseudo-differential operator which we construct as a function of the operator $\partial^2/\partial\theta^2$ of angular variable θ in order to make the problem exactly or approximately equivalent to the original problem.

The exact solution $u(r,\theta)$ for (1)-(3) in B_R^c is given by the following formula with the Hankel function of the first kind of order n which we will denote in this paper by $H^{(1)}(\cdot\,;\,n)(= H_n^{(1)}(\cdot))$:

$$u(r,\theta) = \frac{1}{2\pi} \sum_{n=-\infty}^{\infty} \frac{H^{(1)}(kr;n)}{H^{(1)}(kR;n)} \int_0^{2\pi} u(R,\phi) e^{\mathrm{i}n(\theta-\phi)} d\phi.$$

Using this expression, we introduce the exact Dirichlet-to-Neumann (DtN) mapping as:

$$M_{exact}u(\theta) \equiv \frac{k}{2\pi} \sum_{n=-\infty}^{\infty} \frac{H^{(1)'}(kR;n)}{H^{(1)}(kR;n)} \int_0^{2\pi} u(R,\phi) e^{\mathrm{i}n(\theta-\phi)} d\phi, \tag{7}$$

which relates the Dirichlet data of the solution u on the artificial boundary Γ_R to the Neumann data on the same boundary. If we put $M = M_{exact}$ in (6), the problem (4)-(6) is equivalent to the original problem (1)-(3). The DtN mapping can also be expressed as the function of the elliptic operator D^2 as:

$$M_{exact} = M_{DtN}(D^2) = k\frac{H^{(1)'}(kR;\sqrt{D^2})}{H^{(1)}(kR;\sqrt{D^2})}, \qquad D \equiv -\mathrm{i}\partial/\partial\theta. \tag{8}$$

1.2 Radiation Boundary Conditions

There have been many studies related to the analytical as well as numerical approximations of the DtN mapping. Among them, Engquist and Majda

[1977], Engquist and Majda [1979] introduced a series of non-local approximate radiation boundary condition such as:

$$\begin{aligned} M_1(D^2) &= \frac{\mathbf{i}}{R}\sqrt{k^2R^2 - D^2}, \\ M_2(D^2) &= M_1(D^2) - \frac{1}{2R}\frac{k^2R^2}{(k^2R^2 - D^2)} \end{aligned}$$

and so forth. Some local approximate radiation boundary conditions are well used and they are given as

$$\begin{aligned} M_{1,1}(D^2) &= \mathbf{i}k, \\ M_{2,1}(D^2) &= \mathbf{i}k - \frac{1}{2R}. \end{aligned}$$

Those are derived directly from the Sommerfeld radiation condition. Feng [1983] introduced a series of local type operators such as

$$\begin{aligned} F_3(D^2) &= \mathbf{i}k - \frac{1}{2R} + \frac{\mathbf{i}}{8kR^2} - \frac{\mathbf{i}}{2kR^2}D^2, \\ F_4(D^2) &= \mathbf{i}k - \frac{1}{2R} + \frac{\mathbf{i}}{8kR^2} + \frac{1}{8k^2R^3} - \left(\frac{\mathbf{i}}{2kR^2} + \frac{1}{2k^2R^3}\right)D^2. \end{aligned}$$

Bayliss and Turkel [1980] also introduced in a systematic way a hierarchy of local operators:

$$L_n = \prod_{j=1}^{n}\left(\frac{\partial}{\partial r} - \mathbf{i}k + \frac{4j-3}{2r}\right), n \geq 1.$$

On the other hand, related to the finite element method, Kako and Kano [1999] proposed a non-local approximation by a bounded operator as a higher order correction:

$$M_{LK}(D^2) = \mathbf{i}k - \frac{3}{2R} + \frac{1}{R}\left[1 + \frac{1}{2\mathbf{i}kR}\left(\frac{1}{4} - D^2\right)\right]^{-1}.$$

According to our numerical experiments, the exact DtN operator is the best one when it is combined with the appropriate discrete approximation. In the next subsection, we will briefly review the recent results of Nasir et al. [2003] based on a mixed method approximation.

1.3 Finite Element Approximation

To formulate the finite element method for the Helmholtz problem, we introduce a function spaces $V \equiv H^1(\Omega_R)$. Then the weak formulation of the problem is to find $u \in V$ such that

$$a(u,v) - \langle u,v\rangle_M = (g,v)_{\partial\Omega} \qquad \forall v \in V \tag{9}$$

with

$$a(u,v) = \int_{\Omega_R} \left(\frac{\partial u}{\partial r}\frac{\partial \bar{v}}{\partial r} + \frac{1}{r^2}\frac{\partial u}{\partial \theta}\frac{\partial \bar{v}}{\partial \theta} - k^2 u\bar{v} \right) r dr d\theta,$$

$$\langle u,v\rangle_M = \int_0^{2\pi} (Mu)\bar{v}\, R d\theta, \qquad (f,g)_{\partial\Omega} = \int_{\partial\Omega} f\bar{g} d\sigma,$$

where M is one of the operators appeared in the previous section. Let us introduce a finite dimensional subspace V_h of V. Then the finite element approximation is to find $u_h \in V_h$ such that

$$a(u_h,v_h) - \langle u_h,v_h\rangle_M = (g,v_h)_{\partial\Omega}, \qquad \forall v_h \in V_h. \tag{10}$$

In the following, we will introduce the fictitious domain method combined with a fast direct method. For this purpose, we firstly treat a special problem with annulus region.

1.4 Fast Direct Method

In case that Ω_R is an annulus region, we can make a separation of variables with respect to the radial and angular coordinates, and by dividing the intervals into n_r subintervals in radial direction and into n_θ in angular direction, the finite element method gives a linear system:

$$\mathbf{BU} = \mathbf{F} \tag{11}$$

with a separable matrix $\mathbf{B} = (b_{IJ})$, a given vector $\mathbf{f} = (f_I)$ and a solution vector $\mathbf{u} = (u_J)$ where $b_{IJ} = a(\Phi_J, \Phi_I) - \langle \Phi_J, \Phi_I\rangle_M$ and $f_I = (g, \Phi_I)_{\partial\Omega}$. The matrix $\mathbf{B}$ is given by a tensor product form:

$$\mathbf{B} = \mathbf{R}_2 \otimes \mathbf{T}_1 + \mathbf{R}_1 \otimes \mathbf{T}_2 - k^2\mathbf{R}_1 \otimes \mathbf{T}_1 - \mathbf{e}_{n_r}^{n_r}\mathbf{e}_{n_r}^{n_r T} \otimes \mathbf{M}$$

with a tri-diagonal matrices $\mathbf{R}_i \in C^{n_r\times n_r}$ $(i = 1,2)$ and circulant matrices $\mathbf{T}_i$ $(i = 1,2)$ and $\mathbf{M} \in C^{n_\theta \times n_\theta}$. The matrix $\mathbf{M}$ corresponds to the radiation boundary condition and $\mathbf{e}_j^n$ denotes the usual jth canonical basis vector of R^n. Explicit forms for the matrices for $\mathbf{R}_i, \mathbf{T}_i$ and $\mathbf{M}$ can be found, for example, in Ernst [1996].

To solve the system (11), a fast direct solution method based on separation of variables can be used by diagonalizing the circulant matrices. This leads to

$$\begin{aligned}(\mathbf{I}_{n_r} \otimes \mathbf{Q}^H)(\mathbf{R}_2 \otimes \mathbf{\Lambda}_1 + \mathbf{R}_1 \otimes \mathbf{\Lambda}_2 - k^2\mathbf{R}_1 \otimes \mathbf{\Lambda}_1 \\ -\mathbf{e}_{n_r}^{n_r}\mathbf{e}_{n_r}^{n_r T} \otimes \mathbf{\Lambda}_M)(\mathbf{I}_{n_r} \otimes \mathbf{Q})\mathbf{u} = \mathbf{f},\end{aligned}$$

where $\mathbf{\Lambda}_i, i = 1,2$ and $\mathbf{\Lambda}_M$ are diagonal matrices consisting of eigenvalues of the corresponding circulant matrices and $\mathbf{Q}$ is the Fourier matrix with

ijth component given by $e^{2\pi \mathrm{i} ij/n_\theta}$. $\mathbf{Q}^H$ is the Hermitian conjugate of $\mathbf{Q}$. The column vectors of $\mathbf{Q}$ are the eigenvectors of the circulant matrices and $\mathbf{I}_{n_r}$ is the identity matrix of size n_r. Then, the system reduces to n_θ independent tri-diagonal systems

$$(\lambda_1^j \mathbf{R}_2 + \lambda_2^j \mathbf{R}_1 - k^2 \lambda_1^j \mathbf{R}_1 - \lambda_M^j \mathbf{e}_{n_r}^{n_r} \mathbf{e}_{n_r}^{n_r T}) \hat{\mathbf{u}}^j = \hat{\mathbf{f}}^j,$$
$$j = 0, 1, \cdots, n_\theta - 1$$

where $\hat{\mathbf{u}}^j$ and $\hat{\mathbf{f}}^j$ are the vectors composed of the jth components from the n_r blocks of the discrete Fourier transform of $\mathbf{u}$ and $\mathbf{f}$ respectively. The discrete Fourier transforms can be performed efficiently by using FFT and the solution vector $\mathbf{u}$ is obtained by inverse FFT. Here, one needs only the knowledge of the eigenvalues of the matrices $\mathbf{T}_\mathrm{i}$, $i = 1, 2$ and $\mathbf{M}$. These are circulant matrices for which simple analytical expressions of the eigenvalues can be obtained. For example, for a circulant matrix of the form $[c_0, c_1, \cdots, c_{n-1}]$, the jth eigenvalue is given by $\lambda^j = \sum_{k=0}^{n-1} c_k e^{-2\pi \mathrm{i} jk/n}$.

For a general non-circular obstacle, we use the fictitious domain method to solve the original problem where the matrix B in this subsection is used as a preconditioner (see for details Heikkola et al. [1998] and Nasir [2003]).

1.5 A Mixed-Type Method

As we will see later, to apply the standard finite element method to the exact DtN mapping, we need to compute an infinite sum which should be truncated in some way. In this section, we introduce an alternative way to approximate the DtN mapping by the idea of mixed method.

Let $a(.,.)$ be a bounded V-elliptic sesqui-linear form defined in a Hilbert space V with inner-product $(.,.)_V$ and let A be a corresponding operator from V into V given as $a(u,v) = (Au, v)_V$. Let $b(u,v)$ be another bounded sesqui-linear form in V and let B be defines as $b(u,v) = (Bu, v)_V$. Assume that B is one-to-one and positive. Let L be $L = B^{-1}A$ with $\mathcal{D}(L) = \{u \mid Au \in R(B)\}$. Consider the problem to find $u \in V$ for a given $f \in V$ such that

$$L^n u = f, \tag{12}$$

where n is a non-negative integer. By introducing intermediate unknowns $p_j, j = 1, 2, \cdots, n-1$, the problem (12) can be equivalently written as

$$Ap_{n-1} = Bf, \;\; Ap_{n-2} - Bp_{n-1} = 0, \cdots, Ap_1 - Bp_2 = 0, \;\; Au - Bp_1 = 0.$$

The weak formulation of this problem is to find $p_j \in V, j = 0, 1, \cdots, n-1$ such that for each $j = 0, 1, \cdots, n-1$,

$$a(p_j, q_j) - b(p_{j+1}, q_j) = 0, \qquad \forall q_j \in V, \tag{13}$$

where $u = p_0 \in V$ and $p_n = f \in V$. When we pose the problem in a finite dimensional subspace V_h of V and use a given basis $\{\psi_i\}_0^{N_h}$, the approximate problem is given by the block matrix equation in $C^{N_h}, N_h = \dim V_h$ of the form

$$[A]_h P_j - [B]_h P_{j+1} = 0, \qquad j = 0, 1, \cdots, n-1, \tag{14}$$

where the matrices $[A]_h$ and $[B]_h$ are given by

$$([A]_h)_{mn} = a(\psi_n, \psi_m), \qquad ([B]_h)_{mn} = b(\psi_n, \psi_m)$$

and $P_j = [P_{j1}, \cdots, P_{jN_h}]^T$ are the vector of coefficients in the representation of $p_{j,h} \in V_h$ with respect to the basis: $p_{j,h} = \sum_{k=0}^{N_h} P_{jk}\psi_k$.

Eliminating the block components $P_j, j = 1, 2, \ldots n-1$, we have

$$[A]_h([B]_h^{-1}[A]_h)^{n-1}U = ([A]_h[B]_h^{-1})^n[B]_h U = F$$

where $U = P_0$ and $F = P_n$. The mixed-type approximation matrix of L^n is then given by $([A]_h[B]_h^{-1})^n[B]_h$.

From these observations, we introduce a mixed-type approximation for a function $M(L)$ of operator L as follows.

Definition (A Mixed-Type Method) Let V_h be a finite dimensional subspace of V. Then, a mixed-type approximation matrix for the sesqui-linear form $\langle u, v\rangle_M = b(M(L)u, v)$, denoted by $[M(L)]_h$, with respect to a given basis of V_h is defined as

$$[M(L)]_h = M([A]_h[B]_h^{-1})[B]_h. \tag{15}$$

1.6 Application to Radiation Boundary Condition

Let S be the unit circle. We choose the space $V = H^1(S)$ with the inner-product $(u, v)_V \equiv (u', v') + (u, v)$ where $(.,.)$ is the L_2-inner-product: $(u, v) \equiv \int_0^{2\pi} u\bar{v}d\theta$. Let us define sesqui-linear forms as

$$a(u, v) \equiv (u', v') \quad \text{and} \quad b(u, v) \equiv (u, v).$$

Then the operator L becomes D^2.

We denote by $\mathrm{Circ}(a, b, c)$ a circulant matrix for which the main diagonal is formed by b and lower and upper diagonals are formed by a and c respectively. For the mixed-type approximation matrix given by piecewise linear continuous Lagrangian basis functions defined on a uniform partition with interval size $h_\theta = 2\pi/n_\theta$, we have

$$[A]_h = \frac{1}{h_\theta}\mathrm{Circ}(-1, 2, -1) \quad \text{and} \quad [B]_h = \frac{h_\theta}{6}\mathrm{Circ}(1, 4, 1).$$

The eigenvalues of $[B]_h$ and $[A]_h$ are then given by

$$\lambda_B^j = \frac{h_\theta}{3}(2 + \cos jh_\theta) \text{ and } \lambda_A^j = \frac{2}{h_\theta}(1 - \cos jh_\theta).$$

The eigenvalues of $[M(L)]_h = [M(D^2)]_h$ is hence given as

$$\lambda_M^j = M(\nu_j^2)\lambda_B^j \quad \text{with} \quad \nu_j^2 = \frac{\lambda_A^j}{\lambda_B^j} = \frac{1}{h_\theta^2}\frac{6(1-\cos jh_\theta)}{2+\cos jh_\theta}. \tag{16}$$

For the exact DtN mapping (8) and the corresponding bilinear form $\langle u, v\rangle_M$, the eigenvalues of the mixed-type approximation matrix is

$$\lambda_{DtN}^j = kR\frac{H'(kR;\nu_j)}{H(kR;\nu_j)}\lambda_B^j. \tag{17}$$

This can be directly computed without the Hankel function and its derivative by using the continued fraction(see Nasir [2003])

$$x\frac{H^{(1)'}(x;\nu)}{H^{(1)}(x;\nu)} = \mathbf{i}x - \frac{1}{2} + \mathbf{i}\frac{(1/2)^2-\nu^2}{2(x+\mathbf{i})+}\frac{(3/2)^2-\nu^2}{2(x+2\mathbf{i})+}\cdots. \tag{18}$$

where $x = kR$.

We can prove the theoretical convergence and the practical computational efficiency of this type of mixed method in the case of the DtN mapping (see Nasir et al. [2003]).

Here we remark the standard finite element method for the DtN mapping (see for details Nasir [2003]). When we use piecewise linear continuous functions $\{\phi_i\}_0^{n_\theta-1}$ as the basis for the finite element approximation, the ijth entry in the finite element matrix is given by

$$\langle \phi_j, \phi_i\rangle_M = \frac{1}{2\pi}\sum_{n=-\infty}^{\infty} RM(n^2)(\phi_j, e_n)\overline{(\phi_i, e_n)}$$

$$= \frac{RM(0)h_\theta^2}{2\pi} + \frac{1}{\pi}\sum_{n=1}^{\infty} RM(n^2)\frac{4(1-\cos nh_\theta)^2}{n^4 h_\theta^2}\cos nh_\theta(j-i).$$

One advantage of the mixed method is that there is no infinite sum in the method and hence there is no ambiguity how to truncate the infinite series which exists in the standard finite element method. Furthermore, the computational cost of the mixed method is smaller than the one for the standard finite element method, although the performance is similar to the standard one which is known to be the best among all the methods using artificial boundary conditions.

2 Numerical Simulation of Voice Generation

The basic mathematical models of voice generation are the standard wave equation in unbounded domain and the corresponding Helmholtz equation for the frequency response problem.

As for the boundary conditions of the problem, we impose the inhomogeneous Dirichlet or Neumann boundary condition on the vocal cord part as the sound source and the homogeneous Neumann condition on the rest part of the vocal tract surface which can be modeled as the rigid wall. Imposing the artificial Dirichlet to Neumann type boundary condition on the common artificial boundary between the interior and exterior regions, we can reduce the problem into the one inside the artificial boundary.

In the voice generation process of vowels, vocal cord vibrates nonlinearly and makes a (linear) sound wave with a fixed pitch (or a basic time frequency) having a characteristic Fourier spectrum. Articulation organs like tongue and jaw make a modification of the Fourier spectrum as a kind filter and produce different kind of vowels depending on different shapes of oral cavities (see Figure 1 for the two-dimensional model shape of Japanese and also Russian vowel $/i/$).

Formants are the peaks of the frequency response function. The cognition of different vowels are made by detecting the position of formants according to the experiments. Especially the lowest two or three formants are very important for the cognition.

For the general background of the voice generation problem in human speech, see Flanagan [1972] and Furui [1989].

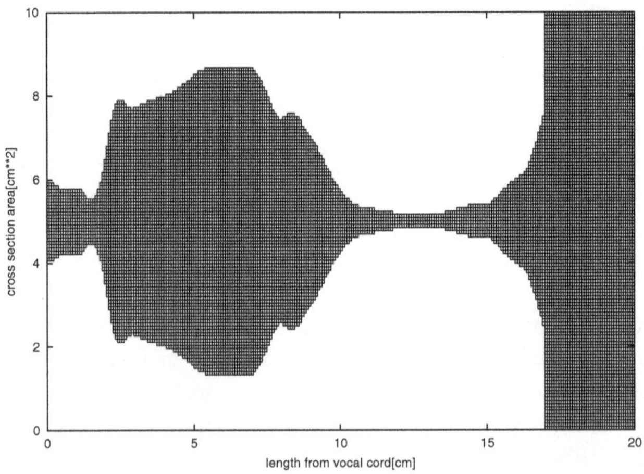

Fig. 1. Two-dimensional shape of vocal tract for vowel /i/

2.1 Mathematical Model of Voice Generation

Let Ω be a domain in R^n consisting of the bounded part Ω_i which corresponds to the vocal tract and the exterior unbounded part Ω_e:

$$\overline{\Omega} = \overline{\Omega_i} \cup \overline{\Omega_e}.$$

The sound pressure $u(t,x)$ in Ω with the source terms f, g satisfies the following wave equation:

$$\left(\frac{\partial^2}{\partial t^2} - c^2 \Delta\right) u(t,x) = f(t,x) \quad \text{in } R \times \Omega, \quad \Delta = \sum_{i=1}^{2} \frac{\partial^2}{\partial x_i^2}, \tag{19}$$

$$\frac{\partial u}{\partial n}(t,x) = g(t,x) \quad \text{on } R \times \partial\Omega, \tag{20}$$

with sound velocity c, and $\partial/\partial n$ denotes the outward unit normal on the boundary $\partial\Omega$ of Ω.

The time harmonic problem associated with this equation with the time frequency ω is formulated as the reduced wave equation or the Helmholtz equation for the stationary pressure $u(x)$:

$$(-\Delta - k^2) u(x) = f(x) \ \text{ in } \Omega, \quad k = \omega/c \tag{21}$$

$$\frac{\partial u}{\partial n}(x) = g(x) \ \text{ on } \partial\Omega. \tag{22}$$

We have to impose an appropriate radiation condition at infinity that exclude unphysical incoming waves. In the following, we treat the problem in one or two dimensional case.

2.2 One and Two Dimensional Vocal Tract Models and DtN Mapping

To reduce the problem in unbounded domain into the one in a bounded domain Ω_i, we introduce an artificial boundary condition on the common boundary Γ of the exterior and the interior domains: $\Gamma \equiv \overline{\Omega_i} \cap \overline{\Omega_e}$.

For this purpose, we firstly give the Dirichlet data $u|_\Gamma$ on Γ and solve the exterior problem, then we have the directional derivative $\partial u/\partial n$ of the solution along the exterior normal. This mapping form $u|_\Gamma$ to $\partial u/\partial n \equiv Mu|_\Gamma$ is the Dirichlet to Neumann mapping in this case.

The original problem then becomes equivalent to boundary value problem in the interior domain:

$$(-\Delta - k^2) u = 0 \ \text{ in } \Omega_i \tag{23}$$

$$\frac{\partial u}{\partial n} = g \ \text{ on } \ \partial\Omega \cap \overline{\Omega_i} \tag{24}$$

$$\frac{\partial u}{\partial n} = Mu|_\Gamma \ \text{ on } \Gamma.$$

In the tubular exterior infinite domain case (see Kako and Kano [1999]), we have

$$M u|_\Gamma = \sum_{n=0}^{\infty} \zeta_n C_n(u|_\Gamma) c_n(y),$$

$$C_n(u|_\Gamma) = \int_0^{y_0} u(L, y) c_n(y)\, dy \quad (n \geq 0),$$

$$c_n(y) = \begin{cases} \sqrt{\frac{1}{y_0}} & (n = 0) \\ \sqrt{\frac{2}{y_0}} \cos(\frac{n\pi}{y_0} y) & (n \geq 1), \end{cases}$$

$$\zeta_n = \begin{cases} i\xi_n, & \xi_n = \{k^2 - (\frac{n\pi}{y_0})^2\}^{1/2}, \quad 0 \leq n < \frac{y_0}{\pi} k \\ -\eta_n, & \eta_n = \{(\frac{n\pi}{y_0})^2 - k^2\}^{1/2}, \quad \frac{y_0}{\pi} k \leq n. \end{cases}$$

On the other hand, the following one dimensional simpler mathematical model is also used for the voice simulation problem and it is called the Webster's horn equation which is derived from the conservation law:

$$-\frac{\partial v}{\partial t} = \frac{A(x)}{\rho} \frac{\partial u}{\partial x}, \quad -\frac{\partial u}{\partial t} = \frac{\rho c^2}{A(x)} \frac{\partial v}{\partial x},$$

as

$$\frac{\partial^2 u}{\partial t^2} - \frac{1}{A(x)} c^2 \frac{\partial}{\partial x}(A(x) \frac{\partial u}{\partial x}) = 0,$$

here $A(x)$ denotes the cross sectional area of the vocal tract. The corresponding time harmonic problem for each wave number $k = \omega/c$ is given as

$$-\frac{1}{A(x)} \frac{d}{dx}(A(x) \frac{du}{dx}) - k^2 u = 0,$$

$$\frac{du}{dx}(0) = 1, \quad \frac{du}{dx}(L) = iku(L).$$

The finite element method can be applied in the similar way as in the case of obstacle problem presented in Section 1, and we can obtain some theoretical results for the convergence of the numerical solution to the exact one in case of the standard method (see Kako and Kano [1999]). We can also use the mixed type method in this case.

2.3 Complex Resonance Eigenvalue and Variational Formula

For the numerical simulation of voice generation, we change the frequency ω and hence the wave number k, and calculate the amplification factor computed as the sound pressure at some observation point for the given unit volume velocity of sound at the vocal cord position where the sound originates. The observation point should be located outside of the vocal tract region, and, in the following examples, we choose the point just on the artificial boundary.

This correspondence between the frequency and the amplification factor is the frequency response function or the formant curve. The peaks of this curve are called formants. A typical example of the frequency response function is shown in Figure 2 in the case of vowel $/i/$ for one and two dimensional cases. In two-dimensional computation based on DtN mapping, we used the shape in Figure 1, and applied the finite element method with 392×196 background rectangular mesh and corresponding triangulation. We used the usual piecewise linear continuous basis functions for discretization. One-dimensional computation is performed similarly for the area function $A(x)$ corresponding to Figure 1. We plotted the results for the frequency range from 0 to 4.7 KHz and the pressure u is shown in the logarithmic scale.

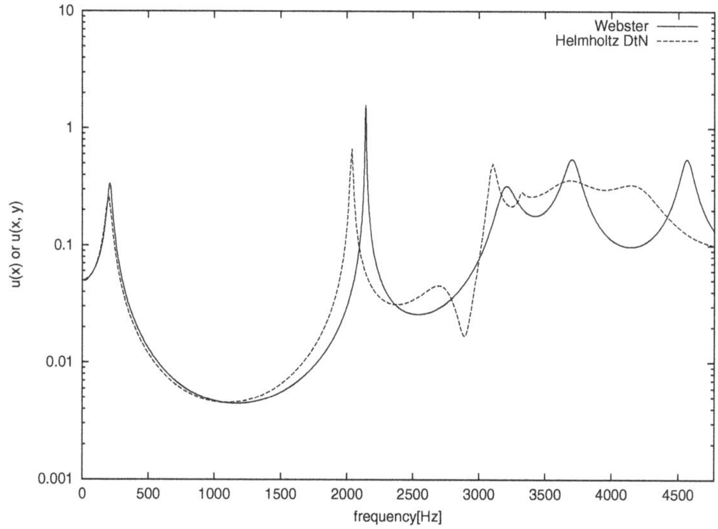

Fig. 2. Frequency response function for vowel /i/

If we change the shape of the vocal tract or the cross section area, the frequency response varies (see Kent and Read [1992] for the heuristic perturbation theory for vocal tract shape and formant curve). The frequency response curve is determined through the complex eigenvalues and its eigenfunction of the Helmholtz equation with the complex frequency ω extended into the lower half complex plane.

We show one-dimensional numerical example for this correspondence between the complex poles and the frequency response function in Figure 3 where we moved the shape of vocal tract from the neutral shape with constant cross sectional area to the one of $/i/$ as is illustrated in Figure 4. These complex poles are calculated by the iteration process of line search in real and imaginary directions alternatively. We took the starting values of the iteration to be the eigenvalues for the previous shape.

As shown in Figure 4, the formants are closely related to the resonant complex eigenvalues. Especially, if the real part of a resonance complex eigenvalue moves, the position of the formant moves correspondingly to the same direction. Furthermore, the imaginary part of the complex eigenvalue corresponds to the strength of the formant. Hence, to control the positions of formants, it would be an efficient way to vary the positions of the resonant complex eigenvalues. As for the mathematical treatments of resonance poles and related topics, refer the works by Lenoir et al. [1992] and Poisson [1995].

In this respect, the variational formula for the resonant complex eigenvalues will be useful to design the format curve to simulate the natural formant curve of human beings. Under the assumption that the eigenvalues and the eigenfunctions move smoothly with respect to the variation of shape of the vocal tract, we can obtain the variational formula as follows.

In one-dimensional case, let k and $u(x;k)$ satisfy the one-dimensional homogeneous Webster's horn equation which is the complex eigenvalue problem for k and u:

$$-\frac{d}{dx}(A(x)\frac{du}{dx}) - k^2 A(x)u(x) = 0,$$

$$\frac{du}{dx}(0) = 0, \quad \frac{du}{dx}(L) = iku(L).$$

Then the variational formula is given as follows with respect to the variation $(\delta A)(x)$ of the cross section area $A(x)$ of the vocal tract:

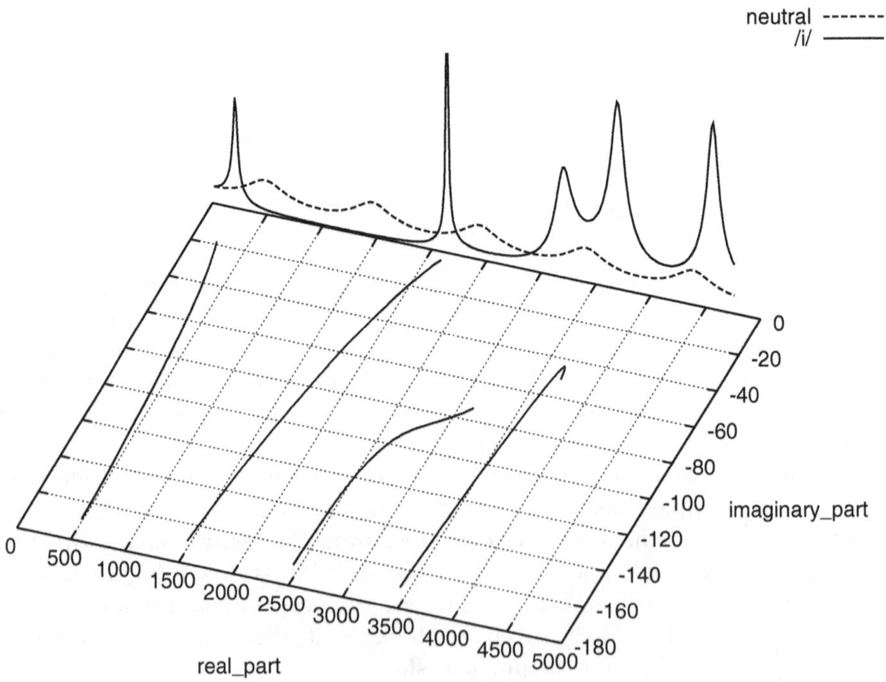

Fig. 3. Correspondence between complex eigenvalues and frequency response

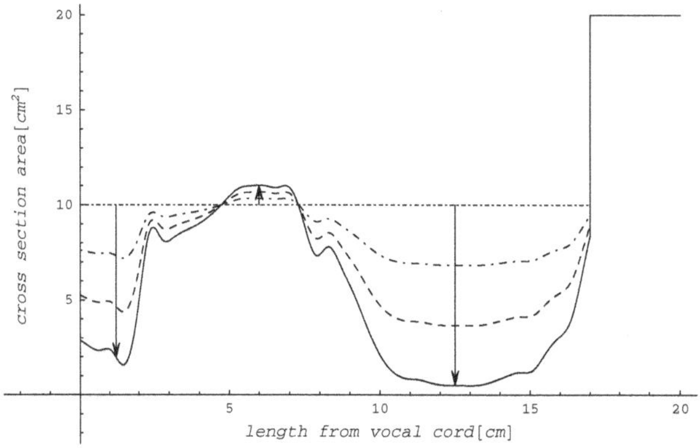

Fig. 4. Changing of area from the neutral shape to /i/

$$\delta k = \frac{\int_0^L (\delta A)(x)\{(\frac{du}{dx}(x))^2 - k^2 u(x)^2\}\mathrm{d}x - ik(\delta A)(L)u(L)^2}{2k\int_0^L A(x)u(x)^2\mathrm{d}x + iA(L)u(L)^2}.$$

We can prove this formula multiplying the equation by the solution and doing the integration by parts.

We show in Figure 5 a numerical example which corresponds to the third formants and the variational formula for the corresponding complex eigenvalue. There is a good coincidence between the tangent directions and the values computed by the variational formula which are illustrated by the vectors with arrow on the eigenvalue trajectory.

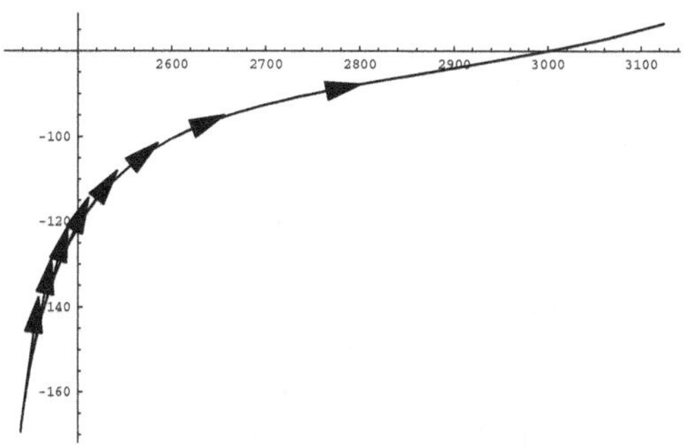

Fig. 5. Complex vectors by variational formula on eigenvalue trajectory

For two dimensional problem, let n be the outward normal and $\psi(x)$ be a function on the boundary, and move the boundary to the direction of the normal as $\epsilon\psi n$.

Firstly we consider the case of completely elastic boundary: $u = 0$ on $\partial\Omega$. The variational derivative of the resonant complex eigenvalue is then given as

$$\frac{\delta k}{\delta\psi} = \frac{-\int_{\partial\Omega}(\frac{\partial u}{\partial n})^2\psi \, d\sigma}{2k\int_{\Omega} u(x)^2 \, dx + \int_{\Gamma}((\frac{\delta M}{\delta k})u) \, u \, d\theta}.$$

Next we consider the case of rigid boundary: $\partial u/\partial n =$ on $\partial\Omega$. In this case the variational formula is given as:

$$\frac{\delta k}{\delta\psi} = \frac{\int_{\partial\Omega}\{n_1^2 \frac{\partial^2 u}{\partial x_1^2} + 2n_1 n_2 \frac{\partial^2 u}{\partial x_1 \partial x_2} + n_2^2 \frac{\partial^2 u}{\partial x_2^2}\} \, \psi \, u \, d\sigma}{2k\int_{\Omega} u(x)^2 \, dx + \int_{\Gamma}((\delta M/\delta k)u) \, u \, d\theta}$$

$$= \frac{\int_{\partial\Omega}\{(\frac{\partial u}{\partial\sigma})^2 - k^2u^2\} \, \psi \, u \, d\sigma + \int_{\partial\Omega}\frac{\partial\phi}{\partial\sigma} \, u \, \frac{\partial u}{\partial\sigma} \, d\sigma}{2k\int_{\Omega} u(x)^2 \, dx + \int_{\Gamma}((\delta M/\delta k)u) \, u \, d\theta}.$$

Now, we modify the Dirichlet to Neumann mapping on the radiation boundary rather artificially multiplying the homotopy parameter $\alpha \in [0,1]$: $\frac{\partial u}{\partial n} = \alpha M u$. Then, if $\alpha = 0$, the boundary condition becomes the homogeneous Neumann condition, and hence the eigenvalues are all real and discrete. If we move α from 0 to 1, we may establish the relationship between real eigenvalues and the resonant complex eigenvalues. In the case of $\alpha = 0$, the variation becomes pure imaginary, and resonance pole moves into the lower complex plane vertically to the real line.

3 Concluding Remarks

We formulated the stationary wave propagation phenomena in unbounded region outside an obstacle via the Helmholtz equation with appropriate boundary conditions. We introduced two kinds of finite element approximation based on the standard method and a mixed type one. The later has better numerical performance in the sense that it needs less computational cost and smaller memory requirement. From the theoretical error analysis, we can conclude the convergence of the numerical solution to the exact one.

We computed one dimensional as well as two dimensional voice generation problem which give reasonable frequency response curves. We then introduced variational formulas for the complex eigenvalues or the resonance poles for two dimensional as well as one dimensional problems. We suggested a procedure to design the frequency response curves based on the variational formula and presented a numerical example to show the validity of our method.

References

A. Bayliss and E. Turkel. Radiation boundary conditions for wave-like equations. *Comm. Pure and Appl. Math.*, 33(6):707–725, 1980.

B. Engquist and A. Majda. Absorbing boundary conditions for the numerical simulation of waves. *Math. Comp.*, 31(139):629–651, 1977.

B. Engquist and A. Majda. Radiation boundary conditions for acoustic and elastic wave calculations. *Comm. Pure Appl. Math.*, 32(3):313–357, 1979.

O. G. Ernst. A finite-element capacitance matrix method for exterior helmholtz problems. *Numer. Math.*, 75(2):175–204, 1996.

K. Feng. Finite element method and natural boundary reduction. In *Proceeding of the International Congress of Mathematicians*, Warsaw, Poland, 1983.

J. Flanagan. *Speech analysis, synthesis, and perception.* Springer, Berlin-New York, 1972.

S. Furui. *Digital speech processing, synthesis, and recognition.* Marcel Dekker, 1989.

E. Heikkola, Y. A. Kuznetsov, P. Neittaanmaki, and J. Toivanen. Fictitious domain methods for the numerical solution of two-dimesional scattering problems. *J. Comput. Phys.*, 145:89–109, 1998.

T. Kako and T. Kano. Finite element method for the helmholtz equation and numerical simulation of the wave propagation in vocal tract. In *GAKUTO Int. Series Math. Sci. and Appli.*, volume 12, pages 55–63, 1999.

R. Kent and C. Read. *The acoustic analysis of speech.* Singular Publ. Group, San Diego, 1992.

M. Lenoir, M. Vullierme-Ledard, and C. Hazard. Variational formulations for the determination of resonant states in scattering problems. *SIAM J. Math. Anal.*, 23(3):579–608, 1992.

H. Nasir. *A mixed type finite element method for radiation and scattering problems with applications to structural-acoustic coupling problem in unbounded region.* PhD thesis, The University of Electro-Communications, Japan, 2003.

H. Nasir, T. Kako, and D. Koyama. A mixed-type finite element approximation for radiation problems using fictitious domain method. *J. Comput. Appl. Math.*, 52:377–392, 2003.

O. Poisson. Étude numérique des pôles de résonance associés à la diffraction d'ondes acoustiques et élastiques par un obstacle en dimension 2. *M^2AN*, 29(2):819–855, 1995.

Selecting Constraints in Dual-Primal FETI Methods for Elasticity in Three Dimensions

Axel Klawonn[1] and Olof B. Widlund[2]

[1] Universität Duisburg-Essen, Campus Essen, Fachbereich Mathematik, (http://www.uni-essen.de/ingmath/Axel.Klawonn/)
[2] Courant Institute of Mathematical Sciences, New York University (http://cs.nyu.edu/cs/faculty/widlund/)

Summary. Iterative substructuring methods with Lagrange multipliers for the elliptic system of linear elasticity are considered. The algorithms belong to the family of dual-primal FETI methods which was introduced for linear elasticity problems in the plane by Farhat et al. [2001] and then extended to three dimensional elasticity problems by Farhat et al. [2000]. In dual-primal FETI methods, some continuity constraints on primal displacement variables are required to hold throughout the iterations, as in primal iterative substructuring methods, while most of the constraints are enforced by the use of dual Lagrange multipliers, as in the older one-level FETI algorithms. The primal constraints should be chosen so that the local problems become invertible. They also provide a coarse problem and they should be chosen so that the iterative method converges rapidly.

Recently, the family of algorithms for scalar elliptic problems in three dimensions was extended and a theory was provided in Klawonn et al. [2002a,b]. It was shown that the condition number of the dual-primal FETI methods can be bounded polylogarithmically as a function of the dimension of the individual subregion problems and that the bounds can otherwise be made independent of the number of subdomains, the mesh size, and jumps in the coefficients.

In the case of the elliptic system of partial differential equations arising from linear elasticity, essential changes in the selection of the primal constraints have to be made in order to obtain the same quality bounds for elasticity problems as in the scalar case. Special emphasis is given to developing robust condition number estimates with bounds which are independent of arbitrarily large jumps of the material coefficients. For benign coefficients, without large jumps, selecting an appropriate set of edge averages as primal constraints are sufficient to obtain good bounds, whereas for arbitrary coefficient distributions, additional primal first order moments are also required.

1 The equations of linear elasticity

The equations of linear elasticity model the displacement of a linear elastic material under the action of external and internal forces. The elastic body occupies a domain $\Omega \subset \mathbb{R}^3$, which we assume to be bounded and polyhedral. We denote its boundary by $\partial\Omega$ and assume that one part of it, $\partial\Omega_D$, is clamped, i.e., with homogeneous Dirichlet boundary conditions, and that the rest, $\partial\Omega_N := \partial\Omega \backslash \partial\Omega_D$, is subject to a surface force $\mathbf{g}$, i.e., a natural boundary condition. We can also introduce a body force $\mathbf{f}$, e.g., gravity. Using the notation $\mathbf{H}^1(\Omega) := (H^1(\Omega))^3$, the appropriate space for a variational formulation is then the Sobolev space $\mathbf{H}_0^1(\Omega, \partial\Omega_D) := \{\mathbf{v} \in \mathbf{H}^1(\Omega) : \mathbf{v} = \mathbf{0} \text{ on } \partial\Omega_D\}$. The linear elasticity problem consists in finding the displacement $\mathbf{u} \in \mathbf{H}_0^1(\Omega, \partial\Omega_D)$ of the elastic body Ω, such that

$$a(\mathbf{u}, \mathbf{v}) = \langle \mathbf{F}, \mathbf{v} \rangle \quad \forall \mathbf{v} \in \mathbf{H}_0^1(\Omega, \partial\Omega_D) \tag{1}$$

where $\langle \mathbf{F}, \mathbf{v} \rangle := \int_\Omega \mathbf{f}^T \mathbf{v} \, d\mathbf{x} \; + \; \int_{\partial\Omega_N} \mathbf{g}^T \mathbf{v} \, d\sigma$ and

$$a(\mathbf{u}, \mathbf{v}) = \int_\Omega G(\mathbf{x}) \varepsilon(\mathbf{u}) : \varepsilon(\mathbf{v}) d\mathbf{x} + \int_\Omega G(\mathbf{x}) \, \beta(\mathbf{x}) \operatorname{div} \mathbf{u} \operatorname{div} \mathbf{v} \, d\mathbf{x}. \tag{2}$$

Here G and β are material parameters which depend on Young's modulus $E > 0$ and the Poisson ratio $\nu \in (0, 1/2]$; we have $G = E/(1+\nu)$ and $\beta = \nu/(1-2\nu)$. In this contribution, we only consider the case of compressible elasticity, which means that the Poisson ratio ν is bounded away from $1/2$. Furthermore, $\varepsilon_{ij}(\mathbf{u}) := \frac{1}{2}(\frac{\partial u_i}{\partial x_j} + \frac{\partial u_j}{\partial x_i})$ is the linearized strain tensor, and

$$\varepsilon(\mathbf{u}) : \varepsilon(\mathbf{v}) = \sum_{i,j=1}^{3} \varepsilon_{ij}(\mathbf{u}) \varepsilon_{ij}(\mathbf{v}).$$

For convenience, we also introduce the notation

$$(\varepsilon(\mathbf{u}), \varepsilon(\mathbf{v}))_{L_2(\Omega)} := \int_\Omega \varepsilon(\mathbf{u}) : \varepsilon(\mathbf{v}) d\mathbf{x}.$$

The bilinear form associated with linear elasticity is then

$$a(\mathbf{u}, \mathbf{v}) = (G \, \varepsilon(\mathbf{u}), \varepsilon(\mathbf{v}))_{L_2(\Omega)} \; + \; (G \, \beta \operatorname{div}, \mathbf{u}, \operatorname{div}, \mathbf{v})_{L_2(\Omega)}.$$

We will also use the standard Sobolev space norm

$$\|\mathbf{u}\|_{H^1(\Omega)} := \left(|\mathbf{u}|_{H^1(\Omega)}^2 + \|\mathbf{u}\|_{L_2(\Omega)}^2 \right)^{1/2}$$

with $\|\mathbf{u}\|_{L_2(\Omega)}^2 := \sum_{i=1}^{3} \int_\Omega |u_i|^2 d\mathbf{x}$, and $|\mathbf{u}|_{H^1(\Omega)}^2 := \sum_{i=1}^{3} \|\nabla u_i\|_{L_2(\Omega)}^2$. It is clear that the bilinear form $a(\cdot, \cdot)$ is continuous with respect to $\|\cdot\|_{H^1(\Omega)}$, although the bound depends on the Lamé parameters. Proving ellipticity is far less trivial but it can be established from a Korn inequality; see, e.g., [Ciarlet, 1988, pp. 292-295]. The wellposedness of the linear system (1) follows immediately from the continuity and ellipticity of the bilinear form $a(\cdot, \cdot)$. This makes it possible to use many technical tools, previously developed for scalar second order elliptic problems, in the analysis of domain decomposition methods for the system of linear elasticity.

The null space $\mathbf{ker}\,(\varepsilon)$ is the space of rigid body motions. Thus, the linearized strain tensor of $\mathbf{u}$ and its divergence vanish only for the elements of the space spanned by the three translations

$$\mathbf{r}_1 := \begin{bmatrix} 1 \\ 0 \\ 0 \end{bmatrix}, \mathbf{r}_2 := \begin{bmatrix} 0 \\ 1 \\ 0 \end{bmatrix}, \mathbf{r}_3 := \begin{bmatrix} 0 \\ 0 \\ 1 \end{bmatrix},$$

and the three rotations

$$\mathbf{r}_4 := \frac{1}{H} \begin{bmatrix} x_2 - \hat{x}_2 \\ -x_1 + \hat{x}_1 \\ 0 \end{bmatrix}, \mathbf{r}_5 := \frac{1}{H} \begin{bmatrix} -x_3 + \hat{x}_3 \\ 0 \\ x_1 - \hat{x}_1 \end{bmatrix}, \mathbf{r}_6 := \frac{1}{H} \begin{bmatrix} 0 \\ x_3 - \hat{x}_3 \\ -x_2 + \hat{x}_2 \end{bmatrix},$$

where $\hat{\mathbf{x}} \in \hat{\Omega}$ and H denotes the diameter of an appropriate region $\hat{\Omega}$. The scaling and shifting of the rotational rigid body modes make the $L_2(\hat{\Omega})-$norms of these six functions scale similarly with H.

2 Finite elements and geometry

We will only consider compressible elastic materials. Since the problem is well posed in $\mathbf{H}^1(\Omega)$, it is sufficient to discretize our elliptic problem (1) by low order, conforming finite elements, e.g., linear or trilinear elements.

We introduce a triangulation τ^h of Ω which is shape regular and has a typical diameter of h. We denote by $\mathbf{W}^h := \mathbf{W}^h(\Omega) \subset \mathbf{H}_0^1(\Omega, \partial\Omega_D)$ the corresponding conforming finite element space of finite element functions. The corresponding discrete problem is finding $\mathbf{u}_h \in \mathbf{W}^h$ such that,

$$a(\mathbf{u}_h, \mathbf{v}_h) = \langle \mathbf{F}, \mathbf{v}_h \rangle \quad \forall \mathbf{v}_h \in \mathbf{W}^h. \tag{3}$$

When there is no risk of confusion, we will drop the subscript h.

Let the domain $\Omega \subset \mathbb{R}^3$ be decomposed into nonoverlapping subdomains $\Omega_i, i = 1, \ldots, N$, each of which is the union of finite elements with matching finite element nodes, on the boundaries of neighboring subdomains, across the interface Γ. The interface Γ is the union of subdomain faces, edges, and vertices, all of them regarded as open sets. We denote the faces of Ω_i by $\mathcal{F}^{ij}$, its edges by $\mathcal{E}^{ik}$, and its vertices by $\mathcal{V}^{il}$. Faces are sets which are shared by two subregions, edges by more than two subregions, and vertices are endpoints of edges. Subdomain vertices that lie on $\partial\Omega_N$ are part of Γ, while subdomain faces that are part of $\partial\Omega_N$ are not; the nodes on those faces will always be treated as interior. If Γ intersects $\partial\Omega_N$ along an edge common to the boundaries of only two subdomains, we will regard it as part of the face common to this pair of subdomains; otherwise it will be regarded as an edge of Γ. We note that any subdomain that does not intersect $\partial\Omega_D$ is a floating subdomain, i.e., a subdomain on which only natural boundary conditions are imposed.

Let us denote the sets of nodes on $\partial\Omega, \partial\Omega_i$, and Γ by $\partial\Omega_h, \partial\Omega_{i,h}$, and Γ_h, respectively. For any interface point $x \in \Gamma_h$, we define

$$\mathcal{N}_x := \{j \in \{1, \ldots, N\} : x \in \partial\Omega_j\},$$

i.e., $\mathcal{N}_x$ is the index set of all subdomains with x on their boundaries. We note that we can characterize individual faces, edges, and vertices of the interface in terms of an equivalence relation defined in terms of these index sets.

In our theoretical analysis, we assume that each subregion Ω_i is the union of a number of shape regular tetrahedral coarse elements and that the number of such tetrahedra is uniformly bounded for each subdomain. Therefore, the subregions are not very thin and we can also easily show that the diameters of any pair of neighboring subdomains are comparable. We also assume that the material parameters are constant in each subdomain.

We denote the standard finite element space of continuous, piecewise linear functions on Ω_i by $\mathbf{W}^h(\Omega_i)$; we always assume that these functions vanish on $\partial\Omega_i \cap \partial\Omega_D$. To simplify the theory, we will assume that the triangulation of each subdomain is quasi uniform. The diameter of Ω_i is H_i, or generically, H. We denote the corresponding trace spaces by $\mathbf{W}^{(i)} := \mathbf{W}^h(\partial\Omega_i \cap \Gamma), i = 1, \ldots, N$, and by $\mathbf{W} := \prod_{i=1}^N \mathbf{W}^{(i)}$ the associated product space. We will often consider elements of $\mathbf{W}$ which are discontinuous across the interface.

For each subdomain Ω_i, we define the local stiffness matrix $K^{(i)}$ which we view as an operator on $\mathbf{W}^h(\Omega_i)$. On the product space $\prod_{i=1}^N \mathbf{W}^h(\Omega_i)$, we define the operator K as the direct sum of the local stiffness operators $K^{(i)}$, i.e.,

$$K := \bigoplus_{i=1}^N K^{(i)}. \tag{4}$$

In an implementation, K corresponds to a block diagonal matrix since, so far, there is no coupling across the interface. The finite element approximation of the elliptic problem is continuous across Γ and we denote the corresponding subspace of $\mathbf{W}$ by $\widehat{\mathbf{W}}$. We note that while the stiffness matrix K and its Schur complement, which corresponds to the product space $\mathbf{W}$, generally are singular, restricted to $\widehat{\mathbf{W}}$ they are not.

In the present study, as in others on dual-primal FETI methods, we also work with subspaces $\widetilde{\mathbf{W}} \subset \mathbf{W}$ for which sufficiently many constraints are enforced so that the resulting leading diagonal block matrix of the FETI saddle point problem, to be introduced in (10), though no longer block diagonal, is strictly positive definite. These constraints are called primal and usually consist of certain edge averages and moments, which have common values across the interface of neighboring subdomains, and possibly of well chosen subdomain vertices (or other nodes), for which a partial subassembly is carried out. One of the benefits of working in $\widetilde{\mathbf{W}}$, rather than in $\mathbf{W}$, is that certain related Schur complements, $\widetilde{S}_\varepsilon$ and S_ε, are strictly positive definite; cf. (6) and (8).

We further introduce two subspaces, $\widehat{\mathbf{W}}_\Pi \subset \widehat{\mathbf{W}}$ and $\widetilde{\mathbf{W}}_\Delta$, corresponding to a primal and a dual part of the space $\widetilde{\mathbf{W}}$. These subspaces play an important role in the description and analysis of our iterative method. The direct sum of these spaces equals $\widetilde{\mathbf{W}}$, i.e.,

$$\widetilde{\mathbf{W}} = \widehat{\mathbf{W}}_\Pi \oplus \widetilde{\mathbf{W}}_\Delta. \tag{5}$$

The second subspace, $\widetilde{\mathbf{W}}_\Delta$, is the direct sum of local subspaces $\widetilde{\mathbf{W}}_\Delta^{(i)}$ of $\widetilde{\mathbf{W}}$, where each subdomain Ω_i contributes a subspace $\widetilde{\mathbf{W}}_\Delta^{(i)}$; only its i-th component in the sense of the product space $\widetilde{\mathbf{W}}$ is non trivial.

We note that the dual subspaces will be associated with Lagrange multipliers to control jumps across the interface, jumps which will only vanish at convergence of our iterative methods. The constraints associated with the degrees of freedom of the primal subspace, on the other hand, will be satisfied throughout the iteration.

We now define certain Schur complements by using a variational formulation. We first define Schur complements $S_\varepsilon^{(i)}, i = 1, \ldots, N$, operating on $\mathbf{W}^{(i)}$ by

$$\langle S_\varepsilon^{(i)} \mathbf{w}^{(i)}, \mathbf{w}^{(i)} \rangle = \min \langle K^{(i)} \mathbf{v}^{(i)}, \mathbf{v}^{(i)} \rangle, \ \forall \mathbf{w}^{(i)} \in \mathbf{W}^{(i)}, \tag{6}$$

where we take the minimum over all $\mathbf{v}^{(i)} \in \mathbf{W}^h(\Omega_i)$ with $\mathbf{v}^{(i)}{}_{|\partial\Omega_i \cap \Gamma} = \mathbf{w}^{(i)}$. We can now define the Schur complement S_ε operating on $\mathbf{W}$ by the direct sum of the local Schur complements

$$S_\varepsilon := \bigoplus_{i=1}^{N} S_\varepsilon^{(i)}. \tag{7}$$

We next introduce a Schur complement $\widetilde{S}_\varepsilon$, operating on $\widetilde{\mathbf{W}}_\Delta$, by a variational problem: for all $\mathbf{w}_\Delta \in \widetilde{\mathbf{W}}_\Delta$,

$$\langle \widetilde{S}_\varepsilon \mathbf{w}_\Delta, \mathbf{w}_\Delta \rangle = \min_{\mathbf{w}_\Pi \in \widehat{\mathbf{W}}_\Pi} \langle S_\varepsilon (\mathbf{w}_\Delta + \mathbf{w}_\Pi), \mathbf{w}_\Delta + \mathbf{w}_\Pi \rangle. \tag{8}$$

We will always assume that we have enough primal constraints, i.e., a large enough primal subspace $\widehat{\mathbf{W}}_\Pi$, to make $\widetilde{S}_\varepsilon$ invertible. We note that any Schur complement of a positive semi-definite, symmetric matrix is always associated with such a variational problem. We also obtain, analogously, a reduced right hand side $\tilde{\mathbf{f}}_\Delta$, from the load vectors associated with the individual subdomains.

3 The dual-primal FETI method

We reformulate the original finite element problem, reduced to the degrees of freedom of the second subspace $\widetilde{\mathbf{W}}_\Delta$, as a minimization problem with constraints given by the requirement of continuity across all of Γ_h: find $\mathbf{u}_\Delta \in \widetilde{\mathbf{W}}_\Delta$, such that

$$\left.\begin{array}{c} J(\mathbf{u}_\Delta) := \frac{1}{2}\langle \widetilde{S}_\varepsilon \mathbf{u}_\Delta, \mathbf{u}_\Delta\rangle - \langle \tilde{\mathbf{f}}_\Delta, \mathbf{u}_\Delta\rangle \rightarrow \min \\ B_\Delta \mathbf{u}_\Delta = 0 \end{array}\right\}. \tag{9}$$

The jump operator B_Δ, with elements from $\{0, 1, -1\}$, operates on $\widetilde{\mathbf{W}}$ and enforces pointwise continuity at the dual displacement degrees of freedom. At possible primal vertices, continuity is already enforced by subassembly and the jump operator applied to a function from $\widetilde{\mathbf{W}}$ is automatically zero at these special degrees of freedom.

By introducing a set of Lagrange multipliers $\boldsymbol{\lambda} \in \mathbf{V} := \mathbf{range}\,(B_\Delta)$, to enforce the constraints $B_\Delta \mathbf{u}_\Delta = 0$, we obtain a saddle point formulation of (9)

$$\begin{bmatrix} \widetilde{S}_\varepsilon & B_\Delta^T \\ B_\Delta & O \end{bmatrix} \begin{bmatrix} \mathbf{u}_\Delta \\ \boldsymbol{\lambda} \end{bmatrix} = \begin{bmatrix} \tilde{\mathbf{f}}_\Delta \\ \mathbf{0} \end{bmatrix}. \tag{10}$$

We note that we can add any element from $\mathbf{ker}\,(B_\Delta^T)$ to $\boldsymbol{\lambda}$ without changing the displacement solution $\mathbf{u}_\Delta$.

Since $\widetilde{S}_\varepsilon$ is invertible, we can eliminate $\mathbf{u}_\Delta$ and obtain the following system for the Lagrange multiplier variables:

$$F\boldsymbol{\lambda} = \mathbf{d}. \tag{11}$$

Here, our new system matrix F is defined by

$$F := B_\Delta \widetilde{S}_\varepsilon^{-1} B_\Delta^T \tag{12}$$

and the new right hand side by $\mathbf{d} := B_\Delta \widetilde{S}_\varepsilon^{-1} \tilde{\mathbf{f}}_\Delta$. Algorithmically, $\widetilde{S}_\varepsilon$ is only needed in terms of $\widetilde{S}_\varepsilon^{-1}$ times a vector $\mathbf{w}_\Delta \in \widetilde{\mathbf{W}}_\Delta$ and such a product can be computed relatively inexpensively although it involves a small subproblem that is global. The operator F will obviously depend on the choice of the subspaces $\widehat{\mathbf{W}}_\Pi$ and $\widetilde{\mathbf{W}}_\Delta$.

The dual-primal FETI Dirichlet preconditioner is defined in terms of certain scale factors $\delta_j^\dagger(x)$. They depend on one of the Lamé parameters. We first define a set of functions $\delta_j(x)$, one for each $\partial\Omega_j$, by

$$\delta_j(x) := \frac{\sum_{i \in \mathcal{N}_x} G_i^\gamma(x)}{G_j^\gamma(x)}, \quad x \in \partial\Omega_{j,h} \cap \Gamma_h, \tag{13}$$

where $\gamma \in [1/2, \infty)$. Here, as before, $\mathcal{N}_x$ is the set of indices of the subregions which have x on its boundary; any $x \in \Gamma_h$ belongs to at least two subdomains. The pseudo inverses $\delta_j^\dagger$ are defined as

$$\delta_j^\dagger(x) = \delta_j^{-1}(x), \quad x \in \partial\Omega_{j,h} \cap \Gamma_h. \tag{14}$$

The scaled jump operators for the dual-primal FETI Dirichlet preconditioner is defined by

$$B_{D,\Delta} := [B_{D,\Delta}^{(1)}, \ldots, B_{D,\Delta}^{(N)}].$$

Here, the $B_{D,\Delta}^{(i)}$ are defined as follows: each row of $B_{\Delta}^{(i)}$ with a nonzero entry corresponds to a Lagrange multiplier connecting the subdomain Ω_i with a neighboring subdomain Ω_j at a point $x \in \partial\Omega_{i,h} \cap \partial\Omega_{j,h}$. Multiplying this row with $\delta_j^\dagger(x)$ and doing so for all rows with nonzero entries gives us $B_{D,\Delta}^{(i)}$.

As in [Klawonn and Widlund, 2001, section 5], we solve the dual system (11) using the preconditioned conjugate gradient algorithm with the preconditioner

$$M^{-1} := P B_{D,\Delta} S_\varepsilon B_{D,\Delta}^T P^T, \tag{15}$$

where P is the ℓ_2-orthogonal projection from $\mathbf{range}\,(B_{D,\Delta})$ onto $\mathbf{V} = \mathbf{range}\,(B_\Delta)$, i.e., P removes the component from $\mathbf{ker}\,(B_\Delta^T)$ of any element in $\mathbf{range}\,(B_{D,\Delta})$. We note that, in the present context, P and P^T are only needed for the theoretical analysis to guarantee that the preconditioned residuals will belong to $\mathbf{V}$; they can be dropped in the implementation.

The dual-primal FETI method is the standard preconditioned conjugate gradient algorithm for solving the preconditioned system

$$M^{-1}F\boldsymbol{\lambda} = M^{-1}\mathbf{d}.$$

We can see that we can drop the projection operator P and its transpose by the following argument. Applying $B_{D,\Delta} S_\varepsilon B_{D,\Delta}^T$ to an element from $\mathbf{V}$ results in a vector μ which can be written as a sum $\mu = \mu_0 + \mu_1$ of components $\mu_0 \in \mathbf{ker}\,(B_\Delta^T)$ and $\mu_1 \in \mathbf{V} = \mathbf{range}\,(B_\Delta)$. When F is applied to μ, the component $F\mu_0$ disappears and we also have $F\mu \in \mathbf{V}$. Examining the standard pcg algorithm, we see that dropping P and P^T only affects the computed Lagrange multiplier solution but not the computed displacements.

Our definition of M^{-1} clearly depends on the choice of the subspaces $\widehat{\mathbf{W}}_\Pi$ and $\widetilde{\mathbf{W}}_\Delta$. We can show that M^{-1} is invertible if $\widetilde{S}$ is, i.e., if the subspace $\widehat{\mathbf{W}}_\Pi$ is large enough; cf. Klawonn and Widlund [2004]

4 Selection of constraints

In order to control the rigid body modes of a subregion, we need at least six constraints. To get an understanding of the type of primal constraints that are required for our preconditioner, it is useful to examine two special cases.

In the first, we assume that we have two subdomains made of the same material, which have a face in common and are surrounded by subdomains made of a material with much smaller coefficients. Such a problem will clearly have six low energy modes corresponding to the rigid body modes of the union of the two special subdomains. Any preconditioner that has less than six linearly independent primal constraints across that face will have at least seven low energy modes and is likely to be poor.

In the second case, we again consider two subdomains surrounded by subdomains with much smaller coefficients. We now assume that the two special subdomains share only an edge. In this case, there are seven low energy modes of the finite element model corresponding to the same rigid body modes as before and an additional one. The new mode corresponds to a relative rotation of the two subdomains around their common edge. We conclude that in such a case, we should introduce five linearly independent primal constraints related to the special edge. Such edges will be called fully primal in our discussion.

In the convergence theory presented in Section 5, we will assume that the requirement of the first special case is met for each face; we will select at least six linearly independent edge constraints for each face of the interface. We note that such a constraint will serve as a constraint for every face adjacent to the edge in question. Nevertheless, it is likely that in many cases we will be able to use fewer constraints and still maintain a good rate of convergence of our algorithm; we plan to return to these questions in future work. We also note that using only face constraints can be inadequate; see Klawonn et al. [2002b, 2003].

For coefficient distributions with only modest jumps across the interface Γ and for some special decompositions, we are able to exclusively work with edge averages; cf. Section 5.1. To be able to treat general coefficient distributions with arbitrarily large jumps, we also need first order moments in addition to the averages on certain edges as in the second special case discussed above. We will also introduce the concept of an acceptable path; cf. Section 5.3.

In our theory, we will work with sets of constraints associated with all the faces of the interface and with the edges designated as fully primal. For a fully primal edge only five constraints are required; cf. the discussion above of the second special case. This is related to the fact that one rotational rigid body mode is invisible on the edge. This can be easily seen by a change of coordinates such that the chosen edge coincides with an axis of the Cartesian coordinate system. Without restriction of generality, we assume that it is the third rotation $\mathbf{r}_6$. This motivates the following definition, where $p = 6$ relates to the case of pure edge averages and $p = 5$ to edge averages and first order moments used on a single edge.

Definition 1. *Let $\mathcal{F}^{ij}$ be a face and $5 \leq p \leq 6$. A set $f_m, m = 1, \ldots, p$, of linearly independent linear functionals on $\mathbf{W}^{(i)}$ is called a set of primal constraint functionals if it has the following properties:*

1. $|f_m(\mathbf{w}^{(i)})|^2 \leq C\, H_i^{-1}(1 + \log(H_i/h_i))\{|\mathbf{w}^{(i)}|^2_{H^{1/2}(\mathcal{F}^{ij})} + \frac{1}{H_i}\|\mathbf{w}^{(i)}\|^2_{L_2(\mathcal{F}^{ij})}\}$
2. $f_m(\mathbf{r}_l) = \delta_{ml} \quad \forall m, l = 1, \ldots, p, \quad \mathbf{r}_l \in \mathbf{ker}\,(\varepsilon).$

We note that these bounds will allow us to prove almost uniform bounds for the condition number of our algorithms. If point constraints were to replace the edge constraints, this would not be possible. We note that while we will work with functionals which are not uniformly bounded, the growth of these bounds is quite modest when H/h grows. These growth factors will appear in the main theorem as is customary for many domain decomposition methods. We also note that the logarithmic factors cannot be eliminated if we wish to obtain a result which is uniform with respect to arbitrary variations of the Lamé parameters.

Let us now first consider the case of six functionals, i.e., $p = 6$. As an example of functionals f_m, we can use appropriately chosen linear combinations of certain edge averages, of components of the displacement,

$$g_m(\mathbf{w}^{(i)}) := \frac{\int_{\mathcal{E}^{ik}} w_\ell^{(i)} dx}{\int_{\mathcal{E}^{ik}} 1 dx}$$

for a function $\mathbf{w}^{(i)} \in \mathbf{W}^{(i)}$. Using a Cauchy-Schwarz inequality, we obtain

$$|g_m(\mathbf{w}^{(i)})|^2 \leq C\, H_i^{-1}\, \|\mathbf{w}^{(i)}\|^2_{L_2(\mathcal{E}^{ik})}.$$

We can show, by using standard tools, that

$$\|\mathbf{w}^{(i)}\|^2_{L_2(\mathcal{E}^{ik})} \leq C\,(1 + \log(H_i/h_i))\,(|\mathbf{w}^{(i)}|^2_{H^{1/2}(\mathcal{F}^{ij})} + \frac{1}{H_i}\|\mathbf{w}^{(i)}\|^2_{L_2(\mathcal{F}^{ij})}).$$

Thus, the first requirement of Definition 1 is satisfied. In order to obtain six linearly independent linear functionals associated with a face $\mathcal{F}^{ij}$, we have to choose a total of six averages on at least three different edges $\mathcal{E}^{ik}$.

The functionals $g_1, \ldots, g_6$, provide a basis of the dual space $(\mathbf{ker}\,(\varepsilon))'$. There always exists a dual basis of $(\mathbf{ker}\,(\varepsilon))'$, which we denote by $f_1, \ldots, f_6$, defined by $f_m(\mathbf{r}_l) = \delta_{ml}, m, l = 1, \ldots, 6$. Obviously, there exist $\beta_{lk} \in \mathbb{R}, l, k = 1, \ldots, 6$, such that for $\mathbf{w} \in \mathbf{ker}\,(\varepsilon)$, we have

$$f_m(\mathbf{w}) = \sum_{n=1}^{6} \beta_{mn} g_n(\mathbf{w}), \quad m = 1, \ldots, 6.$$

We next consider the case of $p = 5$ in Definition 1. Let us introduce the following definition.

Definition 2. *An edge is said to be* **fully primal** *if we use five linearly independent constraints, the averages over the three displacement components and two first order moments.*

Thus, we can define the functionals f_m as

$$f_m(\mathbf{w}^{(i)}) := \frac{(\mathbf{w}^{(i)}, \mathbf{r}_m)_{L_2(\mathcal{E}^{ik})}}{(\mathbf{r}_m, \mathbf{r}_m)_{L_2(\mathcal{E}^{ik})}}, \quad m \in \{1, \ldots, 5\}. \tag{16}$$

Obviously, the second requirement of Definition 1 is satisfied.

Using a Cauchy-Schwarz inequality, we obtain

$$|f_m(\mathbf{w}^{(i)})|^2 \leq \frac{\|\mathbf{w}^{(i)}\|^2_{L_2(\mathcal{E}^{ik})}}{\|\mathbf{r}_m\|^2_{L_2(\mathcal{E}^{ik})}}$$

and the first requirement of Definition 1 follows again by using standard tools.

We also need to introduce the concept of acceptable paths.

Definition 3. *Let us first consider an arbitrary pair of subdomains (Ω_i, Ω_k) which has a face or an edge in common. An* **acceptable path** *is a path $\{\Omega_i, \Omega_{j_1}, \ldots, \Omega_{j_n}, \Omega_k\}$ from Ω_i to Ω_k, possibly via a uniformly bounded number of other subdomains $\Omega_{j_q}, q = 1, \ldots, n$, such that the associated coefficients G_{j_q} satisfy the condition*

$$TOL * G_{j_q} \geq \min(G_i, G_k) \quad q = 1, \ldots, n, \tag{17}$$

for some tolerance TOL. We can pass from one subdomain to another either through a face or through a fully primal edge, if the next subdomain has a coefficient satisfying (17); cf. Figure 1. We also need to consider all vertices and all pairs of substructures which only have a vertex, but not a face or an edge in common. Then, if the vertex is not primal, there must be an acceptable path, of the same nature as before, with the only difference that here we can be more lenient and only insist on

$$TOL * G_{j_q} \geq \frac{h_i}{H_i} \min(G_i, G_k) \quad q = 1, \ldots, n. \tag{18}$$

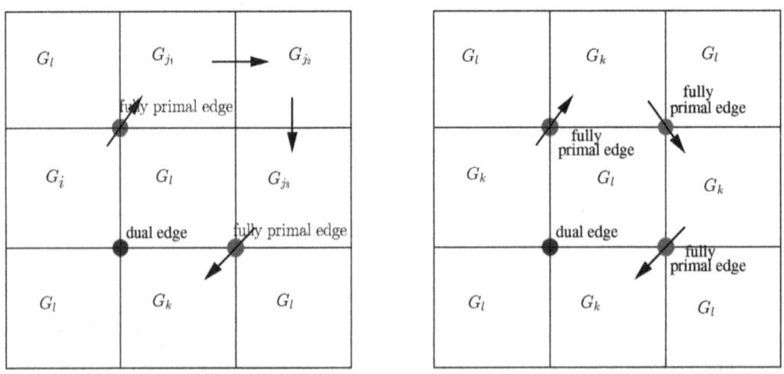

Fig. 1. Acceptable paths, through edges and faces, left, and only through edges, right, (planar cut)

We will assume that for each pair (Ω_i, Ω_k), which has a face, an edge, or a vertex in common, there exists an acceptable path as defined in Definition 3 with a modest tolerance TOL and that the path does not exceed a prescribed length. If TOL becomes too large for a certain edge or vertex or if the length of the acceptable path exceeds a given uniform bound, we can make the edge fully primal or the vertex primal; cf. Figure 2 for an example where certain vertices should be made primal.

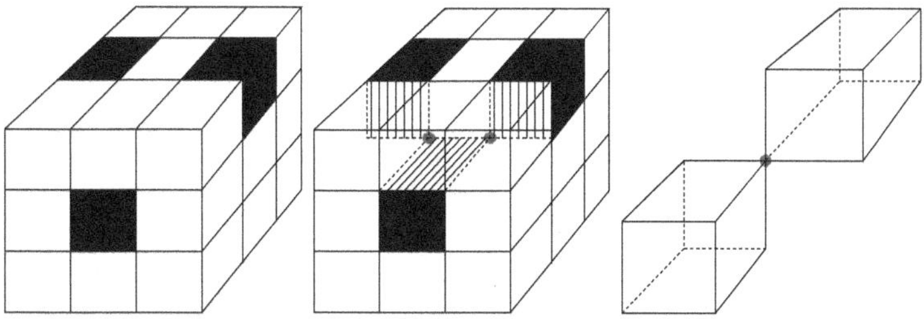

Fig. 2. Example of a decomposition where no acceptable path exists for the vertices which connect the black subdomains, which have much larger coefficients than those of the white. These vertices should be made primal

Finally, we define the spaces $\widehat{\mathbf{W}}_{\Pi}$ and $\widetilde{\mathbf{W}}_{\Delta} = \bigoplus_{i=1}^{N} \widetilde{\mathbf{W}}_{\Delta}^{(i)}$. The first space, $\widehat{\mathbf{W}}_{\Pi}$, is spanned by the nodal finite element functions which are associated with primal vertices and by averages and possibly first order moments, which belong to primal and fully primal edges, respectively. Each such primal constraint is associated with a basis element of $\widehat{\mathbf{W}}_{\Pi}$; all these functions are continuous across the interface Γ. For each subdomain Ω_i, we then define a subspace $\widetilde{\mathbf{W}}_{\Delta}^{(i)}$ by those functions in $\mathbf{W}^{(i)}$ which are zero at primal vertices and have zero averages or first order moments on primal and fully primal edges, respectively.

5 Convergence analysis

As in Klawonn et al. [2002a], the two different Schur complements, $\widetilde{S}_{\varepsilon}$ and S_{ε}, introduced in section 3, play an important role in the analysis of the dual–primal iterative algorithm. Both operate on the second subspace $\widetilde{\mathbf{W}}_{\Delta}$ and we also recall that $\widetilde{S}_{\varepsilon}$ represents a global problem while S_{ε} does not.

We recall that $\mathbf{V} := \mathbf{range}\,(B_{\Delta})$ is the space of Lagrange multipliers. As in [Klawonn and Widlund, 2001, section 5], we introduce a projection

$$P_{\Delta} := B_{D,\Delta}^{T} B_{\Delta}.$$

A simple computation shows, see [Klawonn and Widlund, 2001, Lemma 4.2], that P_{Δ} preserves the jump of any function $\mathbf{u}_{\Delta} \in \widetilde{\mathbf{W}}_{\Delta}$, i.e.,

$$B_\Delta P_\Delta \mathbf{u}_\Delta = B_\Delta \mathbf{u}_\Delta \tag{19}$$

and we therefore have $P_\Delta \mathbf{u} = 0 \;\; \forall \mathbf{u} \in \widehat{\mathbf{W}}$.

In our proof of Theorem 1, we will use representation formulas for F and M which allow us to carry out our analysis in the space of displacement variables. The representation formula for F is given in the next lemma; see also [Klawonn et al., 2002a, p. 175] or Mandel and Tezaur [2001].

Lemma 1. *For any $\boldsymbol{\lambda} \in \mathbf{V}$, we have*

$$\langle F\boldsymbol{\lambda}, \boldsymbol{\lambda}\rangle = \sup_{0 \neq \mathbf{v} \in \widetilde{\mathbf{W}}} \frac{\langle \boldsymbol{\lambda}, B_\Delta \mathbf{v}\rangle^2}{|\mathbf{v}|^2_{S_\varepsilon}}.$$

A similar formula holds for M; it only differs in the denominator from the one for F. For a proof, see Klawonn and Widlund [2004].

Lemma 2. *For any $\boldsymbol{\lambda} \in \mathbf{V}$, we have*

$$\langle M\boldsymbol{\lambda}, \boldsymbol{\lambda}\rangle = \sup_{0 \neq \mathbf{v} \in \widetilde{\mathbf{W}}} \frac{\langle \boldsymbol{\lambda}, B_\Delta \mathbf{v}\rangle^2}{|P_\Delta \mathbf{v}|^2_{S_\varepsilon}}.$$

For a proof of the lower bound in our main theorem, we will use the following lemma; cf. Klawonn and Widlund [2004].

Lemma 3. *For any $\boldsymbol{\mu} \in \mathbf{V}$, there exists a $\mathbf{w}_\Delta \in \widetilde{\mathbf{W}}_\Delta$ such that $\boldsymbol{\mu} = B_\Delta \mathbf{w}_\Delta$ and $(I - P_\Delta)\mathbf{w}_\Delta \in \widehat{\mathbf{W}}_\Pi$. In addition, $\mathbf{z}_w = P_\Delta \mathbf{w}_\Delta \in \widetilde{\mathbf{W}}$ and $\boldsymbol{\mu} = B_\Delta \mathbf{z}_w$.*

We now require P_Δ to satisfy a stability condition which is discussed for different cases in subsections 5.1, 5.2, and 5.3 and, with full details, in Klawonn and Widlund [2004].

Condition 1 *For all $\mathbf{w} \in \widetilde{\mathbf{W}}$, we have,*

$$|P_\Delta \mathbf{w}|^2_{S_\varepsilon} \leq C \, \max(1, TOL) \, (1 + \log(H/h))^2 |\mathbf{w}|^2_{S_\varepsilon}.$$

We note that this bound can be developed for individual subdomains and their next neighbors, one by one. Using Condition 1 and the three previous lemmas, we can now prove our condition number estimate.

Theorem 1. *Assume that Condition 1 holds. Then, the condition number satisfies*

$$\kappa(M^{-1}F) \leq C \, \max(1, TOL) \, (1 + \log(H/h))^2.$$

Here, C is independent of h, H, γ, and the values of the G_i.

Proof. We have to estimate the smallest eigenvalue $\lambda_{min}(M^{-1}F)$ from below and the largest eigenvalue $\lambda_{max}(M^{-1}F)$ from above. We will show that

$$\langle M\boldsymbol{\lambda}, \boldsymbol{\lambda}\rangle \leq \langle F\boldsymbol{\lambda}, \boldsymbol{\lambda}\rangle \leq C \, \max(1, TOL) \, (1+\log(H/h))^2 \langle M\boldsymbol{\lambda}, \boldsymbol{\lambda}\rangle \;\; \forall \boldsymbol{\lambda} \in \mathbf{V}. \tag{20}$$

Lower bound: The lower bound follows by using Lemmas 1, 2, and 3: for all $\boldsymbol{\lambda} \in \mathbf{V}$, we have

$$\langle M\boldsymbol{\lambda}, \boldsymbol{\lambda} \rangle = \sup_{\mathbf{w} \in \widetilde{\mathbf{W}}} \frac{\langle \boldsymbol{\lambda}, B_{\Delta}\mathbf{w} \rangle^2}{|P_{\Delta}\mathbf{w}|^2_{S_\varepsilon}} = \sup_{\mathbf{w} \in \widetilde{\mathbf{W}}} \frac{\langle \boldsymbol{\lambda}, B_{\Delta}\mathbf{z}_w \rangle^2}{|\mathbf{z}_w|^2_{S_\varepsilon}} \leq \sup_{\mathbf{z} \in \widetilde{\mathbf{W}}} \frac{\langle \boldsymbol{\lambda}, B_{\Delta}\mathbf{z} \rangle^2}{|\mathbf{z}|^2_{S_\varepsilon}} = \langle F\boldsymbol{\lambda}, \boldsymbol{\lambda} \rangle$$

Upper bound: Using Condition 1 and Lemmas 1 and 2, we obtain for all $\boldsymbol{\lambda} \in \mathbf{V}$

$$\begin{aligned} \langle F\boldsymbol{\lambda}, \boldsymbol{\lambda} \rangle &= \sup_{0 \neq \mathbf{w} \in \widetilde{\mathbf{W}}} \frac{\langle \boldsymbol{\lambda}, B_{\Delta}\mathbf{w} \rangle^2}{|\mathbf{w}|^2_{S_\varepsilon}} \\ &\leq C \max(1, TOL)\,(1 + \log(H/h))^2 \sup_{0 \neq \mathbf{w} \in \widetilde{\mathbf{W}}} \frac{\langle \boldsymbol{\lambda}, B_{\Delta}\mathbf{w} \rangle^2}{|P_{\Delta}\mathbf{w}|^2_{S_\varepsilon}} \\ &= C \max(1, TOL)\,(1 + \log(H/h))^2 \langle M\boldsymbol{\lambda}, \boldsymbol{\lambda} \rangle. \end{aligned}$$

We will now discuss Condition 1 successively for different cases.

5.1 First case

Let us first consider a decomposition of Ω, where no more than three subdomains are common to any edge and where each of the subdomains shares a face with each of the other two. We further assume that all vertices are primal and that for each face $\mathcal{F}^{ij}$ which is shared by two subdomains Ω_i and Ω_j, we have six linear functionals $f_m(\cdot)$ which satisfy Definition 1 and the property $f_m(\mathbf{w}^{(i)}) = f_m(\mathbf{w}^{(j)}) \ \ \forall \mathbf{w}^{(i)} \in \widetilde{\mathbf{W}}^{(i)}, \mathbf{w}^{(j)} \in \widetilde{\mathbf{W}}^{(j)}$. As mentioned before, cf. the first example after Definition 1, we can define our functionals f_i as properly chosen linear combinations of certain edge averages, over components of the displacement, of the form

$$g_m(\mathbf{w}^{(i)}) = \frac{\int_{\mathcal{E}^{ik}} w_\ell^{(i)}\, dx}{\int_{\mathcal{E}^{ik}} 1 dx},$$

where the $\mathcal{E}^{ik} \subset \partial\mathcal{F}^{ij}$ are appropriately chosen edges. Let us note that for a square face, we would have to work with three different edges to satisfy Definition 1.ii. For this case, we are able to prove Condition 1 with $TOL = 1$; see Klawonn and Widlund [2004] for a proof.

Lemma 4. *For all* $\mathbf{w} \in \widetilde{\mathbf{W}}$, *we have,*

$$|P_{\Delta}\mathbf{w}|^2_{S_\varepsilon} \leq C\,(1 + \log(H/h))^2 |\mathbf{w}|^2_{S_\varepsilon}.$$

Remark 1. The result of Lemma 4 still holds with an additional factor of $\max(1, TOL)$ for decompositions where more than three subdomains have a single edge in common and an acceptable path through the faces of those subdomains exists; cf. Definition 3. This is the case, e.g., if four subdomains

$\Omega_i, \Omega_j, \Omega_k$, and Ω_l have an edge in common as in Figure 3 (left) and the corresponding coefficients G_i, G_j, G_k, and G_l satisfy the condition

$$\min(G_i, G_k) \leq TOL \max(G_j, G_l)$$
$$\min(G_j, G_l) \leq TOL \max(G_i, G_k)$$

with a modest constant $TOL > 0$ independent of H, h, and the values of the G_i. This condition can be easily generalized to more than four subdomains meeting at an edge. Assuming that ε can become arbitrarily large or small, then this condition still rules out a checkerboard distribution as in Figure 3 (middle), but allows coefficient distributions as in Figure 3 (right).

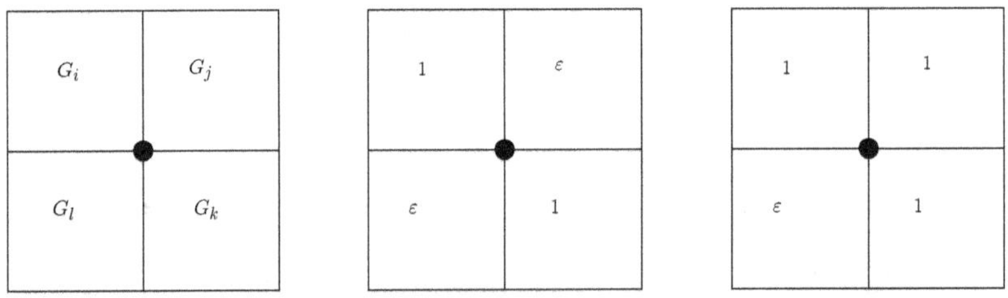

Fig. 3. Planar cut of four subdomains meeting at an edge

5.2 Second case

We again assume that all vertices are primal and also that any edge which is common to more than three subdomains is fully primal; cf. Definition 2. For such an edge, we have five linear functionals $f_m(\cdot)$ which satisfy Definition 1 and have the property $f_m(\mathbf{w}^{(i)}) = f_m(\mathbf{w}^{(k)})\ \ \forall \mathbf{w}^{(i)} \in \mathbf{W}^{(i)}, \mathbf{w}^{(k)} \in \mathbf{W}^{(k)}$. Here, Ω_i and Ω_k is any arbitrary pair of subdomains with such an edge $\mathcal{E}^{ik}$ in common. The functionals $f_m(\cdot), m = 1, \ldots, 5$, are defined in (16). For this case, as in Subsection 5.1, we are able to establish Condition 1 with $TOL = 1$; see Klawonn and Widlund [2004] for a proof.

Lemma 5. *For all* $\mathbf{w} \in \widetilde{\mathbf{W}}$, *we have*

$$|P_\Delta \mathbf{w}|^2_{S_\varepsilon} \leq C\,(1 + \log(H/h))^2 |\mathbf{w}|^2_{S_\varepsilon}.$$

5.3 Third case

Finally, we show that it is often possible to use a smaller number of fully primal edges and to have fewer primal vertices. The next lemma is proven in Klawonn and Widlund [2004] under the assumptions that there are at least six linearly independent edge constraints across any face of the interface and that there is an acceptable path for each pair of subdomains that share an edge or vertex. We have,

Lemma 6. *For all* $\mathbf{w} \in \widetilde{\mathbf{W}}$, *we have,*

$$|P_{\Delta}\mathbf{w}|_{S_\varepsilon}^2 \leq C \max(1, TOL)\,(1 + \log(H/h))^2 |\mathbf{w}|_{S_\varepsilon}^2.$$

References

P. G. Ciarlet. *Mathematical Elasticity Volume I: Three–Dimensional Elasticity.* North-Holland, 1988.

C. Farhat, M. Lesoinne, P. LeTallec, K. Pierson, and D. Rixen. FETI-DP: A dual-primal unified FETI method - part i: A faster alternative to the two-level FETI method. *Int. J. Numer. Meth. Engrg.*, 50:1523–1544, 2001.

C. Farhat, M. Lesoinne, and K. Pierson. A scalable dual-primal domain decomposition method. *Numer. Lin. Alg. Appl.*, 7:687–714, 2000.

A. Klawonn, O. Rheinbach, and O. B. Widlund. Some computational results for Dual-Primal FETI methods for three dimensional elliptic problems. In R. Kornhuber, R. Hoppe, D. Keyes, J. Périaux, O. Pironneau, and J. Xu, editors, *Domain Decomposition Methods.* Springer-Verlag, Lecture Notes in Computational Science and Engineering, 2003. Proceedings of the 15th International Conference on Domain Decomposition Methods, Berlin, July 21-25, 2003.

A. Klawonn and O. Widlund. Dual-Primal FETI methods for linear elasticity. Technical report, Courant Institute of Mathematical Sciences, Department of Computer Science, 2004. In preparation.

A. Klawonn and O. B. Widlund. FETI and Neumann–Neumann iterative substructuring methods: Connections and new results. *Comm. Pure Appl. Math.*, 54:57–90, January 2001.

A. Klawonn, O. B. Widlund, and M. Dryja. Dual-Primal FETI methods for three-dimensional elliptic problems with heterogeneous coefficients. *SIAM J.Numer.Anal.*, 40, 159-179 2002a.

A. Klawonn, O. B. Widlund, and M. Dryja. Dual-Primal FETI methods with face constraints. In L. F. Pavarino and A. Toselli, editors, *Recent developments in domain decomposition methods*, pages 27–40. Springer-Verlag, Lecture Notes in Computational Science and Engineering, Volume 23, 2002b.

J. Mandel and R. Tezaur. On the convergence of a dual-primal substructuring method. *Numer. Math.*, 88:543–558, 2001.

Coupled Boundary and Finite Element Tearing and Interconnecting Methods

Ulrich Langer[1] and Olaf Steinbach[2]

[1] Johannes Kepler University Linz, Institute of Computational Mathematics (http://www.numa.uni-linz.ac.at/Staff/langer/)
[2] University of Stuttgart, Institute for Applied Analysis and Numerical Simulation (http://www.ians.uni-stuttgart/LstAngMath/Steinbach/)

Summary. We have recently introduced the Boundary Element Tearing and Interconnecting (BETI) methods as boundary element counterparts of the well-established Finite Element Tearing and Interconnecting (FETI) methods. Since Finite Element Methods (FEM) and Boundary Element Methods (BEM) have certain complementary properties, it is sometimes very useful to couple these discretization techniques and to benefit from both worlds. Combining our BETI techniques with the FETI methods gives new, quite attractive tearing and interconnecting parallel solvers for large scale coupled boundary and finite element equations. There is an unified framework for coupling, handling, and analyzing both methods. In particular, the FETI methods can benefit from preconditioning components constructed by boundary element techniques. This is especially true for sparse versions of the boundary element method such as the fast multipole method which avoid fully populated matrices arising in classical boundary element methods.

1 Introduction

In Langer and Steinbach [2003] we have recently introduced the Boundary Element Tearing and Interconnecting (BETI) methods as boundary element counterparts of the well-established Finite Element Tearing and Interconnecting (FETI) methods. The first FETI methods were introduced by Farhat and Roux [1991] (see also Farhat and Roux [1994] for a more detailed description by the same authors). Since then the FETI methods have successfully been applied to different engineering problems, new FETI versions have been proposed (see Farhat et al. [2000]), and the analysis has been developed as well (see Mandel and Tezaur [1996, 2001], Klawonn and Widlund [2001], Klawonn et al. [2002], Brenner [2003]). Nowadays, the FETI method is one of the most widely used domain decomposition (DD) methods in parallel codes including commercial codes. This success of the FETI methods is certainly related to the wide applicability of the FETI methods, the possibility of the use of standard components in the solution process, the moderate dependence of the iteration number on the complexity of the problem (see Mandel and Tezaur [1996], Klawonn and Widlund [2001], Brenner [2003]), the scalability (see, e.g., Stefanica [2001]) and, last but not least, the robustness (see Klawonn and Widlund [2001], Klawonn et al. [2002], Brenner [2003]). These facts are true for the BETI methods as well (see Langer and Steinbach [2003]).

In this paper we generalize the tearing and interconnecting technique to symmetric coupled boundary and finite element equations. Since Finite Element Methods (FEM) and Boundary Element Methods (BEM) have certain complementary properties, it is sometimes very useful to couple these discretization techniques and to benefit from the advantages of both worlds. This concerns not only the treatment of unbounded domains (BEM), but also the right handling of singularities (BEM), moving parts (BEM), air regions in electromagnetics (BEM), source terms (FEM), non-linearities (FEM) etc. The symmetric coupling of BEM and FEM goes back to Costabel [1987]. During the last decade iterative substructuring solvers for symmetric coupled boundary and finite element equations have been developed by Langer [1994], Haase et al. [1998], Hsiao et al. [2000], Steinbach [2003] for elliptic boundary value problems in bounded and unbounded, two and three-dimensional domains, and have been successfully applied to real-life problems. Parallel implementations showed high performance on several platforms (see Haase et al. [1998]). Especially in 3D, the preconditioning of the global (assembled) boundary and finite element Schur complement, living on the skeleton of the domain decomposition, is the crucial point for constructing an efficient iterative substructuring method. This weak point of iterative substructuring methods can be avoided by the dual approach. Indeed, combining our BETI techniques with the FETI methods gives new, quite attractive tearing and interconnecting parallel solvers for large scale coupled boundary and finite element equations. Moreover, there is an unified framework for coupling, handling, and analyzing both methods. In particular, the FETI methods can benefit from preconditioning components constructed by boundary element techniques. This is especially true for sparse versions of the boundary element method. Sparse approximation techniques such as the fast multipole method (see Greengard and Rokhlin [1987]) avoid fully populated matrices arising in classical boundary element methods. Our sparse hypersingular BETI/FETI-preconditioner that is based on symmetry and kernel preserving fast multipole techniques requires only $O((H/h)^{d-1}(\log(H/h))^2)$ arithmetical operations in a parallel regime, where d is the dimension of our computational domain ($d = 2$, or 3), and H and h denote the usual scaling of the subdomains and the elements, respectively. Similar to the FETI methods, the relative spectral condition number grows only like $O((1 + \log(H/h))^2)$ and is independent of the jumps in the coefficients of the partial differential equation.

The rest of the paper is organized as follows. In Section 2, we introduce the coupled BETI/FETI techniques for solving large scale coupled boundary and finite element DD equations, and discuss some algorithmical aspects. Section 3 is devoted to the preconditioning and analysis of the combined BETI/FETI solver. Finally, in Section 4, we draw some conclusions for using and developing the tearing and interconnecting technique in both the boundary and finite element worlds.

2 Formulation of BETI/FETI

For a bounded domain $\Omega \subset \mathbb{R}^d$ $(d = 2,3)$ with Lipschitz boundary $\Gamma = \partial\Omega$ we consider the Dirichlet boundary value problem

$$-\operatorname{div}\left[\alpha(x)\nabla u(x)\right] = f(x) \quad \text{for } x \in \Omega, \quad u(x) = g(x) \quad \text{for } x \in \Gamma. \tag{1}$$

We assume that there is given a non–overlapping quasi regular domain decomposition,

$$\overline{\Omega} = \bigcup_{i=1}^{p} \overline{\Omega}_i, \Omega_i \cap \Omega_j = \emptyset \text{ for } i \neq j, \Gamma_i = \partial\Omega_i, \Gamma_{ij} = \Gamma_i \cap \Gamma_j, \Gamma_S = \bigcup_{i=1}^{p} \Gamma_i.$$

Moreover, we assume that the coefficient function α is piecewise constant,

$$\alpha(x) = \alpha_i \quad \text{for } x \in \Omega_i, i = 1, \ldots, p.$$

Instead of the global boundary value problem (1) we now consider the local problems

$$-\alpha_i \Delta u_i(x) = f(x) \quad \text{for } x \in \Omega_i, \quad u_i(x) = g(x) \quad \text{for } x \in \Gamma_i \cap \Gamma \tag{2}$$

together with the transmission conditions

$$u_i(x) = u_j(x), \quad \alpha_i \frac{\partial}{\partial n_i} u_i(x) + \alpha_j \frac{\partial}{\partial n_j} u_j(x) = 0 \quad \text{for } x \in \Gamma_{ij}. \tag{3}$$

The solution of the local Dirichlet boundary value problems

$$-\alpha_i \Delta u_i(x) = f(x) \quad \text{for } x \in \Omega_i, \quad u_i(x) = g_i(x) \quad \text{for } x \in \Gamma_i \tag{4}$$

defines the Dirichlet–Neumann map

$$t_i(x) := \alpha_i \frac{\partial}{\partial n_i} u_i(x) = (S_i u_i)(x) - (N_i f)(x) \quad \text{for } x \in \Gamma_i \tag{5}$$

with the Steklov–Poincaré operator $S_i : H^{1/2}(\Gamma_i) \to H^{-1/2}(\Gamma_i)$ and with some Newton potential $N_i : \widetilde{H}^{-1}(\Omega_i) \to H^{-1/2}(\Gamma_i)$, see, e.g., Steinbach [2003]. The coupled boundary value problem (2)–(3) is therefore equivalent to find $(u_i, t_i) \in H^{1/2}(\Gamma_i) \times H^{-1/2}(\Gamma_i)$ for $i = 1, \ldots, p$ such that

$$\begin{aligned}
t_i(x) &= (S_i u_i)(x) - (N_i f)(x) && \text{for } x \in \Gamma_i, \\
u_i(x) &= g(x) && \text{for } x \in \Gamma_i \cap \Gamma, \\
u_i(x) &= u_j(x) && \text{for } x \in \Gamma_{ij}, \\
0 &= t_i(x) + t_j(x) && \text{for } x \in \Gamma_{ij}.
\end{aligned}$$

Let $\widetilde{g} \in H^1(\Omega)$ be some arbitrary but fixed bounded extension of the given Dirichlet data $g \in H^{1/2}(\Gamma)$. Introducing the trace space $H^{1/2}(\Gamma_S)$ on the skeleton Γ_S and the subspace

$$H_0^{1/2}(\Gamma_S, \Gamma) := \left\{ v \in H^{1/2}(\Gamma_S) \, : \, v(x) = 0 \quad \text{for } x \in \Gamma \right\},$$

we arrive at the skeleton problem: find a function $u_0 \in H_0^{1/2}(\Gamma_S, \Gamma)$ such that

$$(S_i u_i)(x) + (S_j u_j)(x) = (N_i f)(x) + (N_j f)(x) \quad \text{for } x \in \Gamma_{ij}$$

is satisfied on all local coupling boundaries Γ_{ij} and where $u_i(x) := u_0(x) + \widetilde{g}(x)$ for $x \in \Gamma_i$. The resulting variational problem is to find $u_0 \in H_0^{1/2}(\Gamma_S, \Gamma)$ such that

$$\sum_{i=1}^{p} \int_{\Gamma_i} (S_i u_0)(x) v(x) ds_x = \sum_{i=1}^{p} \int_{\Gamma_i} [(N_i f)(x) - (S_i \widetilde{g})(x)] v(x) ds_x \tag{6}$$

is satisfied for all $v \in H_0^{1/2}(\Gamma_S, \Gamma)$.

Let

$$S_h^1(\Gamma_S) = \text{span}\{\varphi_k^1\}_{k=1}^M \subset H_0^{1/2}(\Gamma_S, \Gamma)$$

be a conformal finite dimensional trial space of piecewise linear continuous basis functions φ_k^1. By

$$S_h^1(\Gamma_i) := S_h^1(\Gamma_S)_{|\Gamma_i} = \text{span}\{\varphi_{k,i}^1\}_{k=1}^{M_i} \subset H_0^{1/2}(\Gamma_i, \Gamma)$$

we denote the restriction of the global trial space $S_h^1(\Gamma_S)$ onto the local subdomain boundaries Γ_i. For a global vector $\underline{v} \in \mathbb{R}^M$ and the corresponding finite element function $v_h \in S_h^1(\Gamma_S)$ ($\underline{v} \in \mathbb{R}^M \leftrightarrow v_h \in S_h^1(\Gamma_S)$), we consider the restriction $v_{h,i} := v_{h|\Gamma_i} \in S_h^1(\Gamma_i) \leftrightarrow \underline{v}_i \in \mathbb{R}^{M_i}$. Using connectivity matrices $A_i \in \mathbb{R}^{M_i \times M}$ this can be written as $\underline{v}_i = A_i \underline{v}$. Due to the implicit definition of the local Dirichlet–Neumann map (5) it is in general not possible to discretize the variational problem (6) in an exact manner. Hence we have to approximate the local Dirichlet problems which are involved in the definition of the local Dirichlet to Neumann map. This can be done either by finite or boundary elements, see Steinbach [2003].

We start to consider a finite element approximation of the local Steklov–Poincaré operators S_i to realize the Dirichlet to Neumann map in the subdomains Ω_i, $i = 1, \ldots, q \leq p$. Let

$$S_h^1(\Omega_i) := \text{span}\{\phi_{\kappa,i}^1\}_{\kappa=1}^{\bar{M}_i} \subset H_0^1(\Omega_i)$$

be the local finite element spaces of piecewise linear and continuous basis functions $\phi_{\kappa,i}^1$ which vanish on the subdomain boundary Γ_i. The finite element discretization of the local Dirichlet boundary value problems (4) for given Dirichlet data $\underline{u}_{C,i}$ then leads to the linear systems

$$\begin{pmatrix} K_{II,i} & K_{CI,i} \\ K_{CI,i}^\top & K_{CC,i} \end{pmatrix} \begin{pmatrix} \underline{u}_{I,i} \\ \underline{u}_{C,i} \end{pmatrix} = \begin{pmatrix} \underline{f}_{I,i} \\ \underline{f}_{C,i} \end{pmatrix} \tag{7}$$

with block matrices defined by

$$K_{CC,i}[\ell, k] := \int_{\Omega_i} \alpha_i \nabla \varphi_{k,i}^1(x) \nabla \varphi_{\ell,1}^1(x) dx,$$

$$K_{II,i}[\lambda, \kappa] := \int_{\Omega_i} \alpha_i \nabla \phi_{\kappa,i}^1(x) \nabla \phi_{\lambda,i}^1(x) dx,$$

$$K_{CI}[\lambda, k] := \int_{\Omega_i} \alpha_i \nabla \varphi_{k,i}^1(x) \nabla \phi_{\lambda,i}^1(x) dx$$

and the right-hand side

$$f_{I,i,\lambda} := \int_{\Omega_i} f(x) \phi_{\lambda,i}^1(x) dx - \int_{\Omega_i} \alpha_i \nabla \widetilde{g}(x) \nabla \phi_{\lambda,i}^1(x) dx,$$

$$f_{C,i,\ell} := \int_{\Omega_i} f(x) \varphi_{\ell,i}^1(x) dx - \int_{\Omega_i} \alpha_i \nabla \widetilde{g}(x) \nabla \varphi_{\ell,i}^1(x) dx$$

for all $k, \ell = 1, \ldots, M_i$ and $\kappa, \lambda = 1, \ldots, \bar{M}_i$. Eliminating $\underline{u}_{I,i}$, we now obtain the finite element approximation

$$S_{i,h}^{\text{FEM}} \underline{u}_{C,i} = \left[K_{CC,i} - K_{CI,i}^\top K_{II,i}^{-1} K_{CI,i} \right] \underline{u}_{C,i} = \underline{f}_{C,i} - K_{CI,i}^\top K_{II,i}^{-1} \underline{f}_{I,i} = \underline{f}_i^{\text{FEM}}$$

of the local Dirichlet to Neumann map (5). In the finite element subdomains the coefficients α_i can, of course, depend on x, but should not vary too much on Ω_i for $i = 1, \ldots, q$.

In the remaining subdomains Ω_i, $i = q+1, \ldots, p$ we assume $f(x) = 0$ for $x \in \Omega_i$. Hence we may use a symmetric boundary element method to approximate the local Steklov–Poincaré operators S_i. The fundamental solution of the Laplace operator is given by

$$U^*(x, y) = \begin{cases} -\dfrac{1}{2\pi} \log |x - y| & \text{for } d = 2, \\ \dfrac{1}{4\pi} \dfrac{1}{|x - y|} & \text{for } d = 3. \end{cases}$$

The relation between the local Cauchy data $[t_i, u_i]$ can then be described by the system of boundary integral equations (Calderón projection),

$$\begin{pmatrix} u_i \\ t_i \end{pmatrix} = \begin{pmatrix} \frac{1}{2} I - K_i & V_i \\ D_i & \frac{1}{2} I + K_i' \end{pmatrix} \begin{pmatrix} u_i \\ t_i \end{pmatrix},$$

where we used the standard notations for the single layer potential operator V_i, for the double layer potential operator K_i and its adjoint K_i' and for the hypersingular integral operator D_i defined by

$$(V_i t_i)(x) := \alpha_i \int_{\Gamma_i} U^*(x,y) t_i(y) ds_y \quad \text{for } x \in \Gamma_i,$$

$$(K_i u_i)(x) := \alpha_i \int_{\Gamma_i} \frac{\partial}{\partial n_y} U^*(x,y) u_i(y) ds_y \quad \text{for } x \in \Gamma_i,$$

$$(K_i' t_i)(x) := \alpha_i \int_{\Gamma_i} \frac{\partial}{\partial n_x} U^*(x,y) t_i(y) ds_y \quad \text{for } x \in \Gamma_i,$$

$$(D_i u_i)(x) := -\alpha_i \frac{\partial}{\partial n_x} \int_{\Gamma_i} \frac{\partial}{\partial n_y} U^*(x,y) u_i(y) ds_y \quad \text{for } x \in \Gamma_i,$$

respectively. The mapping properties of all of these boundary integral operators are well known (see Costabel [1988]), in particular, the local single layer potential $V_i : H^{-1/2}(\Gamma_i) \to H^{1/2}(\Gamma_i)$ is $H^{-1/2}(\Gamma_i)$–elliptic and therefore invertible (Hsiao and Wendland [1977]); for $d = 2$ we assume $\operatorname{diam} \Omega_i < 1$ that can be always obtained by scaling the computational domain. Then we obtain a symmetric boundary integral operator representation of the Steklov–Poincaré operator $S_i : H^{1/2}(\Gamma_i) \to H^{-1/2}(\Gamma_i)$,

$$(S_i u_i)(x) = \left[D_i + (\frac{1}{2} I + K_i') V_i^{-1} (\frac{1}{2} I + K_i) \right] u_i(x) \quad \text{for } x \in \Gamma_i.$$

Let

$$S_h^0(\Gamma_i) = \operatorname{span}\{\psi_{\kappa,i}^0\}_{\kappa=1}^{N_i} \subset H^{-1/2}(\Gamma_i) \tag{8}$$

be the trial space of piecewise constant basis functions $\psi_{\kappa,i}^0$ to approximate the local Neumann data $t_i \in H^{-1/2}(\Gamma_i)$. This Galerkin approximation of the local Steklov–Poincaré operator S_i gives the matrix representation

$$S_{i,h}^{\mathrm{BEM}} := D_{i,h} + (\frac{1}{2} M_{i,h}^\top + K_{i,h}^\top) V_{i,h}^{-1} (\frac{1}{2} M_{i,h} + K_{i,h})$$

with

$$\begin{aligned} D_{i,h}[\ell, k] &= \langle D_i \varphi_{k,i}^1, \varphi_{\ell,i}^1 \rangle_{L_2(\Gamma_i)}, \\ V_{i,h}[\lambda, \kappa] &= \langle V_i \psi_{\kappa,i}^0, \psi_{\lambda,i}^0 \rangle_{L_2(\Gamma_i)}, \\ K_{i,h}[\lambda, k] &= \langle K_i \varphi_{k,i}^1, \psi_{\lambda,i}^0 \rangle_{L_2(\Gamma_i)}, \\ M_{i,h}[\lambda, k] &= \langle \varphi_{k,i}^1, \psi_{\lambda,i}^0 \rangle_{L_2(\Gamma_i)} \end{aligned}$$

for all $k, \ell = 1, \ldots, M_i$ and $\kappa, \lambda = 1, \ldots, N_i$.

Note that both finite and boundary element approximations $S_{i,h}^{\text{BEM/FEM}}$ of the local Steklov–Poincaré operators S_i are symmetric and spectrally equivalent to the exact Galerkin matrices $S_{i,h}$. This holds true for an almost arbitrary choice of the local trial spaces $S_h^1(\Omega_i)$ and $S_h^0(\Gamma_i)$, respectively, see Steinbach [2003]. Moreover, the error $S_{i,h} - S_{i,h}^{\text{BEM/FEM}}$ of the approximate Steklov–Poincaré operators can be controlled by the approximation properties of the local trial spaces $S_h^1(\Omega_i)$ and $S_h^0(\Gamma_i)$, respectively.

The Galerkin discretization of the variational problem (6) with the boundary and finite element approximations of the local Dirichlet problems discussed above leads now to the linear system

$$\sum_{i=1}^{q} A_i^\top S_{i,h}^{\text{FEM}} A_i \underline{u} + \sum_{i=q+1}^{p} A_i^\top S_{i,h}^{\text{BEM}} A_i \underline{u} = \sum_{i=1}^{q} A_i^\top \underline{f}_i^{\text{FEM}} - \sum_{i=q+1}^{p} A_i^\top S_{i,h}^{\text{BEM}} A_i \underline{g} \quad (9)$$

which is uniquely solvable due to the positive definiteness of the assembled stiffness matrix. Discretization error estimates are given in Steinbach [2003]. The aim of tearing and interconnecting domain decomposition methods is to design efficient solution strategies to solve the global linear system (9). When introducing local vectors $\underline{u}_i = A_i \underline{u}$ the continuity of the primal variables across the interfaces can be written by the constraint

$$\sum_{i=1}^{p} B_i \underline{u}_i = \underline{0}$$

where $B_i \in \mathbb{R}^{M \times M_i}$. Each row of the matrix $B = (B_1, \dots, B_p)$ is connected with a pair of matching nodes across the interface. The entries of such a row are 1 and -1 for the indices corresponding to the matching nodes and 0 otherwise. By introducing the Lagrange multiplier $\underline{\lambda} \in \mathbb{R}^M$ we have to solve the linear system

$$\begin{pmatrix} S_{1,h}^{\text{FEM}} & & & & & & B_1^\top \\ & \ddots & & & & & \vdots \\ & & S_{q,h}^{\text{FEM}} & & & & B_q^\top \\ & & & S_{q+1,h}^{\text{BEM}} & & & B_{q+1}^\top \\ & & & & \ddots & & \vdots \\ & & & & & S_{p,h}^{\text{BEM}} & B_p^\top \\ B_1 & \cdots & B_q & B_{q+1} & \cdots & B_p & 0 \end{pmatrix} \begin{pmatrix} \underline{u}_1 \\ \vdots \\ \underline{u}_q \\ \underline{u}_{q+1} \\ \vdots \\ \underline{u}_p \\ \underline{\lambda} \end{pmatrix} = \begin{pmatrix} \underline{f}_1^{\text{FEM}} \\ \vdots \\ \underline{f}_q^{\text{FEM}} \\ \underline{f}_{q+1}^{\text{BEM}} \\ \vdots \\ \underline{f}_p^{\text{BEM}} \\ \underline{0} \end{pmatrix} \quad (10)$$

with $\underline{f}_i^{\text{BEM}} := -S_{i,h}^{\text{BEM}} A_i \underline{g}$ for $i = q+1, \dots, p$. For $i = 1, \dots, p$ we now consider the solvability of the local systems

$$S_{i,h}^{\text{FEM/BEM}} \underline{u}_i = \underline{f}_i^{\text{FEM/BEM}} - B_i^\top \underline{\lambda}. \quad (11)$$

For a unique framework we define the modified matrices

$$\widetilde{S}_{i,h}^{\mathrm{BEM/FEM}} := S_{i,h}^{\mathrm{BEM/FEM}} + \beta_i \underline{e}_i \underline{e}_i^\top$$

with $\beta_i = 0$ for non–floating subdomains Ω_i, i.e., the subdomain boundary $\Gamma_i = \partial\Omega_i$ contains some part of the Dirichlet boundary $\Gamma = \partial\Omega$, and some suitable chosen $\beta_i > 0$ for floating subdomains Ω_i with $\Gamma_i \cap \Gamma = \emptyset$. When requiring the solvability condition

$$\underline{e}_i^\top \left[\underline{f}_i^{\mathrm{BEM/FEM}} - B_i^\top \underline{\lambda} \right] = 0, \tag{12}$$

the local linear systems (11) are equivalent to the modified systems

$$\widetilde{S}_{i,h}^{\mathrm{FEM/BEM}} \underline{u}_i = \underline{f}_i^{\mathrm{FEM/BEM}} - B_i^\top \underline{\lambda} \tag{13}$$

which are now unique solvable. However, for floating subdomains we have to incorporate the rigid body motions. Hence the general solutions of the modified linear systems (13) are given by

$$\underline{u}_i = \left[\widetilde{S}_{i,h}^{\mathrm{BEM/FEM}} \right]^{-1} \left[\underline{f}_i^{\mathrm{BEM/FEM}} - B_i^\top \underline{\lambda} \right] + \gamma_i \underline{e}_i \tag{14}$$

with $\gamma_i = 0$ for all non–floating subdomains. Inserting these local solutions into the last equation of (10) we obtain the Schur complement system

$$\sum_{i=1}^p B_i \left[\widetilde{S}_{i,h}^{\mathrm{BEM/FEM}} \right]^{-1} B_i^\top \underline{\lambda} - \sum_{i=1}^p \gamma_i B_i \underline{e}_i = \sum_{i=1}^p B_i \left[\widetilde{S}_{i,h}^{\mathrm{BEM/FEM}} \right]^{-1} \underline{f}_i^{\mathrm{BEM/FEM}}$$

where the compatibility condition (12) is to be assumed for all floating subdomains. Hence we have to solve the linear system

$$\begin{pmatrix} F & -G \\ G^\top & \end{pmatrix} \begin{pmatrix} \underline{\lambda} \\ \underline{\gamma} \end{pmatrix} = \begin{pmatrix} \underline{d} \\ \underline{e} \end{pmatrix} \tag{15}$$

with

$$F := \sum_{i=1}^p B_i \left[\widetilde{S}_{i,h}^{\mathrm{BEM/FEM}} \right]^{-1} B_i^\top, \quad G := (B_i \underline{e}_i)_{i:\Gamma_i \cap \Gamma = \emptyset}$$

and

$$\underline{d} := \sum_{i=1}^p B_i \left[\widetilde{S}_{i,h}^{\mathrm{BEM/FEM}} \right]^{-1} \underline{f}_i^{\mathrm{BEM/FEM}}, \quad \underline{e} := \left(\underline{e}_i^\top \underline{f}_i^{\mathrm{FEM/BEM}} \right)_{i:\Gamma_i \cap \Gamma = \emptyset}.$$

Defining now the orthogonal projection

$$P := I - G(G^\top G)^{-1} G^\top : \Lambda := \mathbb{R}^M \to \Lambda_0 := \ker G^\top = (\operatorname{range} G)^\perp$$

with respect to the Euclidean scalar product, we can split the computation of λ from the definition of $\underline{\gamma}$. Indeed, applying P to the first equation in (15) gives the equation

$$P F \underline{\lambda} = \underline{P} \underline{d} \tag{16}$$

since $PG\underline{\gamma} = \underline{0}$. Once λ is determined by solving (16), we obtain

$$\gamma := (G^\top G)^{-1} G^\top (F\underline{\lambda} - \underline{d}).$$

Finally, we get the vectors $\underline{u}_i$ from (14). Let us mention that in the case of jumping coefficients the scalar product in Λ has to be changed according to the proposal made by Klawonn and Widlund [2001] on pages 63 and 75 (see also Brenner [2003]).

The dual problem (16) is solved by a preconditioned conjugate gradient subspace iteration. The matrix-by-vector multiplication with the stiffness matrix F involves the application of the inverse modified discrete Steklov–Poincaré operators $[\widetilde{S}_{i,h}^{\text{BEM/FEM}}]^{-1}$ to some vector $B_i^\top \underline{\lambda}$ resulting in some $\underline{w}_i = [\widetilde{S}_{i,h}^{\text{BEM/FEM}}]^{-1} B_i^\top \underline{\lambda}$. This can be done by solving directly extended systems for the local boundary and finite element Neumann problems. This is the standard technique in the FETI methods (see Langer and Steinbach [2003] for BETI). Khoromskij et al. [2004] propose the application of the $\mathcal{H}$-matrix technique for an approximate inversion of the boundary and finite element Schur complements resulting in a sparse representation of the approximate inverse Schur complements in $\mathcal{H}$-matrix formate. This representation allows us to perform this matrix-by-vector multiplication with almost optimal complexity. Other approaches are discussed by Langer and Steinbach [2003].

3 Preconditioners and Analysis

In this section we will describe and analyze an efficient solution of the linear system (16) by some projected preconditioned conjugate gradient method. The preconditioning matrix C to be used in the PCG algorithm should be spectrally equivalent to the matrix F on the subspace $\Lambda_0 = \ker G^\top$, i.e.

$$c_1\,(C\underline{\lambda}, \underline{\lambda}) \le (F\underline{\lambda}, \underline{\lambda}) \le c_2\,(C\underline{\lambda}, \underline{\lambda}) \quad \text{for all } \underline{\lambda} \in \Lambda_0 \tag{17}$$

with positive spectral equivalence constants c_1 and c_2 such that the relative spectral condition number $\kappa(PC^{-1}P^\top P^\top FP)$ respectively its bound c_2/c_1 is as small as possible and the application of the preconditioner is as cheap as possible.

Following the FETI approach a first preconditioner is built from the local Schur complements $S_{i,h}^{\text{FEM/BEM}}$,

$$C_{\text{FETI}}^{-1} := (BC_\alpha^{-1}B^\top)^{-1} BC_\alpha^{-1} \left[\sum_{i=1}^{p} B_i S_{i,h}^{\text{FEM/BEM}} B_i^\top \right] C_\alpha^{-1} B^\top (BC_\alpha^{-1}B^\top)^{-1}$$

where $C_\alpha = \operatorname{diag}(C_{\alpha,i})_{i=1:p}$ and $C_{\alpha,i} = \operatorname{diag}(c_\ell^i)_{\ell=1:M_i}$ are diagonal matrices with appropriately chosen weights c_ℓ^i, e.g. as proposed in Klawonn and Widlund [2001], see also Brenner [2003].

The proof of the spectral equivalence inequalities (17) is essentially based on the spectral equivalence inequalities of the approximated Steklov–Poincaré operators $S_{i,h}^{\text{FEM/BEM}}$ with the exact Galerkin approximation $S_{i,h}$. The application C_{FETI}^{-1} of the preconditioning matrix C_{FETI} mainly consists in the application of the local approximate Steklov–Poincaré operators $S_{i,h}^{\text{FEM/BEM}}$, i.e. the solution of local Dirichlet boundary value problems by either finite or boundary element methods. Here we will propose a more efficient preconditioning strategy when replacing the approximate Steklov–Poincaré operators $S_{i,h}^{\text{FEM/BEM}}$ by discrete hypersingular integral operators $D_{i,h}$ which are defined with respect to all subdomain boundaries Γ_i and $i = 1, \ldots, p$.

Lemma 1. *The local boundary element Schur complement matrix $S_{i,h}^{BEM}$ and the local finite element Schur complement matrix $S_{i,h}^{FEM}$ are spectrally equivalent to the exact Galerkin matrix $S_{i,h}$ of the local Steklov–Poincaré operator S_i and to the boundary element matrix $D_{i,h}$ of the local hypersingular boundary integral operator D_i, i.e.*

$$S_{i,h}^{BEM} \simeq S_{i,h}^{FEM} \simeq S_{i,h} \simeq D_{i,h}$$

for all $i = 1, \ldots, p$, where $A \simeq B$ means that the matrices A and B are spectrally equivalent.

Proof. It is well known (see, e.g., theorem 3.5, p. 64 in Steinbach [2003]), that the finite element Schur complement is spectrally equivalent to the $H^{-1/2}(\Gamma_i)$–semi–norm squared, i.e., there exist universal positive constants c_1 and c_2 such that

$$c_1 \, |v_{i,h}|^2_{H^{1/2}(\Gamma_i)} \leq (S_{i,h}^{\text{FEM}} \underline{v}_i, \underline{v}_i) \leq c_2 \, |v_{i,h}|^2_{H^{1/2}(\Gamma_i)}$$

is satisfied for all $v_{i,h} \in S_h^1(\Gamma_i) \leftrightarrow \underline{v}_i \in \mathbb{R}^{M_i}$. On the other hand, the boundary element Schur complement $S_{i,h}^{\text{BEM}}$ is spectrally equivalent to the Galerkin matrix $D_{i,h}$ of the local hypersingular boundary integral operator $D_{i,h}$ see lemma 3.1 in Langer and Steinbach [2003]. Since the energy of the local hypersingular integral operator D_i is also equivalent to the $H^{1/2}(\Gamma_i)$–semi–norm squared, the proof is completed. □

The resulting scaled hypersingular BETI preconditioner is now given by

$$C_{\text{BETI}}^{-1} := (BC_\alpha^{-1}B^\top)^{-1}BC_\alpha^{-1}\left[\sum_{i=1}^{p} B_i D_{i,h} B_i^\top\right] C_\alpha^{-1}B^\top(BC_\alpha^{-1}B^\top)^{-1}. \quad (18)$$

Theorem 1 (Theorem 3.1 in Langer and Steinbach [2003]). *For the scaled hypersingular BETI preconditioner (18), the condition estimate*

$$\kappa(PC_{BETI}^{-1}P^\top P^\top FP) \leq c\left(1 + \log\frac{H}{h}\right)^2$$

holds, where the positive constant c is independent of the local mesh size h, the average subdomain size H, the number p of subdomains and of the coefficients

α_i (coefficient jumps). The matrix by vector operation $D_{i,h}\underline{v}_i$ that is the most expensive operation in the preconditioning step costs ops$(D_{i,h}\underline{v}_i) = \mathcal{O}((H/h)^2)$ and ops$(D_{i,h}\underline{v}_i) = \mathcal{O}((H/h)^4)$ arithmetical operations for $d = 2$ and for $d = 3$, respectively.

To obtain a more efficient preconditioning strategy we may use some fast boundary element method such as the fast multipole method to realize the local matrix by vector multiplication with $D_{i,h}$. The resulting sparse version of the scaled hypersingular BETI preconditioner then reads

$$C_{\text{sBETI}}^{-1} := (BC_\alpha^{-1}B^\top)^{-1}BC_\alpha^{-1}\left[\sum_{i=1}^{p} B_i\widetilde{D}_{i,h}B_i^\top\right]C_\alpha^{-1}B^\top(BC_\alpha^{-1}B^\top)^{-1}. \quad (19)$$

We start the analysis of the sparse hypersingular BETI preconditioner (19) with some considerations of the local sparse approximations $\widetilde{D}_{i,h}$. Using integration by parts (see Nédélec [1982]) the bilinear form of the local hypersingular boundary integral operator D_i can be rewritten as

$$\langle D_i u_i, v_i\rangle_{L_2(\Gamma_i)} = -\frac{\alpha_i}{2\pi}\int_{\Gamma_i}\dot{v}(x)\int_{\Gamma_i}\log|x-y|\dot{u}(y)ds_y ds_x$$

for $d = 2$ where $\dot{u}$ means the derivative with respect to the arc length. Similarly, for $d = 3$ we have

$$\langle D_i u_i, v_i\rangle_{L_2(\Gamma_i)} = \frac{\alpha_i}{4\pi}\int_{\Gamma_i}\int_{\Gamma_i}\frac{\text{curl}_{\Gamma_i}u_i(y)\cdot\text{curl}_{\Gamma_i}v_i(x)}{|x-y|}ds_y ds_x,$$

where

$$\text{curl}_{\Gamma_i}u_i(x) := n_i(x)\times\nabla_x u_i^*(x) \quad \text{for } x\in\Gamma_i$$

is the surface curl and u_i^* is an extension of u_i into a neighborhood of Γ_i. When using an interface triangulation of plane triangles and piecewise linear continuous basis functions φ_k^1, $\text{curl}_{\Gamma_i}\varphi_k^1 \in \mathbb{R}^3$ is piecewise constant. Then the local Galerkin matrix $D_{i,h}$ can be represented in the form ($d = 3$)

$$D_{i,h} = C_{i,h}^\top\begin{pmatrix} V_{i,h} & & \\ & V_{i,h} & \\ & & V_{i,h}\end{pmatrix}C_{i,h},$$

where $V_{i,h}$ is the local Galerkin matrix of the related single layer potential with piecewise constant basis functions. Moreover, $C_{i,h}$ is an appropriate $3N_i \times M_i$ matrix which describes the transformation of the coefficient vector $\underline{v}_i \in \mathbb{R}^{M_i}$ of $v_{h,i} \in S_h^1(\Gamma_i)$ to the piecewise constant vector–valued result in $\mathbb{R}^{3N_i}$ of $\text{curl}_{\Gamma_i}v_{h,i}$. A fast realization $\widetilde{D}_{i,h}$ of the discrete hypersingular integral operator is now reduced to three fast applications $\widetilde{V}_{i,h}$ of the discrete single layer potential,

$$\widetilde{D}_{i,h} = C_{i,h}^{\top} \begin{pmatrix} \widetilde{V}_{i,h} & & \\ & \widetilde{V}_{i,h} & \\ & & \widetilde{V}_{i,h} \end{pmatrix} C_{i,h}. \tag{20}$$

Since the curl of a constant function vanishes we conclude $C_{i,h}\underline{e}_i = \underline{0}$ and therefore $\ker \widetilde{D}_{i,h} = \ker D_{i,h}$, i.e., this approach is kernel–preserving for any possible fast application $\widetilde{V}_{i,h}$ of the approximate discrete single layer potential. Let $V_{i,h}$ be the Galerkin matrix of the local single layer potential operator V_i when using piecewise constant basis functions $\psi^0_{\kappa,i} \in S^0_h(\Gamma_i)$. The matrix by vector product $\underline{v}_i = V_{i,h}\underline{w}_i$ then reads ($d = 3$)

$$v_{i,\lambda} = \sum_{\kappa=1}^{N_i} V_{i,h}[\lambda,\kappa]w_{i,\kappa} = \frac{\alpha_i}{4\pi}\sum_{\kappa=1}^{N_i} w_{i,\kappa} \int_{\tau_\lambda}\int_{\tau_\kappa} \frac{1}{|x-y|} ds_y ds_x.$$

For a fixed boundary element τ_λ we consider the collection of all boundary elements τ_κ, which are in farfield of τ_λ satisfying the admissibility condition

$$\operatorname{dist}(\tau_\kappa, \tau_\lambda) \geq \eta \max\{\operatorname{diam}\tau_\kappa, \operatorname{diam}\tau_\lambda\}$$

with some appropriately chosen parameter $\eta > 1$. The remaining boundary elements τ_κ are called to be in the nearfield of τ_λ. Using some numerical integration scheme in the farfield, the matrix by vector product can be rewritten as

$$v_{i,\lambda} = \sum_{\text{nearfield}} V_{i,h}[\lambda,\kappa]w_{i,\kappa} + \sum_{\text{farfield}} w_{i,\kappa} \sum_{m=1}^{N_{G,\kappa}} \sum_{n=1}^{N_{G,\lambda}} \frac{\omega_{\kappa,m}\omega_{\lambda,n}}{|x_{\kappa,m} - x_{\lambda,n}|}$$

where $x_{\kappa,m}$ are suitable chosen integration nodes and $\omega_{\kappa,m}$ are related integration weights, respectively. The evaluation of

$$v_{i,\lambda,n} = \sum_{\text{farfield}} w_{i,\kappa} \sum_{m=1}^{N_{G,\kappa}} \frac{\omega_{\kappa,m}}{|x_{\kappa,m} - x_{\lambda,n}|}$$

corresponds exactly to the fast multiple particle simulation algorithm as described in Greengard and Rokhlin [1987] and can be implemented efficiently (Of [2001]). This defines a fast multipole approximation $\widetilde{V}_{i,h}$ of the discrete local single layer potential $V_{i,h}$. In fact, the matrix by vector multiplication with the discrete single layer potential by means of the fast multipole method costs $\operatorname{ops}(\widetilde{V}_{i,h}\underline{w}_i) = \mathcal{O}(N_i \log^2 N_i)$ arithmetical operations ($d = 3$). Choosing both the numerical integration scheme in the farfield and the multipole parameters in an appropriate way, we obtain corresponding error estimates for the perturbed single layer potential $\widetilde{V}_{i,h}$ (Of et al. [2004]). In fact, the approximated single layer potential $\widetilde{V}_{i,h}$ turns out to be $H^{-1/2}(\Gamma_i)$–elliptic, i.e.

$$(\widetilde{V}_{i,h}\underline{w}_i, \underline{w}_i) \geq \frac{1}{2} c_1^V \, \|w_{i,h}\|^2_{H^{-1/2}(\Gamma_i)} \tag{21}$$

for all $w_{i,h} \in S_h^0(\Gamma_i) \leftrightarrow \underline{w}_i \in \mathbb{R}^{N_i}$ where c_1^V is the ellipticity constant of the local single layer potential operator V_i. Combining the discrete ellipticity of the approximated single layer potential $\widetilde{V}_{i,h}$ with the representation (20) of the discrete hypersingular integral operator we get the following result.

Lemma 2. *The sparse representation $\widetilde{D}_{i,h}$ as given in (20) is symmetric and spectrally equivalent to the Galerkin matrix $D_{i,h}$ of the local hypersingular integral operator, i.e., there hold the spectral equivalence inequalities*

$$c_1 \, (D_{i,h}\underline{v}_i, \underline{v}_i) \leq (\widetilde{D}_{i,h}\underline{v}_i, \underline{v}_i) \leq c_2 \, (D_{i,h}\underline{v}_i, \underline{v}_i) \quad \textit{for all } \underline{v}_i \in \mathbb{R}^{M_i}.$$

Let us mention that the numerical integration in the farfield and the multipole approximation of the single layer potential only have to ensure corresponding spectral equivalence inequalities, that means basically the discrete ellipticity estimate (21) of $\widetilde{V}_{i,h}$. In fact this cost much less than the stronger requirement of meeting the accuracy given by the discretization error of the Galerkin scheme.

From Lemma 1 and Lemma 2 we now conclude the spectral equivalence of the finite and boundary element Schur complements $S_{i,h}^{\text{FEM/BEM}}$ with the sparse representation $\widetilde{D}_{i,h}$ of the local hypersingular integral operator D_i. Hence we can reformulate Theorem 1 for the sparse version of the scaled hypersingular BETI preconditioners as defined in (19).

Theorem 2. *For the sparse version of the scaled hypersingular BETI preconditioner (19), the condition estimate*

$$\kappa(PC_{sBETI}^{-1}P^\top P^\top FP) \leq c\left(1 + \log\frac{H}{h}\right)^2$$

holds, where the positive constant c is independent of the local mesh size h, the average subdomain size H, the number p of subdomains and of the coefficients α_i (coefficient jumps). The matrix by vector operation $\widetilde{D}_{i,h}\underline{v}_i$ costs $ops(D_{i,h}\underline{v}_i) = \mathcal{O}((H/h)^2 \log^2(H/h))$ arithmetical operations ($d = 3$).

4 Concluding Remarks

In this paper we presented the BETI/FETI technique for solving large scale coupled boundary and finite element equations arising from the non-overlapping domain decomposition. Our BETI/FETI preconditioner was constructed from the discrete hypersingular operator that is especially efficient in its sparse version. In the latter case the complexity of the preconditioning operation is almost proportional to the number of unknowns living on the skeleton of our domain decomposition. Our analysis showed that the

BETI/FETI method has the same nice numerical and practical properties as they are known from the well-established FETI methods. In Langer and Steinbach [2003] we report on the first numerical experiments with our BETI solver that shows the same numerical behaviour as it is typical for the FETI methods.

Klawonn and Widlund [2000] proposed a FETI version with inexact solvers. This technique avoids the exact solution of the local Neumann and Dirichlet problems in the FETI methods and works only with the corresponding preconditioners. In a forthcoming paper we will develop sparse inexact BETI versions the total complexity of which is basically proportional to the number of the subdomain boundary unknowns. The coupling of both inexact techniques will lead to sparse inexact BETI/FETI methods of almost optimal total complexity.

Acknowledgement. This work has been supported by the Austrian Science Fund 'Fonds zur Förderung der wissenschaftlichen Forschung (FWF)' the under grant SFB F013 'Numerical and Symbolic Scientific Computing' and the German Research Foundation 'Deutsche Forschungsgemeinschaft (DFG)' under the grant SFB 404 'Multifield Problems in Continuum Mechanics'.

References

S. Brenner. An additive Schwarz preconditioner for the FETI method. *Numer. Math.*, 94(1):1–31, 2003.

M. Costabel. Symmetric methods for the coupling of finite elements and boundary elements. In C. Brebbia, W. Wendland, and G. Kuhn, editors, *Boundary Elements IX*, pages 411–420, Berlin, Heidelberg, New York, 1987. Springer.

M. Costabel. Boundary integral operators on Lipschitz domains: Elementary results. *SIAM J. Math. Anal.*, 19:613–626, 1988.

C. Farhat, M. Lesoinne, and K. Pierson. A scalable dual-primal domain decomposition method. *Numer. Linear Algebra Appl.*, 7(7-8):687–714, 2000.

C. Farhat and F.-X. Roux. A method of finite element tearing and interconnecting and its parallel solution algorithm. *Int. J. Numer. Meth. Engrg.*, 32:1205–1227, 1991.

C. Farhat and F.-X. Roux. Implicite parallel processing in structural mechanics. In J. Oden, editor, *Comput. Mech. Adv.*, pages 1–124, Amsterdam, 1994. North-Holland.

L. Greengard and V. Rokhlin. A fast algorithm for particle simulations. *J. Comput. Physics*, 73:325–348, 1987.

G. Haase, B. Heise, M. Kuhn, and U. Langer. Adaptive domain decompostion methods for finite and boundary element equations. In *Boundary Element Topics (ed. by W. Wendland)*, pages 121–147, Berlin, 1998. Springer-Verlag.

G. C. Hsiao, O. Steinbach, and W. L. Wendland. Domain decomposition methods via boundary integral equations. *J. Comput. Appl. Math.*, 125: 521–537, 2000.

G. C. Hsiao and W. L. Wendland. A finite element method for some integral equations of the first kind. *J. Math. Anal. Appl.*, 58:449–481, 1977.

B. Khoromskij, W. Hackbusch, and R. Kriemann. Direct Schur complement method by hierarchical matrix techniques. In R. K. et al, editor, *15th International Conference on Domain Decomposition Methods, Berlin, August 20 - 25, 2003*, Lecture Notes in Computational Science and Engineering (LNCSE), Heidelberg, 2004. Springer.

A. Klawonn and O. Widlund. A domain decomposition method with Lagrange multipliers and inexact solvers for linear elasticity. *SIAM J. Sci. Comput.*, 22(4):1199–1219, 2000.

A. Klawonn and O. Widlund. FETI and Neumann–Neumann Iterative Substructuring Methods: Connections and New Results. *Comm. Pure Appl. Math.*, 54:57–90, January 2001.

A. Klawonn, O. Widlund, and M. Dryja. Dual-primal FETI methods for three-dimensional elliptic problems with heterogenous coefficients. *SIAM J. Numer. Anal.*, 40(1):159–179, 2002.

U. Langer. Parallel iterative solution of symmetric coupled FE/BE- equations via domain decomposition. *Contemp. Math.*, 157:335–344, 1994.

U. Langer and O. Steinbach. Boundary element tearing and interconnecting method. *Computing*, 71:205–228, 2003.

J. Mandel and R. Tezaur. Convergence of a substrucuring method with Lagrange multipliers. *Numer. Math.*, 73(4):473–487, 1996.

J. Mandel and R. Tezaur. On the convergence of a dual-primal substructuring method. *Numer. Math.*, 88:543–558, 2001.

J.-C. Nédélec. Integral equations with non–integrable kernels. *Int. Eq. Operat. Th.*, 5:562–572, 1982.

G. Of. Die Multipolmethode für Randintegralgleichungen. Diplomarbeit, Universität Stuttgart, Mathematisches Institut A, 2001.

G. Of, O. Steinbach, and W. L. Wendland. The fast multipole method for the symmetric boundary integral formulation. 2004. in preparation.

D. Stefanica. A numerical study of FETI algorithms for mortar finite element methods. *SIAM J. Sci. Comput.*, 23(4):1135–1160, 2001.

O. Steinbach. *Stability estimates for hybrid coupled domain decomposition methods*, volume 1809 of *Lecture Notes in Mathematics.* Springer, Heidelberg, 2003.

Parallel Simulation of Multiphase/Multicomponent Flow Models

Erlend Øian, Magne S. Espedal, I. Garrido, and G. E. Fladmark

University of Bergen, Center for Integrated Petroleum Research (CIPR)
`http://www.mi.uib.no/~erlendo/`

Summary. The simulation of flow in porous media is a computationally demanding task. Thermodynamical equilibrium calculations and complex, heterogeneous geological structures normally gives a multiphysics/multidomain problem to solve. Thus, efficient solution methods are needed. The research simulator ATHENA is a 3D, multiphase, multicomponent, porous media flow simulator. A parallel version of the simulator was developed based on a non-overlapping domain decomposition strategy, where the domains are defined a-priori from e.g. geological data. Selected domains are refined with locally matching grids, giving a globally non-matching, unstructured grid. In addition to the space domain, novel algorithms for parallel processing in time based on a predictor-corrector strategy has been successfully implemented.

We discuss how the domain decomposition framework can be used to include different physical and numerical models in selected sub-domains. Also we comment on how the two-level solver relates to multiphase upscaling techniques.

Adding communication functionality enables the original serial version to run on each sub-domain in parallel. Motivated by the need for larger time steps, an implicit formulation of the mass transport equations has been formulated and implemented in the existing parallel framework. Further, as the Message Passing Interface (MPI) is used for communication, the simulator is highly portable. Through benchmark experiments, we test the new formulation on platforms ranging from commercial super-computers to heterogeneous networks of workstations.

1 Introduction

The simulation of flow in porous media is a computationally demanding task. Thermodynamical equilibrium calculations and complex, heterogeneous geological structures normally gives a multiphysics/multidomain problem to solve. When studying e.g. various faulted and fractured porous media, important features that can have a large impact on the flow characteristics are localized in space and exist on a much smaller scale than the characteristic length scale of the domain of interest.

In order to give a full three dimensional description of geometrical and physical properties of such a case, efficient numerical tools are necessary. A natural approach to resolve the geometrical details, are local grid refinement (LGR) techniques. The goal of such methods is to reduce the overall size of the problem while retaining a fairly good numerical resolution. A domain decomposition based LGR technique was implemented in an in-house, 3D, research simulator for porous media flow called ATHENA, see Reme and Øye [1999]. By adding communication functionality, the original serial version was extended to run on each sub-domain in parallel, see Øye and Reme [1999]. The communication is enabled through an object oriented, C++ library called OOMPI. This library is based on the Message Passing Interface (MPI) standard.

The framework included in the ATHENA simulator allows various aspects of domain decomposition strategies to be explored. In the space domain different models and discretizations can be used within the total domain. In a similar way, the time domain can be split and solved in parallel. This is achieved through a predictor-corrector strategy in which a coarse time step simulation (predictor) provides initial values for solving fine sub-intervals in parallel (corrector).

As an application, we will show how the domain decomposition framework can be used for modeling flow in fractured porous media. Specifically, we suggest applying a discrete fracture network model in selected domains. Such a model is a flexible and accurate tool to describe the complex geometries of fractures, but at the cost of larger systems of equations. This problem can be solved by using parallel computations and upscaling.

The mathematical model describing multiphase porous media flow includes equations for the mass transport. Previously, these equations were solved by a forward Euler time stepping scheme. Motivated by the need for larger time steps, an implicit formulation of the mass transport equations was formulated. Here we will describe how the implicit formulation is included in the framework of a parallel version of the ATHENA simulator.

We compare the implicit formulation on platforms ranging from commercial super-computers to heterogeneous networks of PC workstations.

In Sect. 2 we recall the mathematical model of porous media flow with multiple phases and thermal effects. Then, in Sect. 3, we present the domain decomposition and local grid refinement framework. The approach which combines fracture modeling and domain decomposition is given in Sect. 4 and a parallel implementation of implicit mass transport formulation in Sect. 5. Section 6 includes an example that combines aspects of the framework presented above. We end with a summary and conclusion in Sect. 7.

2 Mathematical Model

The mathematical model describing multiphase flow in porous media with multiple components and thermal effects constitutes a complex set of coupled equations. These equations involve a set of primary variables and additional constraints imposed by secondary variables. Since the model we use has already been described in detail in e.g. Reme et al. [2000], we will only briefly present the equations for the primary variables. The $2+n_c$ primary variables are the *temperature*, T, the *water pressure*, p^w, and the *molar masses*, N_ν, of each component. Here $\nu = 1, 2, \ldots, n_c$, is the component index, and n_c is the number of components. Further, we let V denote a finite control volume of a porous medium having the closed surface, S. In order to simplify notation we leave the summation index unspecified when summing over phases, i.e.

$$\sum_{\ell} \equiv \sum_{\ell=\mathrm{w,o,g}} .$$

The temperature within a control volume, V, is governed by a heat flow equation. This equation expresses conservation of energy by relating the temperature gradient, ∇T, the heat capacity, ρu, convective flux, $h\rho\mathbf{u}$, and heat sinks/sources q:

$$\frac{\partial}{\partial t}\int_{\mathrm{V}}(\rho u)dV - \int_{\mathrm{S}}\left(\underset{\sim}{k}\nabla T - h\rho\mathbf{u}\right)\cdot d\mathbf{S} = \int_{\mathrm{V}} q dV. \tag{1}$$

An equation for the water pressure is derived by requiring that the pores are totally filled, i.e. that the residual pore volume, $R(t) = 0$, $\forall t$. A Taylor's expansion of $R(t + \Delta t)$ then gives:

$$\frac{\partial R}{\partial p^w}\frac{\partial p^w}{\partial t} + \sum_{\nu=1}^{n_c}\frac{\partial R}{\partial N_\nu}\frac{\partial N_\nu}{\partial t} = -\frac{R}{\Delta t} - \frac{\partial R}{\partial W}\frac{\partial W}{\partial t}. \tag{2}$$

The overburden pressure $W = \sigma + p$ is the sum of effective stress, σ, and pore pressure, p. This derivation of the pressure equation gives a sequential formulation of the mathematical model.

Finally, we have n_c equations expressing conservation of the molar mass of component ν:

$$\frac{\partial}{\partial t}\int_{\mathrm{V}}\left(\phi_p\sum_{\ell} C_\nu^\ell \xi^\ell S^\ell\right)dV = -\int_{\mathrm{S}}\left(\sum_{\ell} C_\nu^\ell \xi^\ell \mathbf{v}^\ell\right)\cdot d\mathbf{S} + \int_{\mathrm{V}} q_\nu dV. \tag{3}$$

Here, ϕ_p is the rock porosity, C_ν^ℓ is the fraction of component ν in phase ℓ, and ξ^ℓ, S^ℓ and $\mathbf{v}^\ell$ are the corresponding molar density, phase saturation and generalized Darcy velocity respectively.

In order to solve these equations numerically, we use a standard, cell centered, piecewise constant finite volume discretization in space. Details on how the resulting systems of equations are solved in each time step are given in the next section.

3 A Two-level Solver

Here we present an iterative solver which we use to solve the linear systems. The solver is equivalent to the so called Fast Adaptive Composite method, see e.g. Teigland [1998], Briggs et al. [2000]. It can also be viewed as a two-level domain decomposition method, see Smith et al. [1996] and references therein.

Assume that a matching, but possibly non-regular, *coarse grid*, $\hat{\Omega}$ is defined. Further, let a subset, Ω_f, of coarse cells be refined. Each of the coarse cells in this subset defines a sub-domain, Ω_{f_i}, that is refined independently of the other sub-domains with a locally matching, non-regular grid. The resulting *composite grid*, Ω, is generally non-regular and non-matching and consists of some (or no) true coarse cells, Ω_c, and several (or all) refined sub-domains

$$\Omega_f = \bigcup_{i=1}^{p} \Omega_{f_i}, \tag{4}$$

where p is the number of refined sub-domains.

Let the number of composite grid cells in the composite grid be N. The underlying coarse grid has $\hat{N}$ cells. The *index set* of fine cells within coarse cell number $\hat{i}$ is denoted $\mathrm{M}_{\hat{i}}$. Further, we associate to this cell a *basis vector* $\psi_{\hat{i}} \in \mathbb{R}^N$ defined as

$$\psi_{\hat{i}} = \{\psi_k^{\hat{i}}\}. \tag{5}$$

The vector components, $\psi_k^{\hat{i}}$, have value one for refinement cell, k, in coarse cell/sub-domain number $\hat{i}$, i.e.

$$\psi_k^{\hat{i}} = \begin{cases} 1, & \forall\, k \in \mathrm{M}_{\hat{i}}, \\ 0, & \text{otherwise}. \end{cases} \tag{6}$$

Then, let $\mathcal{R}^T \in \mathbb{R}^{N \times \hat{N}}$ be the matrix representation of interpolation from the coarse grid, $\hat{\Omega}$, to the composite grid, Ω. The columns of this operator consist of the basis vectors $\psi_{\hat{i}}$. Thus representing constant interpolation. Correspondingly, the restriction operator is $\mathcal{R} \in \mathbb{R}^{\hat{N} \times N}$. Further, the restriction operator $\mathcal{S}_i \in \mathbb{R}^{N_i \times N}$, returns the vector coefficients defined on sub-domain Ω_i, i.e.

$$\mathbf{x}_{\Omega_i} = \mathcal{S}_i \mathbf{x}, \qquad \mathbf{x}_{\Omega_i} \in \mathbb{R}^{N_i}, \quad \mathbf{x} \in \mathbb{R}^N. \tag{7}$$

Finally, the combination of $\mathcal{R}^T$ and $\mathcal{S}_i$ provides a mapping $\mathcal{R}_i^T \in \mathbb{R}^{N_i \times \hat{N}}$ from the coarse grid to sub-domain Ω_i:

$$\mathcal{R}_i^T = \mathcal{S}_i \mathcal{R}^T. \tag{8}$$

The numerical solution of the equations in the preceding section entails solving linear systems of the form $\mathcal{A}\mathbf{x} = \mathbf{b}$. This can be written in block matrix form,

$$\begin{bmatrix} \mathcal{A}_{cc} & \mathcal{A}_{cf} \\ \mathcal{A}_{fc} & \mathcal{A}_{ff} \end{bmatrix} \begin{bmatrix} \mathbf{x}_c \\ \mathbf{x}_f \end{bmatrix} = \begin{bmatrix} \mathbf{b}_c \\ \mathbf{b}_f \end{bmatrix} \tag{9}$$

where the system matrix is decomposed according to the domains of the grid. We use the various transfer operators defined above to define a two-level solution algorithm for the system in Eq. (9) defined on the composite grid.

We proceed by introducing the two-level iterated solution $\mathbf{x}^{(s)}$ of Eq. (9). Let an updated/improved solution be defined by

$$\mathbf{x}^{(s+1/2)} = \mathbf{x}^{(s)} + \mathcal{R}^T \hat{\mathbf{d}}^{(s)}, \tag{10}$$

where $\hat{\mathbf{d}}^{(s)} \in \mathbb{R}^{\hat{N}}$ is a coarse grid correction. Substituting $\mathbf{x}^{(s)}$ for the update $\mathbf{x}^{(s+1/2)}$ and restricting to the coarse grid, we get

$$\mathcal{R}\mathcal{A}\mathcal{R}^T \hat{\mathbf{d}}^{(s)} = \mathcal{R}\big(\mathbf{b} - \mathcal{A}\mathbf{x}^{(s)}\big). \tag{11}$$

The solution of this equation gives the next step, iterated solution for the non-refined coarse cells, and an intermediate update of the refined sub-domain solutions defined by

$$\mathbf{x}_c^{(s+1)} = \mathbf{x}_c^{(s+1/2)} = \mathbf{x}_c^{(s)} + \mathcal{R}_c^T \hat{\mathbf{d}}^{(s)}, \tag{12}$$

$$\mathbf{x}_r^{(s+1/2)} = \mathbf{x}_r^{(s)} + \mathcal{R}_r^T \hat{\mathbf{d}}^{(s)}, \quad r = 1, \ldots, p. \tag{13}$$

The intermediate solutions $\mathbf{x}_r^{(s+1/2)}$ enable the composite problem in Eq. (9) to be split into independent sub-domain problems for all the sub-domains $q = 1, \ldots, p$

$$\mathcal{A}_{qq}\mathbf{x}_q^{(s+1)} = \mathbf{b}_q - \sum_{r \neq q} \mathcal{A}_{qr}\mathbf{x}_r^{(s+1/2)}, \tag{14}$$

where summation index r also includes the set of true coarse cells. The iteration proceeds until the scaled difference between two consecutive iterations is below some prescribed tolerance.

Our group is also working on applying the two-level scheme above to the time domain, following a parallel technique proposed in Baffico et al. [2002]. A coarse time step solution acts as a predictor by providing boundary values for each sub time interval of the coarse step. Then, each sub time step problem is solved, determining a correction to the coarse solution for the next step of the iteration, see Garrido et al..

4 Using the Domain Decomposition Framework

The domain decomposition based local grid refinement strategy we have implemented is similar to what is know as multiblock reservoir simulation, see e.g. Lee et al. [2002] and Lu et al. [2002]. Multiblock grids allows different gridding in each block, i.e. coarse domain, with possibly non-matching gridlines at block boundaries, Lee et al. [2003]. The two-level solver for composite grids in Sect. 3 can be described as a multiblock method. An important advantage of this approach is that it combines the simplicity of globally structured grids with the flexibility of fully unstructured discretizations. Further, since the coarse (Galerkin) operator of Eq. (11) is defined simply as a summation of the fine scale operators, it can be constructed independently of the type of mathematical model which is used on each sub-domain.

A basic requirement in the cell centered finite volume method, is that flux is continuous across interfaces of the grid. In our case, we want the mass to be conserved, i.e. we want continuity of mass flux. For composite interfaces, we introduce ghost cells and calculate the composite interface fluxes by interpolating real cell pressure (potential) values onto the ghost cells. Currently, we only use a constant interpolation.

The so called mortar methods provide a general framework that treats composite interfaces in a systematic way, see e.g. Ewing et al. [2000] et al. for a finite volume element variant and references therein.

Due to our relatively crude composite interface approximation, we currently require that the coarse block interfaces are located away from large gradients and boundaries between regions of different physical properties/models. This implies that the transition zone between e.g. a single phase and a multiphase flow region should be included within a multiphase model coarse block. Another case is flow through faulted porous media. We use such an example to illustrate the use of our framework in Sect. 6 below.

4.1 Discrete Fracture Network Model

As an application of the multiblock/multiphysics and parallel processing capabilities of our simulator, we consider flow in fractured porous media. In particular, we are working on combining structured, Cartesian discretizations with discrete network models in selected domains, Øian [2004]. This applies to fractured porous media simulations, where the fractures occur in localized swarms and the traditional dual continuum approach might not be appropriate. In Karimi-Fard et al. [2003] and Karimi-Fard and Firoozabadi [2003] a discrete fracture model is presented. It is based on unstructured grids and allows for both two and three-dimensional systems. An important aspect of this method is that control volumes at the intersection of fractures can be removed, thus relaxing the restrictions on the stable time step in the simulations. Due to the flexibility of this method in modeling complex geometries, fine scale effects are resolved accurately. This method might easily introduce too many details, though, if it is used on the global domain. The multiblock framework enables us to localize this method to selected domains based on the geological description. In the rest of the domain, we can use traditional, less expensive discretizations. This is illustrated in Fig. 1, where only the upper right domain is discretized with the discrete fracture network model.

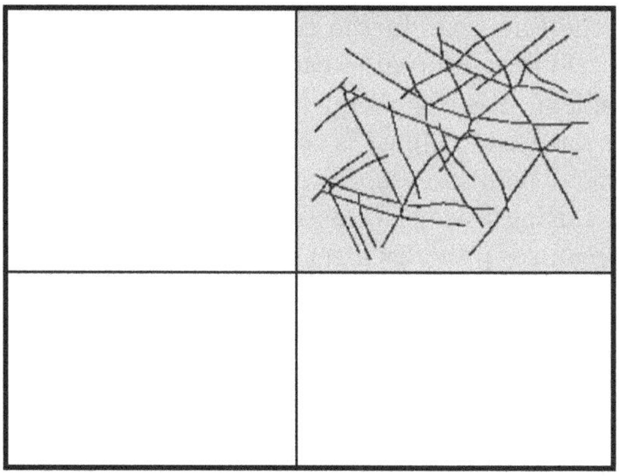

Fig. 1. The figure illustrates using a discrete fracture model in a single domain

4.2 Upscaling Issues

Since geological models of porous media are highly detailed, direct simulations on such models are typically not efficient or feasible. The standard approach to tackle this scale discrepancy problem, is to use various methods of coarsening or upscaling. For single phase flow simulations it is common to calculate an upscaled, effective absolute permeability through either analytical averaging techniques or local numerical methods. Such (quasi-)local methods, may not be adequate in many situations. Typically, these methods can be sensitive to the boundary conditions used. In contrast, global methods use full fine scale simulations to determine the coarse scale parameters and are thus better at capturing coarse scale flow features, but at the expense of more computations. These methods all focus on upscaling of the permeability. Other variants calculate transmissibilities directly.

When we turn to multiphase flow problems, the effect of relative permeabilities and capillary pressures must also be considered. It is common to upscale the various parameters independently. The problem, though, is that combining these upscaled parameters might not give a coarse model which captures the coarse scale features of the flow. Several authors have suggested that due to the complexity and uncertainty in such an approach, upscaling of the solution variables directly should be considered.

The coarse solver found in many preconditioning techniques/multilevel iterative methods directly incorporates all the fine scale information. Multigrid-upscaling methods are based on the idea/observation that these coarse operators might be good approximations of the coarse scale effect of the fine scale differential operator. Various work on upscaling within a multigrid context are given in e.g. Moulton et al. [1998] and Knapek [1998, 1999].

The two level multigrid solver that is implemented in the ATHENA simulator fits this framework. In Aarnes et al. and Reme et al. [2002] a Galerkin-based upscaling procedure was presented. The main idea is to use existing information on the fine scale in the coarse averaging system, i.e. to capture some of the fine grid flow internal to each coarse grid block without solving the full fine grid problem. It is important to note that by using piecewise constant interpolation we ensure that mass is conserved. By solving an inverse problem based on the coarse solution, effective parameters can be calculated. In the single phase case, using Darcy's law on the coarse scale with volume averaged velocities and pressures provides a set of equations for the components of the permeability tensor.

Based on some upscaled parameters, a coarse solve is performed. The coarse solution provides boundary conditions for local fine scale simulations in order to determine domains for further refinement. As this process is only performed once (or a few times) during the course of a simulation, the overhead of the fine scale simulations is permissible. We propose an algorithm which includes both a local simulation approach and multigrid upscaling. The local solve includes improved boundary conditions stemming form a global coarse solve. This is similar to the coupled local-global approach in Chen et al. [2003]. This could serve as a background permeability for a full multiphase simulation. But, more research is needed in order to fully understand the implications of such an approach.

5 Implicit Molar Mass Formulation

Discretizing e.g. fractured domains using locally refined grids, leads to small spatial scales. Previously the mass transport equations were integrated using an explicit, forward Euler scheme. To avoid the severe time step restrictions given by the CFL-condition, an implicit formulation using a backward Euler scheme has been formulated, see Chaib et al. [2002], Øian et al. [2003], Garrido et al..

We start by noting that the molar mass of component ν is equal to the integral on the left hand side of Eq. (3). We let V^ℓ denote the volume of phase ℓ and introduce the volume factor $a^\ell = 1/V^\ell$. Further, we define the molar mass of component ν in phase ℓ as $N_\nu^\ell = C_\nu^\ell N^\ell = C_\nu^\ell V^\ell \xi^\ell$. Then, integrating Eq. (3) over a control volume V_i, with surface S_i in a numerical grid and using up-stream weighting, we get

$$\frac{\partial N_{\nu_i}}{\partial t} + \sum_{\mathrm{is}\in\mathrm{S}_i} \left(\sum_\ell \left(a^\ell N_\nu^\ell\right)_{\mathrm{in}} \theta_{\mathrm{is}}^\ell \right) = Q_{\nu_i}, \tag{15}$$

where $\theta_{\mathrm{is}}^\ell = \mathbf{v}_{\mathrm{is}}^\ell \cdot \mathbf{n}_{\mathrm{is}} A_{\mathrm{is}}$ is the Darcy volume flux. Here $\mathbf{n}_{\mathrm{is}}$ and A_{is} is the outward normal and area of subsurface "is" of S_i respectively. Subscript "in" indicates evaluation in the upstream cell and is phase dependent.

After a Newton-Raphson linearization step, we get

$$\frac{\delta N_{\nu_i}^{(t+1)}}{\Delta t} + \sum_{\mathrm{is}\in\mathrm{S}_i} \sum_\ell \left[\left(a^{\ell^n} \sum_\mu \left(\frac{\partial N_\nu^\ell}{\partial N_\mu} \right)^{(t)} \delta N_\mu^{(t+1)} \right)_{\mathrm{in}} \theta_{\mathrm{is}}^{\ell^n} \right] = \beta_{\nu_i}^{(t)}. \tag{16}$$

The right hand side is given by

$$\beta_{\nu_i}^{(t)} = Q_{\nu_i}^n - \frac{N_{\nu_i}^{(t)} - N_{\nu_i}^n}{\Delta t} - \sum_{\mathrm{is} \in \mathrm{S}_i} \sum_{\ell} \left[\left(a^{\ell^n} N_\nu^{\ell(t)} \right)_{\mathrm{in}} \theta_{\mathrm{is}}^{\ell^n} \right].$$

In matrix notation the molar mass equation for component ν can be expressed as

$$\sum_{\mu} \mathcal{J}_{\nu,\mu}^{(t)} \delta \mathbf{N}_\mu^{(t+1)} = \mathbf{b}_\nu^{(t)}, \qquad \nu, \mu = \mathrm{w}, \mathrm{o}, \mathrm{g}. \tag{17}$$

As a simplifying step, we continue by neglecting off-diagonal blocks representing coupling between different components, i.e.

$$\frac{\partial N_\nu^\ell}{\partial N_\mu} = 0, \qquad \nu \neq \mu. \tag{18}$$

We get then the following decoupled systems for each component (we drop the $\cdot_{\nu,\nu}$ sub-script on the Jacobian matrix)

$$\mathcal{J}^{(t)} \delta \mathbf{N}_\nu^{(t+1)} = \mathbf{b}^{(t)}. \tag{19}$$

In order to solve the linear system in Eq. (19) we use the domain decomposition based iterative method presented in Sect. 3. Thus, the two-level iterated solution is now $\mathbf{N}_\nu^{(s)(t)}$ and the incremented solution is defined as

$$\delta \mathbf{N}_\nu^{(s)(t)} = \mathbf{N}_\nu^{(s)(t)} - \mathbf{N}_\nu^{(t)}. \tag{20}$$

Following Eq. (10) the updated/improved solution is then

$$\mathbf{N}_\nu^{(s+1/2)(t)} = \mathbf{N}_\nu^{(s)(t)} + \mathcal{R}^T \hat{\mathbf{d}}^{(s)}, \tag{21}$$

and the coarse grid equation for the molar mass is

$$\mathcal{R} \mathcal{J}^{(t)} \mathcal{R}^T \hat{\mathbf{d}}^{(s)} = \mathcal{R} \big(\mathbf{b}^{(t)} - \mathcal{J}^{(t)} \delta \mathbf{N}_\nu^{(s)(t)} \big). \tag{22}$$

The independent equations for the sub-domains $q = 1, \ldots, p$, are

$$\mathcal{J}_{qq}^{(t)} \delta \mathbf{N}_{\nu_q}^{(s+1)(t)} = \mathbf{b}_{\nu_q}^{(t)} - \sum_{r \neq q} \mathcal{J}_{qr}^{(t)} \delta \mathbf{N}_{\nu_r}^{(s+1/2)(t)}, \tag{23}$$

where summation index r includes the set of true coarse cells. The right hand side terms involving intermediate solutions $\mathbf{N}_{\nu_r}^{(s+1/2)(t)}$ are given by

$$\mathbf{N}_{\nu_c}^{(s+1)(t)} = \mathbf{N}_{\nu_c}^{(s+1/2)(t)} = \mathbf{N}_{\nu_c}^{(s)(t)} + \mathcal{R}_c^T \hat{\mathbf{d}}^{(s)}, \tag{24}$$

$$\mathbf{N}_{\nu_r}^{(s+1/2)(t)} = \mathbf{N}_{\nu_r}^{(s)(t)} + \mathcal{R}_r^T \hat{\mathbf{d}}^{(s)}, \quad r = 1, \ldots, p. \tag{25}$$

5.1 Parallel Implementation

The parallel version of the ATHENA code is based on the concept of a "simulator parallel model". The idea is that the original serial simulator is used on each sub-domain. To incorporate the solver described in Sect. 3, the main modifications consist of adding functionality for communicating between the domains. This is implemented through a Communicator class, which has the various objects to be communicated as member classes. We use the object oriented MPI library OOMPI, Squyres et al. [2003], to achieve this. OOMPI is a thin layer on top of the MPI, see e.g. Snir et al. [1996], which enables easy creation and communication of user defined objects. Since OOMPI is a fairly lightweight library and introduces little overhead, we have chosen to continue with this in the implementation of the implicit mass transport solver. In Skjellum et al. [2001] a comparison of different design strategies for an object oriented interface to MPI is given. Due to the success of MPI in defining a standard for distributed, parallel programming, the simulator is highly portable.

Following the existing framework and based on the mathematical model, we have introduced two classes inherited from OOMPI_Datatype. These are the RefinedMM class, which contains the values needed for upstream evaluation of the Jacobian matrix terms

$$a^{\ell^n}\left(\frac{\partial N_\nu^\ell}{\partial N_\nu}\right)^{(t)} \quad \text{and} \quad a^{\ell^n} N_\nu^{\ell(t)}, \tag{26}$$

and the CoarseMM class, which contains the sub-domain terms contributing to the system Jacobian and right hand side given in Eq. (11).

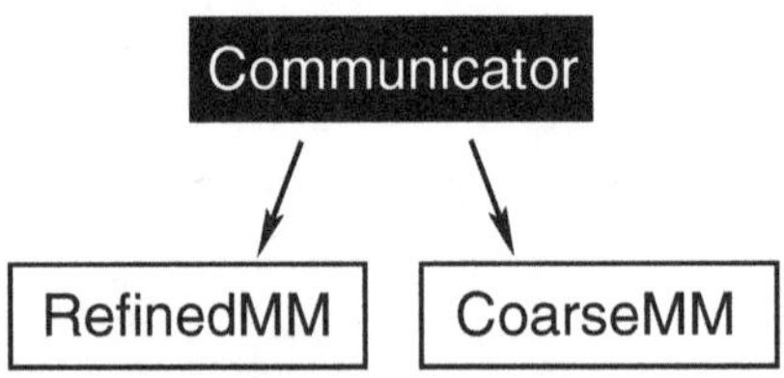

Fig. 2. The figure shows the collaboration of the Communicator class and the classes storing data between refinements, RefinedMM, and data for the coarse solve, CoarseMM

5.2 Numerical Experiments

We perform a numerical experiment using the implicit mass transport formulation. The main goal of the experiment is to evaluate the parallel performance on various platforms. At the current stage of implementation, we map domains onto CPUs in a one-to-one fashion. In the case where other considerations than just getting an even number of grid cells dictate the choice of domain decomposition, this approach is not optimal. Work on allowing a more flexible load balancing of the simulator is in progress.

We use SGI and IBM SP2 super computers, a dedicated IBM Linux cluster and on a network of PC workstations running Linux. Even though high-speed super computers are an important target platform for the simulator, it is also interesting to see how the code performs on lower bandwidth networks of regular workstations, as this is available to a lot of users. Consequently, we have run the simulations on three super-computing platforms and on two commodity off-the-shelf hardware platforms, see Table 1. To avoid the complicating effect of calculating equilibrium in the beginning of the simulation, we do our experiments using restarts at times when equilibrium (hydrostatic) is established.

Table 1. The table lists various hardware platforms and compilers

index	platform	C++ compiler	MPI implementation
I	SGI Origin 3800	MIPSpro	Irix
II	IBM p690	AIX	IBM
III	IBM Linux cluster	Intel	LAM
IV	Linux cluster	GNU	MPICH
V	Linux workst. netw.	GNU	MPICH

The test case domain is depicted in Fig. 3. In this case the porous medium is initially saturated with water. The simulation domain is 366m × 671m × 52m in x-, y- and z-direction respectively and is decomposed into six domains with an equal number of fine cells. Oil and gas phases migrate from one corner of the domain. For a fixed simulation time interval, we have measured the

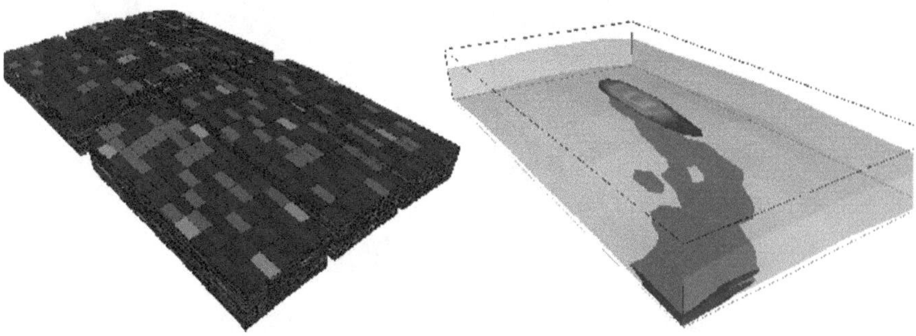

Fig. 3. The figure shows the domain decomposition (left) and gas saturation at a given time (right)

Table 2. Timing (seconds) of simulations on various platforms

	platform I	platform II	platform III	platform IV	platform V
total	2908	764	2472	1056	1801
pressure	711	195	502	252	489
mass	2081	497	1778	725	1194

total CPU time and CPU times for *pressure solution* and *molar mass solve* separately. The results are given in Table 2. More detailed timing results, including a conceptual fault zone case, will be presented in Øian [2004].

6 A Geometrically Complex Case

We demonstrate the simulator framework in use through a multiphase flow case in a faulted porous medium, see Fig. 4. Oil and gas phases migrate through a water saturated layer and into a high permeable fault zone. The

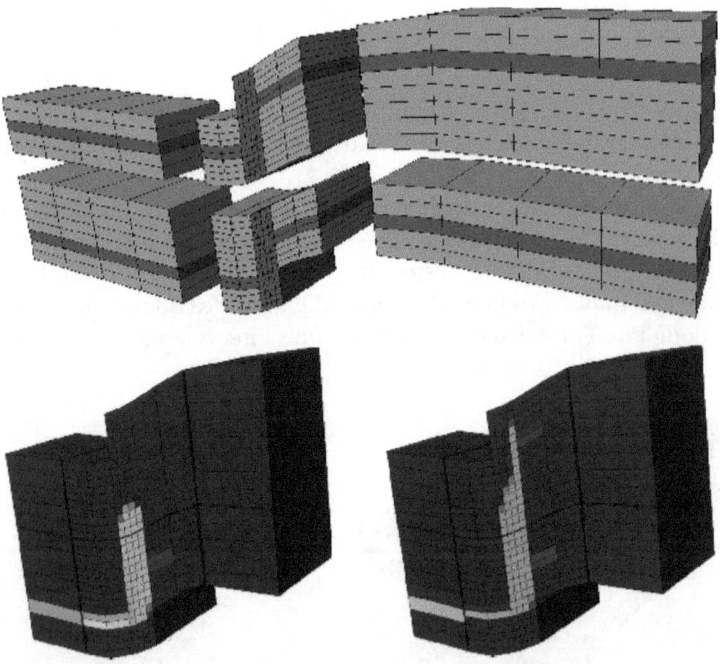

Fig. 4. The top part of the figure shows the domain decomposition including horizontal and vertical flow regions. Red color indicates high permeability. Oil and gas saturations are plotted at the same time level in the bottom left and right parts of the figure respectively, where only a middle section of the grid is visualized

background lithology is a low permeable shale. Due to the density differences of water, oil and gas, a segregation process occurs in the vertical fault zone accompanied by much higher flow rates than in the horizontal layers.

As mentioned in Sect. 4 above, we place the transition between the horizontal and vertical flow regions inside the coarse blocks. The implicit mass transport formulation improves the allowed time step within the vertical flow region. Also, the two-level solver enables the mass transport equations to be solved in parallel, which is important if we want to make a more refined discretization. In Øian [2004] further computational results will be given.

7 Conclusion

We have presented a domain decomposition framework which is implemented in a parallel version of the flow simulator ATHENA. The two-level, iterative solver allows multiblock/multiphysics domains to be built. This is because the coarse operator is based on an algebraic combination of the fine scale operators (Galerkin) rather than an explicit coarse scale discretization. Consequently, by defining the domain decomposition a-priori, we can allow different models or discretizations in different domains. Work is in progress on applying this method on modeling flow in fractured porous media.

Another issue is that the Galerkin coarse scale operator can be viewed as an upscaled model. Traditional upscaling methods typically treat each parameter separately. Combining these into an upscaled model might be incorrect. In the multiphase case, the Galerkin operator gives a combined averaged representation by directly incorporating the fine scale, nonlinear processes caused by changes in absolute and relative permeability and capillary forces.

A sequential, implicit formulation of the mass transport equations has been implemented within the existing object oriented parallel framework. A timing experiment illustrated that the code performs better on high performance, shared memory supercomputers than distributed memory systems. This is the typical behavior for domain decomposition based methods. As we have a sequential, i.e. decoupled, solution procedure for solving the pressure and molar mass equations, the number of iterations to achieve convergence in either, depend on the time steps. The implicit molar mass equations allows larger time steps, but might introduce more iterations in the pressure solve. This will have an effect on the observed parallel efficiency since communication is involved in each iteration. Work is in progress to implement the simultaneous solution of pressure and masses, i.e. fully implicit.

A flexible load balancing scheme has not yet been implemented in the simulator. The effect of this is apparent on networks with low bandwidth connections and will also influence the scaling properties on supercomputers. With a queue system, e.g. running parallel jobs at night time, a network of workstations would still be a valuable parallel platform. Specially for development purposes and setting up simulation cases, this opportunity is useful.

Acknowledgement. The first author would like to thank Prof. L. J. Durlofsky for the opportunity to visit Stanford University, Petroleum Engineering Department, and initiating work on discrete fracture network models with Dr. Mohammad Karimi-Fard.

References

J. Aarnes, H. Reme, and M. S. Espedal. A least-squares approach for upscaling and the acceleration of a galerkin technique. Presented at the Upscaling Downunder Conference, Melbourne, Australia, 7 - 10 February, 2000.

L. Baffico, S. Bernard, Y. Maday, G. Turinici, and G. Zerah. Parallel in time molecular dynamics simulations. *Phys. Rev. E.*, 66:057701, 2002.

W. L. Briggs, V. E. Henson, and S. F. McCormick. *A Multigrid Tutorial.* Society for Industrial and Applied Mathematics, Philadelphia, PA, USA, 2000.

M. Chaib, G. E. Fladmark, and M. S. Espedal. Implicit treatment of molar mass equations in secondary oil migration. *Computing and Visualization in Science*, 4(3):191–196, February 2002.

Y. Chen, L. J. Durlofsky, M. Gerritsen, and X. H. Wen. A coupled local-global upscaling approach for simulating flow in highly heterogeneous formations. *Advances in Water Resources*, (26):1041–1060, 2003.

R. Ewing, R. Lazarov, T. Lin, and Y. Lin. The mortar finite volume element methods and domain decomposition. *East-West J. Numer. Math.*, 8:93–110, 2000.

I. Garrido, M. S. Espedal, and G. E. Fladmark. A convergent algorithm for time parallelization applied to reservoir simulation. Presented at DD15, 2003.

M. Karimi-Fard, L. J. Durlofsky, and K. Aziz. An efficient discrete fracture model applicable for general purpose reservoir simulators. *SPE 79699*, 2003.

M. Karimi-Fard and A. Firoozabadi. Numerical simulation of water injection in fractured media using the discrete-fracture model and the Galerkin method. *SPE Reservoir Evaluation & Engineering*, 6(2):117–126, April 2003.

S. Knapek. Upscaling techniques based on subspace correction and coarse-grid approximation. *InSitu*, 22(1):35–58, 1998. Special issue on reservoir simulation.

S. Knapek. Matrix-dependent multigrid homogenization for diffusion problems. *SIAM J. Sci. Comp.*, 20(2):515–533, 1999.

S. H. Lee, P. Jenny, and H. A. Tchelepi. A finite-volume method with hexahedral multiblock grids for modeling flow in porous media. *Computational Geosciences*, 6:353–379, 2002.

S. H. Lee, C. Wolfsteiner, L. J. Durlofsky, P. Jenny, and H. A. Tchelepi. New developments in multiblock reservoir simulation: Black oil modeling, nonmatching subdomains and near-well upscaling. *Society of Petroleum Engineers*, (SPE 79682), 2003.

Q. Lu, M. Peszynska, and M. F. Wheeler. A parallel multiblock black-oil model in multimodel implementation. *SPE Journal*, 7(3):278–287, September 2002. SPE 79535.

J. D. Moulton, J. E. Dendy, and J. M. Hyman. The black box multigrid numerical homogenization algorithm. *J. Comput. Phys.*, (142):80–108, 1998. Article No. CP985911.

E. Øian. *Modeling Flow in Fractured and Faulted Media.* Dr.Scient. thesis, in preparation, University of Bergen, 2004.

E. Øian, I. Garrido, M. Chaib, G. E. Fladmark, and M. S. Espedal. Modeling fractured and faulted regions: Local grid refinement methods for implicit solvers. *Computing and Visualization in Science*, 2003. Accepted for Publication.

G. Å. Øye and H. Reme. Parallelization of a compositional simulator with a galerkin coarse/fine method. In P. A. et al., editor, *Lecture Notes in Computer Science*, pages 586–594. Springer-Verlag, Berlin, 1999. LNCS 1685.

H. Reme, M. Espedal, and G. E. Fladmark. *A Preconditioning Technique as an Upscaling Procedure*, volume 131 of *The IMA Volumes in Mathematics and its Applications*, pages 283–297. Springer Verlag, Heidelberg, 2002.

H. Reme and G. Å. Øye. Use of local grid refinement and a galerkin technique to study secondary migration in fractured and faulted regions. *Computing and Visualization in Science*, 2:153–162, 1999.

H. Reme, G. Å. Øye, M. S. Espedal, and G. E. Fladmark. Parallelization of a compositional reservoir simulator. In Z.-C. S. Z. Chen, R. E. Ewing, editor, *Numerical Treatment of Multiphase Flows in Porous Media*, number 552 in Lecture Notes in Physics, pages 244–267. Springer-Verlag, Berlin, 2000.

A. Skjellum, D. G. Wooley, Z. Lu, M. Wolf, P. V. Bangalore, A. Lumsdaine, J. M. Squyres, and B. McCandless. Object-oriented analysis and design of the message passing interface. *Concurrency and Computation: Practice and Experience*, 13:245–292, 2001. (DOI: 10.1002/cpe.556).

B. F. Smith, P. E. Bjørstad, and W. Gropp. *Domain Decomposition: Parallel Multilevel Methods for Elliptic Partial Differential Equations.* Cambridge University Press, 1996.

M. Snir, S. Otto, S. Huss-Lederman, D. Walker, and J. Dongarra. *MPI: The Complete Reference.* MIT Press, 1996.

J. M. Squyres, J. Willcock, B. C. McCandless, P. W. Rijks, and A. Lumsdaine. *Object Oriented MPI (OOMPI): A C++ Class Library for MPI.* Open Systems Laboratory Pervasive Technologies Labs, Indiana University, September 3 2003. `http://www.osl.iu.edu/research/oompi/`.

R. Teigland. On some variational acceleration techniques and related methods for local refinement. *Int. J. Numer. Meth. Fluids*, (28):945–960, 1998.

Uncoupling-Coupling Techniques for Metastable Dynamical Systems*

Christof Schütte, Ralf Forster, Eike Meerbach, and Alexander Fischer

Institute of Mathematics II, Free University Berlin, Arnimallee 2–6, 14195 Berlin, Germany
URL: `http://page.mi.fu-berlin.de/~biocomp`

Summary. We shortly review the uncoupling-coupling method, a Markov chain Monte Carlo based approach to compute statistical properties of systems like medium-sized biomolecules. This technique has recently been proposed for the efficient computation of biomolecular conformations. One crucial step of UC is the decomposition of reversible nearly uncoupled Markov chains into rapidly mixing subchains. We show how the underlying scheme of uncoupling-coupling can also be applied to stochastic differential equations where it can be translated into a domain decomposition technique for partial differential equations.

AMS classification: 15A18; 15A51; 60J35; 65C40; 65N55

Keywords: Biomolecules; Reversible Markov chains; Stationary distribution; Fokker-Planck operator; Uncoupling-coupling; Stochastic Differential Equations

1 Introduction

Application of Markov chain Monte Carlo (MCMC) to biomolecular systems has to tackle the *trapping problem*, i.e., the Markov chain remains for a very long time in one part of the state space before it moves on to another part. Such undesirable behavior of the Markov chain is caused by *metastable sets* (also called *modes* or *conformations*) in the state space, between which transitions are extremely rare. There exists a huge body of literature addressing this notoriously difficult problem (Ferguson et al. [1999], Liu [2001]).

We herein review a novel approach to overcome the trapping problem, the uncoupling-coupling scheme (UC), which has recently been introduced by one of the authors (AF) (Fischer [2003], Fischer et al. [2002]). UC combines statistical reweighting techniques with a *hierarchical decomposition* of the state

* Supported by the DFG research center "Mathematics for key technologies" (FZT 86) in Berlin.

space into metastable sets. The key idea is to regard metastable sets as *almost invariant sets* w.r.t. the Markov chain.

It has been shown recently that these metastable sets are strongly connected to the spectral structure of the Markov propagation operator associated with the Markov chain (Schütte et al. [1999]), and that it is even possible for a wide range of problem classes to identify metastable sets by computing dominant eigenvalues of this operator (Deuflhard et al. [2000]). Once n dominant metastable sets are identified, significantly improved convergence properties are achieved by *uncoupling*, i.e., by parallel simulation of n independent chains, each one restricted to one of the metastable sets. Subsequently, information lost in the uncoupling step, i.e., the weighting factors between the n metastable sets, is reconstructed by means of the stationary distribution of an appropriate $(n \times n)$ coupling matrix. We present the uncoupling step in the context of MCMC in Sect. 2, and the underlying uncoupling-coupling scheme in its simplest form in Sect. 3.

In its final section this article is devoted to the demonstration that the UC idea can be translated into a domain decomposition technique for eigenvalue problems of specific partial differential operators. The translation is possible since Markov propagation operators of specific Markov processes (e.g., those governed by stochastic differential equations) are generated by partial differential operators. In such cases the decomposition of state space into metastable sets is strongly connected to the dominant eigenmodes of the generator, and the UC scheme for the propagation operator can be translated into an analogous scheme for the generator. This relation may make uncoupling-coupling techniques as used in biomolecular simulations accessible for research in the direction of domain decomposition.

2 Uncoupling-Coupling Markov Chain Monte Carlo

For biomolecular simulations, MCMC is the method of choice for the task of drawing samples from the canonical distribution. In the presence of strong metastabilities slow convergence can be avoided by uncoupling-coupling, where the state space is decomposed into metastable sets.

Metropolis Algorithm

The Metropolis (or Metropolis-Hastings) algorithm is the most widely used form of MCMC and essentially builds upon Markov chain theory (Brémaud [1999], Liu [2001]).

Suppose that we are interested in a distribution given by a density function $f > 0$ with values in $\Omega \subseteq \mathbb{R}^d$, from which it is practically impossible to draw independent samples (e.g., the canonical distribution of medium-sized biomolecules, where d is in the range of 50 to 500). Usually, f is defined in terms of an *unnormalized density* $\hat{f}$ via

$$f(x) = \frac{\hat{f}(x)}{Z_{\hat{f}}} \quad \text{with} \quad Z_{\hat{f}} = \int_{\Omega} \hat{f}(x)\, dx, \tag{1}$$

where $Z_{\hat{f}}$ denotes the *normalizing constant* of $\hat{f}$. In most applications $\hat{f}$ is the canonical or Boltzmann density $\hat{f} = \exp(-\beta V)$ with inverse temperature β for some potential energy function $V : \Omega \to \mathbb{R}$.

The goal is to obtain expectations of some function g with respect to f, i.e., computing the expectation

$$I_f(g) = \int g(x) f(x)\, dx.$$

The Metropolis algorithm realizes a Markov chain $\mathcal{X} = X^{(1)}, X^{(2)}, X^{(3)}, \ldots$ having f as its invariant density. A sample $\mathbf{x} = (x^{(1)}, \ldots, x^{(n)})$ of $\mathcal{X}$ is obtained by accepting a proposal step $x_{\text{prop}}^{(k+1)}$ with a probability that only depends on the ratio of $f(x_{\text{prop}}^{(k+1)})/f(x^{(k)})$, thereby avoiding a computation of the unknown normalizing constant (which in its integral representation is typically hard to evaluate). The generated (dependent) random sample $\mathbf{x}$ then enables us to estimate the integral I_f by

$$\hat{I}_f(g) = \frac{1}{n} \sum_{k=1}^{n} g(x^{(k)}). \tag{2}$$

The evolution of a Markov chain $\mathcal{X} = (X_k)$ with state space Ω is defined by a *stochastic transition function* $K : \Omega \times \Omega \to \mathbb{R}$, where $K(x, A)$ is the probability density to move from x to the set A in one step (Meyn and Tweedie [1993]). We call f an *invariant density* of the Markov chain given by K, if

$$f(y) = \int_{\Omega} K(x, y) f(x)\, dx \tag{3}$$

holds for all $y \in \Omega$.

In the Metropolis-Hastings algorithm a transition function K which satisfies (3) is realized by first defining an arbitrary but *irreducible* transition kernel $q(x, y)$ together with the *acceptance function*

$$\alpha(x, y) = \begin{cases} \min\left(1, \frac{q(y,x)\, f(y)}{q(x,y)\, f(x)}\right) & \text{for } q(x, y) > 0 \\ 1 & \text{otherwise} \end{cases}. \tag{4}$$

The computation of α requires ratios of the form $f(y)/f(x)$ only, which is feasible even if the normalizing constant $Z_{\hat{f}}$ is unknown.

Based on q and α we define K as the sum of two contributions,

$$K(x, y) = k(x, y) + r(x)\delta(x - y),$$

where the absolutely continuous part k is given by

$$k(x,y) = \begin{cases} q(x,y)\alpha(x,y) & \text{if } x \neq y \\ 0 & \text{otherwise} \end{cases}$$

and the singular component by $r(x) = 1 - \int k(x,y)\,dy$.

With this K one step in the realization of the Markov chain from the state $X_k = x$ consists of: *a*) propose some y distributed according to $q(x,y)$, *b*) accept this step by setting $X_{k+1} = y$ with probability $\alpha(x,y)$ or *c*) reject the proposal leaving $X_{k+1} = x$.

The construction of K guarantees that the associated Markov chain $\mathcal{X}$ is irreducible—provided that q is irreducible—and that for all $x, y \in \Omega$ the detailed balance condition

$$f(x)\,k(x,y) = f(y)\,k(y,x) \tag{5}$$

holds (for details, see e.g.(Tierney [1994])). Due to (5) K is called *reversible* w.r.t. f. If we further assume that $\mathcal{X}$ is aperiodic—which is guaranteed whenever $r > 0$—we can state that f is the unique invariant density of $\mathcal{X}$.

Markov Operator

In the following we want to understand the global behavior of a Markov chain via the eigenmodes of its associated *Markov operator* P. This operator is defined in terms of the transition function K by

$$Pu(y) = \int_{\Omega} k(x,y)u(x)\,dx + r(y)u(y). \tag{6}$$

P describes the propagation of a phase space density with one step of the Markov chain. One can show that the reversibility of K w.r.t. f implies that its spectrum $\sigma(P)$ is real-valued. More exactly, we have $\sigma(P) \subseteq [-1,1]$, and the largest eigenvalue is $\lambda_1 = 1$. We have $Pf = f$, and under some ergodicity conditions f is the unique eigenfunction associated with $\lambda_1 = 1$ (up to normalization). Some additional ergodicity typically is sufficient to guarantee that the essential spectrum $\sigma_{ess}(P)$ of P is bounded such that $\sigma_{ess}(P) \subset [-r,r]$ (see Schütte and Huisinga [2003]) and one typically can assume that there are several discrete eigenvalues λ with $|\lambda| > r$. If this is the case then these eigenvalues subsequently are assumed to be ordered due to their absolute value, i.e., such that $\lambda_1 > \lambda_2 \geq \lambda_3 \geq \ldots$.

Discretization

Identification and restricted sampling is carried out by discretizing the operator in essential degrees of freedom. To that end, let for two sets $A, B \subseteq \Omega$ the *transition probability* between A and B within an ensemble distributed w.r.t. the density f and during one step of the Markov chain be given by

$$\kappa(A, B) = \frac{1}{\int_A f(x)\,dx} \int_A \int_B K(x, y) f(x)\,dx\,dy. \tag{7}$$

Then, discretization is done by *coarse graining* with an arbitrary box decomposition of the phase space Ω into m disjoint sets $B_1, \ldots, B_m \subset \Omega$ with $\bigcup B_j = \Omega$. Based on this box decomposition, we introduce the new finite state space $S = \{B_1, \ldots, B_m\}$ and define the transition function $\tilde{K}$ on S via

$$\tilde{K}(B_k, B_l) \;=\; \kappa(B_k, B_l). \tag{8}$$

The finite dimensional Markov chain defined by $\tilde{K}$ again is reversible w.r.t. its invariant density $\tilde{f}$ given by $\tilde{f}(B_k) = \int_{B_k} f(x)dx$. Whenever f is unique for K, $\tilde{f}$ is also unique for $\tilde{K}$. Then the phase space is finite and the Markov operator P becomes an $(m \times m)$-transition matrix $\mathbf{P}$ which simply is the stochastic matrix with entries $p_{kl} = \tilde{K}(B_k, B_l) = \kappa(B_k, B_l)$.

Metastability

If λ_2 is close to $\lambda_1 = 1$, we often find that the reason for the undesirably slow convergence is that the Markov chain remains for a long time in a *metastable set* (or *conformation*) of the state space, before it moves on to another one. We will call a set A *metastable* under our Markov chain, if the transition probability from A to itself is close to one, i.e., if $\kappa(A, A) \approx 1$.

For an algorithmic exploitation of metastability the following observation is of importance: If there are n eigenvalues close to $\lambda_1 = 1$ (including λ_1 itself) and a significant spectral gap to all remaining eigenvalues, then there also are n disjoint metastable subsets and vice versa (Meyer [1989], Schütte et al. [2001]). If this is the case, the chain is rapidly mixing *within* the corresponding metastable subsets and the undesirably slow overall convergence results from the rareness of transitions between these metastable sets.

The close connection between a separated cluster of dominant eigenvalues and the existence of metastable subsets has another very important algorithmic consequence: it has been shown that one can *identify* the n metastable subsets only on basis of the *eigenvectors* associated with the n dominant eigenvalues (Schütte et al. [1999, 2001]). This insight leads to a significantly general identification algorithm (Deuflhard et al. [2000]) used for the detection of biomolecular conformations.

Restriction

Assume that we know the n disjoint metastable sets $A_1, \ldots, A_n$ of our Markov chain, and that we now want to sample separately in each A_l, for $l = 1, \ldots, n$. Then, for each l we define a restricted Markov kernel K_l from K on A_l by setting

$$K_l(x, y) = k_l(x, y) + r_l(x)\delta(x - y) \tag{9}$$

with

$$k_l(x,y) = \begin{cases} q(x,y)\alpha(x,y) & \text{if } x \neq y \text{ and } y \in A_l \\ 0 & \text{otherwise} \end{cases}$$

and

$$r_l(x) = 1 - \int k_l(x,y)\,dy.$$

Clearly, detailed balance still holds, so that K_l is again a reversible Markov kernel. Now, let $\hat{f}_l = \mathbf{1}_{A_l}\hat{f}$ be the restricted unnormalized density on A_l, with $\mathbf{1}_A$ denoting the indicator function on A, i.e., $\mathbf{1}_A(x) = 1$ if $x \in A$ and $\mathbf{1}_A(x) = 0$ otherwise. Then, under the assumption, that K_l is irreducible, $f_l = \hat{f}_l / Z_{\hat{f}_l}$ is the unique invariant density of K_l.

We denote by P_l the corresponding propagator of K_l. If we assume that A_l is metastable and that it cannot be subdivided further into two or more almost invariant sets, then we can state the following: The second largest eigenvalue λ_2 of P_l is substantially less than 1, otherwise there would exist a decomposition into two or more metastable subsets. As a consequence, due to $\lambda_2 \ll 1$, the corresponding Markov chain $\mathcal{X}_l$ is rapidly mixing.

For the restricted Markov kernel K_l the detailed balance condition (5) still holds for all $x, y \in A_l$; therefore the density f_l is a scalar multiple of the correct global density f of the unrestricted Markov chain. Thus, we can regain the global density via

$$f = \sum_{l=1}^{n} \xi_l f_l \tag{10}$$

in terms of the local densities f_l. Only the scalar coupling factors ξ_l, $l = 1, \ldots, n$, are unknowns which represent the neglected coupling between the sets A_l. Apparently, the coupling factors need to be ratios of normalizing constants of the form $\xi_l = Z_{\hat{f}_l}/Z_{\hat{f}}$, since then we can reconstruct f from the f_l's:

$$\sum_{l=0}^{n} \xi_l f_l = \sum_{l=0}^{n} \frac{Z_{\hat{f}_l}}{Z_{\hat{f}}} \frac{\hat{f}_k}{Z_{\hat{f}_k}} = \frac{\hat{f}}{Z_{\hat{f}}} = f. \tag{11}$$

Hierarchical Uncoupling-Coupling

Restricted sampling alone does not directly provide the necessary coupling vector $\boldsymbol{\xi} = (\xi_1, \ldots, \xi_k)$ and also raises the question of how to decompose the state space. Yet, it is possible to overcome these problems by embedding some Metropolis sampler into a hierarchical annealing structure. For a detailed presentation of this approach we refer to the Uncoupling-coupling Monte Carlo method presented in Fischer [2003], Fischer et al. [2002].

The hierarchical annealing structure is a crucial part for the algorithmic concept in the context of biomolecular simulations with far reaching consequences for the coupling step. However, since it is of lesser importance in the PDE context, we focus on the basic uncoupling-coupling scheme in the following.

3 Basic Uncoupling-Coupling Scheme

We provide some theoretical aspects of uncoupling-coupling, especially the impact of uncoupling on the spectra of restricted Markov operators. Throughout this section we consider a finite state space. Therefore, let $\mathbf{P}$ be an irreducible, aperiodic and reversible stochastic $(m \times m)$-matrix, which might be obtained by a box discretization of a Markov operator P as defined in (8). In this case the state space Ω reduces to $S = \{B_1, \ldots, B_m\}$, the entries p_{ij} of $\mathbf{P}$ are transition probabilities $\kappa(B_k, B_l)$ of the Markov operator between the boxes B_i and B_j, and the global density f becomes a stochastic vector $\boldsymbol{\pi}$.

$\mathbf{P}$ is called a nearly uncoupled Markov chain, if there is a permutation of the state space such that $\mathbf{P}$ becomes diagonal block-dominant, i.e.

$$\mathbf{P} = \widetilde{\mathbf{P}} + \mathbf{E} = \begin{pmatrix} \mathbf{P}_{11} & \mathbf{E}_{12} & \cdots & \mathbf{E}_{1n} \\ \mathbf{E}_{21} & \mathbf{P}_{22} & \ddots & \vdots \\ \vdots & \ddots & \ddots & \mathbf{E}_{n-1,n} \\ \mathbf{E}_{n1} & \cdots & \mathbf{E}_{n,n-1} & \mathbf{P}_{nn} \end{pmatrix}, \tag{12}$$

where each sub-matrix is quadratic and entries in $\mathbf{E}$ are small. In the context of metastability this corresponds to the existence of n metastable sets $S_1, \ldots, S_n$. Computation of an appropriate permutation is by no means trivial, but can be done by the identification algorithm already mentioned in the previous section (Deuflhard et al. [2000]). There are different ways to measure the smallness of $\mathbf{E}$, e.g. by the maximum row sum norm $\|\cdot\|_\infty$ or by some $\boldsymbol{\pi}$-weighted norm.

Reversibility of $\mathbf{P}$ is equivalent to the detailed balance condition

$$\pi_i p_{ij} = \pi_j p_{ji} \tag{13}$$

for all $1 \le i, j \le m$. From (13) it easily follows, that $\mathbf{P}$ is self-adjoint w.r.t. the weighted inner product

$$\langle \mathbf{x}, \mathbf{y} \rangle_{\boldsymbol{\pi}} = x_1 y_1 \pi_1 + \ldots + x_m y_m \pi_m. \tag{14}$$

Therefore all eigenvalues of $\mathbf{P}$ are real and contained in the interval $(0, 1]$. Since the diagonal blocks in (12) are nearly stochastic, continuity of the eigenvalues guarantees the existence of n eigenvalues close to 1. We assume that the other eigenvalues are reasonable bounded away from 1, which corresponds to the assumption that the Markov chain is fast-mixing within each metastable set.

The matrix

$$\mathbf{P}^{\mathrm{rest}} = \begin{pmatrix} \mathbf{R}_{11} & \mathbf{0} & \cdots & \mathbf{0} \\ \mathbf{0} & \mathbf{R}_{22} & \ddots & \vdots \\ \vdots & \ddots & \ddots & \mathbf{0} \\ \mathbf{0} & \cdots & \mathbf{0} & \mathbf{R}_{nn} \end{pmatrix}, \tag{15}$$

where $\mathbf{R}_{ii} = \mathbf{P}_{ii} + \text{diag}(\mathbf{e}_i - \mathbf{P}_{ii}\mathbf{e}_i)$ and $\mathbf{e}_i = (1, \ldots, 1)$ is a vector of size $|S_i|$, is called the restriction matrix of $\mathbf{P}$ along a given partition. $\mathbf{P}^{\text{rest}}$ is obtained from $\mathbf{P}$ by setting all off-diagonal blocks in $\mathbf{P}$ to zero and adding the sum of the deleted entries of the i-th row to p_{ii}, which is the discrete counterpart of (9).

Since the diagonal blocks of $\mathbf{P}^{\text{rest}}$ are stochastic we can conclude that the dominant eigenvalue 1 is n-fold, while due to continuity all other eigenvalues should be well separated from 1. If the smallness of $\mathbf{E}$ is measured by the ∞-norm, we can state a quantitative bound on this phenomena (Meerbach et al. [2003]):

Theorem 1. *Let* $\mathbf{P}$ *be a reversible stochastic matrix partitioned according to (12) and* $\mathbf{P}^{\text{rest}}$ *the restricted matrix, as in (15). Then*

$$\lambda_j(\mathbf{P}^{\text{rest}}) \leq \lambda_j(\mathbf{P}) + 2\|\mathbf{E}\|_\infty \tag{16}$$

holds for each $j = 1, \ldots, m$.

Therefore, by transition from $\mathbf{P}$ to $\mathbf{P}^{\text{rest}}$, we obtain n uncoupled Markov chains restricted to the sets $S_1, \ldots, S_n$, whereby for a metastable decomposition each chain is fast mixing.

The following theorem summarizes some facts about $\mathbf{R}$ and reveals that the behavior of the uncoupled chains is closely related to that of the original chain (Meerbach et al. [2003], Meyer [1989]).

Theorem 2. *Let* $\mathbf{P}$ *be an irreducible and reversible stochastic matrix partitioned as in (12). Furthermore, let all* $\mathbf{P}_{ii}$ *be irreducible (substochastic) matrices. Then,*

(a) all $\mathbf{R}_{ii}$ *are irreducible,*
(b) $\mathbf{R}$ *is stochastic with an n-fold dominant eigenvalue* 1,
(c) if the (unique) stationary distribution $\boldsymbol{\pi}$ *of* $\mathbf{P}$ *is partitioned according to* $\mathbf{P}$,

$$\boldsymbol{\pi} = (\boldsymbol{\pi}^{(1)}, \boldsymbol{\pi}^{(2)}, \ldots, \boldsymbol{\pi}^{(n)}),$$

then for each $i = 1, \ldots, n$ *the unique stationary distribution* $\mathbf{r}^{(i)}$ *of the restriction* $\mathbf{R}_{ii}$ *is identical to* $\xi_i^{-1}\boldsymbol{\pi}^{(i)}$, *where* $\xi_i = \sum_h \pi_h^{(i)}$ *is a constant factor,*
(d) the coupling vector $\boldsymbol{\xi} = (\xi_1, \ldots, \xi_n)$ *is the unique stationary distribution of the irreducible and stochastic* coupling matrix $\mathbf{C} = (c_{ij})$, *with*

$$c_{ij} := \mathbf{r}^{(i)}\mathbf{P}_{ij}\mathbf{e}. \tag{17}$$

Theorem 2 states that coupling factors, which are needed to reweight the restricted stationary distributions of the $\mathbf{R}_{ii}$'s in order to obtain the stationary distribution of $\mathbf{P}$, can be derived via the coupling matrix $\mathbf{C}$ containing the transition probabilities between the metastable sets.

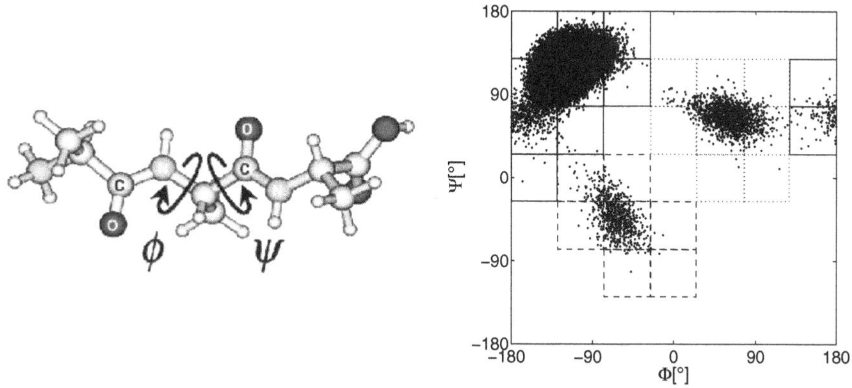

Fig. 1. Left: The trialanine molecule shown in ball-and-stick representation. The overall structure of trialanine is primarily determined by the two torsion angles Φ and Ψ. Right: Plotting Φ versus Ψ results in a so-called Ramachandran plot. The discretization boxes are plotted with different edge lines indicating the different metastable sets they were allocated to

Theorem 1 is closely related to the theory of stochastic complementation (Meyer [1989]), where an analogous result is stated for the general case of non-reversible matrices. Associated with stochastic complementation are so-called aggregation/disaggregation techniques (Cho and Meyer [1999], Stewart and Wu [1992]), which aim for a given stochastic matrix at a fast computation of the stationary distribution by decomposing the state space into stochastic complements. Yet, due to non-reversibility the setup of stochastic complements is much more intricate than restriction. This is also the reason why for biomolecular simulations, the technique of stochastic complementation does not enable to set up restricted Markov operators for MCMC sampling on a continuous state space. However, for reversible problems, we can utilize aggregation/disaggregation techniques combined with restriction in Section 4 to solve eigenvector problems for discretized differential operators.

Note that, since we do not treat the embedding into a hierarchical annealing structure here, the coupling matrix $\mathbf{C}$ in Theorem 2, although it shares the same characteristics, is different from the one actually employed in the UC algorithm.

3.1 Trialanine Simulation

As an example how UC is employed in biomolecular simulations we consider trialanine, a small peptide composed of three alanine amino acid residues. Although the continuous state space Ω is high-dimensional, the structural and dynamical properties of trialanine are primarily determined by the two torsion angles Φ and Ψ, as shown in Fig. 1. We herein only illustrate the initial uncoupling step of UC, which starts with a high-temperature MCMC

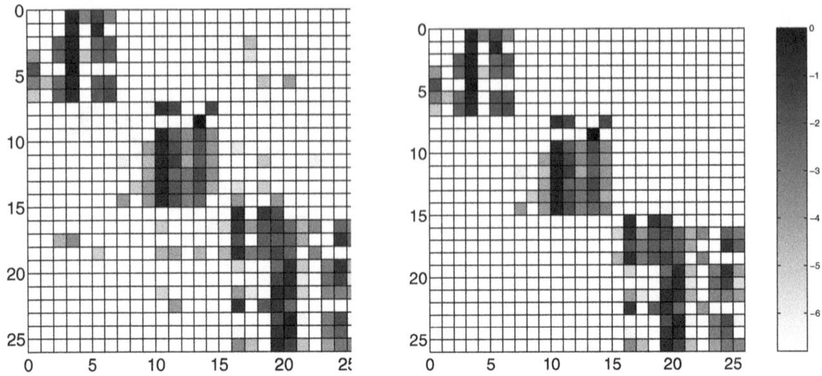

Fig. 2. Left: The permuted transition matrix **P** clearly has a block dominant structure. Right: In the resulting restricted matrix **R** all off-diagonal entries are set to zero. The intensity of the boxes is chosen due to the logarithmic scale on the far right

simulation. More precisely, we used the Hybrid Monte Carlo method (Brass et al. [1993]) to sample at a temperature of 650 K and stored the torsion angles for each simulation step. Discretization of each torsion angle domain $\mathcal{D} = (-180\,°, 180\,°]$ into 7 equidistant intervals resulted in 26 non-empty boxes $(B_1, \ldots, B_{26})$ in $\mathcal{D}^2$, see Fig. 1. On these boxes we set up a transition matrix $\mathbf{P} = (p_{ij})$ (i.e., the discretized Markov operator), where transition probabilities p_{ij} are obtained by counting the number of transitions between boxes B_i and B_j during simulation.

The first eigenvalues of the resulting (26×26)-transition matrix are

j	1	2	3	4	5	
$\lambda_j(\mathbf{P})$	1	0.9952	0.9941	0.5692	0.1425	$\ldots$,

indicating a slow mixing Markov chain with three metastable sets. For the identification of these metastable sets we used the previously mentioned spectral approach (Deuflhard et al. [2000], Weber [2003], Deuflhard and Weber [2003]). In Fig. 1 the identified metastable sets are indicated by different line styles. A corresponding permutation of the transition matrix confirms the computation in that it reveals an obvious block dominant structure, see Fig. 2. Calculating the subdominant eigenvalues of the restrictions $\mathbf{R}_{ii}$ for $i = 1, 2, 3$ results in

$\lambda_2(\mathbf{R}_{11})$	$\lambda_2(\mathbf{R}_{22})$	$\lambda_2(\mathbf{R}_{33})$
0.1376	0.1482	0.5855 ,

which shows that this metastable decomposition in fact leads to three fast mixing restricted Markov chains.

4 UC and Domain Decomposition

In this section, we will pursue the goal of connecting the UC scheme for computing canonical distributions with the world of PDEs. In order to do so, we will first introduce a reversible Markov process in continuous time and discuss its biomolecular background and then will focus on its connection to domain decomposition for PDEs.

Molecular Dynamics

In order to specify what we herein consider as molecular dynamics let q denote the position and p the momenta of a single molecular system consisting of N atoms in state $x = (q,p) \in \mathbb{R}^{3N} \times \mathbb{R}^{3N}$. $V = V(q) : \mathbb{R}^{3N} \to \mathbb{R}$ describes the potential energy, which we assume to be differentiable. The statistics of molecular systems in state space is given by the well-known *canonical density* $f = \frac{1}{Z}\exp(-\beta V(x))$. Let μ be the measure induced by f (for a more detailed description of the biophysical background, cf. Schütte and Huisinga [2003]).

There exist several models of molecular dynamics; we will focus on the *Smoluchowski equation*

$$\dot{q} = -\frac{1}{\gamma}\nabla_q V(q) + \frac{\sigma}{\gamma}\dot{W}, \tag{18}$$

which is an approximation of the well-known *Langevin equation* in case of high friction γ, where $\dot{W}$ is given by a standard $3N$-dimensional Brownian motion W. The continuous time Markov process $(X_t)_{t\geq 0}$ defined by (18) leaves the canonical measure μ invariant and is reversible. Furthermore, $(X_t)_{t\geq 0}$ defines an absolutely continuous *stochastic transition function* $p(t,x,y)$ that describes the probability that the process if started in x at time $t = 0$ is being found in y at time t (for details see Schütte and Huisinga [2003]).

Markov Operator and Generator

The family of Markov operators $(P_t)_{t\geq 0}$ associated with $(X_t)_{t\geq 0}$ is defined analogously to (6) for every $t > 0$ with $k(x,y) = p(t,x,y)$ and $r \equiv 0$. The family $(P_t)_{t\geq 0}$ forms a strongly continuous semigroup such that the *infinitesimal generator*

$$\mathcal{A}y = \lim_{t\to 0}\frac{P_t y - y}{t}$$

is defined and acts on the domain $\mathrm{dom}(\mathcal{A})=\{y \in Y : \lim_{t\to 0}(P_t y - y)/t \text{ exists}\}$. In the following we will simply express the relation between P_t and $\mathcal{A}$ by $P_t = \exp(t\mathcal{A})$ (for details see Huisinga et al. [to appear 2004]).

The reversibility of the underlying Markov process $(X_t)_{t\geq 0}$ has the additional implication that all P_t and the generator $\mathcal{A}$ are *self-adjoint* operators on the Hilbert space $L^2(\mu)$ equipped with the scalar product $\langle u,v\rangle_\mu = \int u(x)\bar{v}(x)\mu(dx)$, cf. Schütte and Huisinga [2003].

More insight into the process and into the form of the generator is available if we consider the evolution of a function under the dynamics given by $(X_t)_{t\geq 0}$. This evolution is governed by the *Fokker-Planck equation*

$$\partial_t\, u = \Bigg(\underbrace{\frac{\sigma^2}{2\gamma^2}\Delta_q + \frac{1}{\gamma}\nabla_q V(q)\cdot\nabla_q + \frac{1}{\gamma}\Delta_q V(q)}_{\mathcal{A}} \Bigg)\, u \tag{19}$$

on some suitable subspace of $L^1(dq)$. In this formula $\mathcal{A}$ is the infinitesimal generator of the semigroup $P_t : L^1(dq) \to L^1(dq)$. That is, for twice differentiable functions u the generator $\mathcal{A}$ is the elliptic partial differential operator given by the RHS of (19).

Finally, the invariance of canonical density μ under the process $(X_t)_{t\geq 0}$ gives us

$$P_t f = f \qquad \text{and} \qquad \mathcal{A}f = 0. \tag{20}$$

Therefore, the computation of the canonical density f can be reduced to the computation of the dominant eigenvector of the generator $\mathcal{A}$. If the potential V satisfies some growth and regularity conditions then, both, the spectra of P_t and $\mathcal{A}$ are discrete in $L^2(\mu)$ and satisfy $\sigma(P_t) = \exp(t\sigma(\mathcal{A}))$. Then, metastability can be discussed via the dominant eigenvalues of $\mathcal{A}$ (i.e., those close to the largest one $\lambda_1 = 0$).

Discretization and Uncoupling-Coupling

Discretizing the operator P_t in position and time, one obtains a Markov chain with transition matrix $\mathbf{P}(t)$. Due to the reversibility, one can apply the UC scheme which was presented in Section 3. Yet another way is to work with the generator $\mathcal{A}$ instead. In the remaining part of this section, it will be shown to what extent the operator $\mathcal{A}$ is related to P_t and how one can extract the required information from $\mathcal{A}$ in almost the same manner as from P_t. For simplicity in what follows we consider a bounded system: The potential V is smooth and takes infinite values at the boundary and outside of a compact domain Ω with sufficiently smooth boundary $\partial\Omega$.

The *discretization* of $\mathcal{A}$, acting on $L^1(dq)$, by means of Finite Elements or Finite Differences is well known and produces a large sparse matrix $\mathbf{A}$ with row sum 0 at the interior nodes. Due to the condition on V, we have Dirichlet boundary conditions equal 0 on $\partial\Omega$ and thus at the boundary nodes. If we assume that there is a Perron cluster of eigenvalues of $\mathcal{A}$ close to its largest eigenvalue $\lambda = 0$ then we have the same spectral property for $\mathbf{A}$ (if the discretization grid is fine enough). If the number of nodes is large efficient numerical solution of the eigenvalue problem for $\mathbf{A}$ will therefore have to apply advanced numerical techniques like subspace oriented multigrid solvers (Friese et al. [1999]), appropriate domain decomposition techniques, or suitably preconditioned linear algebra solvers. Alternatively, we can exploit that the connection between the discretized operators $\mathbf{A}$ and $(\mathbf{P}(t))$ is rather close (Forster [2003]). This allows us to transfer the idea of UC to the generator $\mathbf{A}$. More precisely: decompose the state space into the metastable subsets, restrict $\mathbf{A}$ to these subsets, solve the eigenvector problem locally, and couple the local solutions with weighting factors obtained by some coupling matrix. The *restriction step* is based on the restricted discretized generator (cp. (15))

$$\mathbf{A}^{\text{rest}} = \begin{pmatrix} \mathbf{A}^{\text{rest}}_{(11)} & \mathbf{0} & \cdots & \mathbf{0} \\ \mathbf{0} & \mathbf{A}^{\text{rest}}_{(22)} & \ddots & \vdots \\ \vdots & \ddots & \ddots & \mathbf{0} \\ \mathbf{0} & \cdots & \mathbf{0} & \mathbf{A}^{\text{rest}}_{(nn)} \end{pmatrix} \tag{21}$$

with $\mathbf{A}^{\text{rest}}_{(ii)} = \mathbf{A}^{(ii)} + \text{diag}(\sum_{j,j\neq i} \mathbf{A}^{(ij)}\mathbf{e}_j)$ and blocks $\mathbf{A}^{(ii)}$ of $\mathbf{A}$ according to a decomposition in metastable subsets S_i. This Neumann-like boundary condition is imposed directly on the discretization level. Neumann boundary conditions make sense since the invariant measure should have local minima at the boundaries between the metastable sets. However, these boundary conditions have no simple continuous analogue. Nonetheless, $\mathbf{A}^{\text{rest}}$ is the requested object, since the local solutions of $\mathbf{A}^{\text{rest}}_{(ii)}\mu^{(i)} = 0$ form—apart from the weighting factors (ξ_i)—the canonical density f. That follows from the fact that the detailed balance condition $\mu_k \mathbf{A}_{kl} = \mu_l \mathbf{A}_{lk}$ (which holds since $\mathcal{A}$ is self-adjoint) implies $\mu^T \mathbf{A}^{\text{rest}} = \mu^T \mathbf{A} = 0$ as follows:

$$\begin{aligned}
(\mu^T \mathbf{A}^{\text{rest}})_j &= \mu_j \mathbf{A}^{\text{rest}}_{jj} + \sum_{k\notin S_i} \mu_k \mathbf{A}^{\text{rest}}_{kj} + \sum_{k\in S_i, k\neq j} \mu_k \mathbf{A}^{\text{rest}}_{kj} \\
&= \mu_j \left(\mathbf{A}_{jj} + \sum_{k\notin S_i} \mathbf{A}_{jk} \right) + \sum_{k\in S_i, k\neq j} \mu_k \mathbf{A}_{kj} \\
&= \mu_j \mathbf{A}_{jj} + \sum_{k\notin S_i} \mu_k \mathbf{A}_{kj} + \sum_{k\in S_i, k\neq j} \mu_k \mathbf{A}_{kj} = (\mu^T \mathbf{A})_j,
\end{aligned}$$

where j is a node in S_i and furthermore μ_k and $\mathbf{A}_{kj}$ denote the entries of the large vector $\mu = (\mu^{(1)}, \ldots, \mu^{(n)})$ and the matrix $\mathbf{A}$, respectively. The restricted operator $\mathbf{A}^{\text{rest}}$ is not the generator of $\mathbf{P}(t)^{\text{rest}}$, since $(\mathbf{P}(t)^{\text{rest}})_{t\geq 0}$ does no longer form a semigroup; for the same reason, in general it is $(e^{t\mathbf{A}})^{\text{rest}} \neq e^{t\mathbf{A}^{\text{rest}}}$.

In the *coupling step* the weighting factors can be obtained (Forster [2003]) by the *coupling matrix* $\mathbf{C} = (c_{ij})$ with entries

$$c_{ij} = \langle \mu_i, \mathbf{A}^T \mu_j \rangle_2.$$

The matrix $\mathbf{C}$ arises from a Galerkin discretization of $\mathcal{A}$ on the ansatz space $\mathcal{V} = \text{span}\{\mu_1, \mu_2, \ldots, \mu_n\}$ and inherits the structure of $\mathcal{A}$. The factors $\boldsymbol{\xi} = (\xi_i)$ are the solution of the low-dimensional equation $\mathbf{C}\boldsymbol{\xi} = 0$. Even more can be achieved: under certain conditions, it is also possible to generate the eigenvectors $v_2, \ldots, v_n$ of the other dominant eigenvalues $\lambda_2, \ldots, \lambda_n$ close to 1 by means of (μ_i) and $\mathbf{C}$. More precisely: the solutions $\nu^{(i)}$ of $(\mathbf{CD})\nu^{(i)} = \hat{\lambda}_i \nu^{(i)}$ with $\mathbf{D} = \mathrm{diag}(\xi_1, \ldots, \xi_n)$ allow to define approximations $\sum_k \nu_k^{(i)} \mu_k$ for the eigenvectors v_i. For a detailed description see Forster [2003].

The efficiency of the entire approach critically depends on the underlying decomposition of state space into metastable subsets. If the dynamics given by (18) is rapidly mixing within each of these metastable subsets then the second eigenvalues of all the diagonal blocks $\mathbf{A}^{\mathrm{rest}}_{(ii)}$ of the discretized restricted generator matrix $\mathbf{A}^{\mathrm{rest}}$ will be separated from the largest eigenvalues $\lambda = 0$ by some significant gap such that iterative eigenvalue solvers can be used to compute the eigenvectors μ_i to $\lambda = 0$ for all blocks $\mathbf{A}^{\mathrm{rest}}_{(ii)}$ efficiently. Thus, in order to construct a fully efficient algorithm one has to integrate annealing strategies and grid refinement into some carefully controlled hierarchical approach. This is still under investigation.

References

A. Brass, B. J. Pendleton, Y. Chen, and B. Robson. Hybrid Monte Carlo simulations theory and initial comparison with molecular dynamics. *Biopolymers*, 33:1307–1315, 1993.

P. Brémaud. *Markov Chains: Gibbs Fields, Monte Carlo Simulation, and Queues.* Springer, New York, 1999.

G. E. Cho and C. D. Meyer. Aggregation/disaggregation methods for nearly uncoupled Markov chains. Technical Report NCSU #041600-0400, North Carolina State University, November 1999.

P. Deuflhard, W. Huisinga, A. Fischer, and C. Schütte. Identification of almost invariant aggregates in reversible nearly uncoupled Markov chains. *Lin. Alg. Appl.*, 315:39–59, 2000.

P. Deuflhard and M. Weber. Robust Perron cluster analysis in conformation dynamics. ZIB-Report 03-19, Konrad-Zuse-Zentrum, Berlin, 2003.

D. M. Ferguson, J. I. Siepmann, and D. G. Truhlar, editors. *Monte Carlo Methods in Chemical Physics*, volume 105 of *Advances in Chemical Physics.* Wiley, New York, 1999.

A. Fischer. *An Uncoupling-Coupling Method for Markov Chain Monte Carlo Simulations with an Application to Biomolecules.* PhD thesis, Freie Universität Berlin, 2003.

A. Fischer, C. Schütte, P. Deuflhard, and F. Cordes. Hierarchical uncoupling-coupling of metastable conformations. In T. Schlick and H. H. Gan, editors, *Computational Methods for Macromolecules: Challenges and Applications, Proceedings of the 3rd International Workshop on Algorithms for Macromolecular Modeling, New York, Oct. 12–14, 2000*, volume 24 of *Lecture Notes in Computational Science and Engineering*, Berlin, 2002. Springer.

R. Forster. Ein Algorithmus zur Berechnung invarianter Dichten in metastabilen Systemen. Diploma thesis, Freie Universität Berlin, 2003.

T. Friese, P. Deuflhard, and F. Schmidt. A multigrid method for the complex Helmholtz eigenvalue problem. In C.-H. Lai, P. E. Bjørstad, M. Cross, and O. B. Widlund, editors, *Domain Decomposition Methods in Sciences and Engineering*, DDM-org, pages 18–26, New York, 1999.

W. Huisinga, S. Meyn, and C. Schütte. Phase transitions and metastability in Markovian and molecular systems. *Ann. Appl. Probab.*, to appear 2004.

J. S. Liu. *Monte Carlo Strategies in Scientific Computing.* Springer, New York, 2001.

E. Meerbach, A. Fischer, and C. Schütte. Eigenvalue bounds on restrictions of reversible nearly uncoupled Markov chains. Preprint, 2003.

C. D. Meyer. Stochastic complementation, uncoupling Markov chains, and the theory of nearly reducible systems. *SIAM Rev.*, 31:240–272, 1989.

S. P. Meyn and R. L. Tweedie. *Markov Chains and Stochastic Stability.* Springer, Berlin, 1993.

C. Schütte, A. Fischer, W. Huisinga, and P. Deuflhard. A direct approach to conformational dynamics based on hybrid Monte Carlo. *J. Comput. Phys.*, 151:146–168, 1999.

C. Schütte and W. Huisinga. Biomolecular conformations can be identified as metastable sets of molecular dynamics. In P. G. Ciaret and J.-L. Lions, editors, *Handbook of Numerical Analysis*, volume Computational Chemistry. North-Holland, 2003.

C. Schütte, W. Huisinga, and P. Deuflhard. Transfer operator approach to conformational dynamics in biomolecular systems. In B. Fiedler, editor, *Ergodic Theory, Analysis, and Efficient Simulation of Dynamical Systems.* Springer, 2001.

W. J. Stewart and W. Wu. Numerical experiments with iteration and aggregation for Markov Chains. *ORSA Journal on Computing*, 4(3):336–350, 1992.

L. Tierney. Markov chains for exploring posterior distributions (with discussion). *Ann. Statist.*, 22:1701–1762, 1994.

M. Weber. Improved Perron cluster analysis. ZIB-Report 03-04, Konrad-Zuse-Zentrum, Berlin, 2003.

Part II

Minisymposium: Domain Decomposition Methods for Wave Propagation in Unbounded Media

On the Construction of Approximate Boundary Conditions for Solving the Interior Problem of the Acoustic Scattering Transmission Problem

X. Antoine[1] and H. Barucq[2]

[1] Mathématiques pour l'Industrie et la Physique, MIP, UMR CNRS 5640, UFR MIG, Université P. Sabatier, 118, route de Narbonne, 31062 Toulouse cedex 4, France <antoine@mip.ups-tlse.fr> (http://mip.ups-tlse.fr/~antoine/)

[2] Laboratoire de Mathématiques Appliquées pour l'Industrie, Université de Pau et des Pays de l'Adour, Avenue de l'Université, 64000 Pau, France <Helene.Barucq@univ-pau.fr>

Summary. The construction of accurate generalized impedance boundary conditions for the three-dimensional acoustic scattering problem by a homogeneous dissipative medium is analyzed. The technique relies on an explicit computation of the symbolic asymptotic expansion of the exact impedance operator in the interior domain. An efficient pseudolocalization of this operator based on Padé approximants is then proposed. The condition can be easily integrated in an iterative finite element solver without modifying its performances since the pseudolocal implementation preserves the sparse structure of the linear system. Numerical results are given to illustrate the method.

1 Introduction

The penetration of an acoustic field into a given medium can be approximately modeled through a Fourier-Robin-type (also called impedance) boundary condition (see for instance Senior and Volakis [1995]). To have both a larger application range and a gain of accuracy of the model, a possible approach consists in designing higher-order *generalized impedance boundary conditions*. These conditions are often defined by a differential operator which describes with some finer informations the behaviour of the transmitted acoustic field. We present some new generalized impedance boundary conditions which extend the validity domain of the usual differential conditions for the scattering problem of an acoustic wave by a three-dimensional homogeneous isotropic scatterer. The proposed conditions have also the interest of not increasing the total cost of a resolution by an iterative finite element solver (or possibly an integral equation procedure). All these points are developed below.

2 The acoustic transmission boundary value problem

Let Ω_1 be a regular bounded domain embedded in $\mathbb{R}^3$ with a $\mathcal{C}^\infty$ boundary Γ. We set Ω_2 as the associated infinite domain defined by $\Omega_1 = \mathbb{R}^3/\overline{\Omega_2}$. We assume that both media Ω_j, $j = 1, 2$, are homogeneous and isotropic. Each one is characterized by two positive real constants: the density ρ_j and the sound velocity c_j. We moreover suppose that Ω_1 may be dissipative. This aspect is modeled by the introduction of a damping parameter $\delta \geq 0$.

Consider now an incident wave u_0 defined in the vicinity of Γ and which satisfies the Helmholtz equation: $\Delta u_0 + k_2^2 u_0 = 0$. We make the assumption that the solution has a time-harmonic dependence of the form $e^{-ik_2 t}$, where $k_2 = \varpi/c_2$ is the wave number in the unbounded domain of propagation, setting ϖ as the frequency of the signal. We can then define the (possibly complex) wave number k_1 in Ω_1 by: $k_1^2 = \varpi^2/c_1^{-2}(1 + i\delta/\varpi)$. Two parameters are usually introduced: the complex refraction index $N = c_r^{-1}(1 + i\delta/\varpi)^{1/2}$ and the complex contrast coefficient $\alpha = \rho_r^{-1}(1 + i\delta/\varpi)^{-1}$, where c_r and ρ_r designate respectively the relative velocity and density. Finally, if z is a complex number, we set $z^{1/2}$ as the principal determination of the square root with branch cut along the negative real axis.

We consider now the scattering problem of the wave u_0 by Ω_1 which consists in computing the field v solution to the transmission problem

$$\begin{cases} \Delta v_2 + k_2^2 v_2 = 0, \text{ in } \Omega_2, \\ \Delta v_1 + k_1^2 v_1 = k_2^2(1 - N^2)u_0, \text{ in } \Omega_1, \\ [v] = 0 \text{ and } [\chi \partial_{\mathbf{n}} v] = -[\chi \partial_{\mathbf{n}} u_0], \text{ on } \Gamma, \\ \lim\limits_{|x| \to +\infty} |x|(\nabla v_2 \cdot \dfrac{x}{|x|} - ik_2 v_2) = 0, \end{cases} \tag{1}$$

where χ is the piecewise constant function defined by $\chi = 1$ in Ω_2 and $\chi = \alpha$ in Ω_1. The vector $\mathbf{n}$ stands for the outward unit normal vector to Ω_1. The restriction of the field v to Ω_j, $j = 1, 2$, is denoted by $v_j = v_{|\Omega_j}$; the jump between the exterior and interior traces is given by: $[v] = v_{1|\Gamma} - v_{2|\Gamma}$. The inner product of two complex vector fields $\mathbf{a}$ and $\mathbf{b}$ of $\mathbb{C}^3$ is: $\mathbf{a} \cdot \mathbf{b} = \sum_{\mathrm{j}=1}^{3} \mathrm{a_j} \overline{\mathrm{b_j}}$. The operator ∇ is the gradient operator of a complex-valued vector field and the Laplacian operator is defined by: $\Delta = \nabla^2$. The last equation of (1) is the so-called Sommerfeld radiation condition at infinity which leads to the uniqueness of the solution to the boundary value problem. We denote by SRC the associated operator. The existence and uniqueness of the solution to (1) can be proved in an adequate functional setting.

3 Generalized impedance boundary conditions

When the interior wave number has a sufficiently large modulus $|k_1|$, a reduction of the computational complexity in the practical solution of the boundary value problem (1) can be achieved by approximately modeling the penetration of the wave into the interior domain by a boundary condition set on Γ. This approach is well-known in electromagnetism under the name of *generalized impedance boundary condition* method. The ideas have been introduced during the second world war for modeling the interaction of an electromagnetic field with an irregular terrain (see e.g. Senior and Volakis [1995]). The resulting boundary condition for a given problem takes the form of a generalized mixed boundary condition defined by a differential or a pseudodifferential operator. We give here an outline of the application of the theory of pseudodifferential operators to derive a family of accurate boundary conditions for the transmission problem.

The first point consists in considering the total field formulation of system (1) setting $u = v + u_0$. Therefore, we are led to compute u such that

$$\begin{cases} \Delta u_2 + k_2^2 u_2 = -f, \text{ in } \Omega_2, \\ \Delta u_1 + k_2^2 N^2 u_1 = 0, \text{ in } \Omega_1, \\ [u] = 0 \text{ and } [\chi \partial_{\mathbf{n}} u] = 0, \text{ on } \Gamma, \\ SRC(u_2 - u_0) = 0, \end{cases}$$

for an explicit source term f. Let us now assume that we can construct the Dirichlet-Neumann (DN) operator for the interior problem

$$\begin{cases} \widetilde{\mathcal{Z}^-} : H^{1/2}(\Gamma) \to H^{-1/2}(\Gamma) \\ \qquad u_1 \mapsto \partial_{\mathbf{n}} u_1 = \widetilde{\mathcal{Z}^-} u_1 \end{cases} .$$

This operator, also called the Steklov-Poincaré operator, is a first-order pseudodifferential operator. The determination of this operator yields an *a priori* integro-differential computation of the internal solution from its Cauchy data. Using the transmission conditions at the interface and considering the scattered field formulation, we have to solve the exterior non-standard impedance boundary value problem: find v_2 such that

$$\begin{cases} \Delta v_2 + k_2^2 v_2 = 0, \text{ in } \Omega_2, \\ (\partial_{\mathbf{n}} - \alpha \widetilde{\mathcal{Z}^-}) v_2 = g, \text{on } \Gamma, \\ SRC(v_2) = 0, \end{cases}$$

with $g = -(\partial_{\mathbf{n}} - \alpha \widetilde{\mathcal{Z}^-}) u_0$. In the above system, the operator $\alpha \widetilde{\mathcal{Z}^-}$ is generally called the Exact Impedance Boundary Operator (EIBO).

To achieve an explicit computation of a non-local approximation of the DN operator for an arbitrarily-shaped surface, we rewrite the Helmholtz equation in a generalized coordinates system associated to the surface and next we compute the two first terms of its asymptotic expansion in homogeneous complex symbols. To this end, let us define the wave operator in the interior domain: $L_1 = (\Delta - \partial_t^2)$, where the exponential time dependence of the solution is

$e^{-ik_1 t}$. As a consequence, the multiplication by k_1 must be understood as the action of the first-order time derivative where we take the attenuation effects into account. A calculation in the one-dimensional case in space naturally imposes the choice of k_1. Furthermore, we can notice that we consider the same asymptotic parameter as Senior and Volakis [1995] but without assuming a particular analytical asymptotic form of the interior field. This hypothesis is unnecessary here and yields some more accurate pseudodifferential approximations of the EIBO.

Let us express the operator L_1 in a tubular neighborhood of Γ. Since Γ is a compact submanifold of $\mathbb{R}^3$, we can choose a local coordinates system at any point x_0 of Γ. Let us designate by $s = (s_1, s_2)$ the tangential variable and by r the radial variable along the unit normal vector $\mathbf{n}$ at x_0. Then, a point x near the surface can be locally rewritten under the form: $x = x_0 + r\mathbf{n}(x_0)$, with $x_0 \in \Gamma$. Let us introduce Γ_r as the surface defined for a fixed value of r and let us choose an orthogonal coordinates system on Γ. The covariant basis $(\boldsymbol{\tau}_1, \boldsymbol{\tau}_2)$ of the tangent plane $T_{x_0}(\Gamma)$ which is compatible with the orientation of $\mathbf{n}(x_0)$ is better known as the principal basis. Vectors $\boldsymbol{\tau}_1$ and $\boldsymbol{\tau}_2$ are the principal directions of the curvatures to the surface. If we set $\mathcal{R}$ as the curvature tensor of the tangent plane at a given point of the surface, then the diagonalization of $\mathcal{R}$ yields the determination of the principal curvatures κ_1 and κ_2 of Γ which fulfill: $\mathcal{R}\boldsymbol{\tau}_\beta = \kappa_\beta \boldsymbol{\tau}_\beta$ for $\beta = 1, 2$, and the mean curvature $\mathcal{H} = (\kappa_1 + \kappa_2)/2$. Let $h_\beta = 1 + r\kappa_\beta$, $\beta = 1, 2$. After a few calculations, we find the expression of the Helmholtz operator in generalized coordinates

$$L_1 = \partial_r^2 + 2\mathcal{H}_r \partial_r + h_1^{-1} h_2^{-1} \partial_s \cdot (h_2 h_1^{-1} \partial_{s_1}, h_1 h_2^{-1} \partial_{s_2}) - \partial_t^2,$$

setting $\mathcal{H}_r = (h_1^{-2}\kappa_1 + h_2^{-2}\kappa_2)/2$.

To construct the approximation of the EIBO, we have to introduce some tools available from the theory of pseudodifferential operators. Let $A = A(x, D_x)$ be a pseudodifferential operator of OPS^j, $j \in \mathbb{Z}$, $\sigma(A) = \sigma(A)(x, \eta)$ its symbol and $\sigma_j(A)$ its principal symbol. A symbol $\sigma(A)$ admits a symbolic asymptotic expansion in homogeneous symbols if it can be written on the form $\sigma(A) \sim \sum_{m=-j}^{+\infty} \sigma_{-m}(A)$, where functions $\sigma_{-m}(A)$ are some homogeneous functions of degree $-m$ with respect to η, with $m \geq -j$, which continuously depend on x. The above equality holds in the sense of pseudodifferential operators (see Treves [1980]). The partial symbol $\mathcal{L}_1$ of L_1, according to $s = (s_1, s_2)$ and t and their respective covariables $\xi = (\xi_1, \xi_2)$ and $N\omega$, smoothly depends on r. This symbol can be expressed as

$$\mathcal{L}_1 = \partial_r^2 + 2\mathcal{H}_r \partial_r - |\xi|^2 + i h_1^{-1} h_2^{-1} (\partial_{s_1} h_2 h_1^{-1}, \partial_{s_2} h_1 h_2^{-1}) \cdot \xi + N^2 \omega^2,$$

where the length of ξ is defined by: $|\xi| = (\sum_{\beta=1}^{2} h_\beta^{-2} \xi_\beta^2)^{1/2}$. Since N is a complex number, $\mathcal{L}_1$ is a complex symbol. Therefore, the operator L_1 can be factorized since its characteristic equation: $z^2 + N^2\omega^2 - |\xi|^2 = 0$ admits two distinct complex conjugate roots. These two solutions $z_1^{\pm} = \pm i(N^2\omega^2 - |\xi|^2)^{1/2}$ are first-order homogeneous complex functions according to $(\xi, N\omega)$. For a dissipative medium, we remark that: $\Re z_1^- > 0$ and $\Re z_1^+ < 0$.

According to Antoine et al. [2001], the following proposition holds.

Proposition 1. *There exist two classical pseudodifferential operators Z^- and Z^+ of OPS^1, which continuously depend on r and such that*

$$L_1 = (\partial_r - Z^+)(\partial_r - Z^-) \mod C^\infty,$$

with $\sigma_1(Z^\pm) = z_1^\pm$. Moreover, the uniqueness of the decomposition is satisfied by the following characterization. Let $z^\pm$ be the symbol of $Z^\pm$. From the definition of pseudodifferential operators in OPS^1, symbols $z^\pm$ are some elements of the symbol class S^1 and admit the following asymptotic expansion $z^\pm \sim \sum_{j=-1}^{+\infty} z_{-j}^\pm$, where $z_{-j}^\pm$ are some homogeneous complex valued functions of degree $-j$ with respect to $(\xi, N\omega)$.

In the case of a non-dissipative medium with $\Im N = 0$, it can be proved that the factorization is only valid in the cone of propagation $\{(\xi, N\omega), z_1^+ z_1^- > 0\}$.

Using the calculus rules of classical pseudodifferential operators, one can obtain an explicit recursive and constructive algorithm to compute each homogeneous symbol. We refer to Antoine et al. [2001] for further details. We restrict ourselves to the presentation of the effect of the first two terms of the asymptotic expansion ($m = 0$), taking more terms leading to more complicate formulations. The first symbol $z_1^- = -i(N^2\omega^2 - |\xi|^2)^{1/2}$ has already been computed. Concerning the zeroth-order symbol, one gets the explicit expression $z_0^- = -\mathcal{H} - \sum_{l=1}^{2} \kappa_l \xi_l^2 / (2(z_1^-)^2)$. From the analysis developed in Antoine et al. [2001], the EIBO can be suitably approximated by the following generalized Fourier-Robin boundary condition

$$(\partial_{\mathbf{n}} - \alpha \sum_{j=-1}^{m} \widetilde{Z_{-j}^-}) v_2 = \tilde{g} \equiv -(\partial_{\mathbf{n}} - \alpha \sum_{j=-1}^{m} \widetilde{Z_{-j}^-}) u_0, \tag{2}$$

with the classical pseudodifferential operator: $\widetilde{Z_{-j}^-} = \mathrm{Op}(z^-_{-j|r=0})$.

The resulting approximate boundary condition (2) is not yet completely satisfactory for a numerical treatment. Indeed, the condition is still defined by a non-local pseudodifferential operator. If we approach the numerical solution by a volume finite element method, then we have to consider the following variational formulation: find $v_2 \in H^1(\Omega_b)$ such that

$$\begin{aligned} \int_{\Omega_b} \nabla v_2 \cdot \nabla \varphi - k_2^2 v_2 \varphi d\Omega_b + \int_\Sigma \mathcal{M} v_2 \varphi d\Sigma + \alpha \int_\Gamma \sum_{j=-1}^{0} \widetilde{Z_{-j}^-} v_2 \varphi d\Gamma \\ = - \int_\Gamma \tilde{g} \varphi d\Gamma. \end{aligned} \tag{3}$$

In the above formulation, the unbounded domain has been truncated by the introduction of a non-reflecting boundary condition of the form: $\partial_{\mathbf{n}} v_2 + \mathcal{M} v_2 = 0$, where $\mathcal{M}$ is a local differential operator defined on a fictitious boundary Σ

enclosing the scatterer. The resulting finite domain of computation is denoted here by Ω_b with a boundary $\partial\Omega_b := \Gamma \cup \Sigma$. Generally, such a linear system is solved by an iterative solver (see e.g. Tezaur et al. [2002]). Therefore, we can assume that v_2 is a given entry at the k-th step of the algorithm and we want to evaluate the action of the operator defined by the left-hand side of Eq.(3). The first two terms are actually classical to compute (Antoine [2001], Tezaur et al. [2002]). This is not the case of the third one which involves two pseudodifferential operators leading to a high computational cost similar to the one involved in an integral equation approach. If we stop at this level, the method is inefficient. However, one can overcome this problem using some suitable Padé approximants. To fix the ideas, let us consider the first-order homogeneous pseudodifferential operator and let us introduce the classical Padé approximants of the square root with branch cut along the negative real line from $z = -1$: $\sqrt{1+z} \approx R_M(z) = c_0 + \sum_{j=1}^{M} a_j z(1+b_j z)^{-1}$. The coefficients c_0 and $(a_j, b_j)_{j=1,\dots,M}$ are expressed as $c_0 = 1$, $a_j = 2/(2M + 1)\sin^2(j\pi/(2M+1))$ and $b_j = \cos^2(j\pi/(2M+1))$, for $j = 1, \dots, M$. Then, the evaluation of $\widetilde{Z_1^-}$ applied to a given surface field v_2 is realized by first computing the solution ϕ_j to the surface PDE

$$\int_\Gamma b_j \nabla_\Gamma \phi_j \cdot \nabla_\Gamma \psi - k_1^2 \phi_j \psi d\Gamma = \int_\Gamma v_2 \psi d\Gamma, \quad \text{for } j = 1, \dots, M,$$

and then evaluating

$$\int_\Gamma \widetilde{Z_1^-} v_2 \psi d\Gamma = -ik_1 \int_\Gamma v_2 \psi d\Gamma + ik_1 \sum_{j=1}^{M} a_j \int_\Gamma \nabla_\Gamma \phi_j \cdot \nabla_\Gamma \psi d\Gamma,$$

for any test function ψ in $H^1(\Gamma)$. The operator ∇_Γ is the surfacic gradient operator of a scalar surface field. If the interior medium is weakly dissipative or non-dissipative, the approximation of the square root can require a large number M of PDEs to solve. A modified version of the square root approximation should be preferred as for instance by using the rotating branch cut approximation of Milinazzo et al. [1997]. This new approximation has been introduced within the context of underwater acoustic wave propagation problems resolved by the wide-angle parabolic equations approach. The technique consists of replacing the usual coefficients by the new ones $C_0 = e^{i\theta/2} R_M(e^{-i\theta} - 1)$, $A_j = e^{-i\theta/2} a_j((1 + b_j(e^{-i\theta} - 1))^{-2}$ and $B_j = e^{-i\theta} b_j(1 + b_j(e^{-i\theta} - 1))^{-1}$,for $j = 1, \dots, M$. An optimal experimental value for the free rotation angle is $\theta = \pi/4$ and $M = 4$ for the number of equations (to *a priori* choose with respect to the interior frequency).

4 Numerical performance

To evaluate the efficiency of the pseudolocal impedance boundary condition, we represent both the surface field and the far field pattern which is given by

$RCS(\vartheta) = 10\log_{10}(\lim_{r\to+\infty} 2\pi r|v_2(r,\vartheta)|^2)$ (db). We compare it to the local impedance boundary condition developed in Antoine et al. [2001] using some second-order Taylor expansions of the first four symbols. This latter condition has a wider validity domain than the usual Fourier-Robin condition. We consider an incident plane wave of frequency $k_2 = 25$ and with a null incidence angle illuminating the unit circular cylinder. The physical parameters are $\rho_r = 1.3$, $c_r = 1.05$ and $\delta = 5$ ($|N| = 0.96$ and $\Im k_1 = 2.3$). As it can be seen on Fig. 1, the surface field is accurately computed with the new condition compared to the second-order condition. This remark also holds for the bistatic RCS. A more complete analysis shows that it is always preferable to use the Padé approximation than the second-order Taylor expansion without affecting the total computational cost of the iterative procedure.

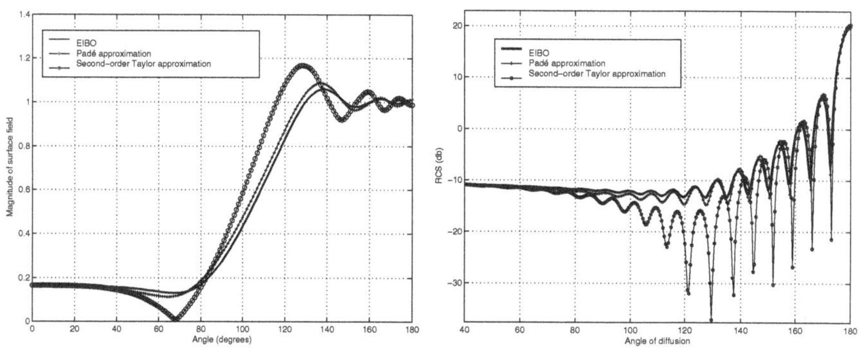

Fig. 1. Surface fields and bistatic RCS computations for the proposed test case

References

X. Antoine. Fast approximate computation of a time-harmonic scattered field using the on-surface radiation condition method. *IMA J. Appl. Math.*, pages 83–110, 2001.

X. Antoine, H. Barucq, and L. Vernhet. High-frequency asymptotic analysis of a dissipative transmission problem resulting in generalized impedance boundary conditions. *Asympt. Anal.*, 26(3-4):257–283, 2001.

F. Milinazzo, C. Zala, and G. Brooke. Rational square-root approximations for parabolic equation algorithms. *J. Acoust. Soc. Am.*, 101(2):760–766, 1997.

T. Senior and J. Volakis. *Approximate Boundary Conditions in Electromagnetics.* IEE Electromagnetic Waves Series, Serie 41, London, 1995.

R. Tezaur, A. Macedo, C. Farhat, and R. Djellouli. Three-dimensional finite element calculations in acoustic scattering using arbitrarily shaped convex artificial boundaries. *J. Numer. Methods Engrg.*, pages 1461–1476, 2002.

F. Treves. *Introduction to Pseudodifferential Operators and Fourier Integral Operators.* Plenum Press, New York and London, 1980.

Approximation and Fast Calculation of Non-local Boundary Conditions for the Time-dependent Schrödinger Equation

Anton Arnold[1], Matthias Ehrhardt[2], and Ivan Sofronov[3]

[1] Universität Münster, Institut für Numerische Mathematik, Einsteinstr. 62, D-48149 Münster, Germany (`http://www.math.uni-muenster.de/u/arnold/`)
[2] Technische Universität Berlin, Institut für Mathematik, Str. des 17. Juni 136, D-10623 Berlin, Germany (`http://www.math.tu-berlin.de/~ehrhardt/`)
[3] Keldysh Institute of Applied Mathematics, Russian Academy of Sciences, Moscow, Russia (`sofronov@spp.keldysh.ru`)

Summary. We present a way to efficiently treat the well-known transparent boundary conditions for the Schrödinger equation. Our approach is based on two ideas: firstly, to derive a discrete transparent boundary condition (DTBC) based on the Crank-Nicolson finite difference scheme for the governing equation. And, secondly, to approximate the discrete convolution kernel of DTBC by sum-of-exponentials for a rapid recursive calculation of the convolution. We illustrate the efficiency of the proposed method on several examples.

A much more detailed version of this article can be found in Arnold et al. [2003].

1 Introduction

Discrete transparent boundary conditions for the discrete 1D–Schrödinger equation

$$-iR(\psi_{j,n+1} - \psi_{j,n}) = \Delta^2 (\psi_{j,n+1} + \psi_{j,n}) - wV_{j,n+\frac{1}{2}} (\psi_{j,n+1} + \psi_{j,n}), \quad (1)$$

where $\Delta^2\psi_j = \psi_{j+1} - 2\psi_j + \psi_{j-1}$, $R = 4\Delta x^2/\Delta t$, $w = 2\Delta x^2$, $V_{j,n+\frac{1}{2}} := V(x_j, t_{n+\frac{1}{2}})$, $x_j = j\Delta x$, $j \in \mathbb{Z}$; and $V(x,t) = V_- =$ const. for $x \leq 0$; $V(x,t) = V_+ =$ const. for $x \geq X$, $t \geq 0$, $\psi(x,0) = \psi^I(x)$, with supp $\psi^I \subset [0, X]$, were introduced in Arnold [1998]. The DTBC at e.g. the left boundary point $j = 0$ reads, cf. Theorem 3.8 in Ehrhardt and Arnold [2001]:

$$\psi_{1,n} - s_0\psi_{0,n} = \textstyle\sum_{k=1}^{n-1} s_{n-k}\psi_{0,k} - \psi_{1,n-1}, \quad n \geq 1. \quad (2)$$

The convolution kernel $\{s_n\}$ can be obtained by explicitly calculating the inverse Z–transform of the function $\hat{s}(z) := \frac{z+1}{z}\hat{\ell}_0(z)$, where $\hat{\ell}_0(z) = 1 - i\zeta \pm \sqrt{-\zeta(\zeta + 2i)}$, $\zeta = \frac{R}{2}\frac{z-1}{z+1} + i\Delta x^2 V_-$ (choose sign such that $|\hat{\ell}_0(z)| > 1$).

Using (2) in numerical simulations permits to avoid any boundary reflections and it renders the fully discrete scheme unconditionally stable, like the Crank-Nicolson scheme (1) for the whole-space problem. However, the numerical effort to evaluate the DTBC increases linearly in t and it can sharply raise the total computational costs. A strategy to overcome this drawback is the key issue of this paper.

2 Approximation by Sums of Exponentials

The convolution coefficients s_n appearing in the DTBC (2) can either be obtained from (lengthy) explicit formulas or evaluated numerically: $s_n \approx \rho^n N^{-1} \sum_{k=0}^{N-1} \hat{s}(\rho e^{i\varphi_k})\, e^{in\varphi_k}$, $n = 0, 1, \dots, N-1$. Here $\varphi_k = 2\pi k/N$, and $\rho > 1$ is a regularization parameter. For the question of choosing ρ we refer the reader to Arnold et al. [2003] and the references therein.

Our fast method to calculate the discrete convolution in (2) is based on approximating these coefficients s_n by the following ansatz (sum of exponentials):

$$s_n \approx \tilde{s}_n := \begin{cases} s_n, & n = 0, \dots, \nu - 1, \\ \sum_{l=1}^{L} b_l q_l^{-n}, & n = \nu, \nu + 1, \dots, \end{cases} \tag{3}$$

where $L, \nu \in \mathbb{N}$ are fixed numbers. In order to find the appropriate constants $\{b_l, q_l\}$, we fix L and ν in (3) (e.g. $\nu = 2$), and consider the Padé approximation $\frac{P_{L-1}(x)}{Q_L(x)}$ for the formal power series: $f(x) := s_\nu + s_{\nu+1}x + s_{\nu+2}x^2 + \dots$, $|x| \le 1$.

Theorem 1. *Let the polynomial $Q_L(x)$ have L simple roots q_l with $|q_l| > 1$, $l = 1, \dots, L$. Then*

$$\tilde{s}_n = \sum_{l=1}^{L} b_l q_l^{-n}, \qquad n = \nu, \nu + 1, \dots, \tag{4}$$

where

$$b_l := -\frac{P_{L-1}(q_l)}{Q_L'(q_l)} q_l^{\nu-1} \neq 0, \qquad l = 1, \dots, L. \tag{5}$$

Remark 1. All our practical calculations confirm that the assumption of Theorem 1 holds for any desired L, although we cannot prove this.

Remark 2. According to the definition of the Padé algorithm the first $2L+\nu-1$ coefficients are reproduced exactly: $\tilde{s}_n = s_n$ for $n = \nu, \nu+1, \dots, 2L+\nu-1$. For the remaining $\tilde{s}_n$ with $n > 2L + \nu - 1$, the following estimate holds: $|s_n - \tilde{s}_n| = \mathcal{O}(n^{-\frac{3}{2}})$. A typical graph of $|s_n|$ and $|s_n - \tilde{s}_n|$ versus n for $L = 20$ is shown in Fig. 1 (note the different scaling for both graphs).

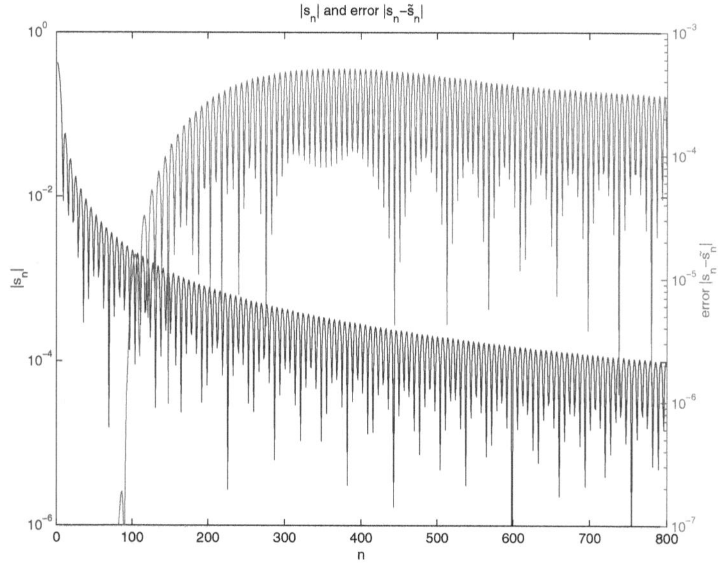

Fig. 1. Convolution coefficients s_n (left axis, dashed line) and error $|s_n - \tilde{s}_n|$ of the convolution coefficients (right axis); $\Delta x = 1/160$, $\Delta t = 2 \cdot 10^{-5}$, $V \equiv 0$ ($L = 20$)

3 The Transformation Rule

A nice property of the considered approach consists of the following: once the approximate convolution coefficients $\{\tilde{s}_n\}$ are calculated for particular discretization parameters $\{\Delta x,\, \Delta t,\, V\}$, it is easy to transform them into appropriate coefficients for any other discretization. We shall confine this discussion to the case $\nu = 2$:

Transformation rule 3.1 *For $\nu = 2$, let the rational function*

$$\hat{\tilde{s}}(z) = s_0 + \frac{s_1}{z} + \sum_{l=1}^{L} \frac{b_l}{q_l z - 1} \frac{1}{q_l z} \tag{6}$$

be the Z-transform of the convolution kernel $\{\tilde{s}_n\}_{n=0}^{\infty}$ from (3), where $\{\tilde{s}_n\}$ is assumed to be an approximation to a DTBC for the equation (1) with a given set $\{\Delta x,\, \Delta t,\, V\}$.
Then, for another set $\{\Delta x_\star,\, \Delta t_\star,\, V_\star\}$, one can take the approximation

$$\hat{\tilde{s}}^\star(z) := s_0^\star + \frac{s_1^\star}{z} + \sum_{l=1}^{L} \frac{b_l^\star}{q_l^\star z - 1} \frac{1}{q_l^\star z}, \tag{7}$$

where

$$q_l^\star := \frac{q_l \bar{a} - \bar{b}}{a - q_l b}, \quad b_l^\star := b_l q_l \frac{a\bar{a} - b\bar{b}}{(a - q_l b)(q_l \bar{a} - \bar{b})} \frac{1 + q_l^\star}{1 + q_l}, \tag{8}$$

$$a := 2\frac{\Delta x^2}{\Delta t} + 2\frac{\Delta x_\star^2}{\Delta t_\star} + i(\Delta x^2 V - \Delta x_\star^2 V_\star), \tag{9}$$

$$b := 2\frac{\Delta x^2}{\Delta t} - 2\frac{\Delta x_\star^2}{\Delta t_\star} - i(\Delta x^2 V - \Delta x_\star^2 V_\star). \tag{10}$$

$s_0^\star$, $s_1^\star$ *are the exact convolution coefficients for the parameters* $\{\Delta x_\star, \Delta t_\star, V_\star\}$.

While the Padé–algorithm provides a method to calculate approximate convolution coefficients $\tilde{s}_n$ for *fixed parameters* $\{\Delta x,\ \Delta t,\ V\}$, the Transformation rule yields the natural link between *different parameter sets* $\{\Delta x_\star,\ \Delta t_\star,\ V_\star\}$ (and L fixed).

Example 1. For $L = 10$ we calculated the coefficients $\{b_l, q_l\}$ with the parameters $\Delta x = 1$, $\Delta t = 1$, $V = 0$ and then used the Transformation 3.1 to calculate the coefficients $\{b_l^*, q_l^*\}$ for the parameters $\Delta x_* = 1/160$, $\Delta t_* = 2\cdot 10^{-5}, V_* = 4500$. Fig. 2 shows that the resulting convolution coefficients $\tilde{s}_n^*$ are in this example even better approximations to the exact coefficients s_n than the coefficients $\tilde{s}_n$, which are obtained directly from the Padé algorithm discussed in Theorem 1. Hence, the numerical solution of the corresponding Schrödinger equation is also more accurate (cf. Fig. 5).

The Maple code that was used to calculate the coefficients q_l, b_l in the approximation (3) including the explicit formulas in Transformation rule 3.1 can be downloaded from the authors' homepages.

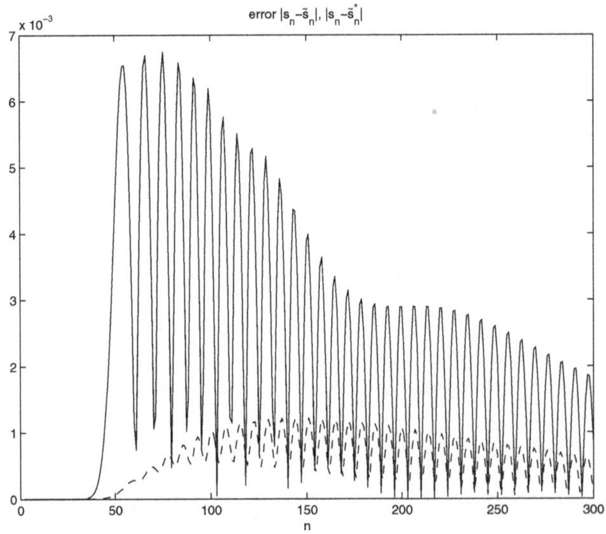

Fig. 2. Approximation error of the approximate convolution coefficients for $\nu = 2$, $\Delta x = 1/160$, $\Delta t = 2\cdot 10^{-5}, V = 4500$: The error of $\tilde{s}_n^*$ (- - -) obtained from the transformation rule and the error of $\tilde{s}_n$ (—) obtained from a direct Padé approximation of the exact coefficients s_n

4 Fast Evaluation of the Discrete Convolution

Given the approximation (3) of the discrete convolution kernel appearing in the DTBC (2), the convolution

$$C^{(n)}(u) := \sum_{k=1}^{n-\nu} u_k \tilde{s}_{n-k}, \quad \tilde{s}_n = \sum_{l=1}^{L} b_l q_l^{-n}, \quad |q_l| > 1, \tag{11}$$

of a discrete function u_k, $k = 1, 2, \ldots$, can be calculated efficiently by recurrence formulas, cf. Sofronov [1998]:

Theorem 2. *The function $C^{(n)}(u)$ from (11) for $n \geq \nu + 1$ is represented by*

$$C^{(n)}(u) = \sum_{l=1}^{L} C_l^{(n)}(u), \tag{12}$$

where

$$C_l^{(n)}(u) = q_l^{-1} C_l^{(n-1)}(u) + b_l q_l^{-\nu} u_{n-\nu} \quad \text{for} \quad n \geq \nu + 1, \quad C_l^{(\nu)}(u) \equiv 0. \tag{13}$$

This recursion drastically reduces the computational effort of evaluating DTBCs for long–time computations ($n \gg 1$): $\mathcal{O}(L{*}n)$ instead of $\mathcal{O}(n^2)$ arithmetic operations.

5 Numerical Examples

In this section we shall present two examples to compare the numerical results from using our approach of the approximated DTBC, i.e. the sum-of-exponentials-ansatz (3) (with $\nu = 2$) to the solution using the exact DTBC (2).

Example 2. As an example, we consider (1) on $0 \leq x \leq 1$ with $V_- = V_+ = 0$, and initial data $\psi^I(x) = \exp(i50x - 30(x-0.5)^2)$. The time evolution of the approximate solution $|\psi_a(x,t)|$ using the approximated DTBC with convolution coefficients $\{\tilde{s}_n\}$ and $L = 10$, $L = 20$, respectively, is shown in Fig. 3 (observe the viewing angle).

While one can observe some reflected wave when using the approximated DTBC with $L = 10$, there are almost no reflections visible when using the approximated DTBC with $L = 20$.

The goal is to investigate the long–time stability behaviour of the approximated DTBC with the sum-of-exponentials ansatz. The reference solution ψ_{ref} with $\Delta x = 1/400$, $\Delta t = 2 \cdot 10^{-5}$ is obtained by using exact DTBCs (2) at the ends $x = 0$ and $x = 1$. We vary the parameter $L = 10, 20, 30, 40$ in (3), find the corresponding approximate DTBCs, and show the relative error of the approximate solution, i.e. $\frac{||\psi_a - \psi_{ref}||_{L_2}(t)}{||\psi^I||_{L_2}}$. The result up to $n = 15000$

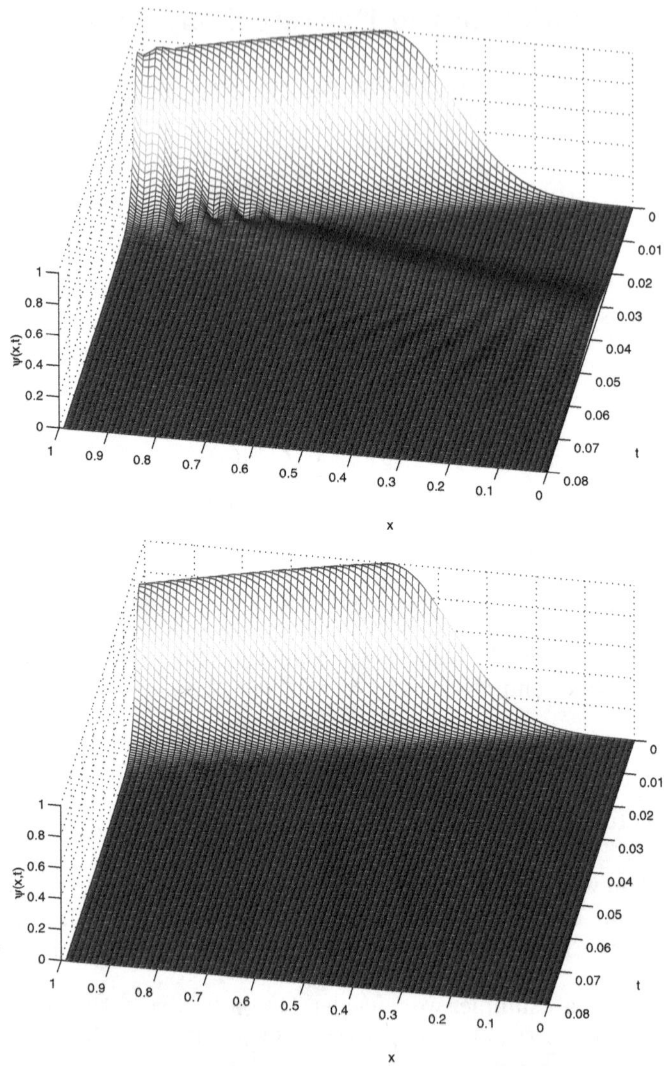

Fig. 3. Time evolution of $|\psi_a(x,t)|$: The approximate convolution coefficients consisting of $L = 10$ discrete exponentials give rise to a reflected wave (upper figure). Using $L = 20$ discrete exponentials make reflections almost invisible (lower figure)

is shown in Fig. 4. Larger values of L clearly yield more accurate coefficients and hence a more accurate solution ψ_a. Fig. 4 also shows the discretization error, i.e. $\frac{\|\psi_{ref}-\psi_{an}\|_{L_2}(t)}{\|\psi^I\|_{L_2}}$, where ψ_{an}, the analytic solution of this example is explicitly computable.

Example 3. The second example considers (1) on $[0,2]$ with zero potential in the interior ($V(x) \equiv 0$ for $0 < x < 2$) and $V(x) \equiv 4500$ outside the computa-

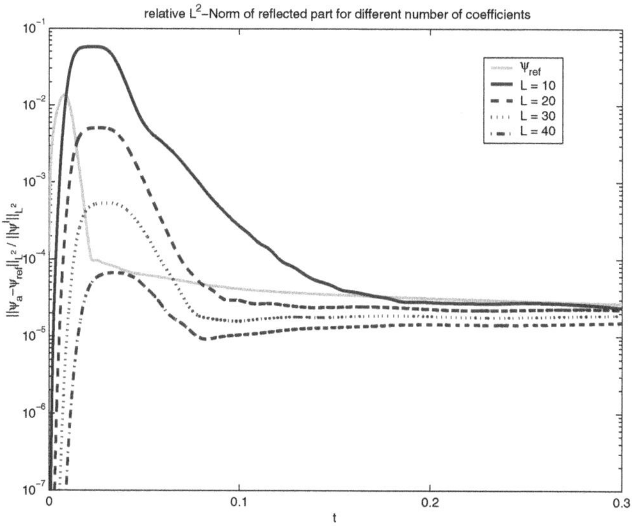

Fig. 4. Error of the approximate solution $\psi_a(t)$ with approximate convolution coefficients consisting of $L = 10, 20, 30, 40$ discrete exponentials and discretization error. ψ_{ref} is the relative error of ψ_{ref}. The error-peak between $t = 0.01$ and $t = 0.02$ corresponds to the first reflected wave

tional domain. The initial data is taken as $[\psi^I(x) = \exp(i100x - 30(x-1)^2)]$, and this wave packet is partially reflected at the boundaries. We use the rather coarse space discretization $\Delta x = 1/160$, the time step $\Delta t = 2 \cdot 10^{-5}$, and the exact DTBC (2). The value of the potential is chosen such that at time $t = 0.08$, i.e. after 4000 time steps 75% of the mass $(\|\psi(.,t)\|_2^2)$ has left the domain. Fig. 5 shows the time decay of the discrete ℓ^2-norm $\|\psi(.,t)\|_2$ and the temporal evolution of the error $\|e_L(.,t)\|_2 := \|\psi_a(.,t) - \psi_{ref}(.,t)\|_2/\|\psi^I\|_{L_2}$ when using an approximated DTBC with $L = 20, 30, 40$. Additionally, we calculated for $L = 20$ the coefficients $\{b_l, q_l\}$ for the "normalized parameters" $\Delta x = 1$, $\Delta t = 1$, $V = 0$ and then used the Transformation rule 3.1 to calculate the coefficients $\{b_l^*, q_l^*\}$ for the desired parameters.

6 Conclusion

For numerical simulations of the Schrödinger equation one has to introduce artificial (preferable *transparent*) boundary conditions in order to confine the calculation to a finite region. Such TBCs are non-local in time (of convolution form). Hence, the numerical costs (just) for evaluating these BCs grow quadratically in time. And for long-time calculations it can easily outweigh the costs for solving the PDE inside the computational domain.

Here, we presented an efficient method to overcome this problem. We construct approximate DTBCs that are of a sum-of-exponential form and hence

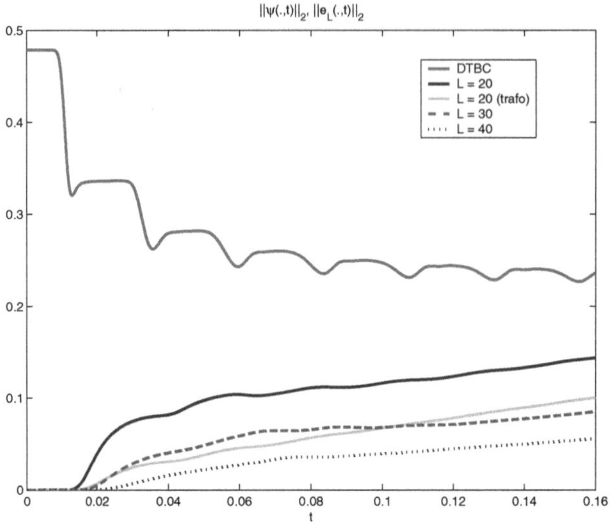

Fig. 5. Time evolution within the potential well ($V = 4500$) of $\|\psi(.,t)\|_2$ and of the errors $\|e_L(.,t)\|_2$ that are due to approximated DTBCs with $L = 20, 30, 40$. "$L = 20$ (trafo)" uses coefficients calculated by the Transformation rule 3.1

only involve a linearly growing numerical effort. Moreover, these BCs yield very accurate solutions, and it was shown in Arnold et al. [2003] that the resulting initial-boundary value scheme is conditionally ℓ^2-stable on $[0, T]$ as $\Delta t \to 0$ (e.g. for $0 < \Delta t < \Delta t_0$ and $\Delta x = \Delta x_0 = \text{const.}$).

Acknowledgement. The first two authors were partially supported by the IHP–Network HPRN-CT-2002-00282 from the EU and the DFG under Grant-No. AR 277/3-1. The second author was supported by the DFG Research Center "Mathematics for key technologies" (FZT 86) in Berlin. The third author was partially supported by RFBR-Grant No. 01-01-00520 and by Saarland University.

References

A. Arnold. Numerically absorbing boundary conditions for quantum evolution equations. *VLSI Design*, 6:313–319, 1998.

A. Arnold, M. Ehrhardt, and I. Sofronov. Discrete transparent boundary conditions for the Schrödinger equation: Fast calculation, approximation, and stability. *Comm. Math. Sci.*, 1:501–556, 2003.

M. Ehrhardt and A. Arnold. Discrete transparent boundary conditions for the Schrödinger equation. *Riv. Mat. Univ. Parma*, 6:57–108, 2001.

I. Sofronov. Artificial boundary conditions of absolute transparency for two- and threedimensional external time–dependent scattering problems. *Euro. J. Appl. Math.*, 9:561–588, 1998.

Domain Decomposition and Additive Schwarz Techniques in the Solution of a TE Model of the Scattering by an Electrically Deep Cavity

Nolwenn Balin[1,2], Abderrahmane Bendali[2,3], and Francis Collino[2]

[1] MBDA-France
[2] CERFACS (`http://www.cerfacs.fr/emc/balin/balin.html`)
[3] MIP

Summary. Two techniques are coupled to solve a model problem relative to the scattering of a 2D time-harmonic electromagnetic wave by an obstacle including an electrically deep cavity. Both of them are based on a boundary element method. The first technique uses a domain decomposition procedure to reduce the contribution of the cavity to a set of equations supported by the aperture. The second one is an additive Schwarz procedure to solve the problem after the reduction of the cavity. Numerical results are reported to give an insight into the approach.

Key words: Scattering, cavity, boundary elements, Schwarz additive method.

1 Introduction

It is a well-known fact for experts in stealth technology that a cavity residing in a scatterer can significantly contribute to the Radar Cross Section (RCS). Because of several difficulties, standard methods cannot be applied to solve this type of problem. Indeed, the size of the problem and the complexity of the involved phenomena (diffraction, resonance, etc.) prevent the use of available methods either direct or fast (like the fast multipole method) or asymptotic (like physical optics or geometrical theory of diffraction).

Several approaches, based on domain decomposition (DD) or hybrid methods have already been proposed: finite element-boundary integral (FE-BI) formulations (Jin [1993], Liu and Jin [2003]), multi-methods (Barka et al. [2000]) based on generalized scattering matrices, etc. However, in our opinion, none of these approaches can be considered as completely satisfactory in general. Some well-known dispersion deficiencies of FE methods can seriously damage the accuracy of the solution. Similarly, the determination of scattering matrices can rapidly become unwieldy.

We have investigated two new directions based on BI formulations to enhance the solution procedure. The first technique consists in exploiting the geometry of the cavity as in (Liu and Jin [2003]) to reduce its contribution to a set of equations supported only by the aperture. However, to avoid the dispersion flaws present in FE schemes, we use a BI formulation as well as a DD method to reduce the computing time and the memory storage. The second one is an additive overlapping Schwarz method for solving the equations on the aperture of the cavity and the rest of the boundary.

2 Nonoverlapping Domain Decomposition Method

2.1 The full problem

The geometrical data of the scattering problem are depicted in Fig. 1. They are related to a 2D model for the scattering of an electromagnetic wave by an open-ended thick cavity, indeed a time-harmonic H_z-wave. The scatterer is endowed with the perfect conducting boundary condition on Γ. The surrounding medium Ω is assumed to be the free-space. The unit normal to Γ inwardly directed to Ω is denoted by $\mathbf{n}$.

Assuming an implicit time dependence in $\mathrm{e}^{-\mathrm{i}\omega t}$, we are led to solve the following boundary-value problem (see e.g., Jin [1993])

$$\begin{cases} \Delta u + k^2 u = 0 \text{ in } \mathbb{R}^2, \\ \partial_{\mathbf{n}} u = 0 \text{ on } \Gamma, \\ \lim\limits_{|x| \to +\infty} |x|^{1/2} \left(\partial_{|x|}(u - u^{\mathrm{inc}}) - \mathrm{i}k(u - u^{\mathrm{inc}}) \right) = 0. \end{cases} \tag{1}$$

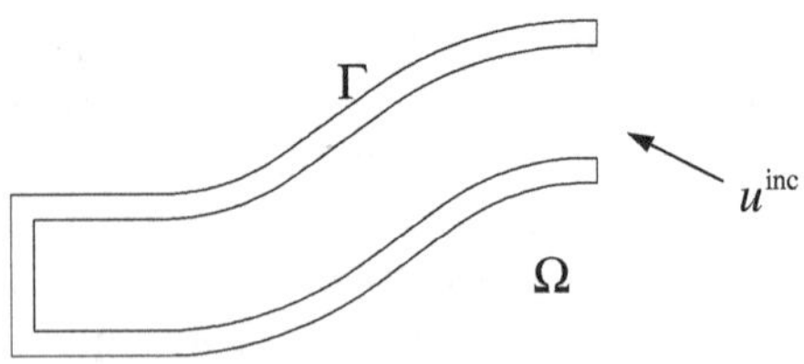

Fig. 1. The full problem

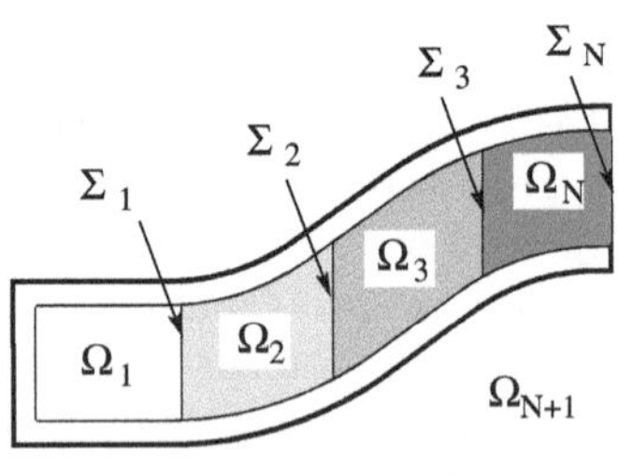

Fig. 2. The decomposition of the cavity

2.2 Domain decomposition and problem formulation

The cavity is sliced into N domains Ω_i $(i = 1, \dots, N)$ as shown in Fig. 2. The unbounded part Ω_{N+1} of this DD of Ω lies outside the cavity. The interfaces

Σ_i between the subdomains are sectional surfaces of the cavity. Finally, the unit normal to the boundary $\partial\Omega_i$ of Ω_i outwardly directed to Ω_i is denoted by $\mathbf{n_i}$.

Denoting by $u_i := u|_{\Omega_i}$ and by Γ_i the part of $\partial\Omega_i$ on Γ, we are led to the following equivalent formulation of problem (1)

$$\begin{cases} \left.\begin{array}{l} \Delta u_i + k^2 u_i = 0 \text{ in } \Omega_i, \\ \partial_{\mathbf{n}} u_i = 0 \text{ on } \Gamma_i, \end{array}\right\} \text{ for } i = 1, \dots, N{+}1, \\ \lim_{|x|\to+\infty} |x|^{1/2} \left(\partial_{|x|} \left(u_{N+1} - u^{\text{inc}}\right) - ik\left(u_{N+1} - u^{\text{inc}}\right)\right)(x) = 0, \end{cases} \tag{2}$$

subject to the following matching conditions

$$u_i = u_{i+1} \text{ and } \partial_{\mathbf{n_i}} u_i + \partial_{\mathbf{n_{i+1}}} u_{i+1} = 0 \text{ on } \Sigma_i, \text{ for } i = 1, \dots, N. \tag{3}$$

The most used BI formulations reduce the determination of u_i in Ω_i to its Cauchy data $\lambda_i := u_i|_{\partial\Omega_i}$ and $p_i := \partial_{\mathbf{n}_i} u_i|_{\partial\Omega_i}$ (e.g. Jin [1993]). Denoting the restriction of these Cauchy data to some part of $\partial\Omega_i$ in an obvious way, we directly obtain the following relations from the above boundary and matching conditions

$$p_i^{\Gamma_i} = 0, \tag{4}$$

$$\lambda_i^{\Sigma_i} = \lambda_{i+1}^{\Sigma_i} \text{ and } p_i^{\Sigma_i} + p_{i+1}^{\Sigma_i} = 0. \tag{5}$$

We use Rumsey's reactions principle to express the boundary and matching conditions variationally with testing functions λ_i' and p_i' subject to the same conditions as Cauchy data (4) and (5):

$$\sum_{i=1}^{N+1} \int_{\partial\Omega_i} \left(\partial_{\mathbf{n_i}} u_i \lambda_i' - u_i p_i'\right) ds = 0. \tag{6}$$

Expressing $u_i|_{\partial\Omega_i}$ and $\partial_{\mathbf{n_i}} u_i|_{\partial\Omega_i}$ through their integral representation in terms of λ_i and p_i, we obtain the following integral equations in a straightforward way

$$\sum_{i=1}^{N+1} \left\{ \lambda_i'^T \; p_i'^T \right\} Z_i \left\{ \begin{array}{c} \lambda_i \\ p_i \end{array} \right\} = \left\{ \lambda_{N+1}'^T \; p_{N+1}'^T \right\} U^{\text{inc}}. \tag{7}$$

The integrodifferential operator Z_i becomes a complex dense matrix representing the interactions between the unknowns related to subdomain Ω_i once λ_i, λ_i', p_i and p_i' have been discretized as in (Bendali and Souilah [1994]) for instance.

Clearly, this variational system has the same structure as the usual ones associated with a substructuring procedure in FE methods. It yields a linear system of the type depicted in Fig. 3. A Schur complement procedure, dealing with one subdomain at a time, can hence be used to reduce the equations relative to the cavity to a matrix coupling the Cauchy data on the aperture $\lambda_{N+1}^{\Sigma_N}$ and $p_{N+1}^{\Sigma_N}$. The procedure saves computing time and storage in a significant way.

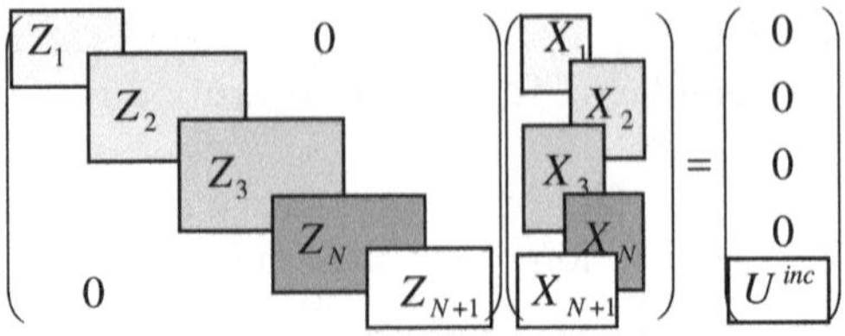

Fig. 3. Linear system

2.3 Numerical results

To give an insight into the performance of the method, we consider the COBRA JINA test case. To examine the effect of the exterior structure, we add a thickness (see Fig. 1) and fix the frequency at 8 GHz.

Figure 4 represents the CPU time necessary to compute the monostatic RCS for 361 incidences and for several decompositions of the cavity. Splitting the domain Ω into only two subdomains reduces this CPU time by a quite good factor of 65%, the optimal number of subdomains being 4 for the case at hand. Meanwhile the memory storage is reduced by a factor of 60%.

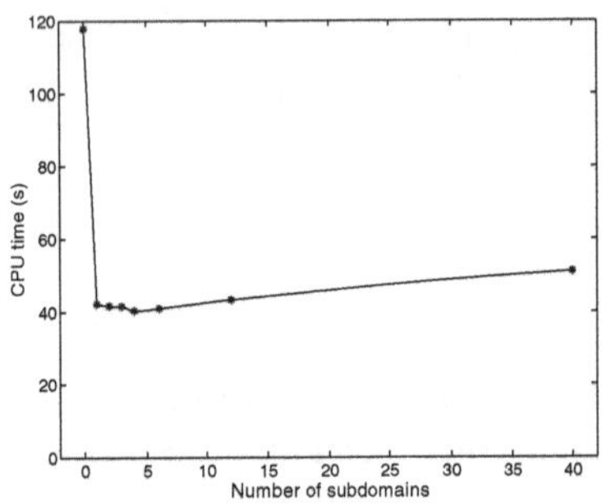

Fig. 4. Number of internal subdomains and CPU time

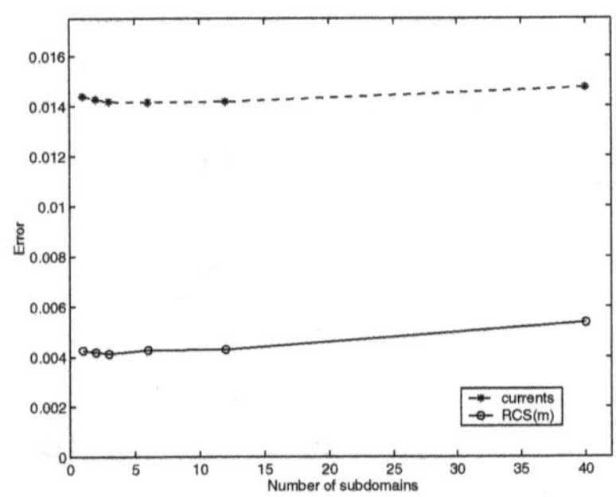

Fig. 5. Error (L^2-norm) on the currents (*dashed line*) and RCS (*solid line*)

The currents on the external boundary Γ_{N+1} and the RCS (in m) are compared to those obtained using the direct solution (Fig. 5). Although the introduction of Cauchy data on the interfaces induces an error on the currents, they remain small and do not increase in a significant way with the number of subdomains.

3 Additive Overlapping Schwarz Method

3.1 Introduction

The Schwarz methods (Lions [1988]) are efficient iterative processes for solving usual boundary value problems. The principle is to solve only small size problems in each subdomain in each iteration. We give an adaptation of the additive version of the Schwarz algorithm (Frommer and Szyld [1999] for example) for the problem set on the boundary of Ω_{N+1}, obtained once the cavity has been reduced, to efficiently deal with its solution.

3.2 Boundary decomposition

We start from a generic problem like the following one

$$X^{'T} B X = X^{'T} U \tag{8}$$

assuming that this system is related to the nodal values X and $X^{'}$ of respectively unknown and test functions defined on the boundary Γ. For the subsequent description, it will be more meaningful to denote the components of X and $X^{'}$ as $X(x)$ and $X^{'}(x)$ respectively, x being a node on Γ.

We consider Γ_i $(i = 1, \ldots, N)$ an overlapping decomposition of Γ (Fig. 6) as well as a partition of unity α_i associated with this covering of Γ.

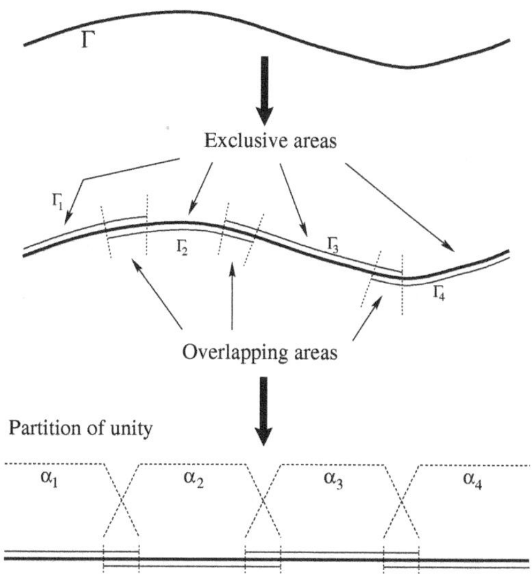

Fig. 6. Domain decomposition and partition of unity

Starting from this decomposition of the boundary, we can decompose X as follows

$$X(x) = \sum_{j=1}^{N} \alpha_j(x) X_j(x), \quad \text{for all node } x, \quad X_j = \mathbb{I}_j X. \tag{9}$$

$\mathbb{I}_j$ is the matrix obtained from the identity matrix $\mathbb{I}$ with the same size than B by removing all the rows corresponding to a node x where $\alpha_j(x) = 0$. Now expressing the vector with the same size as X_j and whose components are $\alpha_j(x)X_j(x)$ by means of a diagonal matrix still denoted by α_j as $\alpha_j X_j$, we can write (9) in the form of a matrix product as follows

$$X = \sum_{j=1}^{N} \mathbb{I}_j^T \alpha_j X_j. \tag{10}$$

Inserting (9) in (8) and testing by $X_i^{'T} \alpha_i \mathbb{I}_i$, we are led to

$$X_i^{'T} B_{ii} \; X_i = X_i^{'T} U_i - X_i^{'T} \sum_{\substack{j=1 \\ j \neq i}}^{N} B_{ij} X_j \tag{11}$$

where

$$B_{ij} = \alpha_i \mathbb{I}_i B \mathbb{I}_j^T \alpha_j, \quad U_i = \alpha_i \mathbb{I}_i U. \tag{12}$$

We can then reconstruct X to show that it solves the following fixed point problem

$$X = \sum_{i=1}^{N} \alpha_i X_i = \sum_{i=1}^{N} \alpha_i \left(B_{ii}^{-1} \Big(U_i - \sum_{j \neq i} B_{ij} X_j \Big) \right). \tag{13}$$

This system corresponds to the classical form of the additive Schwarz algorithm (Frommer and Szyld [1999]). Once derived for a linear system $CX = D$, it can be solved by the GMRES algorithm.

3.3 Numerical results

This method has been tested on the COBRA cavity with thin walls at a frequency of 30 GHz. This is known to be a difficult problem for the convergence of iterative methods.

Distribution of the eigenvalues

Figure 7 depicts the eigenvalues of the matrix of the initial system and those of the matrix obtained by the Schwarz procedure using a decomposition of the boundary into 75 patches. All the eigenvalues of the new matrix lie in the right half plane whereas the initial matrix has an important number of eigenvalues almost uniformly distributed in a circle centered at zero. It is well-known that distributions of the eigenvalues of the latter type are the worst cases relatively to the convergence of iterative methods whereas the former is much more adapted to this convergence.

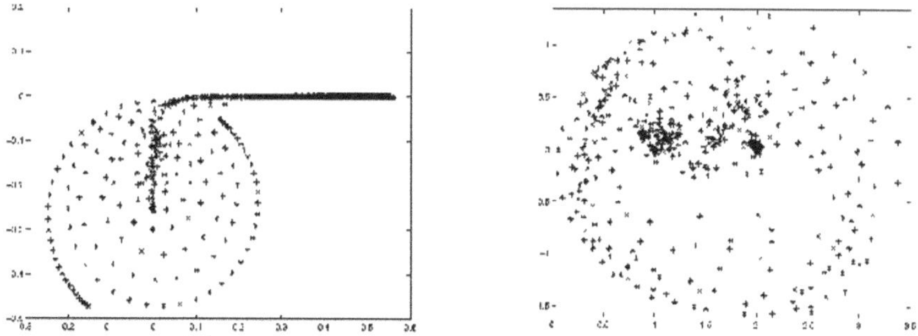

(a) (b)

Fig. 7. Eigenvalues repartition: (**a**) initial matrix; (**b**) matrix resulting from the Schwarz method

Convergence

The results of this method have been compared to those obtained by a SParse Approximate Inverse (SPAI) preconditioning technique. The Krylov method which has been used is the GMRES algorithm with a restart every 20 iterations.

Figure 8 represents the norms of the residuals relative to the initial matrix, without any preconditioning and with a SPAI preconditioner and for the Schwarz procedure. As expected, the Schwarz technique shows a better convergence rate than the method without the preconditioner. Furthermore, the convergence rate is almost the same as that one of the SPAI method. It is worth noting that we have considered a cavity with thin walls, corresponding in fact to an open surface, which is the most unfavourable case for the convergence of the iterative process, to check its robustness.

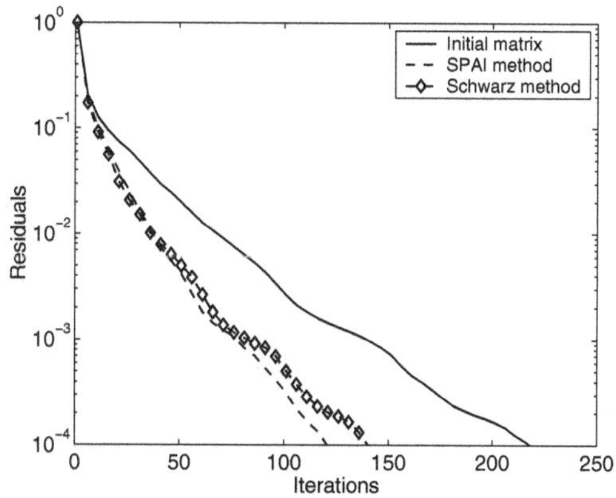

Fig. 8. Convergence rate

4 Conclusion and forthcoming studies

The solution procedure proposed here has been fully validated in the 2D case and has efficiently handled several deep cavity problems. Work on extensions to 3D is currently going on. We have been inspired to do so by a prospective work in 3D case by M. Fares. The authors would like to acknowledge this invaluable information as well as the support of CINES which has provided the possibility in terms of massively parallel platforms to deal with such problems of really huge size.

References

André Barka, Paul Soudais, and D. Volpert. Scattering from 3D cavities with a plug&play numerical scheme combining IE, PDE and modal techniques. *IEEE Trans. Antennas and Propagation*, 2000.

A. Bendali and M. Souilah. Consistency estimates for a double-layer potential and application to the numerical analysis of the boundary-element approximation of acoustic scattering by a penetrable object. *Math. Comp.*, 62 (205):65–91, 1994.

Andreas Frommer and Daniel B. Szyld. Weighted max norms, splittings, and overlapping additive Schwarz iterations. *Numer. Math.*, 83:259–278, 1999. URL `http://www.springer-ny.com/journals/211/`.

Jian-Ming Jin. *The Finite Element Method in Electromagnetism*. Wiley, New-York, 1993.

Pierre-Louis Lions. On the Schwarz alternating method. I. In Roland Glowinski, Gene H. Golub, Gérard A. Meurant, and Jacques Périaux, editors, *First International Symposium on Domain Decomposition Methods for Partial Differential Equations*, pages 1–42, Philadelphia, PA, 1988. SIAM.

J. Liu and Jian-Ming Jin. Scattering analysis of a large body with deep cavities. *IEEE Trans. Antennas and Propagation*, 51(6):1157–1167, June 2003.

Part III

Minisymposium: Parallel Finite Element Software

A Model for Parallel Adaptive Finite Element Software

Krzysztof Banaś

Cracow University of Technology, Section of Applied Mathematics ICM

Summary. The paper presents a conceptual model and details of an implementation for parallel adaptive finite element systems, particularly their computational kernels. The whole methodology is based on domain decomposition while message passing is used as a model of programming. The proposed finite element architecture consist of independent modules, most of them taken from sequential codes. The sequential modules are only slightly modified for parallel execution and three new modules, explicitly aimed at handling parallelism, are added. The most important new module is the domain decomposition manager that performs most tasks related to parallel execution. An example implementation that utilizes 3D prismatic meshes and discontinuous Galerkin approximation is presented. Two numerical examples, the first in which Laplace's equation is approximated using GMRES with multi-grid preconditioning and the second where dynamic adaptivity with load balancing is utilized for simulating linear convection, illustrate capabilities of the approach.

1 Introduction

Parallelization of adaptive finite element systems is a complex and complicated task. There are only few systems (Bastian et al. [1997], Beall and Shephard [1999], Bangerth and Kanschat [1999]) combining adaptivity and parallelism within a comprehensive finite element environment. Comprehensive is understood as offering such capabilities as 2D and 3D meshes of various types, continuous, discontinuous and higher order approximations, multi-level iterative solvers for linear systems, handling of coupled, possibly multi-physics, non-linear problems. On the other hand there is a growing interest in using principles of software engineering and object orientedness for design of scientific codes (Bruaset and Langtangen [1997], Beck et al. [1997]). The goal of the research is to combine flexibility and maintainability of object oriented codes with efficiency of monolithic Fortran or C programs.

The stress in the present paper is put on the modular structure of codes, design of modules' interfaces and fundamental principles for parallelizing adaptive finite element codes. The starting point is a modular finite element framework for sequential computations, designed for extendibility, reusability and efficiency. The goal of the parallelization process is to preserve the modular structure of the framework and add efficient mechanisms for parallel execution. Additionally parallelization is designed to change sequential modules as little as possible and to be conceptually simple. The aim of these last features is to offer an easy way for parallelizing existing legacy finite element programs.

2 A model architecture

It is assumed that the whole code is built of independent modules having their own data structures and communicating through interfaces accessible using the main programming languages of scientific computing (Fortran90, C, C++) (Banaś [2002]). There are four fundamental sequential modules:

- mesh manipulation module - that provides the other modules with all topological data concerning elements and other entities (faces, edges, vertices) present in the mesh; mesh manipulation module performs refinements/derefinements of individual mesh entities
- approximation module - that performs all tasks related to approximation fields defined for finite element meshes; main tasks include numerical integration, finite element interpolation and different kinds of projections
- linear solver module - the module may form a solver itself or serve as an interface for some external solver
- problem dependent module - the rest of the code that includes, among others, submodules specifying the solved PDE problem and driving the process of adaptation (the latter involves error estimation)

The mesh manipulation module is the simplest to design. It possess its own data structure and is organized as a set of services, a library of functions returning data concerning mesh entities. It does not use data from other finite element modules, although it can, and sometimes should, interact with external modules for mesh generation and geometry modeling.

The approximation module handles all tasks related to approximation fields that are defined in terms of finite element shape functions. Since definitions of shape functions are specific to different types of elements, approximation modules are strongly related to particular mesh manipulation modules. Besides the dependence on a specific mesh, the approximation module is also strongly coupled with the problem dependent module. The problem of efficient realization of numerical integration in flexible, multi-purpose codes is solved using an interface with few well defined call-backs (Banaś [2002]).

The mechanism of call-backs is also utilized in the design of the interface between the problem dependent module and the linear solver module. It is assumed that the problem dependent module calls the linear solver to perform basic steps of the solution procedure, but it is the linear solver that gathers all data (on mesh entities, approximation fields and particular entries to the system matrix and the load vector) necessary to perform multi-level solution of linear equations.

2.1 Parallel execution modules

The proposed model of parallelization is based on domain decomposition as an algorithmic foundation and message passing as a programming technique. One of reasons for such a choice is the possibility of reusing much of sequential modules for parallel codes.

It is assumed that sequential modules are included into parallel codes without substantial modifications. There are three new modules added to handle parallel execution. The two first are simple interface modules. The first, named parallel execution interface, gathers the main calls related to parallel execution within the problem dependent module. These calls are then passed to the main parallel module, the domain decomposition manager, or left with no effect in case of sequential runs.

The second simple module connects the finite element program to a parallel execution environment. It consist of a set of generic send/receive and group operations, that have to be implemented for various communication libraries.

The domain decomposition manager performs the following tasks:

- interfacing an external mesh partitioner (and, possibly different, repartitioner)
- distributing the mesh among processors
- creating necessary overlap and managing all requests related to overlap entities
- implementing domain decomposition algorithm
- adapting mesh in parallel
- load balancing and data transfer

The domain decomposition manager is composed of several submodules, responsible for parallel execution of different tasks. In such a way it is possible to parallelize only parts of the code (e.g. linear solver) while the rest remains sequential.

3 Implementation

Based on the proposed architecture (see Fig. 1) a prototype implementation has been created that uses 3D prismatic meshes and discontinuous Galerkin approximation.

The basis for implementation of the domain decomposition manager is formed by the assumption that every mesh entity and every set of approximation data present in the data structure is equipped with a global (interprocessor) identifier (IPID). This identifier can be understood as a substitute for a global address space used in sequential codes and is composed of a processor (subdomain) number and a local (to a given processor) identifier. IPIDs are not known to sequential modules of the code and all situation where the access to non-local data is necessary are handled by the domain decomposition manager.

4 Numerical examples

Two numerical examples showing capabilities of the described approach and the prototype implementation are presented in this section. The example problems are very simple from mathematical point of view. However, they show the effect of practical realization of two important and technically difficult, from implementation point of view, phases of simulation: parallel multilevel solution of linear equations and parallel adaptivity combined with transfer of mesh entities to maintain load balance.

The computational environment for both examples consist of a set of Linux workstations connected using a standard 100 Mbit Ethernet network. The results in tables have been obtained using computers equipped with 1.6 GHz Pentium IV processor and 1 GByte memory.

4.1 Simulating diffusion

The first example is Laplace's equation

$$\Delta u = \Delta u_{ex}$$

where u_{ex} is the known exact solution:

$$u_{ex} = \exp\left(-x^2 - y^2 - z^2\right)$$

The computational domain consist of the box $[0, 0.1] \times [0, 1] \times [0, 10]$ and boundary conditions are chosen to match the exact solution. Discontinuous Galerkin approximation (Oden et al. [1998]) and the preconditioned GMRES method are used for solving the problem.

Table 1 presents results for a series of computations corresponding to the described problem. Two preconditioners are employed, both use the combination of additive Schwarz preconditioning for the whole problem and multiplicative Schwarz within subdomains. The first is single level preconditioner and the second uses three consecutive mesh levels to achieve multigrid preconditioning. For each preconditioner problems of different sizes, corresponding to subsequently uniformly refined meshes, are considered. For each combination preconditioner/problem size results of computations using 1, 2, 4 and 8 workstations are shown. For the largest problem the reference number of processors to compute speed up and efficiency is two, since the problem did not

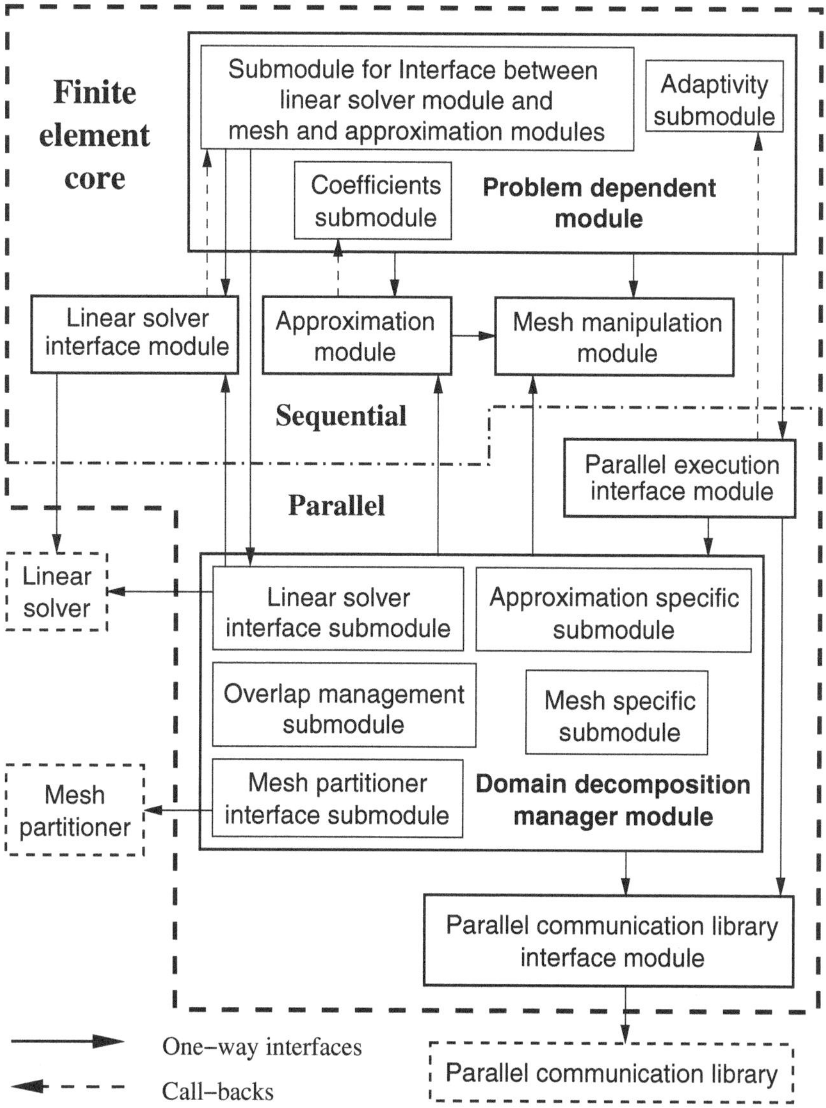

Fig. 1. Diagram of the proposed modular architecture for computational kernels of parallel adaptive finite element codes

fit into a memory of a single computer. For the smallest problem the number of mesh levels was equal two and the only possible preconditioner was single level.

Results are reported for 10 iterations of the preconditioned GMRES method to focus on the efficiency of parallel implementation, not considering the influence of parallelization on the convergence of GMRES (nevertheless the latter is reported for completeness). Subsequent meshes are obtained by uniform refinements and for each mesh N_{DOF} is the number of degrees of freedom. N_{proc} is the number of workstations solving the problem. *Error* is the norm of residual after 10 GMRES iterations (within a single restart) and *Rate* is the total GMRES convergence rate during solution. Execution time *Time* is a wall clock time. Speed-up and efficiency are computed in the standard way.

Table 1. Results for 10 iterations of the preconditioned GMRES method and discontinuous Galerkin approximation used for solving Laplace's equation in a box domain (description in the text)

Single level preconditioner						
N_{DOF}	N_{proc}	Error*10^9	Conv. rate	Exec. time	Speed up	Efficiency
48 896	1	0.288	0.443	2.26	1.00	100%
	2	0.328	0.448	1.16	1.94	97%
	4	0.340	0.450	0.61	3.70	92%
	8	0.361	0.452	0.36	6.28	78%
391 168	1	9.313	0.626	17.85	1.00	100%
	2	10.173	0.632	8.93	1.99	100%
	4	10.252	0.633	4.53	3.94	98%
	8	11.183	0.638	2.34	7.63	95%
3 129 344	2	48.041	0.738	70.76	1.00	100%
	4	47.950	0.738	35.63	1.98	99%
	8	48.748	0.739	17.71	3.99	100%

Three level preconditioner						
N_{DOF}	N_{proc}	Error*10^9	Conv. rate	Exec. time	Speed up	Efficiency
391 168	1	0.018	0.335	26.18	1.00	100%
	2	0.017	0.334	14.18	1.85	92%
	4	0.018	0.335	9.08	2.88	72%
	8	0.024	0.346	7.60	3.44	43%
3 129 344	2	0.027	0.350	111.16	1.00	100%
	4	0.027	0.350	57.76	1.92	96%
	8	0.027	0.348	33.15	3.35	84%

4.2 Simulating convection

The second example is a simple convection problem in the box $[0, 38] \times [0..1000] \times [0..18]$. A rectangular pattern is traveling from left to right (along the y-axis). GMRES with single level Schwarz preconditioning is used, once again with discontinuous Galerkin approximation. The only interesting process for this example, that will be described in more detail, are the subsequent parallel mesh adaptations and load balancing achieved through transfer of mesh entities. There are four workstations used for simulation and the computational domain is divided into four subdomains. Subdomains have two element overlap to enable mesh adaptations and overlapping Schwarz preconditioning. After each time step (in the example run there were 120 time steps) the mesh is adapted in parallel.

After each mesh adaptation, the number of degrees of freedom in each subdomain is checked against the average (it is assumed that processors are of the same speed). If imbalance larger than 10% is encountered, mesh repartitioner is called, to provide new domain decomposition. According to the new assignment of elements to processors and two element overlap requirements, mesh entities are marked respectively, and the transfer between subdomains takes place. To enable clustering, mesh transfers consider always whole element families - initial elements that are marked for a transfer and all their antecedents.

Table 2 presents characteristics of mesh transfers for five subsequent time steps, from 100 to 104. The average number of DOFs in a subdomain remains constant since the same number of elements appears due to refinements and disappears due to derefinements. Since refinements and derefinements takes place in different regions the difference between the subdomain with the greatest number of DOFs and the subdomain with the smallest number of DOFs grows after each time step. The numbers of mesh entities reported in the table concern the total number of entities effectively transferred between all the subdomains. The numbers do not include entities for which IPIDs only are exchanged.

For the whole simulation, the speed up obtained using 4 processors was equal to 2.67, giving the efficiency of 67%. For the overhead that includes mesh repartitioning, mesh transfers and the fact that, according to the overall strategy, the load for processors is not perfectly balanced, the results appear to be reasonable.

5 Conclusions

The new architecture proposed for parallel adaptive finite element codes fulfills the requirement of combining execution efficiency and code modularity. Further improvements of the prototype implementation concerning efficiency and the creation of new specialized modules that would increase code's flexibility are under way.

Acknowledgement. This work has been supported by the Polish State Committee for Scientific Research under grant 7 T11F 014 20

Table 2. Characteristics of mesh transfers during parallel simulation for the convection problem

	Time step number				
	100	101	102	103	104
Average number of DOFs	5086	5086	5086	5086	5086
Maximal number of DOFs	5636	5120	5372	5596	5120
Minimal number of DOFs	4468	5012	4732	4508	4996
Number of transferred vertices	300	0	0	390	0
Number of transferred edges	1212	0	0	1671	0
Number of transferred faces	1284	0	0	1863	0
Number of transferred elements	438	0	0	657	0

References

K. Banaś. Finite element kernel modules for parallel adaptive codes. Report 4/2002, Section of Applied Mathematics ICM, Cracow University of Technology, Warszawska 24, 31-155 Kraków, Poland, 2002. *submitted to Computing and Visualization in Science.*

W. Bangerth and G. Kanschat. Concepts for object-oriented finite element software - the deal.II library. Preprint SFB 359, Universitat Heidelberg, 1999.

P. Bastian, K. Birken, K. Johannsen, S. Lang, N. Neuss, H. Rentz-Reichert, and C. Wieners. UG - a flexible software toolbox for solving partial differential equations. *Computing and Visualization in Science*, 1(1):27–40, 1997.

M. Beall and M. Shephard. An object-oriented framework for reliable numerical simulations. *Engineering with Computers*, 15:61–72, 1999.

R. Beck, B. Erdman, and R. Roitzsch. An object-oriented adaptive finite element code: design issues and application in hyperthermia treatment planning. In E. Arge, A. Bruaset, and e. H.P Langtangen, editors, *Modern software tools for scientific computing*, pages 105–123. Birkhauser Press, 1997.

A. Bruaset and H. Langtangen. A comprehensive set of tools for solving partial differential equations - Diffpack. In M. Daehlen and A. Tveito, editors, *Numerical Methods and Software Tools in Industrial Mathematics*, pages 63–92. Birkhauser, 1997.

J. Oden, I. Babuska, and C. Baumann. A discontinous *hp* finite element method for diffusion problems. *Journal of Computational Physics*, 146: 491–519, 1998.

Towards a Unified Framework for Scientific Computing

Peter Bastian[1], Mark Droske[3], Christian Engwer[1], Robert Klöfkorn[2], Thimo Neubauer[1], Mario Ohlberger[2], and Martin Rumpf[3]

[1] Interdisziplinäres Zentrum für Wissenschaftliches Rechnen, Universität Heidelberg, Im Neuenheimer Feld 368, D-69120 Heidelberg,
[2] Abteilung für Angewandte Mathematik, Universität Freiburg, Hermann-Herder-Str. 10, D-79104 Freiburg,
[3] Fachbereich Mathematik, Universität Duisburg Lotharstr. 63/65, D-47048 Duisburg; DUNE website: `http://hal.iwr.uni-heidelberg.de/dune/`

Summary. Most finite element, or finite volume software is built around a fixed mesh data structure. Therefore, each software package can only be used efficiently for a relatively narrow class of applications. For example, implementations supporting unstructured meshes allow the approximation of complex geometries but are in general much slower and require more memory than implementations using structured meshes. In this paper we show how a generic mesh interface can be defined such that one algorithm, e. g. a discretization scheme, works on different mesh implementations. For a cell centered finite volume scheme we show that the same algorithm runs thirty times faster on a structured mesh implementation than on an unstructured mesh and is only four times slower than a non-generic version for a structured mesh. The generic mesh interface is realized within the *Distributed Unified Numerics Environment* DUNE.

1 Introduction

There exist many simulation packages for the numerical solution of partial differential equations ranging from small codes for particular applications or teaching purposes up to large ones developed over many years which can solve a variety of problems. Each of these packages has a set of features which the designers decided to need to solve their problems. In particular, the codes differ in the kind of meshes they support: (block) structured meshes, unstructured meshes, simplicial meshes, multi-element type meshes, hierarchical meshes, bisection and red-green type refinement, conforming or non-conforming meshes, sequential or parallel mesh data structures are possible.

Using one particular code it may be impossible to have a particular feature (e. g. local mesh refinement in a structured mesh code) or a feature may be very inefficient to use (e. g. structured mesh in unstructured mesh code). If efficiency matters, there will never be one optimal code because the goals

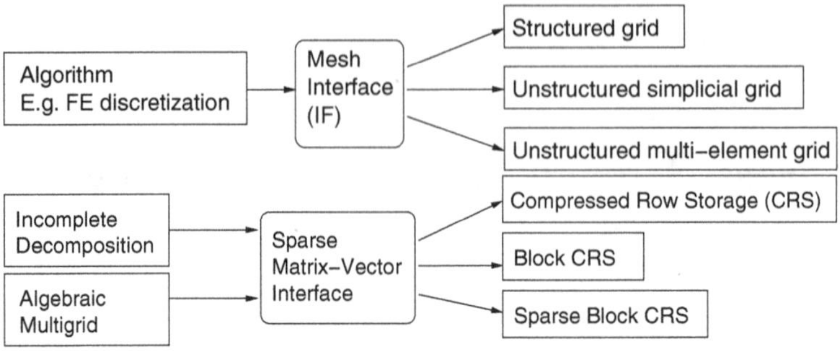

Fig. 1. Encapsulation of data structures with abstract interfaces

are conflicting. Extension of the set of features of a code is often very hard. The reason for this is that most codes are built upon a particular mesh data structure. This fact is well known in computer science (Brooks [1975]).

A solution to this problem is to separate data structures and algorithms by an abstract interface, i. e.

- one writes algorithms based on an abstract interface and
- uses exactly the data structure that fits best to the problem.

Figure 1 shows the application of this concept to two different places in a finite element code: A discretization scheme accesses the mesh data structure through an abstract interface. The interface can be implemented in different ways, each offering a different set of features efficiently. In the second example an algebraic multigrid method accesses a sparse matrix data structure through an abstract interface.

Of course, this principle also has its implications: The set of supported features is built into the abstract interface. Again, it is in general very difficult to change the interface. However, not all implementations need to support the whole interface (efficiently). Therefore, the interface can be made very general. At run-time the user pays only for functionality needed in the particular application.

The paper is organized as follows: The next section describes the *Distributed Unified Numerics Environment* (DUNE) which is based on abstract interfaces and shows how these interfaces can be implemented very efficiently using generic programming in C++. Then, in Section 3, we describe in more detail the abstract interface for a general finite element or finite volume mesh and in Section 4 we evaluate the concept on the basis of a cell centered finite volume scheme for various implementations of the mesh interface.

2 The DUNE Library

Writing algorithms based on abstract interfaces is not a new concept. Classical implementations of this concept in procedural languages use function calls. As an example consider the basic linear algebra subroutines BLAST [2001]. In

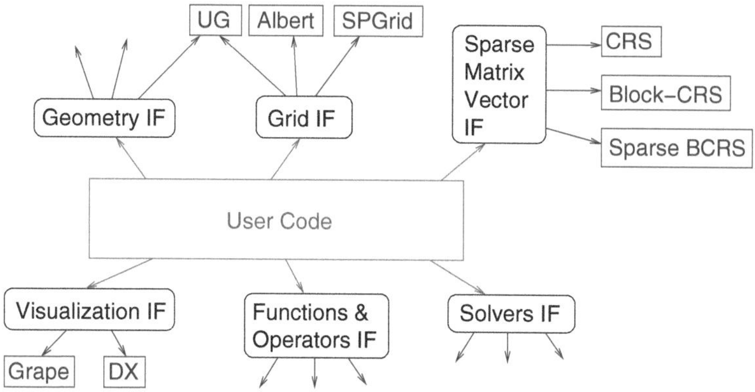

Fig. 2. DUNE module structure

object oriented languages one uses abstract base classes and inheritance to implement polymorphism. E. g., C++ offers virtual functions to implement dynamic polymorphism. The function call itself poses a serious performance penalty in case a function/method in the interface consists only of a few instructions. Therefore, function calls and virtual method invocation can only be used efficiently for interfaces with sufficiently coarse granularity.

However, to utilize the concept of abstract interfaces to full extent one needs interfaces with fine granularity. E. g., in the case of a mesh interface one needs to access coordinates of nodes, normals of faces or evaluate element transformations at individual quadrature points. Generic programming, implemented in the C++ language through templates, offers a possibility to implement interfaces without performance penalty. The abstract algorithm is parameterized by an implementation of the interface (a concrete class) at compile-time. The compiler will then be able to inline small functions and to employ all code optimizations. Basically, the interface is removed completely at compile-time. This technique is also called static (or compile-time) polymorphism and is used extensively in the well-known standard template library STL, see Musser et al. [2001]. Many C++ programming techniques we use are described in Barton and Nackman [1994] and Veldhuizen [2000].

DUNE is a template library for all software components required for the numerical solution of partial differential equations. Figure 2 shows the high level design. User code written in C++ will access geometries, grids, sparse linear algebra, visualization and the finite element functionality through abstract interfaces. Many implementations of one interface are possible and particular implementations are selected at compile-time. It is very important that incorporation of existing codes is very natural within this concept. Moreover, the design can also be used to couple different existing codes in one application.

In the rest of this paper we concentrate on the design of the abstract interface for finite element and finite volume meshes.

3 Design of the Sequential Grid Interface

There are many different types of finite element or finite volume grids. We have selected the features of our grid interface according to the needs of our applications. In particular, we wanted to support grids that

- discretize unions of manifolds (e. g. fracture networks, shell elements),
- consist of elements of different geometric shapes (e. g tetrahedra, prisms, pyramids and hexahedra),
- support local, hierarchical mesh refinement.

In the following we define a grid $\mathcal{T}$ in mathematical terms. It is supposed to discretize a domain $\Omega \subset \mathbb{R}^n$, $n \in \mathbb{N}$, $n > 0$, with piecewise smooth boundary $\partial\Omega$. A grid $\mathcal{T}$ consists of $L+1$ grid levels

$$\mathcal{T} = \{\mathcal{T}_0, \mathcal{T}_1, \ldots, \mathcal{T}_L\}.$$

Each grid level $\mathcal{T}_l$ consists of sets of grid entities $\mathcal{E}_l^c$ of codimension $c \in \{0, 1, \ldots, d\}$ where $d \leq n$ is the dimensionality of the grid:

$$\mathcal{T}_l = \{\mathcal{E}_l^0, \ldots, \mathcal{E}_l^d\}.$$

Each entity set consists of individual grid entities which are denoted by $\Omega_{l,i}^c$:

$$\mathcal{E}_l^c = \left\{\Omega_{l,0}^c, \Omega_{l,1}^c, \ldots, \Omega_{l,N(l,c)-1}^c\right\}.$$

The number of entities of codimension c on level l is $N(l,c)$ and we define a corresponding index set

$$I_l^c = \{0, 1, \ldots, N(l,c) - 1\}.$$

Definition 1. *$\mathcal{T}$ is called a grid if the following conditions hold:*

1. *(Partitioning). The entities of codimension 0 on level 0 define a partitioning of the whole domain:*

$$\bigcup_{i \in I_0^0} \overline{\Omega_{0,i}^0} = \overline{\Omega}, \qquad \forall i \neq j : \Omega_{0,i}^0 \cap \Omega_{0,j}^0 = \emptyset.$$

2. *(Nestedness). Entities of codimension 0 on different levels form a tree structure. We require:*

$$\forall l > 0, i \in I_l^0 : \exists! j \in I_{l-1}^0 : \Omega_{l,i}^0 \subset \Omega_{l-1,j}^0.$$

This $\Omega_{l-1,j}^0$ is called father of $\Omega_{l,i}^0$. For entities with at least one side on the boundary this condition can be relaxed. We define the set of all descendant entities of codimension 0 and level $l \leq L$ of an entity $\Omega_{k,i}^0$ as

$$C_L(\Omega_{k,i}^0) = \{\Omega_{l,j}^0 | \ \Omega_{l,j}^0 \subset \Omega_{k,i}^0, l \leq L\}.$$

3. *(Recursion over codimension). The boundary of a grid entity is composed of grid entities of the next higher codimension, i. e. for $c < d$ we have*

$$\partial\Omega_{l,i}^c = \bigcup_{j \in I_{l,i}^{c+1} \subset I_l^{c+1}} \overline{\Omega_{l,j}^{c+1}}.$$

Grid entities $\Omega_{l,j}^d$ of codimension d are points in $\mathbb{R}^n$.

4. *(Reference elements and dimension). For each grid entity $\Omega_{l,i}^c$ there is a reference element $\omega_{l,i}^c \subset \mathbb{R}^{d-c}$ and a sufficiently smooth map*

$$m_{l,i}^c : \overline{\omega_{l,i}^c} \to \overline{\Omega_{l,i}^c}$$

from the reference element to the actual element. Reference elements are convex polyhedrons in $\mathbb{R}^{d-c}$. The dimension of the grid d is the dimension of the reference elements corresponding to grid entities of codimension 0. For $c = d$ the map $m_{l,i}^d$ simply returns the corresponding point in $\mathbb{R}^n$.

5. *(Nonconformity). Note that we do not require the mesh to be conforming in the sense that the intersection of the closure of two grid entities of codimension c is either zero or a grid entity with codimension greater than c. However, we require that all grid entities in $\mathcal{E}_l^c$ are distinct, i. e. :*

$$\forall i, j, c, l : \quad \Omega_{l,i}^c = \Omega_{l,j}^c \Rightarrow i = j.$$

The set of all neighbors of an entity $\Omega_{l,i}^0$ is represented by the set of all non empty intersections with that entity:

$$\mathcal{I}(\Omega_{l,i}^0) = \{\overline{\Omega_{l,i}^0} \cap \overline{\Omega_{l,j}^0} | \ \overline{\Omega_{l,i}^0} \cap \overline{\Omega_{l,j}^0} \neq \emptyset, i \neq j\}.$$

Classes in the DUNE grid interface

According to the description in Definition 1, the grid interface consists of the following abstract classes:

1. `Grid⟨dim, dimworld, ...⟩`
   This class corresponds to the whole grid $\mathcal{T}$. It is parametrized by the grid dimension d = *dim* and the space dimension n = *dimworld*. The grid class provides iterators for the access to its entities.
2. `Entity⟨codim, dim, dimworld, ...⟩`, `Element⟨dim, dimworld, ...⟩`
   Grid entities $\Omega_{l,i}^c$ of codimension c = *codim* are realized by the classes Entity and Element. The Entity class contains all topological information, while geometrical specifications are provided by the Element class.
3. `LevelIterator⟨codim, dim, dimworld, ...⟩`
   The level iterator gives access to all grid entities on a specified level l. This allows a traversal of the set $\mathcal{E}_l^c$.

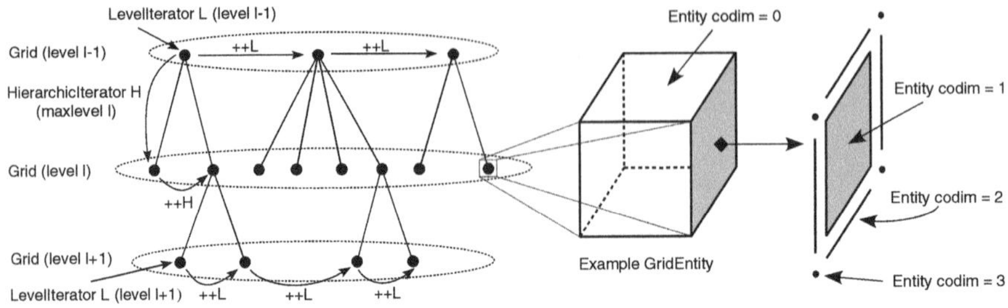

Fig. 3. DUNE Grid walk-trough of an 3 dimensional Grid

4. `HierarchicIterator⟨dim, dimworld, ...⟩`
   Another possibility to access grid entities is provided by the hierarchic iterator. This iterator runs over all descendant entities with level $l \leq L$ of a given entity $\Omega^0_{k,i}$. Therefore, it traverses the set $\mathcal{C}_L(\Omega^0_{k,i})$.
5. `IntersectionIterator⟨dim, dimworld, ...⟩`
   Part of the topological information provided by the Entity class of codimension 0 is realized by the intersection iterator. For a given entity $\Omega^0_{l,i}$ the iterator traverses the set $\mathcal{I}(\Omega^0_{l,i})$.

A specific grid is realized by an implementation of derived classes of these abstract interface classes. The efficiency of the specific implementations is guaranteed by using static polymorphism (see Barton and Nackman [1994]). Figure 3 gives a sketch of the functionality of the grid interface. It shows the access of the grid entities via level, or hierarchic iterators and displays the recursive definition of entities via codimension.

4 Example and Performance Evaluation

In order to assess the performance of the proposed concept we compare different implementations for the numerical solution of the following linear hyperbolic equation:

$$\frac{\partial u}{\partial t} + \nabla \cdot (\mathbf{v}u) = 0 \text{ in } \Omega, \quad u = g \text{ on } \Gamma_{in} = \{\mathbf{x} \in \partial\Omega \mid \mathbf{v}(\mathbf{x}) \cdot \mathbf{n} \leq 0\}.$$

We discretize this equation with a cell-centered finite volume scheme using full upwind and an implicit Euler scheme in time. We note that this is not a particularly good scheme. However, it is very well suited to compare run-times since it is simple but contains all essential features of more complicated schemes as far as the mesh interface is concerned.

Table 1 shows the run-time for assembling the system matrix using various implementations. Implementation A is for a structured mesh. In implementations B and C the discretization is based on the DUNE mesh interface, thus it can be used for any dimension and element type. B uses simplegrid, an implementation of the mesh interface supporting structured meshes of variable dimension with entities of codimension d and 0. C uses an implementation of the

Table 1. Run-time for matrix assembly in a cell-centered finite volume scheme. We used a Pentium IV computer (2.4 GHz) with the Intel C++ compiler `icc` 7.0

Key	Implementation	Grid	Run-time $[s]$
A	structured grid	128^3 hex	0.37
B	DUNE/simplegrid	128^3 hex	1.59
C	DUNE/albertgrid	$6 \cdot 64^3$ tet	4.20
D	Albert	$6 \cdot 64^3$ tet	3.13
E	UG	$6 \cdot 64^3$ tet	2.64
F	UG	128^3 hex	45.60

mesh interface based on the PDE toolbox ALBERT, see Schmidt and Siebert [2000], D is the same test using ALBERT directly without going through the interface. ALBERT supports simplicial elements in two and three space dimensions with bisection refinement. Finally, in E and F, we implemented the discretization scheme within the PDE framework UG (Bastian et al. [1997]) using hexahedral and tetrahedral meshes. Increase of run-time per element from E to F is due to the more costly element transformation for hexahedra. We are currently implementing the DUNE mesh interface based on UG. Run-times of cases E and F can be considered as preliminary results for a DUNE/UG mesh module.

From Table 1 we conclude that performance can be increased by a factor of 30 when we replace the unstructured hexahedral mesh (F) by a structured mesh (B). Memory requirements are reduced by a factor 10. These improvements are achieved *without* changes in the application code (here the discretization). Additional savings by a factor 4 in run-time (A) are only possible at the cost of reduced functionality of the user code. Memory requirements of the DUNE/simplegrid and structured mesh variants are the same.

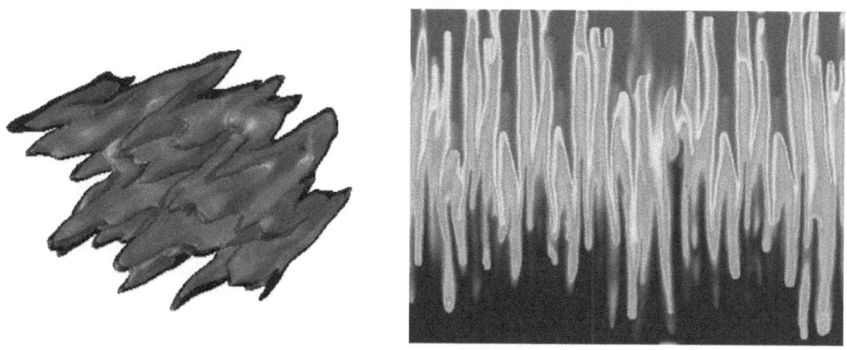

Fig. 4. Large scale parallel three-dimensional simulations. Contaminant transport in a heterogeneous medium (higher order Godunov scheme, $5 \cdot 10^8$ cells, left), density driven flow in a porous medium ($8 \cdot 10^8$ cells in three dimensions, right)

5 Conclusion and Future Prospects

In this paper we presented a new framework for the numerical solution of partial differential equations. The concept consequently separates data structures and algorithms. Algorithms are written in terms of abstract interfaces, the interfaces are implemented efficiently using static polymorphism in C++. We evaluated the performance of this approach for a simple discretization scheme. The run-time can be improved by up to a factor 30 by replacing an unstructured mesh implementation with a structured mesh implementation. The improvement is possible without any changes in the application code.

Currently we are extending the mesh interface by a general parallel data distribution model which will allow the formulation of overlapping and non overlapping domain decomposition methods as well as parallel multigrid methods on the same interface. First results of the parallel implementation are shown in Figure 4. Large scale computations with up to 10^{10} grid cells are possible on a 500 processor Linux cluster with a structured mesh implementation.

For data visualization, DUNE will be linked to the graphics packages GRAPE, Geßner et al. [1999], and AMIRA, Amira [2002].

References

Amira. *Amira 3.0 Visualization Software.* http://www.amiravis.com/, 2002.

J. Barton and L. Nackman. *Scientific and Engineering C++.* Addison-Wesley, 1994.

P. Bastian et al. UG - A flexible software toolbox for solving partial differential equations. *Computing and Visualization in Science*, 1:27–40, 1997.

BLAST. *Basic Linear Algebra Subprograms Technical (BLAST) Forum Standard.* http://www.netlib.org/blas/blast-forum/, 2001.

F. P. Brooks. *The Mythical Man-Month: Essays on Software Engineering.* Addison-Wesley, 1975.

T. Geßner et al. A procedural interface for multiresolutional visualization of general numerical data. Report 28, SFB 256, Bonn, 1999.

D. R. Musser, G. J. Derge, and A. Saini. *STL Tutorial and Reference Guide.* Addison-Wesley, 2001.

A. Schmidt and K. Siebert. ALBERT – An adaptive hierarchical finite element toolbox. Preprint 06/2000 Freiburg, 2000.

T. Veldhuizen. Techniques for scientific C++. Technical Report 542, Indiana University Computer Science, 2000. http://osl.iu.edu/~tveldhui/papers/techniques/.

Distributed Point Objects. A New Concept for Parallel Finite Elements

Christian Wieners

Universität Karlsruhe, Institut für Praktische Mathematik
Englerstraße 2, 76128 Karlsruhe, Germany
(`http://www.mathematik.uni-karlsruhe.de/~ipm`)

Summary. We present a new concept for the realization of finite element computations on parallel machines with distributed memory. The parallel programming model is based on a dynamic data structure addressed by points. All geometric objects (cells, faces, edges) are referenced by their midpoints, and all algebraic data structures (vectors and matrices) are tied to the nodal points of the finite elements. The parallel distribution of all objects is determined by processor lists assigned to the reference points.
Based on this new model for Distributed Point Objects (DPO) a first application to a geotechnical application with Taylor-Hood elements on hexahedra has been presented in Wieners et al. [2004]. Here, we consider the extension to parallel refinement, curved boundaries, and multigrid preconditioners. Finally, we present parallel results for a nonlinear model problem with isoparametric cubic elements.

1 Introduction

Many finite element applications require a fine mesh resolution or a huge number of time steps. Together with the increasing complexity of the considered models, the solution of such problems is only possible with elaborated solvers such as domain decomposition methods or multigrid preconditioners. In order to obtain a reasonable computing time, very often this can be realized only on a parallel machine.

For this purpose, large software developments are available, e. g., *PLTMG* (Bank [1998]), *PETSc* (Balay et al. [2001]), and *UG* (Bastian et al. [1997]). Now, a discussion starts, how such developments can be unified and combined by a general framework Bastian et al. [2004], so that one is not fixed to the underlying programming model of the software.

Here, we introduce a new parallel programming model which is specially designed for the support of nonlinear engineering finite element applications, and which provides a platform for the development of modern solvers and their adaptation to such problems. The main features of the concept are flexibility, transparence, and extensibility. It allows the realization of complex algorithms in a very compact form, and it reduces the implementation time for new applications. In particular, this make the code attractive for educational purposes.

Our concept is based on long experiences with parallel software development and parallel simulations for partial differential equations, cf. Bastian et al. [1998, 1999, 2000], Lang et al. [2002]. It can be understood as an abstraction and simplification of our previous code. In terms of software engineering this study describes the underlying structure of a prototype implementation which provides optimal support for numerical experiments and numerical analysis in the investigation of new models, discretizations, and solvers.

This contribution is organized as follows. In Section 2, we introduce the programming model for Point Objects, which is enhanced in Section 3 to parallel distributed objects. In addition, we define a parallel refinement process leading to a hierarchical mesh structure. This is coupled in Section 4 with a multilevel parallel linear algebra, where parallel multigrid methods can be defined. In Section 5 this is combined with an abstract model for finite elements (including the requirements for isoparametric elements). Finally, in Section 6 the application to a nonlinear model problem is presented.

2 Point Objects

Our geometry model is based on a finite set of points $\mathcal{P} \subset \mathbf{R}^d$. We consider different types of points: corner points, edge midpoints, face midpoints, cell midpoints, and the exception point $P = \infty$. A cell $C = (P_1, ..., P_N)$ is determined by a vector of N different corner points $P_j \in \mathcal{P}$, and the convex hull of the corner points is denoted by $\operatorname{conv}(C) \subset \mathbf{R}^d$. A face $F = (P_{\text{left}}, P_{\text{right}}) \in \mathcal{P}^2$ is determined by a pair of points representing the midpoints of the cells C_{left} and C_{right} of the common face $\operatorname{conv}(C_{\text{left}}) \cap \operatorname{conv}(C_{\text{right}})$; boundary faces are characterized by $P_{\text{right}} = \infty$. An edge $E = (P_{\text{left}}, P_{\text{right}}) \in \mathcal{P}^2$ represents the line from P_{left} to P_{right}. Now, a mesh $\mathcal{M} = (\mathcal{C}, \mathcal{F}, \mathcal{E}, \mathcal{V}, \mathcal{B}, \mathcal{G})$ is given by

- a cell mapping $\mathcal{C}\colon \mathcal{P} \longrightarrow \bigcup_N \mathcal{P}^N$, which assigns every cell midpoint P_C the cell $C = \mathcal{C}(P_C)$ represented by the vector of N corner points;
- a face mapping $\mathcal{F}\colon \mathcal{P} \longrightarrow \mathcal{P}^2$, which assigns every face midpoint P_F the two adjacent element midpoints;
- an edge mapping $\mathcal{E}\colon \mathcal{P} \longrightarrow \mathcal{P}^2$, which assigns every edge midpoint P_E the two adjacent vertices;
- a vertex mapping $\mathcal{V}$ representing a list of vertices;
- a boundary mapping $\mathcal{B}$ representing a list of boundary faces for the assignment of segment numbers specifying boundary conditions;
- a geometry mapping $\mathcal{G}\colon \mathcal{P} \longrightarrow \mathbf{R}^d$, which assigns points P on the polygonal boundary of $\bigcup_C \operatorname{conv}(C)$ the projection onto the (possibly) curved boundary of the computational domain Ω.

All mappings return an empty object for points of wrong type. We use the notation $C \in \mathcal{C}$ if $C \in \mathcal{C}(\mathcal{P})$, and $P \in \mathcal{P}_\mathcal{C}$ for the cell midpoints, etc.

We consider only consistent meshes with admissible triangulations, i. e., we assume that the cells define a polygonal approximation $\bar{\Omega}_\mathcal{C} = \bigcup_{C\in\mathcal{C}} \mathrm{conv}(C)$ of a domain $\Omega \subset \mathbf{R}^d$, such that $\mathrm{conv}(C) \cap \mathrm{conv}(C') = \mathrm{conv}(C \cap C')$ for two cells $C, C' \in \mathcal{C}$, i. e., the intersection is empty, a common vertex in $\mathcal{V}$, a common edge in $\mathcal{E}$, or a common face in $\mathcal{F}$. In general, we assume $\mathcal{G}(P) = P$ for the boundary vertices $P \in \mathcal{P}_\mathcal{V} \cap \partial\Omega$, and $\mathcal{G}(P) \in \partial\Omega$ for the $P \in \mathcal{P}_\mathcal{E} \cap \partial\Omega_\mathcal{C}$ and $P \in \mathcal{P}_\mathcal{F} \cap \partial\Omega_\mathcal{C}$. The geometry mapping is used for the realization of isoparametric elements and for the refinement algorithm (see below).

Example We illustrate the data structure for a mesh with two triangles in $\mathbf{R}^2$. Inserting the first triangle $C_1 = (P_1, P_2, P_3)$ in $\mathcal{M}$ results in the point set $\mathcal{P} = \{P_1, P_2, P_3,\ P_{12} = \frac{1}{2}(P_1 + P_2), P_{13} = \frac{1}{2}(P_1 + P_3), P_{23} = \frac{1}{2}(P_2 + P_3), P_{123} = \frac{1}{3}(P_1 + P_2 + P_3)\}$, the edges $\mathcal{E}(P_{12}) = (P_1, P_2)$, $\mathcal{E}(P_{23}) = (P_2, P_3)$, $\mathcal{E}(P_{13}) = (P_1, P_3)$, and the faces $\mathcal{F}(P_{12}) = (P_{123}, \infty)$, $\mathcal{F}(P_{23}) = (P_{123}, \infty)$, $\mathcal{F}(P_{13}) = (P_{123}, \infty)$.

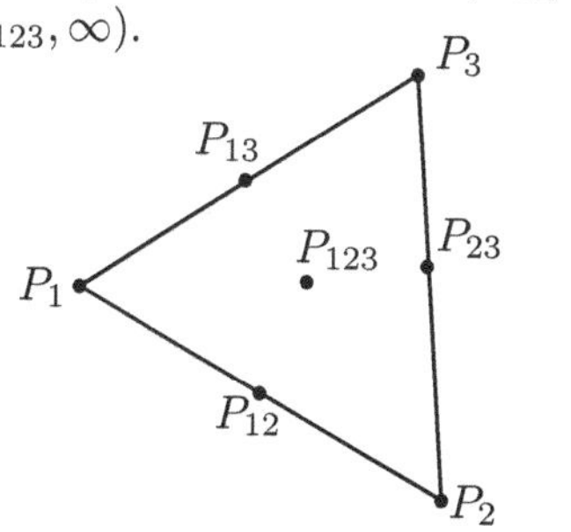

Now, the second triangle $C_2 = (P_2, P_4, P_3)$ is inserted, which adds the new points $P_{24} = \frac{1}{2}(P_2 + P_4)$, $P_{34} = \frac{1}{2}(P_3 + P_4)$, $P_{234} = \frac{1}{3}(P_2 + P_3 + P_4)$ to the points set $\mathcal{P}$, new edges $\mathcal{E}(P_{24}) = (P_2, P_4)$, $\mathcal{E}(P_{34}) = (P_3, P_4)$, and the new faces $\mathcal{F}(P_{24}) = (P_{234}, \infty)$, $\mathcal{F}(P_{34}) = (P_{234}, \infty)$; the face $\mathcal{F}(P_{23}) := (P_{123}, P_{234})$ is now updated.

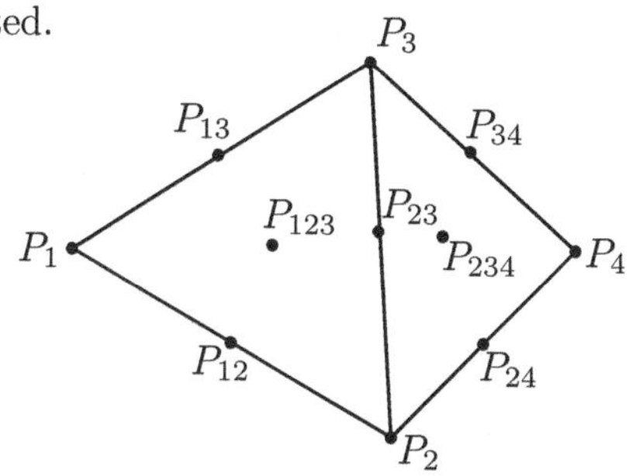

After inserting all cells, we can identify the neighborhood relationship by the faces, where a boundary face can be identified by testing for $P_{\mathrm{right}} = \infty$. Then, for curved boundaries the projection of edge midpoints and face midpoints onto the boundary can be computed. In general, this requires an additional data structure for the boundary definition. In our application, where the boundary is defined by a periodic cubic spline which is uniquely defined by the corner vertices, this can be realized without further geometry information.

3 Distributed Objects

A parallel distribution is determined by a partition map

$$\pi\colon \mathcal{P} \longrightarrow 2^{\{1,2,\dots,N_{\text{procs}}\}}$$

assigning to every point $P \in \mathcal{P}$ the subset $\pi(P) \subset \{1,2,\dots,N_{\text{procs}}\}$ of processors, where this point is represented. This defines also a unique master processor $\mu(P) = \min \pi(P)$ for every point, and it determines an overlapping partition

$$\mathcal{P} = \mathcal{P}_1 \cup \dots \cup \mathcal{P}_{N_{\text{procs}}}, \qquad \mathcal{P}_q = \{P \in \mathcal{P}\colon q \in \pi(P)\}.$$

We obtain the local mesh $\mathcal{M}_q = (\mathcal{C}_q, \mathcal{F}_q, \mathcal{E}_q, \mathcal{V}_q, \mathcal{B}_q, \pi_q)$ on processor q by restricting all mappings to $\mathcal{P}_q$. Then, the parallel distribution is completely determined by the partition map π. An admissible parallel distribution requires that every cell can be represented at least on one processor q, i. e., for $C = (P_1, \dots, P_N) \in \mathcal{C}_q$ we require $P \in \mathcal{P}_q$. Moreover, for every face $F \in \mathcal{F}$ we require $F \in \bigcup_{p,q \in \pi(P_F)} \mathcal{P}_p \times \mathcal{P}_q$.

For the determination of an admissible distribution of the mesh $\mathcal{M}$ onto N_{procs} processors, we assign a destination processor $\text{dest}(C) \in \{1,2,\dots,N_{\text{procs}}\}$ to every cell $C \in \mathcal{C}$, defining a disjoint partition

$$\mathcal{C} = \mathcal{C}_1 \cup \dots \cup \mathcal{C}_{N_{\text{procs}}}, \qquad \mathcal{C}_q = \{C \in \mathcal{C}\colon\ \text{dest}(C) = q\}$$

and a domain decomposition

$$\bar{\Omega}_{\mathcal{C}} = \bar{\Omega}_1 \cup \dots \cup \bar{\Omega}_{N_{\text{procs}}}, \qquad \bar{\Omega}_q = \bigcup_{C \in \mathcal{C}_q} \text{conv}(C).$$

A corresponding compatible partition map is defined by

$$\begin{aligned}
\pi(P_C) &= \{\text{dest}(C)\}, & C &\in \mathcal{C},\\
\pi(P_F) &= \{\text{dest}(C)\colon P_C \in F\}, & F &\in \mathcal{F},\\
\pi(P_E) &= \{\text{dest}(C)\colon E \subset C\}, & E &\in \mathcal{E},\\
\pi(P) &= \{\text{dest}(C)\colon P \in C\}, & P &\in \mathcal{V}.
\end{aligned}$$

Thus, the partition map can be computed in advance before the realization of the parallel distribution.

Example (continued) The parallel distribution with $\text{dest}(C_1) = 1$ and $\text{dest}(C_2) = 2$ results in $\pi(P_1) = \pi(P_{12}) = \pi(P_{13}) = \pi(P_{123}) = \{1\}$, $\pi(P_2) = \pi(P_3) = \pi(P_{23}) = \{1,2\}$, and $\pi(P_4) = \pi(P_{24}) = \pi(P_{34}) = \pi(P_{234}) = \{2\}$.

Parallel Refinement

A uniform refinement of a cell $C = (P_1, ..., P_N)$ in $\mathcal{M}$ is defined by a refinement rule $\mathcal{R} = \{r_{ij}: i = 1, ..., N,\ j = 1, ..., 2^d\}$: let $(P_1, ..., P_M) = (\mathcal{V}_C, \mathcal{E}_C, \mathcal{F}_C, P_C)$ be the vector of cell vertices, edge midpoints, face midpoints, and the cell midpoint. Then, insert the cells $C_j = (\mathcal{G}(P_{r_{1j}}), ..., \mathcal{G}(P_{r_{Nj}}))$, $j = 1, ..., 2^d$ in the new mesh $\mathcal{N}$.

The new partition map is determined independently and can be computed in advance before the refinement of the cells is realized:

$$\begin{aligned}
&\pi_{\mathcal{N}}(P) = \pi_{\mathcal{M}}(P) \quad \text{for all} \quad P \in \mathcal{P}_{\mathcal{M}},\\
&\pi_{\mathcal{N}}(\tfrac{1}{2}P_{\text{left}} + \tfrac{1}{2}\mathcal{G}(P_E)) = \pi_{\mathcal{N}}(\tfrac{1}{2}\mathcal{G}(P_E) + \tfrac{1}{2}P_{\text{right}}) = \pi_{\mathcal{M}}(P_E)\\
&\qquad\qquad \text{for} \quad E = (P_{\text{left}}, P_{\text{right}}) \in \mathcal{E}_{\mathcal{M}},\\
&\pi_{\mathcal{N}}(\mathcal{G}(P_{F_j})) = \pi_{\mathcal{M}}(P_F) \quad \text{for} \quad F \in \mathcal{F}_{\mathcal{M}},\ j = 1, .., 2^{d-1}\\
&\pi_{\mathcal{N}}(P_{C_j}) = \pi_{\mathcal{N}}(P_{E_k}) = \pi_{\mathcal{M}}(P_C) \text{ for } C \in \mathcal{C}_{\mathcal{M}},\ j = 1, .., 2^d \text{ and inner edges,}
\end{aligned}$$

where P_{F_j} and P_{C_j} are the midpoints of the refined face F or cell C.

Example (continued) The cell (P_1, P_2, P_3) is refined (in case of no boundary projections) to four cells $C_1 = (P_1, P_{12}, P_{31})$, $C_2 = (P_2, P_{23}, P_{12})$, $C_3 = (P_3, P_{31}, P_{23})$, $C_4 = (P_{12}, P_{23}, P_{31})$, and we set $\pi_{\mathcal{N}}(P_j) = \pi_{\mathcal{M}}(P_j)$, $\pi_{\mathcal{N}}(P_{jk}) = \pi_{\mathcal{N}}(\frac{1}{2}P_j + \frac{1}{2}P_{jk}) = \pi_{\mathcal{N}}(\frac{1}{2}P_{jk} + \frac{1}{2}P_k) = \pi_{\mathcal{M}}(P_{jk})$, and finally $\pi_{\mathcal{N}}(P_{C_1}) = \pi_{\mathcal{N}}(P_{C_2}) = \pi_{\mathcal{N}}(P_{C_3}) = \pi_{\mathcal{N}}(P_{C_4}) = \pi_{\mathcal{N}}(\frac{1}{2}P_{ij} + \frac{1}{2}P_{jk}) = \pi_{\mathcal{M}}(P_C)$.

4 Parallel Linear Algebra

We assign to every point $P \in \mathcal{P}$ the number of degrees of freedom $N_P \geq 0$, where the point set $\mathcal{P}$ may be extended by nodal points of the finite element discretization. Let $N_{\mathcal{P}} = \sum_{P \in \mathcal{P}} N_P$ be the total number of unknowns.
A vector $\underline{u} \in \mathbf{R}^{N_{\mathcal{P}}} \simeq \prod_{P \in \mathcal{P}} \mathbf{R}^{N_P}$ maps a point P to the vector $\underline{u}[P] \in \mathbf{R}^{N_P}$ of unknowns associated to the point $P \in \mathcal{P}$.
We use two different representation of distributed vectors (cf. Bastian [1996]):

- Solution vectors and correction vectors $\underline{u}$ are represented consistently in

  $$V[\mathcal{M}] := \{(\underline{u}_1, ..., \underline{u}_{N_{\text{procs}}}) \in \prod \mathbf{R}^{N_{\mathcal{P}_q}} : \underline{u}_p[P] = \underline{u}_q[P], \quad p, q \in \pi(P)\}.$$

  This defines a global vector $\underline{u}$ by $\underline{u}[P] = \underline{u}_q[P]$ for any $q \in \pi(P)$.
- Right-hand side vectors and residual vectors $\underline{r}$ are represented additively, i. e., any additive vector $(\underline{r}_q) \in \prod \mathbf{R}^{N_{\mathcal{P}_q}}$ represents at $P \in \mathcal{P}$ the value $r[P] = \sum_{q \in \pi(P)} \underline{r}_q[P]$. Collecting the distributed values at the master points

and replacing $(\underline{r}_q)$ by $r_q[P] = r[P]$ for $q = \mu(P)$ and $\underline{r}_q[P] = 0$ else results in a unique representation in

$$V(\mathcal{M}) := \{(\underline{r}_1, ..., \underline{r}_{N_{\text{procs}}}) \in \prod \mathbf{R}^{N_{\mathcal{P}_q}} : \underline{r}_q[P] = \underline{0}, \quad q \neq \mu(P)\}.$$

This allows for the parallel evaluation of the norm $\sqrt{\underline{r}^T\underline{r}} = \sqrt{\sum \underline{r}_q^T \underline{r}_q}$.

In our programming model we define the following parallel operators:

- The stiffness matrix corresponds to an operator $\underline{A}\colon V[\mathcal{M}] \longrightarrow V(\mathcal{M})$. In particular, for the solution vector $(\underline{u}_q) \in V[\mathcal{M}]$ and the right-hand side $(\underline{b}_q) \in V(\mathcal{M})$ the parallel residual $(\underline{b}_q - \underline{A}_q\underline{u}_q) \in V(\mathcal{M})$ is independent of the distribution. For any additive matrix $(\underline{A}_q) \in \prod \mathbf{R}^{N_{\mathcal{P}_q}} \times \mathbf{R}^{N_{\mathcal{P}_q}}$ the application of the parallel operator consists of two steps: compute an additive representation $(\underline{A}_q\underline{u}_q) \in \prod \mathbf{R}^{N_{\mathcal{P}_q}}$ by local matrix-vector multiplication without parallel communication, and then collect the values on $\pi(P)$ for all points P to obtain a vector in $V(\mathcal{M})$.
- A parallel preconditioner corresponds to an operator $\underline{B}\colon V(\mathcal{M}) \longrightarrow V[\mathcal{M}]$. E. g., $\underline{B}[P,P] = \Big(\sum_{q\in\pi(P)} \underline{A}_q[P,P] \in \mathbf{R}^{N_P\times N_P}\Big)^{-1}$ defines the block-Jacobi preconditioner. For the application, in the first step the local computation results in an additive result $(\underline{B}_q\underline{r}_q)$ without parallel communication. Then, this is accumulated on all processors in $\pi(P)$ to obtain a vector in $V[\mathcal{M}]$.
- We consider transfer operators $\underline{I} = \underline{I}_{\mathcal{M}}^{\mathcal{N}}\colon V[\mathcal{M}] \longrightarrow V[\mathcal{N}]$ prolongating values on a coarse mesh $\mathcal{M}$ to a fine mesh $\mathcal{N}$. For the application, in the first step the local application results in $(\underline{I}_q\underline{u}_q)$ without parallel communication. For conforming discretizations and uniform parallel refinement described in the previous section, the result is consistent in $V[\mathcal{N}]$; in general, this requires local communication.
  The adjoined operator $\underline{I}^T\colon V(\mathcal{N}) \longrightarrow V(\mathcal{M})$ restricts values by first computing an additive result $(\underline{I}_q^T\underline{r}_q)$ without parallel communication and then collecting the values on $\pi(P)$ for all points P to obtain a vector in $V(\mathcal{M})$.

Together with the dual pairing in $V[\mathcal{M}] \times V(\mathcal{M})$, this allows for a general formulation of parallel iterative solvers such as Krylov methods with multigrid preconditioner, where global communication is restricted to the inner products and the application of the parallel operators.

5 Parallel Finite Elements

Corresponding to cell based finite element discretizations we define the cell nodal points by $\mathcal{P}_C := \{P \in \mathcal{P} \cap \operatorname{conv}(C)\colon N_P > 0\}$, the cell vector $\underline{u}[C] = (x[P])_{P\in\mathcal{P}_C} \in \mathbf{R}^{N_C}$ with $N_C = \sum_{P\in\mathcal{P}_C} N_P$, and the cell matrix $A[C] = (A[P,Q])_{P,Q\in\mathcal{P}_C} \in \mathbf{R}^{N_C,N_C}$. In the case of cubic triangular elements we have $N_C = 10$. Again, the isoparametric transformation is determined by the application of the geometry mapping $\mathcal{G}$ to the nodal points.

A nonlinear finite element problem is given by the following assembling routines D_C, R_C, J_C, running in parallel over all cells $C \in \mathcal{C}_q$ of a mesh $\mathcal{M}$:

- Essential boundary conditions are assigned by a *Dirichlet* routine $D_C(\underline{u}[C])$ to the corresponding indices $i = 1, ..., N_P$, $P \in \mathcal{P}_C$; after the assembling it has to be guaranteed that this results in parallel consistent Dirichlet values respecting $(\underline{u}_q) \in V[\mathcal{M}]$.
- The additive *residual* $\underline{r}[C] = R_C(\underline{u}[C])$ is computed depending on the actual solution vector $\underline{u}[C]$; after the assembling the residual values at the distributed points P have to be collected on $\mu(P)$ to obtain a unique additive representation in $V(\mathcal{M})$.
- If the residual norm is not small enough, the *Jacobian* $A[C] = J_C(\underline{u}[C])$ is assembled additively; in every nonlinear step a consistent correction vector $\underline{c} \in V[\mathcal{M}]$ is computed by solving the linear equation $\underline{Ac} = \underline{r}$ iteratively up to a given accuracy, and the solution vector is updated by $\underline{u} := \underline{u} + \underline{c}$.

6 A Numerical Experiment

We present a numerical approximation of the quasi-linear model problem

$$u \in H_0^1(\Omega): \qquad -\Delta u = \lambda \exp(u)$$

(Gelfand-Bratu problem) with cubic isoparametric finite elements. It is well known that this equation admits at least two solutions for $\lambda \in (0, \lambda^*)$, with a turning point at the critical parameter λ^* (see, e. g., Lions [1982]).

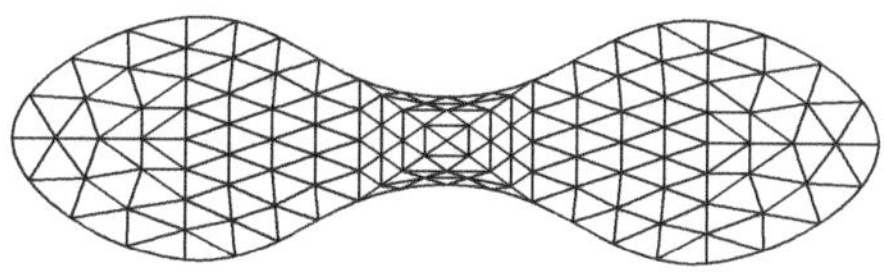

Fig. 1. A nonconvex domain Ω, where the boundary $\partial\Omega$ is of class $C^{2,1}$ (represented by a cubic spline)

While analytical tools provide the existence of only two solutions, it could be shown by a computer assisted existence proof in Plum and Wieners [2002] for a special nonconvex domain, that a further (symmetry breaking) solution branch exists. Such a solution is illustrated in Fig. 2. For this method (showing the existence of a continuous solution via Schauders fixpoint theorem), extremely smooth and accurate approximations are required.

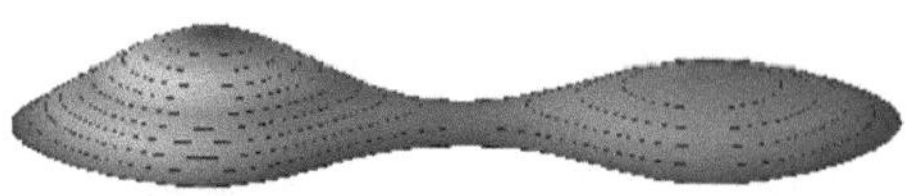

Fig. 2. Symmetry breaking solution of the Gelfand problem for $\lambda = 0.45$, computed in parallel with isoparametric cubic elements after 6 geometry preserving refinement steps of the mesh in Fig. 1

This example is realized (by the extended use of the standard template library) within ca. 10000 lines of code (and ca. 100 lines for the definition of the boundary value problem). The solution is computed on a Linux cluster with 36 processors in less than 5 minutes.

References

S. Balay, W. D. Gropp, L. C. McInnes, and B. F. Smith. PETSc users manual. Technical Report ANL-95/11 - Revision 2.1.1, Argonne National Laboratory, 2001.

R. E. Bank. *PLTMG: A Software Package for Solving Elliptic Partial DifferentiaEquations, Users' Guide 8.0l*, volume 5 of *Software, Environments and Tools*. SIAM, Philadelphia, 1998.

P. Bastian. *Parallele adaptive Mehrgitterverfahren.* Teubner Skripten zur Numerik. Teubner, Stuttgart, 1996.

P. Bastian, K. Birken, K. Johannsen, S. Lang, N. Neuß, H. Rentz-Reichert, and C. Wieners. UG – a flexible software toolbox for solving partial differential equations. *Comp. Vis. Sci.*, 1:27–40, 1997.

P. Bastian, K. Birken, K. Johannsen, S. Lang, V. Reichenberger, H. Rentz-Reichert, C. Wieners, and G. Wittum. A parallel software-platform for solving problems of partial differential equations using unstructured grids and adaptive multigrid methods. In E. Krause and W. Jäger, editors, *High Performance Computing in Science and Engineering '98*, pages 326–339, Berlin, 1998. Springer-Verlag.

P. Bastian, M. Droske, C. Engwer, R. Klöfkorn, T. Neubauer, M. Ohlberger, and M. Rumpf. Towards a unified framework for scientific computing. In R. Kornhuber, O. Pironneau, R. Hoppe, J. Périaux, D. Keyes, and J. Xu, editors, *Proc. Int. Conf. on Domain Decomposition Methods DD15*, 2004.

P. Bastian, K. Johannsen, S. Lang, S. Nägele, V. Reichenberger, C. Wieners, G. Wittum, and C. Wrobel. Advances in high-performance computing: Multigrid methods for partial differential equations and its applications. In E. Krause and W. Jäger, editors, *High Performance Computing in Science and Engineering '99*, pages 506–519, Berlin, 1999. Springer-Verlag.

P. Bastian, K. Johannsen, S. Lang, V. Reichenberger, C. Wieners, G. Wittum, and C. Wrobel. Parallel solutions of partial differential equations with adaptive multigrid methods on unstructured grids. In E. Krause and W. Jäger, editors, *High Performance Computing in Science and Engineering '00*, pages 496–508, Berlin, 2000. Springer-Verlag.

S. Lang, C. Wieners, and G. Wittum. The application of adaptive parallel multigrid methods to problems in nonlinear solid mechanics. In E. Stein, editor, *Error-Controlled Adaptive Finite Element Methods in Solid Mechanics*, pages 347–384, New-York, 2002. Wiley.

P. L. Lions. On the existence of positive solutions of semilinear elliptic equations. *SIAM Review*, 24:441–467, 1982.

M. Plum and C. Wieners. New solutions of the Gelfand problem. *J. Math. Anal. Appl.*, 269:588–606, 2002.

C. Wieners, M. Ammann, and W. Ehlers. Distributed point objects: A new concept for parallel finite elements applied to a geomechanical problem. *Future Generation Computer Systems*, 2004. to appear.

Part IV

Minisymposium: Collaborating Subdomains for Multi-Scale Multi-Physics Modelling

Local Defect Correction Techniques Applied to a Combustion Problem

Martijn Anthonissen

Technische Universiteit Eindhoven, Scientific Computing Group
m.j.h.anthonissen@tue.nl

Summary. The standard local defect correction (LDC) method has been extended to include multilevel adaptive gridding, domain decomposition, and regridding. The domain decomposition algorithm provides a natural route for parallelization by employing many small tensor-product grids, rather than a single large unstructured grid. The algorithm is applied to a laminar Bunsen flame with one-step chemistry.

1 Introduction

Partial differential equations (PDEs) with solutions that have highly localized properties appear in many application areas, e.g. combustion, shock hydrodynamics, and transport in porous media. Such problems require a fine grid only in the region(s) of high activity, whereas elsewhere a coarser grid suffices. We consider a discretization method for elliptic boundary value problems introduced by Hackbusch [1984]. In this technique, the *local defect correction* (LDC) method, the discretization on the composite grid is based on a combination of standard discretizations on uniform grids with different spacings that cover different parts of the domain. The coarse grid must cover the entire domain, and its spacing is chosen in agreement with the relatively smooth behavior of the solution outside the high activity areas. Apart from this global coarse grid, one or more local fine grids are used that are also uniform, each of which covers only a (small) part of the domain and contains a high activity region. The grid spacings of the local grids are chosen in agreement with the behavior of the continuous solution in that part of the domain.

The LDC method is closely related to the *fast adaptive composite grid* (FAC) method (McCormick [1984a], McCormick and Thomas [1986]). An important difference with LDC is that an explicit discretization scheme for the composite grid is proposed, in which special difference stars near the grid interfaces are used. The resulting discrete system is solved by an iterative method which may take advantage of the composite grid structure. This is a crucial difference with the LDC method, which combines standard discretizations on uniform grids only and does not use an *a priori* given composite grid discretization. For the FAC method in a variational setting, convergence results have been given by McCormick [1984b]. The variational theory is extended to the finite volume element method in McCormick and Rüde [1994]. Xu [1992] presents an abstract framework and general convergence theory for a wide range of iterative methods, among which domain decomposition, multigrid and multilevel methods. He groups the algorithms in *parallel* and *successive subspace correction* methods. Xu shows in particular that FAC is equivalent to classic multigrid with smoothing in the area of refinement only.

The idea to approximate low frequency components on a coarse grid and high frequency components on a (local) fine grid forms the basis of multigrid. McCormick and Ruge [1986] present *unigrid*, an algorithm based on these principles and especially suited for testing the feasibility of using multigrid in a given application. Xu and Zhou [1999, 2000, 2001] use the fact that the global behavior is dominated by low frequencies and the local behavior by high frequencies to design discretization schemes in a finite element context. They study elliptic boundary value problems and prove error estimates for the finite element solution. Based on these estimates, they develop several algorithms. The simplest one is to solve a global problem on a locally refined grid. This algorithm is improved by using a residual correction technique, in which a global coarse grid problem is solved first. Next the coarse grid residual is corrected by solving the problem on one or more locally refined grids that cover the whole domain but are very coarse outside the area of refinement. Note that the global coarse grid problem needs to be solved only once and is not coupled with the local problems. This is different from the LDC method, in which the local problems take artificial boundary conditions from the coarse grid solution and cover part of the domain only. Xu and Zhou [1999, 2000] present parallel versions of their algorithms by subdividing the domain in disjoint subdomains. In Xu and Zhou [2000], further algorithms are developed by ignoring the lower order terms of the PDE on the local grids, which can be done because the symmetric positive definite part dominates the high frequency components. Xu and Zhou [2001] study a solution technique for nonlinear elliptic PDEs. The full nonlinear problem is first discretized by a standard finite element technique on a global coarse grid. Next, the residual is corrected using linearized discretizations on fine grids.

This paper deals with some extensions to the standard LDC method; we add adaptivity, multilevel refinement, domain decomposition and regridding. We apply the new algorithm to a Bunsen flame problem previously treated by Bennett and Smooke [1998].

2 Formulation of the LDC method

Before presenting our extensions to the LDC method, we begin by describing the standard LDC method. We consider the elliptic boundary value problem

$$\begin{cases} Lu = f, & \text{in } \Omega, \\ u = g, & \text{on } \partial\Omega. \end{cases} \tag{1}$$

In (1), L is a linear elliptic differential operator, and f and g are the source term and Dirichlet boundary condition, respectively. Other types of boundary conditions can be used as well, but for ease of presentation we formulate the method for (1). To discretize (1), we first choose a global coarse grid (grid spacing H), which we denote by Ω^H. An initial approximation u_0^H on Ω^H can be found by solving the system

$$L^H u_0^H = f^H, \tag{2}$$

which is a discretization of (1). In (2), the right-hand side f^H incorporates the source term f as well as the Dirichlet boundary condition g. We assume L^H to be invertible.

Assume that the continuous solution u of (1) has a high activity region in some (small) part of the domain. We select a subdomain $\Omega_l \subset \Omega$ such that the high activity region of u is contained in Ω_l. In Ω_l, we choose a local fine grid (grid spacing h), which we denote by Ω_l^h, such that grid points of the global coarse grid that lie in the area of refinement also belong to the local fine grid. In order to formulate a discrete problem on Ω_l^h, we define artificial boundary conditions on Γ, the interface between Ω_l and $\Omega \setminus \Omega_l$. We apply an interpolation operator $P^{h,H}$ that maps function values at coarse grid points on the interface to function values at fine grid points on the interface. In the numerical simulations we use linear interpolation. In this way, we find the following approximation $u_{l,i}^h$, iteration $i = 0$, on Ω_l^h:

$$L_l^h u_{l,i}^h = f_l^h - B_{l,\Gamma}^h P^{h,H}\left(u_i^H|_\Gamma\right). \tag{3}$$

In (3), matrix L_l^h (assumed to be invertible) is a discrete approximation to L on the subdomain Ω_l. The first term on the right-hand side incorporates the source term f as well as the Dirichlet boundary condition g on $\partial\Omega_l \setminus \Gamma$ given in (1). In the second term, the operator $B_{l,\Gamma}^h$ represents the dependence of the fine grid points on the coarse grid solution at the artificial boundary Γ.

We will now use the local fine grid solution to update the coarse grid approximation. If we were able to substitute the projection on Ω^H of the exact solution u of boundary value problem (1) into the coarse grid discretization (2), we would find the local discretization error or *local defect* d^H, given by $L^H\left(u|_{\Omega^H}\right) = f^H + d^H$. We could then use d^H within the right-hand side of (2) to find a better approximation on the coarse grid. However, as we do not know u, we instead use the fine grid approximation $u_{l,0}^h$ to estimate d^H at the coarse grid points inside the area of refinement $(x,y) \in \Omega_l^H := \Omega^H \cap \Omega_l$. We define w_0^H as the global coarse grid function of best approximations so far:

$$w_0^H(x,y) := \begin{cases} u_{l,0}^h(x,y), \; (x,y) \in \Omega_l^H, \\ u_0^H(x,y), \; (x,y) \in \Omega^H \setminus \Omega_l^H, \end{cases}$$

and estimate the defect by $d^H = L^H\left(u|_{\Omega^H}\right) - f^H \approx L^H w_0^H - f^H =: d_0^H$. Assuming that the stencil at grid point (x,y) involves (at most) function values at $(x + iH, y + jH)$ with $i, j \in \{-1, 0, 1\}$, d_0^H provides an estimate of the local discretization error of the coarse grid discretization at all points of Ω_l^H. We apply the *coarse grid correction step* to find u_i^H, $i = 1$:

$$L^H u_{i+1}^H = \begin{cases} f^H(x,y) + d_i^H(x,y), & (x,y) \in \Omega_l^H, \\ f^H(x,y), & (x,y) \in \Omega^H \setminus \Omega_l^H. \end{cases} \tag{4}$$

Because (4) incorporates estimates of the local discretization error of the coarse grid discretization, u_1^H is assumed to be more accurate than u_0^H. Hence it provides a better boundary condition on Γ, and a better solution on the local fine grid can be found by solving (3) with $i = 1$. This leads to an iterative method: we can solve a new updated coarse grid problem.

Often one or two LDC iterations will suffice to obtain a satisfactory approximation on the composite grid due to the high rate of convergence of the method. Typically, iteration errors are reduced by a factor of 10 to 1,000 in each iteration step (*cf.* Ferket and Reusken [1996], Hackbusch [1984], Nefedov and Mattheij [2002], Anthonissen [2001]). A detailed analysis of the convergence behavior for diffusion equations is given in Anthonissen et al. [2003b].

3 Extensions to the LDC method

We now extend the LDC algorithm by adding adaptivity, multilevel refinement, domain decomposition, and regridding. The result will be a technique for discretizing and solving (1) on a composite grid found by adaptive grid refinement, given a code for solving boundary value problem (1) on a tensor-product grid in a rectangular domain.

Adaptive multilevel refinement

We assume that the continuous solution u has one area of high activity; it is straightforward to generalize the algorithm to the case where there is more than one area of high activity. We assume that the initial coarse grid is given by its x- and y-coordinates x_i, y_j, and define the boxes B_{ij} formed by grid points and points on the boundary, *viz.* $B_{ij} = (x_i, x_{i+1}) \times (y_j, y_{j+1})$. In order to determine which boxes require refinement, we introduce the positive weight function of Bennett and Smooke [1998, 1999] as an indicator for solution roughness. As detailed in Anthonissen et al. [2003a], the weight function assigns a value to each box B_{ij} and points will be added in regions where the weight function is large, so it should measure the rapidity of change of u. Xu and Zhou [2000] give theoretical justification why we may use a global *a posteriori* error estimate for equi-distribution of the error, as we do here. Apart from high activity boxes, we also flag their neighbors for refinement in order to prevent the solution from being artificially trapped at interfaces between coarse and fine grids, which can happen if high activity areas move during recalculation on the finer grid.

For the area of refinement Ω_l, we choose the smallest rectangle that encloses all flagged boxes; more efficient choices are discussed in the next section.

In Ω_l, we choose a local fine grid Ω_l^h by uniform refinement. The integer refinement factor σ is typically set to 2. Ideally, σ should be chosen largest at places where the weight function is largest. However, this approach would lead to an unstructured composite grid, which we want to avoid. Therefore, we will use multiple refinement levels.

After adding successive levels of refinement, the fine grid approximations are used to improve the coarse grid approximations via coarse grid correction steps. Once we have returned to the base grid, we will solve discrete problems on finer levels again.

Domain decomposition

In the previous section, we determined the smallest rectangle enclosing all flagged boxes and chose to refine this rectangle entirely. However, this approach may refine many boxes that have not been flagged for refinement, especially when an area of high activity is not aligned with the grid directions. To remedy this inefficiency as well as to prevent the grids from becoming too large, we combine the multilevel LDC algorithm with domain decomposition, in which we use a set of rectangles to cover all flagged boxes. We require each flagged box to be enclosed in at least one rectangle, and we want the rectangles to be overlapping. The overlap of the rectangles is necessary in situations where interfaces between rectangles intersect high activity zones. We remedy large errors at these interfaces by performing a number of domain decomposition iterations via a standard multiplicative Schwarz procedure.

To find a set of rectangles satisfying the conditions just stated, a cost function is defined that states how expensive using a certain set is. The algorithm evaluates the cost of using a single rectangle (as we did in the previous section), splitting it horizontally in two smaller rectangles or splitting vertically. This procedure is performed recursively on the smaller rectangles if splitting has occurred; see Anthonissen et al. [2003a] for details. The algorithm, including the Schwarz alternating procedure, is shown in Figure 1.

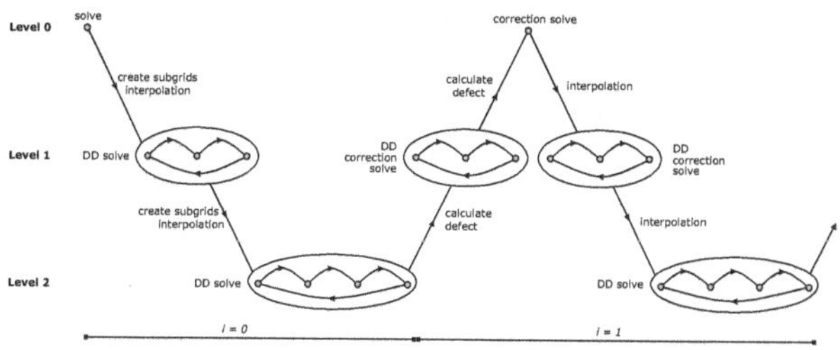

Fig. 1. Solution procedure with domain decomposition

Regridding

Refining a grid and solving the boundary value problem on the new composite grid may cause the region(s) of high solution activity to move. Therefore, we apply the regridding procedure from Bennett and Smooke [1998] before proceeding from Level l to Level $l+1$.

4 Application to a combustion problem

We now turn our attention to the axisymmetric laminar Bunsen flame with one-step chemistry. This problem was previously presented by Bennett and Smooke [1998]. Because almost all of the dependent variables in the Bunsen flame problem have large gradients in a very small region of the computational domain, adaptive gridding is a must for this simulation. The physical configuration for the Bunsen flame is shown in Figure 2. A mixture of methane and air flows up from a central jet, which is surrounded by a coflowing air stream. A steady conical flame forms at the mouth of the cylindrical burner.

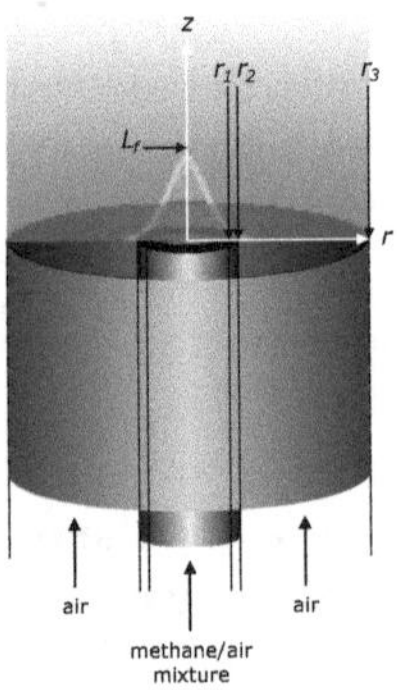

Fig. 2. Physical configuration for the axisymmetric Bunsen flame

The chemical model we consider has five species: methane, oxygen, water, carbon dioxide, and the abundant inert, nitrogen. There are nine dependent variables in the Bunsen flame problem: radial velocity, axial velocity, vorticity, temperature, and five mass fractions. These variables satisfy a set of strongly coupled nonlinear PDEs, see Anthonissen et al. [2003a] for details. The initial coarse grid is chosen to be more finely spaced in the region above the inner jet, because it is known that the flame forms in that area. The exact r- and z-coordinates of the initial grid are given in Anthonissen et al. [2003a].

Due to the nature of the LDC method, the PDEs need only be discretized on tensor-product grids. We apply standard finite difference stencils at interior points. First-order upwinding is used on convective terms. Details can be found in Anthonissen et al. [2003a]. The discretized governing equations and boundary conditions form a system of equations, that is linearized by a damped, modified Newton's method (Deuflhard [1974], Smooke [1983]) with a nested Bi-CGSTAB linear algebra solver; the latter is preconditioned using a block Gauss-Seidel preconditioner.

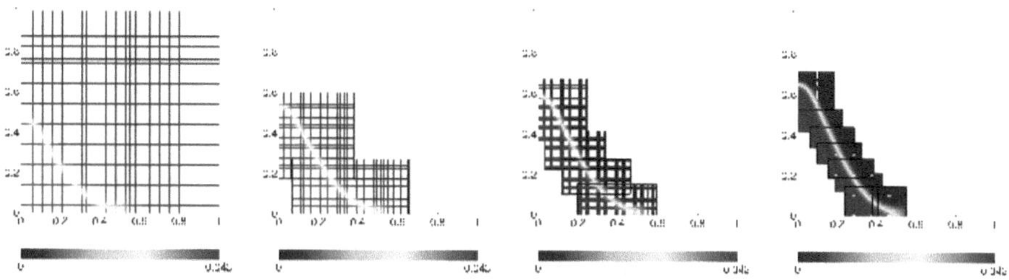

(a) $l_{\max} = 0$. (b) $l_{\max} = 1$. (c) $l_{\max} = 2$. (d) $l_{\max} = 3$.

Fig. 3. Plots of the methane mass fraction on the finest level for the LDC simulations with various values of the maximum level of refinement $l_{\max}$

Four different LDC simulations have been carried out on a 175 MHz SGI Octane with 1 GB of RAM. Each simulation starts from an already converged solution on the base tensor-product grid. For each first solve on a local fine grid, we use as an initial guess the approximation found by interpolating the approximation on the parent grid. At each level with more than one grid, five domain decomposition iterations are done to improve the boundary conditions at the internal interfaces. The flagging of high activity boxes is based on the methane mass fraction. The refinement factor is 2.

Figure 3 shows the projection of the methane mass fraction on LDC composite grids with increasingly fine resolution. Although the flame structure is similar in each plot, the flame length increases, with the largest increase occurring when the first refinement level is added. In Anthonissen et al. [2003a], the LDC results are shown to have excellent agreement with both the local rectangular refinement (LRR) results for the same problem as well as with results found on equivalent tensor-product (ETP) grids with the same resolution presented by Bennett and Smooke [1998]. In the LRR method, an unstructured grid is constructed from an initial tensor-product grid by flagging and refining high activity boxes individually. Unlike tensor-product grids, grid lines in an LRR grid are not required to extend from one domain boundary to the other. A Newton solver is subsequently applied to the discretized PDE system on the complete unstructured grid.

In the LDC simulations, however, the Newton solver is applied to many small grids individually rather than to one large grid. In the LDC simulation for $l_{\max} = 3$, the composite grid consists of the initial tensor-product grid with three additional refinement levels, that have two, four, and eight subgrids, respectively. The biggest tensor-product grid in this hierarchy has only 16,653 points — a substantial memory savings over the ETP grid of 312,872 points.

5 Conclusions

In this paper, we have extended the standard LDC method by including multilevel adaptive gridding, domain decomposition, and regridding. We have successfully applied this method to a lean axisymmetric laminar Bunsen flame with one-step chemistry. In the future, we would like to investigate the inclusion of more sophisticated domain decomposition techniques within the method, so that fewer iterations will be required among grids at a given level. We would also be interested in the possibility to solve different problems on the global coarse and local fine grid, using e.g. chemical equilibrium; *cf.* Xu and Zhou [2000, 2001] in which only the symmetric positive definite part or a linearized discretization is solved locally. As there are virtually no conceptual hurdles in expanding the approach to higher dimensions, our ultimate goal is to apply the extended LDC method to three-dimensional combustion problems, for which its low memory usage and parallelization opportunities will play an important role.

References

M. J. H. Anthonissen. *Local defect correction techniques: analysis and application to combustion.* PhD thesis, Eindhoven University of Technology, Eindhoven, 2001.

M. J. H. Anthonissen, B. A. V. Bennett, and M. D. Smooke. An adaptive multilevel local defect correction technique with application to combustion. Technical Report RANA 03-24, Eindhoven University of Technology, Eindhoven, Oct. 2003a.

M. J. H. Anthonissen, R. M. M. Mattheij, and J. H. M. ten Thije Boonkkamp. Convergence analysis of the local defect correction method for diffusion equations. *Numerische Mathematik*, 95(3):401–425, 2003b.

B. A. V. Bennett and M. D. Smooke. Local rectangular refinement with application to axisymmetric laminar flames. *Combust. Theory Modelling*, 2:221–258, 1998.

B. A. V. Bennett and M. D. Smooke. Local rectangular refinement with application to nonreacting and reacting fluid flow problems. *J. Comput. Phys.*, 151:684–727, 1999.

P. Deuflhard. A modified Newton method for the solution of ill-conditioned systems of nonlinear equations with application to muliple shooting. *Num. Math.*, 22:289–315, 1974.

P. J. J. Ferket and A. A. Reusken. Further analysis of the local defect correction method. *Computing*, 56:117–139, 1996.

W. Hackbusch. Local defect correction and domain decomposition techniques. In K. Böhmer and H. J. Stetter, editors, *Defect Correction Methods. Theory and Applications, Computing, Suppl. 5*, pages 89–113, Wien, New York, 1984. Springer.

S. McCormick. Fast adaptive composite grid (FAC) methods. In K. Böhmer and H. J. Stetter, editors, *Defect Correction Methods: Theory and Applications*, pages 115–121. Computing Supplementum 5, Springer-Verlag, Wien, 1984a.

S. McCormick. Fast adaptive composite grid (FAC) methods. In K. Böhmer and H. J. Stetter, editors, *Defect Correction Methods: Theory and Applications*, pages 115–121. Computing Supplementum 5, Springer-Verlag, Wien, 1984b.

S. McCormick and U. Rüde. A finite volume convergence theory for the fast adaptive composite grid methods. *Appl. Numer. Math.*, 14:91–103, 1994.

S. McCormick and J. Ruge. Unigrid for multigrid simulation. *Math. Comp.*, 41:43–62, 1986.

S. McCormick and J. Thomas. The fast adaptive composite grid (FAC) method for elliptic equations. *Math. Comp.*, 46(174):439–456, 1986.

V. Nefedov and R. M. M. Mattheij. Local defect correction with different grid types. *Numerical Methods for Partial Differential Equations*, 18:454–468, 2002.

M. D. Smooke. Error estimate for the modified Newton method with applications to the solution of nonlinear, two-point boundary value problems. *J. Optim. Theory Appl.*, 39:489–511, 1983.

J. Xu. Iterative methods by space decomposition and subspace correction. *SIAM Review*, 34(4): 581–613, December 1992.

J. Xu and A. Zhou. Some local and parallel properties of finite element discretizations. In C.-H. Lai, P. E. Bjørstad, M. Cross, and O. B. Widlund, editors, *Eleventh International Conference on Domain Decomposition Methods*, pages 140–147, Bergen, 1999. Domain Decomposition Press.

J. Xu and A. Zhou. Local and parallel finite element algorithms based on two-grid discretizations. *Mathematics of Computation*, 69(231):881–909, 2000.

J. Xu and A. Zhou. Local and parallel finite element algorithms based on two-grid discretizations for nonlinear problems. *Advances in Computational Mathematics*, 14(4):293–327, 2001.

Electronic Packaging and Reduction in Modelling Time Using Domain Decomposition

Peter Chow[1] and Choi-Hong Lai[2]

[1] Fujitsu Laboratories of Europe Ltd, Physical and Life Sciences (P.Chow@fle.fujitsu.com)
[2] School of Computing and Mathematical Sciences, University of Greenwich (C.H.Lai@gre.ac.uk)

Summary. The domain decomposition method is directed to electronic packaging simulation in this article. The objective is to address the entire simulation process chain, to alleviate user interactions where they are heavy to mechanization by component approach to streamline the model simulation process.

1 Introduction

Small is exquisite and cool to the consumers of electronic products, but it has enormous technical challenges that need to be overcome by designers and engineers. Elements such as health and safety compliance, power and heat management, and usability are commonly top on the list of issues. The primary technical challenges (Chow and Addison [2002]) are: 1) High density of components leads to an increase of model complexity that needs to address, including any interactions and interferences between the processes. 2) Creating highly intricate and detailed geometry and mesh models with parts of dissimilar scales in appropriate time, for examples, the entire electronic components in a laptop computer and exposure analysis of electronic devices on an entire human body and tissues. 3) Size of computational demands million plus cells/elements models are common in industrial simulations and models with tens of millions of elements are becoming more frequent. 4) The market demands add to the ever-increasing pressures on engineers to speed up the modelling and design cycles.

Whilst domain decomposition has been widely applied to areas such as parallel solvers and preconditioners, coupling of different numerical methods and physical models, it has not been considered for the entire simulation process chain in order to achieve a comprehensive reduction in modelling time. An early concept of using domain decomposition methods in the reduction of modelling time can be found in Chow and Addison [2002]. This paper gives a rigorous approach of the concept and uses the framework of a defect equation in the coupling (C.-H. Lai and Pericleous [1997]). Algorithms developed in sections below concentrate on problems with geometrical multi-scale at the macroscopic mathematical models. Numerical experiments, including monophase and multi-phase, are examined with efficiency of the algorithm being linked to the overall modelling time.

2 Component Approach and Concept

A basic conventional simulation process chain would look something like: Geometry-Meshing-Analysis-Visualization. An alteration to the model at any stage means going back to the Geometry stage and repeating the procedure. For large and complex models re-do it all is time consuming. A smart way is to take a component based approach where only the altered components are being re-done and thus significantly more efficient. This is generally not possible due to conformal mesh constraint, one continuous volume mesh, in mesh generation and analysis stages. In instances where it is possible, it is frequently for special purposes and non-standard in nature. While in user interaction intensity, geometry creation and meshing are the most user intensive stages, whereas analysis stage is the least, and visualization is interpreting the solution for specific requirements. The concept detailed below is component based with each component independently created, meshed and solved. The model solution is reached when interface conditions between components agrees, this is attained by iterations with the domain decomposition method.

The component meshing and gluing (CMG) approach(Chow and Addison [2002]), takes an approach the same as manufacturing products are assembled A product is a collection of assembled components or parts, connected and bonded together, and commonly, the parts themselves are products produced and marketed by others. This component nesting is the base of CMG and let existing models to be reused for other models. This kind of model assembly is probably most suited to applications where models are constructed from a few basic shapes such as multi-chip module models in electronic packaging. Here, it can be realized by a database of components with simple tools that uses parametric to define basic objects relative to parameters such as length, thickness, mesh density, etc., for rapid primitive component creation.

The model of assembled components is then glue together by either merging components to create a single knitted mesh model or collaborating components using an iterative domain decomposition method. The former methodology requires unstructured meshes or the use of polyhedral type elements to combine into one mesh model, and we will refer to this as the CMG-Knitted strategy in this paper. One disadvantage of the knitted strategy is that it does not apply to all solvers, for example, structured mesh solvers. Only solvers with polyhedral element capability can be considered. The latter methodology requires the domain decomposition method (DDM) method to attain the model solution through the exchange of boundary conditions between common interfaces that the components shared. We will refer to this as the CMG-DDM strategy in this paper. The solution of each component may be obtained by means of existing fast solvers. This is more universally applicable to all types of solvers, but one known disadvantage is that the computing time to achieved a converged solution is longer. Fast iterative methods in domain decomposition can significantly shorten the time to solution but it is unlikely to match the knitted mesh case. For appropriate solvers, a combination of the two gluing strategies is possible.

The significant benefit of CMG is it virtually removes all the difficulties commonly associated with model creation and mesh generation which made the two processes extremely manpower intensive in the process chain. And with the volume-mesh generation element no longer called, a considerable saving in time and computing resources. Perhaps the only meshing related element that may need some manpower input are the interface regions where the component meets. This is not envisage, but if needed, it is a surface meshing problem and not a volume one which is one-degree of dimension less in complexity and requires significantly less computing to do. This gain needs to be summed with the increase computing times in the analysis stage expected in the CMG-DDM strategy to give an account of profit or lost balance. When the balance is at a significant lost we do have the parallel processing option to address the problem. Compared to model creation and mesh generation, parallel processing has drastically cut the time for the analysis element and the trend is still downward.

3 Numerical Algorithms

Provided that the solvers can take polyhedral elements, the CMG-Knitted strategy does not require extra effort to put into the solvers. It is the mesh-model that needs to be knitted at the finite-element mesh topology level, gluing the interfaces of the mesh components. In the CMG-DDM strategy, the domain decomposition method (DDM) (Chow and Addison [2002]) is ideally suited for the assembled-component model, with the non-overlapping class the most appropriate. A non-overlapping approach allows flexibility in the mesh processing, the methods of numerical solution, the handling of different physics, and the adoption of numerical solvers in each of the model components. This choice also makes the defect equation technique as developed in C.-H. Lai and Pericleous [1997] an ideal method for CMG-DDM.

Let $\mathcal{L}u = f$ be defined in the domain Ω and $u = g$ on $\partial\Omega$, where $\mathcal{L}$ may be a nonlinear operator that depends on u, and g is a known function. The domain Ω is partitioned into M non-overlapped sub-domains such that $\bigcup_{i=1}^{M} \Omega_i = \Omega$ and $\Omega_i \bigcap \Omega_j =$, for $i \neq j$. Each sub-domain is associated with a sub-model defined by $\mathcal{L}_i u_i = f_i$. The boundary of each sub-domain, $\partial\Omega_i$, subtracting the part of boundary which overlaps with the boundary of the entire problem is in essence a part of the interface. Therefore the interface, which attached to

Ω_i, may be defined as $\gamma_i = \partial\Omega_i\partial\Omega$. The boundary conditions defined on γ_i may be denoted by u_γ and it satisfies a defect equation, such as $D(u_\gamma) = 0$ (C.-H. Lai and Pericleous [1997]). Using superscripts to denote the number of gluing process, the CMG-DDM algorithm may be written as follows.

Initial values: $n = 0$; $u_i^{(0)}$, $i = 1, ..., M$ are given.
Repeat { $n := n + 1$;
  Do $i = 1, ..., M$
    $u_i^{(n)} :=$ { Solve $\mathcal{L}_i u_i^{(n)} = f_i$ in Ω_i };
      subject to:
      $u_i^{(n)} = g$ on $\partial\Omega \bigcap \partial\Omega_i$ and $u_i^{(n)}|_{\gamma_i} = u_{\gamma_i}$;
  End-Do
  Solve $D(u_\gamma) = 0$; }
Until $||D||_2 < \epsilon$

When the model consists of a single domain (meshed component) then the For loop and the defect calculation, $D(u_\gamma) = 0$, are redundant. The CMG-Coupled cases are performed in this way. From the above algorithm the For loop may be run in parallel and on homogeneous computing systems the solution are identical between parallel and scalar computations.

4 Numerical Experiments

The particular problem to be considered in this paper is governed by the 2-D energy equation, limited to conduction only, in temperature u. The variables in (2) are density (ρ), specific heat (c), thermal conductivity (k), time (t) and the source term (S).

$$\rho c \frac{\partial u}{\partial t} = \nabla \cdot (k \nabla u) + S(u) \tag{1}$$

The nonlinearity is introduced in the form of a material phase-change in the source term. For solidification using the enthalpy source-based method this is given by

$$S(u) = L\rho \frac{\partial f(u)}{\partial t} \tag{2}$$

where L is the latent heat and f is the liquid fraction. The algorithm for solving these kinds of problems may be found in papers by Chow and Cross [1992] and Voller and Swaminathan [1991] and is not discussed in this paper. Readers interested in obtaining more information are directed to these references. In this study, the numerical stable method of Voller and Prakash [1987] solidification algorithm is used.

A nonlinear problem with phase-change occurring inside the domain, geometry as that of Fig. 1, was used to conduct the numerical experiments and investigations. Three experiments conducted were: 1) A steady state heat transfer (no phase-change) where the top surface is at a temperature of $10°C$

and bottom surface of 100°C. The left side of the model is symmetry and for all other boundaries, a convective heat boundary condition of ambient temperature of 25°C with a heat transfer coefficient of 10.0 W/m^2C. 2) A transient heat transfer problem that has the same boundary conditions as the first experiment with an initial temperature of 100°C. The time step size taken was 10 seconds interval and simulation time end at 120 seconds. 3) The final experiment is a heat transfer with the small solder bumps (Ch4 Solder Bump in Fig. 1 undergoing solidification. The boundary condition is essentially the same as the previous two experiments with both top and bottom surfaces now have the convective heat boundary conditions. The initial temperature is at 183°C with time step size of 2 seconds interval and simulation time end at 600 seconds.

Table 1 shows the dimension of the components and Table 2 shows the material properties data used in the experiments. For convenient, the Si-Chip, MCM-L and motherboard take on the material property of the Board dataset, and both the Ch4 and BGA solder bumps take on the Solder dataset. For the third experiment, only the Ch4 solder bumps are solidifying, the liquidus and solidus temperatures for BGA solder bumps are set above that given thus no solidification occurs.

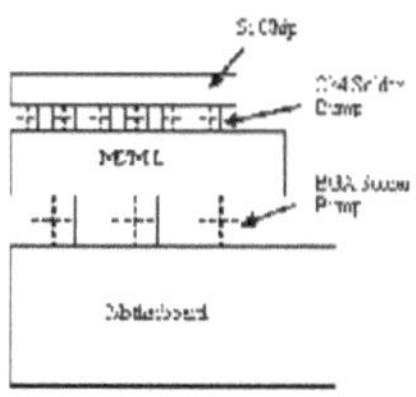

Fig. 1. An example of multiple chip model geometry

Table 1. Geometric dimension of components in the test model

	Length (mm)	Height (mm)	Gap Interval (mm)
Si Chip	10.5	1.5	
Ch4 Solder Bump	1.0	1.0	1.0
MCM-4	15.5	2.0	
BGA Solder Bump	2.0	2.0	3.0
Motherboard	19.5	3.0	

Fig. 2 shows two different meshes used in present experiments. Figures 3 to 5 show the cell invariant temperature distribution of the CMG-Knitted computation and CMG-DDM for the three experiments. The temperature profile on the two meshes (conformal and non-conformal) is virtually the same. Ta-

Table 2. Material properties

	Density kg/m^3	Specific Heat J/kg C	Conductivity W/m C	Liquidus Temp. C	Solidus Temp. C	Latent Heat J/kg
Board	1400	838	0.18			
Solder	8400	171	50.6	183	180	3700

ble 3 give the total energy in the system domain and computing costs for the simulation, together with iteration numbers required. The knitted conformal mesh result is used as the reference guide towards measuring accuracy and computing performance. In the transient problems, the iteration numbers shows the first time step has the highest iteration counts, this is obvious due to the cold starting the simulation, whereas lowest is found in time steps towards end of simulation. The CMG-knitted computation for linear problems, Experiments 1 and 2, require 2 iterations for both steady state and per time step in transient to achieve convergence on temperature. The largest deviation of the solution from the referenced data is under 0.2 % in Experiment 3, and under 0.07% and 0.01% respectively for Experiments 2 and 1. Computing times for CMG-DDM for the two meshes (conformal and non-conformal) are 72.1 and 65.0 in Experiment 1, 44.5 and 40.2 in Experiment 2, and 2.1 and 2.0 in Experiment 3, times more expensive respective to the referenced knitted conformal mesh cases.

Based on these results the CMG-DDM approach for linear problem is not competitive, but non-linear problem is a different proposition. Assuming 25% of overall time is used for analysis, this implies the projected total modelling time of 197 seconds (= 49.296 / 25%) for coupled computation in Experiment 3, which suggests that the CMG-DDM approach to be competitive in non-linear phase-change problems in electronic packaging.

5 Summary

Numerical experiments conducted indicates potential advantages of the CMG method in electronic packaging for non-linear solder solidification based on the enthalpy method (Chow and Cross [1992], Voller and Prakash [1987]) simulation of multiple chip modules. The success of the method is the mechanization by component approach to streamline the model simulation process at model creation and mesh generation stages which are the most manpower intensive.

References

A. C. C.-H. Lai and K. Pericleous. A defect equation approach for the coupling of subdomains in domain decomposition methods. *Computers Math.*

Applic., 6:81–94, 1997.
P. Chow and C. Addison. Putting domain decomposition at the heart of a mesh-based simulation process. *Int. J. Numer. Meth. Fluids*, 40:1471–1484, 2002.
P. Chow and M. Cross. An enthalpy control volume-unstructured mesh (cv-um) algorithm for solidification by conduction only. *Int. J. Numer. Meth. in Engg.*, 35:1849–1870, 1992.
V. Voller and C. Prakash. A fixed grid numerical modelling methodology for convection-diffusion mushy region phase change problems. *Int. J. of Heat and Mass Transfer*, 30:1709–1719, 1987.
V. Voller and C. Swaminathan. General source-based methods for solidificationn phase change. *Numerical Heat Transfer*, 19:175–190, 1991.

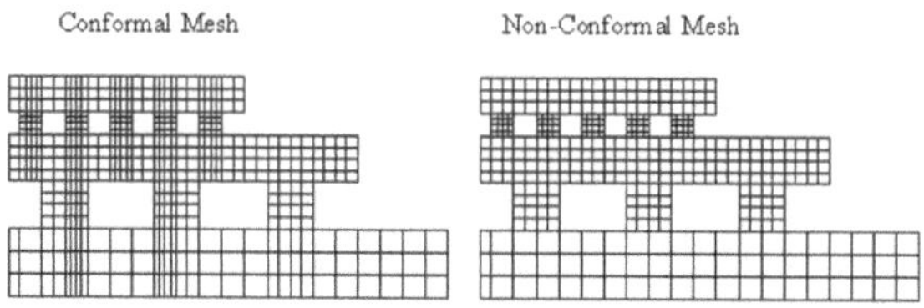

Fig. 2. Two different meshes used in experiments

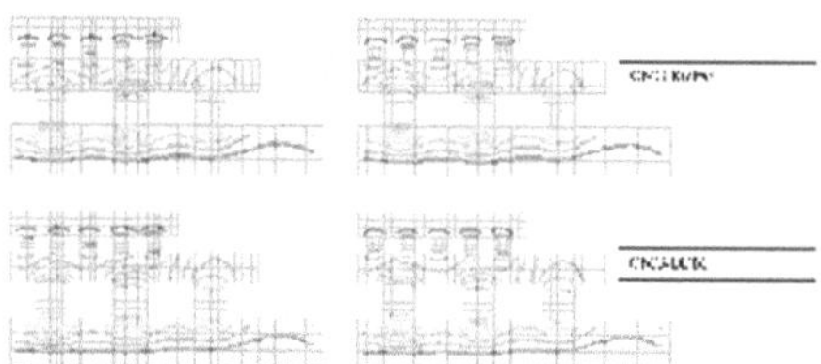

Fig. 3. Comparison of the temperature distribution of steady state results

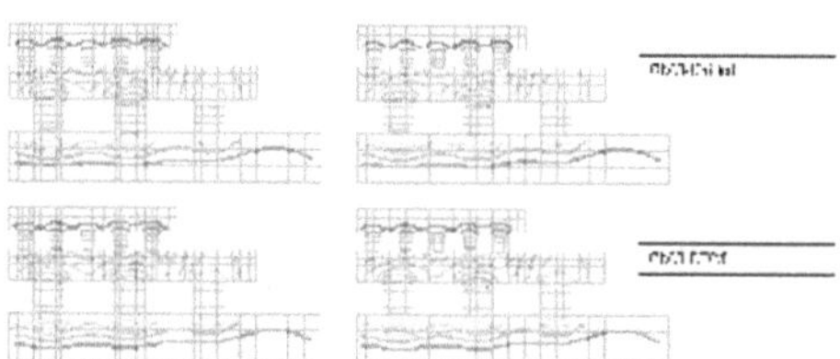

Fig. 4. Comparison of the temperature distribution of transient results

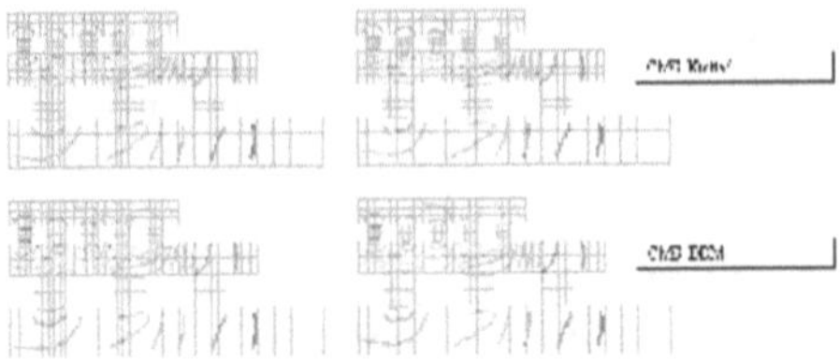

Fig. 5. Comparison of the temperature distribution of solidification results

Table 3. Total energy in system and computing times (Computing platform: P4, 2GHz, 1GB RAM)

Steady state results ($\|\|D\|\|_2 < 10^{-6}$)				
	Conformal mesh		Non-conformal mesh	
	CMG-Knitted	CMG-DDM	CMG-Knitted	CMG-DDM
Total energy	8.998910166	8.998910109	8.998117691	8.998120956
Relative error		$6.334100x10^{-9}$	$8.806344x10^{-5}$	$8.770062x10^{-5}$
Computing time	0.031	2.234	0.015	2.015
Iteration number	2	230	2	232

Transient results ($\|\|D\|\|_2 < 10^{-6}$)				
	Conformal mesh		Non-conformal mesh	
	CMG-Knitted	CMG-DDM	CMG-Knitted	CMG-DDM
Total energy	9.378820792	9.378820836	9.384847314	9.384851563
Relative error		$4.691421x10^{-9}$	$6.425671x10^{-4}$	$6.430202x10^{-4}$
Computing time	0.172	7.640	0.156	6.906
Max. Iterations	2	75	2	75
Min. Iterations	2	61	2	59

Solidification results ($\|\|D\|\|_2 < 10^{-6}$)				
	Conformal mesh		Non-conformal mesh	
	CMG-Knitted	CMG-DDM	CMG-Knitted	CMG-DDM
Total energy	15.214174680	15.214174700	15.185239930	15.185216610
Relative error		$1.314564x10^{-9}$	$1.901828x10^{-3}$	$1.903361x10^{-4}$
Computing time	49.296	101.563	35.094	94.547
Max. Iterations	78	79	77	78
Min. Iterations	19	23	18	23

Improving Robustness and Parallel Scalability of Newton Method Through Nonlinear Preconditioning*

Feng-Nan Hwang[1] and Xiao-Chuan Cai[2]

[1] Department of Applied Mathematics, University of Colorado at Boulder, Boulder, CO 80309, USA (hwangf@colorado.edu)
[2] Department of Computer Science, University of Colorado at Boulder, Boulder, CO 80309, USA (cai@cs.colorado.edu,http://www.cs.colorado.edu/~cai/)

Summary. Inexact Newton method with backtracking is one of the most popular techniques for solving large sparse nonlinear systems of equations. The method is easy to implement, and converges well for many practical problems. However, the method is not robust. More precisely speaking, the convergence may stagnate for no obvious reason. In this paper, we extend the recent work of Tuminaro, Walker and Shadid [2002] on detecting the stagnation of Newton method using the angle between the Newton direction and the steepest descent direction. We also study a nonlinear additive Schwarz preconditioned inexact Newton method, and show that it is numerically more robust. Our discussion will be based on parallel numerical experiments on solving some high Reynolds numbers steady-state incompressible Navier-Stokes equations in the velocity-pressure formulation.

1 Introduction

Many computational science and engineering problems require the numerical solution of large, sparse nonlinear systems of equations. Several classes of approaches are available, including Newton type methods, multigrid type methods, and continuation type methods. However, for some difficult problems, such as incompressible flows with high Reynolds number (Re), none of the methods works well, except the continuation methods, e.g. parameter continuation Gunzburger [1989] and pseudo time stepping Kelley and Keyes [1998], which are often too slow to be considered practical. In general, nonlinear iterative methods are fragile. They may converge rapidly for a well-selected set of parameters (for example, certain initial guesses, certain range of Re), but

* The work was partially supported by the Department of Energy, DE-FC02-01ER25479, and by the National Science Foundation, CCR-0219190, ACI-0072089 and ACI-0305666.

diverge if we slightly change some of the parameters. They may converge well at the beginning of the iterations, then suddenly stall for no apparent reason. In this paper we develop some techniques for detecting the bad behavior of Newton method, and focus on a class of nonlinear preconditioning methods that make Newton more robust; i.e., not too sensitive to some of the unfriendly parameters such as large *Re*. The preconditioner is constructed using the nonlinear additive Schwarz method, which not only increases the robustness of Newton, but also maintains the parallel scalability of the algorithm.

2 A brief review of inexact Newton method

Solving a nonlinear system of equations,

$$F(x) = 0, \tag{1}$$

using inexact Newton with backtracking (INB) Eisenstat and Walker [1996] can be described briefly as

$$x^{(k+1)} = x^{(k)} - \lambda^{(k)} s^{(k)},$$

where $\lambda^{(k)}$ is the step length computed using a linesearch technique, and $s^{(k)}$ is a good search direction if a non-zero $\lambda^{(k)}$ can be found. $s^{(k)}$ is computed, often from a *linearly preconditioned* Jacobian equation

$$M_k^{-1} J s^{(k)} = M_k^{-1} F(x^{(k)}),$$

where J is a Jacobian of F and M_k^{-1} is a linear preconditioner. It has been known for a long time that, even with global strategies, INB often stagnates for many problems. A recent study Tuminaro et al. [2002] shows that this is likely because the angle between the Newton direction and the steepest descent direction is too close to $\pi/2$. In this case, the Newton direction becomes only a weak descent direction. As a result, only extremely small steps can be accepted by linesearch. More precisely, let θ be the angle between $s^{(k)}$ and the negative gradient direction of $||F||$ at $x^{(k)}$. Then, according to Tuminaro et al. [2002], in the worst case,

$$\frac{1}{\kappa(J)} \le \cos(\theta) \le \frac{2}{\kappa(J)}, \tag{2}$$

where $\kappa(J)$ is the condition number of J. This means that the Newton direction can be nearly orthogonal to the gradient of $||F||$ when $\kappa(J)$ is large. In the incompressible Navier-Stokes equations, $\kappa(J)$ becomes very large when *Re* is high or when the mesh size is fine. Estimate (2) also suggests that sometimes solving the Jacobian system too accurately is not a good idea, even without considering the issue of computational cost. It might be better to stop the Jacobian iteration earlier. The following stopping conditions were suggested in Eisenstat and Walker [1996],

$$||F(x^{(k)}) - J s^{(k)}||_2 \le \eta_k ||F(x^{(k)})||_2$$

- Choice 0: η_k is a constant (not too small)
- Choice 1:
$$\eta_k = \frac{\left| ||F(x^{(k)})||_2 - ||F(x^{(k-1)}) - Js^{(k-1)}||_2 \right|}{||F(x^{(k-1)})||_2}$$
- Choice 2:
$$\eta_k = \gamma \left(\frac{||F(x^{(k)})||_2}{||F(x^{(k-1)})||_2} \right)^{\alpha}, \gamma \in [0,1], \alpha \in (1,2].$$

INB with these forcing terms is more robust, but is still not enough to solve the Navier-Stokes equations for a large range of *Re* because the parameters in the "choices" are too problem-dependent Shadid et al. [1997]. A closer look at (2) and its proof in Tuminaro et al. [2002] shows that the linear preconditioner M_k^{-1} does not appear in the estimate (2), which means that even though the linear preconditioning may speed up the solution algorithm for the Jacobian system, it does not help improve the quality of the search direction. Therefore, to enhance the robustness of Newton method by finding a better search direction we believe that the preconditioner has to be *nonlinear*. An alternative approach to improve the quality of the search direction is based on the affine invariant Newton methods Deuflhard [1991] using the natural monotonicity test for highly nonlinear systems.

3 Nonlinear additive Schwarz preconditioning

This section describes a nonlinearly preconditioned inexact Newton algorithm (ASPIN) Cai and Keyes [2002], Hwang and Cai [2003]. Suppose that $F(x) = 0$ is a nonlinear system of equations arising from a finite element discretization. The finite element mesh on Ω is partitioned into non-overlapping subdomains Ω_i, $i = 1, \ldots, N$, then, each subdomain is extended into a larger overlapping subdomain Ω_i'. Let R_i be a restriction operator on Ω_i', we define the subdomain nonlinear function

$$F_i = R_i F.$$

For any given $x \in R^n$, $T_i(x)$ is defined as the solution of the subspace nonlinear systems,

$$F_i\left(x - R_i^T T_i(x)\right) = 0, \text{ for } = 1, ..., N. \tag{3}$$

Using the subdomain functions, we introduce a new global nonlinear system

$$\mathcal{F}(x) = \sum_{i=1}^{N} R_i^T T_i(x) = 0, \tag{4}$$

which we refer to as the nonlinear additive Schwarz preconditioned system. Then, ASPIN algorithm is defined as: find a solution of (1) by solving

$$\mathcal{F}(x) = 0,$$

with INB, starting with an initial guess $x^{(0)}$. As shown in Cai and Keyes [2002], Hwang and Cai [2003], an approximation of the Jacobian of $\mathcal{F}$ takes the form $\sum_{i=1}^{N} J_i^{-1} J$. Through nonlinear preconditioning, we have:

- an improved angle estimate

$$\frac{1}{\kappa(\sum_{i=1}^{N} J_i^{-1} J)} \leq \cos(\theta) \leq \frac{2}{\kappa(\sum_{i=1}^{N} J_i^{-1} J)}; \text{ and}$$

- an improved conditioning of the Jacobian system

$$\left(\sum_{i=1}^{N} J_i^{-1} J \right) s^{(k)} = \mathcal{F}(x^{(k)}); \text{ and}$$

- an improved merit function $\|\mathcal{F}\|^2/2$ for the linesearch.

4 Stabilized finite element method for incompressible Navier-Stokes equations in the primitive variable

Consider two-dimensional steady-state incompressible Navier-Stokes equations in the primitive variable form Gunzburger [1989], Reddy and Gartling [2000]:

$$\begin{cases} \boldsymbol{u} \cdot \nabla \boldsymbol{u} - 2\nu \nabla \cdot \epsilon(\boldsymbol{u}) + \nabla p = 0 & \text{in} \quad \Omega, \\ \nabla \cdot \boldsymbol{u} = 0 & \text{in} \quad \Omega, \\ \boldsymbol{u} = \boldsymbol{g} & \text{on} \quad \Gamma, \end{cases} \tag{5}$$

where $\boldsymbol{u}$ is the velocity, p is the pressure, $\nu = 1/Re$ is the dynamic viscosity, and $\epsilon(\boldsymbol{u}) = 1/2(\nabla \boldsymbol{u} + (\nabla \boldsymbol{u})^T)$ is the symmetric part of the velocity gradient. The pressure p is determined up to a constant. To make p unique, we impose an additional condition $\int_\Omega p\, dx = 0$.

To discretize (5), we use a stabilized $Q_1 - Q_1$ finite element method (Franca and Frey [1992]). For simplicity, we consider only rectangular bilinear mesh $\mathcal{T}_h = \{K\}$. Let V^h and P^h be a pair of finite element spaces for the velocity and pressure, given by

$$V^h = \{\boldsymbol{v} \in (C^0(\Omega) \cap H^1(\Omega))^2 : \ \boldsymbol{v}|_K \in Q_1(K)^2, \ K \in \mathcal{T}_h \}$$
$$P^h = \{p \in C^0(\Omega) \cap L^2(\Omega) : \ p|_K \in Q_1(K), \ K \in \mathcal{T}_h\}.$$

The weighting and trial velocity function spaces V_0^h and V_g^h are

$$V_0^h = \{\boldsymbol{v} \in V^h : \boldsymbol{v} = 0 \text{ on } \Gamma\} \text{ and } V_g^h = \{\boldsymbol{v} \in V^h : \ \boldsymbol{v} = g \text{ on } \Gamma\}.$$

Similarly, let the finite element space P_0^h be both the weighting and trial pressure function spaces:

$$P_0^h = \left\{ p \in P^h : \int_\Omega p \, dx = 0 \right\}.$$

Following Franca and Frey [1992], the stabilized finite element method for steady-state incompressible Navier-Stokes equations reads: Find $\boldsymbol{u}^h \in V_g^h$ and $p^h \in P_0^h$, such that

$$B\left(\boldsymbol{u}^h, p^h; \boldsymbol{v}, q\right) = 0 \qquad \forall (\boldsymbol{v}, q) \in V_0^h \times P_0^h \tag{6}$$

with

$$B(\boldsymbol{u}, p; \boldsymbol{v}, q) = ((\nabla \boldsymbol{u}) \cdot \boldsymbol{u}, \boldsymbol{v}) + (2\nu\epsilon(\boldsymbol{u}), \epsilon(\boldsymbol{v})) - (\nabla \cdot \boldsymbol{v}, p) - (\nabla \cdot \boldsymbol{u}, q) + \sum_{K \in \mathcal{T}_h} ((\nabla \boldsymbol{u}) \cdot \boldsymbol{u} + \nabla p, \tau((\nabla \boldsymbol{v}) \cdot \boldsymbol{v} - \nabla q))_K + (\nabla \cdot \boldsymbol{u}, \delta \nabla \cdot \boldsymbol{v})$$

We use the stability parameters δ and τ suggested in Franca and Frey [1992]. The stabilized finite element formulation (6) can be written as a nonlinear algebraic system

$$F(x) = 0, \tag{7}$$

which is often large, sparse, and highly nonlinear when the value of Reynolds number is large. A vector x corresponds to the nodal values of $\boldsymbol{u}^h = (u_1^h, u_2^h)$ and p^h in (6). Now, we define the subdomain velocity space as

$$V_i^h = \left\{ v^h \in V^h \cap \left(H^1(\Omega_i')\right)^2 : v^h = 0 \text{ on } \partial\Omega_i' \right\}$$

and the subdomain pressure space as

$$P_i^h = \left\{ p^h \in P^h \cap L^2(\Omega_i') : p^h = 0 \text{ on } \partial\Omega_i' \backslash \Gamma \right\}.$$

Using these subspaces we can define subspace nonlinear problems as in (3). Note that, implicitly defined in the subspaces V_i^h and P_i^h, we impose Dirichlet conditions according to the original equations (5) on the physical boundaries, and on artificial boundaries, we assume both $\boldsymbol{u} = 0$ and $p = 0$. This is similar to the conditions used in Klawonn and Pavarino [1998].

5 Experimental results

To show the convergence properties of ASPIN and its robustness with respect to high Reynolds numbers, in this section we consider a lid-driven cavity flow problem described by (5) on the unit square. We also compare the results with those obtained using a standard Newton-Krylov-Schwarz algorithm Cai et al. [1998], which is here referred to as INB. GMRES is used for solving Jacobian systems. A zero initial guess is used for all test cases, and a constant nonlinear tolerance 10^{-6} is used for ASPIN and INB. Other parameters to be studied are described briefly as follows. Two meshes of size 64×64 and 128×128

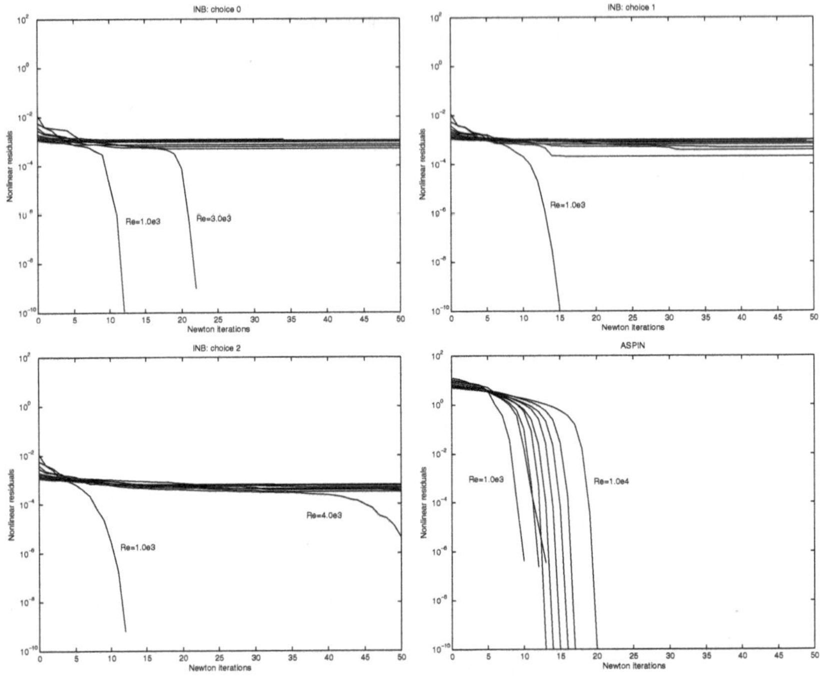

Fig. 1. Nonlinear residual curves of ASPIN and INB with three different forcing terms. *Re* ranges from 10^3 to 10^4

are considered. Reynolds numbers range from 10^3 to 10^4. The subdomains are obtained by partitioning the mesh uniformly into either a 2×2 or a 4×4 partition. The number of processors is the same as the number of subdomains. Our parallel software is developed using PETSc of Argonne National Laboratory Balay et al. [2002]. More implementation details and numerical results are available in Hwang and Cai [2003].

Figure 1 compares the nonlinear residual history of ASPIN with those of INB with three different choices of forcing terms as described in Section 2. Ten tests are run for Reynolds numbers ranging from 10^3 to 10^4, with an increment of 10^3. All results are obtained by on a 128×128 mesh using 16(=4×4) processors. We see that nonlinear residuals of INB with all choices of forcing terms behave similarly. Except for a few cases with low Reynolds numbers, INB nonlinear residuals stagnate around 10^{-3} without any progress after about the first 15 iterations. Different choices of forcing terms do not help much. On the other hand, ASPIN converges for the whole range of Reynolds numbers. Furthermore, ASPIN preserves the local quadratic convergence of Newton when the intermediate solution is near the desired solution.

To understand the robustness of ASPIN and INB, we next compare the minimum values of $\cos(\theta)$ for ASPIN and INB with different forcing terms in

Table 1. Comparison of the minimum values of $\cos(\theta)$ for ASPIN and INB

	$Re = 10^3$	$Re = 5 \times 10^3$	$Re = 10^4$
	Mesh size: 64×64		
Choice 0	1.68e-03	8.50e-12*	6.70e-11*
Choice 1	4.21e-03	6.22e-08*	1.09e-04*
Choice 2	4.80e-03	4.91e-05*	1.54e-04
ASPIN	7.37e-03	1.74e-03	1.82e-03
	Mesh size: 128×128		
Choice 0	8.65e-04	1.97e-07*	3.31e-07*
Choice 1	3.78e-03	3.30e-05*	1.82e-08*
Choice 2	3.33e-03	1.20e-04*	9.27e-05*
ASPIN	2.98e-03	2.94e-03	3.90e-03

Table 1. The values marked with asterisks in the table indicate that INB fails to converge either after 150 nonlinear iterations, or the backtracking step fails. For INB, the minimum value of $\cos(\theta)$ is tiny when INB fails. This agrees well with estimate (2), since $\kappa(J)$ is expected to be very large for this high Re. On the other hand, the minimum value of $\cos(\theta)$ for ASPIN is always away from zero and is not sensitive to the change of Re as well as the refinement of the mesh.

Scalability is an important issue in parallel computing and the issue becomes significant when we solve large scale problems with many processors. Table 2 shows that the number of ASPIN iterations does not change much, while the average number of GMRES iterations increases when the number of processors increases from 4 to 16 on a fixed 128×128 mesh. The increase of GMRES iteration numbers is not unexpected since we do not have a coarse space in the preconditioner. The number of GMRES iterations can be kept near a constant if a multilevel ASPIN is used Cai et al. [2002], Marcinkowski and Cai [2003].

Table 2. Varying the number of processors and the Reynolds number on a 128×128 mesh

np	$Re = 10^3$	$Re = 5 \times 10^3$	$Re = 10^4$
	ASPIN iterations		
$2 \times 2 = 4$	11	13	19
$4 \times 4 = 16$	14	13	20
	Average GMRES iterations		
$2 \times 2 = 4$	67	71	74
$4 \times 4 = 16$	128	132	140

References

S. Balay, K. Buschelman, W. Gropp, D. Kaushik, M. Knepley, L. McInnes, B. Smith, and H. Zhang. PETSc users manual. Technical Report ANL-95/11 - Revision 2.1.5, Argonne National Laboratory, 2002.

X.-C. Cai, W. Gropp, D. Keyes, R. Melvin, and D. Young. Parallel Newton-Krylov Schwarz algorithms for the transonic full potential equation. *SIAM J. Sci. Comput.*, 19:246–265, 1998.

X.-C. Cai and D. Keyes. Nonlinearly preconditioned inexact Newton algorithms. *SIAM J. Sci. Comput.*, 24:183–200, 2002.

X.-C. Cai, D. Keyes, and L. Marcinkowski. Nonlinear additive Schwarz preconditioners and applications in computational fluid dynamics. *Int. J. Numer. Meth. Fluids*, 40:1463–1470, 2002.

P. Deuflhard. Global inexact Newton methods for very large scale nonlinear problems. *IMPACT Comp. Sci. Eng.*, 3:366–393, 1991.

S. Eisenstat and H. Walker. Choosing the forcing terms in an inexact Newton method. *SIAM J. Sci. Comput.*, 17:16–32, 1996.

L. Franca and S. Frey. Stabilized finite element method: II. The incompressible Navier-Stokes equation. *Comput. Methods Appl. Mech. Engrg.*, 99:209–233, 1992.

M. Gunzburger. *Finite Element Methods for Viscous Incompressible Flows.* Academics Press, New York, 1989.

F.-N. Hwang and X.-C. Cai. A parallel nonlinear additive Schwarz preconditioned inexact Newton algorithm for incompressible Navier-Stokes equations. Technical report, 2003.

C. Kelley and D. Keyes. Convergence analysis of pseudo-transient continuation. *J. Sci. Comput.*, 35:508–523, 1998.

A. Klawonn and L. Pavarino. Overlapping Schwarz method for mixed linear elasticity and Stokes problems. *Comput. Methods Appl. Mech. Engrg.*, 165: 233–245, 1998.

L. Marcinkowski and X.-C. Cai. Parallel performance of some two-level ASPIN algorithms. In *this proceeding*, 2003.

J. Reddy and D. Gartling. *The Finite Element Method in Heat Transfer and Fluid Dynamics.* CRC Press, Florida, 2000.

J. Shadid, R. Tuminaro, and H. Walker. An inexact Newton method for fully coupled solution of the Navier-Stokes equations with heat transport. *J. Comput. Phys.*, 137:155–185, 1997.

R. Tuminaro, H. Walker, and J. Shadid. On backtracking failure in Newton-GMRES methods with a demonstration for the Navier-Stokes equations. *J. Comput. Phys.*, 180:549–558, 2002.

Iterative Substructuring Methods for Indoor Air Flow Simulation

Tobias Knopp[1], Gert Lube[1], Ralf Gritzki[2], and Markus Rösler[2]

[1] Georg-August-University of Göttingen, Math. Department, NAM
[2] Dresden University of Technology, Faculty of Mech. Engrg., TGA

Summary. The numerical simulation of turbulent indoor-air flows is performed using iterative substructuring methods. We present a framework for coupling eddy-viscosity turbulence models based on the non-stationary, incompressible, non-isothermal Navier-Stokes problem with non-isothermal near-wall models; this approach covers the k/ϵ model with an improved wall function concept. The iterative process requires the fast solution of linearized Navier-Stokes problems and of advection-diffusion-reaction problems. These subproblems are discretized using stabilized FEM together with a shock-capturing technique. For the linearized problems we apply an iterative substructuring technique which couples the subdomain problems via Robin-type transmission conditions. The method is applied to a benchmark problem, including comparison with experimental data by Tian and Karayiannis [2000] and to realistic ventilation problems.

1 A full-overlapping DDM for wall-bounded flows

Let $\Omega \subset \mathbf{R}^d$, $d = 2, 3$ be a bounded Lipschitz domain. As the basic mathematical model we consider the (non-dimensional) incompressible, non-isothermal Navier-Stokes equations with an eddy-viscosity model to be specified later and the Boussinesq approximation for buoyancy forces. We seek a velocity field $\mathbf{u}$, pressure p, and temperature θ as solutions of

$$\begin{aligned} \partial_t \mathbf{u} - \boldsymbol{\nabla} \cdot (2\nu_e \mathbb{S}(\mathbf{u})) + (\mathbf{u} \cdot \boldsymbol{\nabla})\mathbf{u} + \boldsymbol{\nabla} p &= -\beta\theta\mathbf{g} \\ \boldsymbol{\nabla} \cdot \mathbf{u} &= 0 \\ \partial_t \theta + (\mathbf{u} \cdot \boldsymbol{\nabla})\theta - \boldsymbol{\nabla} \cdot (a_e \boldsymbol{\nabla}\theta) &= \dot{q}^V c_p^{-1} \end{aligned} \qquad (1)$$

with $\mathbb{S}(\mathbf{u}) := \frac{1}{2}(\boldsymbol{\nabla}\mathbf{u} + \boldsymbol{\nabla}\mathbf{u}^T)$, isobaric volume expansion coefficient β, gravitational acceleration $\mathbf{g}$, volumetric heat source $\dot{q}^V$, and specific heat capacity (at constant pressure) c_p. Moreover, we introduce effective viscosities $\nu_e = \nu + \nu_t$ and $a_e = a + a_t$ with kinematic viscosity ν, turbulent viscosity ν_t, thermal diffusivity $a = \nu Pr^{-1}$ and turbulent thermal diffusivity $a_t = \nu_t Pr_t^{-1}$ with Prandtl numbers $Pr = 0.7$ and $Pr_t = 0.9$. Therein, the non-constant ν_t and

a_t are supposed to model turbulent effects and are considered in detail later. Depending on the sign of $\mathbf{u} \cdot \mathbf{n}$, the boundary $\partial\Omega$ is divided into wall zones $\Gamma_0 \equiv \Gamma_W$, inlet zones Γ_- and outlet zones Γ_+. We impose

$$\sigma(\mathbf{u},p)\mathbf{n} = \tau_n \mathbf{n} \quad \text{on } \Gamma_- \cup \Gamma_+ \,, \qquad \mathbf{u} = \mathbf{0} \quad \text{on } \Gamma_0 \tag{2}$$

with $\sigma(\mathbf{u},p) = 2\nu_e \mathbb{S}(\mathbf{u}) - p\mathbb{I}$. For θ we require

$$\theta = \theta_{in} \quad \text{on } \Gamma_- \,, \quad a_e \nabla\theta \cdot \mathbf{n} = 0 \quad \text{on } \Gamma_+ \,, \qquad \theta = \theta_w \quad \text{on } \Gamma_0. \tag{3}$$

In an outer loop, for the semidiscretization in time we apply the implicit Euler scheme which leads to a sequence of coupled non-linear problems to be solved from time step to time step. Denote $\tilde{\partial}_t \phi = (\phi - \phi^{old})/(\Delta t)$ the backward-difference quotient in time for a certain variable ϕ with time-step Δt.

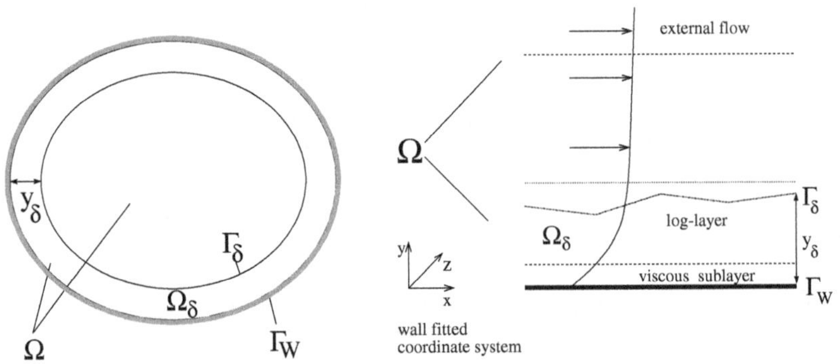

Fig. 1. Domain decomposition in the boundary layer region

Near Γ_W, the solutions for $\mathbf{u}$ and θ often exhibit strong gradients. As an illustration, Fig. 1 (right) shows the typical near-wall profile of the streamwise component of $\mathbf{u}$. The aim is to circumvent an anisotropic grid refinement in the near-wall region, which is computationally very expensive. For this purpose we study an overlapping domain-decomposition method which is presented in the sequel, see also Fig. 1 (left). For clarity of the presentation we assume $\partial\Omega = \Gamma_0 \equiv \Gamma_W$; for the general case we refer to Knopp et al. [2002] and Knopp [2003]. We start with the global problem with modified boundary conditions on Γ_W compared to (2), (3):

$$\begin{aligned}
\tilde{\partial}_t \mathbf{u} - \nabla \cdot (\nu_e \nabla \mathbf{u}) + (\mathbf{u} \cdot \nabla)\mathbf{u} + \nabla p &= -\beta\theta\mathbf{g} && \text{in } \Omega \\
\nabla \cdot \mathbf{u} &= 0 && \text{in } \Omega \\
\mathbf{u} \cdot \mathbf{n} = 0 \,, \qquad (\mathbb{I} - \mathbf{n} \otimes \mathbf{n})\sigma(\mathbf{u},p)\mathbf{n} &= \boldsymbol{\tau}_t(\mathbf{u}, \mathbf{u}^{BL}, \theta^{BL}) && \text{on } \Gamma_W \\
\tilde{\partial}_t \theta + (\mathbf{u} \cdot \nabla)\theta - \nabla \cdot (a_e \nabla\theta) &= \dot{q}^V c_p^{-1} && \text{in } \Omega \\
a_e \nabla\theta \cdot \mathbf{n} &= \dot{q}(\mathbf{u}^{BL}, \theta^{BL}) c_p^{-1} && \text{on } \Gamma_W
\end{aligned} \tag{4}$$

where the r.h.s. data $\boldsymbol{\tau}_t(\mathbf{u}, \mathbf{u}^{BL}, \theta^{BL})$, $\dot{q}(\mathbf{u}^{BL}, \theta^{BL})$ are determined from

$$\begin{aligned}
\tilde{\partial}_t \mathbf{u}^{BL} - \boldsymbol{\nabla} \cdot (\nu_e^{BL} \boldsymbol{\nabla} \mathbf{u}^{BL}) + (\mathbf{u}^{BL} \cdot \boldsymbol{\nabla}) \mathbf{u}^{BL} + \boldsymbol{\nabla} p^{BL} &= \mathbf{f} && \text{in } \Omega_\delta \\
\boldsymbol{\nabla} \cdot \mathbf{u}^{BL} &= 0 && \text{in } \Omega_\delta \\
\mathbf{u}^{BL} = \mathbf{0} \quad \text{on } \Gamma_W, \qquad \mathbf{u}^{BL} &= \mathbf{u} && \text{on } \Gamma_\delta \quad (5) \\
\tilde{\partial}_t \theta^{BL} + (\mathbf{u}^{BL} \cdot \boldsymbol{\nabla}) \theta^{BL} - \boldsymbol{\nabla} \cdot (a_e^{BL} \boldsymbol{\nabla} \theta^{BL}) &= \dot{q}^V c_p^{-1} && \text{in } \Omega_\delta \\
\theta^{BL} = \theta_w \quad \text{on } \Gamma_W, \qquad \theta^{BL} &= \theta && \text{on } \Gamma_\delta .
\end{aligned}$$

Now we specify ν_t, $\boldsymbol{\tau}_t(\mathbf{u}, \mathbf{u}^{BL}, \theta^{BL})$, $\dot{q}(\mathbf{u}^{BL}, \theta^{BL})$ in (4), and we modify (5).
(I) Global turbulence model in Ω: In (4), as a particular but successful choice for indoor-air flow simulation, we apply the k/ϵ model for ν_t (see, e.g., Codina and Soto [1999]) using the formula $\nu_t = c_\mu k^2 \epsilon^{-1}$ ($c_\mu = 0.09$) with turbulent kinetic energy k and turbulent dissipation ϵ being the solution of

$$\begin{aligned}
\tilde{\partial}_t k + (\mathbf{u} \cdot \boldsymbol{\nabla}) k - \boldsymbol{\nabla} \cdot (\nu_k \boldsymbol{\nabla} k) &= P_k + G - \epsilon \qquad (6) \\
\tilde{\partial}_t \epsilon + (\mathbf{u} \cdot \boldsymbol{\nabla}) \epsilon - \boldsymbol{\nabla} \cdot (\nu_\epsilon \boldsymbol{\nabla} \epsilon) + C_2 \epsilon^2 k^{-1} &= C_1 \epsilon k^{-1} (P_k + G)
\end{aligned}$$

with constants $C_1 = 1.44, C_2 = 1.92, Pr_k = 1.0, Pr_\epsilon = 1.3$, effective viscosities $\nu_k = \nu + \nu_t Pr_k^{-1}, \nu_\epsilon = \nu + \nu_t Pr_\epsilon^{-1}$, production and buoyancy terms

$$P_k := 2\nu_t |\mathbb{S}(\mathbf{u})|^2, \quad G := C_t \beta a_t \mathbf{g} \cdot \boldsymbol{\nabla} \theta \ , \quad C_t = 0.8.$$

The k/ϵ-equations (6) are solved in $\Omega \backslash \Omega_\delta$ with the following boundary conditions (with $\kappa = 0.41$ and $U_* = |\boldsymbol{\tau}_t|^{1/2}$)

$$k = c_\mu^{-1/2} U_*^2, \qquad \epsilon = U_*^3 / (\kappa y) \quad \text{on } \Gamma_\delta .$$

Alternatively to (6), we can use an eddy-viscosity-based LES model for ν_t in Ω, e.g., the non-isothermal *Smagorinsky model* with Eidson's modification

$$\nu_t = (C_S \Delta)^2 \left(\max\{ \, 0 \, ; \, ||\mathbb{S}(\mathbf{u})||_F^2 + \frac{\beta}{Pr_t} \mathbf{g} \cdot \boldsymbol{\nabla} \theta \} \right)^{1/2} , \; a_t = \frac{\nu_t}{Pr_q}$$

with $C_S = 0.21$ and $Pr_q = 0.04$.
(II) Boundary layer model in Ω_δ: Denote x, y, z the streamwise, wall-normal and spanwise direction resp. in a wall-fitted coordinate system, see Fig. 1 (right). We simplify (5) in Ω_δ under standard assumptions in Prandtl's boundary layer theory (cf. Knopp [2003]) and using modified effective viscosities in Ω_δ

$$\nu_e^{BL} = \nu \max \left(1; \frac{Re}{Re_{min}} \right), \quad a_e^{BL} = \frac{\nu}{Pr} \max \left(1; \frac{Pr}{Pr_t^{BL}} \frac{Re}{Re_{min}} \right) \qquad (7)$$

with $Re(x, y, z) = |\mathbf{u}^{BL}(x, y, z)| y / \nu$, $Pr_t^{BL} = 1.16$ and with the following empirical formula which accounts for effects of thermal stratification in the boundary layer, see Knopp [2003] and references therein, viz.,

$$Re_{min} = R_0 \min[\exp(-K_s \dot{q} Pr \nu U_*^{-4} \mathbf{g}\cdot\mathbf{n}); 70], \; R_0 = 20.0, \; K_s = 25.0. \quad (8)$$

Then, instead of a set of partial differential equations (5), in Ω_δ we solve

$$\begin{aligned} -\frac{d}{dy}\left(\nu_e^{BL}\frac{du_x^{BL}}{dy}\right) &= -\beta\theta^{BL} g_x, \\ -\frac{d}{dy}\left(a_e^{BL}\frac{d\theta^{BL}}{dy}\right) &= 0, \\ u_x^{BL}|_{y=0} = 0\,, \qquad \theta^{BL}|_{y=0} &= \theta_w, \end{aligned} \quad (9)$$

with g_x being the streamwise component of $\mathbf{g}$ and matching conditions

$$u_x^{BL}|_{y=y_\delta} = u_x(y_\delta), \qquad \theta^{BL}|_{y=y_\delta} = \theta(y_\delta). \quad (10)$$

Now we decouple and linearize the model (I), (II) within each time step:

(A) First update ν_t, a_t. Then update $\boldsymbol{\tau}_t$, $\dot{q}$: Given u_x, θ on Γ_δ from the previous iteration cycle, we replace the boundary condition (10) with

$$\nu_e \frac{du_x^{BL}}{dy}|_{y=0} = R, \qquad a_e \frac{d\theta^{BL}}{dy}|_{y=0} = S. \quad (11)$$

and solve the initial value problem (7),(8),(9),(11) using a shooting method for (R,S) until the conditions (10) are fulfilled. Then we find the r.h.s. $\boldsymbol{\tau}_t = -U_*^2\mathbf{u}/||\mathbf{u}||$ and $\dot{q}$ in (4) by setting $U_*^2 = R$ and $\dot{q} = c_p S$.

(B) We solve (4) and, if the k/ϵ model is used for ν_t, additionally (6), using a block Gauss-Seidel method.

(C) If a certain stopping-criterion is not yet fulfilled, then goto step (A). Otherwise goto next time step.

Step (B) requires the solution of two basic problems. First, the linearized equations for θ, k and ϵ are *advection-diffusion-reaction* (ADR) problems with non-constant viscosity of the general form (skipping the restriction $\partial\Omega = \Gamma_0$):

$$\begin{aligned} Lu \equiv -\nabla\cdot(\nu\nabla u) + (\mathbf{b}\cdot\nabla)u + cu &= f && \text{in } \tilde{\Omega} \\ u &= g && \text{on } \tilde{\Gamma}_D \\ \nu\nabla u\cdot\mathbf{n} &= h && \text{on } \tilde{\Gamma}_N. \end{aligned} \quad (12)$$

Secondly, the linearized Navier-Stokes equations are of *Oseen*-type with a positive reaction term and non-constant viscosity:

$$\begin{aligned} L_O(\mathbf{a},\mathbf{u},p) \equiv -\nabla\cdot(2\nu\mathbb{S}(\mathbf{u})) + (\mathbf{a}\cdot\nabla)\mathbf{u} + c\mathbf{u} + \nabla p &= \mathbf{f} && \text{in } \Omega \\ \nabla\cdot\mathbf{u} &= 0 && \text{in } \Omega \\ \sigma(\mathbf{u},p)\mathbf{n} &= \tau_n\mathbf{n} && \text{on } \Gamma_-\cup\Gamma_+ \\ (\mathbb{I}-\mathbf{n}\otimes\mathbf{n})\sigma(\mathbf{u},p)\mathbf{n} = \boldsymbol{\tau}_t, \qquad \mathbf{u}\cdot\mathbf{n} &= 0 && \text{on } \Gamma_0. \end{aligned} \quad (13)$$

For the finite element discretization of (12)-(13) we assume an admissible triangulation $\mathcal{T}_h = \{K\}$ of Ω and define discrete subspaces $X_h^l \equiv \{v \in C(\overline{\Omega}) \mid v|_K \in \Pi_l(K) \ \forall K \in \mathcal{T}_h\}$, $l \in \mathbf{N}$.
For the ADR-problem (12), for simplicity with $g = 0$ on Γ_D, we apply the Galerkin-FEM with SUPG-stabilization:

$$\text{Find } u \in V_h = \{v \in X_h^l \mid v|_{\Gamma_D} = 0\} \text{ s.t.} : \ b^s(u,v) = l^s(v) \quad \forall v \in V_h \,, \tag{14}$$

$$b^s(u,v) = (\nu \nabla u, \nabla v)_\Omega + ((\mathbf{b}\cdot\nabla)u + cu, v)_\Omega + \sum_{T\in\mathcal{T}_h} (\delta_T L u, (\mathbf{b}\cdot\nabla)v)_T$$
$$l^s(v) = (f,v)_\Omega \ + \ (h,v)_{\Gamma_N} + \sum_{T\in\mathcal{T}_h} (\delta_T f, (\mathbf{b}\cdot\nabla)v))_T$$

where $(\cdot,\cdot)_S$ denotes the inner product on some S and with an appropriate parameter set $\{\delta_T\}_T$, see Knopp et al. [2002]. Additionally, we use a (nonlinear) shock-capturing method, see Knopp et al. [2002].
For the Oseen problem (13), we define the discrete spaces $\mathbf{V}_h \times Q_h = (X_h^r)^d \times X_h^s$ with $r, s \in \mathbf{N}$. The Galerkin FEM reads:

$$\text{Find } U = (\mathbf{u},p) \in \mathbf{V}_h \times Q_h, \text{ s.t. } \mathcal{A}(U,V) = \mathcal{L}(V) \ \forall\, V = (\mathbf{v},q) \in \mathbf{V}_h \times Q_h \tag{15}$$

with the (bi)linear forms

$$\mathcal{A}(U,V) = a(\mathbf{u},\mathbf{v}) + b(\mathbf{v},p) - b(\mathbf{u},q)\,, \qquad b(\mathbf{v},p) = -(p, \nabla\cdot\mathbf{v}),$$
$$a(\mathbf{u},\mathbf{v}) = (2\nu\mathbb{S}(\mathbf{u}), \nabla\mathbf{v})_\Omega + ((\mathbf{a}\cdot\nabla)\mathbf{u} + c\mathbf{u}, \mathbf{v})_\Omega - (\mathbf{n}\otimes\mathbf{n}\sigma(\mathbf{u},p)\mathbf{n}, \mathbf{v})_{\Gamma_0}$$
$$\mathcal{L}(V) = (\mathbf{f},\mathbf{v})_\Omega + (\tau_n \mathbf{n}, \mathbf{v})_{\Gamma_- \cup \Gamma_+} + (\boldsymbol{\tau}_t, \mathbf{v})_{\Gamma_0}.$$

Here we use an equal-order ansatz in $\mathbf{V}_h \times Q_h$ ($r = s = 1$); thus the discrete inf-sup condition is not satisfied. As a remedy we apply a pressure stabilization (PSPG) together with divergence and SUPG stabilizations, cf. Knopp et al. [2002].

2 Domain decomposition of the linearized problems

A nonoverlapping domain decomposition method with Robin interface conditions is applied to the basic linearized problems (12), (13). Consider a nonoverlapping partition of Ω (which, for simplicity, is assumed to be stripwise) into convex, polyhedral subdomains being aligned with the FE mesh, i.e.

$$\overline{\Omega} = \cup_{k=1}^N \overline{\Omega}_k, \quad \Omega_k \cap \Omega_j = \emptyset \quad \forall k \neq j\,, \quad \forall K \in \mathcal{T}_h \ \exists k \ : \ K \subset \Omega_k.$$

Moreover, we set $\Gamma_{jk} := \partial\Omega_j \cap \partial\Omega_k$, $j \neq k$, with $\Gamma_{kj} \equiv \Gamma_{jk}$.
For the (continuous) ADR-problem (12) the DDM reads: for given u_k^n from iteration step n on each Ω_k, seek (in parallel) for u_k^{n+1}

$$\begin{aligned} Lu_k^{n+1} &= f && \text{in } \Omega_k \\ u_k^{n+1} &= 0 && \text{on } \Gamma_D \cap \partial\Omega_k \\ \nu \boldsymbol{\nabla} u_k^{n+1} \cdot \mathbf{n}_k &= h && \text{on } \Gamma_N \cap \partial\Omega_k \end{aligned} \tag{16}$$

together with the interface conditions (with a relaxation parameter $\theta \in (0,1]$)

$$\begin{aligned} &\Phi_k(u_k^{n+1}) = \theta\Phi_k(u_j^n) + (1-\theta)\Phi_k(u_k^n) \quad \text{on } \Gamma_{jk},\ j = 1,\dots,N,\ j \neq k \\ &\Phi_k(u) = \nu\boldsymbol{\nabla} u \cdot \mathbf{n}_k + (-\frac{1}{2}\mathbf{b}\cdot\mathbf{n}_k + z_k)u. \end{aligned} \tag{17}$$

Let $V_{k,h}$, b_k^s, and l_k^s denote the restrictions of V_h, b^s, and l^s to a subdomain Ω_k. Moreover, $W_{kj,h}$ is the restriction of V_h to the interface part Γ_{kj}. The inner product in $L^2(\Gamma_{kj})$ or, whenever needed, the dual product in $(W_{kj,h})^* \times W_{kj,h}$ is denoted by $\langle\cdot,\cdot\rangle_{\Gamma_{kj}}$.

The *fully discretized* DDM reads for $k = 1,\dots,N$ and given u_k^n, Λ_{jk}^n:

Parallel computation step: find $u_k^{n+1} \in V_{k,h}$ s.t. $\forall\, v_k \in V_{k,h}$

$$b_k^s(u_k^{n+1}, v_k) + \langle(-\frac{1}{2}\mathbf{b}\cdot\mathbf{n}_k + z_k)u_k^{n+1}, v_k\rangle_{\Gamma_{kj}} = l_k^s(v_k) + \sum_{j(\neq k)} \langle \Lambda_{jk}^n, v_k\rangle_{\Gamma_{kj}}.$$

Communication step: for all $j \neq k$, update the Lagrangian multipliers

$$\langle \Lambda_{kj}^{n+1}, \phi\rangle_{\Gamma_{kj}} = \langle \theta(z_k + z_j)u_k^{n+1} - \theta\Lambda_{jk}^n + (1-\theta)\Lambda_{kj}^n, \phi\rangle_{\Gamma_{kj}} \quad \forall \phi \in W_{kj,h}.$$

In Knopp et al. [2002], the analysis of the method is resumed and the following design of the interface function is proposed (motivated by an *a-posteriori estimate*)

$$\begin{aligned} z_k &= \frac{1}{2}|\mathbf{b}\cdot\mathbf{n}_k| + R_k(H), \\ R_k(H) &\sim \frac{\nu_{min}}{H}\left[1 + H\sqrt{\frac{c_{max}}{\nu_{min}}} + \min\left(\frac{H\|\mathbf{b}\|_{max}}{\nu_{min}}; \frac{\|\mathbf{b}\|_{max}}{\sqrt{(\nu c)_{min}}}\right)\right], \end{aligned} \tag{18}$$

with H being the diameter of the interface. A further improvement is achieved with a multilevel type approach with appropriate change of $R_k(\cdot)$ corresponding to higher frequencies of the error, for details see Lube et al. [2003].

For the Oseen problem (13) we proceed similar to the method (16) for the ADR problem. We use the interface conditions

$$\Phi_k(\mathbf{u}_k^{n+1}, p_k^{n+1}) = \theta\Phi_k(\mathbf{u}_j^n, p_j^n) + (1-\theta)\Phi_k(\mathbf{u}_k^n, p_k^n) \quad \text{on } \Gamma_{jk}.$$

with relaxation parameter $\theta \in (0,1]$ and the interface function

$$\Phi_k(u,p) = \nu\boldsymbol{\nabla}\mathbf{u}\cdot\mathbf{n}_k - p\mathbf{n}_k + (-\frac{1}{2}\mathbf{a}\cdot\mathbf{n}_k + z_k)\mathbf{u} \tag{19}$$

with acceleration parameter z_k which has the same structure as in (18). Concerning the corresponding parallel algorithm (in weak form), its analysis and further details, we refer to Knopp et al. [2002] and references therein.

3 Application to Indoor Air Flow Simulation

The approach is applied to a standard benchmark test case for indoor-air flow simulation, viz., turbulent natural convection in an air-filled square cavity as sketched in Fig. 2 (left), using the research code *Parallel NS*. Let a tilde denote dimensional quantities. Denote $\tilde{\Omega} = (0, \tilde{H})^3$ with $\tilde{H} = 0.75m$. We impose $\tilde{\theta}_w = 323.15K$ on Γ_h and $\tilde{\theta}_w = 283.15K$ on Γ_c. On $\Gamma_b \cup \Gamma_t$, alternatively, *(i)* we impose θ_w using the experimental data given in Tian and Karayiannis [2000] or *(ii)* we simply require that $a_e \nabla\theta \cdot \mathbf{n} = 0$. Moreover, we have $\tilde{\nu} = 1.53 \times 10^{-5} m^2 s^{-1}$, $\tilde{\beta} = 3.192 \times 10^{-3} K^{-1}$, $\tilde{g} = 9.81 ms^{-2}$, thus giving a Rayleigh number $Ra = \tilde{g}\tilde{\beta}(\tilde{\theta}_h - \tilde{\theta}_c)\tilde{H}^3 Pr/\tilde{\nu} = 1.58 \times 10^9$. We used a structured mesh

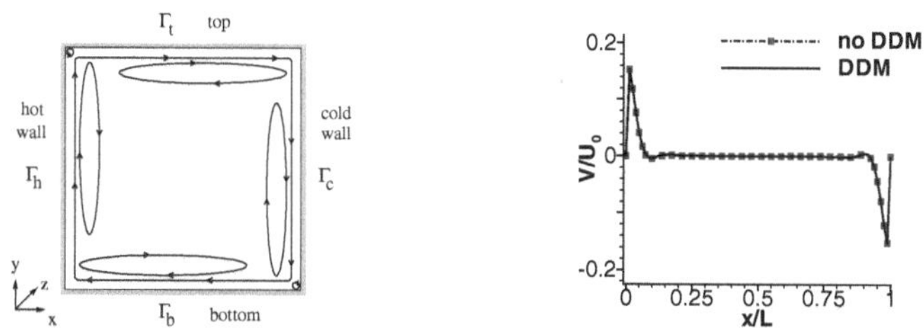

Fig. 2. Sketch of cavity and flow (left) and prediction for V/U_0 at $y/H = 0.5$ (right)

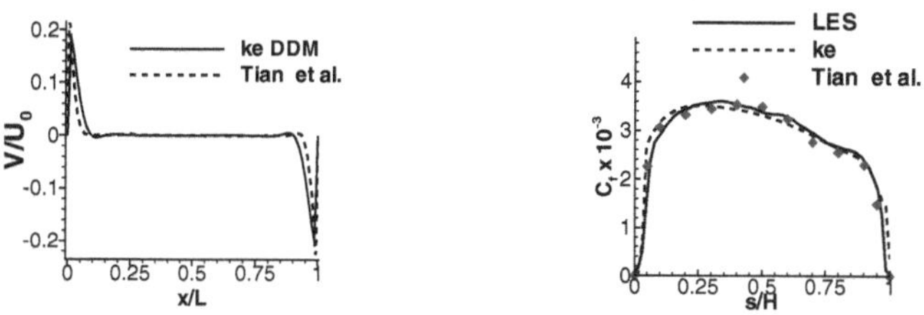

Fig. 3. V/U_0 at $y/H = 0.5$ (left) and $C_f = 2U_*^2/U_0^2$ (right) for variant (i)

with $81 \times 65 \times 29$ grid points being equidistantly distributed in each coordinate direction and we use $\Delta\tilde{t} = 1.0$ for the time step. Computations were performed on a cluster of 4 COMPAQ Professional Workstations XP1000 (667 MHz) connected by Ethernet. Parallelization is accomplished using a master/slave paradigm in the PVM configuration. No coarse-grid solver is used so far.
First, the agreement of the solution with DDM (using a coarse-granular $2 \times 2 \times 1$ partition of Ω) and without DDM (for variant *(ii)*) is obvious, see Fig. 2 (right). Therein, V denotes the streamwise component of $\mathbf{u}$ and $\tilde{U}_0 = 0.9692$. The parallel speed-up achieved was 3.7. The accuracy of the approach (for variant *(i)*) is validated by reference to the experimental data by Tian and Karayiannis [2000]. Fig. 3 (left) shows the k/ϵ model prediction (with DDM) for V. Fig. 3 (right) gives the predictions for $C_f \equiv 2U_*^2/\tilde{U}_0$ on Γ_h with $s \equiv y$ (k/ϵ with DDM for a $2 \times 2 \times 1$ partition, LES model (7) without DDM).

The method is applied at Dresden University as an analysis tool for the design and investigation of natural ventilation systems, see Richter et al. [2003]. Note that for the simulation presented in Fig. 4, the DDM described in Sect. 2 is applied where one subdomain is used for the room and one for the surrounding air with the interface being located in the window.

Summarizing, in this paper we combined two DD strategies for turbulent flows, one for near-wall modelling and one for parallel computation of the linearized problems. For this approach, we demonstrated both the accuracy for a benchmark problem and the applicability to a real-life problem.

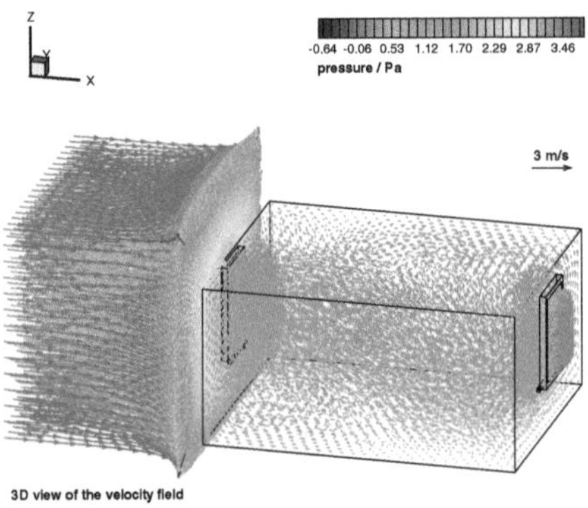

Fig. 4. Indoor-air flow simulation for natural building ventilation

References

R. Codina and O. Soto. Finite element implementations of two-equation and algebraic stress turbulence models for steady incompressible flow. *Intern. J. Numer. Meths. Fluids*, 90(3):309–334, 1999.

T. Knopp. *Finite-element simulation of buoyancy-driven turbulent flows*. PhD thesis, Universität Göttingen, 2003.

T. Knopp, G. Lube, R. Gritzki, and M. Rösler. Iterative substructuring techniques for incompressible nonisothermal flows and its application to indoor air flow simulation. *Intern. J. Numer. Meths. Fluids*, 40:1527–1538, 2002.

G. Lube, T. Knopp, and G. Rapin. Acceleration of an iterative substructuring method for singularly perturbed elliptic problems. Technical report, Universität Göttingen, 2003.

W. Richter, J. Seifert, R. Gritzki, and M. Rösler. Bestimmung des realen Luftwechsels bei Fensterlüftung aus energetischer und bauphysikalischer Sicht. Technical report, Dresden University of Technology, 2003.

Y. Tian and T. Karayiannis. Low turbulence natural convection in an air filled square cavity, part i: The thermal and fluid flow field. *Int. J. Heat Mass Transfer*, 43:849–866, 2000.

Fluid-Structure Interaction Using Nonconforming Finite Element Methods

Edward Swim and Padmanabhan Seshaiyer

Texas Tech University, Mathematics and Statistics (eswim@math.ttu.edu)

Summary. Direct numerical solution of the highly nonlinear equations governing even the most simplified models of fluid-structure interaction requires that both the flow field and the domain shape be determined as part of the solution since neither is known *a priori*. To accomplish this, previous algorithms have decoupled the solid and fluid mechanics, solving for each separately and converging iteratively to a solution which satisfies both. In this paper, we describe a nonconforming finite element method which solves the problem of interaction between a viscous incompressible fluid and a structure whose deformation defines the interface between the two simultaneously. A general methodology is described for the model 2D problem and the algorithm is validated computationally for a one-dimensional example.

1 Introduction

Many applications from engineering and biological sciences, such as blood flow through arteries, require detailed simulation of an interaction between a fluid and an elastic membrane surrounding it. However, meshes generated for the purpose of analyzing the two materials may be incompatible and the cost of producing matching grids may be prohibitive. Much work has been done to build efficient numerical schemes using nonconforming finite element methods (Seshaiyer [2003], Seshaiyer and Suri [2000b], Seshaiyer and Suri [2000a], Belgacem et al. [2000], and references therein.) Thus far, these methods have focused on model problems where the governing equations on each sub-domain are the same, *e.g.*, interaction between fluids (Chilton and Seshaiyer [2002]) or interaction between structures (Belgacem et al. [2003].)

Furthermore, it has been shown (Wan et al. [2003]) that even one-dimensional models can be useful in predicting important characteristics of blood flow despite their simplicity. Our purpose here is to present a nonconforming finite element method for fluid-structure interaction problems which allows for both mesh refinement and degree enhancement independently on each component. The methodology is described for a two-dimensional model problem and the algorithm is computationally validated for a one-dimensional model.

2 Governing equations

Let $\mathbf{x} = (x_1, x_2)$. We consider a rectangular domain Ω which is divided into two sub-domains, $\Omega_F(t)$ and $\Omega_S(t)$, at any time t, as illustrated in Figure 1. We assume that a viscous incompressible fluid occupies Ω_F while an elastic solid material occupies Ω_S. Initially, assume that $\Omega_F^0 \equiv \Omega_F(0) = I \times I$, where $I = [0, 1]$, and that $\Omega_S^0 \equiv \Omega_S(0) = [1, 2] \times I$. Let $\gamma(t, x_2)$ represent the interface between the two sub-domains.

We model the velocity $\mathbf{u} \in \mathbb{R}^2$ and pressure p of the fluid using the Navier-Stokes equations,

$$\sum_{i=1}^{2} \frac{\partial}{\partial x_i} \left[\mu_F \left(\frac{\partial u_j}{\partial x_i} + \frac{\partial u_i}{\partial x_j} \right) \right] - \frac{\partial p}{\partial x_j} = \rho_F \left(\frac{\partial u_j}{\partial t} + \nabla \cdot (u_j \mathbf{u}) - f_j \right), \quad (1)$$

$j = 1, 2$, $\forall \mathbf{x} \in \Omega_F(t)$, $t > 0$. Here, μ_F is the dynamic viscosity, ρ_F is the fluid density, and $\mathbf{f} = (f_1, f_2)$ is the applied force. Moreover, due to the incompressible nature of the fluid, the velocity must satisfy $\nabla \cdot \mathbf{u} = 0$. Additionally, we model the displacement $\mathbf{d}$ of the solid from its initial position at time $t = 0$ using the Navier-space equations,

$$\sum_{i=1}^{2} \frac{\partial}{\partial x_i} \left[\mu_S \left(\frac{\partial d_j}{\partial x_i} + \frac{\partial d_i}{\partial x_j} \right) \right] + \lambda \frac{\partial}{\partial x_j} (\nabla \cdot \mathbf{d}) = \rho_S \left(\frac{\partial^2 d_j}{\partial t^2} - g_j \right), \quad (2)$$

$j = 1, 2$, $\forall \mathbf{x} \in \Omega_S^0$. Here, λ and μ_S are the Lamè coefficients, ρ_S is the solid density, and $\mathbf{g} = (g_1, g_2)$ is the applied load on the structure. We impose homogeneous Dirichlet boundary conditions on $\partial\Omega_F \setminus \gamma$ and $\partial\Omega_S \setminus \gamma$ except on Γ_{DW}, the downwind boundary of Ω_F. On this boundary, we only assume that the velocity $\mathbf{u}$ is known. In particular, we let $\mathbf{u}|_{\Gamma_{DW}} = (0, \tilde{u})$.

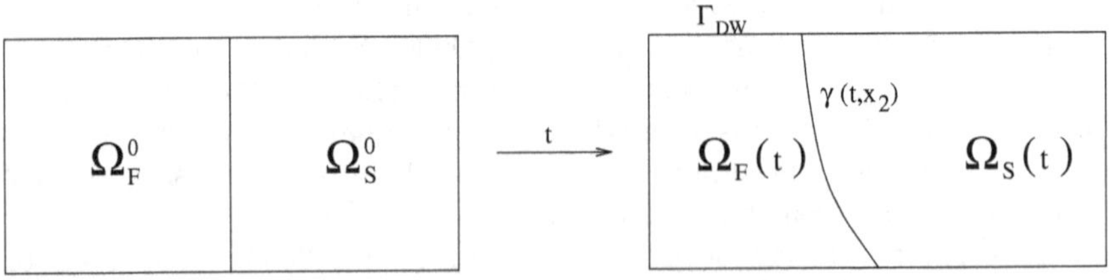

Fig. 1. Deformation of the fluid and solid sub-domains over time

Letting $\nu = \dfrac{\mu_F}{\rho_F}$, the kinematic viscosity of the fluid, and using incompressibility of the fluid, (1) reduces to

$$\frac{\partial u_j}{\partial t} - \nu \Delta u_j + \mathbf{u} \cdot \nabla u_j + \frac{\partial p}{\partial x_j} = f_j, \tag{3}$$

$j = 1, 2$, $\forall \mathbf{x} \in \Omega_F(t)$, $t > 0$. Similarly, if we let $\mu = \mu_S$ and $\varepsilon = \dfrac{\mu + \lambda}{\rho_S}$ then (2) becomes

$$\frac{\partial^2 d_j}{\partial t^2} - \mu \Delta d_j - \varepsilon \frac{\partial}{\partial x_j}(\nabla \cdot \mathbf{d}) = g_j. \tag{4}$$

$j = 1, 2$, $\forall \mathbf{x} \in \Omega_S^0$.

The weak formulation for (3) is to find u_j satisfying

$$\int_{\Omega_F(t)} (\partial_t u_j) w_F^{(j)} dA + \nu \int_{\Omega_F(t)} \nabla u_j \cdot \nabla w_F^{(j)} dA - \nu \int_{\partial\Omega_F(t)} \frac{\partial u_j}{\partial \mathbf{n}} w_F^{(j)} ds$$
$$+ \int_{\Omega_F(t)} (\mathbf{u} \cdot \nabla u_j) w_F^{(j)} dA + \int_{\Omega_F(t)} (\partial_{x_j} p) w_F^{(j)} dA = \int_{\Omega_F(t)} f_j w_F^{(j)} dA,$$

for appropriate $w_F^{(j)} \in H^1(\Omega_F(t))$, $j = 1, 2$. Thus,

$$a_F(\mathbf{u}, \mathbf{w_F}) + b_F(p, \mathbf{w_F}) + \tilde{c}(\mathbf{u}, \mathbf{w_F}) + B_F(\mathbf{\Lambda}_F, \mathbf{w_F}) = F_F(\mathbf{w_F}), \tag{5}$$

where

$$a_F(\mathbf{u}, \mathbf{w_F}) = \nu \int_{\Omega_F} \sum_{j=1}^{2} (\nabla u_j \cdot \nabla w_F^{(j)}) dA,$$

$$b_F(p, \mathbf{w_F}) = -\int_{\Omega_F} p (\nabla \cdot \mathbf{w_F}) dA,$$

$$\tilde{c}(\mathbf{u}, \mathbf{w_F}) = \int_{\Omega_F} \sum_{j=1}^{2} (\partial_t u_j + \mathbf{u} \cdot \nabla u_j) w_F^{(j)} dA,$$

$$B_F(\mathbf{\Lambda}_F, \mathbf{w_F}) = -\int_{\gamma} (\mathbf{\Lambda}_F \cdot \mathbf{w_F}) ds,$$

$$\mathbf{\Lambda}_F{}^j = \nu (\nabla u_j \cdot \mathbf{n}) - p n_j,$$

and

$$F_F(\mathbf{w_F}) = \int_{\Omega_F} \mathbf{f} \cdot \mathbf{w_F} dA + \int_{\Gamma_{DW}} \tilde{u} w_F^{(2)} ds.$$

Next, note that

$$b_F(v, \mathbf{u}) = 0 \qquad \forall v \in H^1(\Omega_F(t)). \tag{6}$$

And finally, $\forall w_S^{(j)} \in H^1(\Omega_S^0)$,

$$\int_{\Omega_S^0} (\partial_{tt} d_j) w_S^{(j)} dA + \mu \int_{\Omega_S^0} \nabla d_j \cdot \nabla w_S^{(j)} dA - \mu \int_{\partial\Omega_S^0} \frac{\partial d_j}{\partial \mathbf{n}} w_S^{(j)} ds$$
$$- \varepsilon \int_{\Omega_S^0} [\partial_{x_j} (\nabla \cdot \mathbf{d})] w_S^{(j)} dA = \int_{\Omega_S^0} g_j w_S^{(j)} dA,$$

$j = 1, 2$. Thus,

$$a_S(\mathbf{d}, \mathbf{w_s}) + c_S(\mathbf{d}, \mathbf{w_s}) + B_S(\mathbf{\Lambda}_S, \mathbf{w_s}) = F_S(\mathbf{w_s}), \tag{7}$$

where

$$a_S(\mathbf{d}, \mathbf{w_s}) = \int_{\Omega_S} \mu \sum_{j-1}^{2} (\nabla d_j \cdot \nabla w_S^{(j)}) + \varepsilon(\nabla \cdot \mathbf{d})(\nabla \cdot \mathbf{w_s}) dA,$$

$$c_S(\mathbf{d}, \mathbf{w_s}) = \int_{\Omega_S} (\partial_{tt}\mathbf{d}) \cdot \mathbf{w_s} dA,$$

$$B_S(\mathbf{\Lambda}_S, \mathbf{w_s}) = -\int_{\gamma} \mathbf{\Lambda}_S \cdot \mathbf{w_s} ds,$$

$$\mathbf{\Lambda}_S{}^j = \mu(\nabla d_j \cdot \mathbf{n}) + \varepsilon(\nabla \cdot \mathbf{d}) n_j,$$

and

$$F_S(\mathbf{w_s}) = \int_{\Omega_S} \mathbf{g} \cdot \mathbf{w_s} dA.$$

Now, on the interface

$$\gamma(t, x_2) = 1 + d_1(t, 1, x_2),$$

where $\mathbf{d} = (d_1, d_2)$, we wish to enforce continuity of the velocities,

$$\mathbf{u}(t, \gamma(t, x_2), x_2) = \frac{\partial \mathbf{d}}{\partial t}(t, 1, x_2),$$

and continuity of the flux,

$$(\mathbf{\Lambda}_F)|_{\gamma(t,x_2)} + (\mathbf{\Lambda}_S)|_{\gamma(0,x_2)} = \mathbf{0}.$$

To construct a two-field method, we set $\mathbf{\Lambda}_N = (\mathbf{\Lambda}_F)|_{\gamma(t,x_2)} = -(\mathbf{\Lambda}_S)|_{\gamma(0,x_2)}$. Then the global weak formulation is to find

$$(\mathbf{u}, p, \mathbf{d}, \mathbf{\Lambda}_N) \in X_N \equiv [H^1(\Omega_F)]^2 \times H^1(\Omega_F) \times [H^1(\Omega_S)]^2 \times [H^{-\frac{1}{2}}(\gamma)]^2$$

such that $\forall (\mathbf{w_F}, v, \mathbf{w_s}, \mathbf{\Psi}) \in X_N$,

$$a_F(\mathbf{u}, \mathbf{w_F}) + b_F(p, \mathbf{w_F}) + \tilde{c}(\mathbf{u}, \mathbf{w_F}) + B_F(\mathbf{\Lambda}_N, \mathbf{w_F}) = F_F(\mathbf{w_F}), \tag{8}$$

$$a_S(\mathbf{d}, \mathbf{w_s}) + c_S(\mathbf{d}, \mathbf{w_s}) - B_S(\mathbf{\Lambda}_N, \mathbf{w_s}) = F_S(\mathbf{w_s}), \tag{9}$$

$$b_F(v, \mathbf{u}) = 0, \tag{10}$$

and

$$G(\mathbf{u} - \frac{\partial}{\partial t}\mathbf{d}, \mathbf{\Psi}) = 0, \tag{11}$$

where

$$G(\mathbf{u}, \mathbf{\Psi}) = \int_{\gamma} \mathbf{u} \cdot \mathbf{\Psi} ds.$$

To solve a fully coupled fluid-structure interaction problem simultaneously for both $\mathbf{u}$ and $\mathbf{d}$, one must compute the solution to (8)-(11) and then extrapolate a piecewise linear approximation for the interface γ at each time step.

3 A one-dimensional model problem

Imposing (possibly nonconforming) meshes on Ω_F^0 and Ω_S^0 and assuming a piecewise linear interface $\gamma(t, x_2)$ corresponding to the mesh for $\Omega_F(t)$ as in Figure 2, the spatially discrete problem is closely related to a collection of one-dimensional problems of the following form.

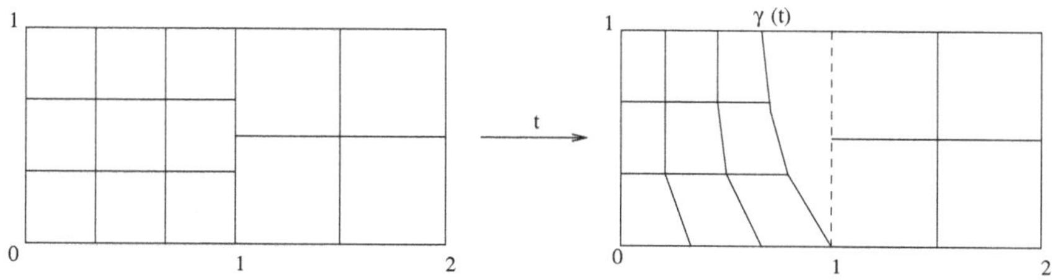

Fig. 2. Evolution of a nonconforming mesh for Ω

We consider a coupled system (Grandmont et al. [2001]) for a velocity u satisfying a modified Burgers' equation and displacement d satisfying a wave equation, *e.g.*,

$$\frac{\partial u}{\partial t} - \nu \frac{\partial^2 u}{\partial x^2} + \frac{3}{2} u \frac{\partial u}{\partial x} = f, \qquad x \in (0, \gamma(t)), \tag{12}$$

where $u(t, 0) = 0$, and

$$\frac{\partial^2 d}{\partial t^2} - \mu \frac{\partial^2 d}{\partial x^2} = g, \qquad x \in (1, 2), \tag{13}$$

where $d(t, 2) = 0$ and $d(0, x) = 0$, $x \in (1, 2)$, with interface $\gamma(t) = 1 + d(t, 1)$, as illustrated in Figure 3. Again, we enforce continuity of velocities, *i.e.*,

$$u(t, \gamma(t)) = \frac{\partial d}{\partial t}(t, 1), \tag{14}$$

and continuity of flux,

$$\nu \frac{\partial u}{\partial x}(t, \gamma(t)) = \mu \frac{\partial d}{\partial x}(t, 1). \tag{15}$$

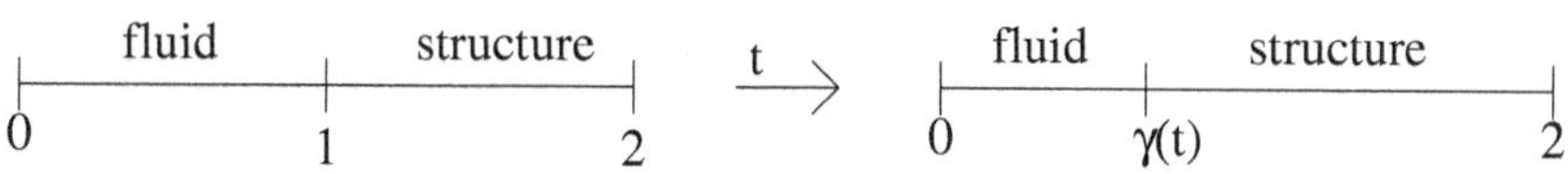

Fig. 3. Evolution of a one-dimensional domain

Employing an arbitrary Lagrangian-Eulerian (ALE) formulation for the fluid equation and an *implicit* formulation for the interface position, the discrete variational problem is to find $\bar{u}^{n+1}$, d^{n+1}, and Γ^{n+1} satisfying

$$\begin{aligned}(\bar{u}^{n+1}, \phi)_n & - \Delta t \nu \phi(\Gamma^n) \partial_x \bar{u}^{n+1}(\Gamma^n) + \Delta t \nu (\partial_x \bar{u}^{n+1}, \partial_x \phi)_n \\ & + \Delta t((u^n - w^n)\partial_x \bar{u}^{n+1}, \phi)_n + \tfrac{1}{2}\Delta t(\bar{u}^{n+1}\partial_x u^n, \phi)_n \\ & = (u^n, \phi)_n + \Delta t(f^{n+1}, \phi)_n,\end{aligned}$$

$$\begin{aligned}\frac{1}{\Delta t}(d^{n+1}, \psi) & + \Delta t \mu \psi(1) \partial_x d^{n+1}(1) + \Delta t \mu (\partial_x d^{n+1}, \partial_x \psi) \\ & = \frac{1}{\Delta t}(2d^n - d^{n-1}, \psi) + \Delta t(g^{n+1}, \psi),\end{aligned}$$

$$\bar{u}^{n+1}(\Gamma^n) = \frac{1}{\Delta t}(d^{n+1} - d^n)(1),$$

$$\nu \partial_x \bar{u}^{n+1}(\Gamma^n) = \mu \partial_x d^{n+1}(1),$$

and

$$\Gamma^{n+1} = 1 + D^{n+1}(1),$$

where $\bar{u}^{n+1}(x) = u^{n+1}(x + w^n(x)\Delta t)$ and $w^n(x) = \frac{x}{\Gamma^n} u^n(\Gamma^n)$ for $n = 1, 2, ...$, Γ^n is the time-discrete approximation of $\gamma(t^n)$, $(\,\cdot\,,\,\cdot\,)_n$ is the scalar product on $L^2(0, \Gamma^n)$, and $(\,\cdot\,,\,\cdot\,)$ is the scalar product on $L^2(1, 2)$.

4 Computational experiments

Using hierarchic basis functions (Szabo and Babuska [1991]), we construct finite element approximations $\bar{U}^n \approx \bar{u}^n$ and $D^n \approx d^n$ and solve the resulting linear system for $\bar{U}^{n+1}$, D^{n+1}, and Γ^{n+1}.

For our experiments, we let $\mu = 2$ and consider an exact solution for (12)-(15) given by

$$u(t, x) = \frac{-2\pi x^2 \cos(\pi t)}{(2 - \sin(\pi t))^2},$$

$$d(t, x) = \frac{1}{2} x(x - 2)\sin(\pi x t),$$

and

$$\gamma(t) = 1 - \frac{1}{2}\sin(\pi t),$$

where $\nu(t) = \frac{1}{2}\mu t(1 - \frac{1}{2}\sin(\pi t))$. Note that both u and d are nonlinear.

Assuming a uniform grid and linear basis functions, let M be the number of subintervals allowed on the interval $(0, \Gamma^n)$ and let N be the number of subintervals on $(1, 2)$. Figure 4 compares the relative error of our method whenever $M = N$ to the case $M = N - 1$ as the degrees of freedom increase.

The relative error under the L^2 norm is plotted against the total degrees of freedom. Panel (a) shows the results for the fluid velocity and panel (b) shows the results for the structural displacement. Furthermore, it was previously shown (Grandmont et al. [2001]) that the consistency error for both $u(t,x)$ and $d(t,x)$ is of order Δt. Table 1 illustrates this property for our test problem.

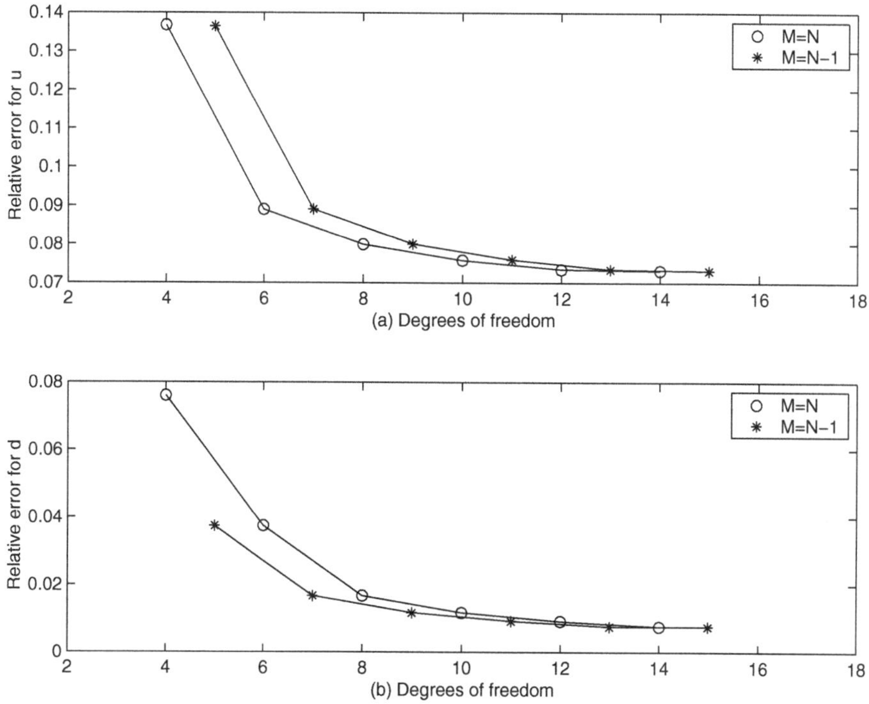

Fig. 4. Convergence of relative error on a uniform mesh

Table 1. Relative L^2 error for fluid velocity and structure displacement as Δt decreases

Δt	$\|\|u-U\|\|_{L^2(0,\Gamma^n)}/\|\|u\|\|_{L^2(0,\Gamma^n)}$	$\|\|d-D\|\|_{L^2(1,2)}/\|\|d\|\|_{L^2(1,2)}$
0.1	0.445914	0.169221
0.05	0.243088	0.070792
0.025	0.170837	0.031750
0.00625	0.133612	0.009507

In conclusion, our numerical results provide confidence that this method will be successful when extended to higher-dimensional problems. To do this, one only needs to employ the ALE formulation along a given line segment $\{(x_1, x_2) : 0 \leq x_1 \leq \gamma(t, x_2)\}$. We intend to present an implementation of this method for 2D problems in the *hp* context in future work.

Acknowledgement. This work was sponsored in part by the ARCS Foundation of Lubbock and by the National Science Foundation under grant DMS 0207327.

References

F. B. Belgacem, L. Chilton, and P. Seshaiyer. The *hp* mortar finite element method for the mixed elasticity and stokes problems. *Comp. Math. App.*, 46:35–55, 2003.

F. B. Belgacem, P. Seshaiyer, and M. Suri. Optimal convergence rates of *hp* mortar finite element methods for second-order elliptic problems. *RAIRO Math. Mod. Numer. Anal.*, 34:591–608, 2000.

L. Chilton and P. Seshaiyer. The *hp* mortar domain decomposition method for problems in fluid mechanics. *Int. Jour. Numer. Meth. Fluids*, 40:1561–1570, 2002.

C. Grandmont, V. Guimet, and Y. Maday. Numerical analysis of some decoupling techniques for the approximation of the unsteady fluid structure interaction. *Math. Mod. Meth. App. Sci.*, 11:1349–1377, 2001.

P. Seshaiyer. Stability and convergence of nonconforming *hp* finite-element methods. *Comp. Math. App.*, 46:165–182, 2003.

P. Seshaiyer and M. Suri. *hp* submeshing via non-conforming finite element methods. *Comp. Meth. Appl. Mech. Engrg.*, 189:1011–1030, 2000a.

P. Seshaiyer and M. Suri. Uniform *hp* convergence results for the mortar finite element method. *Math. Comp.*, 69:521–546, 2000b.

B. Szabo and I. Babuska. *Finite element analysis.* Wiley. New York, 1991.

J. Wan, B. Steele, S. Spicer, S. Strohband, G. Feijoo, T. Hughes, and C. Taylor. A one-dimensional finite element method for simulation-based medical planning for cardiovascular disease. *Comp. Meth. Biomech. Biomed. Engrg.*, 2003. Submitted.

Interaction Laws in Viscous-Inviscid Coupling

Arthur E. P. Veldman[1] and Edith G.M. Coenen[2]

[1] University of Groningen, Institute of Mathematics and Computing Science (http://www.math.rug.nl/~veldman/)
[2] Present address: KPN, Den Haag

Summary. Viscous-inviscid coupling methods for the computation of aerodynamic boundary layers are discussed, with emphasis on the quasi-simultaneous method. Its interaction law is analysed using matrix theory and reduced to its essentials. The redesigned interaction law is tested for separated airfoil flow at maximum lift.

1 Introduction

The accurate and fast prediction of viscous flow over two- and three-dimensional surfaces is an important problem in aero- and hydrodynamics. The continuing advances in efficiency of numerical algorithms, together with the increasing speed and memory size of computers, are enabling viscous flows to be calculated by methods that solve the full (Reynolds-averaged) Navier–Stokes equations. Whilst Navier–Stokes simulation potentially offers generality, its computational requirements currently limit its use for practical applications, especially within a design optimisation environment.

An alternative is to use the older technique of viscous-inviscid interaction (VII), where an inviscid-flow solver is coupled to a viscous boundary-layer calculation method (Figure 1). VII methods have shown to be very efficient and robust (Lock and Williams [1987]). For many cases of aerodynamic interest the coupled solution matches experimental data as well as Navier–Stokes simulation, and this at much lower computational cost.

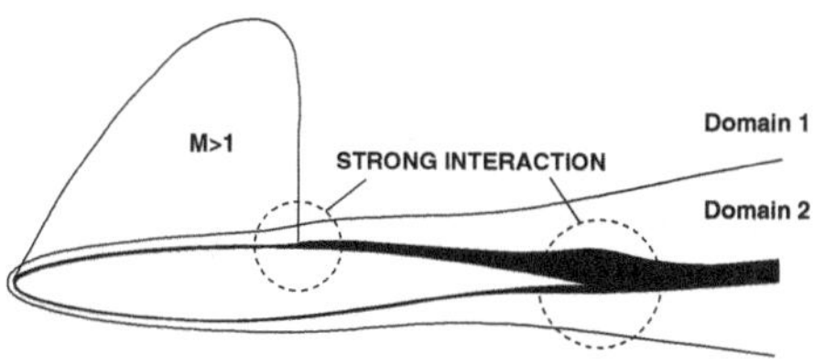

Fig. 1. Decomposition of flow field into boundary layer and inviscid flow

Since Prandtl [1905] introduced his boundary-layer concept in 1904, several VII methods have been developed. The oldest technique is known as the direct method, which solves the boundary layer with a prescribed velocity (pressure) distribution u_e. In turn, the boundary layer expresses its presence in terms of a virtually thickened profile, called displacement thickness δ^*.

For situations with attached flow the direct method works well. However, when regions of reversed flow are present the direct boundary-layer calculation breaks down. In 1948, Goldstein [1948] presented an extensive study of the breakdown, but no definitive conclusion about its origin could be given. Since then the singularity at flow separation bears his name.

It was not until 1966 that the first clue on how to prevent this singularity was given, when Catherall and Mangler [1966] presented calculations of a boundary layer with prescribed δ^*. They were able to pass the critical point of flow separation, but ran into numerical difficulties somewhat further on in the reversed-flow region. A further clue was provided by the asymptotic triple-deck theory introduced by Stewartson [1969], indicating a change in hierarchy between the inviscid-flow region and the boundary layer.

Led by these ideas, in the late 70's alternatives for the direct VII method were introduced. An obvious choice is to reverse the order of information exchange. This inverse method survives in flow separation, but its convergence is extremely slow. Henceforth, both methods were mixed into the semi-inverse method (LeBalleur [1978]), where both flow regions are solved with prescribed δ^*, which is then updated based on the difference in the respective u_e distributions. The latter update requires careful tuning of relaxation parameters.

Another idea was to avoid any hierarchy in the treatment of both flow regions. However, a fully simultaneous approach would require both sets of flow equations to be merged into one big system, which is quite complicated in software terms and defies any flexibility in flow modeling. Hence, an attempt was made to *approximate* such a simultaneous approach. Thus the concept of the interaction law came up: a simple, yet powerful, description of the inviscid flow, which can easily be solved simultaneously with the boundary-layer equations (Veldman [1981]). In this paper we will analyse this quasi-simultaneous approach and simplify the interaction law. We end up with a method which is very close to the direct method, and yet has no problems with reversed flow, as demonstrated with a calculation of airfoil flow.

2 Quasi-Simultaneous VII and the Interaction Law

In an abstract setting, the coupled VII problem can be written as

$$\left.\begin{array}{ll}\text{external inviscid flow:} & u_e = E\delta^* \\ \text{boundary-layer flow:} & u_e = -V\delta^*\end{array}\right\} \Rightarrow (V+E)\delta^* = 0\,. \qquad (1)$$

The classical way of solving these equations is to prescribe δ^* to the external inviscid flow, and then to return u_e to the boundary layer. As the boundary-layer step breaks down when flow separation occurs, the quasi-simultaneous method tries to avoid the iterative hierarchy involved. The basic idea is to inform the boundary layer instantaneously how the external flow will react on changes inside the boundary layer. Hereto, a sufficiently accurate, yet simple, approximation of the external inviscid flow is introduced, denoted as $u_e = I\delta^*$. This interaction law is to be solved simultaneously with the boundary-layer equations, i.e.

$$\left.\begin{aligned} u_e^{(n)} - I\delta^{*(n)} &= (E - I)\delta^{*(n-1)} \\ u_e^{(n)} + V\delta^{*(n)} &= 0 \end{aligned}\right\} \Rightarrow (V + I)\delta^{*(n)} = (I - E)\delta^{*(n-1)} \, . \quad (2)$$

Note that the VII iterations 'only' need to account for the difference between the external flow E and its approximation I.

Next the question arises how to choose the interaction law. A fair description of how an inviscid flow reacts on displacement effects is delivered by thin-airfoil theory, in its simplest form given by

$$u_e(x) = u_{e0}(x) + \frac{1}{\pi} \int_\Gamma \frac{\mathrm{d}\delta^*}{\mathrm{d}\xi} \frac{\mathrm{d}\xi}{x - \xi} \, , \quad (3)$$

where u_{e0} is the edge velocity without displacement effects. Also triple-deck theory provides this type of approximation, which makes (3) a good candidate as an interaction law. In fact, this interaction law (describing thickness effects) together with its anti-symmetric counterpart (describing camber effects) has been used successfully in subsonic and transonic(!) airfoil/wake calculations (Veldman et al. [1990]). As these thin-airfoil expressions are somewhat complicated it is worthwhile to try to simplify the interaction law, yet retaining a robust and efficient VII algorithm. Thus, the question to be answered is

How 'small' can I be chosen?

As a reminder, the direct choice $I = 0$ blows up in Goldstein's singularity. We will first address this question from a theoretical point of view. Thereafter, the usefulness of the theory will be demonstrated on realistic flow problems.

3 A Model Problem

As a model problem to shape the theory, the flow past an indented plate (Figure 2) is studied for which the external flow will be described by the thin-airfoil expression (3). It is our aim to construct a simple interaction law for this case. Let us first collect some properties of the operators E and V.

External Flow. The integral (3) is discretized on a uniform grid with mesh size h. The displacement thickness δ^* is interpolated by a piece-wise linear function; only on the two intervals adjacent to the Cauchy principal value a quadratic is used:

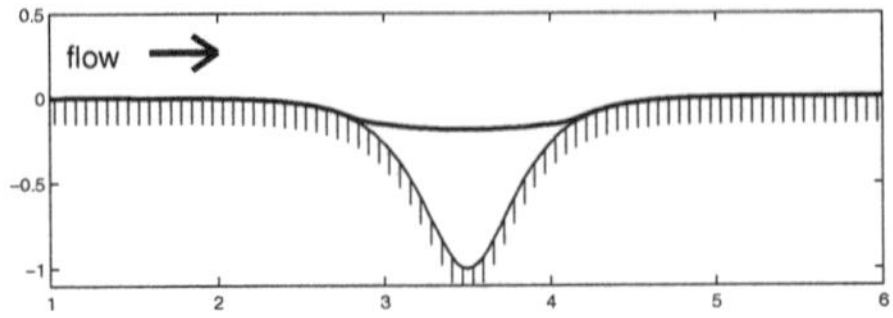

Fig. 2. Geometry sketch of indented plate geometry

$$\frac{1}{\pi}\int_{\Gamma}\frac{\mathrm{d}\delta^*}{\mathrm{d}\xi}\frac{\mathrm{d}\xi}{x_i-\xi}=\frac{1}{\pi}\left\{\int_{x_{i-1}}^{x_{i+1}}+\Sigma_{j\neq i-1,i}\frac{1}{\pi}\int_{x_j}^{x_{j+1}}\right\}\frac{\mathrm{d}\delta^*}{\mathrm{d}\xi}\frac{\mathrm{d}\xi}{x_i-\xi}$$
$$\approx -\frac{2h}{\pi}\left.\frac{\mathrm{d}^2\delta^*}{\mathrm{d}\xi^2}\right|_i+\frac{h}{\pi}\Sigma_{j\neq i-1,i}\left.\frac{\mathrm{d}\delta^*}{\mathrm{d}\xi}\right|_{j+1/2}\ln\left|\frac{i-j}{i-j-1}\right| .$$

The corresponding discrete matrix $\mathbf{E}$ is symmetric, positive definite with diagonal $4/\pi h$, and with non-positive off-diagonal entries.

Boundary-Layer Flow. Referring e.g. to Veldman [1984], a discrete boundary-layer operator typically is lower diagonal, with positive diagonal entries for attached flow and (slightly) negative diagonal entries for reversed flow.

4 Theory of Viscous-Inviscid Interaction

Throughout our mathematical analysis the following assumption is made:

> **Assumption** The matrix $\mathbf{V}+\mathbf{E}$ is assumed to be an M-matrix, i.e. it has positive diagonal entries, non-positive off-diagonal entries and all eigenvalues have positive real part.

The location of the eigenvalues corresponds with steady flow; the sign of the matrix entries allows theory, and will hold approximately in practice.

VII Iterations. An interaction law (2) corresponds with a splitting $\mathbf{V}+\mathbf{E}=(\mathbf{V}+\mathbf{I})-(\mathbf{I}-\mathbf{E})$. When $\mathbf{I}\geq\mathbf{E}$ and when also $\mathbf{V}+\mathbf{I}$ is an M-matrix, then $(\mathbf{V}+\mathbf{E})^{-1}\geq(\mathbf{V}+\mathbf{I})^{-1}\geq 0$ [Horn and Johnson, 1991, p. 117 & 127] and the splitting is regular [Varga, 1962, p. 88]. As the off-diagonals of $\mathbf{E}$ are non-positive, this suggests to construct $\mathbf{I}$ from $\mathbf{E}$ by dropping one or more off-diagonals. Subsequently, the comparison theorem on regular splittings [Varga, 1962, p. 90] implies that the convergence of the VII iterations deteriorates monotonously with the number of dropped off-diagonals in $\mathbf{I}$.

Boundary-Layer Iterations. In each VII iteration a boundary-layer computation has to be performed, in which (2) is to be solved. This is done by repeated marching through the boundary layer, starting near the stagnation point and proceeding in downstream direction. Thus, a Gauss–Seidel type of iteration is performed. This method 'only' has to iterate on the upper triangular part of the matrix $\mathbf{V}+\mathbf{I}$, which here consists of entries from $\mathbf{I}$. Hence it may be expected that a 'small' upper-diagonal part will speed up convergence. And, indeed, under the above assumption it can be proven that the Gauss-Seidel convergence improves monotonously with the number of dropped off-diagonals in **I** (Coenen [2001]). This is opposite to the behaviour of the VII iterations, hence a trade-off is opportune (see below). Further, it is remarked that an interaction law which only consists of a main diagonal does not require boundary-layer iterations.

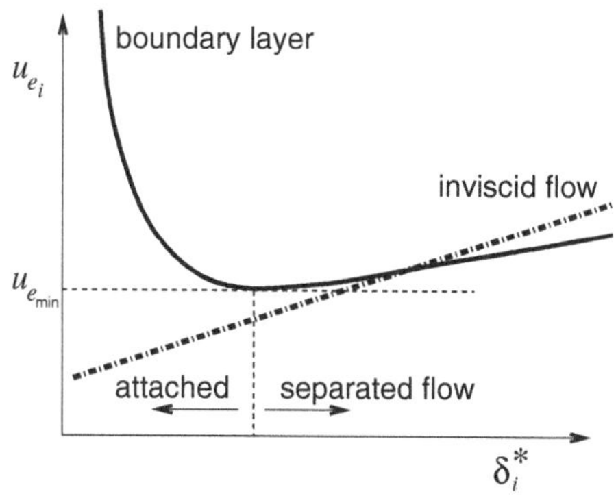

Fig. 3. A sketch of the boundary-layer behaviour, combined with an inviscid-flow relation

Robustness. The boundary-layer formulation is highly nonlinear. A sketch of the situation is given in Figure 3. Here, at a fixed boundary-layer station, the dependence between the edge velocity and the displacement thickness is shown (Veldman [1984]). This sketch shows clearly that below a certain value for a prescribed u_e no solution can be found anymore; this non-existence of a boundary-layer solution induces the breakdown! Prescribing a linear combination of u_e and δ^*, as is the case when an interaction law is applied, should be useful provided the coefficient of δ^* stays away from zero sufficiently. Also during the iterations the process should not break down, hence the eigenvalues of the interaction matrix **I**, and herewith the eigenvalues of $\mathbf{V}+\mathbf{I}$, should stay sufficiently far from the imaginary axis.

Again theory can be developed. With the above assumption, $\mathbf{V}+\mathbf{I}$ can be written as a constant positive diagonal matrix minus a non-negative matrix. For the latter type of matrices the largest eigenvalue grows monotonously with the matrix entries (the Perron-Frobenius theorem [Varga, 1962, p. 30]). Thus for $\mathbf{V}+\mathbf{I}$ this dependency holds for the smallest eigenvalue. With $\mathbf{I} \geq \mathbf{E}$, this eigenvalue is located in the stable (positive) half plane. Further, it grows with the number of dropped diagonals, herewith increasing the robustness of the boundary-layer calculation. Also in this respect an interaction law consisting of only the main diagonal of **E** scores best.

5 Viscous-Inviscid Interaction in Practice

Indented Plate

The applicability of the theory will first be investigated with the above indented plate model. The length of the domain is five units, with a dent up to one unit deep (which is very deep in comparison with the boundary-layer thickness, but gives a severe testcase for the VII algorithm). The Reynolds number based on unit length is 10^8. The boundary layer is modelled with Head's entrainment method; see Coenen [2001].

The interaction law **I** is chosen by simply dropping off-diagonals in the 'exact' inviscid flow matrix **E**. Figure 4 gives the number of VII iterations as a function of the number of retained diagonals (including the main diagonal). Three flow situations have been distinguished: one with attached flow (when the dent is very shallow), one with mild separation, and one with severe separation (as in Figure 2).

For all choices of **I** the VII iterations are found to converge. Moreover, according to theory, the convergence of the VII iterations improves monotonously with the number of diagonals retained in **I**. The limit number of iterations is 2-3. For attached flow this can be compared with the direct method which also requires 3 iterations to converge (in the separated-flow cases the direct method breaks down).

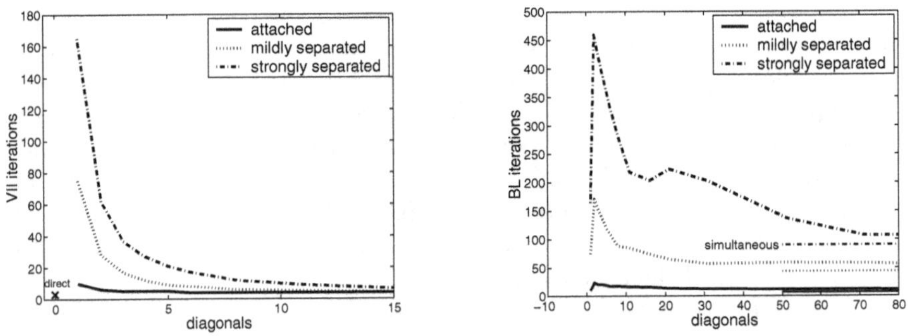

Fig. 4. Number of VII iterations (*left*) and total number of boundary-layer sweeps (*right*) as a function of the number of retained diagonals

Further, the number of VII iterations drops fast when the number of retained diagonals increases. On the other hand this leads to slower convergence of the boundary-layer iterations, and therefore the total number of boundary-layer sweeps has also been monitored in Figure 4. A local minimum exists when the interaction law only consists of the main diagonal; remember that in this case only one boundary-layer sweep per VII iteration is required. When one off-diagonal is added, the number of Gauss–Seidel sweeps per VII iteration increases, such that the total number of boundary-layer sweeps also increases, even though the number of VII iterations has decreased. Adding more off-diagonals the decrease in VII iterations becomes dominant. A minimum number of boundary-layer sweeps is found in the limit $\mathbf{I} = \mathbf{E}$. Here, the number of boundary-layer sweeps should be equal to the number required for a fully simultaneous treatment, i.e. when the system (1) is solved by Gauss–Seidel. Indeed, this is found to be the case.

Thus for the interaction law two interesting choices exist. One option is to choose it according to the 'full' external flow; the other is to choose it equal to only the diagonal $4/\pi h$ of the external-flow matrix. As the first option is against our quest for simplicity, below we will test the second option on a realistic problem of boundary-layer flow past a two-dimensional airfoil.

Subsonic Airfoil Flow

The above ideas on simplifying the interaction law will now be tested for aerodynamic flow past a NACA0012 airfoil (at $Re = 9 \cdot 10^6$, and $M_\infty = 0$); experimental data is available. The inviscid flow is modelled by potential theory, and computed by means of a panel method. Boundary layer plus wake are modelled with Head's entrainment method (for more information we refer to Coenen [2001]). They are solved together with the interaction law $\mathbf{I} = \text{diag}(4/\pi h)$. We stress that this interaction law is unaware of the Kutta condition and its effect on the global circulation; it only accounts for the local VII physics – but this turns out to be sufficient.

A large part of the lift polar has been computed. The calculations turn out to be highly robust. It appears that even for separated-flow cases beyond maximum lift the calculations converge without any need for a good initial guess; they can be started from scratch. The number of VII iterations typically is less than 100 at zero lift upto 1000 around maximum lift (computing times count in seconds on an average PC). Only for larger angles of attack the computations break down; using a good initial guess obtained at a slightly smaller angle does not help.

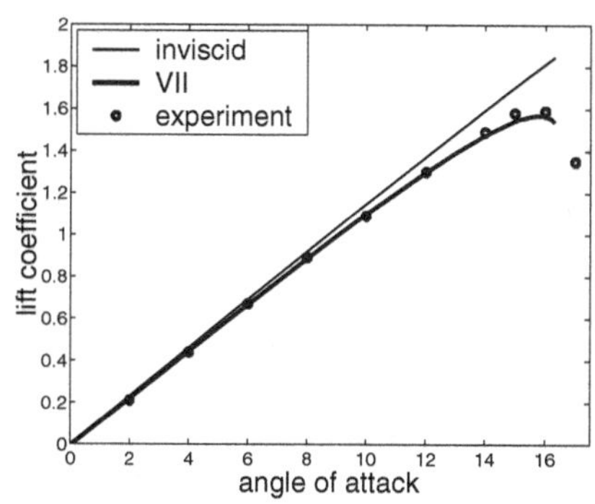

Fig. 5. Lift polar for NACA0012 airfoil: calculation versus experiment

6 Conclusion

A short overview of viscous-inviscid interaction has been presented, starting with Prandtl one century ago, and encountering Goldstein's analysis half a century later. Today, after a whole century of boundary-layer research, it has been found that Goldstein's singularity can be prevented by changing the 'classical' boundary condition of prescribed edge velocity into

$$\left(u_e + \frac{4}{\pi h}\,\delta^*\right)^{(\text{new})} = \left(u_e + \frac{4}{\pi h}\,\delta^*\right)^{(\text{old})} .$$

This slight change, unlikely to be further simplified, results in a highly robust calculation method, applicable to airfoil calculations beyond maximum lift.

References

D. Catherall and K. Mangler. The integration of the two-dimensional laminar boundary-layer equations past the point of vanishing skin friction. *J. Fluid Mech.*, 26:163–182, 1966.

E. Coenen. *Viscous-Inviscid Interaction with the Quasi-Simultaneous Method for 2D and 3D Airfoil Flow.* PhD thesis, Groningen, 2001.

S. Goldstein. On laminar boundary layer flow near a point of separation. *Quart. J. Mech. Appl. Math.*, 1:43–69, 1948.

R. Horn and C. Johnson. *Topics in Matrix Analysis.* CUP, 1991.

J. LeBalleur. Couplage visqueux-non visqueux: méthode numérique et applications aux écoulements bidimensionnels transsoniques et supersoniques. *La Recherche Aérospatiale*, 183:65–76, 1978.

R. Lock and B. Williams. Viscous-inviscid interactions in external aerodynamics. *Prog. Aerospace Science*, 24:51–171, 1987.

L. Prandtl. Ueber Fluessigkeitsbewegung mit kleiner Reibung. In *Verhandlungen des dritten internationalen Mathematischen Kongresses, Heidelberg*, pages 484–491, Leipzig, 1905. Teubner Verlag.

K. Stewartson. On the flow near the trailing edge of a flat plate II. *Mathematika*, 16:106–121, 1969.

R. Varga. *Matrix Iterative Analysis.* Prentice-Hall, 1962.

A. Veldman. New, quasi-simultaneous method to calculate interacting boundary layers. *AIAA J.*, 19:79–85, 1981.

A. Veldman. A numerical view on strong viscous-inviscid interaction. In W. Habashi, editor, *Computational Methods in Viscous Flows*, pages 343–363, Southampton, 1984. Pineridge Press.

A. Veldman, J. Lindhout, E. de Boer, and M. Somers. Vistrafs: a simulation method for strongly-interacting viscous transonic flow. In T. Cebeci, editor, *Numerical and Physical Aspects of Aerodynamic Flow IV*, pages 37–51, Berlin, 1990. Springer Verlag.

Part V

Minisymposium: Recent Developments for Schwarz Methods

Comparison of the Dirichlet-Neumann and Optimal Schwarz Method on the Sphere

J. Côté[1,*], M. J. Gander[2], L. Laayouni[3], and S. Loisel[4]

[1] Recherche en prévision numérique, Meteorological Service of Canada, jean.cote@ec.gc.ca
[2] Department of Mathematics and Statistics, McGill University, Montreal, mgander@math.mcgill.ca
[3] Department of Mathematics and Statistics, McGill University, Montreal, laayouni@math.mcgill.ca
[4] Department of Mathematics and Statistics, McGill University, Montreal, loisel@math.mcgill.ca

Summary. We investigate the performance of domain decomposition methods for solving the Poisson equation on the surface of the sphere. This equation arises in a global weather model as a consequence of an implicit time discretization. We consider two different types of algorithms: the Dirichlet-Neumann algorithm and the optimal Schwarz method. We show that both algorithms applied to a simple two subdomain decomposition of the surface of the sphere converge in two iterations. While the Dirichlet-Neumann algorithm achieves this with local transmission conditions, the optimal Schwarz algorithm needs non-local transmission conditions. This seems to be a disadvantage of the optimal Schwarz method. We then show however that for more than two subdomains or overlapping subdomains, both the optimal Schwarz algorithm and the Dirichlet Neumann algorithm need non-local interface conditions to converge in a finite number of steps. Hence the apparent advantage of Dirichlet-Neumann over optimal Schwarz is only an artifact of the special two subdomain decomposition.

1 Introduction

Numerical efficiency is very important when modeling the atmosphere, see Côté et al. [1998]. This is particularly true of operational weather forecasts that must be run in real-time during a given time window, and weather being a global phenomenon, one must use a global model to accurately forecast or analyze data. Furthermore, fast waves, which carry little energy, propagate many times faster than the local wind speed, by a factor three or more

* The author was partly supported by the Office of Science (BER), U.S. Department of Energy, Grant No. DE-FG02-01ER63199

depending on the application, and these waves restrict the time-step of explicit Eulerian integration schemes. The restrictions are particularly severe for global finite-difference models, due to the convergence of the meridians at the poles. This motivates the use of an implicit (or semi-implicit) time treatment of the terms that govern the propagation of these oscillations in order to greatly retard their propagation and permit a much larger time-step. This approach results in the need to solve an elliptic problem on the sphere. For a time-implicit scheme to be computationally advantageous, it must be possible to integrate with a sufficiently-large time-step to offset the overhead of solving the elliptic-boundary-value problem. This is often the case, even for non-hydrostatic flows, as discussed in Skamarock et al. [1997].

Meteorological operational centers have recently acquired new high-performance significantly-parallel computers. In order to reap the benefits afforded by those systems, parallel algorithms need to be designed for solving the models used in numerical-weather-prediction and data assimilation systems. This motivates the present study, where the parallel solution based on domain-decomposition methods on the surface of the sphere is analyzed. We investigate two domain decomposition methods in this paper: the Dirichlet-Neumann and the optimal Schwarz method. The Dirichlet-Neumann method is a well studied method on the plane, see for example Bjørstad and Widlund [1986], Bramble et al. [1986], Marini and Quarteroni [1989], and references therein. The choice of the optimal relaxation parameter in the Dirichlet-Neumann method on the plane is also well understood: for the case of two subdomains with special symmetry, it is $\frac{1}{2}$. In more general situations, the parameter of relaxation needs to be in a specific interval to obtain a fast algorithm. The key idea underlying the optimal Schwarz method has been introduced in Hagstrom et al. [1988] in the context of non-linear problems. A new class of Schwarz methods based on this idea was then introduced in Charton et al. [1991] and further analyzed in Nataf and Rogier [1995] and Japhet [1998] for convection diffusion problems. For the case of the Poisson equation, see Gander et al. [2001], where also the terms optimal and optimized Schwarz were introduced. Optimal Schwarz methods have in general non-local transmission conditions at the interfaces between subdomains, and are therefore not as easy to use as classical Schwarz methods. Optimized Schwarz methods use local approximations of the optimal, non-local transmission conditions at the interfaces and are therefore as easy to use as the classical Schwarz method, but have a greatly enhanced performance.

In Section 2, we introduce the Poisson equation on the sphere and the tools of Fourier analysis, on which our results are based. In Section 3, we present the Dirichlet-Neumann algorithm for the Poisson equation on the surface of the sphere with possible overlap. We show that convergence in two iterations can be achieved with an appropriate choice of the relaxation parameter. In the case of two subdomains without overlap, this optimal parameter is a constant, but with overlap, and in the case of three subdomains, convergence in a finite number of steps is only possible with a non-local convolution relaxation parameter. In Section 4, we present the optimal Schwarz algorithm for the same configuration. We prove convergence in two iterations for the two subdomain case and in three iterations for the three subdomain case, in both cases with non-local transmission conditions. In Section 5 we illustrate our findings with numerical experiments.

2 The Poisson Equation on the Sphere

We consider the solution of the Poisson problem

$$\mathcal{L}u = \Delta u = f, \qquad \text{in} \quad S \subset \mathbb{R}^3, \tag{1}$$

where S is the unit sphere centered at the origin. Using spherical coordinates, the equation (1) can be rewritten in the form

$$\mathcal{L}u = \frac{1}{r^2}\frac{\partial}{\partial r}\left(r^2\frac{\partial u}{\partial r}\right) + \frac{1}{r^2 sin^2\psi}\frac{\partial^2 u}{\partial\theta^2} + \frac{1}{r^2 sin\psi}\frac{\partial}{\partial\psi}\left(\sin\psi\frac{\partial u}{\partial\psi}\right) = f, \tag{2}$$

where ψ stands for the colatitude, with 0 being the north pole and π being the south pole, and θ is the longitude. For our case on the surface of the unit sphere, we consider solutions independent of r, which simplifies (2) to

$$\mathcal{L}u = \frac{1}{sin^2\psi}\frac{\partial^2 u}{\partial\theta^2} + \frac{1}{sin\psi}\frac{\partial}{\partial\psi}\left(\sin\psi\frac{\partial u}{\partial\psi}\right) = f. \tag{3}$$

Our results are based on Fourier analysis. Because u is periodic in θ, it can be expanded in a Fourier series,

$$u(\psi,\theta) = \sum_{m=-\infty}^{\infty} \hat{u}(\psi,m)e^{im\theta}, \quad \hat{u}(\psi,m) = \frac{1}{2\pi}\int_0^{2\pi} e^{-im\theta}u(\psi,\theta)d\theta.$$

Equation (3) then becomes a family of ordinary differential equations; for any positive or negative integer m, we have

$$-\frac{m^2}{sin^2\psi}\hat{u}(\psi,m) + \frac{1}{sin\psi}\frac{\partial}{\partial\psi}\left(sin\psi\frac{\partial\hat{u}(\psi,m)}{\partial\psi}\right) = \hat{f}(\psi,m). \tag{4}$$

For m fixed, the homogeneous problem, $\hat{f}(\psi,m) = 0$ in (4), has the two fundamental solutions

$$g_{\pm}(\psi,m) = \left(\frac{\sin(\psi)}{\cos(\psi)+1}\right)^{\pm|m|}. \tag{5}$$

Remark 1. g_+ has a singularity at the south pole and g_- has a singularity at the north pole.

Remark 2. The function $\sin x/(\cos x + 1)$ is monotonically increasing on the interval $[0,\pi)$, which can be seen by taking a derivative.

3 The Dirichlet-Neumann Algorithm

We first decompose the surface of the sphere into two overlapping subdomains as shown in Figure 1-(i), where $a \leq b$. The Dirichlet-Neumann method to

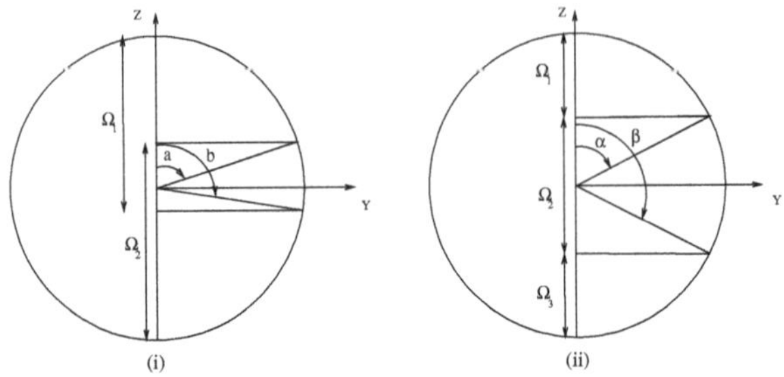

Fig. 1. (i) Two overlapping subdomains (ii) Three nonoverlapping subdomains

solve (1) solves iteratively the Poisson equation on Ω_1 and Ω_2 and exchanges Dirichlet and Neumann conditions respectively on the interfaces a and b. In its classical form, the algorithm is defined without overlap, but here we are also interested in the influence of overlap on the algorithm. By linearity, it suffices to consider only the homogeneous case in the convergence analysis, $f = 0$. For an initial guess $\lambda^0(\theta)$ on the interface b, the algorithm then performs the iteration

$$\begin{array}{ll} \mathcal{L}u_1^{k+1} = 0 \text{ in } \Omega_1, & u_1^{k+1}(b,\theta) = \lambda^k(\theta), \\ \mathcal{L}u_2^{k+1} = 0 \text{ in } \Omega_2, & \frac{\partial}{\partial\psi}u_2^{k+1}(a,\theta) = \frac{\partial}{\partial\psi}u_1^{k+1}(a,\theta), \end{array} \tag{6}$$

where the new function $\lambda^{k+1}(\theta)$ is defined by the linear combination

$$\lambda^{k+1}(\theta) := \gamma u_2^{k+1}(b,\theta) + (1-\gamma)\lambda^k(\theta). \tag{7}$$

Here γ is a relaxation parameter which is assumed to be non-negative. Expanding the iterates in a Fourier series and taking into account Remark 1 for bounded solutions on the subdomains, we obtain the subdomain solutions

$$\begin{array}{l} \hat{u}_1^{k+1}(\psi,m) = \hat{\lambda}^k(m)\frac{g_+(\psi,m)}{g_+(b,m)}, \\ \hat{u}_2^{k+1}(\psi,m) = -\hat{\lambda}^k(m)\frac{g_+(a,m)}{g_+(b,m)g_-(a,m)}g_-(\psi,m). \end{array} \tag{8}$$

Letting $\sigma(m) = g_+(a,m)g_-(b,m)$, the iteration on λ in (7) becomes

$$\hat{\lambda}^{k+1}(m) = (1-\hat{\gamma}(1+\sigma^2(m)))\hat{\lambda}^k(m). \tag{9}$$

Proposition 1. *The Dirichlet-Neumann iteration (6) with the two overlapping subdomains on the surface of the sphere converges in two iterations, provided the relaxation parameter is*

$$\hat{\gamma} = \hat{\gamma}_{opt}(m) := \frac{1}{1+\sigma^2(m)}. \tag{10}$$

Remark 3. The optimal relaxation parameter depends on m, which implies a non-local convolution operation in real space. Without overlap however, $a = b$, we have $\sigma(m) = 1$, and the optimal relaxation parameter becomes $\gamma_{opt} = 1/2$, which is now independent of m and thus a simple local operation in real space.

To see if the optimal result for the non-overlapping two subdomain case can be generalized to more subdomains, we consider a decomposition of the surface of the sphere into three non-overlapping subdomains, see Figure 1-(ii), where $\alpha < \beta$. In this case the Dirichlet-Neumann algorithm for (1) with initial guesses $\lambda_1^0(\theta)$ and $\lambda_2^0(\theta)$ is

$$\begin{array}{llrl} \mathcal{L}u_1^{k+1} = 0 & \text{in } \Omega_1, & u_1^{k+1}(\alpha,\theta) &= \lambda_1^k(\theta), \\ \mathcal{L}u_2^{k+1} = 0 & \text{in } \Omega_2, & \frac{\partial}{\partial\psi}u_2^{k+1}(\alpha,\theta) &= \frac{\partial}{\partial\psi}u_1^{k+1}(\alpha,\theta), \\ & & u_2^{k+1}(\beta,\theta) &= \lambda_2^k(\theta), \\ \mathcal{L}u_3^{k+1} = 0 & \text{in } \Omega_3, & \frac{\partial}{\partial\psi}u_3^{k+1}(\beta,\theta) &= \frac{\partial}{\partial\psi}u_2^{k+1}(\beta,\theta), \end{array} \tag{11}$$

where the new functions $\lambda_j^k(\theta)$, $j = 1, 2$, are defined by

$$\begin{aligned} \lambda_1^{k+1}(\theta) &= \gamma_1 u_2^{k+1}(\alpha,\theta) + (1-\gamma_1)\lambda_1^k(\theta), \\ \lambda_2^{k+1}(\theta) &= \gamma_2 u_3^{k+1}(\beta,\theta) + (1-\gamma_2)\lambda_2^k(\theta), \end{aligned} \tag{12}$$

and γ_1 and γ_2 are non-negative relaxation parameters. Using Fourier series as before, we arrive at the matrix iteration

$$\begin{bmatrix} \hat{\lambda}_1^{k+1} \\ \hat{\lambda}_2^{k+1} \end{bmatrix} = A \begin{bmatrix} \hat{\lambda}_1^{k} \\ \hat{\lambda}_2^{k} \end{bmatrix}, \quad A := \begin{bmatrix} \hat{\gamma}_1 \frac{\Delta_-}{\Delta_+} + (1-\hat{\gamma}_1) & 2\frac{\hat{\gamma}_1}{\Delta_+} \\ -2\frac{\hat{\gamma}_2}{\Delta_+} & \hat{\gamma}_2 \frac{\Delta_-}{\Delta_+} + (1-\hat{\gamma}_2) \end{bmatrix}, \tag{13}$$

where $\Delta_\pm := \eta(m) \pm \eta^{-1}(m)$ and $\eta(m) := g_+(\alpha, m)/g_+(\beta, m)$.

Proposition 2. *The Dirichlet-Neumann iteration with three non-overlapping subdomains on the surface of the sphere converges in three iterations, provided the relaxation parameters are*

$$\begin{aligned} \hat{\gamma}_1 = \hat{\gamma}_{1,opt}(m) &:= \tfrac{1}{2} + \tfrac{1}{2}\eta^2(m) - \tfrac{1}{2}\eta(m)\sqrt{1+\eta^2(m)}, \\ \hat{\gamma}_2 = \hat{\gamma}_{2,opt}(m) &:= \tfrac{1}{2} + \tfrac{1}{2}\eta^2(m) + \tfrac{1}{2}\eta(m)\sqrt{1+\eta^2(m)}. \end{aligned} \tag{14}$$

This result shows that for more than two subdomains, convergence in a finite number of steps can only be achieved with non-local convolution relaxation parameters in the Dirichlet-Neumann algorithm.

4 Schwarz Algorithms

We decompose the surface of the sphere into two overlapping domains as shown in Figure (1)-(i). The classical Schwarz algorithm is given by

$$\begin{aligned} \mathcal{L}u_1^{k+1} &= 0 \text{ in } \Omega_1, \quad u_1^{k+1}(b,\theta) = u_2^k(b,\theta), \\ \mathcal{L}u_2^{k+1} &= 0, \text{ in } \Omega_2, \quad u_2^{k+1}(a,\theta) = u_1^{k+1}(a,\theta). \end{aligned} \tag{15}$$

Using a Fourier series expansion as before, we find

$$\begin{aligned} \hat{u}_1^{k+1}(a,m) &= \rho \hat{u}_1^k(a,m) \\ \hat{u}_2^{k+1}(b,m) &= \rho \hat{u}_2^k(b,m) \end{aligned}, \quad \rho := \frac{g_+(a,m)}{g_+(b,m)} \frac{g_-(b,m)}{g_-(a,m)}. \tag{16}$$

Because of Remark 2, the fractions are less than one and this process is a contraction and hence convergent. We have proved the following

Proposition 3. *For each $m \neq 0$, the Schwarz iteration on the surface of the sphere partitioned along two colatitudes $a < b$ converges linearly with the convergence factor*

$$\left(\frac{\sin(a)}{\cos(a)+1}\right)^{2|m|} \left(\frac{\sin(b)}{\cos(b)+1}\right)^{-2|m|} < 1. \tag{17}$$

This shows that for small values of m the speed of convergence is very poor, since the convergence factor in (17) is nearly one. Following the approach in Gander et al. [2001], we introduce the following new transmission conditions:

$$\begin{aligned} (1 + \hat{p}(m)\tfrac{\partial}{\partial\psi})\hat{u}_1^{k+1}(b,m) &= (1 + \hat{p}(m)\tfrac{\partial}{\partial\psi})\hat{u}_2^k(b,m), \\ (1 + \hat{q}(m)\tfrac{\partial}{\partial\psi})\hat{u}_2^{k+1}(a,m) &= (1 + \hat{q}(m)\tfrac{\partial}{\partial\psi})\hat{u}_1^{k+1}(a,m), \end{aligned} \tag{18}$$

where $\hat{p}$ and $\hat{q}$ are functions we can use to optimize the performance.

Proposition 4. *If, for each $m \neq 0$, $\hat{p}(m) = \sin(b)/|m|$ and $\hat{q}(m) = -\sin(a)/|m|$, then the new Schwarz algorithm with transmission conditions (18) converges in two iterations, even without overlap, $a = b$.*

Proof. Using $\hat{u}_2^1(\psi,m) = C_2 g_-(\psi,m)$ and $\hat{u}_1^1(\psi,m) = C_1 g_+(\psi,m)$, where C_2 is a coefficient to be determined by the method and C_1 is given through the initial guess, and substituting into the transmission condition, yields

$$C_2 g_-(a,m)\left(1 - \hat{q}(m)\frac{|m|}{\sin(a)}\right) = C_1 g_+(a,m)\left(1 + \hat{q}(m)\frac{|m|}{\sin(a)}\right) = 0.$$

Hence $C_2 = 0$ and the iteration has converged for Ω_2. A similar argument shows that in the second step, the iteration converges on Ω_1 as well.

To see if this result generalizes to more than two subdomains, we consider the Schwarz algorithm with three subdomains,

$$\begin{aligned} \mathcal{L}u_1^{k+1} &= 0 \text{ in } \Omega_1, & (1 + p\tfrac{\partial}{\partial\psi})u_1^{k+1}(\alpha,\theta) &= (1 + p\tfrac{\partial}{\partial\psi})u_2^k(\alpha,\theta), \\ \mathcal{L}u_2^{k+1} &= 0 \text{ in } \Omega_2, & (1 + q_1\tfrac{\partial}{\partial\psi})u_2^{k+1}(\alpha,\theta) &= (1 + q_1\tfrac{\partial}{\partial\psi})u_1^{k+1}(\alpha,\theta), \\ & & (1 + q_2\tfrac{\partial}{\partial\psi})u_2^{k+1}(\beta,\theta) &= (1 + q_2\tfrac{\partial}{\partial\psi})u_3^k(\beta,\theta), \\ \mathcal{L}u_3^{k+1} &= 0 \text{ in } \Omega_3, & (1 + r\tfrac{\partial}{\partial\psi})u_3^{k+1}(\beta,\theta) &= (1 + r\tfrac{\partial}{\partial\psi})u_2^{k+1}(\beta,\theta), \end{aligned} \tag{19}$$

where p, q_1, q_2 and r are convolution operators in θ with Fourier symbol $\hat{p}$, $\hat{q}_1$, $\hat{q}_2$ and $\hat{r}$ respectively.

Proposition 5. *If, for each* $m \neq 0$, $\hat{p}(m) = \sin(\alpha)/|m|$, $\hat{q}_1(m) = -\sin(\alpha)/|m|$, $\hat{q}_2(m) = \sin(\beta)/|m|$, $\hat{r}(m) = -\sin(\beta)/|m|$, *then the new Schwarz algorithm for three subdomains (19) converges in three iterations.*

The proof of this last result is similar to the proof for the two subdomain case.

Remark 4. The choice $\hat{p}(m) = \sin(\alpha)/|m|$ and $\hat{q}_2(m) = \sin(\beta)/|m|$ is not necessary in this Gauss-Seidel form of the optimal Schwarz method: $\hat{p}(m)$ and $\hat{q}_2(m)$ can be any real number, except $-\sin(\alpha)/|m|$ and $-\sin(\beta)/|m|$ respectively, and the stated results still hold (the situation is similar for the two subdomain case). In the more parallel Jacobi form of the algorithms however the given choice is necessary to obtain the convergence results stated.

5 Numerical experiments

We used a spectral method in the longitude with 20 modes, and a finite difference method in the colatitude with discretization parameter $h = \pi/3000$. In the first set of experiments, we used two subdomains, once with overlap $[\frac{9}{20}\pi, \frac{11}{20}\pi]$, and once without overlap. A comparison of the convergence behavior of the algorithms is shown in Figure 2 on the left. While the classical

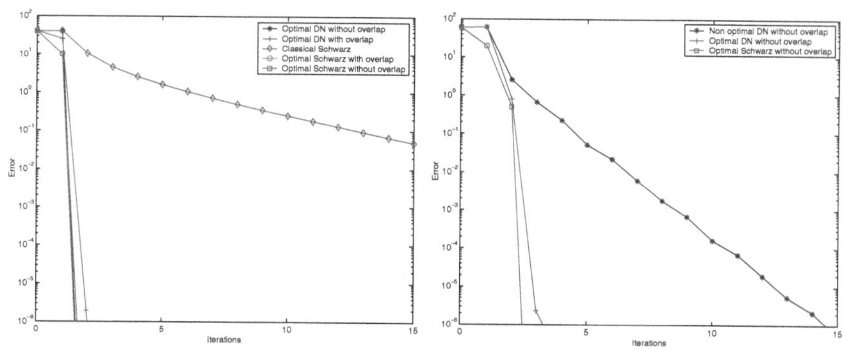

Fig. 2. Convergence behavior for the methods analyzed: the two subdomain case on the left and the three subdomain case on the right

Schwarz algorithm converges very slowly, both the optimal Schwarz and the Dirichlet-Neumann algorithm converge in two steps with and without overlap, as predicted by the analysis.

In the second set of experiments, we use three non-overlapping subdomains. In Figure 2 on the right, one can see that the optimal Schwarz and Dirichlet-Neumann algorithms converge in three steps, as predicted by the analysis, whereas the Dirichlet-Neumann algorithm with the constant relaxation parameter 1/2, which was optimal for the two subdomain case without overlap, is now much slower.

6 Conclusion

The Dirichlet-Neumann algorithm converges with local relaxation parameter in a finite number of steps only in the special case of two subdomains. To obtain convergence in a finite number of steps for more than two subdomains or if overlap is used, non-local relaxation parameters are needed, like for the optimal Schwarz method. These non-local transmission conditions will serve as a guiding principle to develop local approximations which lead to fast algorithms.

References

P. E. Bjørstad and O. B. Widlund. Iterative methods for the solution of elliptic problems on regions partitioned into substructures. *SIAM J. Numer. Anal.*, 23(6):1093–1120, 1986.

J. H. Bramble, J. E. Pasciak, and A. H. Schatz. An iterative method for elliptic problems on regions partitioned into substructures. *Math. Comp.*, 46(173):361–369, 1986.

P. Charton, F. Nataf, and F. Rogier. Méthode de décomposition de domaine pour l'équation d'advection-diffusion. *C. R. Acad. Sci.*, 313(9):623–626, 1991.

J. Côté, S. Gravel, A. Méthot, A. Patoine, M. Roch, and A. Staniforth. The operational CMC-MRB global environmental multiscale (GEM) model: Part I - design considerations and formulation. *Mon. Wea. Rev.*, 126:1373–1395, 1998.

M. J. Gander, L. Halpern, and F. Nataf. Optimized Schwarz methods. In T. Chan, T. Kako, H. Kawarada, and O. Pironneau, editors, *Twelfth International Conference on Domain Decomposition Methods, Chiba, Japan*, pages 15–28, Bergen, 2001. Domain Decomposition Press.

T. Hagstrom, R. P. Tewarson, and A. Jazcilevich. Numerical experiments on a domain decomposition algorithm for nonlinear elliptic boundary value problems. *Appl. Math. Lett.*, 1(3), 1988.

C. Japhet. Optimized Krylov-Ventcell method. Application to convection-diffusion problems. In P. E. Bjørstad, M. S. Espedal, and D. E. Keyes, editors, *Proceedings of the 9th international conference on domain decomposition methods*, pages 382–389. ddm.org, 1998.

L. D. Marini and A. Quarteroni. A relaxation procedure for domain decomposition methods using finite elements. *Numer. Math*, (5):575–598, 1989.

F. Nataf and F. Rogier. Factorization of the convection-diffusion operator and the Schwarz algorithm. M^3AS, 5(1):67–93, 1995.

W. Skamarock, P. Smolarkiewicz, and J. Klemp. Preconditioned conjugate-residual solvers for Helmholtz equations in nonhydrostatic models. *Mon. Wea. Rev.*, 125:587–599, 1997.

Finite Volume Methods on Non-Matching Grids with Arbitrary Interface Conditions and Highly Heterogeneous Media

I. Faille[1], F. Nataf[2], L. Saas[1,2], and F. Willien[1]

[1] IFP, 1 et 4 Avenue Bois-Préau, 92852 Rueil-Malmaison cedex, France
Isabelle.Faille@ifp.fr, Laurent.Saas@ifp.fr, Francoise.Willien@ifp.fr

[2] Ecole Polytechnique, CMAP, 91128 Palaiseau Cedex, France
http://www.cmap.polytechnique.fr/~nataf/, saas@cmapx.polytechnique.fr

Summary. We are interested in a robust and accurate domain decomposition method with arbitrary interface conditions on non-matching grids using a finite volume discretization. We introduce transmission operators to take into account the non-matching grids. Under compatibility assumptions, we have the well-posedness of the global problem and of the local subproblems with a new discretization of the arbitrary interface conditions. Then, we give two error estimates in the discrete H^1 norm: the first one is in $O(h^{1/2})$ with L^2 orthogonal projections onto piecewise functions along the interface and the second one in $O(h)$ with transmission conditions based on a linear rebuilding along the interface. Finally, numerical results confirm the theory. Particular attention is paid to the situation with non matching grids and highly heterogeneous coefficients both across and inside subdomains. The addition of a third very thin subdomain between geological blocks is necessary to ensure a good accuracy.

1 Introduction

The aim of basin modelling is to simulate maturation of source rocks and migration of oil in sedimentary basins in order to provide quantitative prediction about phenomena leading to oil accumulations. A sedimentary basin is divided by faults in several blocks, which are themselves composed of several layers of different lithology. In order to account for these heterogeneities, the mesh used in each block follows the stratigraphic layers. Blocks displacement along faults results in sliding and therefore leads to non matching grids between two adjacent blocks (eventually between two adjacent layers). Our objective is to develop numerical methods based on finite volume discretization (as it is well adapted to multiphase flow modelling), and to handle efficiently non-matching grids. We work in the context of domain decomposition techniques which offer a general framework to handle non matching grids.

As a first simplified model, we consider the following problem in Ω, bounded polygonal subset of $\mathbb{R}^d$ ($d = 2, 3$):

$$(\eta - \Delta)(p) = f \text{ in } \Omega \text{ and } p = 0 \text{ on } \partial\Omega \tag{1}$$

where $\eta > 0$. For the sake of simplicity, we assume that the domain Ω is divided in two non overlapping subdomains Ω_i ($i = 1, 2$), with grids that do not match on the interface.
Previous works have shown that Robin or more general interface conditions in domain decomposition methods ensure robustness and efficiency of the iterative domain decomposition Faille et al. [2000], Achdou et al. [1999]. A continuous domain decomposition formulation of (1) reads:

$$\begin{array}{l} (\eta - \Delta)(p_i^{n+1}) = f \text{ in } \Omega_i \text{ and } p_i^{n+1} = 0 \text{ on } \partial\Omega \cap \partial\Omega_i \\ \frac{\partial p_i^{n+1}}{\partial n_i} + \alpha_j p_i^{n+1} = -\frac{\partial p_j^n}{\partial n_j} + \alpha_j p_j^n \text{ on } \partial\Omega_j \cap \partial\Omega_i, \; i, j = 1, 2 \text{ and } i \neq j \end{array} \tag{2}$$

where $\alpha_j > 0$. Our aim is to combine this domain decomposition algorithm with a cell centered finite volume discretization, while satisfying the following properties. First, the method should be robust enough (at least existence and uniqueness of the discrete solution). Then it should allow a wide range of values for Robin coefficients or even more general interface conditions and it should be accurate enough as our ultimate goal is to consider grids that do not match between layers. Finally, as sliding blocks are considered, the discretization in one block should not depend on the grid of the adjacent block. In the framework of finite volume or mixed finite element method, several discretization methods for non-matching grids have been developed Arbogast et al. [1996], Ewing et al. [1991], Achdou et al. [2002],Cautrés et al. [2000], Aavatsmark et al. [2001] but these methods do not use Robin conditions or loose finite volume accuracy.

The rest of the paper is organized as follows. In the next section, we describe the finite volume discretization inside a subdomain. In Section 3, we introduce the transmission operators used to match the unknowns. In Section 5, error estimates are given. In Section 6, numerical results are shown. In Section 7, discontinuous coefficients are taken into account.

2 Finite volume discretization

We consider a finite volume admissible mesh $\mathcal{T}_i$ associated with each subdomain Ω_i Eymard et al. [2000] which is a set of closed polygonal subsets of Ω_i such that $\Omega_i = \cup_{K \in \mathcal{T}_i} K$ and $\mathcal{E}_{\Omega_i}$ is the set of faces of $\mathcal{T}_i$. We shall use the following notations: Let ϵ_i be a face of $\mathcal{E}_{\Omega_i}$ located on the boundary of Ω_i, $K(\epsilon_i)$ denotes the control cell $K \in \mathcal{T}_i$ such that $\epsilon_i \in K$, $\mathcal{E}_i$ is the set of faces of domain Ω_i located on the interface, $\mathcal{E}(K)$ is the set of faces of $K \in \mathcal{T}_i$, $\mathcal{E}_i(K)$ is the set of faces of $K \in \mathcal{T}_i$ which are on the interface,

$\mathcal{N}_i(K) = \{K' \in \mathcal{T}_i : K \cap K' \in \mathcal{E}_{\Omega_i}\}$ is the set of the control cells adjacent to K and $[K, K']$ denotes the face $K \cap K'$.
We introduce p^i_K an approximation of $p(x_K)$ (where x_K is a point inside the control cell K), p^i_ϵ an approximation of $p(y_\epsilon)$ (where y_ϵ is the center of the face $\epsilon \in \mathcal{E}_i$) and u^i_ϵ an approximation of the flux $\frac{\partial p_i}{\partial n_i}(y_\epsilon)$ outward Ω_i through $\mathcal{E}_i$. Then, not taking into account the Dirichlet boundary condition, a finite volume scheme for (1) can be defined by the set of equations Eymard et al. [2000].

$$\eta p^i_K m(K) - \sum_{K' \in \mathcal{N}_i(K)} \frac{p^i_{K'} - p^i_K}{d(x_{K'}, x_K)} m([K, K']) - \sum_{\epsilon \in \mathcal{E}_i(K)} u^i_\epsilon m(\epsilon) = F^i_K \quad (3)$$

$$\text{with } u^i_\epsilon = \frac{p^i_\epsilon - p^i_K}{d(y_\epsilon, x_K)} \quad \text{for } \epsilon \in \mathcal{E}_i \quad (4)$$

for all control cells K of $\mathcal{T}_i$ and where $m(A)$ is the measure of $A \subset \Omega$. Discretized Robin interface conditions or more general interface conditions on $\mathcal{E}_i$ are introduced in the next section.

3 Transmission operators

We introduce the operators $Q_i : P^0(\mathcal{E}_j) \mapsto P^0(\mathcal{E}_i)$ $(i, j = 1, 2\ i \neq j)$ where $P^0(\mathcal{E}_i)$ is the space of piecewise constant functions on $\mathcal{E}_i$.
Assumption 1 Operators Q_1 and Q_2 are transposed of each other for the standard L^2 scalar product.

Method Constant The first type of transmission operators that we consider are the restrictions on $P^0(\mathcal{E}_j)$ of P^c_i the L^2 orthogonal projection onto $P^0(\mathcal{E}_i)$. They satisfy Assumption 1.

Method Linear The second type of transmission operators uses a linear rebuilding to ensure a more accurate transmission than P^c_i. We introduce for $i = 1, 2$

- the interface grid: $\mathcal{E}^2_i$ coarsening by a factor 2 of $\mathcal{E}_i$
- $P^1_d(\mathcal{E}^2_i)$ discontinuous piecewise linear functions on $\mathcal{E}^2_i$.
- interpolation operator $I_i : P^0(\mathcal{E}_i) \longmapsto P^1_d(\mathcal{E}^2_i)$ and its transpose I^t_i (w.r.t. the scalar product $L^2(\Gamma)$, $\forall u \in P^0(\epsilon_i)$ and $\forall v \in P^1(\epsilon^2_i)$ $< I_i(u), v >_{L^2(\Gamma)} = < u, I^t_i(v) >_{L^2(\Gamma)}$).
- P^L_i L^2 orthogonal projection on $P^1_d(\mathcal{E}^2_i)$

The definitions of the transmission operators are inspired by previous works Arbogast et al. [1996] in mixed finite element method:

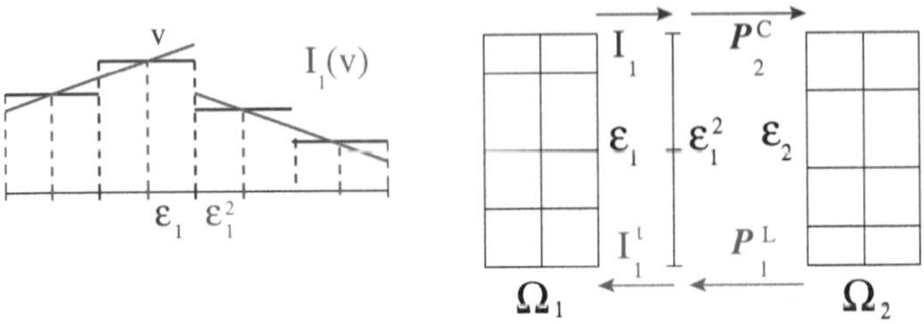

Fig. 1. Linear rebuilding and transmission operators

$$Q_1 = I_1^t P_1^L \qquad (5)$$
$$Q_2 = P_2^C I_1$$

They satisfy Assumption 1 but are not projections.

4 Interface Conditions

In analogy with Bernardi et al. [1994], transmission operators are used to write discrete matching conditions ensuring continuity of the solution and of its normal derivative on the interface:

$$p_2 = Q_2(p_1) \text{ on } \mathcal{E}_2 \text{ and } u_1 = Q_1(-u_2) \text{ on } \mathcal{E}_1 \qquad (6)$$

where $p_i \in P^0(\mathcal{E}_i)$ is the approximate pressure on $\mathcal{E}_i$ and $u_i \in P^0(\mathcal{E}_i)$ is the approximate flux outward Ω_i on $\mathcal{E}_i$ ($p_i = (p_i^\epsilon)_{\epsilon \in \mathcal{E}_i}$ and $u_i = (u_i^\epsilon)_{\epsilon \in \mathcal{E}_i}$). In mortar terminology Bernardi et al. [1994], domain Ω_1 is called the master because it imposes the pressure and Ω_2 is called the slave.
These matching conditions are made compatible with arbitrary interface conditions defined via operators $S_i : P^0(\mathcal{E}_i) \longmapsto P^0(\mathcal{E}_i)$ which satisfy
Assumption 2 S_i is positive definite
The corresponding interface conditions read:

$$Q_1(S_2(Q_2(p_1))) + u_1 = Q_1(S_2(p_2) - u_2) \qquad (7)$$
$$p_2 + Q_2(S_1^{-1}(Q_1(u_2))) = Q_2(p_1 - S_1^{-1}(u_1)) \qquad (8)$$

Examples of interface conditions are:

- Discrete Steklov-Poincaré operator ($S_i = (DtN_i)_h$)
- Robin interface conditions de $S_i = diag(\alpha_\epsilon^i)$, $S_i = diag(\alpha_{opt}^i)$
- optimized of order 1 or 2 (S_i tridiagonal)

Lemma 1. *Under Assumptions 1 and 2, mortar matching conditions (6) and arbitrary interface conditions (7)-(8) are equivalent.*

5 Error Estimates

It is proved in Saas et al. [2002] that under Assumptions 1 and 2,

- the global problem defined by the set of equations (3)-(4)-(6) is well-posed and stable.
- the local problem defined in Ω_1 by the set of equations (3)-(4)-(7) and the local problem defined in Ω_2 by the set of equations (3)-(4)-(8) are both well-posed and stable.

Under Assumptions 1 and 2 and additional assumptions on the mesh:
Assumption 3 $\exists C > 0$ such that $\forall \epsilon \in \mathcal{E}_i$, $diam(\epsilon) \leq Cd(x_{K(\epsilon)}, y_\epsilon)^{1/2}$
error estimates can be derived

Theorem 1. *The H^1 discrete norm of the error is in $O(h^{1/2})$ when the piecewise constant projections are used ($Q_i = P_i^c$).*
The H^1 discrete norm of the error is in $O(h)$ when the linear rebuilding (5) is used.

6 Numerical Results in the homogeneous case

Numerical tests have been done with the equation in four subdomains:

$$p - \Delta p = x^3y^2 - 6x^2y^2 - 2x^3 + (1 + x^2 + y^2)sin(xy) \text{ in } \Omega$$
$$p = p_0 \text{ on } \partial\Omega$$

This results have been compared to the analytical solution which is $p(x, y) = x^3y^2 + sin(xy)$. The domain decomposition method is reformulated with a substructuring method and is solved with a GMRES algorithm. For asymptotic study, we use an initial non conforming mesh which we refine successively by a factor 2. We compare different methods TPFA (Two point flux approximation, see Cautrés et al. [2000]), Ceres (like TPFA but a linear interpolation is performed in order to have a consistent flux approximation on the interface, see Faille et al. [1994]), New Cement (Achdou et al. [2002]), Constant and Linear (Section 3). For all these methods, we take different values for $S_i = diag(\alpha)$ with $\alpha = 1$ or $\alpha = 1/h$ or $\alpha = \alpha_{opt} = 1/h^{1/2}$. The numerical solution depends on the choice of S_i only for the New Cement method. Accuracy is given in figure 2 and iteration counts of the GMRES algorithm in figure 3.

7 Numerical Results in the heterogeneous case

We consider now the problem with discontinuous coefficients

$$\eta p - div(\kappa \nabla p) = f \text{ in } \Omega \text{ and } p = 0 \text{ on } \partial\Omega \tag{9}$$

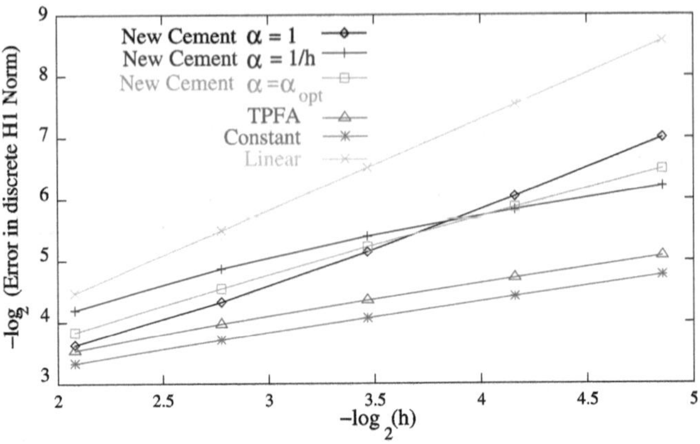

Fig. 2. Slopes: Linear $\simeq 1.3$; New Cement: ($S_i = cte$: $\simeq 1.3$), ($S_i = 1/h^{1/2}$: $\simeq 0.9$), ($S_i = 1/h$: $\simeq 0.6$), Constant: $\simeq 0.5$, TPFA: $\simeq 0.5$

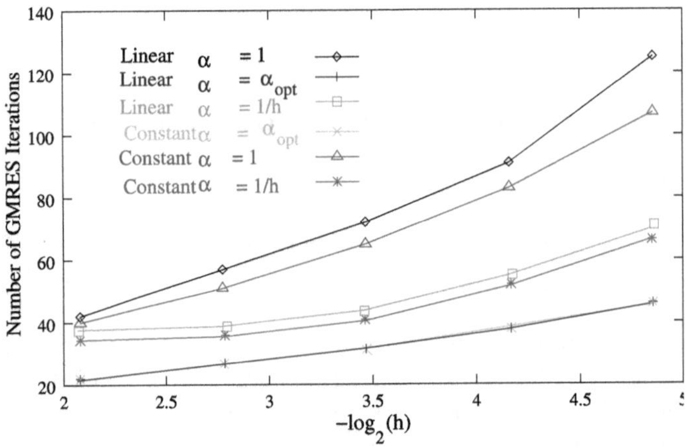

Fig. 3. Iteration counts for the GMRES algorithm

where $\eta > 0$ and κ are highly discontinuous, typically two or three orders of magnitude, see figure 4. For Test 2 for instance, with a very coarse grid methods Constant and Linear (see Section 3) work very poorly especially compared to TPFA and Ceres methods. Typically we have the following relative errors: Linear: 60%, Constant 10%, TPFA 3.2% and Ceres 2%. The errors are computed thanks to a computation on a very fine mesh since we don't have analytic solutions in these cases. Poor results for methods Linear and Constant are due to the fact that the flux on the interface is a very discontinuous function whose jumps are located on the jumps of the coefficients on both blocks. In Ceres and TPFA, a subgrid containing all locations of the jumps of the coefficients on the interface is involved which is not the case for methods Constant and Linear. In order to remedy this situation, a very thin

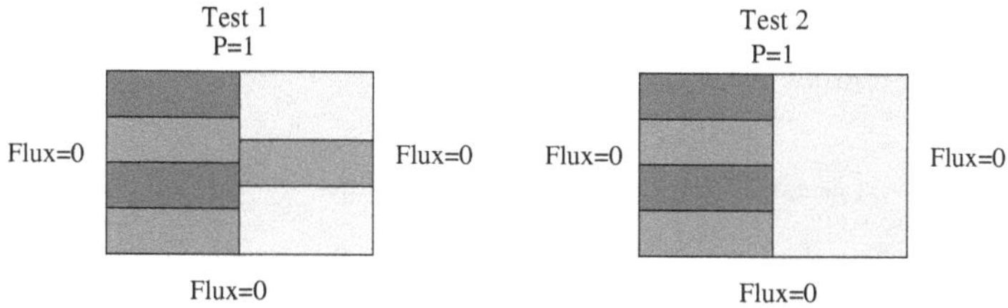

Fig. 4. Heterogeneous media

third subdomain is introduced between the blocks. The mesh of this additional subdomain along the interface is the intersection of the grid interfaces of the neighboring blocks, see figure 5. Methods Constant and Linear are then applied to this three subdomains case. The improvement is dramatic. The relative errors are then: Linear: 1.6%, Constant 1.6% (compared to respectively 60% and 10% in the two-subdomain case).

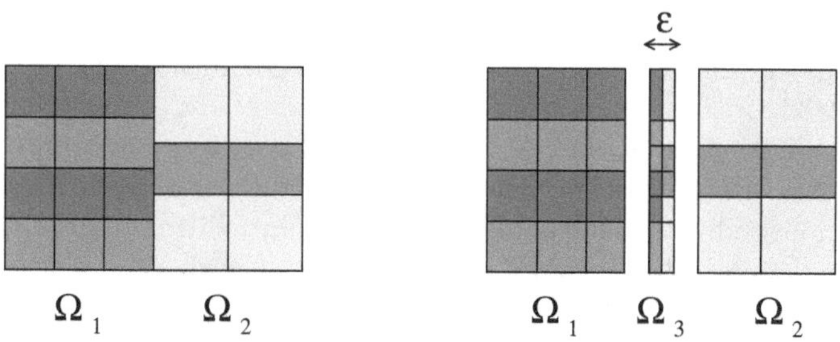

Fig. 5. Addition of a third subdomain

8 Conclusion

We have introduced matching operators to take into account the non-matching grids. Under compatibility assumptions, we have the well-posedness of the global problem and of the local subproblems with a new discretization of the arbitrary interface conditions. We give two error estimates in the discrete H^1 norm: the first one is in $O(h^{1/2})$ with L^2 orthogonal projections onto piecewise functions along the interface and the second one in $O(h)$ with transmission conditions based on a linear rebuilding along the interface. The error estimates depend only on the transmission operators, see Section 3. But, the numerical solutions are independent of the interface conditions whose discretizations are given by (7)-(8). Particular attention was paid to the situation with non matching grids and highly heterogeneous coefficients both across and inside subdomains. The addition of a third very thin subdomain between geological blocks is necessary to ensure a good accuracy. Extension to a finite element discretization would be interesting.

References

I. Aavatsmark, E. Reiso, and R. Teigland. Control-volume discretization method for quadrilateral grids with faults and local refinements. *Computational Geosciences*, pages 1–23, 2001.

Y. Achdou, C. Japhet, Y. Maday, and F. Nataf. A new cement to glue non-conforming grids with Robin interface conditions: the finite volume case. *Numer. Math.*, 92(4):593–620, 2002.

Y. Achdou, C. Japhet, P. L. Tallec, F. Nataf, F. Rogier, and M. Vidrascu. Domain decomposition methods for non-symmetric problems. In C.-H. Lai, P. E. Bjørstad, M. Cross, and O. B. Widlund, editors, *Eleventh International Conference on Domain Decomposition Methods*, pages 3–17, Bergen, 1999. Domain Decomposition Press.

T. Arbogast, L. C. Cowsar, M. F. Wheeler, and I. Yotov. Mixed finite element methods on non-matching multiblock grids. *SIAM J. Numer. Anal.*, 1996. submitted.

C. Bernardi, Y. Maday, and A. T. Patera. A new non conforming approach to domain decomposition: The mortar element method. In H. Brezis and J.-L. Lions, editors, *Collège de France Seminar*. Pitman, 1994. This paper appeared as a technical report about five years earlier.

R. Cautrés, R. Herbin, and F. Hubert. Non matching finite volume grids and the non overlapping schwarz algorithm. In N. Debit, M. Garbey, R. Hoppe, J. Periaux, D. Keyes, and Y. Kuznetsov, editors, *13th International Conference on Domain Decomposition Methods, Lyon*, pages 213–219, 2000.

R. E. Ewing, R. D. Lazarov, and P. S. Vassilevski. Local refinement techniques for elliptic problems on cell-centered grids. i: Error analysis. *Math. Comput.*, 56(194):437–461, 1991.

R. Eymard, T. Gallouët, and R. Herbin. The finite volume method. In P. Ciarlet and J.-L. Lions, editors, *Handbook of Numerical Analysis*, pages 713–1020. North Holland, 2000. This paper appeared as a technical report four years ago.

I. Faille, E. Flauraud, F. Nataf, F. Schneider, and F. Willien. Optimized interface conditions for sedimentary basin modeling. In I. Herrera, D. E. Keyes, O. B. Widlund, and R. Yates, editors, *13th International Conference on Domain Decomposition Methods, Lyon*, 2000.

I. Faille, S. Wolf, and F. Schneider. Aspect numérique de la modèlisation de bassins sédimentaires. Technical Report 41 237, IFP, Mars 1994.

L. Saas, I. Faille, F. Nataf, and F. Willien. Décomposition de domaine non conforme avec conditions de robin optimisées à l'interface et volumes finis. Technical Report 56947, IFP, 2002.

Nonlinear Advection Problems and Overlapping Schwarz Waveform Relaxation

Martin J. Gander[1] and Christian Rohde[2]

[1] McGill University Montreal, Department of Mathematics and Statistics (http://www.math.mcgill.ca/~mgander/)

[2] Albert-Ludwigs-Universität Freiburg, Mathematisches Institut (http://www.mathematik.uni-freiburg.de/IAM/homepages/chris/chris.html)

Summary. We analyze the convergence behavior of the overlapping Schwarz waveform relaxation algorithm applied to nonlinear advection problems. We show for Burgers' equation that the algorithm converges super-linearly at a rate which is asymptotically comparable to the rate of the algorithm applied to linear advection problems. The convergence rate depends on the overlap and the length of the time interval. We carefully track dependencies on the viscosity parameter and show the robustness of all estimates with respect to this parameter.

1 Introduction

Overlapping Schwarz waveform relaxation algorithms have been applied successfully to many evolution problems. However, a rigorous error analysis is only available in the case of linear and weakly nonlinear problems. For results covering the heat equation, advection-diffusion equations, and problems with nonlinear source terms, we refer to Gander [1997], Giladi and Keller [2002], Gander [1998], Daoud and Gander [2000], Gander and Zhao [2002].

We present here first convergence results for the algorithm applied to a class of strongly nonlinear problems: scalar parabolic conservation laws with nonlinear fluxes. In Section 2 we present the problem and the necessary analytical background. In particular we focus on conservation laws in the advection dominated case when the problem is singularly perturbed. In Section 3 we introduce the Schwarz waveform relaxation algorithm for parabolic conservation laws. In Section 4 the error analysis for the algorithm is presented for the special case of Burgers' equation. We focus on two topics: the comparison of the results for the linear and the nonlinear case and the influence of the diffusion parameter ε on the convergence rate. The paper concludes with a numerical experiment that confirms the theoretical results.

We note that there is a fundamentally different approach to solve nonlinear conservation laws using domain decomposition. One first discretizes the problem uniformly in time using an implicit scheme, and then applies domain decomposition to the steady problems obtained at each time step, see Dolean et al. [2000] and references therein. For a heterogeneous approach, see also Garbey [1996], Garbey and Kaper [1997].

2 Advection Dominated Conservation Laws

We consider for $T > 0$ and a function $u_0 \in W^{1,\infty}(\mathbb{R})$ the initial boundary value problem

$$\frac{\partial u^\varepsilon}{\partial t} + \frac{\partial}{\partial x} f(u^\varepsilon) = \varepsilon \frac{\partial^2 u^\varepsilon}{\partial x^2} \text{ in } \mathbb{R} \times (0,T), \quad u^\varepsilon(.,0) = u_0 \text{ in } \mathbb{R} \tag{1}$$

for the unknown $u^\varepsilon = u^\varepsilon(x,t) : \mathbb{R} \times (0,T) \to \mathbb{R}$. Here $\varepsilon > 0$ is a constant and $f \in C^2(\mathbb{R})$ denotes the possibly nonlinear flux function. The scalar problem (1) is a simple model for nonlinear systems of conservation laws which arise frequently to describe dynamical processes in continuum mechanics. Important examples are the Navier-Stokes equations in fluid mechanics and the system of thermo-elasticity in solid mechanics. An interesting feature of many applications governed by conservation laws is the fact that they are *advection dominated.* On the level of problem (1) this implies that the diffusion parameter ε is small and we have to consider a singularly perturbed problem. In the limit $\varepsilon = 0$ the parabolic equation in (1) changes type and becomes the hyperbolic equation

$$\frac{\partial u^0}{\partial t} + \frac{\partial}{\partial x} f(u^0) = 0 \quad \text{in } \mathbb{R} \times (0,T) \tag{2}$$

for the unknown $u^0 : \mathbb{R} \times (0,T) \to \mathbb{R}$. It is well-known that classical solutions of the initial value problem for (2) do not exist globally in time for all smooth initial data if f is nonlinear, see for example Dafermos [2000]. Singularities called shock waves occur. In the singularly perturbed case with $\varepsilon > 0$, diffusive layers with $\frac{\partial u^\varepsilon}{\partial x} = \mathcal{O}(\varepsilon^{-1})$ take the role of shock waves. The following theorem reflects the relationships between solutions of (1), (2) and summarizes the results we need later from the theory of conservation laws.

Theorem 1. *There exists a unique classical solution $u^\varepsilon \in C^1(0,T;C^2(\mathbb{R}))$ of (1) that satisfies*

$$\inf_{x\in\mathbb{R}}\{u_0(x)\} \le u^\varepsilon(x,t) \le \sup_{x\in\mathbb{R}}\{u_0(x)\}, \quad (x,t) \in \mathbb{R} \times [0,T],$$
$$\varepsilon \|u_x^\varepsilon\|_{L^\infty(\mathbb{R}\times[0,T])} \le C,$$

where the positive constant C does not depend on ε. Furthermore there exists a function $u^0 \in L^\infty(\mathbb{R} \times [0,T])$ such that for each compact set $Q \subset \mathbb{R}$ we have

$$\lim_{\varepsilon\to 0} \|u^0 - u^\varepsilon\|_{L^1(Q\times[0,T])} = 0.$$

The analogous statements hold for the initial boundary value problem with Dirichlet boundary conditions.

The proofs of these results are classical and can be found for instance in Dafermos [2000].

3 Overlapping Schwarz Waveform Relaxation

We approximate the solution of (1) using the overlapping Schwarz waveform relaxation algorithm on the two subdomains $\Omega_1 = (-\infty, L)$ and $\Omega_2 = (0, \infty)$ with overlap $L > 0$. The parameter ε is fixed here, so we therefore drop the index ε in this and the next section to simplify the notation. For iteration index $n \in \mathbb{N}$, the overlapping Schwarz waveform relaxation algorithm is defined by

$$\begin{aligned} \frac{\partial u_1^n}{\partial t} + f'(u_1^n)\frac{\partial u_1^n}{\partial x} &= \varepsilon \frac{\partial^2 u_1^n}{\partial x^2} && \text{in } \Omega_1 \times (0,T), \\ u_1^n(\cdot,0) &= u_0 && \text{in } \Omega_1, \\ u_1^n(L,\cdot) &= u_2^{n-1}(L,\cdot) && \text{on } [0,T], \end{aligned} \tag{3}$$

and

$$\begin{aligned} \frac{\partial u_2^n}{\partial t} + f'(u_2^n)\frac{\partial u_2^n}{\partial x} &= \varepsilon \frac{\partial^2 u_2^n}{\partial x^2} && \text{in } \Omega_2 \times (0,T), \\ u_2^n(\cdot,0) &= u_0 && \text{in } \Omega_2, \\ u_2^n(0,\cdot) &= u_1^{n-1}(0,\cdot) && \text{on } [0,T]. \end{aligned} \tag{4}$$

4 Convergence Analysis

We first review results on the iteration (3), (4) for the linear flux $f(u) = cu, \quad c \in \mathbb{R}$. We define the errors in the Schwarz waveform relaxation iteration by $e_1^n := u - u_1^n$ on the left subdomain and $e_2^n := u - u_2^n$ on the right subdomain for $n \in \mathbb{N}_0$. For $n \in \mathbb{N}$, we find that the error e_1^n satisfies

$$\begin{aligned} \frac{\partial e_1^n}{\partial t} + c\frac{\partial e_1^n}{\partial x} &= \varepsilon \frac{\partial^2 e_1^n}{\partial x^2} && \text{in } \Omega_1 \times (0,T), \\ e_1^n(\cdot,0) &= 0 && \text{in } \Omega_1, \\ e_1^n(L,\cdot) &= e_2^{n-1}(L,\cdot) && \text{on } [0,T], \end{aligned} \tag{5}$$

and the analogous equations hold for e_2^n. The error analysis for (5) has been performed in Gander [1997] and independently in Giladi and Keller [2002]. We cite the final result.

Theorem 2 (Linear Advection Diffusion). *The overlapping Schwarz waveform relaxation algorithm (3), (4) for the advection diffusion problem (1) with $f(u) = cu$ converges super-linearly. For each $T > 0$ and $i = 1, 2$ we have*

$$\sup_{x\in\Omega_i,0\le t\le T}|e_i^{2n}(x,t)|\le C_i\mathrm{erfc}\left(\frac{nL}{\sqrt{\varepsilon T}}\right), \tag{6}$$

where $C_1=\sup_{0\le t\le T}|e_1^0(L,t)|$ *and* $C_2=\sup_{0\le t\le T}|e_2^0(0,t)|$.

Remark 1. If we apply the expansion $\sqrt{\pi}\mathrm{erfc}(z)=e^{-z^2}(z^{-1}+\mathcal{O}(z^{-3}))$ for large values $z>0$ in the estimate (6), we obtain

$$\sup_{x\in\Omega_i,0\le t\le T}|e_i^{2n}(x,t)|\approx\frac{C_i}{\sqrt{\pi}}e^{-\frac{n^2L^2}{\varepsilon T}}\frac{\sqrt{\varepsilon T}}{nL}.$$

For fixed $T,L,\varepsilon>0$ we observe that the algorithm converges super-linearly for $n\to\infty$ and $t\le T$. The error vanishes also for $\varepsilon\to 0$, reflecting the fact that the algorithm applied to the pure advection equation converges in two steps.

We now consider the iteration (3), (4) for the quadratic flux $f(u)=\frac{u^2}{2}$, that is Burgers' equation. For $n\in\mathbb{N}$, we find that the error $e_1^n:=u-u_1^n$ satisfies the equation

$$\begin{aligned}\frac{\partial e_1^n}{\partial t}+u\frac{\partial e_1^n}{\partial x}+\frac{\partial u_1^n}{\partial x}e_1^n&=\varepsilon\frac{\partial^2 e_1^n}{\partial x^2} &&\text{in } \Omega_1\times(0,T),\\ e_1^n(\cdot,0)&=0 &&\text{in } \Omega_1,\\ e_1^n(L,\cdot)&=e_2^{n-1}(L,\cdot) &&\text{on } [0,T],\end{aligned} \tag{7}$$

and an analogous problem for e_2^n. We note that in contrast to the linear equation the error equations for the Burgers case contain an additional source term scaled with the spatial derivative of the iterate. Moreover due to Theorem 1 these terms behave like $\mathcal{O}(\varepsilon^{-1})$ (The estimates in Theorem 1 hold mutatis mutandis also for initial boundary value problems).

For our analysis of the non-linear case we require that the iteration starts with the initial guesses

$$u_i^0(x,t)=\inf_{x'\in\Omega_i}\{u_0(x')\},\quad (x,t)\in\Omega_i\times(0,T),\ i=1,2. \tag{8}$$

Because of this choice and the comparison principle for parabolic differential equations we have for all iterations $n\in\mathbb{N}_0$

$$e_i^n(x,t)\ge 0,\quad (x,t)\in\Omega_i\times(0,T),\ i=1,2.$$

It suffices therefore to derive upper bounds for the errors to obtain a bound on the convergence rate of the overlapping Schwarz waveform relaxation algorithm applied to Burgers' equation. The first step of our analysis is to determine *linear advection diffusion problems* that bound the evolution of the errors. We show the derivation of the linear problems in detail, because it is here where the influence of the viscosity parameter ε needs to be traced carefully.

Lemma 1 (Super-Solutions). *For all $n \in \mathbb{N}$ we have*

$$0 \le e_1^n(x,t) \le \bar{e}_1^n(x,t), \quad \forall (x,t) \in \Omega_1 \times (0,T),$$

where the super-solution $\bar{e}_1^n$ is the solution of the linear, constant coefficient problem

$$\begin{aligned} \frac{\partial \bar{e}_1^n}{\partial t} + a_1 \frac{\partial \bar{e}_1^n}{\partial x} + b_1 \bar{e}_1^n &= \varepsilon \frac{\partial^2 \bar{e}_1^n}{\partial x^2} && \text{in } \Omega_1 \times (0,T), \\ \bar{e}_1^n(\cdot,0) &= 0 && \text{in } \Omega_1, \\ \bar{e}_1^n(L,t) &= \exp(\sigma_1 t) \sup_{0\le\tau\le t} e_2^{n-1}(L,\tau) && t \in [0,T], \end{aligned} \tag{9}$$

with the constants $a_1, b_1, \sigma_1 \in \mathbb{R}$ given by

$$\begin{aligned} a_1 &:= \inf_{(x,t)\in\Omega_1\times(0,T)} \{u(x,t)\}, \\ b_1 &:= \inf_{(x,t)\in\Omega_1\times(0,T)} \left\{ \frac{\partial u_1^n}{\partial x}(x,t) + (u(x,t) - a_1)\frac{a_1}{2\varepsilon} \right\}, \\ \sigma_1 &:= \begin{cases} -\frac{a_1^2}{4\varepsilon} - b_1 & \text{if } -\frac{a_1^2}{4\varepsilon} - b_1 \ge 0, \\ 0 & \text{otherwise.} \end{cases} \end{aligned}$$

The number σ_1 is finite but can be of order $\mathcal{O}(\varepsilon^{-1})$ due to (ii) in Theorem 1.

Remark 2. It is not surprising that the constant coefficient problems contain source terms that are not present in the linear case. Note that the spatial derivatives $\frac{\partial u_i^n}{\partial x}$ are in fact multiplied with the second derivative of the flux $f = f(u)$ which is one for $f(u) = u^2/2$ and vanishes in the linear case.

Proof. (of Lemma 1) We use explicit solutions of the constant-coefficient equation (9) by means of the heat kernel. We define the shifted derivative of the heat kernel by

$$K_{1,x}(x,t) = -\frac{1}{2\sqrt{\pi}} \frac{x-L}{\varepsilon^{1/2} t^{3/2}} \exp\Big(-\frac{(x-L)^2}{4\varepsilon t} \Big). \tag{10}$$

For the linear, constant coefficient problem (9) satisfied by the super-solution, we then have the closed form solution formula

$$\bar{e}_1^n(x,t) = \exp(p_1 x + q_1 t) \int_0^t K_{1,x}(x, t-\tau) g_1(\tau)\, d\tau, \tag{11}$$

where we used the constants

$$p_i = \frac{a_i}{2\varepsilon}, \quad q_i = -\frac{a_i^2}{4\varepsilon} - b_i, \qquad i = 1,2 \tag{12}$$

and the function $g_1 = g_1(t) = \exp(-p_1 L + (\sigma_1 - q_1)t) \sup_{0\le\tau\le t} e_2^{n-1}(L,\tau)$. Note that g_1 is nonnegative due to the non-negativity of the errors, and monotonically increasing because of our choice of σ_1. To show that $\bar{e}_1^n$ is indeed a super-solution, we have to show that

$$d_1^n := \bar{e}_1^n - e_1^n \geq 0. \tag{13}$$

Now the difference function d_1^n satisfies the linear advection diffusion equation

$$\frac{\partial d_1^n}{\partial t} + u\frac{\partial d_1^n}{\partial x} + \frac{\partial u_1^n}{\partial x} d_1^n - \varepsilon \frac{\partial^2 d_1^n}{\partial x^2} = Q_1(x,t), \tag{14}$$

where the source term $Q_1(x,t)$ is given by

$$\begin{aligned}
Q_1(x,t) &= (u(x,t)-a_1)\frac{\partial \bar{e}_1^n}{\partial x} + \left(\frac{\partial u_1^n}{\partial x}(x,t) - b_1\right)\bar{e}_1^n(x,t)\\
&= (u(x,t)-a_1)\frac{e^{(p_1x+q_1t)}}{2\sqrt{\pi}}\int_0^t \frac{e^{\left(-\frac{(x-L)^2}{4\varepsilon(t-\tau)}\right)}}{\varepsilon^{1/2}(t-\tau)^{3/2}}\left[\frac{(x-L)^2}{2\varepsilon(t-\tau)} - 1\right] g_1(\tau)\,d\tau\\
&\quad -\left((u(x,t)-a_1)p_1 + \frac{\partial u_1^n}{\partial x}(x,t) - b_1\right)(x-L)\\
&\quad \times \frac{e^{(p_1x+q_1t)}}{2\sqrt{\pi}}\int_0^t \frac{e^{\left(-\frac{(x-L)^2}{4\varepsilon(t-\tau)}\right)}}{\varepsilon^{1/2}(t-\tau)^{3/2}} g_1(\tau)\,d\tau\\
&=: (u(x,t)-a_1)e^{(p_1x+q_1t)}Q_{11}(x,t)\\
&\quad + \left((u(x,t)-a_1)p_1 + \frac{\partial u_1^n}{\partial x}(x,t) - b_1\right)e^{(p_1x+q_1t)}(L-x)Q_{12}(x,t).
\end{aligned}$$

If we can show that $Q_{11}(x,t)$ and $Q_{12}(x,t)$ are non-negative for all $(x,t) \in \Omega_1 \times (0,T)$, we obtain $Q_1(x,t) \geq 0$ for all $(x,t) \in \Omega_1 \times (0,T)$ by the definition of a_1, b_1, which implies (13) by the maximum principle for (14) with zero initial and boundary data. But Q_{12} is nonnegative since g_1 from (11) is nonnegative. For Q_{11} we observe that it is the x-derivative of the solution w of the heat equation $w_t = \varepsilon w_{xx}$ in $\Omega_1 \times (0,T)$ which satisfies $w(L,.) = g_1$ and $w(.,0) \equiv 0$. Since g_1 is nonnegative and monotonically increasing, Q_{11} must also be nonnegative, which concludes the proof that $\bar{e}_1^n$ is a super-solution of e_1^n.

For the preceding proof we used an explicit solution for the constant coefficient problem (9) which serves to bound the error at a given iteration step. To obtain an upper bound on the error over many iteration steps, one considers then the iterated formula using a similar result on the subdomain Ω_2 for e_2^n. Since the iteration for the bounds is an iteration for linear problems, one can obtain, using similar techniques as the ones used in Gander [1997] or Giladi and Keller [2002], the following result.

Theorem 3 (Burgers' Equation). *The overlapping Schwarz waveform relaxation algorithm (3), (4) for the nonlinear advection problem (1) with $f(u) = u^2/2$ and initial guess (8) converges super-linearly. For each $T > 0$ and $i = 1, 2$ we have*

$$\sup_{x\in\Omega_i, 0\leq\tau\leq T} \{e_i^{2n}(x,t)\} \leq C_i e^{\frac{D(T+L)}{\varepsilon}n} \operatorname{erfc}\left(\frac{nL}{\sqrt{\varepsilon T}}\right), \tag{15}$$

where the constants $C_1 = \sup_{0\le t\le T}\{e_1^0(L,t)\}$, $C_2 = \sup_{0\le t\le T}\{e_2^0(0,t)\}$, *and* D *are independent of* ε, L, T *and* n *(but depends on* C *from Theorem 1).*

Remark 3.

(i) If we apply the expansion for the erfc-function for fixed $T, L, \varepsilon > 0$ as in Remark 1, we observe that the algorithm converges super-linearly for $n \to \infty$ and $t \le T$ at the same asymptotic rate as for the linear advection diffusion equation.

(ii) For Burgers' equation, the error estimate contains in addition the factor $e^{\frac{D(T+L)}{\varepsilon}n}$. Thus there exists a $T^* = T^*(n)$ such that (a) the algorithm converges for $\varepsilon \to 0$ on any time interval $[0,T]$ with $T < T^*(n)$, and (b) the estimate for the error e_1^{2n} does not converge to 0 for $\varepsilon \to 0$ on time intervals with $T > T^*(n)$. This scenario does not happen for the linear advection diffusion equation. Even though our estimate might not be sharp, this factor reflects the fact that in the purely hyperbolic case the Schwarz algorithm converges in a finite number of steps. The number of steps however depends on the nonlinearity and the initial data, see Gander and Rohde [2003].

(iii) Theorem 3 can be extended to the case of multiple subdomains, such that the estimate is independent of the number of subdomains, as in the linear case, see Gander and Rohde [2003].

We conclude the paper with a numerical experiment that illustrates the results of Theorem 3. As initial data we take the continuous function

$$u_0(x,t) = \begin{cases} 1 & : \quad x < 0, \\ 1-2x & : \quad 0 \le x < 1, \\ -1 & : \quad x \ge 1. \end{cases}$$

The hyperbolic limit problem with $\varepsilon = 0$ will develop a (standing) shock at $t = 0.5$. Thus for small but positive values of ε the solution will exhibit

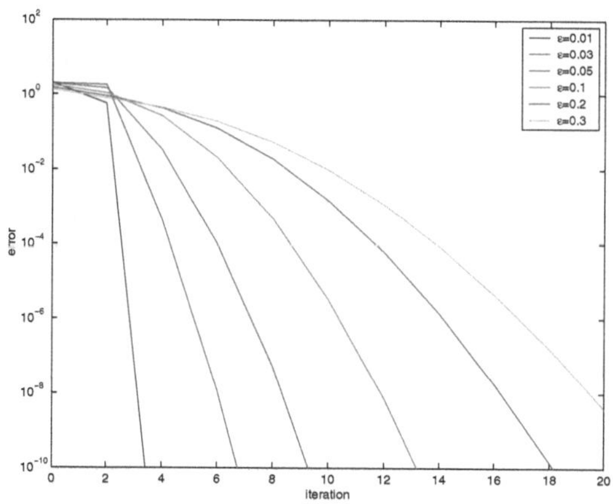

Fig. 1. Convergence rates for various values of the diffusion parameter ε

a sharp layer. For the numerical method we take two bounded subdomains $\Omega_1 = (0, \frac{1}{2} + L)$ and $\Omega_2 = (\frac{1}{2} - L, 1)$ with the overlap parameter $L = 0.1$. We compute the numerical solution up to $T = 0.6$ with a centered finite difference scheme in space, explicit for the nonlinear term and implicit for the Laplacian. The discretization parameters were $\Delta x = 0.01$ and $\Delta t = 0.003$. Figure 1 shows the error on $[0,1] \times [0, 0.6]$ in the L^∞-norm versus the number of iterations at even iteration steps. One can clearly see the super-linear convergence behavior of the overlapping Schwarz waveform relaxation algorithm applied to Burgers' equation with the dependence on ε, as predicted by Theorem 3. One can also see that for ε small, the convergence in a finite number of steps of the hyperbolic limit starts to manifest itself.

References

C. Dafermos. *Hyperbolic conservation laws in continuum physics.* Springer, New York, 2000.

D. S. Daoud and M. J. Gander. Overlapping Schwarz waveform relaxation for convection reaction diffusion problems. In N. D. et al., editor, *13th International Conference on Domain Decomposition Methods*, pages 253–260, 2000.

V. Dolean, S. Lanteri, and F. Nataf. Convergence analysis of a Schwarz type domain decomposition method for the solution of the Euler equations. Technical Report 3916, INRIA, aopr 2000. URL `http://www.inria.fr/RRRT/RR-3916.html`.

M. J. Gander. *Analysis of Parallel Algorithms for Time Dependent Partial Differential Equations.* PhD thesis, Stanford University, Stanford, CA 94305, USA, September 1997.

M. J. Gander. A waveform relaxation algorithm with overlapping splitting for reaction diffusion equations. *Numerical Linear Algebra with Applications*, 6:125–145, 1998.

M. J. Gander and C. Rohde. Overlapping Schwarz waveform relaxation for convection dominated nonlinear conservation laws. Technical Report 12, Mathematisches Institut, Albert-Ludwigs-Universität Freiburg, 2003.

M. J. Gander and H. Zhao. Overlapping Schwarz waveform relaxation for the heat equation in n-dimensions. *BIT*, 42(4):779–795, 2002.

M. Garbey. A Schwarz alternating procedure for singular perturbation problems. *SIAM J. Sci. Comput.*, 17:1175–1201, 1996.

M. Garbey and H. G. Kaper. Heterogeneous domain decomposition for singularly perturbed boundary problems. *SIAM J. Numer. Anal.*, 34:1513–1544, 1997.

E. Giladi and H. B. Keller. Heterogeneous domain decomposition for singularly perturbed boundary problems. *Numer. Math.*, 93(2):279–313, 2002.

A New Cement to Glue Nonconforming Grids with Robin Interface Conditions: The Finite Element Case

Martin J. Gander[1], Caroline Japhet[2], Yvon Maday[3], and Frédéric Nataf[4]

[1] McGill University, Dept. of Mathematics and Statistics, Montreal (`http://www.math.mcgill.ca/mgander/`)
[2] Université Paris 13, Laboratoire d'Analyse, Géométrie et Applications
[3] Université Pierre et Marie Curie, Laboratoire Jacques Louis Lions (`http://www.ann.jussieu.fr/~maday/`)
[4] CNRS, UMR 7641, CMAP, Ecole Polytechnique, 91128 Palaiseau (`http://www.cmap.polytechnique.fr/~nataf/`)

Summary. We present and analyze a new nonconforming domain decomposition method based on a Schwarz method with Robin transmission conditions. We prove that the method is well posed and convergent. Our error analysis is valid in two dimensions for piecewise polynomials of low and high order and also in three dimensions for P_1 elements. We further present an efficient algorithm in two dimensions to perform the required projections between arbitrary grids. We finally illustrate the new method with numerical results.

1 Introduction

We propose a domain decomposition method based on the Schwarz algorithm that permits the use of optimized interface conditions on nonconforming grids. Such interface conditions have been shown to be a key ingredient for efficient domain decomposition methods in the case of conforming approximations (see Després [1991], Nataf et al. [1995], Japhet [1998], Chevalier and Nataf [1998]). Our goal is to use these interface conditions on nonconforming grids, because this simplifies greatly the parallel generation and adaptation of meshes per subdomain. The mortar method, first introduced in Bernardi et al. [1994], also permits the use of nonconforming grids, and it is well suited to the use of "Dirichlet-Neumann" (Gastaldi et al. [1996]) or "Neumann-Neumann" methods applied to the Schur complement matrix. But the mortar method can not be used easily with optimized transmission conditions in the framework of Schwarz methods. In Achdou et al. [2002], the case of finite volume discretizations has been introduced and analyzed. This paper is a first step in the finite element case; we consider only interface conditions of order 0 here.

2 Definition of the method and the iterative solver

We consider the model problem

$$\begin{aligned}(Id - \Delta)u &= f \quad \text{in } \Omega,\\ u &= 0 \quad \text{on } \partial\Omega,\end{aligned} \tag{1}$$

where f is given in $L^2(\Omega)$ and Ω is a $\mathcal{C}^{1,1}$ (or convex) domain in $\mathbb{R}^d$, $d = 2$ or 3. We assume that it is decomposed into K non-overlapping subdomains $\overline{\Omega} = \cup_{k=1}^{K} \overline{\Omega}^k$, where Ω_k, $1 \le k \le K$ are $\mathcal{C}^{1,1}$ or convex polygons in two or polyhedrons in three dimensions. We also assume that this domain decomposition is conforming. Let $\mathbf{n}_k$ be the unit outward normal for Ω^k and $\Gamma^{k,\ell} = \partial\Omega^k \cap \partial\Omega^\ell$.

The variational statement of problem (1) consists of writing the problem as follows: Find $u \in H_0^1(\Omega)$ such that

$$\int_\Omega (\nabla u \nabla v + uv)\, dx = \int_\Omega f v dx, \quad \forall v \in H_0^1(\Omega). \tag{2}$$

We introduce now the space $H_*^1(\Omega^k) = \{\varphi \in H^1(\Omega^k),\, \varphi = 0 \text{ over } \partial\Omega \cap \partial\Omega^k\}$, and the constrained space

$$\mathcal{V} = \{(\mathbf{v},\mathbf{q}) \in (\prod_{k=1}^{K} H_*^1(\Omega^k)) \times (\prod_{k=1}^{K} H^{-1/2}(\partial\Omega^k)),\ v_k = v_\ell \text{ and } q_k = -q_\ell \text{ on } \Gamma^{k,\ell}\}.$$

Problem (2) is then equivalent to the following: Find $(\mathbf{u},\mathbf{p}) \in \mathcal{V}$ such that

$$\sum_{k=1}^{K} \int_{\Omega^k} (\nabla u_k \nabla v_k + u_k v_k)\, dx - \sum_{k=1}^{K} {}_{H^{-1/2}(\partial\Omega^k)} < p_k, v_k >_{H^{1/2}(\partial\Omega^k)} = \sum_{k=1}^{K} \int_{\Omega^k} f_k v_k dx, \quad \forall \mathbf{v} \in \prod_{k=1}^{K} H_*^1(\Omega^k).$$

Being equivalent with the original problem, where $p_k = \frac{\partial u}{\partial \mathbf{n}_k}$ over $\partial\Omega^k$, this problem is naturally well posed. We now describe the iterative procedure in the continuous case, and then its discrete, non-conforming analog.

2.1 The continuous case

We introduce for $\alpha \in \mathbb{R}$, $\alpha > 0$, the zeroth order transmission condition

$$p_k + \alpha u_k = -p_\ell + \alpha u_\ell \quad \text{over } \Gamma^{k,\ell}$$

and the following algorithm: let $(u_k^n, p_k^n) \in H_*^1(\Omega^k) \times H^{-1/2}(\partial\Omega^k)$ be an approximation of (u,p) in Ω^k at step n. Then, (u_k^{n+1}, p_k^{n+1}) is the solution in $H_*^1(\Omega^k) \times H^{-1/2}(\partial\Omega^k)$ of

$$\int_{\Omega^k} \left(\nabla u_k^{n+1} \nabla v_k + u_k^{n+1} v_k\right) dx -_{H^{-1/2}(\partial\Omega^k)} < p_k^{n+1}, v_k >_{H^{1/2}(\partial\Omega^k)}$$
$$= \int_{\Omega^k} f_k v_k dx, \quad \forall v_k \in H^1_*(\Omega^k), \quad (3)$$
$$< p_k^{n+1} + \alpha u_k^{n+1}, v_k >_{\Gamma^{k,\ell}} = < -p_\ell^n + \alpha u_\ell^n, v_k >_{\Gamma^{k,\ell}}, \quad \forall v_k \in H_{00}^{1/2}(\Gamma^{k,\ell}).$$

Convergence of this algorithm is shown in Després [1991] using energy estimates and summarized in the following

Theorem 1. *Assume that f is in $L^2(\Omega)$ and $(p_k^0)_{1\le k\le K} \in \prod_\ell H^{1/2}(\Gamma^{k,\ell})$. Then, algorithm (3) converges in the sense that*

$$\lim_{n\longrightarrow\infty} \left(\|u_k^n - u_k\|_{H^1(\Omega^k)} + \|p_k^n - p_k\|_{H^{-1/2}(\partial\Omega^k)}\right) = 0, \quad \text{for } 1 \le k \le K,$$

where u solves (1), $u_k = u_{|\Omega^k}$, $p_k = \frac{\partial u_k}{\partial \mathbf{n}_k}$ on $\partial\Omega^k$ for $1 \le k \le K$.

2.2 The discrete case

We introduce now the discrete spaces: each Ω_k is provided with its own mesh $\mathcal{T}_h^k$, $1 \le k \le K$, such that $\overline{\Omega}_k = \cup_{T\in\mathcal{T}_h^k} T$. For $T \in \mathcal{T}_h^k$, let h_T be the diameter of T and h the discretization parameter, $h = \max_{1\le k\le K}(\max_{T\in\mathcal{T}_h^k} h_T)$. Let ρ_T be the diameter of the circle in two dimensions or sphere in three dimensions inscribed in T. We suppose that $\mathcal{T}_h^k$ is uniformly regular: there exists σ and τ independent of h such that $\forall T \in \mathcal{T}_h^k$, $\sigma_T \le \sigma$ and $\tau h \le h_T$. We consider that the sets belonging to the meshes are of simplicial type (triangles or tetrahedra), but the following analysis can be applied as well for quadrangular or hexahedral meshes. Let $\mathcal{P}_M(T)$ denote the space of all polynomials defined over T of total degree less than or equal to M for our Lagrangian finite elements. Then, we define over each subdomain two conforming spaces Y_h^k and X_h^k by

$$\begin{aligned} Y_h^k &= \{v_{h,k} \in \mathcal{C}^0(\overline{\Omega}_k),\ v_{h,k|T} \in \mathcal{P}_M(T),\ \forall T \in \mathcal{T}_h^k\}, \\ X_h^k &= \{v_{h,k} \in Y_h^k,\ v_{h,k|\partial\Omega_k\cap\partial\Omega} = 0\}. \end{aligned} \quad (4)$$

The space of traces over each $\Gamma^{k,\ell}$ of elements of Y_h^k is denoted by $\mathcal{Y}_h^{k,\ell}$. In the sequel we assume for the sake of simplicity that referring to a pair (k,ℓ) implies that $\Gamma^{k,\ell}$ is not empty. With each such interface we associate a subspace $\tilde{W}_h^{k,\ell}$ of $\mathcal{Y}_h^{k,\ell}$ like in the mortar element method; for two dimensions, see Bernardi et al. [1994], and for three dimensions see Belgacem and Maday [1997] and Braess and Dahmen [1998]. To be more specific, we recall the situation in two dimensions: if the space X_h^k consists of continuous piecewise polynomials of degree $\le M$, then it is readily noticed that the restriction of X_h^k to $\Gamma^{k,\ell}$ consists of finite element functions adapted to the (possibly curved) side $\Gamma^{k,\ell}$ of piecewise polynomials of degree $\le M$. This side has two end points which we denote by $x_0^{k,\ell}$ and $x_n^{k,\ell}$ and which belong to the set of vertices of the corresponding triangulation of $\Gamma^{k,\ell}$: $x_0^{k,\ell}, x_1^{k,\ell}, \dots, x_{n-1}^{k,\ell}, x_n^{k,\ell}$. The space $\tilde{W}_h^{k,\ell}$ is then the subspace of those elements of $\mathcal{Y}_h^{k,\ell}$ that are polynomials of degree $\leq M-1$ over both $[x_0^{k,\ell}, x_1^{k,\ell}]$ and $[x_{n-1}^{k,\ell}, x_n^{k,\ell}]$. As before, the space $\tilde{W}_h^k$ is the product space of the $\tilde{W}_h^{k,\ell}$ over each ℓ such that $\Gamma^{k,\ell} \neq \emptyset$.

The discrete constrained space is then defined by

$$\mathcal{V}_h = \{(\mathbf{u}_h, \mathbf{p}_h) \in (\prod_{k=1}^{K} X_h^k) \times (\prod_{k=1}^{K} \tilde{W}_h^k),$$

$$\int_{\Gamma^{k,\ell}} ((p_{h,k} + \alpha u_{h,k}) - (-p_{h,\ell} + \alpha u_{h,\ell}))\psi_{h,k,\ell} = 0, \ \forall \psi_{h,k,\ell} \in \tilde{W}_h^{k,\ell}\},$$

and the discrete problem is the following: Find $(\mathbf{u}_h, \mathbf{p}_h) \in \mathcal{V}_h$ such that $\forall \mathbf{v}_h = (v_{h,1}, \dots v_{h,K}) \in \prod_{k=1}^K X_h^k$,

$$\sum_{k=1}^{K} \int_{\Omega^k} (\nabla u_{h,k} \nabla v_{h,k} + u_{h,k} v_{h,k})\, dx - \sum_{k=1}^{K} \int_{\partial\Omega^k} p_{h,k} v_{h,k} ds = \sum_{k=1}^{K} \int_{\Omega^k} f_k v_{h,k} dx. \quad (5)$$

The discrete algorithm is then as follows: let $(u_{h,k}^n, p_{h,k}^n) \in X_h^k \times \tilde{W}_h^k$ be a discrete approximation of $(\mathbf{u}, \mathbf{p})$ in Ω^k at step n. Then, $(u_{h,k}^{n+1}, p_{h,k}^{n+1})$ is the solution in $X_h^k \times \tilde{W}_h^k$ of

$$\int_{\Omega^k} \left(\nabla u_{h,k}^{n+1} \nabla v_{h,k} + u_{h,k}^{n+1} v_{h,k} \right) dx - \int_{\partial\Omega^k} p_{h,k}^{n+1} v_{h,k} ds = \int_{\Omega^k} f_k v_{h,k} dx, \ \forall v_{h,k} \in X_h^k, \quad (6)$$

$$\int_{\Gamma^{k,\ell}} (p_{h,k}^{n+1} + \alpha u_{h,k}^{n+1})\psi_{h,k,\ell} = \int_{\Gamma^{k,\ell}} (-p_{h,\ell}^n + \alpha u_{h,\ell}^n)\psi_{h,k,\ell}, \ \forall \psi_{h,k,\ell} \in \tilde{W}_h^{k,\ell}. \quad (7)$$

Remark 1. Let $\pi_{k,\ell}$ denote the orthogonal projection operator from $L^2(\Gamma^{k,\ell})$ onto $\tilde{W}_h^{k,\ell}$. Then (7) corresponds to

$$p_{h,k}^{n+1} + \alpha \pi_{k,\ell}(u_{h,k}^{n+1}) = \pi_{k,\ell}(-p_{h,\ell}^n + \alpha u_{h,\ell}^n) \quad \text{over } \Gamma^{k,\ell}. \quad (8)$$

Remark 2. A fundamental difference between this method and the original mortar method in Bernardi et al. [1994] is that the interface conditions are chosen in a symmetric way: there is no master and no slave, see also Gander et al. [2001]. Equation (8) is the transmission condition on $\Gamma^{k,\ell}$ for Ω^k, and the transmission condition on $\Gamma^{k,\ell}$ for Ω^ℓ is

$$p_{h,\ell}^{n+1} + \alpha \pi_{\ell,k}(u_{h,\ell}^{n+1}) = \pi_{\ell,k}(-p_{h,k}^n + \alpha u_{h,k}^n) \quad \text{over } \Gamma^{k,\ell}. \quad (9)$$

In order to analyze the convergence of this iterative scheme, we define for any $\mathbf{p}$ in $\prod_{k=1}^K L^2(\partial\Omega_k)$ the norm

$$\|\mathbf{p}\|_{-\frac{1}{2},*} = (\sum_{k=1}^{K} \sum_{\substack{\ell=1 \\ \ell \neq k}}^{K} \|p_k\|^2_{H_*^{-\frac{1}{2}}(\Gamma^{k,\ell})})^{\frac{1}{2}},$$

where $\|.\|_{H_*^{-\frac{1}{2}}(\Gamma^{k,\ell})}$ stands for the dual norm of $H_{00}^{\frac{1}{2}}(\Gamma^{k,\ell})$. Convergence of the algorithm (6)-(7) can be shown again using an energy estimate, see Japhet et al. [2003].

Theorem 2. *Assume that $\alpha h \leq c$ for some constant c small enough. Then, the discrete problem (5) has a unique solution $(\mathbf{u}_h, \mathbf{p}_h) \in \mathcal{V}_h$. The algorithm (6)-(7) is well posed and converges in the sense that*

$$\lim_{n\longrightarrow\infty}(\|u_{h,k}^n - u_{h,k}\|_{H^1(\Omega^k)} + \sum_{\ell\neq k}\|p_{h,k}^n - p_{h,k}\|_{H_*^{-\frac{1}{2}}(\Gamma^{k,\ell})}) = 0, \text{ for } 1 \leq k \leq K.$$

3 Best approximation properties

In this part we give best approximation results of $(\mathbf{u}, \mathbf{p})$ by elements in $\mathcal{V}_h$. The proofs can be found in Japhet et al. [2003] for the two dimensional case with the degree of the finite element approximations $M \leq 13$ and in three dimensions for first order approximations.

Theorem 3. *Assume that the solution $\mathbf{u}$ of (1) is in $H^2(\Omega) \cap H_0^1(\Omega)$ and $u_k = \mathbf{u}_{|\Omega^k} \in H^{2+m}(\Omega^k)$ with $M-1 \geq m \geq 0$, and let $p_{k,\ell} = \frac{\partial u}{\partial \mathbf{n}_k}$ over each $\Gamma^{k,\ell}$. Then, there exists a constant c independent of h and α such that*

$$\|\mathbf{u}_h - \mathbf{u}\|_* + \|\mathbf{p}_h - \mathbf{p}\|_{-\frac{1}{2},*} \leq c(\alpha h^{2+m} + h^{1+m})\sum_{k=1}^{K}\|\mathbf{u}\|_{H^{2+m}(\Omega_k)}$$
$$+ c(\frac{h^m}{\alpha} + h^{1+m})\sum_{k=1}^{K}\sum_{\ell}\|p_{k,\ell}\|_{H^{\frac{1}{2}+m}(\Gamma^{k,\ell})}.$$

Assuming more regularity on the normal derivatives on the interfaces, we have

Theorem 4. *Assume that the solution $\mathbf{u}$ of (1) is in $H^2(\Omega)\cap H_0^1(\Omega)$ and $u_k = \mathbf{u}_{|\Omega^k} \in H^{2+m}(\Omega^k)$ with $M-1 \geq m \geq 0$, and $p_{k,\ell} = \frac{\partial u}{\partial \mathbf{n}_k}$ is in $H^{\frac{3}{2}+m}(\Gamma_{k,\ell})$. Then there exists a constant c independent of h and α such that*

$$\|\mathbf{u}_h - \mathbf{u}\|_* + \|\mathbf{p}_h - \mathbf{p}\|_{-\frac{1}{2},*} \leq c(\alpha h^{2+m} + h^{1+m})\sum_{k=1}^{K}\|\mathbf{u}\|_{H^{2+m}(\Omega_k)}$$
$$+ c(\frac{h^{1+m}}{\alpha} + h^{2+m})(\log h)^{\beta(m)}\sum_{k=1}^{K}\sum_{\ell}\|p_{k,\ell}\|_{H^{\frac{3}{2}+m}(\Gamma^{k,\ell})}.$$

Remark 3. The Robin parameter α can depend on h in the previous theorems, like the optimal Robin parameter α_{opt} in Section 5.

4 Efficient projection algorithm

The projection (8) between non conforming grids is not an easy task in an algorithm, already for two dimensional problems, since one needs to find the intersections of corresponding arbitrary grid cells. A short and efficient algorithm has been proposed in Gander et al. [2001] in the finite volume case with projections on piecewise constant functions. In our case, we denote by n the dimension of $W_h^{k,\ell}$, and we introduce the shape functions $\{\psi_i^{k,\ell}\}_{1\leq i\leq n}$ of $W_h^{k,\ell}$. Then, to compute the right hand side in (7), we need to compute the interface matrix

$$M = (\int_{\Gamma^{k,\ell}} \psi_i^{k,\ell}\psi_j^{\ell,k})_{1\leq i,j\leq n}.$$

In the same spirit as in Gander et al. [2001], the following short algorithm in Matlab computes the interface matrix M for non-matching grids in one pass.

```
function M=InterfaceMatrix(ta,tb);
n=length(tb);
m=length(ta);
ta(m)=tb(n);                          % must be numerically equal
j=1;
M=zeros(n,length(ta));
for i=1:n-1,
  tm=tb(i);
  while ta(j+1)<tb(i+1),
    M(i:i+1,j:j+1)=M(i:i+1,j:j+1)+intMortar(ta(j),ta(j+1),...
      tb(i),tb(i+1),tm,ta(j+1),j==1|j==m-1,i==1|i==n-1);
    j=j+1;
    tm=ta(j);
  end;
  M(i:i+1,j:j+1)=M(i:i+1,j:j+1)+intMortar(ta(j),ta(j+1),...
    tb(i),tb(i+1),tm,tb(i+1),j==1|j==m-1,i==1|i==n-1);
end;
```

It takes two vectors `ta` and `tb` with ordered entries, which represent two non-matching grids at the interface, with `ta(1)=tb(1)`, `ta(end)=tb(end)`, and computes the matrix `M(i,j)=`$\int_{\Gamma^{k,\ell}} b^i a^j$, where b^i is the hat function for the node `tb(i)` and a^j is the hat function for the node `ta(j)`. The mortar condition of constant shape functions at the corners is taken into account, and from the resulting matrix M the first and last row and column needs to be removed. This algorithm has linear complexity; it does a single pass without any special cases or any additional grid. It advances automatically on whatever side the next cell boundary is coming and handles any possible cases of non-matching grids at a one dimensional interface.

5 Numerical results

On the unit square $\Omega = (0,1) \times (0,1)$ we consider the problem

$$(Id - \Delta)u(x,y) = x^3(y^2-2) - 6xy^2 + (1+x^2+y^2)sin(xy), \quad (x,y) \in \Omega,$$
$$u = x^3y^2 + sin(xy), \quad (x,y) \in \partial\Omega,$$

whose exact solution is $u(x,y) = x^3y^2 + sin(xy)$. We decompose the unit square into four non-overlapping subdomains with meshes generated in an independent manner, as shown in Figure 1 on the left. The computed solution

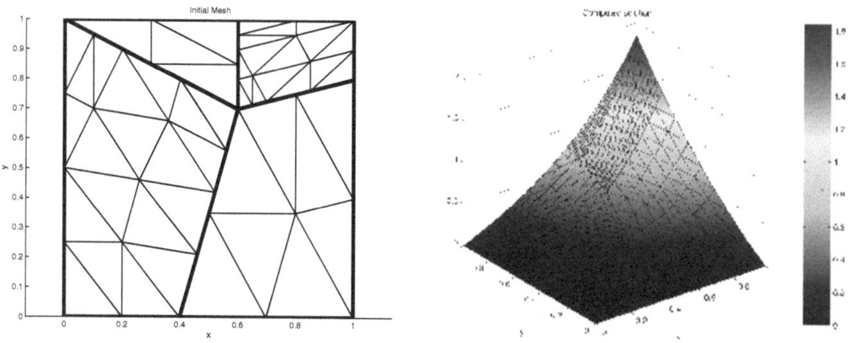

Fig. 1. Initial mesh and computed solution after two refinements

is the solution at convergence of the discrete algorithm (6)-(7), with stopping criterion $\max_{k,\ell/\Gamma^{k,\ell}\neq\emptyset}\left(\int_{\Gamma^{k,\ell}}((p_{h,k}+\alpha u_{h,k})-(-p_{h,\ell}+\alpha u_{h,\ell}))\psi_{k,\ell}\right) < 10^{-8}$, and $\alpha = 10$. On Figure 1 on the right, we show the computed solution.

Figure 2 on the left corresponds to the best approximation error of Theorem 4. On the right, we compare in the case of two subdomains the optimal

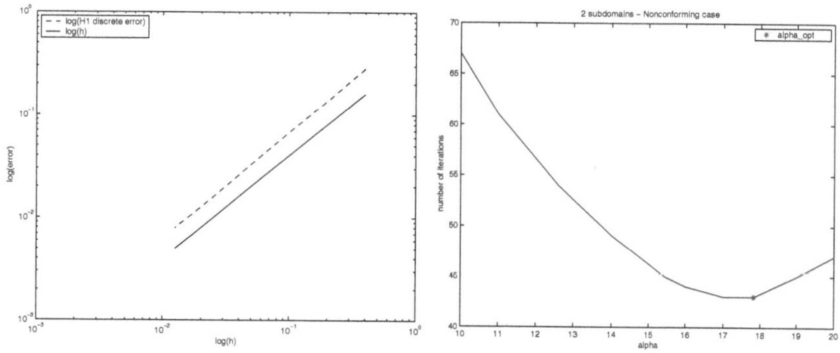

Fig. 2. H^1 error versus h on the left and number of iterations versus α on the right

numerical α to the theoretical value, which minimizes the convergence rate at the continuous level: $\alpha_{opt} = [(\pi^2+1)((\frac{\pi}{h_{min}})^2+1)]^{\frac{1}{4}}$. The nonconforming meshes have 289 and 561 nodes respectively, and the discretization parameters are $h_1 = 0.065$ and $h_2 = 0.032$. We observe that the optimal numerical α is very close to α_{opt}.

References

Y. Achdou, C. Japhet, Y. Maday, and F. Nataf. A new cement to glue non-conforming grids with Robin interface conditions: the finite volume case. *Numer. Math.*, 92(4):593–620, 2002.

F. B. Belgacem and Y. Maday. The mortar element method for three dimensional finite elements. *RAIRO Mathematical Modelling and Numerical Analysis*, 31(2):289–302, 1997.

C. Bernardi, Y. Maday, and A. T. Patera. A new non conforming approach to domain decomposition: The mortar element method. In H. Brezis and J.-L. Lions, editors, *Collège de France Seminar*. Pitman, 1994. This paper appeared as a technical report about five years earlier.

D. Braess and W. Dahmen. Stability estimates of the mortar finite element method for 3-dimensional problems. *East-West J. Numer. Math.*, 6(4):249–264, 1998.

P. Chevalier and F. Nataf. Symmetrized method with optimized second-order conditions for the Helmholtz equation. In *Domain decomposition methods, 10 (Boulder, CO, 1997)*, pages 400–407. Amer. Math. Soc., Providence, RI, 1998.

B. Després. Domain decomposition method and the helmholtz problem. In SIAM, editor, *Mathematical and Numerical aspects of wave propagation phenomena*, pages 44–52. Philadelphia PA, 1991.

M. J. Gander, L. Halpern, and F. Nataf. Optimal Schwarz waveform relaxation for the one dimensional wave equation. Technical Report 469, CMAP, Ecole Polytechnique, September 2001.

F. Gastaldi, L. Gastaldi, and A. Quarteroni. Adaptive domain decomposition methods for advection dominated equations. *East-West J. Numer. Math.*, 4:165–206, 1996.

C. Japhet. Optimized Krylov-Ventcell method. Application to convection-diffusion problems. In P. E. Bjørstad, M. S. Espedal, and D. E. Keyes, editors, *Proceedings of the 9th international conference on domain decomposition methods*, pages 382–389. ddm.org, 1998.

C. Japhet, Y. Maday, and F. Nataf. A new cement to glue nonconforming grids with robin interface conditions: The finite element case. to be submitted, 2003.

F. Nataf, F. Rogier, and E. de Sturler. Domain decomposition methods for fluid dynamics, Navier-Stokes equations and related nonlinear analysis. *Edited by A. Sequeira, Plenum Press Corporation*, pages 367–376, 1995.

Acceleration of a Domain Decomposition Method for Advection-Diffusion Problems

Gert Lube[1], Tobias Knopp[2], and Gerd Rapin[2]

[1] University of Göttingen, Institute of Numerical and Applied Mathematics (http://www.num.math.uni-goettingen.de/lube/)
[2] University of Göttingen, Institute of Numerical and Applied Mathematics

Summary. For advection-diffusion problems we show that a non-overlapping domain decomposition method with interface conditions of Robin type can be accelerated by using a critical parameter of the transmission condition in a cyclic way.

1 Introduction

We consider a non-overlapping domain decomposition method (DDM) with Robin transmission conditions for the advection-diffusion-reaction model

$$Lu := -\epsilon \Delta u + (\mathbf{b} \cdot \boldsymbol{\nabla})u + cu = f \qquad \text{in } \Omega \tag{1}$$

$$u = 0 \qquad \text{on } \partial\Omega \tag{2}$$

in a bounded polyhedral domain $\Omega \subset \mathbf{R}^d$ with a Lipschitz boundary $\partial\Omega$ and $0 < \epsilon \leq 1, \mathbf{b} \in [H^1(\Omega) \cap L^\infty(\Omega)]^d,\ c \in L^\infty(\Omega), f \in L^2(\Omega),\ c - \frac{1}{2}\nabla \cdot \mathbf{b} \geq 0$.

Let $\{\Omega_k\}$ be a non-overlapping macro partition with $\overline{\Omega} = \cup_{k=1}^N \overline{\Omega}_k$. The goal of the well-known DDM of Robin type (Lions [1990]) is to enforce (in appropriate trace spaces) continuity of the solution u and of the diffusive and advective fluxes $\epsilon \nabla u \cdot \mathbf{n}_{kj}$ resp. $-\frac{1}{2}(\mathbf{b} \cdot \mathbf{n}_{kj})u$ on the interfaces $\Gamma_{kj} := \partial\Omega_k \cap \partial\Omega_j$. The algorithm reads in strong form:

For given u_k^n, $n \in \mathbf{N}_0$, in each Ω_k, find (in parallel) u_k^{n+1}, such that

$$Lu_k^{n+1} = f \qquad \text{in } \Omega_k \tag{3}$$

$$u_k^{n+1} = 0 \qquad \text{on } \partial\Omega_k \cap \partial\Omega \tag{4}$$

$$\Phi_k(u_k^{n+1}) = \Phi_k(u_j^n) \qquad \text{on } \Gamma_{kj} \tag{5}$$

with $\Phi_k(u) = \epsilon \boldsymbol{\nabla} u \cdot \mathbf{n}_{kj} + (z_k - \frac{1}{2}\mathbf{b} \cdot \mathbf{n}_{kj})u$ on $\Gamma_{kj},\ k \neq j$ and the outer normal vector $\mathbf{n}_{kj}$ on $\partial\Omega_k \cap \partial\Omega_j$. By determining the interface parameter $z_k > 0$ in a proper way the convergence of the method (3)-(5) can be accelerated.

Let $\mathcal{T}_h$ be an admissible, quasi-uniform triangulation of Ω being aligned with the macro partition. $V_h := \{v \in H_0^1(\Omega) \mid v|_K \in \mathcal{P}_l(K)\ \forall K \in \mathcal{T}_h\}$ denotes a conforming finite element (FE) space of order l. The stabilized FE method

$$Find\ u_h \in V_h,\ such\ that: \quad a^s(u_h, v_h) = l^s(v_h) \qquad \forall v_h \in V_h\ , \tag{6}$$

$$a^s(u,v) = (\epsilon \nabla u, \nabla v)_\Omega + (\mathbf{b} \cdot \nabla u + c\,u,\ v)_\Omega + \sum_{T \in \mathcal{T}_h} (\delta_T L u,\ \mathbf{b} \cdot \nabla v)_T,$$

$$l^s(v) = (f,\ v)_\Omega + \sum_{T \in \mathcal{T}_h} (\delta_T f,\ \mathbf{b} \cdot \nabla v)_T, \qquad \delta_T \sim h_T^2(\epsilon + h_T \|\mathbf{b}\|_{\infty,T})^{-1}$$

with residual stabilization provides improved stability and accuracy (w.r.t. the streamline derivative $\mathbf{b} \cdot \nabla u$) in the hyperbolic limit $\epsilon \to 0$ of (1)-(2).

Assume, for simplicity, that the macro partition has no interior cross-points. Then we restrict the bilinear and linear forms a^s and l^s to each subdomain by $a_k^s = a^s|_{\Omega_k}$ and $l_k^s = l^s|_{\Omega_k}$. Moreover, we define $V_{k,h} = V_h|_{\Omega_k}$ and the trace space $W_{kj,h} = V_h|_{\Gamma_{kj}}$. Finally, we denote by $\langle \cdot, \cdot \rangle_{\Gamma_{kj}}$ the dual product on $(W_{kj,h})^* \times W_{kj,h}$. Then the weak formulation of the fully discretized DDM is given by

(I) Parallel computation step :
For $k = 1, \ldots, N$ *find* $u_{h,k}^{n+1} \in V_{k,h}$ *such that for all* $v_k \in V_{k,h}$:

$$a_k^s(u_{h,k}^{n+1}, v_k) + \langle z_k u_{h,k}^{n+1}, v_k \rangle_{\Gamma_k} = l_k^s(v_k) + \sum_{j(\neq k)} \langle \Lambda_{jk}^n, v_k \rangle_{\Gamma_{kj}}. \tag{7}$$

(II) Communication step :
For all $k \neq j$, *update Lagrange multipliers* $\Lambda_{kj}^{n+1} \in W_{kj,h}^*$ *such that:*

$$\langle \Lambda_{kj}^{n+1}, \phi \rangle_{\Gamma_{kj}} = \langle (z_k + z_j) u_{h,k}^{n+1} - \Lambda_{jk}^n, \phi \rangle_{\Gamma_{kj}}, \qquad \forall \phi \in W_{kj,h}. \tag{8}$$

This method is very easy to implement. It is used as a building block in a parallelized flow solver for the thermally driven incompressible Navier-Stokes problem, cf. Knopp et al. [2002]. A fast convergence of the DDM is desirable, in particular for time-dependent problems.

It is well-known that the algorithm (7)-(8) is well-posed if $z_k = z_j > 0$. Moreover, the sequence $\{u_k^n\}_{n \in \mathbf{N}}$ is strongly convergent according to

$$\lim_{n \to \infty} |||\ u_{h,k}^n - u_h|_{\Omega_k} |||_{\Omega_k} = 0 \tag{9}$$

where $|||v|||_{\Omega_k} := \sqrt{a_k^s(v,v)}$, Lube et al. [2000]. The convergence speed depends in a sensitive way on the parameters z_k which have to be optimized. In Sect. 2 we review an *a priori* optimization introduced by Nataf and Gander. Sect. 3 is devoted to an *a posteriori based* approach.

2 A-priori optimization of the interface condition

The convergence of the Robin DDM (3)-(5) can be described in simple cases using a Fourier analysis. Nataf and Gander proposed a semi-continuous *a priori optimization* of the interface parameter z over a relevant range $S = (s_{min}, s_{max})$ of Fourier modes. An optimization is important for highly oscillatory solutions, e.g. for the Helmholtz equation (1)-(2) with $\mathbf{b} \equiv \mathbf{0}$ and $c/\epsilon \ll -1$. An improved idea is a *cyclic* change of z for appropriate frequency ranges. For the *symmetric* case with $\mathbf{b} = \mathbf{0}$ and $\epsilon = 1$, Gander and Golub [2002] proposed the following variant of the DDM (3)-(5) in $\Omega = \mathbf{R}^2$, $\Omega_1 = \mathbf{R}^+ \times \mathbf{R}$, $\Omega_2 = \mathbf{R}^- \times \mathbf{R}$ with the *cyclic* Robin condition

$$\nabla u_1^{n+1} \cdot \mathbf{n}_1 + z^{n\, mod(m)} u_1^{n+1} = \nabla u_2^n \cdot \mathbf{n}_1 + z^{n\, mod(m)} u_2^n \tag{10}$$

and similarly for u_2^{n+1} on $\Gamma = \{0\} \times \mathbf{R}$ for $m = 2^l$ and appropriate chosen $z^0, \ldots, z^{m-1}$. For $l = 0$, Gander and Golub [2002] obtain the following contraction rate

$$\rho(s, z) = \left(\frac{\sqrt{c + s^2} - z^0}{\sqrt{c + s^2} + z^0} \right)^2$$

for the $s-$th Fourier mode. For the mesh parameter h, an optimization over $S = (s_{min}, \pi/h)$ gives $\min_{z^0 \geq 0} \left(\max_{s \in S} \rho(s, z^0) \right) = 1 - \mathcal{O}(\sqrt{h})$. In the cyclic case $m = 2^l$ one gets the rate

$$\rho(m, s, z) = \prod_{j=1}^{m} \left(\frac{\sqrt{c + s^2} - z^{n\, mod(m)}}{\sqrt{c + s^2} - z^{n\, mod(m)}} \right)^{2/m}$$

and the optimized result

$$\min_{z \geq 0} \left(\max_{s \in S} \rho(m, s, z) \right) \approx 1 - \frac{4}{m} \left[\frac{(c + s_{min}^2) h^2}{16 \pi^2} \right]^{\frac{1}{4m}}, \qquad h \to +0.$$

This result is useful for meshes with $\frac{(c + s_{min}^2) h^2}{16 \pi^2} \leq 1$, but this estimate deteriorates in the singularly perturbed case, i.e. for fixed h and $c \to +\infty$.

For the *non-symmetric case*, the optimization of the Schwarz method corresponding to $l = 0$ can be found in Nataf [2001]. The extension to a cyclic Schwarz method with $l \geq 1$ is open.

3 A posteriori based design of the interface condition

As an alternative to the a priori based design of the interface parameter we propose an approach based on an *a posteriori* estimate. Consider a simplified situation with $\overline{\Omega} = \overline{\Omega}_1 \cup \overline{\Omega}_2 \subset \mathbf{R}^2$ with $\text{meas}_1(\partial\Omega \cap \partial\Omega_i) > 0$ and straight interface $\Gamma = \partial\Omega_1 \cap \partial\Omega_2$ of size $H = \text{meas}\ (\Gamma) \sim \text{diam}(\Omega_i)$, $i = 1, 2$. Set $W = H_{00}^{\frac{1}{2}}(\Gamma)$. We assume constant data ϵ, $\mathbf{b}$, c. In Lube et al. [2000] we proved

Theorem 1. *Let u_h be the solution of (6). The DDM-subdomain error $e_{h,k}^n = u_{h,k}^n - u_h|_{\Omega_k}$, $k \in \{1, 2\}$, can be controlled via (computable) interface data:*

$$|||e_{h,k}^{n+1}|||_{\Omega_k} \leq A_j \ \|u_{h,k}^n - u_{h,j}^{n+1}\|_W + B_j \left| z_k - \frac{\mathbf{b} \cdot \mathbf{n}_k}{2} \right| \|u_{h,k}^n - u_{h,j}^{n+1}\|_{L^2(\Gamma)} \tag{11}$$

for $j = 3 - k$ and with data-dependent constants

$$A_j = \sqrt{\epsilon}\left(1 + \sqrt{\frac{c}{\epsilon}}H + \min\left[\frac{\|\mathbf{b}\|_\infty H}{\epsilon}; \frac{\|\mathbf{b}\|_\infty}{\sqrt{c\epsilon}}\right]\right), \qquad B_j = \sqrt{\frac{H}{\epsilon}}. \tag{12}$$

This result motivates to equilibrate the two right-hand side terms in (11) in order to obtain information about the design of the interface parameter z_k. In Lube et al. [2000] we considered the estimate

$$|||e_{h,k}^{n+1}|||_{s,\Omega_k} \leq \max\left(A_j;\ B_j C\sqrt{H}\left|z_k - \frac{1}{2}\mathbf{b}\cdot\mathbf{n}_k\right|\right)\|u_{h,k}^n - u_{h,j}^{n+1}\|_W \tag{13}$$

using the continuous embedding result $\|\phi\|_{L^2(\Gamma)} \leq C\sqrt{H}\|\phi\|_W$ for all $\phi \in W$. On the other hand, an inverse estimate in (11) leads to

$$|||e_{h,k}^{n+1}|||_{s,\Omega_k} \leq \max\left(CA_j h^{-\frac{1}{2}}; B_j\left|z_k - \frac{1}{2}\mathbf{b}\cdot\mathbf{n}_k\right|\right)\|u_{h,k}^n - u_{h,j}^{n+1}\|_{L^2(\Gamma)}. \tag{14}$$

In the *symmetric* case $\mathbf{b} = \mathbf{0}$ we get from (13) and (14)

$$z_k \sim \frac{\epsilon}{H}\sqrt{\frac{H}{L}}\left(1 + H\sqrt{\frac{c}{\epsilon}}\right), \qquad L \in \{h, H\}. \tag{15}$$

In the *non-symmetric case* $\mathbf{b} \neq \mathbf{0}$, the design of z_k has to match the hyperbolic limit of the Robin condition, i.e.

$$0 = \lim_{\epsilon\to 0}\Phi_k(u) = (-\frac{1}{2}\mathbf{b}\cdot\mathbf{n}_k + \lim_{\epsilon\to 0} z_k)u \ \text{ if } \mathbf{b}\cdot\mathbf{n}_k \geq 0.$$

By extending this condition to the inflow part of $\partial\Omega_k$ with $\mathbf{b}\cdot\mathbf{n}_k < 0$, we obtain from (13)-(14) as a reasonable choice

$$z_k = \frac{1}{2}|\mathbf{b}\cdot\mathbf{n}_k| + R_k(L), \qquad L \in \{h, H\}, \tag{16}$$

$$R_k(L) \sim \frac{\epsilon}{H}\sqrt{\frac{H}{L}}\left(1 + H\sqrt{\frac{c}{\epsilon}} + \min\left[\frac{H\|\mathbf{b}\|}{\epsilon}; \frac{\|\mathbf{b}\|}{\sqrt{c\epsilon}}\right]\right). \tag{17}$$

Inserting (16), (17) with $L = H$ in (13) and applying an inverse inequality, we obtain the *optimized a posteriori* estimates

$$|||e_{h,k}^{n+1}|||_{\Omega_k} \leq A_j\|u_{h,k}^n - u_{h,j}^{n+1}\|_W \leq CA_j h^{-\frac{1}{2}}\|u_{h,k}^n - u_{h,j}^{n+1}\|_{L^2(\Gamma)}. \tag{18}$$

The last estimate also follows directly by inserting (16), (17) with $L = h$ in (14). Therefore we propose to extend the condition (16) to $L \in [h, H]$.

In Lube et al. [2000] we considered the case $L = H$. This choice usually allows a fast error reduction down to the discretization error level if the solution has no highly oscillatory behaviour. Fortunately, the latter case is rare for problem (1)-(2) with $c - \frac{1}{2}\nabla\cdot\mathbf{b} \geq 0$.

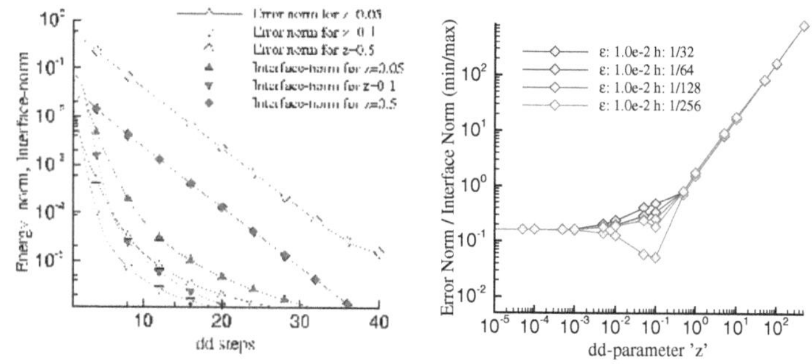

Fig. 1. Reliability of the a posteriori estimate for $h = \frac{1}{128}$ (left), Control of $\max_n / \min_n \sum_{i=1}^{4} |||e_{h,k}^{n+1}|||_{\Omega_i} / \sum_{i \neq j} \|u_{h,i}^{n+1} - u_{h,j}^{n}\|_{L^2(\Gamma_{ij})}$ vs. z (right)

Example 1. Consider the problem (1)-(2) with $\mathbf{b} \equiv 0$, $\epsilon = 10^{-2}$, $c = 1$ in $\Omega = (0,1)^2$. The exact (smooth) solution is $u = x_1(1-x_1)x_2(1-x_2)e^{x_1 x_2}$. We denote the solution of (6) with $\mathcal{P}_1$-elements and $h = \frac{1}{128}$ by $u_k = u_h|_{\Omega_k}$. The DDM on an equidistant 2×2 macro partition with an initial guess $\Lambda_{jk}^0 = 0$ leads to the sequence $u_{h,k}^n$. The stopping criterion $\sum_k |||u_{h,k}^n - u_h|_{\Omega_k}|||_{\Omega_k} \leq 10^{-6}$ has a tolerance beyond the discretization error level.

Fig. 1 (left) shows that the subdomain error $|||\cdot|||_{\Omega_k}$ is clearly controlled by the $L^2(\Gamma)$ interface error according to Theorem 1. Moreover, the convergence of the DD-iteration depends strongly on z. The fast error reduction in the first phase corresponds to a fast reduction of "low" frequencies; but then a (very) slow reduction of "higher" modes can be seen. In Fig. 1 (right) we control the maximal/minimal (w.r.t. to the number n of DD steps) ratio between the subdomain and interface errors for varying h. The value $z_k \sim \frac{1}{10}$ corresponding to the minimum of this ratio for $h = \frac{1}{256}$ is in agreement with the value predicted by (15) with $L = H$. As predicted by Theorem 1, we observe a linear dependence of the error on z for increasing z. □

Obviously, the results of Example 1 with the optimized value of z according to (16), (17) with $L = H$ depend only on the data of the problem (1)-(2) and not on h. We want to check this result for other typical cases.

Example 2. Let be Ω and the solution u as in Example 1. The FEM solution u_h of (6) is computed with $\mathcal{P}_1$-elements on a fine mesh with $h = \frac{1}{256}$ and with SUPG stabilization in advection-dominated cases for

A: Symmetric case: $\mathbf{b} = (0,0)$, $c = 1$, DDM with 4 subdomains,
B: Case $|\mathbf{b} \cdot \mathbf{n}_i| > 0$: $\mathbf{b} = (2,1)$, $c = 1$, DDM with 2 subdomains,
C: Case $|\mathbf{b} \cdot \mathbf{n}_i| \equiv 0$: $\mathbf{b} = (0,1)$, $c = 1$, DDM with 2 subdomains .

The initial guess for the Lagrange multipliers is $\Lambda_{ij}^0 = 0$. The stopping criterion for the error between the discrete solutions with and without DDM is $\sum_k \|u_{h,k}^n - u_h|_{\Omega_k}\|_{L^2(\Omega_k)} \leq 10^{-6}$. The convergence in this range is predicted

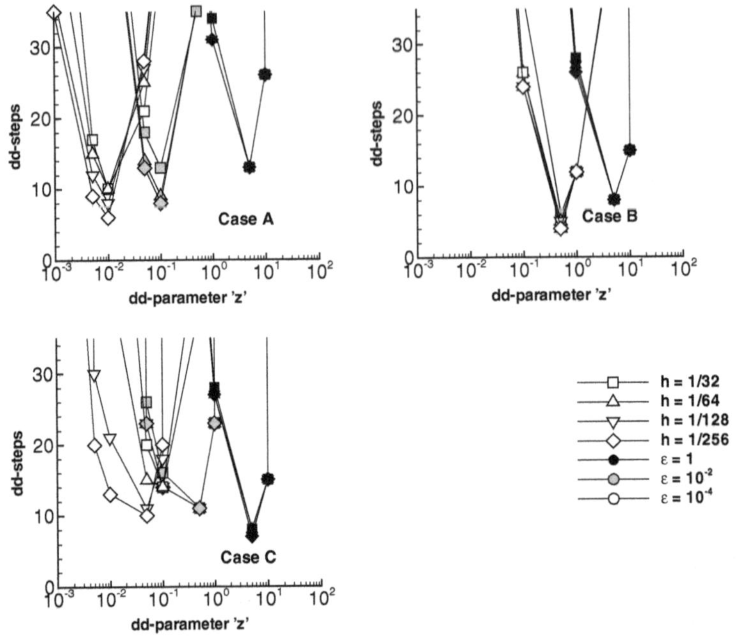

Fig. 2. Optimization of the interface parameter z with one-level approach

by the data of (1)-(2) and is h-independent. The optimal values of z_k are predicted by the optimized z_k from (16), (17) with $L = H$, see Fig. 2. □

The nice convergence behaviour can be explained by the smoothness of the solutions and of the initial guess Λ_{ij}^0. Moreover, in our experiments we never found problems for singularly perturbed problems with sharp layers.

Nevertheless, the convergence behaviour of the Robin-DDM is not satisfactory beyond the discretization error level. Moreover, regarding our application to flow problems (Knopp et al. [2002]), in the turbulent case the solution usually has high-frequent components which may not be efficiently damped in our previous approach. As a remedy we propose a combination of the a posteriori control of the interface error with a *cyclic multi-level* version of the DDM:

1: (optionally) Apply (7)-(8) with the optimized z_k from (16)-(17) with $L = H$ until reduction of the interface error down to discretization error level, e.g. $\|u_{h,i}^n - u_{h,j}^{n+1}\|_{L^2(\Gamma)} \leq \kappa h^{l+1/2}$ for $\mathcal{P}_l$ elements.

2: Apply (7)-(8) in a cyclic way with p levels (see below) using (16)-(17) with $z_k^1, ..., z_k^p$ related to $L = H$ (for z_k^1) and $l = h$ (for z_k^p), resp., and an even number (to our experience, 4 or 6 are sufficient) of DD steps per level until $\|u_{h,i}^n - u_{h,j}^{n+1}\|_{L^2(\Gamma)} \leq TOL$.

Let us discuss this approach for some cases of Example 2. First of all, we have to fix the number p of levels. Assume a dyadic representation of the coarse and of the fine mesh of the domain $\Omega = (0,1)^2$ with $H = 2^{-s}$, $h = 2^{-t}$, $s,t \in \mathbf{N}$. From (16)-(17) we obtain $R_k(L) \leq R_k(h) \sim \sqrt{H/h}$, i.e. a mild dependence on $\sqrt{H/h}$. We propose the following rule: For $2^p < \sqrt{H/h} \leq 2^{p+1}$, take p levels. Thus we obtain for a very fine mesh width $h = 2^{-10}$ a number of two levels for a coarse grid width $H = 2^{-5}$ and of four levels for $H = 2^{-1}$.

Example 3. Consider the situation of Example 2 with a 2×2 macro partition with $H = \frac{1}{2}$ and a fine mesh with $h = 2^{-6}$. This leads to $p = 2$ levels. We start with the symmetric case of (1)-(2) with $\epsilon = 1, \mathbf{b} = \mathbf{0},\ c = 1$. The fast error reduction within the first steps is followed by a very slow reduction in the one-level case, cf. Fig. 3 (left). Here u^h and u^h_{seq} denote the solutions with and without DDM. The two-level method with 6 DD-steps per level leads to a dramatic acceleration, cf. Fig. 3 (right).

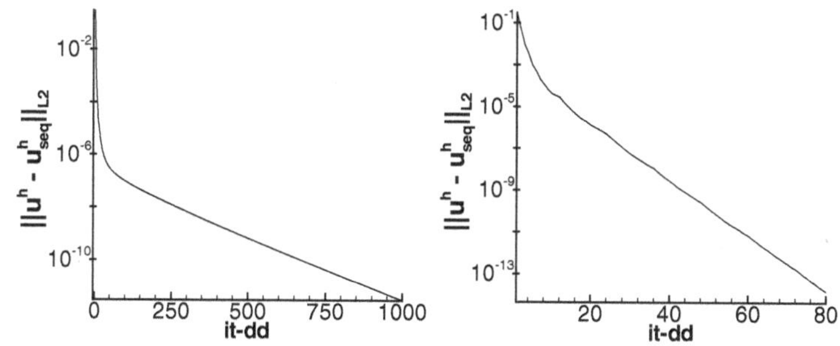

Fig. 3. Error reduction for the *symmetric* case: $p = 1$ (left), $p = 2$ (right)

Consider now the non-symmetric and advection-dominated case with $\epsilon = 10^{-5}, \mathbf{b} = \frac{(1,2)^T}{\sqrt{5}},\ c = 0$. In Fig. 4 we observe a similar behaviour of the proposed approach with $p = 1$ (left) and $p = 2$ (right) levels, although the acceleration is not so dramatic as in the symmetric case. □

Finally, let us note an observation of Gander and Golub [2002] for the symmetric case: The quality of the cyclic DDM (3)-(5) with an optimized condition (10) as a solver increases with the number of levels such that no improvement can be found with Krylov acceleration. A similar behaviour is very likely in the non-symmetric case.

4 Summary

Considerable progress has been reached for Schwarz methods with (a priori) optimized transmission conditions. We propose an approach based on a refined a posteriori error estimate for a DDM with transmission conditions of Robin type. For the one-level variant, this condition can be optimized in such a

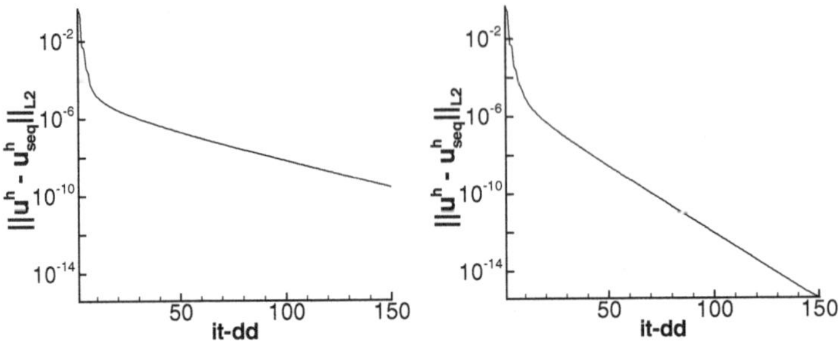

Fig. 4. Error reduction for the *non-symmetric* case: $p = 1$ (left), $p = 2$ (right)

way that the convergence is very reasonable down to the discretization error level; but then one observes a rapid slow-down of error reduction for higher error modes. This is valid for "smooth" solutions and is in contrast to highly oscillatory solutions typically appearing, e.g., for turbulent flows.

A multilevel-type method with optimized interface parameters allows a strong acceleration of the convergence. The approach is motivated by theoretical results, but more efforts are necessary to improve its present state. An advantage of the method over a priori optimized methods is the control of the convergence within the iteration. Moreover, a combination with adaptive mesh refinement is possible. It remains open whether the method is linearly convergent. Moreover, a genuinely multilevel-type implementation might be possible. Finally, the extension to incompressible flows has to be done.

References

M. J. Gander and G. Golub. A non-overlapping optimized Schwarz method which converges with arbitrarily weak dependence on h. In I. H. et.al., editor, *Proc. Fourteenth Intern. Conf. on Domain Decomposition Methods*, pages 281–288. DDM.org, 2002.

T. Knopp, G. Lube, R. Gritzki, and M. Rösler. Iterative substructuring techniques for incompressible nonisothermal flows and its application to indoor air flow simulation. *Intern. J. Numer. Meths. Fluids*, 40:1527–1538, 2002.

P. L. Lions. On the Schwarz alternating method III: A variant for non-overlapping domains. In Chan et al., editor, *Proc. Third Intern. Symp. on Domain Decomposition Methods*, pages 202–223. SIAM, 1990.

G. Lube, L. Müller, and F. Otto. A non-overlapping domain decomposition method for the advection-diffusion problem. *Computing*, 64:49–68, 2000.

F. Nataf. Interface connections in domain decomposition methods. In *NATO Advanced Study Institute, Modern Methods in Scientific Computing and Applications*. NATO Science Ser.II., vol. 75, 2001.

A Stabilized Three-Field Formulation and its Decoupling for Advection-Diffusion Problems

Gerd Rapin[1] and Gert Lube[1]

Georg-August-Universität Göttingen, Institut für Numerische und Angewandte Mathematik (`http://www.num.math.uni-goettingen.de/`)

Summary. We propose a new stabilized three-field formulation applied to the advection-diffusion equation. Using finite elements with SUPG stabilization in the interior of the subdomains our approach enables us to use almost arbitrary discrete function spaces. They need not to satisfy the inf-sup conditions of the standard three-field formulation. The scheme is stable and satisfies an optimal a priori estimate. Furthermore, we show how the scheme can be solved efficiently in parallel by an adapted Schur complement equation and an alternating Schwarz algorithm. Finally some numerical experiments confirm our theoretical results.

1 Introduction

In an bounded Lipschitz domain $\Omega \subset \mathbb{R}^d$, $d = 2, 3$, we consider the problem

$$Lu := -\epsilon \triangle u + \mathbf{b} \cdot \nabla u + cu = f \quad \text{in } \Omega, \qquad u = 0 \quad \text{on } \partial\Omega \tag{1}$$

with $\epsilon > 0$, $\mathbf{b} \in (W^{1,\infty}(\Omega))^d$, $c \in L^\infty(\Omega)$, and $f \in L^2(\Omega)$. Especially the singularly perturbed case $\epsilon << 1$ is of interest, since there the solution can possess sharp layers. Moreover, it is well known, that simple numerical methods fail, since spurious oscillations of the numerical solution may occur.

The three-field formulation was introduced by Baiocchi et al. [1992] (see also Brezzi and Marini [2001]). Decomposing the domain into non-overlapping subdomains the method allows different discretization techniques in different subdomains. Especially, the treatment of non-matching grids is possible. In the discrete case the corresponding function spaces must satisfy two inf-sup conditions. This is quite restrictive for the choice of the discrete spaces. In our stabilized scheme we circumvent these conditions by appending additional terms. The latter terms are well-adapted to the hyperbolic limit $\epsilon = 0$.

2 The three-field formulation

First let us tackle the global problem (1). The weak formulation reads:

$$\text{Find } w \in H_0^1(\Omega) \mid a_\Omega(w,v) = l_\Omega(v), \quad \forall v \in H_0^1(\Omega) \tag{2}$$

with

$$a_G(w,v) = \epsilon \int_G \nabla w \cdot \nabla v dx + \int_G (\mathbf{b} \cdot \nabla w + cw)v dx, \quad l_G(v) = \int_G f v dx$$

for all $w, v \in H^1(\Omega)$ and a domain $G \subset \Omega$. To ensure the well-posedness of (2) we assume the existence of a constant $c_0 > 0$ such that $\tilde{c} := c - \frac{1}{2}\nabla \cdot \mathbf{b} \geq c_0$. Then the Lemma of Lax-Milgram shows that (2) possesses a unique solution.

In order to introduce the three-field formulation, we divide the domain Ω into N non-overlapping subdomains Ω_i with sufficiently smooth boundaries, i.e. $\overline{\Omega} = \bigcup_{i=1}^N \overline{\Omega}_i, \quad \Omega_i \cap \Omega_j = \emptyset, i \neq j$. Moreover, we define local interfaces $\Gamma_i := \partial\Omega_i \setminus \partial\Omega$ and the skeleton $\Gamma := \left(\bigcup_{i=1}^N \partial\Omega_i\right) \setminus \partial\Omega$. For the three-field formulation three different function spaces are introduced. The first function space $\mathbf{V} := \prod_{i=1}^N V^i$ with $V^i := \{v^i \in H^1(\Omega_i) \mid v|_{\partial\Omega \cap \partial\Omega_i} = 0\}$ is defined on the subdomains. Furthermore, we need a space of Lagrange multipliers Λ^i on each local interface Γ_i. The local Lagrange multiplier space Λ^i is given by the dual space of $H_{00}^{\frac{1}{2}}(\Gamma_i)$. We denote the dual product on $\langle \Lambda^i, H_{00}^{\frac{1}{2}}(\Gamma_i) \rangle$ by $\langle \cdot, \cdot \rangle_i$. The global space is given by $\boldsymbol{\Lambda} := \prod_{i=1}^N \Lambda^i$. The third function space is defined on Γ by $\Phi := \{\varphi \in L^2(\Gamma) \ : \ \text{there exists } u \in H_0^1(\Omega), \ u = \varphi \text{ on } \Gamma \ \}$.

Now we formulate the following three-field formulation (cf. Bertoluzza and Kunoth [2000]): Find $\mathbf{u} \in \mathbf{V}$, $\boldsymbol{\lambda} \in \boldsymbol{\Lambda}$ and $\psi \in \Phi$, such that

$$\begin{aligned}
&i) \sum_{i=1}^N a_{\Omega_i}(u^i, v^i) - \sum_{i=1}^N \epsilon\langle \lambda^i, v^i \rangle_i = \sum_{i=1}^N l_{\Omega_i}(v^i) && \forall \mathbf{v} \in \mathbf{V} \\
&ii) \sum_{i=1}^N \epsilon\langle \nu^i, \psi - u^i \rangle_i = 0 && \forall \boldsymbol{\nu} \in \boldsymbol{\Lambda} \\
&iii) \sum_{i=1}^N \epsilon\langle \lambda^i, \phi \rangle_i = 0 && \forall \phi \in \Phi.
\end{aligned} \tag{3}$$

It can be shown that the three-field formulation (3) possesses a unique solution $(\mathbf{u}, \boldsymbol{\lambda}, \varphi) \in \mathbf{V} \times \boldsymbol{\Lambda} \times \Phi$. If the solution $w \in H_0^1(\Omega)$ of (2) is sufficiently regular, i.e. $\triangle w \in L^2(\Omega_i)$, $i = 1, \ldots N$, the solution can be represented by

$$\mathbf{u} = (w|_{\Omega_1}, \ldots, w|_{\Omega_N}), \quad \boldsymbol{\lambda} = \left(\frac{\partial w}{\partial \mathbf{n}_1}|_{\Gamma_1}, \ldots, \frac{\partial w}{\partial \mathbf{n}_N}|_{\Gamma_N}\right), \quad \varphi = w|_\Gamma \tag{4}$$

where $\mathbf{n}_i$ is the outward normal of Ω_i (cf. Baiocchi et al. [1992]).

3 A stabilized three-field formulation

Now the three-field formulation (3) is discretized by linear finite elements. To this end we introduce quasi-uniform meshes $\mathcal{T}_u^i$, $\mathcal{T}_\lambda^i$ and $\mathcal{T}_\varphi$ on Ω_i, Γ_i and Γ. The meshes can be non-matching. But for simplicity we assume that all meshes have the same global mesh size h. Moreover, we need the notation $\partial\mathcal{T}_u^i$ for the restriction of $\mathcal{T}_u^i$ onto the local interface Γ_i. Thus we obtain discrete spaces $\mathbf{V}_h \subset \mathbf{V}$, $\quad \boldsymbol{\Lambda}_h \subset \boldsymbol{\Lambda}$ and $\Phi_h \subset \Phi$.

Replacing the continuous function spaces by discrete subspaces, the well-posedness of the discrete scheme of (3) requires two certain inf-sup conditions. One idea to guarantee the conditions is proposed in Brezzi and Marini [2001]. They enrich the space $\boldsymbol{V}_h$ by bubble functions. Here, we avoid the constraints by adding some stabilization terms modifying ideas of Baiocchi et al. [1992].

In the advection dominated case further problems occur. Using a standard discretization it is well known that there may arise spurious oscillations of the computed solution. Therefore we use the SUPG method and define

$$a_{\Omega_i}^{SD}(u_h^i, v_h^i) := a_{\Omega_i}(u_h^i, v_h^i) + \sum\nolimits_{T\in\mathcal{T}_u^i} \delta_T (Lu_h^i, \mathbf{b}\cdot\nabla v_h^i)_T,$$
$$l_{\Omega_i}^{SD}(v_h^i) := l_{\Omega_i}(v_h^i) + \sum\nolimits_{T\in\mathcal{T}_u^i} \delta_T (f, \mathbf{b}\cdot\nabla v_h^i)_T$$

for $i = 1, \ldots, N$. The parameter δ_T is defined by $\delta_T := \delta_0 h_T \|\mathbf{b}\|_{L^\infty(T)}^{-1}$ in the advection dominated regime for $Pe_T := \frac{1}{2} h_T \epsilon^{-1} \|\mathbf{b}\|_{L^\infty(T)} > 1$ and by $\delta_T := \frac{1}{2}\delta_0 h_T^2 \epsilon^{-1}$ else. Now the error in the interior of the subdomains can be controlled by the streamline diffusion norm

$$|||v_h|||_{SD,\Omega_i}^2 := \epsilon |v_h|_{1,\Omega_i}^2 + \|\sqrt{\tilde{c}} v_h\|_{0,T}^2 + \sum\nolimits_{T\in\mathcal{T}_u^i} \delta_T \|\mathbf{b}\cdot\nabla v_h\|_{0,T}^2$$

which gives us additional control in the streamline direction. Taking all the mentioned problems into account we end up with the following stabilized three-field formulation: Find $\mathbf{u}_h \in \mathbf{V}_h, \boldsymbol{\lambda}_h \in \boldsymbol{\Lambda}_h$ and $\varphi_h \in \Phi_h$, such that

$$\begin{aligned}
&i) \sum\nolimits_{i=1}^N \left\{ a_{\Omega_i}^{SD}(u_h^i, v_h^i) - l_{\Omega_i}^{SD}(v_h^i) - \epsilon\langle \lambda_h^i, v_h^i\rangle_i + f_i^-(u_h^i - \varphi_h, v_h^i) \right\} = 0 \\
&ii) \sum\nolimits_{i=1}^N \left\{ \epsilon\langle \nu_h^i, u_h^i - \varphi_h\rangle_i - \sum\nolimits_{E\in\partial\mathcal{T}_u^i} \beta_E \left(\frac{\partial u_h^i}{\partial \mathbf{n}_i} - \lambda_h^i, \nu_h^i \right)_E \right\} = 0 \qquad (5) \\
&iii) \sum\nolimits_{i=1}^N \left\{ \epsilon\langle \lambda_h^i, \phi_h\rangle_i - f_i^+(u_h^i - \varphi_h, \phi_h) \right\} = 0
\end{aligned}$$

for all $\mathbf{v}_h \in \mathbf{V}_h, \boldsymbol{\nu}_h \in \boldsymbol{\Lambda}_h, \phi_h \in \Phi_h$. We have used the notation

$$f_i^{\pm}(u_h^i - \varphi_h, \psi) := \sum\nolimits_{E\in\partial\mathcal{T}_u^i} \int_E (\alpha_E + (\mathbf{b}\cdot\mathbf{n}_i)^{\pm})(u_h^i - \varphi_h)\psi\, ds, \quad \psi \in H_*^{\frac{1}{2}}(\Gamma_i)$$

with $(\mathbf{b}\cdot\mathbf{n}_i)^{\pm} := \frac{1}{2}|\mathbf{b}\cdot\mathbf{n}_i| \pm \frac{1}{2}(\mathbf{b}\cdot\mathbf{n}_i)$. Thus $(\mathbf{b}\cdot\mathbf{n}_i)^-$ acts only on the inflow part $\Gamma_i^- := \{x \in \Gamma_i \mid \mathbf{b}(x)\cdot\mathbf{n}_i(x) < 0\}$ and $(\mathbf{b}\cdot\mathbf{n}_i)^+$ only on the outflow part. The parameters $\alpha_E, \beta_E \geq 0$ will be specified later.

Let us shortly explain, why we have added the different stabilization terms. $f_i^{\pm}(\cdot,\cdot)$, which are added to the first resp. third line of (5), couple the local spaces V_h^i and the space Φ_h . They give additional control in stream-wise direction, especially in the hyperbolic limit $\epsilon \to 0$. By $\sum_{E\in\partial\mathcal{T}_u^i} \beta_E \left(\frac{\partial u_h^i}{\partial \mathbf{n}_i} - \lambda_h^i, \nu_h^i \right)_E$

the spaces Φ_h and Λ_h are coupled. Theses couplings enable us to ignore the mentioned inf-sup conditions.

We can prove the following a priori estimate (for the rather technical proof cf. Rapin and Lube [2003b], Theorems 1 and 2). Here, $a \preceq b$ means, that there exists a constant $C > 0$ independent of h and ϵ such that $a \leq Cb$.

Theorem 1. *Assume for the stabilization parameters the inequalities*

$$\min\{\epsilon/h, \epsilon^2/h^2\} \preceq \alpha_E \preceq \max\{1, \epsilon/h\}, \quad \min\{\epsilon^2, h\epsilon\} \preceq \beta_E \preceq h\max\{h^2, \epsilon\}.$$

Then there exists a unique solution $\mathbf{u}_h \in \mathbf{V}_h, \boldsymbol{\lambda}_h \in \boldsymbol{\Lambda}_h$ *and* $\varphi_h \in \Phi_h$ *of (5) and the error is bounded by*

$$|||(\mathbf{u}, \boldsymbol{\lambda}, \varphi) - (\mathbf{u_h}, \boldsymbol{\lambda}_h, \varphi_h)||| \preceq \left(\epsilon^{\frac{1}{2}} + h^{\frac{1}{2}}\right) h \sum_{i=1}^{N} |u^i|_{2,\Omega_i} \tag{6}$$

for a solution $\mathbf{u} \in \mathbf{V} \cap H^2(\Omega)$. *The norm is given by*

$$|||(\mathbf{u}_h, \boldsymbol{\lambda}_h, \varphi_h)|||^2 := \sum_{i=1}^{N} (1-\beta_0)|||u_h^i|||^2_{SD,\Omega_i}$$
$$+ \sum_{i=1}^{N} \sum_{E \in \partial \mathcal{T}_u^i} \int_E \left[(2\alpha_E + |\mathbf{b}\cdot\mathbf{n}_i|)(u_h^i - \varphi_h)^2 + \beta_E(\lambda_h^i)^2\right] ds.$$

If we insert a sufficiently regular solution $(\mathbf{u}, \boldsymbol{\lambda}, \varphi)$ of (3) into the stabilized formulation (5), all additional terms vanish. In this sense (5) is consistent.

There is some degree of freedom for the choice of the stabilization parameters in the advection dominated regime. In the diffusion dominated case we obtain the well known choice of the discontinuous Galerkin method $\alpha_E \sim \epsilon/h_E$ (and $\beta_E \sim \epsilon h_E$). Using suitable global constants $0 < \alpha_0, \beta_0 < 1$ we determine

$$\alpha_E = \alpha_0 \begin{cases} \epsilon/h_E, & \epsilon \geq h_E^2 \\ \epsilon^2/h_E^3, & \epsilon < h_E^2 \end{cases}, \qquad \beta_E = \beta_0 \begin{cases} \epsilon h_E, & \epsilon \geq h_E^2 \\ h_E^3, & \epsilon < h_E^2 \end{cases}. \tag{7}$$

By (7) we mainly enforce boundary conditions in a weak sense on the inflow part of the subdomains, even for $\epsilon = 0$.

Remark 1. For given $\varphi_h \in \Phi_h$ and right hand side $f \in L^2(\Omega)$ the equations (5,i), (5,ii) are discretizations of the local Dirichlet problems

$$Lw^i = f \text{ in } \Omega_i \qquad w^i = \varphi_h \text{ on } \partial\Omega_i, \qquad w^i = 0 \text{ on } \partial\Omega \setminus \partial\Omega_i.$$

These problems are well-posed (cf. Rapin and Lube [2003a]).

4 A Schur complement method

Now we derive the corresponding Schur complement equation for our stabilized scheme. Then the solution of (3) can be obtained by solving local problems. Computing the local problems can be done completely in parallel.

Recall that for given $f \in L^2(\Omega)$, $\varphi_h \in \Phi_h$ the first two lines of (5) are local Dirichlet problems (cf. Remark 1). Denoting the local solutions by $(\mathbf{z}_h(f,\varphi_h), \gamma_h(f,\varphi_h)) \in \boldsymbol{V}_h \times \boldsymbol{\Lambda}_h$ we see

$$(\mathbf{z}_h(f,\varphi_h), \gamma_h(f,\varphi_h)) = (\mathbf{z}_h(f,0), \gamma_h(f,0)) + (\mathbf{z}_h(0,\varphi_h), \gamma_h(0,\varphi_h))$$

due to the linearity of the scheme. Inserting this in the third line of (5) yields the Schur complement equation for our scheme: Find $\varphi_h \in \Phi_h$, such that

$$\langle S_h\varphi_h, \psi_h\rangle = \sum_{i=1}^{N} \left\{-\epsilon\langle\gamma_h^i(f,0), \psi_h\rangle_i + f_i^+(z_h^i(f,0), \psi_h)\right\}, \quad \forall\psi_h \in \Phi_h \quad (8)$$

where the discrete Steklov-Poincaré operator S_h is defined by

$$\langle S_h\varphi_h, \psi_h\rangle := \sum_{i=1}^{N} \left\{\epsilon\langle\gamma_h^i(0,\varphi_h), \psi_h\rangle_i - f_i^+(z_h^i(0,\varphi_h) - \varphi_h), \psi_h)\right\}.$$

Theorem 2. *The discrete Schur complement equation (8) possesses a unique solution. Moreover,* $(\mathbf{z}_h(f,\varphi_h), \gamma_h(f,\varphi_h), \varphi_h)$ *is the solution of (5).*

Proof. cf. Rapin and Lube [2003b], Lemma 3.

5 An alternating Schwarz method

Tallec and Sassi [1995] describe a non-conforming discretization for the Poisson problem. We extend the algorithm to the advection-diffusion problem using the additional stabilization terms $f_i^{\pm}(\cdot,\cdot)$ of (5). Starting with an initial guess $(\psi_h)_0 \in \Phi_h$, $(\lambda_h)_0 \in \boldsymbol{\Lambda}_h$, we obtain the algorithm:

1. Find $(\mathbf{u}_h)_{k+1} \in \mathbf{V}_h$ such that

$$a_{\Omega_i}^{SD}((u_h^i)_{k+1}, v_h^i) + f_i^-((u_h^i)_{k+1} - (\psi_h)_k, v_h^i) = l_{\Omega_i}^{SD}(v_h^i) + \epsilon\langle(\lambda_h^i)_k, v_h^i\rangle_{\Gamma_i}, \quad \forall v_h^i \in V_h^i.$$

2. Compute $(\lambda_h^i)_{k+\frac{1}{2}} \in \Lambda_h^i$ by

$$\epsilon\langle(\lambda_h^i)_{k+\frac{1}{2}}, \mu_h^i\rangle_{\Gamma_i} = \epsilon\langle(\lambda_h^i)_k, \mu_h^i\rangle_{\Gamma_i} - f_i^-((u_h^i)_{k+1} - (\psi_h)_k, \mu_h^i), \ \forall\mu_h^i \in \Lambda_h^i.$$

3. Find $(\psi_h)_{k+1} \in \Phi_h$ such that there holds for all $\phi_h \in \Phi_h$

$$\sum_{i=1}^{N} \left\{\epsilon\langle(\lambda_h^i)_{k+\frac{1}{2}}, \phi_h\rangle_{\Gamma_i} - f_i^+((u_h^i)_{k+1} - (\psi_h)_{k+1}, \phi_h)\right\} = 0.$$

4. Compute $(\lambda_h^i)_{k+1} \in \Lambda_h^i$ such that there holds for all $\mu_h^i \in \Lambda_h^i$

$$\epsilon\left\langle(\lambda_h^i)_{k+1}, \mu_h^i\right\rangle_{\Gamma_i} = \epsilon\langle(\lambda_h^i)_{k+\frac{1}{2}}, \mu_h^i\rangle_{\Gamma_i} - f_i^+((u_h^i)_{k+1} - (\psi_h)_{k+1}), \mu_h^i).$$

It can be proved that the algorithm is well-posed. In step 1 local problems with Robin conditions on the interface are solved. The algorithm is quite similar to the algorithm proposed by Lube et al. [2003]. The Robin values on the inflow part of the local problems are mainly determined by the Robin values of the neighbouring subdomains, computed in the previous step.

A convergence proof of this algorithm is still an open problem. But the numerical results are very promising (cf. Rapin [2003]).

6 Numerical experiments

The main focus of our algorithm is the application to the advection dominated case. Especially the case of nontrivial flows is of interest. To demonstrate the power of our approach we consider the following (quite hard) example.

Example 1. We search for a solution $Lu = f$ in $\Omega = (0,1)^2$ with boundary conditions $u = -0.5$ on $\gamma_1 := \{(x_1, x_2) \in \partial\Omega \mid x_2 = 0\}$, $u = 0.5$ on $\gamma_2 := \{(x_1, x_2) \in \partial\Omega \mid x_2 = 1\}$, and $u = 0$ on the remainder $\partial\Omega \setminus (\gamma_1 \cup \gamma_2)$ of the boundary. The flow is given by

$$\mathbf{b}(x_1, x_2) := \left((2x_2 - 1)(1 - (2x_1 - 1)^2), 4x_2(2x_1 - 1)(x_2 - 1)\right)^T .$$

$\mathbf{b}$ is a rotational flow with a center in $(\frac{1}{2}, \frac{1}{2})$ and $\nabla \cdot \mathbf{b} = 0$.

We decompose the unit square Ω into (6×6) squares. In the context of domain decomposition this example is particularly interesting. In the advection dominated case the solution is almost constant in the interior of Ω. The constant is given by the mean value of the Dirichlet data on the boundary. Now any discretization has to find this value by mixing the boundary information.

For a global mesh size h_{int} the local meshes are chosen by a checkerboard pattern with local mesh sizes $h_u = h_{int} \setminus 2h_{int}$, $h_\lambda = (1/3)h_{int} \setminus (2/3)h_{int}$, $h_\varphi = h_{int}$. For all computations we have chosen $\alpha_0 = 1$ and $\beta_0 = 1$. The result for $\epsilon = 10^{-6}$ is plotted in Figure 1 (a). It coincides quite well with the solution of the one-domain case. Although the SUPG stabilization technique is used, typically crosswind wiggles of the finite element solution appear.

The purpose of the next example is to numerically validate the a priori estimate of Theorem 1.

Example 2. For $-\epsilon\Delta u + (-1, -1)^T \cdot \nabla u = f$ in Ω and $u = g$ on $\partial\Omega$ we distinguish two cases. (a) We choose f, g in such a way that $u(x, y) = x\cos(\pi y)$ becomes the exact solution. In the second case (b) with $f = 1$ and $g = 0$ strong boundary layers appear in the singularly perturbed case.

We consider Example 2 (a). Using a decomposition of Ω into (6×6) subrectangles we alter the mesh size for $\epsilon = 1,\ 0.1,\ 10^{-4}$. The results are plotted in Figure 1 (b) and agree with Theorem 1. If we choose the nonsmooth Example 2 (b) with layers, we obtain a convergence rate of $1/2$ in the $L^2(\Omega)$ norm as in the SUPG case without domain decomposition, since the layers are not resolved. Moreover, we obtain the optimal rates on subdomains $\Omega' \subset \Omega$ away from the layers (cf. Rapin [2003]).

Next, we study the effect of the stabilization on the discrete Schur complement equation (8) and the alternating Schwarz algorithm.

We start with the Schur complement equation (8) applied to Example 2 (a). The equation is solved by the GMRES method. In Table 1 (a) we observe that the number of iteration steps is independent of the mesh size for the singularly perturbed case ($\epsilon = 10^{-4},\ 10^{-6}$). In the diffusion dominated regime

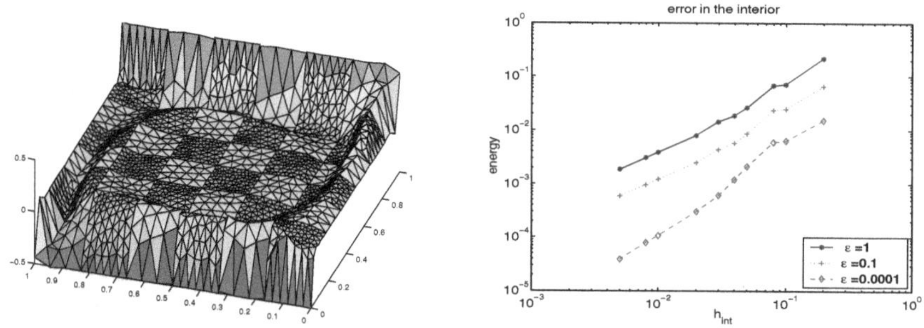

Fig. 1. (a) Plot of Example 1; (b) error in the energy norm $\sum_{i=1}^{N}(\epsilon|\cdot|^2_{1,\Omega_i}+\|\cdot\|^2_{0,\Omega_i})^{\frac{1}{2}}$ for Example 2 (a)

($\epsilon = 1,\ 0.1$) the number of iteration steps increases for smaller mesh sizes. Therefore in this case we have to introduce a preconditioner. First experiments with a generalized Neumann–Neumann preconditioner can be found in Rapin [2003]. As expected, in Table 1 (b) we observe an increase of the number of iteration steps for more subdomains.

Please note, that, in general, the local solutions of the first iteration steps possess sharp boundary layers on the outflow part, although the reference solution is smooth. The layers become smaller within the convergence process. Therefore, we obtain the same results for Example 2 (b).

Now we consider the alternating Schwarz algorithm. In our numerical experiments we compare the discrete solution with the reference solution of the continuous problem. In Figure 2 we see that the discretization error is reached within a few steps for the singularly perturbed case. In the diffusion dominated case the convergence is quite slow, but can be accelerated by an adaptive choice of the parameter α_E (cf. Lube et al. [2003]).

Summarized one can state that both methods work well both in the diffusion dominated case and the singularly perturbed case. But we suggest to

Table 1. Number of iteration steps of the GMRES algorithm, which is needed to reduce the initial residuum by the factor 10^{-8} for Example 2 (a). The initial guess is always 0. In (a) we consider different mesh sizes h_{int} and diffusion coefficients ϵ for a (4×3) partition. In (b) the domain is decomposed into $(n \times n)$ subdomains for mesh size $h_{int} = 0.01$

$\epsilon \setminus h_{int}$	0.05	0.02	0.01	0.005
1	21	32	46	66
10^{-1}	19	31	43	59
10^{-4}	19	20	19	18
10^{-6}	19	20	19	19

(a)

$\epsilon \setminus n$	2	4	6	8	10	12
1	25	51	65	78	81	97
10^{-1}	26	21	26	30	34	38
10^{-4}	16	21	26	30	34	38
10^{-6}	6	21	26	30	34	38

(b)

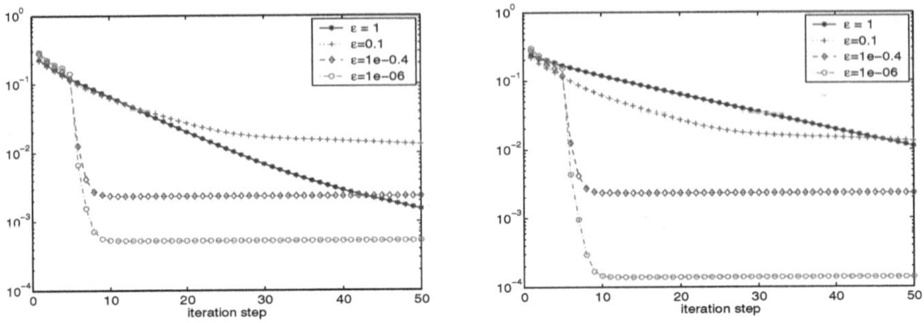

Fig. 2. Convergence behavior of the alternating Schwarz algorithm in the $L^2(\Omega)$ norm for a (4×3) decomposition. On the left hand side the mesh size is chosen by $h_{int} = 0.02$ and on the right hand side by $h_{int} = 0.01$

use the Schur complement method in the diffusion dominated case and the alternating Schwarz method in the advection dominated case.

References

C. Baiocchi, F. Brezzi, and L. Marini. Stabilization of Galerkin methods and applications to domain decomposition. In A. Bensoussan and J. Verjus, editors, *Future Tendencies in Computer Science, Control and Applied Mathematics*, pages 345–355, Berlin-Heidelberg-New York, 1992. Springer-Verlag.

S. Bertoluzza and A. Kunoth. Wavelet Stabilization and Preconditioning for Domain Decomposition. *IMA J. Numer. Anal.*, 20:533–559, 2000.

F. Brezzi and D. Marini. Error Estimates for the three-field formulation with bubble stabilization. *Math. Comp.*, 70:911–934, 2001.

G. Lube, T. Knopp, and G. Rapin. Acceleration of a non-overlapping Schwarz method for advection diffusion problems. Technical report, Universität Göttingen, 2003. DD 15 Preprint.

G. Rapin. *The Three-field Formulation for Elliptic Equations: Stabilization and Decoupling Strategies.* PhD thesis, Universität Göttingen, 2003.

G. Rapin and G. Lube. A stabilized scheme for the Lagrange multiplier method for advection-diffusion equations. Technical report, Universität Göttingen, 2003a. appears in M^3AS.

G. Rapin and G. Lube. A stabilized scheme of the three-field approach for advection-diffusion equations. Technical report, Universität Göttingen, 2003b. submitted.

P. L. Tallec and T. Sassi. Domain Decomposition with nonmatching grids: Augmented Lagrangian Approach. *Math. Comp.*, 64:1367–1396, 1995.

Approximation of Optimal Interface Boundary Conditions for Two-Lagrange Multiplier FETI Method

F.-X. Roux, F. Magoulès, L. Series, and Y. Boubendir

ONERA, 29 av. de la Division Leclerc, BP72, 92322 Châtillon, France,
<roux@onera.fr>, <series@onera.fr>, <boubendir@onera.fr>
Univ. Henri Poincaré, BP239, 54506 Vandoeuvre-les-Nancy, France,
<magoules@iecn.u-nancy.fr>

Summary. Interface boundary conditions are the key ingredient to design efficient domain decomposition methods. However, convergence cannot be obtained for any method in a number of iterations less than the number of subdomains minus one in the case of a one-way splitting. This optimal convergence can be obtained with generalized Robin type boundary conditions associated with an operator equal to the Schur complement of the outer domain. Since the Schur complement is too expensive to compute exactly, a new approach based on the computation of the exact Schur complement for a small patch around each interface node is presented for the two-Lagrange multiplier FETI method.

1 Introduction

Interface boundary conditions are the key ingredient to design efficient domain decomposition methods, see Chevalier and Nataf [1998], Benamou and Després [1997], Gander et al. [2002]. However, convergence cannot be obtained for any method in a number of iterations less than the number of subdomains minus one in the case of a one-way splitting. For the two-Lagrange multiplier FETI method, this optimal convergence can be obtained with generalized Robin type boundary conditions associated with an operator equal to the Schur complement of the outer domain, see Roux et al. [2002]. In practice this optimal condition cannot be implemented since the Schur complement is too expensive to compute exactly. Furthermore, the Schur complement is a dense matrix on each interface and even if it were computed, using it would create a very large increase of the bandwidth of the local subproblem matrix. Hence the issue is how to build a sparse approximation of the Schur complement that is not expensive to compute and that leads to good convergence properties of the two-Lagrange multiplier FETI iterative method.

Different approaches based on approximate factorization or inverse computation of the subproblem matrix have been tested, see Roux et al. [2002]. Here, a new approach based on the computation of the exact Schur complement for a small patch around each interface node appears to be a very efficient method for designing approximations of the complete Schur complement. Furthermore this approach can be easily implemented without any other information than the local matrix in each subdomain.

2 Review of the Two-Lagrange Multiplier FETI Method

2.1 Introduction of Two-Lagrange Multiplier on the Interface

Consider a splitting of the domain Ω as in Figure 1 and note by subscripts i and p the degrees of freedom located inside subdomain $\Omega^{(s)}$, $s = 1, 2$, and on the interface Γ. Then, the contribution of subdomain $\Omega^{(s)}$, $s = 1, 2$ to the matrix and the right-hand side of a finite element discretization of a linear partial differential equation on Ω can be written as follows:

$$K^{(s)} = \begin{bmatrix} K_{ii}^{(s)} & K_{ip}^{(s)} \\ K_{pi}^{(s)} & K_{pp}^{(s)} \end{bmatrix}, \quad b^{(s)} = \begin{bmatrix} b_i^{(s)} \\ b_p^{(s)} \end{bmatrix}$$

where $K_{pp}^{(1)}$ and $K_{pp}^{(2)}$ represent the interaction matrices between the nodes on the interface obtained by integration on $\Omega^{(1)}$ and on $\Omega^{(2)}$. The global problem is a block system obtained by assembling local contribution of each subdomain:

$$\begin{bmatrix} K_{ii}^{(1)} & 0 & K_{ip}^{(1)} \\ 0 & K_{ii}^{(2)} & K_{ip}^{(2)} \\ K_{pi}^{(1)} & K_{pi}^{(2)} & K_{pp} \end{bmatrix} \begin{bmatrix} x_i^{(1)} \\ x_i^{(2)} \\ x_p \end{bmatrix} = \begin{bmatrix} b_i^{(1)} \\ b_i^{(2)} \\ b_p \end{bmatrix}. \tag{1}$$

The block K_{pp} is the sum of the two blocks $K_{pp}^{(1)}$ and $K_{pp}^{(2)}$. In the same way, $b_p = b_p^{(1)} + b_p^{(2)}$ is obtained by local integration in each subdomain and sum on the interface.

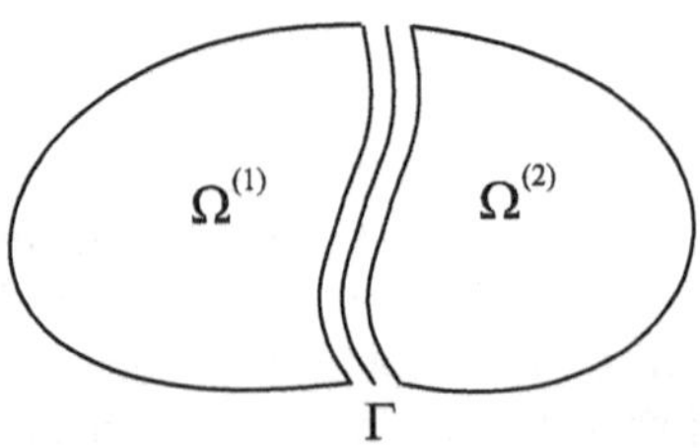

Fig. 1. Non-overlapping domain splitting

The two-Lagrange multiplier FETI method, see Farhat et al. [2000], is an iterative based domain decomposition method which consists to determine the solution of the following coupled problem:

$$\begin{bmatrix} K_{ii}^{(1)} & K_{ip}^{(1)} \\ K_{pi}^{(1)} & K_{pp}^{(1)} + A^{(1)} \end{bmatrix} \begin{bmatrix} x_i^{(1)} \\ x_p^{(1)} \end{bmatrix} = \begin{bmatrix} b_i^{(1)} \\ b_p^{(1)} + \lambda^{(1)} \end{bmatrix}$$
$$\begin{bmatrix} K_{ii}^{(2)} & K_{ip}^{(2)} \\ K_{pi}^{(2)} & K_{pp}^{(2)} + A^{(2)} \end{bmatrix} \begin{bmatrix} x_i^{(2)} \\ x_p^{(2)} \end{bmatrix} = \begin{bmatrix} b_i^{(2)} \\ b_p^{(2)} + \lambda^{(2)} \end{bmatrix}$$
$$\lambda^{(1)} + \lambda^{(2)} - (A^{(1)} + A^{(2)})x_p^{(1)} = 0$$
$$\lambda^{(1)} + \lambda^{(2)} - (A^{(1)} + A^{(2)})x_p^{(2)} = 0$$

where the free matrices $A^{(1)}$ and $A^{(2)}$ are to be determined for the best performance of the algorithm. It is clear that this coupled problem is equivalent to the global problem (1), see Roux et al. [2002]. The elimination of $x_i^{(s)}$ in favor of $x_p^{(s)}$ in the two first equations and substitution in the two last equations leads to the following linear system upon the variable $\lambda := (\lambda^{(1)}, \lambda^{(2)})^T$:

$$F\lambda = d \tag{2}$$

with F and d the matrix and right hand side defined as:

$$F := \begin{bmatrix} I & I - (A^{(1)} + A^{(2)})[S^{(2)} + A^{(2)}]^{-1} \\ I - (A^{(1)} + A^{(2)})[S^{(1)} + A^{(1)}]^{-1} & I \end{bmatrix}$$
$$d := \begin{bmatrix} (A^{(1)} + A^{(2)})[S^{(2)} + A^{(2)}]^{-1} c_p^{(2)} \\ (A^{(1)} + A^{(2)})[S^{(1)} + A^{(1)}]^{-1} c_p^{(1)} \end{bmatrix}$$

The iterative solution of this system is usually performs with a Krylov method.

2.2 Optimal Interface Boundary Conditions

It is shown in Roux et al. [2002] that the best choice for the free matrices $A^{(s)}$, $s = 1, 2$ corresponds to the complete outer Schur complement, i.e. the discretization of the optimal continuous boundary conditions associated to the Steklov-Poincaré operator, see Ghanemi [1997], Collino et al. [2000] and Boubendir [2002]. An extension of this result in the case of a one way splitting can be obtained in the discrete case, see Roux et al. [2002], and in the continuous case, see Nataf et al. [1994].

Theorem 1. *In a case of a two-domain splitting, the Jacobi iterative algorithm for the two-Lagrange multiplier FETI method with augmented term equal to the complete outer Schur complement converges in one iteration at most.*

Theorem 2. *In a case of a one way splitting, the Jacobi iterative algorithm for the two-Lagrange multiplier FETI method with augmented term equal to the complete outer Schur complement converges in a number of iteration equal to the number of subdomain minus one.*

3 Approximation of Optimal Interface Boundary Conditions

In the previous section, we have recalled that the best choice for the augmented matrix in the case of a one way splitting domain decomposition is the complete outer Schur complement matrix. This choice can not be done in practice since the computational cost of the complete outer Schur complement matrix is too expensive.

3.1 Neighbor Schur Complement

From a physical point of view, the complete outer Schur complement matrix represent the interactions of all the degree of freedom of the subdomains condensed on the interfaces. The restriction of the interactions only with the neighboring subdomains, leads to approximate the complete outer Schur complement with the neighbor Schur complement. The computational cost and the exchange of data are thus reduced to the neighboring subdomains only. Unfortunately, this approach still leads to an expensive computational cost. Hence the issue is how to build a sparse approximation of the Schur complement that is not expensive to compute and that gives good convergence for the two-Lagrange multiplier FETI method.

3.2 Lumped Approximation

We have shown in Roux et al. [2002] that an approximation of the neighbor Schur complement matrix $K_{bb}^{(s)} - K_{bi}^{(s)}[K_{ii}^{(s)}]^{-1}K_{ib}^{(s)}$ with its first term, i.e. with the matrix $K_{bb}^{(s)}$ gives good results. Such an approximation, presents the advantage to be very easy to implement since this matrix is computed by the neighboring subdomain during the assembly procedure and the integration of the contribution of the interface nodes. Only an exchange with the neighboring subdomain is required for this regularization procedure.

3.3 Sparse Approximation based on Overlapping Layers

In this section we present a new approach for the approximation of the neighboring Schur complement with a sparse matrix, which leads to a better approximation than the lumped approximation, as shown in the numerical results. The goal is to obtained a spectral density of the approximated matrix close to the spectral density of the neighbor Schur complement matrix. We first define the following subsets of indexes:

$$\begin{aligned}
V_{\Omega^{(2)}} &= \{\text{indexes of nodes inside the subdomain } \Omega^{(2)}\} \\
V_{\Gamma} &= \{\text{indexes of nodes on the interface } \Gamma\} \\
V_i^l &= \{\text{indexes of the nodes } j \text{ such that the minimum} \\
&\qquad \text{connectivity distance between } i \text{ and } j \text{ is lower or} \\
&\qquad \text{equal than } l, l \in \mathbb{N}\} \\
V_{\Gamma,i}^l &= V_\Gamma \cap V_i^l
\end{aligned}$$

The sparse approximation investigated here consist to define a sparse augmented matrix obtained through an extraction of some coefficients and local condensation along the interface. The complete algorithm to compute the augmented matrix—in the case of a two domain splitting—in subdomain $\Omega^{(1)}$ can be define as:

Algorithm 1. [sparse approximation]

1. construction of the structure of the interface matrix $A_1 \in \mathbb{R}^{dimV_\Gamma \times dimV_\Gamma}$.
2. construction of the sparse structure of the subdomain matrix $K^{(2)} \in \mathbb{R}^{dimV_{\Omega^{(2)}} \times dimV_{\Omega^{(2)}}}$.
3. assembly of the matrix $K^{(2)}$.
4. for all i in V_Γ do
   4.1. extraction of the coefficients K_{mn}, $(m,n) \in V_i^l \times V_i^l$, and construction of the sparse matrix $A_2 \in \mathbb{R}^{dimV_i^l \times dimV_i^l}$ with these coefficients.
   4.2. computation of the dense matrix $A_3 \in \mathbb{R}^{dimV_{\Gamma,i}^1 \times dimV_{\Gamma,i}^1}$ by condensation of the matrix A_2 on the degree of freedom $V_{\Gamma,i}^1$.
   4.2. extraction of the coefficients of the line associated with the node i from the matrix A_3 and insertion inside the matrix A_1 at the line associated with the node i.
5. construction of the symmetric matrix $A_4 = \frac{(A_1^T + A_1)}{2}$.
6. regularization of the matrix $K^{(1)}$ with the matrix A_4.

where l denotes the number of layers considered.

Similar calculation performed in the subdomain $\Omega^{(2)}$ gives the augmented matrix $A^{(2)}$ to add to the subdomain matrix $K^{(2)}$. As an example the regular mesh with $\mathbb{Q}_1$-finite elements presented Figure 2 leads to the subsets of

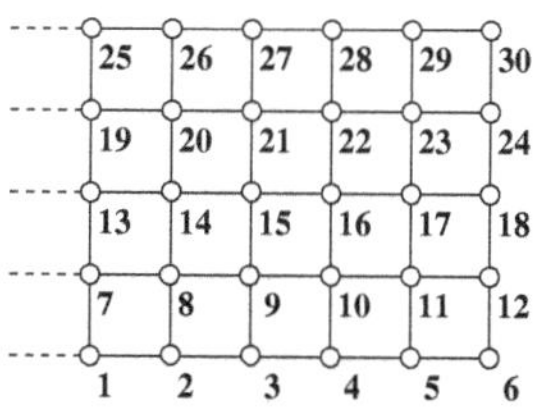

Fig. 2. Numbering of the nodes in subdomain $\Omega^{(2)}$

indexes $V_7^1 = \{1,2,7,8,13,14\}$, $V_7^2 = \{1,2,3,7,8,9,13,14,15,19,20,21\}$ and $V_{\Gamma,7}^1 = \{1,7,13\}$. These subsets correspond to the overlapping layers represented Figure 3.

4 Numerical Results

4.1 The Model Problem

In this section, a two dimensional beam of length L_1 and high L_2 submitted to flexion is analyzed. The Poisson ratio and the Young modulus are respectively $\nu = 0.3$ and $E = 2.0\ 10^5 N/m^2$. Homogeneous Dirichlet boundary conditions

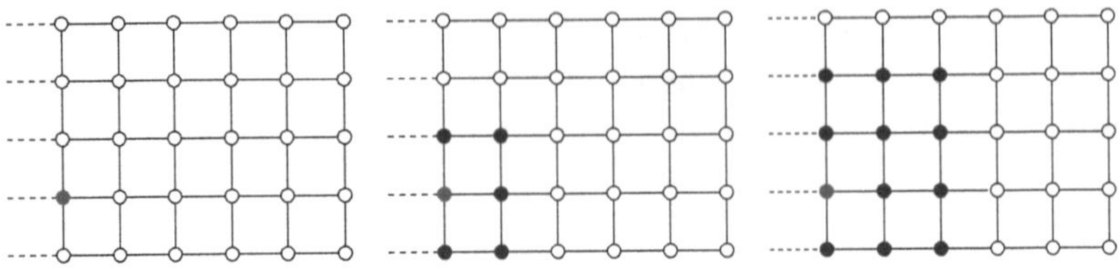

Fig. 3. On the left one interface node, on the middle one interface node with one layer, and on the right one interface node with two layers

are imposed on the left and homogeneous Neumann boundary conditions are set on the top and on the bottom. Loading, model as non homogeneous Neumann boundary condition are imposed on the right of the structure. The beam is meshed with triangular elements and discretized with $\mathbb{P}_1$ finite elements. The domain is then split into two or ten subdomains in a one way splitting and the condensed interface problem is solved iteratively with the ORTHODIR Krylov method. The stopping criterion is set to $||r_n||_2 < 10^{-6}||r_0||_2$, where r_n and r_0 are the nth and initial global residuals.

4.2 Spectral Analysis

Figure 4 represent the spectral density of the eigenvalues of the matrix of the condensed interface problem (2) for different augmented matrices. An augmented matrix equal to the neighbor Schur complement will leads to eigenvalues equal to one which correspond to a spectral density equal to a Dirac function.

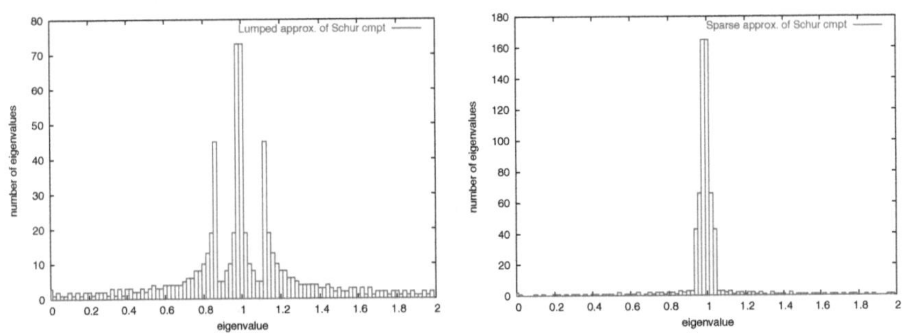

Fig. 4. Spectral density of the condensed interface problem with an augmented matrix issue from the lumped approximation (left) vs from the sparse approximation (right) ($L_1 = 10$, $L_2 = 1$, $h = 1/160$). Case of two subdomains

We can see on Figures 4 that a sparse approximation performed with a number of layers equal to four leads to a spectral density close to a Dirac function. Opposite, a lumped approximation leads to spectrum much more different. This result can be explain by the fact that the sparse approximation is based on local condensation i.e. on local Steklov-Poincaré operators which is not the case of the lumped approximation.

4.3 Asymptotic Analysis

The asymptotic analysis of the proposed methods upon different parameters is now analyzed. The analysis upon the domain size reported Figure 5 show the respective dependence of the methods. The asymptotic behavior of the pro-

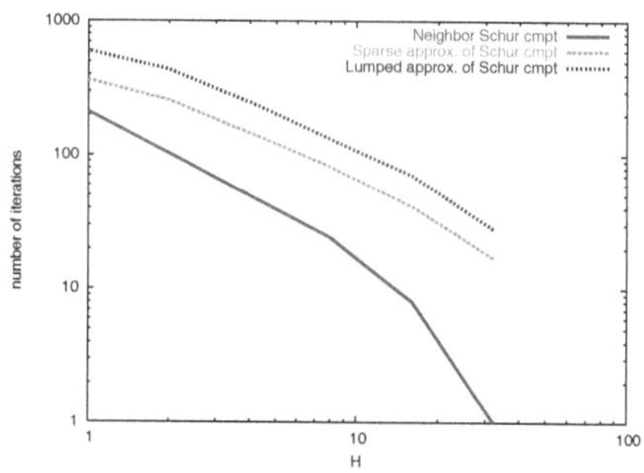

Fig. 5. Asymptotic behavior for different augmented matrices and different subdomain size. ($L_1 = 64$, $L_2 = 1$, $h = 1/20$)

posed methods upon the mesh size is presented Figure 6. On the left picture, four layers are considered for the sparse approximation. A linear dependence upon the mesh size can be noticed for all the methods. On the right picture, the number of layers of the sparse approximation increase proportionally with the mesh size. A linear dependence still occurs for the sparse approximation but the slope of the curve is lower than with a constant number of layers equal to four. The asymptotic results obtained with this last approximation

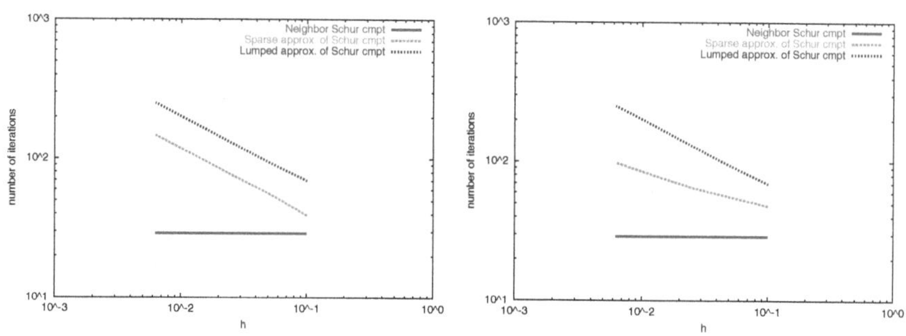

Fig. 6. Asymptotic behavior for different augmented matrices and different mesh size on the left for a constant number of layers, and on the right for a number of layers increasing proportionally with the mesh size. ($L_1 = 10$, $L_2 = 1$). Case of ten subdomains

are still less efficient than those obtained with a continuous approach, see Gander et al. [2002], but the implementation of the previous method doesn't depends on a priori knowledge of the problem to be solved (coefficients of the partial differential equation, mesh size, ...) and thus helps its use as a black box routine!

5 Conclusions

In this paper the principle of the two-Lagrange multiplier FETI method with optimal interface boundary conditions has been remain. A new method for the approximation of these optimal conditions has been introduced. This new method is based on the computation of the exact Schur complement for a small patch around each interface node. This method appears to be a very efficient method for designing approximations of the complete Schur complement that give robust iterative algorithms for solving many different kinds of problems.

References

J.-D. Benamou and B. Després. A domain decomposition method for the Helmholtz equation and related optimal control problems. *J. of Comp. Physics*, 136:68–82, 1997.

Y. Boubendir. *Techniques de décomposition de domaine et méthode d'équations intégrales.* PhD thesis, INSA de Toulouse, Jun 2002.

P. Chevalier and F. Nataf. Symmetrized method with optimized second-order conditions for the Helmholtz equation. *Contemp. Math.*, 218:400–407, 1998.

F. Collino, S. Ghanemi, and P. Joly. Domain decomposition method for harmonic wave propagation: a general presentation. *Computer methods in applied mechanics and engineering*, 184:171–211, 2000.

C. Farhat, A. Macedo, M. Lesoinne, F.-X. Roux, F. Magoulès, and A. de La Bourdonnaye. Two-level domain decomposition methods with lagrange multipliers for the fast iterative solution of acoustic scattering problems. *Comput. Methods in Appl. Mech. and Engrg.*, 184(2):213–240, 2000.

M. Gander, F. Magoulès, and F. Nataf. Optimized schwarz method without overlap for the Helmholtz equation. *SIAM J. Sci. Comput.*, 24(1):38–60, 2002.

S. Ghanemi. A domain decomposition method for Helmholtz scattering problems. In P. E. Bjørstad, M. Espedal, and D. Keyes, editors, *Ninth Int. Conf. on Domain Decomposition Methods*, pages 105–112. ddm.org, 1997.

F. Nataf, F. Rogier, and E. de Sturler. Optimal interface conditions for domain decomposition methods. Technical Report 301, CMAP (Ecole Polytechnique), 1994.

F.-X. Roux, F. Magoulès, S. Salmon, and L. Series. Optimization of interface operator based on algebraic approach. In *Domain Decomposition Methods in Sci. Engrg.*, pages 297–304, 2002.

Optimized Overlapping Schwarz Methods for Parabolic PDEs with Time-Delay

Stefan Vandewalle[1] and Martin J. Gander[2]

[1] Katholieke Universiteit Leuven, Department of Computer Science (http://www.cs.kuleuven.ac.be/~stefan/)

[2] McGill University, Department of Mathematics and Statistics (http://www.math.mcgill.ca/~mgander/)

Summary. We present overlapping Schwarz methods for the numerical solution of two model problems of delay PDEs: the heat equation with a fixed delay term, and the heat equation with a distributed delay in the form of an integral over the past. We first analyze properties of the solutions of these PDEs and find that their dynamics is fundamentally different from that of regular time-dependent PDEs without time delay. We then introduce and study overlapping Schwarz methods of waveform relaxation type for the two model problems. These methods compute the local solution in each subdomain over many time-levels before exchanging interface information to neighboring subdomains. We analyze the effect of the overlap and derive optimized transmission conditions of Robin type. Finally we illustrate the theoretical results and convergence estimates with numerical experiments.

1 Introduction

Delay differential equations model physical systems for which the evolution does not only depend on the present state of the system but also on the past history. Such models are found, for example, in population dynamics and epidemiology, where the delay is due to a gestation or maturation period, or in numerical control, where the delay arises from the processing in the controller feedback loop. Delay differential equations have been studied extensively (and almost exclusively) in the context of ordinary differential equations. An ordinary delay differential equation is an equation of the form

$$\dot{y}(t) = F(t, y(t), y_t), \quad t \in [0, T], \tag{1}$$

where y_t denotes a function segment extending over a time-interval of length τ into the past: $y_t(s) = y(t+s)$, $s \in [-\tau, 0]$. Equation (1) is usually complemented with an initial condition of the type $y_0(s) = g(s)$, where $g(s)$ is a given function over the interval $s \in [-\tau, 0]$. A good starting point to study the analysis and numerical computation of ordinary delay differential equations is Bellen and Zennaro [2003], and the references therein.

Delay PDEs are less well understood. They are typically of the form

$$\frac{\partial}{\partial t}u(t,x) = \mathcal{L}u(t,x,u_{(t,x)}) + f(t,x), \tag{2}$$

where $u_{(t,x)}$ is a function segment, which can extend both in the past and over some region in space: $u_{(t,x)}(v,w) = u(t+v,x+w)$, $(v,w) \in [-\tau,0] \times [-\sigma,\sigma]$. Equation (2) has to be completed with boundary conditions and an initial condition, which, typically, have to be specified over some initial and boundary regions around the domain of definition of the delay PDE. A set of examples, illustrating the wide range of existing delay PDE models can be found in Wu [1996]. A characteristic example from numerical control is the equation

$$\frac{\partial u}{\partial t} = D\frac{\partial^2 u}{\partial x^2} + v(g(u(t-\tau,x)))\frac{\partial u}{\partial x} + c[f(u(t-\tau,x)) - u(t,x)],$$

which models a furnace used to process metal sheets. Here, u is the temperature distribution in a metal sheet, moving at a velocity v and heated by a source specified by the function f; both v and f are dynamically adapted by a controlling device monitoring the current temperature distribution. The finite speed of the controller, however, introduces a fixed delay of length τ. An example from population dynamics is the so-called Britton-model,

$$\frac{\partial u}{\partial t} = D\Delta u + u(1 - g\star u) \quad \text{with} \quad g\star u = \int_{t-\tau}^{t}\int_{\Omega} g(t-s,x-y)u(s,y)\,dy\,ds.$$

Here, $u(t,x)$ denotes a population density, which evolves through random migration (modeled by the diffusion term) and reproduction (modeled by the nonlinear reaction term). The latter involves a convolution operator with a kernel $g(t,x)$, which models the distributed age-structure dependence of the evolution and its dependence on the population levels in the neighborhood.

There is little experience with numerical methods for solving delay PDEs. Zubik-Kowal [2001] and Huang and Vandewalle [2003] analyze the accuracy and stability of spatial and temporal discretization schemes. Zubik-Kowal and Vandewalle [1999] analyze the convergence of a waveform relaxation scheme of Gauss-Seidel and Jacobi type, for solving the discretized problems. In this paper we present a first analysis of domain decomposition based waveform relaxation methods for the solution of two model delay PDEs. Waveform relaxation schemes using domain decomposition in space for parabolic equations without delay were introduced in Gander and Stuart [1998] and independently in Giladi and Keller [2002], and further analyzed in Gander [1998] and Gander and Zhao [2002], see also the references therein. In those papers, it was shown that domain decomposition leads to a fundamentally faster convergence rate than the classical waveform relaxation methods. The performance of these methods can however still be drastically improved using better transmission conditions between subdomains, see Gander et al. [1999]. Our goal is to analyse whether such optimization is also possible in the parabolic delay PDE case.

The structure of the paper is as follows. In §2 we define two characteristic models of delay PDEs, and we analyze the stability of the solution of those problems as a function of the parameters appearing in the model. In §3, we analyze the performance of the classical overlapping Schwarz waveform relaxation method when used as a solver for delay PDEs. An algorithm using optimized Robin type transmission conditions is studied in §4. Finally, in §5, the theoretical results are verified by some numerical experiments.

2 Analysis of Delay PDEs

We consider two representative model problems: a PDE with a constant delay and one with a distributed delay. By analyzing the properties of their solutions, we hope to gain some insight into the behavior of solutions to the more complex problems introduced in §1. The constant delay PDE is given by

$$\frac{\partial u}{\partial t} = \frac{\partial^2 u}{\partial x^2} + au(t-\tau), \quad \text{with} \begin{cases} x \in \mathbb{R}, \ t \in \mathbb{R}^+, \\ a \in \mathbb{R}, \ \tau \in \mathbb{R}^+. \end{cases} \tag{3}$$

Using separation of variables, we arrive at solutions of the form $u(t,x) = e^{\lambda t} \cdot e^{ikx}$. The constants $\lambda \in \mathbb{C}$ and $k \in \mathbb{R}$ satisfy the so-called characteristic equation $\lambda = -k^2 + ae^{-\lambda\tau}$. Separating real and imaginary parts, $\lambda = \eta + i\omega$, we obtain the system of equations

$$\begin{cases} \eta = -k^2 + ae^{-\eta\tau}\cos(\omega\tau), \\ \omega = \qquad\ \ - ae^{-\eta\tau}\sin(\omega\tau). \end{cases} \tag{4}$$

The natural question that arises therefore is for what (a,τ)-pairs the characteristic equation has only solutions with $\eta \leq 0$ and thus solutions of the delay PDE stay bounded for all time. To answer this question of stability, we distinguish two cases. First, we identify the region in the (a,τ) parameter space where unstable solutions exist corresponding to real roots λ; next we treat the unstable, oscillatory solutions case, i.e. corresponding to complex-valued roots with non-vanishing imaginary part ω.

Setting $\omega = 0$ and $\eta > 0$, equation (4) simplifies to $\eta = -k^2 + ae^{-\eta\tau}$. For positive a, and a given k, this equation has a unique solution η, as illustrated in Figure 1 (left). If $k^2 < a$, the corresponding η is positive. Hence, for any $a > 0$ there always exist modes (with k small enough) that grow exponentially. A similar graphical argument shows that, for $a < 0$, any roots η must necessarily be negative. Hence, there are no unstable real modes in that case.

Next, by setting $\omega > 0$ and $\eta = 0$ in (4) we determine the boundary of the (a,τ)-region where unstable oscillatory modes exist. This leads to the conditions $k^2 = a\cos(\omega\tau)$ and $-\omega = a\sin(\omega\tau)$. An analysis of these conditions reveals that they can only be satisfied for $a < 0$ if $\pm\,\omega\tau \in [\pi/2, \pi] + 2\pi n$ with n an arbitrary positive integer; for $a > 0$ the corresponding condition becomes $\pm\,\omega\tau \in [3\pi/2, 2\pi] + 2\pi n$.

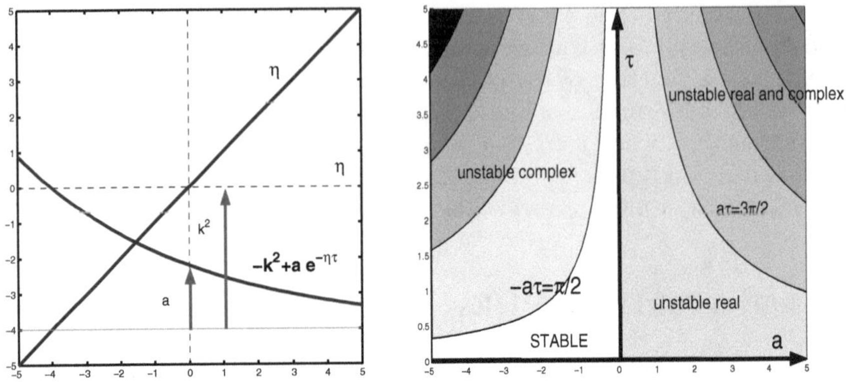

Fig. 1. Left: stability analysis of the constant delay PDE, the case of real roots. Right: stable and unstable regions in the (a, τ) parameter space

The curves $a\tau = -\omega\tau/\sin(\omega\tau)$, for $\omega\tau = \pi/2+\pi n$ are especially important. They determine the (a, τ)-values at which the constant mode, $k=0$, becomes unstable, with an oscillation determined by the corresponding ω. It can be shown that the constant mode is the first mode to become unstable; this leads to the theorem below. In Figure 1 (right) the stability region is shown in white. In the other regions each unstable mode is of a specific multiplicity.

Theorem 1. *The solution to the constant delay PDE* (3) *is stable if* $-\pi/2 \leq a\tau \leq 0$.

The second model problem is a distributed delay PDE,

$$\frac{\partial u}{\partial t} = \frac{\partial^2 u}{\partial x^2} + a\int_{-\tau}^{0} u(t+s)\,ds \quad \text{with} \quad \begin{cases} x \in \mathbb{R},\, t \in \mathbb{R}^+, \\ a \in \mathbb{R},\, \tau \in \mathbb{R}^+. \end{cases} \tag{5}$$

Now, the characteristic equation is given by $\lambda = -k^2+\frac{a}{\lambda}(1-e^{-\lambda\tau})$. Separating real and imaginary parts as we did before, one obtains the system

$$\begin{cases} \eta^2 - \omega^2 + \eta k^2 = a(1-\cos(\omega\tau)e^{-\eta\tau}), \\ 2\eta\omega + \omega k^2 \quad\;\; = a\sin(\omega\tau)e^{-\eta\tau}. \end{cases} \tag{6}$$

We determine the (a, τ)-pairs for which the characteristic equation has only solutions with $\eta \leq 0$. An elementary graphical argument reveals that any positive a admits unstable real roots, i.e., with $\omega = 0$. There are no such roots for $a < 0$. Setting $\eta = 0$ in (6), we arrive at two equations, which can only be satisfied for $a < 0$ and for $\pm\,\omega\tau \in [\pi, 2\pi] + 2\pi n$. As before, the constant mode, with $k=0$, is the stability determining one. The curves $a\tau^2 = -\omega^2\tau^2/(1-\cos(\omega\tau))$ for $\omega\tau = \pi + 2\pi n$ determine the (a, τ)-values where a constant mode instability appears. Figure 2 shows the stability region.

Theorem 2. *The solution to the distributed delay PDE* (5) *is stable if* $-\pi^2/2 \leq a\tau^2 \leq 0$.

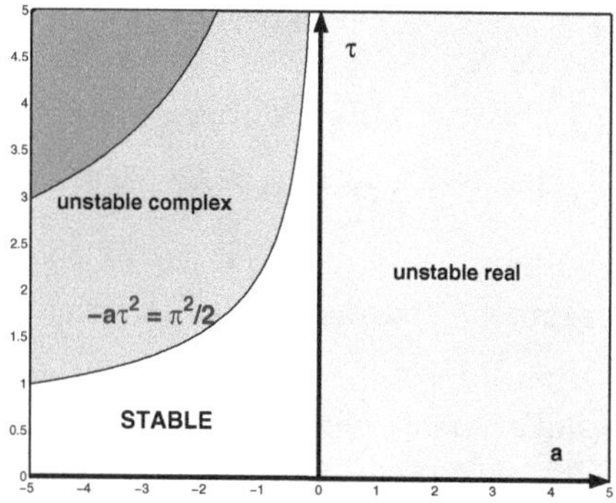

Fig. 2. Stable and unstable regions for the distributed delay PDE

3 Domain Decomposition

The classical Schwarz algorithm. We decompose the domain of PDE (3) into two overlapping subdomains $\Omega_1 = (-\infty, L)$ and $\Omega_2 = (0, \infty)$, with overlap $L > 0$. The classical Schwarz waveform iteration is then given by

$$\begin{cases} \frac{\partial u_1^n}{\partial t} = \frac{\partial^2 u_1^n}{\partial x^2} + a u_1^n(t-\tau) & \text{on } \Omega_1, \ u_1^n(t,L) = u_2^{n-1}(t,L), \\ \frac{\partial u_2^n}{\partial t} = \frac{\partial^2 u_2^n}{\partial x^2} + a u_2^n(t-\tau) & \text{on } \Omega_2, \ u_2^n(t,0) = u_1^{n-1}(t,0), \end{cases} \tag{7}$$

starting with some initial guesses $u_1^0(t,L)$ and $u_2^0(t,0)$. For the analysis we will assume those to be in L^2. Using Laplace transforms, we can rewrite (7) as an iteration in 'frequency space' and explicitly solve for the Laplace transform of the solutions $u_1^n(t,L)$ and $u_2^n(t,0)$. Using arguments very similar to the ones in Gander et al. [1999] we arrive at the following result.

Theorem 3. *Assume a and τ satisfy the stability condition of Theorem 1. Then, the classical Schwarz waveform relaxation algorithm (7) for the constant delay PDE (3) converges linearly, i.e., with $e_1^n = u - u_1^n$ and $e_2^n = u - u_2^n$,*

$$||e_1^n(\cdot, L)||_2 + ||(e_2^n(\cdot, 0)||_2 \leq \rho^n(||e_1^0(\cdot, L)||_2 + ||e_2^0(\cdot, 0)||_2), \tag{8}$$

where $\rho := \rho_{cla} = \sup_{\omega \in \mathbb{R}} \left| e^{-\sqrt{i\omega - a e^{-i\omega\tau}}\, L} \right| < 1.$

The full details of the derivation are given in the companion report Vandewalle and Gander. Using elementary, but very technical arguments, the convergence rate ρ_{cla} can be bounded, as a function of the problem parameters and the size of the overlap.

Corollary 1. *The convergence rate of the classical Schwarz method for problem (3) satisfies $\rho_{cla} \leq e^{-\sqrt{-a\cos(a\tau)/2}\, L}$, provided $-a\tau < 1$.*

Next, we consider the Schwarz algorithm for the distributed delay PDE,

$$\begin{cases} \frac{\partial u_1^n}{\partial t} = \frac{\partial^2 u_1^n}{\partial x^2} + a\int_{-\tau}^{0} u_1(t+s)\,ds & \text{on } \Omega_1, \quad u_1^n(t,L) = u_2^{n-1}(t,L), \\ \frac{\partial u_2^n}{\partial t} = \frac{\partial^2 u_2^n}{\partial x^2} + a\int_{-\tau}^{0} u_2(t+s)\,ds & \text{on } \Omega_2, \quad u_2^n(t,0) = u_1^{n-1}(t,0). \end{cases} \tag{9}$$

With a Laplace transform analysis similar to that for the constant delay case, and some technical arguments, we derive the following theorem and corollary.

Theorem 4. *Assume a and τ satisfy the stability condition of Theorem 2. Then, the classical Schwarz algorithm (9) for delay PDE (5) converges linearly as in (8), where $\rho := \rho_{cla} = \sup_{\omega\in\mathbb{R}} \left| e^{-\sqrt{i\omega + i\frac{a}{\omega}(1-e^{-i\omega\tau})}\,L} \right| < 1$.*

Corollary 2. *The convergence rate of the classical Schwarz method for problem (5) satisfies $\rho \le e^{-\sqrt{\sqrt{-a}\sin(\sqrt{-2a\tau})/2}\,L}$.*

The optimized Schwarz algorithm. We introduce new transmission conditions in (7) and (9), using $\mathcal{B}_+$ and $\mathcal{B}_-$ to denote $(\frac{\partial}{\partial x}+p)$ and $(\frac{\partial}{\partial x}-p)$,

$$\mathcal{B}_+ u_1^n(t,L) = \mathcal{B}_+ u_2^{n-1}(t,L), \quad \mathcal{B}_- u_2^n(t,0) = \mathcal{B}_- u_1^{n-1}(t,0). \tag{10}$$

Theorem 5. *Assume a and τ satisfy the stability condition of Theorem 1. Then, the Schwarz waveform relaxation algorithm (7) with the Robin transmission conditions (10) converges as stated in (8), where*

$$\rho := \rho_{opt}(p) = \sup_{\omega\in\mathbb{R}} \left| \frac{\sqrt{i\omega - ae^{-i\omega\tau}} - p}{\sqrt{i\omega - ae^{-i\omega\tau}} + p} e^{-\sqrt{i\omega - ae^{-i\omega\tau}}\,L} \right| < 1. \tag{11}$$

Defining the curve $\Gamma = \{z : z = \sqrt{i\omega - ae^{-i\omega\tau}}, \omega \in \mathbb{R}\}$, the optimal choice of the parameter p is the value $p^\star$ which solves the min-max problem

$$\min_p \max_{z\in\Gamma} \left| \frac{z-p}{z+p} \cdot e^{-z\,L} \right|. \tag{12}$$

In Figure 3 we graphically depict the curve Γ, together with the contour lines of the function that appears in the min-max problem, for the case $L=0$.

In a numerical computation, $\omega \in [-\omega_{\max}, \omega_{\max}]$, because a numerical grid in time with spacing Δt can not carry arbitrary high frequencies; an estimate of $\omega_{\max}$ is $\omega_{\max} = \frac{\pi}{\Delta t}$. This simplifies the min-max problem (12) to a problem in a bounded domain, but it is still difficult to solve analytically, even for the special case $L = 0$. We therefore propose to solve the min-max problem over the bounding box given in Figure 3 containing the curve. This problem can be solved in closed form for $L = 0$.

Theorem 6. *Let $L = 0$ and set $b = \Re(\sqrt{i(\omega_{\max} + 2\pi/\tau) - ae^{-i\omega_{\max}\tau}})$. Assume $-a\tau \le 1$. If $b \ge -a\cos(a\tau) + 1/\cos(a\tau)$, then the solution of the approximate min-max problem is given by $p^\star = \sqrt{2\cos(a\tau)b + a}$; otherwise it is $p^\star = \sqrt{2a^2\cos^2(a\tau) - a}$.*

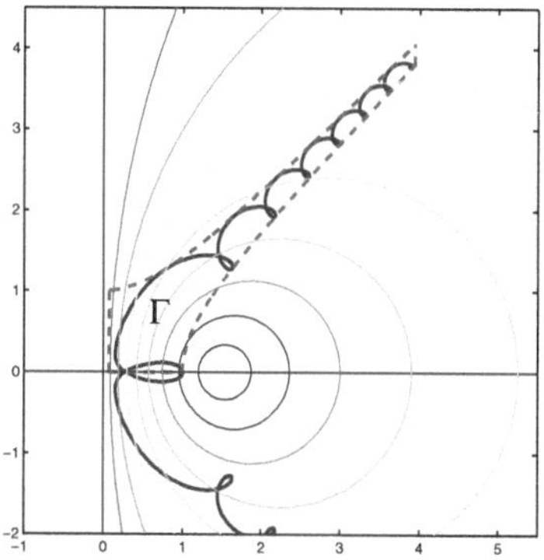

Fig. 3. Curve Γ along which one needs to solve the min-max problem to find the optimal p in the Robin transmission conditions, and level sets of $|z - \frac{3}{2}|/|z + \frac{3}{2}|$

The parameter $p^\star$ guarantees a lower bound on achievable acceleration,

$$\rho_{opt}(p^\star) \leq \sqrt{\frac{(p^\star - q)^2 + q^2 - a}{(p^\star + q)^2 + q^2 - a}} \cdot \rho_{cla}, \quad q = -a\cos(a\tau). \tag{13}$$

A similar analysis can also be done for the distributed delay PDE, leading to a min-max problem as in (12), along a curve

$$\tilde{\Gamma} = \{z : z = \sqrt{i\omega + i\frac{a}{\omega}(1 - e^{-i\omega\tau})}, \omega \in \mathbb{R}\}. \tag{14}$$

One can show that $\tilde{\Gamma}$ belongs to the same bounding box as the curve Γ. Hence, using the value of $p^\star$ from Theorem 6, a similar convergence acceleration will be achieved over the classical algorithm as in (13).

4 Numerical Results

We investigate the influence of the overlap L on the convergence of the classical Schwarz algorithm, and the influence of the parameter p in the Robin transmission conditions, on the convergence of the optimized Schwarz method. The results presented are for the constant delay PDE. We chose the parameters $a = -1.55$, $\tau = 1$, i.e., within the stability region, and $x \in [0, 2]$, $t \in [0, 10]$, $\Delta x = \frac{1}{50}$ and $\Delta t = \frac{1}{50}$. In Figure 4 (left) we show the evolution of the error as a function of the iteration index n, for various values of the overlap. The convergence improvement with increasing overlap is evident. The influence of the parameter in the Robin transmission conditions is shown in Figure 4 (right). Here, a minimal overlap of size $L = \Delta x$ was used.

Our experiments show clearly that the transmission conditions play a very important role for the performance of the algorithm. Compared to the overlap, where an increase corresponds to an increase in the subdomain solution cost, a change in p does not increase the subdomain solution cost.

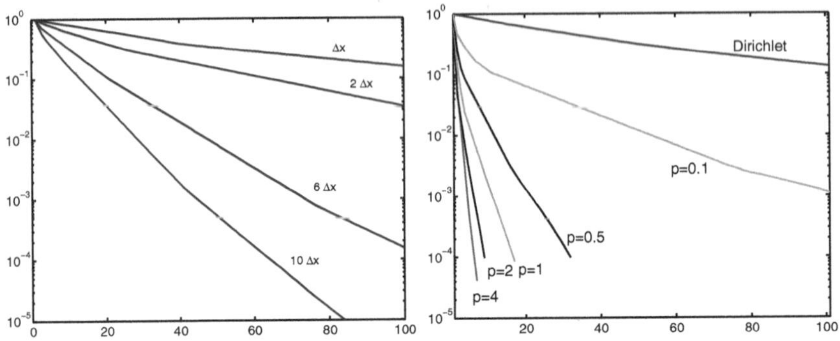

Fig. 4. Left: influence of the overlap on the performance of the classical Schwarz waveform relaxation algorithm for the constant delay PDE problem. Right: influence of the parameter p in the Robin transmission condition on the performance of the optimized Schwarz algorithm

References

A. Bellen and M. Zennaro. *Numerical Methods for Delay Differential Equations.* Oxford University Press, Oxford, U.K., 2003.

M. J. Gander. A waveform relaxation algorithm with overlapping splitting for reaction diffusion equations. *Numerical Linear Algebra with Applications*, 6:125–145, 1998.

M. J. Gander, L. Halpern, and F. Nataf. Optimal convergence for overlapping and non-overlapping Schwarz waveform relaxation. In C.-H. Lai, P. Bjørstad, M. Cross, and O. Widlund, editors, *Eleventh international Conference of Domain Decomposition Methods.* ddm.org, 1999.

M. J. Gander and A. M. Stuart. Space-time continuous analysis of waveform relaxation for the heat equation. *SIAM J. Sci. Comput.*, 19(6):2014–2031, 1998.

M. J. Gander and H. Zhao. Overlapping Schwarz waveform relaxation for the heat equation in n-dimensions. *BIT*, 42(4):779–795, 2002.

E. Giladi and H. B. Keller. Space time domain decomposition for parabolic problems. *Numerische Mathematik*, 93(2):279–313, 2002.

C. Huang and S. Vandewalle. An analysis of delay-dependent stability for ordinary and partial differential equations with fixed and distributed delays. *SIAM J. Sci. Comput.*, 2003. To appear.

S. Vandewalle and M. J. Gander. An analysis of Schwarz methods for delay partial differential equations. Technical report. in preparation.

J. Wu. *Theory and applications of Partial Functional Differential Equations.* Springer-Verlag, New York, 1996.

B. Zubik-Kowal. Stability in the numerical solution of linear parabolic equations with a delay term. *BIT*, 41:191–206, 2001.

B. Zubik-Kowal and S. Vandewalle. Waveform relaxation for functional differential equations. *SIAM J. Sci. Comput.*, 21:207–226, 1999.

Part VI

Minisymposium: Trefftz-Methods

A More General Version of the Hybrid-Trefftz Finite Element Model by Application of TH-Domain Decomposition

Ismael Herrera[1], Martin Diaz[2], and Robert Yates[1]

[1] Instituto de Geofísica, Universidad Nacional Autónoma de México (UNAM), iherrera@servidor.unam.mx
[2] Instituto Mexicano del Petróleo

Summary. In recent years the hybrid-Trefftz finite element (hT-FE) model, which originated in the work by Jirousek and his collaborators and makes use of an independently defined auxiliary inter-element frame, has been considerably improved. It has indeed become a highly efficient computational tool for the solution of difficult boundary value problems In parallel and to a large extent independently, a general and elegant theory of Domain Decomposition Methods (DDM) has been developed by Herrera and his coworkers, which has already produced very significant numerical results. Theirs is a general formulation of DDM, which subsumes and generalizes other standard approaches. In particular, it supplies a natural theoretical framework for Trefftz methods. To clarify further this point, it is important to spell out in greater detail than has been done so far, the relation between Herrera's theory and the procedures studied by researchers working in standard approaches to Trefftz method (Trefftz-Jirousek approach). As a contribution to this end, in this paper the hybrid-Trefftz finite element model is derived in considerable detail, from Herrera's theory of DDM. By so doing, the hT-FE model is generalized to non-symmetric systems (actually, to any linear differential equation, or system of such equations, independently of its type) and to boundary value problems with prescribed jumps. This process also yields some numerical simplifications.

1 Introduction

Trefftz [1926] method was originated by this author. However, the origins of the hybrid-Trefftz (HT) finite element (FE) model are only around twenty five years old, Jirousek and Leon [1977], Jirousek [1978]. Since then it has become a highly efficient computational tool for the solution of difficult boundary value problems, Jirousek and Wroblewski [1996], Qin [2000], with an increasing popularity among researchers and practitioners. In parallel and to a large extent independently, a general and elegant theory of domain decomposition methods (DDM) has been developed by Herrera and coworkers (Herrera et al.

[2002] and Herrera [2003]). This, throughout its different stages of development, has been known by a variety of names; mainly, localized adjoint method (LAM), Trefftz-Herrera method and unified theory of DDM. This is a general formulation, which subsumes and generalizes many other approaches. In particular, it seems to be the natural framework for Trefftz methods and several aspects of that theory have been recognized as fundamental by some of the most conspicuous researchers of these methodologies (Jirousek and Wroblewski [1996], Zielinski [1995] and Jirousek and Zielinski [1997]). However, it is important to spell out in greater detail than thus far, the relation between Herrera's theory and the procedures of Trefftz-Jirousek approach, Jirousek and Wroblewski [1996], which are extensively used by the researchers working in Trefftz method.

In particular, to this end, in the present paper a detailed analysis and comparison of the hybrid-Trefftz finite element model is carried out using Herrera's theory. In this manner, the HT-FE approach is generalized to problems with prescribed jumps and to non-symmetric operators. Also, a manner in which a significant reduction of the number of degrees of freedom involved in the HT-FE global equations is indicated. Although only problems formulated in terms Laplace operator are considered, the results can be extended to very general classes of differential operators using Herrera's general framework, as it will be explained in a paper now being prepared.

2 Notations and auxiliary results

Since the main purpose of this paper is to clarify the relation between Herrera's theory and Trefftz-Jirousek approach, as was stated in the Introduction, the notation that is used follows closely that which is standard is expositions of this latter approach (Qin [2000]). In addition, it is related with that which has been applied in Herrera's theory developments. A domain, Ω, is considered and one of its partitions $\{\Omega_1, ..., \Omega_E\}$, referred as *'the partition'*. In addition to the boundary Γ, of Ω, to be referred as the *'outer boundary'*, one considers the *'internal boundary'* Γ_I, which separates the subdomains from each other. The outer boundary is assumed to be the union of Γ_u and Γ_q. The boundary Γ_e, of every subdomain, Ω_e, of the partition, is assumed to be the union of $\Gamma_{eu} \equiv \Gamma_e \cap \Gamma_u$, $\Gamma_{eq} \equiv \Gamma_e \cap \Gamma_q$ and $\Gamma_{eI} \equiv \Gamma_e \cap \Gamma_I$. Trial and test functions are taken from the same linear space, D, whose members are functions defined in each one of the subdomains and, therefore, are generally discontinuous across Γ_I, together with their derivatives. Borrowing from Herrera's notation, one writes

$$[u] \equiv u_+ - u_- \quad \text{and} \quad \dot{\widehat{u}} \equiv \frac{1}{2}(u_+ + u_-) \tag{1}$$

$[u]$ and $\dot{\widehat{u}}$ are referred as the *'jump'* and the *'average'* of u, respectively. Here, u_+ and u_- are the limits from the positive and negative sides, respectively. The internal boundary is oriented by defining a unit normal vector $\underline{n}$

whose sense is chosen arbitrarily; then, the convention is that $\underline{n}$ points toward the positive side.

Given two functions, $u \in D$ and $w \in D$, the following relation between Jirousek's and Herrera's notations will be applied in the sequel

$$\sum_{e=1}^{E} \int_{\Gamma_{eI}} w \frac{\partial u}{\partial n} d\sigma \equiv - \int_{\Gamma_I} \left[w \frac{\partial u}{\partial n} \right] d\sigma \tag{2}$$

In the left-hand side of this equation, using Jirousek's notation, the normal derivative is taken with respect to the unit normal vector that points outwards of Ω_e. Thus, when Jirousek's notation is used one has two unit normal vectors defined at each point of Γ_I, while in Herrera's notation there is only one.

3 Trefftz-Jirousek Approach

For simplicity, we restrict attention to the case when the differential operator is Laplace's operator and adopt a notation similar to that followed by Jirousek and his collaborators (see for example Qin [2000]). The boundary value problem considered in Trefftz-Jirousek method is

$$\mathcal{L}u \equiv \Delta u = \bar{b}, \quad \text{in} \quad \Omega_e, \quad e = 1, ..., E \tag{3}$$

$$u = \bar{u} \quad \text{on} \quad \Gamma_u \quad \text{and} \quad \frac{\partial u}{\partial n} = \bar{q}_n \quad \text{on} \quad \Gamma_q \tag{4}$$

Together with

$$[u] = \left[\frac{\partial u}{\partial n} \right] = 0 \quad \text{on} \quad \Gamma_I \tag{5}$$

Observe that any function $u \in D$, which satisfies Eq.(3), can be written as

$$u = u_P + u_H \tag{6}$$

Where

$$\Delta u_P = \bar{b}, \quad \Delta u_H = 0, \quad \text{in} \quad \Omega_e, \quad e = 1, ..., E \tag{7}$$

The above equation, which is fulfilled by $u_H \in D$, is homogeneous. Therefore the set of functions that satisfy Eq.(3), constitutes a linear subspace $D_H \subset D$. In addition, $u_P \in D$ is not uniquely determined by Eq.(7). However, once $u_P \in D$ is chosen, $u_H = u - u_P$ is unique. Assuming that a function, $u_P \in D$, fulfilling Eq.(7), has been constructed the search to determine the solution $u \in D$ is carried out in the (affine) subspace $D_P \equiv u_P + D_H \subset D$.

4 Jirousek's Variational Principle

The variational principles to be applied are derived from the functional (see Qin [2000])

$$\Pi_m \equiv \frac{1}{2}\int_{\Omega}\left(q_1^2+q_2^2\right)^2 d\Omega - \int_{\Gamma_u} q_n\bar{u}ds + \int_{\Gamma_q}\left(\bar{q}_n - q_n\right)uds - \sum_{e=1}^{E}\int_{\Gamma_{eI}} q_n\tilde{u}ds \quad (8)$$

Observe that Π_m is a functional of a pair: $(u,\tilde{u})$, where $u \in D$ and $\tilde{u}$, the so called *'displacement frame'*, is a function defined on Γ_I. In Herrera's notation, the above functional is

$$\Pi_m \equiv \frac{1}{2}\int_{\Omega}\nabla u\cdot\nabla u dx - \int_{\Gamma_u}\bar{u}\frac{\partial u}{\partial n}dx + \int_{\Gamma_q}\left(\bar{q}_n - \frac{\partial u}{\partial n}\right)udx - \sum_{e=1}^{E}\int_{\Gamma_{eI}}\frac{\partial u}{\partial n}\tilde{u}dx \quad (9)$$

or, introducing the jumps (see the Notations Section)

$$\Pi_m \equiv \frac{1}{2}\int_{\Omega}\nabla u\cdot\nabla u dx - \int_{\Gamma_u}\bar{u}\frac{\partial u}{\partial n}dx + \int_{\Gamma_q}\left(\bar{q}_n - \frac{\partial u}{\partial n}\right)uds + \int_{\Gamma_I}\tilde{u}\left[\frac{\partial u}{\partial n}\right]dx \quad (10)$$

The system of equations used in Trefftz-Jirousek method is obtained by requiring that the variation of this functional be zero, in the (affine) subspace of functions that fulfill Eq. (3), while no restriction is imposed on the frame, $\tilde{u}$. The weak formulation derived using this functional is $\delta\Pi_m = 0$, which can be written as

$$\begin{aligned}&\int_{\Omega}\nabla u\cdot\nabla w dx - \int_{\Gamma_u}\bar{u}\frac{\partial w}{\partial n}dx + \int_{\Gamma_q}\left\{\left(\bar{q}_n - \frac{\partial u}{\partial n}\right)w - u\frac{\partial w}{\partial n}\right\}dx\\&+\int_{\Gamma_I}\left\{\tilde{w}\left[\frac{\partial u}{\partial n}\right]+\tilde{u}\left[\frac{\partial w}{\partial n}\right]\right\}dx = 0\end{aligned} \quad (11)$$

Here, $w \in D_H$ and $\tilde{w}$ stand for the variation of u and $\tilde{u}$, respectively. This is the form in which it is most frequently applied. However, for our analysis it is more convenient to write it as

$$\begin{aligned}&\int_{\Gamma_u}(u-\bar{u})\frac{\partial w}{\partial n}dx - \int_{\Gamma_q}w\left(\frac{\partial u}{\partial n}-\bar{q}_n\right)dx\\&+\int_{\Gamma_I}\left\{\tilde{w}\left[\frac{\partial u}{\partial n}\right]-[u]\widehat{\frac{\partial w}{\partial n}}^{\cdot}+\left(\tilde{u}-\dot{\widehat{u}}\right)\left[\frac{\partial w}{\partial n}\right]\right\}dx = 0\end{aligned} \quad (12)$$

Therefore, the Euler equations for this variational formulation are the boundary conditions of Eqs. (4) and the continuity conditions for the function and its normal derivative of Eqs. (5), together with

$$\tilde{u} = \dot{\widehat{u}} \equiv u \quad \text{on} \quad \Gamma_I \quad (13)$$

Clearly $\dot{\widehat{u}} \equiv u$ because u is continuous across Γ_I.

In conclusion, a pair $(u,\tilde{u})$, with $u \in D_P$, that satisfies Eq. (12) for every variation $w \in D_H$, has the following properties:

1. u is solution of the BVP, and
2. $\tilde{u} = u$ on Γ_I.

Generally the linear subspace $D_H \subset D$ is infinite dimensional and therefore the search for $u \in D_P$, in the entirety of $D_P \subset D$, is not feasible. In order to make it feasible, a finite-dimensional (affine) subspace $\hat{D}_P \equiv u_P + \hat{D}_H \subset D$ is introduced as follows: a finite family of linearly independent functions $\mathcal{E} \equiv \{w^1, ..., w^N\} \subset D$ is chosen and $\hat{D}_H \subset D_H$ is defined to be

$$\hat{D}_H \equiv \text{span}\{w^1, ..., w^N\} \tag{14}$$

The system $\mathcal{E} \equiv \{w^1, ..., w^N\} \subset \hat{D}_H$, above, is usually referred as a truncated *T-complete system of homogeneous solutions*, Jirousek and Wroblewski [1996]. In addition a system of functions each one of them defined on Γ_I exclusively, $\left\{\tilde{w}^1, ..., \tilde{w}^{\tilde{N}}\right\}$, is introduced. This is referred as the frame basis. Then one approximates $u \in D_P$ and $\tilde{u}$, by

$$\hat{u} = u_P + \hat{u}_H \tag{15}$$

and

$$\widehat{\tilde{u}} = \sum_{\alpha=1}^{\tilde{N}} \tilde{c}_\alpha \tilde{w}^\alpha \tag{16}$$

respectively. Here $\hat{u}_H \in \hat{D}_H$. Clearly, Eq. (14) implies

$$\hat{u}_H = \sum_{i=1}^{N} c_i w^i \tag{17}$$

Above $\{c_1, .., c_N\}$ and $\{\tilde{c}_1, .., \tilde{c}_{\tilde{N}}\}$ are suitable choices of the coefficients. They are determined by application of the variational principle discussed before. Actually, the weak formulations of Eq. (11) or (12) are applied, with u and $\tilde{u}$ replaced by $\hat{u}$ and $\widehat{\tilde{u}}$, respectively. By inspection, it is seen that the Euler equations associated with Eq. (12) yield the following approximate relations

$$\begin{aligned} &\hat{u} \approx \bar{u} \quad \text{on} \quad \Gamma_u, \quad \frac{\partial \hat{u}}{\partial n} \approx \bar{q}_n \quad \text{on} \quad \Gamma_q \\ &[\hat{u}] \approx \left[\frac{\partial \hat{u}}{\partial n}\right] \approx 0 \quad \text{and} \quad \tilde{u} \approx \widehat{\dot{u}} \quad \text{on} \quad \Gamma_I \end{aligned} \tag{18}$$

It is relevant to observe that $\hat{u} \in \hat{D}_P$, as given by Eqs. (15) and (17), is a discontinuous function and, therefore, its average across Γ_I, in Eq. (18), can not be replaced by its value on Γ_I. Also, usually the internal Γ_I boundary is much larger than the external boundary, $\Gamma_u \cup \Gamma_q$, then the relation $N \approx 2\tilde{N}$ is fulfilled approximately and the total number of degrees of freedom is

$$N + \tilde{N} \approx \frac{3}{2} N \tag{19}$$

5 The BVPJ and Herrera's Variational Principles

The boundary value problem considered in Herrera's theory is a boundary value problem with prescribed jumps (BVPJ) in the internal boundary, Γ_I, which is the same as that considered in Section 3, except that Eq. (5) is replaced by

$$[u] = j_{\Sigma}^{0}, \quad \left[\frac{\partial u}{\partial n}\right] = j_{\Sigma}^{1} \quad \text{on } \Gamma_{\mathrm{I}} \tag{20}$$

where j_{Σ}^{0} and j_{Σ}^{1} are given functions defined on Γ_{I}. In Herrera's theory two weak formulations are introduced, Herrera [2001], Herrera [1985]: the *'weak formulation in terms of the data of the problem'*. This is the basis of the direct approach and it yields weak formulations that are quite similar those usually applied by other authors. For the BVPJ here considered, it is

$$\begin{gathered}\langle \mathcal{L}_H u, w\rangle \equiv \\ \int_\Omega w\Delta u dx + \int_{\Gamma_u} u\tfrac{\partial w}{\partial n}dx - \int_{\Gamma_q} w\tfrac{\partial u}{\partial n}dx + \int_{\Gamma_I}\left\{\dot{\widehat{w}}\left[\tfrac{\partial u}{\partial n}\right] - [u]\,\dot{\widehat{\tfrac{\partial w}{\partial n}}}\right\}dx = \\ \int_\Omega w\bar{b}dx + \int_{\Gamma_u} \bar{u}\tfrac{\partial w}{\partial n}dx - \int_{\Gamma_q} w\bar{q}_n dx + \int_{\Gamma_I}\left\{\dot{\widehat{w}}j_\Sigma^1 - j_\Sigma^0\dot{\widehat{\tfrac{\partial w}{\partial n}}}\right\}dx\end{gathered} \tag{21}$$

Which is equivalent to the *'weak formulation in terms of the complementary information'*

$$\begin{gathered}\langle \mathcal{L}_H^* u, w\rangle \equiv \\ \int_\Omega u\Delta w dx + \int_{\Gamma_u} w\tfrac{\partial u}{\partial n}dx - \int_{\Gamma_q} u\tfrac{\partial w}{\partial n}dx + \int_{\Gamma_I}\left\{\dot{\widehat{u}}\left[\tfrac{\partial w}{\partial n}\right] - [w]\,\dot{\widehat{\tfrac{\partial u}{\partial n}}}\right\}dx = \\ \int_\Omega w\bar{b}dx + \int_{\Gamma_u} \bar{u}\tfrac{\partial w}{\partial n}dx - \int_{\Gamma_q} w\bar{q}_n dx + \int_{\Gamma_I}\left\{\dot{\widehat{w}}j_\Sigma^1 - j_\Sigma^0\dot{\widehat{\tfrac{\partial w}{\partial n}}}\right\}dx\end{gathered} \tag{22}$$

Both of these formulations are equivalent, because it can be shown that $\langle \mathcal{L}_H^* u, w\rangle \equiv \langle \mathcal{L}_H u, w\rangle = \langle \mathcal{L}_H w, u\rangle$. Furthermore, they are equivalent to the variational condition $\delta\Pi_H(u) = 0$, if $\Pi_H(u)$ is defined to be

$$\begin{gathered}2\Pi_H(u) \equiv \\ \int_\Omega u\Delta u dx + \int_{\Gamma_u} u\tfrac{\partial u}{\partial n}dx - \int_{\Gamma_q} u\tfrac{\partial u}{\partial n}dx + \int_{\Gamma_I}\left\{\dot{\widehat{u}}\left[\tfrac{\partial u}{\partial n}\right] - [u]\,\dot{\widehat{\tfrac{\partial u}{\partial n}}}\right\}dx - \\ 2\left\{\int_\Omega w\bar{b}dx + \int_{\Gamma_u} \bar{u}\tfrac{\partial w}{\partial n}dx - \int_{\Gamma_q} w\bar{q}_n dx + \int_{\Gamma_I}\left\{\dot{\widehat{w}}j_\Sigma^1 - j_\Sigma^0\dot{\widehat{\tfrac{\partial w}{\partial n}}}\right\}dx\right\}\end{gathered} \tag{23}$$

When u is varied subjected to the restriction $u \in D_P$, so that $\Delta w = 0$, Eqs. (21) and (22) can also be written as

$$\begin{gathered}\int_{\Gamma_u}(u-\bar{u})\tfrac{\partial w}{\partial n}dx - \int_{\Gamma_q} w\left(\tfrac{\partial u}{\partial n} - \bar{q}_n\right)dx \\ + \int_{\Gamma_I}\left\{\dot{\widehat{w}}\left(\left[\tfrac{\partial u}{\partial n}\right] - j_\Sigma^1\right) - \left([u] - j_\Sigma^0\right)\dot{\widehat{\tfrac{\partial w}{\partial n}}}\right\}dx = 0\end{gathered} \tag{24}$$

and

$$\begin{gathered}\int_{\Gamma_u} w\frac{\partial u}{\partial n}dx - \int_{\Gamma_q} u\frac{\partial w}{\partial n}dx + \int_{\Gamma_I}\left\{\dot{\widehat{u}}\left[\frac{\partial w}{\partial n}\right] - [w]\,\dot{\widehat{\frac{\partial u}{\partial n}}}\right\}dx = \\ \int_{\Omega} w\bar{b}dx + \int_{\Gamma_u}\bar{u}\frac{\partial w}{\partial n}dx - \int_{\Gamma_q} w\bar{q}_n dx + \int_{\Gamma_I}\left\{\dot{\widehat{w}}j^1_{\Sigma} - j^0_{\Sigma}\dot{\widehat{\frac{\partial w}{\partial n}}}\right\}dx\end{gathered} \tag{25}$$

respectively. Eq. (24) exhibits the Eqs. (4) and (20), as the Euler equations of *the variational principle in terms of the data* of the BVPJ. However, the use of Eq. (25) is different.

Let $\tilde{u}$, $\tilde{q}_n$, $\tilde{u}_a$ and $\tilde{q}_a$ be functions defined, the first two, on Γ_q and Γ_u respectively, and on Γ_I, the last two. Assume that they satisfy

$$\begin{gathered}\int_{\Gamma_u} w\tilde{q}_n dx - \int_{\Gamma_q}\tilde{u}\frac{\partial w}{\partial n}dx + \int_{\Gamma_I}\left\{\tilde{u}_a\left[\frac{\partial w}{\partial n}\right] - [w]\,\tilde{q}_a\right\}dx = \\ \int_{\Omega} w\bar{b}dx + \int_{\Gamma_u}\bar{u}\frac{\partial w}{\partial n}dx - \int_{\Gamma_q} w\bar{q}_n dx + \int_{\Gamma_I}\left\{\dot{\widehat{w}}j^1_{\Sigma} - j^0_{\Sigma}\dot{\widehat{\frac{\partial w}{\partial n}}}\right\}dx\end{gathered} \tag{26}$$

for every $w \in D_H$, then subtracting Eq. (26) from Eq. (25), it is seen that

$$\begin{gathered}\int_{\Gamma_u} w\left(\frac{\partial u}{\partial n} - \tilde{q}_n\right)dx - \int_{\Gamma_q}(u-\tilde{u})\frac{\partial w}{\partial n}dx \\ +\int_{\Gamma_I}\left\{\left(\dot{\widehat{u}} - \tilde{u}_a\right)\left[\frac{\partial w}{\partial n}\right] - [w]\left(\dot{\widehat{\frac{\partial u}{\partial n}}} - \tilde{q}_a\right)\right\}dx = 0\end{gathered} \tag{27}$$

Eq. (26) is a variational principle whose Euler equations, in view of Eq. (27), are $\tilde{q}_n = \frac{\partial u}{\partial n}$ on Γ_u, $\tilde{u} = u$ on Γ_q and $\tilde{u}_a = \dot{\widehat{u}}$, $\tilde{q}_a = \dot{\widehat{\frac{\partial u}{\partial n}}}$ on Γ_I.

When a truncated *T-complete system of homogeneous solutions*, $\mathcal{E} \equiv \{w^1, ..., w^N\} \subset D_H$, is used to generate a subspace $\hat{D}_H \subset D_H$, and D_H is replaced by $\hat{D}_H$, then these equations are only approximately satisfied. In particular, $\tilde{u}_a$ and $\tilde{q}_a$ are approximations of the averages, across Γ_I, of the function and its normal derivative, respectively. The following systems of *'frames'* are introduced: $\tilde{\mathcal{E}}_u \equiv \left\{\tilde{w}^1_u, ..., \tilde{w}^{\tilde{N}_u}_u\right\}$, $\tilde{\mathcal{E}}_q \equiv \left\{\tilde{w}^1_q, ..., \tilde{w}^{\tilde{N}_q}_q\right\}$, $\tilde{\mathcal{E}}_{ua} \equiv \left\{\tilde{w}^1_{ua}, ..., \tilde{w}^{\tilde{N}_{ua}}_{ua}\right\}$ and $\tilde{\mathcal{E}}_{qa} \equiv \left\{\tilde{w}^1_{qa}, ..., \tilde{w}^{\tilde{N}_{qa}}_{qa}\right\}$. The first two are defined on Γ_q and Γ_u respectively, and the last two on Γ_I. Then the functions are taken to be linear combinations of these bases with suitable coefficients, which are determined by application of the weak formulation of Eq. (26). Of course a necessary condition for this to be possible is that $N = \tilde{N}_u + \tilde{N}_q + \tilde{N}_{ua} + \tilde{N}_{qa}$. The total number of degrees of freedom is N and global matrix associated the system of equations (26) is $N \times N$.

6 Comparisons and Conclusions

Clearly Trefftz-Jirousek and Trefftz-Herrera formulations are closely related. However, this latter approach generalizes Jirousek's since the boundary value problem considered in Section 4 is a particular case of the more general BVPJ treated in Section 5; namely, Trefftz-Jirousek method deals with the particular case of this BVPJ when $j^0_\Sigma = j^1_\Sigma = 0$. Also, according to Eq. (19) in Trefftz-Herrera formulation the number of degrees of freedom is reduced a 33%, in comparison with Trefftz-Jirousek formulation. Indeed, in this latter approach one deals with $\frac{3}{2}N \times \frac{3}{2}N$ global matrices, while these are only $N \times N$ in the former. A more thorough discussion of these points will be presented in a paper now being prepared.

References

I. Herrera. Unified Approach to Numerical Methods. Part 1. Green's Formulas for Operators in Discontinuous Fields. *Numerical Methods for Partial Differential Equations*, 1(1):12–37, 1985.

I. Herrera. On Jirousek Method and its Generalizations. *Computer Assisted. Mech. Eng. Sci.*, 8:325–342, 2001.

I. Herrera. A Unified Theory of Domain Decomposition Methods. In I. Herrera, D. E. Keyes, O. B. Widlund, and R. Yates, editors, *14th International Conference on Domain Decomposition Methods*, pages 243–248, 2003.

I. Herrera, R. Yates, and M. Diaz. General Theory of Domain Decomposition: Indirect Methods. *Numerical Methods for Partial Differential Equations*, 18 (3):296–322, 2002.

J. Jirousek. Basis for development of large finite elements locally satisfying all field equations. *Comp. Meth. Appl. Mech. Eng.*, 14:65–92, 1978.

J. Jirousek and N. Leon. A powerful finite element for plate bending. *Comp. Meth. Appl. Mech. Eng.*, 12:77–96, 1977.

J. Jirousek and Wroblewski. T-elements: State of the Art and Future Trends. *Archives of Computational Methods in Engineering*, 3–4:323–434, 1996.

J. Jirousek and A. P. Zielinski. Survey of Trefftz-Type Element Formulations. *Compu. and Struct.*, 63:225–242, 1997.

Q. H. Qin. *The Trefftz Finite and Boundary Element Method.* WIT Press. Southampton, 2000.

E. Trefftz. Ein Gegenstck zum Ritzschen Verfahren. In *Proceedings 2nd International Congress of Applied mechanics*, pages 131–137, Zurich, 1926.

A. P. Zielinski. On trial functions applied in the generalized Trefftz method. *Advances in Engineering Software*, 24:147–155, 1995.

Part VII

Minisymposium: Domain Decomposition on Nonmatching Grids

Mixed Finite Element Methods for Diffusion Equations on Nonmatching Grids

Yuri Kuznetsov

Department of Mathematics, University of Houston (kuz@math.uh.edu)

Summary. The hybridization technique is applied to replace the macro-hybrid mixed finite element problem for the diffusion equation by the equivalent cell-based formulation. The underlying algebraic system is condensed by eliminating the degrees of freedom which represent the interface flux and cell pressure variables to the system containing the Lagrange multipliers variables. An approach to the numerical solution of the condensed system is briefly discussed.

1 Introduction

In this paper, we consider macro-hybrid mixed finite element method for the diffusion equation on nonmatching grids. The paper is organized as follows. The four-field macro-hybrid mixed formulation for the diffusion equation is given in Sect. 2.

In Sect. 3, we apply the hybridization technique to replace the macro-hybrid formulation by the cell-based formulation and describe the condensation procedure to reduce the underlying algebraic system to the system for the Lagrange multipliers only. In Sect. 4, we briefly discuss an algebraic solution method for the condensed system.

2 Problem formulation

We consider the diffusion problem in the form of a system of the first order differential equations

$$\begin{aligned} K^{-1}\,\bar{u} + \operatorname{grad} p &= 0 \\ \operatorname{div} \bar{u} + \qquad cp &= f \end{aligned} \tag{1}$$

in a bounded connected polygonal (polyhedral) domain Ω in $\mathbb{R}^d$, $d = 2$ ($d = 3$) with homogeneous boundary conditions

$$\begin{aligned} p &= 0 \quad \text{on} \quad \Gamma_D, \\ \bar{u} \cdot \bar{n} &= 0 \quad \text{on} \quad \Gamma_N. \end{aligned} \tag{2}$$

Here Γ_D and Γ_N are the Dirichlet and the Neumann parts of the boundary $\partial\Omega$, $\bar{n}$ is the outward unit normal to $\partial\Omega$, $K = K(x)$ is the diffusion tensor, $K = K^T > 0$, $c = c(x)$ is a nonnegative function, and $f \in L_2(\Omega)$. We assume that Γ_D is a closed subset of $\partial\Omega$ consisting of a finite number of segments (polygons) in the case $d = 2$ ($d = 3$).

For the sake of simplicity, in the paper, we consider only the case $d = 2$. The extension to the three dimensional problem is basically straightforward.

The weak formulation of (1), (2) reads as follows: find

$$\bar{u} \in V \equiv \Big\{\bar{v} : \bar{v} \in H_{\mathrm{div}}(\Omega),\ \int\limits_{\partial\Omega} |\bar{v} \cdot \bar{n}|^2 \,\mathrm{ds} < +\infty,\ \bar{v} \cdot \bar{n} = 0 \text{ on } \Gamma_N\Big\},$$

$p \in Q \equiv L_2(\Omega)$ such that

$$\begin{aligned} \int\limits_\Omega \left(K^{-1}\bar{u}\right) \cdot \bar{v} \,\mathrm{dx} - \int\limits_\Omega p(\nabla \cdot \bar{v}) \,\mathrm{dx} &= 0 \\ \int\limits_\Omega (\nabla \cdot \bar{u}) q \,\mathrm{dx} \quad + \int\limits_\Omega cpq \,\mathrm{dx} \quad &= \int\limits_\Omega fq \,\mathrm{dx} \end{aligned} \tag{3}$$

for all $(\bar{v},\ q) \in V \times Q$.

Let Ω_h be a partitioning of Ω into m nonoverlapping polygonal cells e_k:

$$\Omega_h = \bigcup_{k=1}^{m} e_k, \tag{4}$$

and V_h and Q_h be finite element subspaces of V and Q, respectively. We assume that the partitioning Ω_h is conforming, i.e. the interface Γ_{st} between any adjacent cells e_s and e_t is always a common edge for both cells and the set $\Gamma_N \cap \Gamma_D$ belongs to the set of vertices in Ω_h. If all the cells e_k are triangles then V_h can be chosen as the proper subspace of the lowest Raviart-Thomas finite element space $\mathrm{RT}_0(\Omega_h)$ (see, Brezzi and Fortin [1991]). Otherwise, we can use the new method for the construction of V_h recently invented in Kuznetsov and Repin [2003]. The normal components $\bar{u} \cdot \bar{n}_{st}$ of the flux $\bar{u}$ at the interfaces Γ_{st} between cells e_s and e_t are constants in both choices of V_h. Here $\bar{n}_{st}$ denotes the unit normal to Γ_{st} directed from e_s to e_t.

The mixed finite element approximation to (1), (2) reads as follows: find $(\bar{u}_h,\ p_h) \in V_h \times Q_h$ such that

$$\begin{aligned} \int\limits_\Omega \left(K^{-1}\bar{u}_h\right) \cdot \bar{v} \,\mathrm{dx} - \int\limits_\Omega p_h(\nabla \cdot \bar{v}) \,\mathrm{dx} &= 0 \\ \int\limits_\Omega (\nabla \cdot \bar{u}_h)\, q \,\mathrm{dx} \quad + \int\limits_\Omega cp_h q \,\mathrm{dx} \quad &= \int\limits_\Omega fq \,\mathrm{dx} \end{aligned} \tag{5}$$

for all $(\bar{v},\ q) \in V_h \times Q_h$.

Let Ω be splitted into two nonoverlapping subdomains Ω_1 and Ω_2 with a piece-wise linear simply connected interface boundary Γ. Then, the four-field macro-hybrid mixed formulation of (1), (2) originally proposed in Kuznetsov and Wheeler [1995] reads as follows: find $(\bar{u}_k,\ p_k,\ \lambda_k) \in V_k \times Q_k \times \Lambda_k$, $k = 1, 2$, $\phi \in \Phi$ such that

$$
\begin{aligned}
&a_1(\bar{u}_1,\ \bar{v}_1) + b_1(p_1,\ \bar{v}_1) + c_1(\lambda_1,\ \bar{v}_1) = 0\\
&a_2(\bar{u}_2,\ \bar{v}_2) + b_2(p_2,\ \bar{v}_2) + c_2(\lambda_2,\ \bar{v}_2) = 0\\
&b_1(q_1,\ \bar{u}_1) - \sigma_1(p_1,\ q_1) = l_1(q_1)\\
&b_2(q_2,\ \bar{u}_2) - \sigma_2(p_2,\ q_2) = l_2(q_2)\\
&c_1(\mu_1,\ \bar{u}_1) + d_1(\phi,\ \mu_1) = 0\\
&c_2(\mu_2,\ \bar{u}_2) + d_2(\phi,\ \mu_1) = 0\\
&d_1(\psi,\ \lambda_1) + d_2(\psi,\ \lambda_2) = 0
\end{aligned}
\tag{6}
$$

for all $(\bar{v}_k,\ p_k,\ \mu_k) \in V_k \times Q_k \times \Lambda_k$, $k = 1, 2$, $\psi \in \Psi$.
Here

$$
\begin{aligned}
V_k &= \Big\{\bar{v} : \bar{v} \in H_{\mathrm{div}}(\Omega_k),\ \int\limits_{\partial\Omega_k} (\bar{v}\cdot\bar{n}_k)^2\,\mathrm{ds} < +\infty,\ \bar{v}\cdot\bar{n} = 0 \text{ on } \partial\Omega_k \cap \Gamma_N\Big\},\\
Q_k &= L_2(\Omega_k), \quad \Lambda_k = L_2(\Gamma), \quad k = 1, 2,\\
\Phi &= L_2(\Gamma),
\end{aligned}
\tag{7}
$$

and

$$
\begin{aligned}
a_k(\bar{u}, \bar{v}) &= \int\limits_{\Omega_k} (K^{-1}\bar{u})\cdot\bar{v}\,\mathrm{dx}, \quad \sigma_k(p,\ q) = \int\limits_{\Omega_k} cpq\,\mathrm{dx},\\
b_k(p,\ \bar{v}) &= -\int\limits_{\Omega_k} p\,(\nabla\cdot\bar{v})\,\mathrm{dx}, \quad c_k(\lambda,\ \bar{v}) = (-1)^{k-1}\int\limits_{\Gamma} \lambda\,(\bar{v}\cdot\bar{n}_\Gamma)\,\mathrm{ds},\\
d_k(\phi,\ \mu) &= (-1)^k \int\limits_{\Gamma} \phi\mu\,\mathrm{ds}, \quad l_k(q) = -\int\limits_{\Omega_k} fq\,\mathrm{dx},
\end{aligned}
\tag{8}
$$

$k = 1, 2$, where $\bar{n}_\Gamma$ is the unit normal vector to Γ directed from Ω_1 to Ω_2.

Let $\Omega_{k,h}$ be a partitioning of Ω_k into m_k polygons $e_i^{(k)}$, $k = 1, 2$. We assume that both partitionings are conformal and the set of vertices of Γ belongs to the set of vertices of both partitionings $\Omega_{1,h}$ and $\Omega_{2,h}$. Subspaces $V_{k,h}$ and $Q_{k,h}$ of the spaces V_k and Q_k, respectively, are defined similar to V_h and Q_h in problem (5).

Let $\Gamma_h^{(k)} = \bigcup_{i=1}^{n_k} \gamma_{i,h}^{(k)}$ be the trace of $\Omega_{k,h}$ onto Γ where $\gamma_{i,h}^{(k)}$ are the edges of the cells in $\Omega_{k,h}$ adjacent to Γ, $i = \overline{1,\ n_k}$, $k = 1, 2$. Here n_k is the number of cells' edges in Ω_h belonging to Γ, $k = 1, 2$. We define $\Lambda_{k,h}$ by

$$\Lambda_{k,h} = \{\lambda : \lambda = \text{const on } \gamma_{i,h}^{(k)},\ i = \overline{1,\ n_k}\} \tag{9}$$

$k = 1, 2$, and choose

$$\Phi_h = \Lambda_{1,h}. \tag{10}$$

The finite element approximation to (6)-(8) reads as follows: find $(\bar{u}_{k,h},\ p_{k,h},\ \lambda_{k,h}) \in V_{k,h} \times Q_{k,h} \times \Lambda_{k,h}$, $k = 1, 2$, $\phi_h \in \Phi_h$, such that the equations (6) with $\bar{u}_k = \bar{u}_{k,h}$, $p_k = p_{k,h}$, $\lambda_k = \lambda_{k,h}$, $k = 1, 2$, $\phi = \phi_h$ are satisfied for all $(\bar{v}_k,\ q_k,\ \mu_k) \in V_{k,h} \times Q_{k,h} \times \Lambda_{k,h}$, $k = 1, 2$, $\psi \in \Phi_h$. This approximation results in the system

$$\mathcal{A} \begin{pmatrix} w_1 \\ w_2 \\ \phi \end{pmatrix} = \begin{pmatrix} F_1 \\ F_2 \\ 0 \end{pmatrix} \tag{11}$$

with the matrix

$$\mathcal{A} = \begin{pmatrix} A_1 & 0 & D_1^T \\ 0 & A_2 & D_2^T \\ D_1 & D_2 & 0 \end{pmatrix} \tag{12}$$

where

$$A_k = \begin{pmatrix} M_k & B_k^T & C_k^T \\ B_k & -\Sigma_k & 0 \\ C_k & 0 & 0 \end{pmatrix} \tag{13}$$

are the saddle point matrices, $k = 1, 2$, and

$$w_k = \begin{pmatrix} u_k \\ p_k \\ \lambda_k \end{pmatrix}, \quad F_k = \begin{pmatrix} 0 \\ -f_k \\ 0 \end{pmatrix}, \quad k = 1, 2. \tag{14}$$

Here M_k is a symmetric positive definite matrix, and Σ_k is a symmetric positive definite (or semidefinite) matrix, $k = 1, 2$.

3 Hybridization and condensation

The extension of (6)-(8) to the case of many subdomains is straightforward. We consider the hybrid mixed formulation based on partitionings of $\Omega_{k,h}$ into subdomains/cells $e_i^{(k)}$ used in Sect. 2 for the approximation of the problem (6)-(8).

We introduce new spaces $V_{k,i,h}$ and $Q_{k,i,h}$ to be the restrictions onto $e_i^{(k)}$ of $V_{k,h}$ subject to $\Gamma_N = \emptyset$ and $Q_{k,h}$, respectively, $i = \overline{1, m_k}$, $k = 1, 2$, and define the spaces $\widehat{V}_{k,h}$ and $\widehat{Q}_{k,h}$ as the products of the spaces $V_{k,i,h}$ and $Q_{k,i,h}$, $i = \overline{1,\ m_k}$, respectively, $k = 1, 2$. Then, we introduce spaces $\Lambda_{k,i,h}$ of functions λ defined on $\partial e_i^{(k)}$ which are constants on each interface $\Gamma_{k,i,h}$ between $e_i^{(k)}$ and adjacent cells $e_j^{(k)}$ as well as on the intersections $\partial e_i^{(k)}$ with the linear parts of $\partial\Omega$, $i = \overline{1,\ m_k}$, $k = 1, 2$. The functions in $\Lambda_{k,i,h}$ should vanish on Γ_D. We denote by $\widehat{\Lambda}_{k,h}$ the product of all spaces $\Lambda_{k,i,h}$, $k = 1, 2$. Finally, we preserve the definition for Φ_h from Sect. 2.

The new finite element problem reads as follows: find $(\hat{u}_{k,h},\ \hat{p}_{k,h},\ \hat{\lambda}_{k,h}) \in \widehat{V}_{k,h} \times \widehat{Q}_{k,h} \times \widehat{\Lambda}_{k,h}$, $k = 1, 2$, $\hat{\phi}_h \in \Phi_h$, such that

$$\begin{aligned}
&\hat{a}_1(\hat{u}_{1,h},\ \bar{v}_1) + \hat{b}_1(\hat{p}_{1,h},\ \bar{v}_1) + \hat{c}_1(\hat{\lambda}_{1,h},\ \bar{v}_1) = 0\\
&\hat{a}_2(\hat{u}_{2,h},\ \bar{v}_2) + \hat{b}_2(\hat{p}_{2,h},\ \bar{v}_2) + \hat{c}_2(\hat{\lambda}_{2,h},\ \bar{v}_2) = 0\\
&\hat{b}_1(q_1,\ \hat{u}_{1,h}) - \hat{\sigma}_1(\hat{p}_{1,h},\ q_1) = \hat{l}_1(q_1)\\
&\hat{b}_2(q_2,\ \hat{u}_{2,h}) - \hat{\sigma}_2(\hat{p}_{2,h},\ q_2) = \hat{l}_2(q_2)\\
&\hat{c}_1(\mu_1,\ \hat{u}_{1,h}) + \hat{d}_1(\hat{\phi}_h,\ \mu_1) = 0\\
&\hat{c}_2(\mu_2,\ \hat{u}_{2,h}) + \hat{d}_2(\hat{\phi}_h,\ \mu_1) = 0\\
&\hat{d}_1(\psi,\ \hat{\lambda}_{1,h}) + \hat{d}_2(\psi,\ \hat{\lambda}_{2,h}) = 0
\end{aligned} \tag{15}$$

for all $(\bar{v}_k,\ p_k,\ \mu_k) \in \widehat{V}_{k,h} \times \widehat{Q}_{k,h} \times \widehat{\Lambda}_{k,h}$, $k = 1, 2$, $\psi \in \Phi_h$.
Here,

$$\begin{aligned}
\hat{a}_k(\bar{u}_k,\ \bar{v}_k) &= \sum_{i=1}^{m_k} \int_{e_i^{(k)}} (K^{-1}\bar{u}_{k,i}) \cdot \bar{v}_{k,i}\, \mathrm{dx}, \quad \hat{b}_k(p_k,\ \bar{v}_k) = -\sum_{i=1}^{m_k} \int_{e_i^{(k)}} p_{k,i}\, (\nabla \cdot \bar{v}_{k,i})\, \mathrm{dx},\\
\hat{c}_k(\lambda_k,\ \bar{v}_k) &= \sum_{i=1}^{m_k} \int_{\partial e_i^{(k)} \setminus \Gamma_D} \lambda_{k,i}\, (\bar{v}_{k,i} \cdot \bar{n}_{k,i})\, \mathrm{ds}, \quad \hat{\sigma}_k(p_k,\ q_k) = \sum_{i=1}^{m_k} \int_{e_i^{(k)}} c p_{k,i} q_{k,i}\, \mathrm{dx},\\
\hat{d}_k(\phi,\ \mu_k) &= (-1)^k \sum_{i=1}^{m_k} \int_{\Gamma \cap \partial e_i^{(k)}} \phi \mu_{k,i}\, \mathrm{ds}, \quad \hat{l}_k(q_k) = -\sum_{i=1}^{m_k} \int_{e_i^{(k)}} f q_{k,i}\, \mathrm{dx},
\end{aligned} \tag{16}$$

where $\bar{n}_{k,i}$ is the outward unit normal to $\partial e_i^{(k)}$, $i = \overline{1,\ m_k}$, $k = 1, 2$.

The finite element problem (15), (16) is said to be the hybridization of the finite element problem of the previous Section. It can be proved that the problems are equivalent, i.e. the restrictions of $\bar{u}_{k,h}$ and $p_{k,h}$ onto a cell $e_i^{(k)}$ coincide with $\hat{u}_{k,i,h}$ and $\hat{p}_{k,i,h}$, respectively, $\lambda_{k,h}$ coincides with restriction of $\hat{\lambda}_{k,h}$ onto Γ, and ϕ_h coincides with $\hat{\phi}_h$.

Problem (15), (16) results in the system of linear algebraic equations

$$\mathcal{A}\begin{pmatrix} w_1 \\ \lambda_1 \\ w_2 \\ \lambda_2 \\ \phi \end{pmatrix} = \begin{pmatrix} F_1 \\ 0 \\ F_2 \\ 0 \\ 0 \end{pmatrix} \tag{17}$$

with the 5×5 block matrix

$$\mathcal{A} = \begin{pmatrix} A_1 & C_1^T & 0 & 0 & 0 \\ C_1 & 0 & 0 & 0 & D_1^T \\ 0 & 0 & A_2 & C_2^T & 0 \\ 0 & 0 & C_2 & 0 & D_2^T \\ 0 & D_1 & 0 & D_2 & 0 \end{pmatrix} \tag{18}$$

where A_k is the block diagonal matrix with the diagonal blocks

$$A_{k,i} = \begin{pmatrix} M_{k,i} & B_{k,i}^T \\ B_{k,i} & -\Sigma_{k,i} \end{pmatrix}, \tag{19}$$

$$C_k = \begin{pmatrix} C_{k,1} & \dots & C_{k,m_k} \end{pmatrix}, \tag{20}$$

and

$$F_k = \begin{pmatrix} F_{k,1} \\ \vdots \\ F_{k,m_k} \end{pmatrix}, \qquad F_{k,i} = \begin{pmatrix} 0 \\ -f_{k,i} \end{pmatrix}, \quad i = \overline{1,\, m_k}, \tag{21}$$

$k = 1, 2$. The subvectors w_1 and w_2 can be excluded from the system by the block Gauss elimination method. The reduced system is given by

$$\begin{pmatrix} S_1 & 0 & -D_1^T \\ 0 & S_2 & -D_2^T \\ -D_1 & -D_2 & 0 \end{pmatrix} \begin{pmatrix} \lambda_1 \\ \lambda_2 \\ \phi \end{pmatrix} = \begin{pmatrix} g_1 \\ g_2 \\ 0 \end{pmatrix} \tag{22}$$

where

$$S_k = \sum_{i=1}^{m_k} C_{k,i}\, A_{k,i}^{-1}\, C_{k,i}^T \tag{23}$$

$$g_k = \sum_{i=1}^{m_k} C_{k,i}\, A_{k,i}^{-1}\, F_{k,i}, \tag{24}$$

$k = 1, 2$. The system (22)-(24) is said to be the condensation of the system (17)-(21).

4 Algebraic solvers

The saddle point system (22) can be explicitly reduced to a system with a positive definite matrix. With additional block partitioning

$$\begin{aligned} S_k &= \begin{pmatrix} S_{kk} & S_{k\Gamma} \\ S_{\Gamma k} & S_{\Gamma\Gamma}^{(k)} \end{pmatrix}, \quad D_k = \begin{pmatrix} 0 & D_{k\Gamma} \end{pmatrix}, \\ \lambda_k &= \begin{pmatrix} \lambda_{kk} \\ \lambda_{k\Gamma} \end{pmatrix}, \quad g_k = \begin{pmatrix} g_{k1} \\ g_{k\Gamma} \end{pmatrix} \end{aligned} \tag{25}$$

where the blocks $S_{\Gamma\Gamma}^{(k)}$, $D_{k\Gamma}$, $\lambda_{k\Gamma}$, and $g_{k\Gamma}$ correspond to the degrees of freedom located on the interface Γ, $k = 1, 2$. System (22) can be written in the form of 5×5 block system

$$\begin{pmatrix} S_{11} & S_{1\Gamma} & 0 & 0 & 0 \\ S_{\Gamma 1} & S_{\Gamma\Gamma}^{(1)} & 0 & 0 & -D_{1\Gamma}^T \\ 0 & 0 & S_{22} & S_{2\Gamma} & 0 \\ 0 & 0 & S_{\Gamma 2} & S_{\Gamma\Gamma}^{(2)} & -D_{2\Gamma}^T \\ 0 & -D_{1\Gamma} & 0 & -D_{2\Gamma} & 0 \end{pmatrix} \begin{pmatrix} \lambda_{11} \\ \lambda_{1\Gamma} \\ \lambda_{22} \\ \lambda_{2\Gamma} \\ \phi \end{pmatrix} = \begin{pmatrix} g_{11} \\ g_{1\Gamma} \\ g_{22} \\ g_{2\Gamma} \\ 0 \end{pmatrix}. \tag{26}$$

In this system $D_{1\Gamma}$ is the diagonal matrix. Then, excluding the subvectors $\lambda_{1\Gamma}$ and ϕ by the block Gauss elimination method we get the system

$$R \begin{pmatrix} \lambda_{11} \\ \lambda_{22} \\ \lambda_{2\Gamma} \end{pmatrix} = \begin{pmatrix} g_{11} \\ g_{22} \\ g_{\Gamma} \end{pmatrix} \tag{27}$$

with the symmetric positive definite matrix

$$R = \begin{pmatrix} S_{11} & 0 & R_{1\Gamma}^T \\ 0 & S_{22} & S_{2\Gamma} \\ R_{1\Gamma} & S_{\Gamma 2} & R_{\Gamma\Gamma} \end{pmatrix} \tag{28}$$

where

$$\begin{aligned} R_{\Gamma\Gamma} &= S_{\Gamma\Gamma}^{(2)} + D_{2\Gamma}^T D_{1\Gamma}^{-1} S_{\Gamma\Gamma}^{(1)} D_{1\Gamma}^{-1} D_{2\Gamma}, \\ R_{1\Gamma} &= -D_{2\Gamma}^T D_{1\Gamma}^{-1} S_{\Gamma 1}, \\ g_{\Gamma} &= g_{2\Gamma} - D_{2\Gamma}^T D_{1\Gamma}^{-1} g_{1\Gamma}. \end{aligned} \tag{29}$$

To solve the system (27) we can use iterative techniques developed for algebraic systems with symmetric positive definite matrices. We recall that for the mortar P_1 finite element methods the above explicit reduction is not applicable.

The preconditioned Lanczos method is a good candidate to solve the saddle point system (22). In Kuznetsov [1995] an efficient preconditioner was proposed for the P_1 mortar element method. By coupling the ideas from Kuznetsov [1995], Kuznetsov and Wheeler [1995] with the new results from the recent publication by Kuznetsov [2003] we are able to derive efficient preconditioners for the system matrix in (22) as well. This is a topic for another publication.

Acknowledgement. The author is thankful to Oleg Boyarkin for his help in preparation of the paper.

References

F. Brezzi and M. Fortin. *Mixed and Hybrid Finite Element Methods.* Springer-Verlag, New York – Berlin – Heidelberg, 1991.

Y. A. Kuznetsov. Efficient iterative solvers for elliptic problems on nonmatching grids. *Russ. J. Numer. Anal. Math. Modelling*, 10(3):187–211, 1995.

Y. A. Kuznetsov. Spectrally equivalent preconditioners for mixed hybrid discretizations of diffusion equations on distorted meshes. *J. Numer. Math.*, 11(1):61–74, 2003.

Y. A. Kuznetsov and S. K. Repin. New mixed finite element method on polygonal and polyhedral meshes. *Russ. J. Numer. Anal. Math. Modelling*, 18(3):261–278, 2003.

Y. A. Kuznetsov and M. F. Wheeler. Optimal order substructuring preconditioners for mixed finite element methods on nonmatching grids. *East-West J. Numer. Math.*, 3(2):127–143, 1995.

Mortar Finite Elements with Dual Lagrange Multipliers: Some Applications*

Bishnu P. Lamichhane and Barbara I. Wohlmuth

University of Stuttgart, Institute of Applied Analysis and Numerical Simulation
http://www.ians.uni-stuttgart.de/nmh,{lamichhane,wohlmuth}@mathematik.uni-stuttgart.de

Summary. Domain decomposition techniques provide a powerful tool for the numerical approximation of partial differential equations. We consider mortar techniques with dual Lagrange multiplier spaces to couple different discretization schemes. It is well known that the discretization error for linear mortar finite elements in the energy norm is of order h. Here, we apply these techniques to curvilinear boundaries, nonlinear problems and the coupling of different model equations and discretizations.

1 Introduction

The numerical approximation of partial differential equations is often a challenging task. When different physical models should be used in different subregions, a suitable discretization scheme has to be used in each region. Mortar methods yield efficient and flexible coupling techniques for different discretization schemes. The central idea of mortar methods is to decompose the domain of interest into non-overlapping subdomains and impose a weak continuity condition across the interface by requiring that the jump of the solution is orthogonal to a suitable Lagrange multiplier space, see Bernardi et al. [1993, 1994]. Here, we work with mortar techniques and dual Lagrange multiplier spaces. These non-standard Lagrange multipliers show the same qualitative a priori estimates and quantitative numerical results as the standard ones and yield locally supported basis functions for the constrained space leading to a cheaper numerical realization, see Wohlmuth [2001]. This paper is concerned with applications of mortar methods to couple different physical models in different simulation regions. In the next section, we apply mortar methods to couple compressible and nearly incompressible materials in linear elasticity. In Section 3, the linear Laplace operator is coupled with the non-linear p-Laplace operator. Finally in Section 4, we show an application to an elasto-acoustic problem, and a generalized eigenvalue problem has to be solved. For

* This work was supported in part by the Deutsche Forschungsgemeinschaft, SFB 404, C12.

all our models, we provide numerical results. The weak coupling in terms of dual Lagrange multipliers results in a diagonal matrix on the slave side. As a consequence, the Lagrange multiplier can be eliminated locally, and optimal multigrid methods can be applied to the resulting positive definite system.

2 Compressible and Nearly Incompressible Materials

In this section, we consider a problem in linear elasticity with two different materials in two subdomains, one of them being nearly incompressible. We assume that the domain $\Omega \subset \mathbb{R}^2$ is decomposed into two non-overlapping subdomains Ω_1 and Ω_2 with a common interface $\bar{\Gamma} = \bar{\Omega}_1 \cap \bar{\Omega}_2$, and the subdomain Ω_1 is occupied with a nearly incompressible material having a very large Lamé parameter λ_1. It is well-known that standard low order finite elements for nearly incompressible materials suffer from locking, see Babuška and Suri [1992], and various approaches have been introduced to improve the numerical results. Working with a mixed formulation on Ω_1, see, e.g., Braess [2001], and standard finite element approach on Ω_2, we use mortar techniques with dual Lagrange multipliers to realize the coupling between the two formulations. On each subdomain, we define the space

$$\mathbf{H}^1_*(\Omega_k) := \{\mathbf{v} \in H^1(\Omega_k)^2, \mathbf{v}_{|\partial\Omega \cap \partial\Omega_k} = 0\}, \quad k = 1, 2,$$

and consider the constrained product space

$$\mathbf{V} := \{\mathbf{v} \in \prod_{k=1}^{2} \mathbf{H}^1_*(\Omega_k) \mid \int_\Gamma [\mathbf{v}] \cdot \psi \, d\sigma = 0, \, \psi \in \mathbf{M}\},$$

where $\mathbf{M} := \mathbf{H}^{-\frac{1}{2}}(\Gamma)$ is the Lagrange multiplier space, and $[\mathbf{v}]$ is the jump of $\mathbf{v}$ across Γ. Introducing an additional unknown $p := \lambda_1 \mathrm{div}\mathbf{u}$ in Ω_1, the variational problem is given by: find $[\mathbf{u}, p] \in \mathbf{V} \times L^2(\Omega_1)$ such that

$$\begin{aligned} a(\mathbf{u}, \mathbf{v}) + b(\mathbf{v}, p) &= l(\mathbf{v}), \quad \mathbf{v} \in \mathbf{V}, \\ b(\mathbf{u}, q) - \frac{1}{\lambda_1} c(p, q) &= 0, \quad q \in L^2(\Omega_1), \end{aligned}$$

where $l \in \mathbf{V}'$ and

$$a(\mathbf{u}, \mathbf{v}) := \sum_{i=1}^{2} 2\mu_i \int_{\Omega_i} \varepsilon(\mathbf{u}) : \varepsilon(\mathbf{v}) \, dx + \lambda_2 \int_{\Omega_2} \mathrm{div}\mathbf{u} \, \mathrm{div}\mathbf{v} \, dx,$$

$$b(\mathbf{v}, q) := \int_{\Omega_1} \mathrm{div}\mathbf{v} \, q \, dx, \quad c(p, q) := \int_{\Omega_1} p \, q \, dx, \quad \text{and} \quad l(\mathbf{v}) := \int_\Omega f \cdot \mathbf{v} \, dx.$$

Here, $\varepsilon(\mathbf{u})$ is the linear strain tensor. For our example, the domain $\Omega := \mathrm{conv}\{(0,0), (48,44), (48,60), (0,44)\}$ is decomposed into two subdomains Ω_1

and Ω_2 with $\Omega_1 := \text{conv}\{(12, 20.25), (36, 38.75), (36, 50.25), (12, 38.75)\}$, and $\Omega_2 := \Omega \backslash \bar{\Omega}_1$. Here, conv$\xi$ is the convex hull of the set ξ. The decomposition of the domain and the initial triangulation are shown in the left picture of Figure 1. Here, the left boundary of Ω is fixed and the right boundary is subjected to an in-plane shearing load of 100N along the positive y-direction. The lower and upper boundaries are set free, and we do not apply any volume force. The material parameters are taken to be $E_1 = 250$Pa, $E_2 = 80$Pa, $\nu_1 = 0.4999$, and $\nu_2 = 0.35$ to get a nearly incompressible response in Ω_1, where E_i and ν_i are the Young's modulus and the Poisson ratio on Ω_i, $i = 1, 2$, respectively. The displacement field is discretized with bilinear finite elements, and the pressure in Ω_1 is discretized with piecewise constant functions. The right picture of Figure 1 shows the vertical displacement at $(48, 60)$ versus the number of elements. We compare three different numerical schemes. Using standard conforming finite elements (standard) in Ω does not give satisfying numerical results, whereas the more expensive mixed formulation (mixed) in Ω provides good results. Our numerical results show that the mortar approach (coupled) is almost as good as the mixed formulation and significantly better than the standard one.

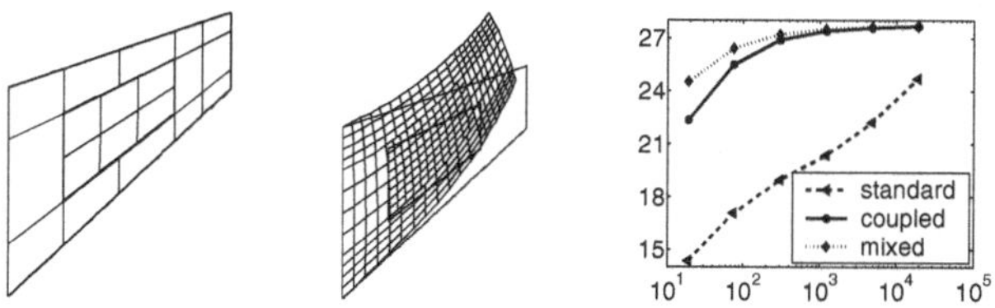

Fig. 1. Decomposition of the domain and initial triangulation (left), distorted grid on level 2 (middle), and vertical tip displacement versus number of elements (right)

3 The Laplace and the p-Laplace Operator

In this section, we consider the coupling of a linear and a non-linear model. The linear model is described by a Poisson equation, and we use the p-Laplacian for the non-linear model. Here, we decompose the domain $\Omega := (-1, 1) \times (-1, 1)$ into four non-overlapping subdomains defined by $\Omega_1 := (-1, 0) \times (-1, 0)$, $\Omega_2 := (0, 1) \times (-1, 0)$, $\Omega_3 := (-1, 0) \times (0, 1)$ and $\Omega_4 := (0, 1) \times (0, 1)$. We have given the decomposition of the domain and the initial triangulation in the left picture of Figure 2. We consider the Poisson equation $-\nabla \cdot (\alpha \nabla) u = f$ in Ω_1 and Ω_4 and the p-Laplacian $-\nabla \cdot (\alpha |\nabla u|^{p-2} \nabla u) = f$ in Ω_2 and Ω_3. The p-Laplace equation occurs in the theory of two-dimensional plasticity under longitudinal shear or in the diffusion problem with non-linear diffusivity, see Atkinson and Champion [1984], and we are considering here different material models in different subdomains. For the regularity of the solutions and error estimates of the p-Laplacian, we refer to Liu and Barret [1993] and Liu and Yan

[2001]. Let $\mathcal{T}_{h_k}$ be a shape regular simplicial triangulation on Ω_k with mesh-

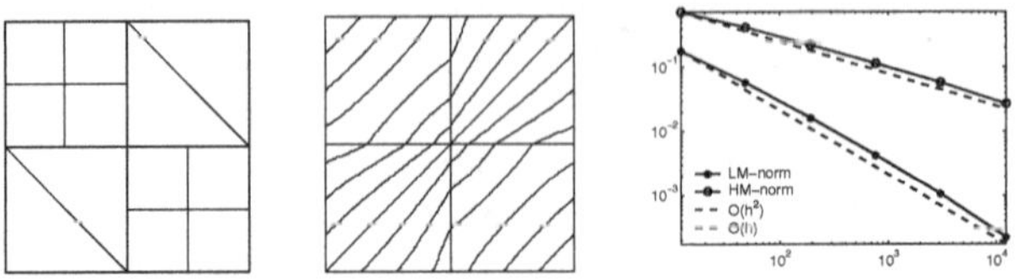

Fig. 2. Decomposition of the domain and initial triangulation (left), isolines of the solution (middle) and discretization errors versus number of elements (right)

sizes bounded by h_k, and $\mathcal{S}(\Omega_k, \mathcal{T}_{h_k})$ stands for the space of linear conforming finite elements in the subdomain Ω_k associated with the triangulation $\mathcal{T}_{h_k}$ satisfying the Dirichlet boundary conditions on $\partial\Omega_k \cap \partial\Omega$, $k = 1, \cdots, 4$. Then, the unconstrained finite element space X_h is given by $X_h := \prod_{k=1}^{4} \mathcal{S}(\Omega_k, \mathcal{T}_{h_k})$. The interface $\Gamma := \{(0, y), -1 < y < 1\} \cup \{(x, 0), -1 < x < 1\}$ inherits its one-dimensional triangulation $\mathcal{S}_\Gamma$ from the mesh on Ω_2 and Ω_3. We recall that $(0, 0)$ is a crosspoint, and M_h does not have any degree of freedom at this point. Now, the Lagrange multiplier space M_h is defined on Γ and is associated with the triangulation $\mathcal{S}_\Gamma$. Assuming $q_1 := 2$, $q_2 := p$, $q_3 := p$, and $q_4 := 2$, we can write the weak formulation of the problem as: find $(u_h, \lambda_h) \in X_h \times M_h$ such that

$$\begin{aligned} a(u_h, v) + b(v, \lambda_h) &= l(v), \qquad v \in X_h, \\ b(u_h, \mu) &= 0, \qquad \mu \in M_h, \end{aligned} \tag{1}$$

where $a(u, v) := \sum_{i=1}^{4} \int_{\Omega_i} \alpha |\nabla u|^{q_i - 2} \nabla u \cdot \nabla v \, dx$, $b(v, \mu) := \int_\Gamma [v] \, \mu \, d\sigma$, and $l(v) := \int_\Omega f \, v \, dx$. If $\alpha > 0$, and the right hand side function f is sufficiently smooth, we can show by monotonicity techniques that the problem (1) has a unique solution, see Liu [1999]. However, the regularity of the solution is not known. Let $u_h := \sum_{k=1}^{n} u_k \phi_k$ and $\lambda_h := \sum_{k=1}^{n_s} \lambda_k \mu_k$, where n and n_s are the dimensions of X_h and M_h, respectively. Suppose $w = (u_1, \cdots, u_n, \lambda_1, \cdots, \lambda_{n_s})$ be a vector. Now, we define $F(w) := (F_1(w), F_2(w))^T$ with

$$F_1(w) := \begin{pmatrix} a(u_h, \phi_1) + b(\phi_1, \lambda_h) - l(\phi_1) \\ \vdots \\ a(u_h, \phi_n) + b(\phi_n, \lambda_h) - l(\phi_n) \end{pmatrix}, \quad F_2(w) := \begin{pmatrix} b(u_h, \mu_1) \\ \vdots \\ b(u_h, \mu_{n_s}) \end{pmatrix}.$$

The system $F(w) = 0$ is a non-linear system if $p \neq 2$, and we apply the Newton method to solve this system. First, we initialize the solution vector w_0 satisfying the given Dirichlet boundary conditions. Then, we iterate until convergence with

$$J_k \Delta w_k = F(w_k),$$

where $\Delta w_k := w_k - w_{k+1}$, and J_k is the Jacobian of F evaluated at w_k. Working with a dual Lagrange multiplier space has the advantage that the

Jacobian matrix J_k has exactly the same form as the saddle point matrix arising from the mortar finite elements with a dual Lagrange multiplier space for the Laplace operator, see Wohlmuth and Krause [2001]. Hence, we can apply the multigrid approach introduced in Wohlmuth and Krause [2001] to solve the linear system on each level. Suppose that $\tilde{\Omega}_l := \Omega_1 \cup \Omega_4$, and $\tilde{\Omega}_p := \Omega_2 \cup \Omega_3$. In our numerical example, we choose $\alpha = 0.5$ in $\tilde{\Omega}_l$, and $\alpha = 1$ in $\tilde{\Omega}_p$, $p = 1.5$, and $f = 0$. For boundary conditions, we set $u(-1,-1) = u(1,1) = 0$, $u(-1,1) = 1$ and $u(1,-1) = -1$, and the Dirichlet boundary condition on $\partial\Omega$ is imposed by taking the linear combination of them in between. Here, we do not have the exact solution. To get the approximation of the discretization errors, we compute a reference solution u_{ref} at a fine level and compare it with the solution u_h at each level until $h_{ref} \leq 2h$. We have given the discretization errors in the LM- and HM- norms defined by

$$\|v\|_{LM} := \|v\|_{L^2(\tilde{\Omega}_l)} + \|v\|_{L^p(\tilde{\Omega}_p)}, \quad \text{and} \quad \|v\|_{HM} := |v|_{W^{1,2}(\tilde{\Omega}_l)} + |v|_{W^{1,p}(\tilde{\Omega}_p)}$$

in the right picture of Figure 2, and the isolines of the solution are given in the middle. Although the regularity of the solution is not known, we get convergence of order h^2 in the LM-norm and of order h in the HM-norm.

4 Application to an Elasto-Acoustic Problem

In this section, we show the application of mortar finite element methods for an elasto-acoustic problem. We consider the situation that the fluid is completely surrounded by the structure. The problem is described by a linear elastic structure occupying a subdomain $\Omega_S \subset \mathbb{R}^2$ coupled with an irrotational fluid in $\Omega_F \subset \mathbb{R}^2$. The interface $\Gamma(= \partial\Omega_S \cap \partial\Omega_F)$ separates the fluid and solid regions. Given the fluid-density ρ_F, the solid-density ρ_S, and the acoustic speed c, we seek the frequency ω, the velocity-field $\mathbf{u}$, and the pressure p such that

$$\begin{aligned}
\nabla p - \omega^2 \rho_F \mathbf{u}_F &= \mathbf{0} \quad \text{in} \quad \Omega_F, \\
p + c^2 \rho_F \nabla \cdot \mathbf{u}_F &= 0 \quad \text{in} \quad \Omega_F, \\
\nabla \cdot \sigma(\mathbf{u}_S) + \omega^2 \rho_S \mathbf{u}_S &= \mathbf{0} \quad \text{in} \quad \Omega_S, \\
\mathbf{u}_S &= \mathbf{0} \quad \text{on} \quad \Gamma_D, \\
\sigma(\mathbf{u}_S) \cdot \mathbf{n}_S &= \mathbf{0} \quad \text{on} \quad \Gamma_N, \\
\sigma_n(\mathbf{u}_S) + p = 0, \quad \sigma_t(\mathbf{u}_S) = \mathbf{0}, &\quad \text{and} \quad (\mathbf{u}_F - \mathbf{u}_S) \cdot \mathbf{n} = 0 \quad \text{on} \quad \Gamma.
\end{aligned}$$

Here, σ is the usual stress tensor from linear elasticity, $\sigma_n = \mathbf{n} \cdot (\sigma \cdot \mathbf{n})$ is the normal stress on Γ, and $\sigma_t = \sigma \cdot \mathbf{n} - \sigma_n \mathbf{n}$ is the tangential traction vector on Γ, where $\mathbf{n}$ is the outward normal to Ω_F on Γ. This problem has become a subject of different papers, see, e.g., Hansbo and Hermansson [2003], Bermúdez and Rodríguez [1994], Alonso et al. [2001]. We introduce the following function spaces to formulate our problem in the weak form

$$X := \mathbf{H}(\mathrm{div}, \Omega_F) \times \mathbf{H}^1_{\Gamma_D}(\Omega_S), \text{ and } \mathbf{V} := \{(\mathbf{u}_F, \mathbf{u}_S) \in X, \ [\mathbf{u}] \cdot \mathbf{n} = 0 \text{ on } \Gamma\},$$

where

$$\begin{aligned} \mathbf{H}(\mathrm{div}, \Omega_F) &:= \{\mathbf{v} \in L^2(\Omega_F)^2, \ \|\nabla \cdot \mathbf{v}\|_{L^2(\Omega_F)} < \infty\}, \\ \mathbf{H}^1_{\Gamma_D}(\Omega_S) &:= \{\mathbf{v} \in H^1(\Omega_S)^2, \ \mathbf{v}_{|\Gamma_D} = 0\}, \quad \text{and} \quad [\mathbf{u}] := (\mathbf{u}_F - \mathbf{u}_S). \end{aligned}$$

The weak form of the continuous problem is: find $\mathbf{u} \in \mathbf{V}$ and $\omega \in \mathbb{R}$ such that

$$\begin{aligned} a(\mathbf{u}, \mathbf{v}) &= \omega^2 m(\mathbf{u}, \mathbf{v}), \quad \mathbf{v} \in \mathbf{V}, \quad \text{where} \\ a(\mathbf{u}, \mathbf{v}) &:= (\rho_F c^2 \nabla \cdot \mathbf{u}_F, \nabla \cdot \mathbf{v}_F)_{\Omega_F} + (\sigma(\mathbf{u}_S), \epsilon(\mathbf{v}_S))_{\Omega_S}, \quad \text{and} \\ m(\mathbf{u}, \mathbf{v}) &:= (\omega^2 \rho_F \mathbf{u}_F, \mathbf{v}_F)_{\Omega_F} + (\omega^2 \rho_S \mathbf{u}_S, \mathbf{v}_S)_{\Omega_S}. \end{aligned}$$

Here, $\epsilon(\mathbf{v}_S)$ is the linear strain tensor and is related to the stress tensor by Hooke's law, i.e., $\sigma_{ij}(\mathbf{v}_S) = 2\mu\epsilon_{ij}(\mathbf{v}_S) + \lambda \sum_{k=1}^2 \epsilon_{kk}(\mathbf{v}_S)\delta_{ij}$, $i, j = 1, 2$. Let $\mathcal{T}_{h_s}$ and $\mathcal{T}_{h_f}$ be shape regular simplicial triangulations on Ω_S and Ω_F, respectively, and Γ inherits its triangulation $\mathcal{S}_\Gamma$ from the side of Ω_F. It is a well-known fact that if standard Lagrangian finite elements are used to discretize the fluid, it will give rise to spurious eigensolutions with positive eigenvalues interspersed among the 'real' ones, and a possible remedy of this problem is to use Raviart-Thomas elements in the fluid domain, see Bermúdez et al. [1995]. Therefore, we discretize the fluid domain with Raviart-Thomas elements of lowest order:

$$RT_0 := \{\mathbf{u} \in \mathbf{H}(\mathrm{div}, \Omega_F) : \mathbf{u}_{|K} = (a + bx, c + by), \ K \in \mathcal{T}_{h_f}, \ \ a, b, c \in \mathbb{R}\},$$

and the solid domain with Lagrangian finite elements of lowest order:

$$W_h^D := S_D(\Omega_S, \mathcal{T}_{h_s}) \times S_D(\Omega_S, \mathcal{T}_{h_s}),$$

where $S_D(\Omega_S, \mathcal{T}_{h_s})$ is the finite element space on Ω_S satisfying the Dirichlet boundary condition on Γ_D. The kinematic constraint can be imposed by piecewise constant Lagrange multipliers yielding a uniform inf-sup condition. Suppose $X_h := RT_0 \times W_h^D$, and $M_h := \{\mu_h \in L^2(\Gamma) : \mu_h|_e \in \mathcal{P}_0(e), e \in \mathcal{S}_\Gamma\}$. Now the finite element space can be written as

$$\mathbf{V}_h := \{(\mathbf{u}_{hF}, \mathbf{u}_{hS}) \in X_h, \ \int_\Gamma [\mathbf{u}_h] \cdot \mathbf{n} \, \mu_h \, d\sigma = 0, \ \mu_h \in M_h\}.$$

The discrete problem reads: find $\mathbf{u}_h \in \mathbf{V}_h$, and $\omega_h \in \mathbb{R}$ such that

$$a(\mathbf{u}_h, \mathbf{v}_h) = \omega_h^2 m(\mathbf{u}_h, \mathbf{v}_h), \quad \mathbf{v}_h \in \mathbf{V}_h.$$

Remark 1. We remark that the Lagrange multiplier λ_h approximates the pressure on the interface Γ. The Lagrange multipliers are associated with the one-dimensional mesh inherited from the triangulation on the fluid domain. Due to the special structure of the support of the nodal basis functions of RT_0 and M_h, the degree of freedom corresponding to the Lagrange multiplier can locally be eliminated by inverting a diagonal mass matrix.

In Alonso et al. [2001], an adaptive finite element scheme is analyzed to solve the fluid-structure vibration problem, where the kinematic constraint is imposed by means of piecewise constant Lagrange multiplier. Following this technique, we arrive at the same mortar setting as we discuss here. Now, we consider the domain $\Omega := \{(x,y) \in \mathbb{R}^2,\, x^2+y^2 < 1\}$ decomposed into two subdomains Ω_S and Ω_F with $\Omega_F := \{(x,y) \in \mathbb{R}^2,\, x^2+y^2 < 0.6\}$, and $\Omega_S := \Omega \backslash \bar{\Omega}_F$. Here, $\Gamma_D = \{(\cos\theta, \sin\theta),\, \frac{5\pi}{4} \le \theta \le \frac{7\pi}{4}\}$. We have used the following parameters in our numerical example: $\rho_F = 1000\text{kg/m}^3$, $c = 1430\text{m/s}$, $\rho_S = 7700\text{kg/m}^3$, $E = 144\text{GPa}$, and $\nu = 0.35$. The first three consecutive eigenmodes along with the pressure in the fluid domain and the distorted grids in the solid domain are shown in Figure 3. We note that Γ defines a curvilinear interface. To evaluate the weak coupling, we commit an additional variational crime by projecting the mesh of the structure side to the mesh on the fluid side.

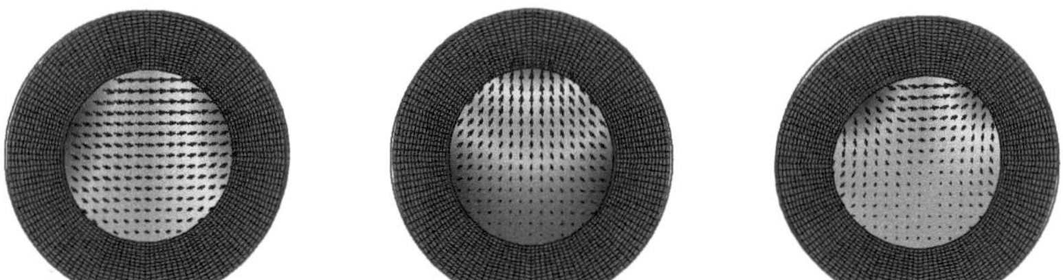

Fig. 3. The first, second and the third eigenmodes corresponding to the eigenvalues 809.1481, 1980.7519 and 3606.3907 (rad/s)

The second numerical example is taken from Bermúdez and Rodríguez [1994]. The domain $\Omega := (0,1.5) \times (0,1.5)$ is decomposed into two subdomains Ω_S and Ω_F with $\Omega_F := (0.25,1.25) \times (0.25,1.25)$, $\Omega_S := \Omega \backslash \bar{\Omega}_F$, and $\Gamma_D = \{(x,0) \in \mathbb{R}^2,\, 0 \le x \le 1.5\}$. We have used the same physical parameters as in the previous example. The computed eigenfrequencies (in rad/s) along with the extrapolated ones referred to as 'Exact' in Bermúdez and Rodríguez [1994] are given in Table 1.

Table 1. The computed eigenfrequencies using mortar techniques compared with the extrapolated eigenfrequencies ('Exact') in Bermúdez and Rodríguez [1994]

Eigenmodes	Computed Eigenfrequencies	'Exact'
1	648.1847	641.837
2	2147.3593	2116.398
3	3419.5020	3201.475
4	3885.9022	3804.124
5	4214.0865	4211.620
6	4699.6782	4687.927

References

A. Alonso, A. D. Russo, C. Otero-Souto, C. Padra, and R. Rodríguez. An adaptive finite element scheme to solve fluid-structure vibration problems on non-matching grids. *Computing and Visualization in Science*, 4:67–78, 2001.

C. Atkinson and C. R. Champion. Some boundary value problems for the equation $\nabla \cdot (|\nabla\phi|^N) = 0$. *Quart. J. Mech. Appl. Math.*, 37:401–419, 1984.

I. Babuška and M. Suri. On locking and robustness in the finite element method. *SIAM J. Numer. Anal.*, 29:1261–1293, 1992.

A. Bermúdez, R. Duran, M. Muschietti, R. Rodríguez, and J. Solomin. Finite element vibration analysis of fluid-solid systems without spurious modes. *SIAM J. Numer. Anal.*, 32:1280–1295, 1995.

A. Bermúdez and R. Rodríguez. Finite element computation of the vibration modes of a a fluid-solid system. *Comput. Meth. in Appl. Mech. and Engrg.*, 119:355–370, 1994.

C. Bernardi, Y. Maday, and A. Patera. Domain decomposition by the mortar element method. In H. K. et al., editor, *Asymptotic and numerical methods for partial differential equations with critical parameters*, pages 269–286. Reidel, Dordrecht, 1993.

C. Bernardi, Y. Maday, and A. Patera. A new nonconforming approach to domain decomposition: the mortar element method. In H. B. et al., editor, *Nonlinear partial differential equations and their applications*, pages 13–51. Paris, 1994.

D. Braess. *Finite Elements. Theory, fast solver, and applications in solid mechanics.* Cambridge Univ. Press, Second Edition, 2001.

P. Hansbo and J. Hermansson. Nitsche's method for coupling non-matching meshes in fluid-structure vibration problems. *Comput. Mech.*, 32:134–139, 2003.

W. Liu. Degenerate quasilinear elliptic equations arising from bimaterial problems in elastic-plastic mechanics. *Nonlinear Analysis*, 35:517–529, 1999.

W. Liu and J. Barret. A remark on the regularity of the solutions of the p-Laplacian and its application. *J. Math. Anal. Appl.*, 178:470–488, 1993.

W. Liu and N. Yan. Quasi-norm local error estimators for p-Laplacian. *SIAM J. Numer. Anal.*, 39:100–127, 2001.

B. Wohlmuth. *Discretization Methods and Iterative Solvers Based on Domain Decomposition*, volume 17 of *LNCS*. Springer Heidelberg, 2001.

B. Wohlmuth and R. Krause. Multigrid methods based on the unconstrained product space arising from mortar finite element discretizations. *SIAM J. Numer. Anal.*, 39:192–213, 2001.

Non-Conforming Finite Element Methods for Nonmatching Grids in Three Dimensions

Wayne McGee and Padmanabhan Seshaiyer*

Texas Tech University, Mathematics and Statistics (padhu@math.ttu.edu)

Summary. In the last decade, non-conforming domain decomposition methods such as the mortar finite element method have been shown to be reliable techniques for several engineering applications that often employ complex finite element design. With this technique, one can conveniently assemble local subcomponents into a global domain without matching the finite element nodes of each subcomponent at the common interface. In this work, we present computational results for the convergence of a mortar finite element technique in three dimensions for a model problem. We employ the mortar finite element formulation in conjunction with higher-order elements, where both mesh refinement and degree enhancement are combined to increase accuracy. Our numerical results demonstrate optimality for the resulting non-conforming method for various discretizations.

1 Introduction

As computational resources are rapidly increasing, numerical modeling of physical processes is being performed on increasingly complex domains. Often an analysis may be performed by decomposing the global domain into several local subdomains, each of which can be modeled independently. The global domain can then be reconstructed by assembling the subdomains appropriately. In the standard conforming method, it is required that the corners of a given element intersect other elements only on their corners, that is, corners must not coincide with edges of other elements. It is often infeasible or inconvenient to coordinate the decomposition and reassembly processes so that the subdomains conform at the common interfaces. The use of a non-conforming method circumvents this difficulty. In practical applications, the non-conforming method has two noteworthy advantages. First, the discretization of the domain can be selectively increased in localized regions, such as around corners or other features where the error in the solution is likely to be

* This work of this author is supported in part by the National Science Foundation under Grant DMS 0207327.

greatest. This allows for greater accuracy in the method without the computational load associated with increasing the discretization of the entire domain.

Another practical benefit of the non-conforming method is that the process may be utilized to connect independently analyzed substructures in a large problem. For example, in the construction of an aircraft, the fuselage and wing structures may have been analyzed independently by different engineers, possibly in different organizations. It is highly unlikely that the independently constructed meshes of each subcomponent would coincide when assembled. Moreover, transition meshing could become highly complex and expensive to achieve. The non-conforming approach eliminates this need.

The mortar finite element method (Bernardi et al. [1993], Belgacem [1999], Seshaiyer and Suri [2000b], D. Braess and Wieners [2000], Wohlmuth [2000] and references therein) is an example of a non-conforming technique. In the last decade, there has been a lot of research on the theoretical and computational aspects of this domain decomposition technique (Seshaiyer and Suri [1998], Seshaiyer [2003], Ewing et al. [2000], Braess et al. [1999]). It has been well-established that the mortar finite element method yields optimal results both in the presence of highly non-quasiuniform meshes and high polynomial degree (Seshaiyer and Suri [2000a]) and also preserves the optimal rates afforded by conforming h, p, and hp discretizations for a variety of applications (Belgacem et al. [2000, 2003]).

In the last few years, the extension of the mortar finite element technique has been analyzed (see Belgacem and Maday [1997], Braess and Dahmen [1998], Kim et al. [2001] and references therein). In Belgacem and Maday [1997] the mortar finite element method was extended for the special case of linear polynomials. However, the method is difficult to generalize for a general mesh of parallelograms for any polynomial degree. To circumvent this, a variant of the mortar method, M_1, was introduced in (Seshaiyer and Suri [2000a], Seshaiyer [2003]) which easily extends the technique to any number of dimensions. The computational performance of this method, however, was not tested which is the focus, herein. In this paper, we computationally validate the convergence behavior for the mortar finite element formulation for a time-dependent model problem in three-dimensions. In particular, we show via numerical experiments that the M_1 mortar method is stable and behaves as well as the conforming finite element method.

2 Model problem and its discretization

Consider the model problem for $\mathbf{x} = (x_1, x_2, x_3) \in \Omega, t > 0$:

$$\frac{\partial u(\mathbf{x},t)}{\partial t} - \nabla \cdot (P(\mathbf{x}) \nabla u(\mathbf{x},t)) + Q(\mathbf{x}) u(\mathbf{x},t) = f(\mathbf{x},t), \tag{1}$$

where P is uniformly positive and Q is a nonnegative function in the bounded domain Ω, with the boundary and initial conditions

$$u(\mathbf{x},t) = 0 \quad \text{for } \mathbf{x} \in \partial\Omega, \tag{2}$$

$$u(\mathbf{x},0) = g(\mathbf{x}). \tag{3}$$

Discretizing time using a backward Euler scheme yields

$$-\nabla \cdot (P_{(0)} \nabla u_n) + Q_{(0)} u_n = f_{n(0)}, \tag{4}$$

where $P_{(0)} = (\Delta t)P, \quad Q_{(0)} = 1 + (\Delta t)Q, \quad$ and $f_{n(0)} = u_{n-1} + (\Delta t) f_n$.

Let us for simplicity, decompose Ω into two geometrically conforming non-overlapping subdomains Ω_1 and Ω_2, which share a common interface Γ (denoted by the dotted line). For each subdomain Ω_i, we consider a regular sequence of geometrically conforming triangulations τ_i. Note that, no compatibility is assumed between meshes in different subdomains, i.e. the meshes of Ω_1 and Ω_2 need not match on Γ. This is illustrated in the figure 1.

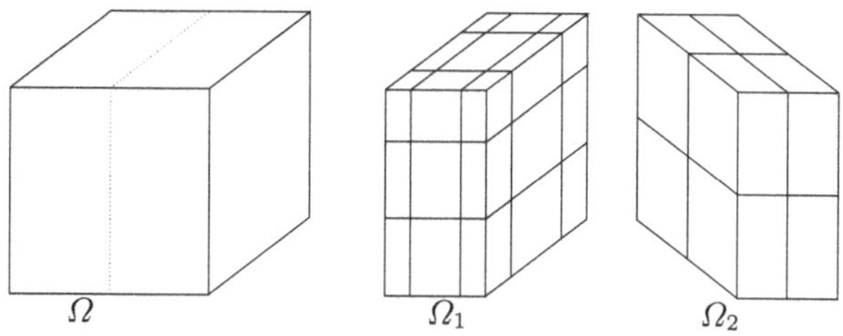

Fig. 1. Geometrically conforming decomposition of Ω partitioned into Ω_1 and Ω_2 with non-conforming meshes

Let $u_n^{(i)}$ denote the interior solution in each Ω_i, which satisfies the global continuity restriction $u_n^{(1)}(\mathbf{x}) = u_n^{(2)}(\mathbf{x})$ for $\mathbf{x} \in \Gamma$. Due to the non-conformity of the grids across Γ, we enforce this continuity in a weak sense as

$$b(u_n, \psi) := \int_\Gamma (u_n^{(1)} - u_n^{(2)})\, \psi \, dx = 0 \quad \forall \psi \in H^{-1/2}(\Gamma). \tag{5}$$

Let us now describe the weak formulation of our model problem (4) as a mixed method formulation, which is a convenient method for implementation. Using standard Sobolev space notation, define $H_D^1(\Omega_i) = \{v \in H^1(\Omega_i) | v = 0 \text{ on } \partial\Omega_i \bigcap \partial\Omega_D\}$. The weak form of (4) then becomes: For $i = 1, 2$, find $u_n^{(i)} \in H_D^1(\Omega_i)$ such that for all $v_i \in H_D^1(\Omega_i)$,

$$\int_{\Omega_i} P_{(0)} \nabla u_n^{(i)} \cdot \nabla v_i \, dx - \int_\Gamma P_{(0)} \frac{\partial u_n^{(i)}}{\partial n} v_i \, ds + \int_{\Omega_i} Q_{(0)} u_n^{(i)} v_i \, dx = \int_{\Omega_i} f_{n(0)} v_i \, dx. \tag{6}$$

Let $\lambda = -P_{(0)} \dfrac{\partial u_n^{(1)}}{\partial n} = P_{(0)} \dfrac{\partial u_n^{(2)}}{\partial n}$. Define the spaces $\tilde{V} = \{v \in L^2(\Omega),\ v|_{\Omega_i} \in H_D^1(\Omega_i)\}$ and $\Lambda = \{\psi \in \mathcal{D}'(\Gamma),\ \psi|_\Gamma \in H^{-\frac{1}{2}}(\Gamma)\}$ (where $\mathcal{D}'$ is the Schwarz set of distributions) equipped with their respective norms.

For each $T \in \tau_i$, denote the set $S_p(T)$ to be all polynomials generated by the *serendipity* (or trunk) space families. Hence $S_2(T)$ has 20 degrees of freedom. We have used these spaces for our computations in the next section. Assume the finite element spaces $V_F^{(i)} = \{u \in H^1(\Omega_i) \mid u|_T \in S_p(T), u = 0 \text{ on } \partial\Omega_i \cap \partial\Omega_D\}$ are given. We can then define the non-conforming space $\tilde{V}_F = \{u \in L_2(\Omega) \mid u|_{\Omega_i} \in V_F^{(i)}\} \subset \tilde{V}$.

To define the finite-dimensional Lagrange multiplier space, let us suppose that the mesh on the interface Γ matches the mesh on Ω_1 (Note that this choice is arbitrary). For $K \subset \mathbb{R}^2$, we denote by $Q_{p,s}(K)$ the set of polynomials on K which is of degree p in x and s in y (so that $Q_{p,p}(K) = Q_p(K)$). Let us denote the rectangles in the mesh on the interface Γ by $K_{ij}, 0 \leq i, j \leq N$. Then the Lagrange multiplier space will be defined as $\Lambda_F = \left\{\chi \in C(\Gamma) : \chi_{|K_{ij}} \in Q_{p-1}(K_{ij})\right\} \subset \Lambda$ (see Figure 2). The

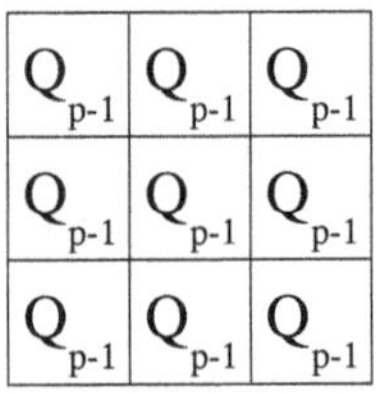

Fig. 2. Lagrange multiplier space for M_1 method

associated mortar method is called the M_1 mortar finite element method, and has been implemented in the next section. It can be shown that this choice of the Lagrange multiplier space leads to optimal results in three-dimensions by extending the arguments of Seshaiyer and Suri [2000a] and Seshaiyer [2003].

Let us now define $\overset{0}{V}{}_F^{\Gamma} = \{u|_\Gamma, u \in V_F^{(i)}\} \cap H_D^1(\Gamma)$. Then for any $z \in L^2(\Gamma)$, we define the space $X_F^\Gamma(z) = \{w \in \overset{0}{V}{}_F^{\Gamma}, \int_\Gamma (w - z)\, \chi\, ds = 0 \;\; \forall\, \chi \in \Lambda_F\}$. Let us now make the following restriction.

Condition I: $X_F^\Gamma(z) \neq \emptyset$ for all $z \in L_2(\Gamma)$.

If Condition I holds, then one can prove that the mixed formulation satisfies the inf-sup condition:

$$\inf_{\substack{\lambda \in \Lambda_F \\ \lambda \neq 0}} \sup_{v \in \tilde{V}_F} \frac{b(v, \lambda)}{||v||_{\tilde{V}}\; ||\lambda||_\Lambda} > 0$$

Let the finite dimensional spaces $V_F^{(i)}$ and Λ_F be spanned by basis functions $\{\Psi_j^{(i)}\}_{j=1}^{N_i}$ and $\{\Phi_j\}_{j=1}^{N_\lambda}$ respectively. Writing $u_n^{(i)} = \sum_{k=1}^{N_i} a_k^{(i)} \Psi_k^{(i)}$ and $\lambda = \sum_{k=1}^{N_\lambda} b_k \Phi_k$ respectively, (6) and (5) yield a discrete system of integral equations, which can be written in block matrix form as:

$$\begin{bmatrix} A_1 & 0 & B_1 \\ 0 & A_2 & B_2 \\ B_1^T & B_2^T & 0 \end{bmatrix} \begin{bmatrix} \mathbf{a}^{(1)} \\ \mathbf{a}^{(2)} \\ \mathbf{b} \end{bmatrix} = \begin{bmatrix} \mathbf{F}_1 \\ \mathbf{F}_2 \\ 0 \end{bmatrix} \tag{7}$$

Here $\mathbf{a}^{(\mathbf{i})} = \{a_1^{(i)}, a_2^{(i)}, \ldots, a_{N_i}^{(i)}\}$, $\mathbf{b} = \{b_1, b_2, \ldots, b_{N_\lambda}\}$ and

$$A_{i(sj)} = \int_{\Omega_i} P_{(0)}^{(i)} \nabla \Psi_s^{(i)} \cdot \nabla \Psi_j^{(i)}\, dx + \int_{\Omega_i} Q_{(0)}^{(i)} \Psi_s^{(i)} \Psi_j^{(i)}\, dx$$

$$B_{1(js)} = \int_\Gamma \Phi_s \Psi_j^{(1)}\, ds \qquad B_{2(js)} = -\int_\Gamma \Phi_s \Psi_j^{(2)}\, ds \qquad F_{i(j)} = \int_{\Omega_i} f_{n(0)}^{(i)} \Psi_j^{(i)}\, dx$$

for $i = 1, 2$. Note that the invertibility of the stiffness matrix in (7) is related to Condition I.

3 Numerical Results

In this section, we demonstrate the performance of the numerical technique described. Our computations were performed for the model problem (1) on the domain $\Omega = (-1,1) \times (-1,1) \times (-1,1)$, and we decompose this domain into $\Omega_1 = (-1,1) \times (-1,0) \times (-1,1)$ and $\Omega_2 = (-1,1) \times (0,1) \times (-1,1)$. We take h_1 subintervals along the x, y, and z axes for Ω_1, and h_2 subintervals for Ω_2. A sample partition of Ω into two subdomains with $h_1 = 3$ and $h_2 = 2$ is shown in Figure 1. Note that the grids do not match on the interface. For our experiments, we consider uniform polynomial degrees p in both subdomains.

Steady-state, constant coefficients

Our initial experiment involves a steady-state ($\frac{\partial u}{\partial t} = 0$) equation with constant coefficients $P = Q = 1$. We choose the right hand side f such that our exact solution is $u(x,y,z) = (1-x^2)(1-y^2)(1-z^2)$.

We consider the h-version for the non-conforming method for the combinations $(h_1, h_2) = \{(3,3), (3,4), (4,4), \ldots, (7,8), (8,8)\}$ with polynomial degrees $p = 1$ and $p = 2$. The results are demonstrated for both the L_2 (Figure 3(a)) and H^1 (Figure 3(b)) errors. For our computations, we have used tensor products of one-dimensional Gauss-Legendre quadratures for numerical integration and the errors have been computed at the Gauss points on the rectangular grids. Due to our simplified geometry with a smooth regular solution, we not only get optimal solutions but one can also observe superconvergence rates. Although not obvious, one may need to perform a detailed analysis for the mixed mortar method, to study this superconvergence behaviour following the details Ewing and Lazarov [1993].

Figure 4 demonstrates the performance of the non-conforming method versus the conforming method for $p = 2$. For this experiment, the L_2 error for

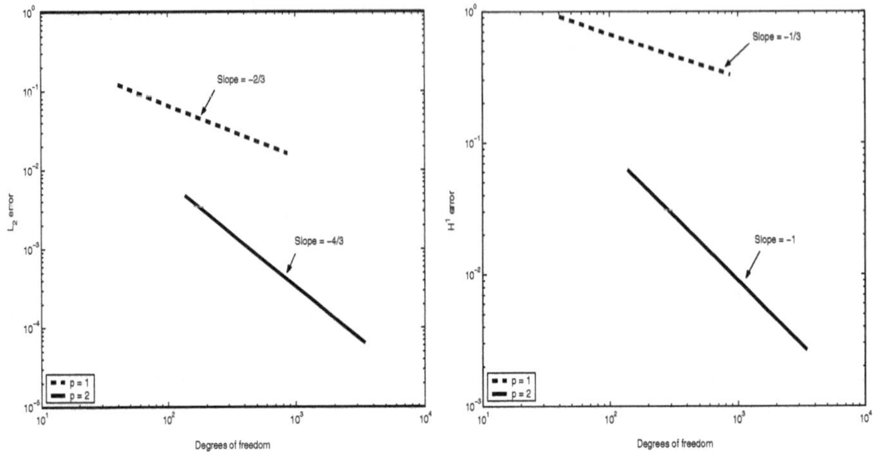

Fig. 3. Steady-state convergence: (a) L^2 error, (b)H^1 error

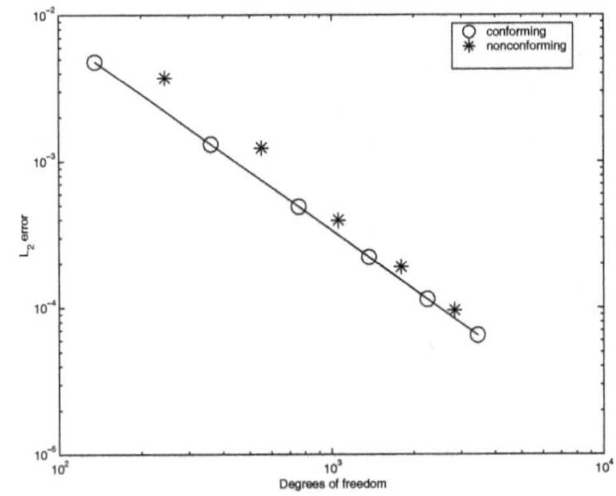

Fig. 4. Conforming versus non-conforming method

the conforming meshes $(h_1, h_2) = \{(3,3), (4,4), \ldots, (8,8)\}$ (circles) is plotted against the non-conforming meshes $(h_1, h_2) = \{(3,4), (4,5), \ldots, (7,8)\}$ (asterisks). The results indicate that the non-conforming method performs no worse than the conforming method in higher dimensions.

Steady-state, varying coefficients

Next, we performed computations for the steady-state problem with $P(y) = \sin y + 2$ and $Q(y) = \cos y + 2$. The results of this experiment are shown in Table 1. We denote by DOF the number of degrees of freedom, which is the size of the stiffness matrix in (7). L_2 and H^1 denote the errors in the respective norms, and $L_2\%$ and $H^1\%$ are the respective relative errors.

Table 1. Nonconstant coefficients, $p = 2$

h_1	h_2	DOF	L_2	$L_2\%$	H^1	$H^1\%$
3	3	136	0.004798	0.435505	0.062567	1.948016
3	4	244	0.003711	0.336868	0.048629	1.514050
4	4	361	0.001307	0.118605	0.024271	0.755660
4	5	553	0.001238	0.112389	0.020820	0.648217
5	5	756	0.000491	0.044614	0.011831	0.368343
5	6	1056	0.000395	0.035888	0.009703	0.302095
6	6	1369	0.000222	0.020174	0.006605	0.205649
6	7	1801	0.000190	0.017269	0.005602	0.174420
7	7	2248	0.000114	0.010382	0.004051	0.126116
7	8	2836	0.000096	0.008694	0.003458	0.107656
8	8	3441	0.000065	0.005862	0.002658	0.082762

Table 2. Convergence in Δt

n	L_2 8:8	H^1	L_2 7:8	H^1	L_2 7:7	H^1
2	0.550825	1.605919	0.550899	1.606458	0.554554	1.629562
4	0.435470	0.802964	0.275490	0.803821	0.279264	0.838348
8	0.137706	0.401490	0.137790	0.402982	0.141801	0.458332
16	0.068853	0.200759	0.068948	0.203510	0.073412	0.290824
32	0.034427	0.100407	0.034545	0.105592	0.039823	0.228157
64	0.017214	0.050257	0.017377	0.059764	0.023938	0.208207

Time-dependent case

Our final experiment confirmed convergence for the unsteady equation. The exact solution was chosen to be $u(x, y, z, t) = t(1 - x^2)(1 - y^2)(1 - z^2)$. We considered several matching and non-matching mesh combinations and the results for the combinations $(h_1, h_2) = \{(8, 8), (7, 8), (7, 7)\}$ for polynomial degree 2, are presented in Table 2. Our computations were run from time $t = 0$ to time $t = 1$, with varying numbers of time steps n. The results not only demonstrate convergence as we refine the time discretization but also suggest that the errors for the non-matching combination $(8, 7)$ are between the matching combinations $(7, 7)$ and $(8, 8)$ as one should expect.

Conclusion

A non-conforming finite element method for non-matching grids in three dimensions was described and implemented. Our numerical results for a model problem clearly demonstrate that the technique performs as well as the standard conforming finite element method in higher dimensions.

References

F. B. Belgacem. The mortar finite element method with lagrange multipliers. *Numer. Math.*, 84:173–197, 1999.

F. B. Belgacem, L. Chilton, and P. Seshaiyer. The *hp*-mortar finite-element method for mixed elasticity and stokes problems. *Comp. Math. Appl.*, 46: 35–56, 2003.

F. B. Belgacem and Y. Maday. The mortar element method for three dimensional finite elements. *RAIRO Math. Mod. Numer. Anal.*, 31:289–302, 1997.

F. B. Belgacem, P. Seshaiyer, and M. Suri. Optimal convergence rates of *hp* mortar finite element methods for second-order elliptic problems. *RAIRO Math. Mod. Numer. Anal.*, 34:591–608, 2000.

C. Bernardi, Y. Maday, and A. Patera. Domain decomposition by the mortar finite element method. *Asymptotic and Numerical Methods for PDEs with Critical Parameters*, 384:269–286, 1993.

D. Braess and W. Dahmen. Stability estimates of the mortar finite element method for 3dimensional problems. *EastWest J. Numer. Math.*, 6:249–263, 1998.

D. Braess, M. Dryja, and W. Hackbusch. Multigrid method for nonconforming finite element discretization with application to non-matching grids. *Comput.*, 63:1–25, 1999.

W. D. D. Braess and C. Wieners. A multigrid algorithm for the mortar finite element method. *SIAM J. Numer. Anal.*, 37:48–69, 2000.

R. Ewing, R. Lazarov, T. Lin, and Y. Lin. Mortar finite volume element methods of second order elliptic problems. *East-West J. Numer. Math.*, 8: 93–110, 2000.

R. E. Ewing and R. D. Lazarov. Superconvergence of the mixed finite element approximations of parabolic problems using rectangular finite elements. *East-West J. Numer. Math.*, 1:199–212, 1993.

C. Kim, R. Lazarov, J. Pasciak, and P. Vassilevski. Multiplier spaces for the mortar finite element method in three dimensions. *SIAM J. Numer. Anal.*, 39:519–538, 2001.

P. Seshaiyer. Stability and convergence of nonconforming *hp* finite-element methods. *Comp. Math. Appl.*, 46:165–182, 2003.

P. Seshaiyer and M. Suri. Convergence results for non-conforming *hp* methods. *Contemp. Math.*, 218:453–459, 1998.

P. Seshaiyer and M. Suri. *hp* submeshing via non-conforming finite element methods. *Comp. Meth. Appl. Mech. Engrg.*, 189:1011–1030, 2000a.

P. Seshaiyer and M. Suri. Uniform *hp* convergence results for the mortar finite element method. *Math. Comp.*, 69:521–546, 2000b.

B. I. Wohlmuth. A mortar finite elment method using dual spaces for the lagrange multiplier. *SIAM J. Numer. Anal.*, 38:989–1012, 2000.

On an Additive Schwarz Preconditioner for the Crouzeix-Raviart Mortar Finite Element

Talal Rahman[1], Xuejun Xu[2], and Ronald H.W. Hoppe[3]

[1] University of Augsburg, Department of Mathematics, Universitätsstr. 14, D-86159 Augsburg, Germany (talal.rahman@math.uni-augsburg.de)
[2] LSEC, Institute of Computational Mathematics, Chinese Academy of Sciences, P.O.Box 2719, Beijing, 100080, People's Republic of China. (xxj@lsec.cc.ac.cn)
[3] Department of Mathematics, University of Houston, Calhoun Street, Houston, TX-77204-3008, U.S.A., (rohop@math.uh.edu) and Institute for Mathematics, University of Augsburg, Universitätsstr. 14, D-86159 Augsburg, Germany. (hoppe@math.uni-augsburg.de)

Summary. We consider an additive Schwarz preconditioner for the algebraic system resulting from the discretization of second order elliptic equations with discontinuous coefficients, using the lowest order Crouzeix-Raviart element on nonmatching meshes. The overall discretization is based on the mortar technique for coupling nonmatching meshes. A convergence analysis of the preconditioner has recently been given in Rahman et al. [2003]. In this paper, we give a matrix formulation of the preconditioner, and discuss some of its numerical properties.

1 Introduction

We consider the Crouzeix-Raviart (CR), or the nonconforming P1 finite element discretization on nonmatching meshes of the following elliptic problem with discontinuous coefficients: Find $u^* \in H_0^1(\Omega)$ such that

$$a(u^*, v) = f(v), \; v \in H_0^1(\Omega), \tag{1}$$

where $\Omega \subset R^2$ is a bounded, simply connected polygonal domain, $a(u,v) = \sum_{i=1}^N \rho_i (\nabla u, \nabla v)_{L^2(\Omega_i)}$ and $f(v) = \sum_{i=1}^N \int_{\Omega_i} fv \, dx$, and $\overline{\Omega} = \cup_{i=1}^N \overline{\Omega}_i$ is the partition of Ω into nonoverlapping polygonal subdomains Ω_i of diameter H_i. The coefficients ρ_i are positive constants with possibly large jumps across subdomain interfaces. Let $X_h(\Omega_i)$ be the nonconforming P1 (Crouzeix-Raviart) finite element space defined on a quasi-uniform triangulation $\mathcal{T}_h(\Omega_i)$ of mesh size h_i, of the subdomain Ω_i, consisting of functions which are piecewise linear in each triangle $\tau \subset \Omega_i$, continuous at the interior edge midpoints $x_k \in \Omega_{ih}^{CR}$, and vanishing at the edge midpoints lying on the boundary $\partial\Omega$.

Since the triangulations on Ω_i and Ω_j may not match on their common interface $\overline{\Gamma}_{ij} = \overline{\Omega}_i \cap \overline{\Omega}_j$, the functions in $X_h(\Omega) = \Pi_{i=1}^N X_h(\Omega_i)$ are discontinuous at edge midpoints along the interface. We use a special technique, known as the mortar technique, cf. Bernardi et al. [1994], for the coupling of nonmatching meshes. An analysis of the mortar technique for the Crouzeix-Raviart element has been given in Marcinkowski [1999].

According to the mortar technique, a weak continuity condition, called the mortar condition, is imposed on the function along the interfaces. A function $u_h = \{u_i\}_{i=1}^N \in X_h$ satisfies the mortar condition on the interface Γ_{ij}, if $Q_m u_i = Q_m u_j$, where $Q_m : L^2(\Gamma_{ij}) \rightarrow M^{h_j}(\delta_{m(j)})$ is the L^2-projection operator defined as: $(Q_m u, \psi)_{L^2(\delta_{m(j)})} = (u, \psi)_{L^2(\delta_{m(j)})}, \quad \forall \psi \in M^{h_j}(\delta_{m(j)})$, where $\delta_{m(j)} \subset \partial\Omega_j$ is the nonmortar side of Γ_{ij}, $M^{h_j}(\delta_{m(j)}) \subset L^2(\Gamma_{ij})$ is the test space of functions which are piecewise constant on the triangulation of $\delta_{m(j)}$, and $(\cdot,\cdot)_{L^2(\delta_{m(j)})}$ denotes the L^2 innerproduct on $L^2(\delta_{m(j)})$. The other side of Γ_{ij}, called the mortar side, is denoted by $\gamma_{m(i)} \subset \partial\Omega_i$. The discrete problem takes the following form: Find $u_h^* = \{u_i\}_{i=1}^N \in V_h$ such that

$$a_h(u_h^*, v_h) = f(v_h), \qquad \forall v_h \in V_h, \tag{2}$$

where $V_h \subset X_h$ is a subspace of functions which satisfy the mortar condition on all interfaces, and $a_h(u,v) = \sum_{i=1}^N \rho_i \sum_{\tau \in \mathcal{T}_h(\Omega_i)} (\nabla u, \nabla v)_{L^2(\tau)} = \sum_{i=1}^N a_i(u,v)$. V_h is a Hilbert space with an inner product defined by $a_h(\cdot,\cdot)$. The problem has a unique solution and a priori error estimates have been provided in Marcinkowski [1999].

Even though, there exists a lot of work concerning the nonconforming P1 element on matching grids, cf., e.g., Brenner [1996], Hoppe and Wohlmuth [1995], Sarkis [1997], the work on nonmatching grids is very limited, cf., e.g., Marcinkowski [1999], Xu and Chen [2001]. Recently, an efficient additive Schwarz method for the nonconforming P1 element on nonmatching grids has been proposed in Rahman et al. [2003]. In this paper, we complement the work by introducing the matrix formulation of the preconditioner, and discuss some of its numerical properties.

2 An additive Schwarz preconditioner

In this section, we describe the additive Schwarz preconditioner of Rahman et al. [2003], for the problem (2), which is based on the idea of solving local subproblems on nonoverlapping subdomains, and coarse problems on specially constructed subspaces of small dimensions. The preconditioner is defined using the general framework for additive Schwarz methods, cf. Smith et al. [1996]. We decompose V_h as $V_h = \sum_\gamma V^\gamma + V^0 + \sum_{i=1}^N V^i$, where the first sum is taken over the set of all mortar sides $\{\gamma\}$. For $i = 1, \cdots, N$, V^i is the restriction of V_h to Ω_i, with functions vanishing at subdomain boundary edge midpoints $\partial\Omega_{ih}^{CR}$ as well as on the remaining subdomains. V^γ is a space of

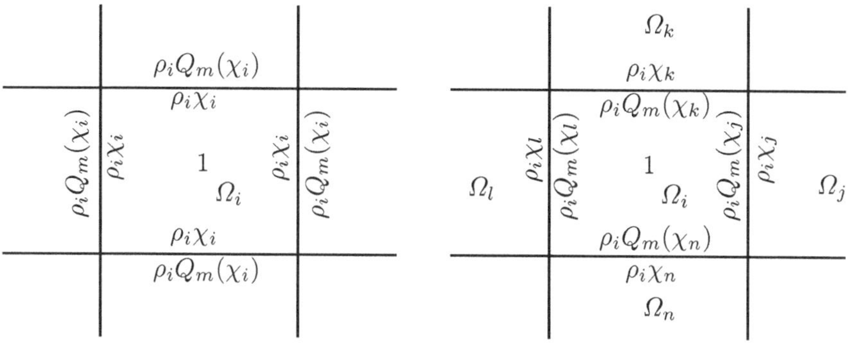

Fig. 1. Examples of Φ_i corresponding to a subdomain Ω_i having only mortar sides (left) or nonmortar sides (right), indicating the nonzero values of the function.

functions given by their values on mortar edge midpoints γ_h^{CR}, $V^\gamma = \{v \in V_h : v(x) = 0,\ x \in \overline{\Omega}_h^{CR} \setminus \gamma_h^{CR}\}$. The coarse space V^0, a special space having a dimension equal to the number of subdomains, is defined using the function $\chi_i \in X_h(\Omega_i)$ associated with the subdomain Ω_i. χ_i is defined by its nodal values as: $\chi_i(x) = 1/\sum_j \rho_j(x)$ at $x \in \overline{\Omega}_{ih}^{CR}$, where the sum is taken over the subdomains which x is common to, V^0 is given as the span of its basis functions, $\Phi_i, i = 1, \cdots, N$, i.e., $V^0 = span\{\Phi_i : i = 1, \cdots, N\}$, where Φ_i associated with Ω_i, is defined as follows (cf. Figure 1).

$$\Phi_i(x) = \begin{cases} 1, & x \in \Omega_{ih}^{CR}, \\ \rho_i \chi_i(x), & x \in \gamma_{m(i)h}^{CR}, \\ \rho_i Q_m(\chi_j)(x), & x \in \delta_{m(i)h}^{CR},\ \delta_{m(i)} = \gamma_{m(j)}, \\ \rho_i Q_m(\chi_i)(x), & x \in \delta_{m(j)h}^{CR},\ \delta_{m(j)} = \gamma_{m(i)}, \\ \rho_i \chi_j(x), & x \in \gamma_{m(j)h}^{CR},\ \gamma_{m(j)} = \delta_{m(i)}, \\ 0, & x \in \partial\Omega_{ih}^{CR} \cap \partial\Omega, \end{cases} \tag{3}$$

and $\Phi_i(x) = 0$ at all other x in $\overline{\Omega}_h^{CR}$. We use exact bilinear forms for all our subproblems. The projection like operators $T^i : V_h \to V^i$ are defined in the standard way, i.e., for $i \in \{\{\gamma\}, 0, \cdots, N\}$ and $u \in V_h$, $T^i u \in V^i$ is the solution of $a_h(T^i u, v) = a_h(u, v),\ v \in V^i$. Let $T = \sum_\gamma T^\gamma + T^0 + T^1 + \cdots + T^N$. The problem (2) is now replaced by the following preconditioned system,

$$Tu_h^* = g, \tag{4}$$

where $g = \sum_\gamma T^\gamma u_h^* + \sum_{i=0}^N T^i u_h^*$. Let c and C represent generic constants independent of the mesh sizes $h = \inf_i h_i$ and $H = \max_i H_i$, and of the jumps of the coefficients ρ_i, then the following result holds.

Theorem 1 (Rahman et al. [2003]). *For all $u \in V_h$,*

$$c\frac{h}{H} a_h(u, u) \le a_h(Tu, u) \le C a_h(u, u). \tag{5}$$

The proof of this theorem is given in Rahman et al. [2003], which uses the general theory for Schwarz methods, cf. Smith et al. [1996]. It follows from the theorem, that the condition number of the operator T is bounded by $c(\frac{H}{h})$.

2.1 Matrix formulation

Our aim is to derive a matrix representation for the preconditioned system (4). The finite element space V^h can be expressed as $V^h = span\{\phi_k\}$, where each basis function ϕ_k is associated with a node x_k which is either a subdomain interior edge midpoint or a mortar edge midpoint. Let $\varphi_k^{(i)}$ denote the standard nodal basis function of $X_h(\Omega_i)$, associated with the edge midpoint x_k. The basis functions are defined as follows.

If $x_k \in \Omega_{ih}^{CR}$, a subdomain interior node, then $\phi_k(x)$ is exactly equal to $\varphi_k(x)$. If $x_k \in \gamma_{m(i)h}^{CR}$, a mortar node, then $\phi_k(x) = \varphi_k(x)$ on $\overline{\Omega}_i$, while on $\overline{\delta}_{m(j)}$, where $\gamma_{m(i)} = \delta_{m(j)}$, $\phi_k(x) = Q_m(\varphi_k)(x)$ at $x \in \delta_{m(j)h}^{CR}$. ϕ_k is zero at the remaining edge midpoints of $\overline{\Omega}_j$, and zero everywhere on the remaining subdomains. Note that there are no basis functions associated with nonmortar edge midpoints. Using these basis functions of V_h, the problem (2) can be rewritten in the matrix form as

$$\mathbf{A}\mathbf{u}^* = \mathbf{f}, \tag{6}$$

where $\mathbf{u}^*$ is a vector of nodal values of u_h^*, and $\mathbf{A}$ is a matrix generated by the bilinear form $a_h(.,.)$ on $V_h \times V_h$. We shall now see how this matrix can be obtained from the local matrices $\hat{\mathbf{E}}_i$ generated by $a_i(.,.)$ on $X_h(\Omega_i) \times X_h(\Omega_i)$.

Observing that $a_h(.,.) = \sum_{i=1}^N a_i(.,.)$, where $a_i(.,.) = a_h(.,.)|_{\Omega_i}$, we can calculate the elements of $\mathbf{A}$ from their local contributions restricted to individual subdomains Ω_i. In order to calculate the local contribution $a_i(.,.)$, we use only those basis functions that have nonzero supports on $\overline{\Omega}_i$. These basis functions are exactly the ones associated with the nodes of Ω_{ih}^{CR}, $\gamma_{m(i)h}^{CR}$ ($\gamma_{m(i)} \subset \partial\Omega_i$), and the set $\gamma_{m(j)h}^{CR}$ ($\gamma_{m(j)} = \delta_{m(i)} \subset \partial\Omega_i$) of neighboring mortar edge midpoints except those on $\partial\Omega$. Let Λ_i be the set of all these nodes.

Let $\mathbf{P}_i$ be the restriction matrix which is a permutation of a rectangular identity matrix, such that $\mathbf{P}_i\mathbf{u}$ returns the vector of all coefficients of $\mathbf{u}$, associated with the nodes of Λ_i. $\mathbf{P}_i^T$ is the corresponding extension matrix. Let $\mathbf{E}_i$, associated with the subdomain Ω_i, be the matrix generated by $a_i(.,.)$ on $span\{\phi_k : x_k \in \Lambda_i\} \times span\{\phi_l : x_l \in \Lambda_i\}$. Using these three types of matrices, we can assemble the global matrix as $\mathbf{A} = \sum_{i=1}^N \mathbf{P}_i^T\mathbf{E}_i\mathbf{P}_i$.

We note that $\mathbf{E}_i = \{a_i(\phi_k, \phi_l)\}$, where $x_k, x_l \in \Lambda_i$, and $\hat{\mathbf{E}}_i = \{a_i(\varphi_k, \varphi_l)\}$, where $x_k, x_l \in \overline{\Omega}_{ih}^{CR}$. If $x_k, x_l \in \Omega_{ih}^{CR} \cup \gamma_{m(i)h}^{CR}$, then $a_i(\phi_k, \phi_l) = a_i(\varphi_k, \varphi_l)$. If $x_k \in \gamma_{m(j)h}^{CR}$, then the calculation of an element of $\mathbf{E}_i$ involving ϕ_k, requires the values of $Q_m(\varphi_k)(x_o)$ at the nodes $x_o \in \delta_{m(i)h}^{CR}$, since, by definition, $\phi_k = \sum_{x_o \in \delta_{m(i)h}^{CR}} Q_m(\varphi_k)(x_o)\varphi_o$ in $\overline{\Omega}_i$. In the following, we derive these coefficients $\{Q_m(\varphi_k)(x_o)\}$ from the mortar condition.

We assume that the subdomain Ω_i has only one nonmortar side $\delta_{m(i)}$, the extension to more than one nonmortar edge is straightforward. Let $\mathbf{M}_{\gamma_{m(j)}} = \{(\varphi_k, \psi_o)_{L^2(\delta_{m(i)})}\}$ and $\mathbf{S}_{\delta_{m(i)}} = \{(\varphi_o, \psi_o)_{L^2(\delta_{m(i)})}\}$, for $x_k \in \gamma_{m(j)h}^{CR}$ and $x_o \in \delta_{m(i)h}^{CR}$, be the master and the slave matrix, respectively. Then

$$\mathbf{Q}_{m(i)} = \mathbf{S}_{\delta_{m(i)}}^{-1} \mathbf{M}_{\gamma_{m(j)}}$$

is the matrix representation of the mortar projection Q_m. The columns of this matrix correspond to the nodes $x_k \in \gamma_{m(j)h}^{CR}$, containing exactly the coefficients $\{Q_m(\varphi_k)(x_o)\}$. We note that $\mathbf{S}_{\delta_{m(i)}}$ is a diagonal matrix containing the lengths of the edges along $\delta_{m(i)}$, as entries.

Now define the matrix $\mathbf{Q}_i = diag(\mathbf{I}, \mathbf{Q}_{m(i)})$, where $\mathbf{I}$ is the identity matrix corresponding to the nodes of Ω_{ih}^{CR} and $\gamma_{m(i)h}^{CR}$, and $\mathbf{Q}_{m(i)}$ is the projection matrix corresponding to the nodes of $\gamma_{m(j)h}^{CR}$. Then it is easy to see that $\mathbf{E}_i = \mathbf{Q}_i^T \hat{\mathbf{E}}_i \mathbf{Q}_i$. Finally, we have $\mathbf{A} = \sum_{i=1}^N \mathbf{P}_i^T \mathbf{Q}_i^T \hat{\mathbf{E}}_i \mathbf{Q}_i \mathbf{P}_i$. In the same way, we get $\mathbf{f} = \sum_{i=1}^N \mathbf{P}_i^T \mathbf{Q}_i^T \hat{\mathbf{f}}_i$.

We follow the standard procedure, cf. Smith et al. [1996], for expressing the preconditioned system in matrix form. Since $V^i \subset V_h$, for $i \in \{\{\gamma\}, 0, 1, \cdots, N\}$, the interpolation operators $I_i : V^i \rightarrow V_h$ are simply the imbedding operators. Let $\mathbf{R}_i^T$ be the matrix representation of I_i. In matrix form, T^i is then given by $\mathbf{T}_i = \mathbf{R}_i^T \mathbf{A}_i^{-1} \mathbf{R}_i \mathbf{A}$, where $\mathbf{A}_i = \mathbf{R}_i \mathbf{A} \mathbf{R}_i^T$. Now, setting $\mathbf{T} = \sum_\gamma \mathbf{T}_\gamma + \sum_{i=0}^N \mathbf{T}_i$, which is the matrix representation of T, the preconditioned system (4) takes the following matrix form.

$$\mathbf{T}\mathbf{u}^* = \mathbf{g}. \tag{7}$$

In this, $\mathbf{T} = \mathbf{B}\mathbf{A}$ and $\mathbf{g} = \mathbf{B}\mathbf{f}$, where $\mathbf{B} = \sum_\gamma \mathbf{R}_\gamma^T \mathbf{A}_\gamma^{-1} \mathbf{R}_\gamma + \sum_{i=0}^N \mathbf{R}_i^T \mathbf{A}_i^{-1} \mathbf{R}_i$ is the preconditioner. The restriction matrices $\mathbf{R}_i$, $i = 1, \cdots, N$, and $\mathbf{R}_\gamma$ are all permutations of rectangular identity matrices such that $\mathbf{R}_i \mathbf{u}$ and $\mathbf{R}_\gamma \mathbf{u}$ return vectors of coefficients of $\mathbf{u}$, associated with the subdomain interior edge midpoints Ω_{ih}^{CR} and the mortar edge midpoints γ_h^{CR}, respectively. The construction of $\mathbf{R}_0$ is different but simple. Let $\mathbf{v}_i$ be the vector of nodal values of Φ_i, then the columns of $\mathbf{R}_0^T$ consist of exactly these vectors, i.e., $\mathbf{v}_i$, $i = 1, \cdots, N$. The matrices $\mathbf{P}_i$ and $\mathbf{R}_i$, $i = 1, \cdots, N$, and $\mathbf{R}_\gamma$ are never formed in practice. Their use in this section has been merely for the representation.

Most iterative methods for solving (7) require the actions of $\mathbf{A}$ and $\mathbf{B}$ on different vectors in each iteration. Once the matrices $\mathbf{Q}_{m(i)}$ and $\hat{\mathbf{E}}_i$, $i = 1, \cdots, N$, and $\mathbf{A}_i$, $i \in \{\{\gamma\}, 0, \cdots, N\}$, are generated, the actions of $\mathbf{A}$ and $\mathbf{B}$ on the vectors can be calculated by multiplying the vectors with the expressions of $\mathbf{A}$ and $\mathbf{B}$, respectively.

3 Numerical Examples

In this section, we present numerical results and discuss some of the properties of the preconditioner presented in the previous section. The Preconditioned

Table 1. Numerical results for varying jumps in the coefficients. $d \times d = 36$ subdomains, each having $2m^2 = 72$ or $2n^2 = 50$ elements, are used for the triangulation

Coefficient jump ρ	Condition number κ_2	CG-iteration counts	L^2-norm of error	H_h^1-seminorm of error
10^0	31.580	35	$0.9516 \cdot 10^{-3}$	$0.4366 \cdot 10^{-1}$
10^2	32.755	39	$0.1099 \cdot 10^{-2}$	$0.4558 \cdot 10^{-1}$
10^4	32.825	39	$0.1104 \cdot 10^{-2}$	$0.4565 \cdot 10^{-1}$
10^6	32.834	39	$0.1104 \cdot 10^{-2}$	$0.4565 \cdot 10^{-1}$

Conjugate Gradients (PCG) method has been used for solving the preconditioned system (4), which stops as the relative norm of the residual drops below the tolerance 10^{-6}. In all our experiments, a uniform triangulation has been used in each subdomain employing, in a checker board order, either $2m^2$ or $2n^2$ triangular elements, where m and n are chosen differently in order to have nonmatching grids with different mesh sizes across subdomain interfaces.

Test I

The objective of this test is to study the effect of coefficient jumps on the convergence. We consider a model problem for which the exact solution is known. The problem is defined on a unit square initially defined as the union of 2×2 nonoverlapping square subregions with coefficients $\rho_1 = \rho,\ \rho_2 = 1,\ \rho_3 = 1$, and $\rho_4 = \rho$ so that the jump in the coefficients across any subregion interface is equal to $\rho > 0$. The function f is chosen such that the exact solution can be given by $u(x,y) = \sin(\pi x)\sin(\pi y)$. Note that $\nabla u \cdot \eta$ vanishes along the subregion interfaces.

Numerical results for varying jumps in the coefficients are presented in Table 1, showing condition number estimate of the preconditioned system, and the L^2-norm and the broken H^1-seminorm (H_h^1) of the error in the numerical solution. The condition number estimates, as shown in the table, remain unchanged as the jump increases, illustrating that the preconditioner is robust with respect to jumps in the coefficients. In Table 2, we present the weighted L^2-norm (L_ρ^2) and the weighted broken H^1-seminorm ($H_{\rho h}^1$) of the error for varying subdomain size and mesh size, where the weights are the coefficients

Table 2. Numerical results for varying subdomain size and mesh size, and fixed $\rho = 10$. H and h of the second row correspond to $d = 4$, $m = 12$ and $n = 11$

Subdomain size	Mesh size	Condition number κ_2	CG-iteration counts	L_ρ^2-norm of error	$H_{\rho h}^1$-seminorm of error
H	$\frac{1}{2}h$	131.27	68	$0.2541 \cdot 10^{-3}$	$0.2914 \cdot 10^{-1}$
H	h	65.08	47	$0.1056 \cdot 10^{-2}$	$0.6057 \cdot 10^{-1}$
$\frac{1}{2}H$	$\frac{1}{2}h$	66.20	53	$0.3778 \cdot 10^{-3}$	$0.3293 \cdot 10^{-1}$

ρ_i. The condition number estimates of the table are in accordance with the theory. The error in the L^2_ρ-norm and $H^1_{\rho h}$-seminorm indicate convergence as $O(h^2)$ and $O(h)$, respectively.

Test II

In our second test, we consider the choice of mortar or nonmortar sides, and the ratio between mesh sizes from neighboring subdomains, and discuss their possible influence on the convergence. The problem is defined on a unit square, with the force function $f(x) = 2\pi^2 \sin(\pi x)\sin(\pi x)$. We assume the domain to be the union of $d \times d$ subregions (subdomains) with coefficients $\rho_i = 1$ or $\rho_i = \rho = 10^4$ distributed in a checkerboard order.

Table 3. Condition number and CG-iteration counts (in parentheses) for two opposite choices of mortar sides and varying mesh size ratio. $m = 12$ for $\rho_i = \rho$, and $n = 11, 6$ for $\rho_i = 1$ giving $h_{\rho_i=\rho}/h_{\rho_i=1} \approx 1, \frac{1}{2}$

Subdomains	Choice I		Choice II	
$d \times d$	$m = 12, n = 11$	$m = 12, n = 6$	$m = 12, n = 11$	$m = 12, n = 6$
6×6	68.63 (57)	68.63 (54)	68.62 (57)	63.28 (49)
9×9	68.79 (59)	68.95 (55)	68.72 (57)	63.40 (49)

Each column in Table 3 and Table 4, corresponds to a fixed pair $\{m, n\}$ representing a fixed $\frac{H}{h}$ ratio. Two opposite choices of mortar sides, called 'Choice I' and 'Choice II', have been chosen for the experiment. 'Choice I' corresponds to choosing the sides with larger coefficients as the mortar sides, and 'Choice II' corresponds to the opposite choice. Under each choice of mortar sides, two sets of results corresponding to different mesh size ratios between neighboring subdomains are presented. The difference between the tables is as follows: In Table 3, we use a finer mesh on subdomains with larger coefficient, i.e. $h_{\rho_i=\rho} < h_{\rho_i=1}$, (this gives a better a priori error, cf. Bernardi and Verfürth [2000]), and in Table 4, we do the opposite.

Table 4. Condition number estimate and iteration counts (in parentheses) for two opposite choices of mortar sides and varying mesh size ratio. $m = 11, 6$ for $\rho_i = \rho$, and $n = 12$ for $\rho_i = 1$ giving $h_{\rho_i=1}/h_{\rho_i=\rho} \approx 1, \frac{1}{2}$

Subdomains	Choice II		Choice I	
$d \times d$	$m = 11, n = 12$	$m = 6, n = 12$	$m = 11, n = 12$	$m = 6, n = 12$
6×6	62.86 (55)	33.78 (45)	62.87 (57)	33.77 (43)
9×9	62.99 (51)	34.04 (42)	62.99 (51)	34.05 (40)

As seen from the tables above, for a particular choice of mortar sides and a fixed $\frac{H}{h}$ ratio, the condition number estimates remain bounded. In fact, the choice of mortar sides show no or only a mild influence on the condition number estimates. The place where this mild influence is seen in the table, is for the mesh size ratio $h_{\rho_i=\rho}/h_{\rho_i=1} = \frac{1}{2}$, cf. Table 3. It has however been observed through experiments that the percentile difference between the estimates reduces gradually with the mesh size. We close this section by making a final remark on Table 4. As seen from the table, the condition number estimates for the ratio $h_{\rho_i=1}/h_{\rho_i=\rho} = \frac{1}{2}$ are approximately half of those for $h_{\rho_i=1}/h_{\rho_i=\rho} = 1$. This is due to the minimum eigenvalue. It is not difficult to show, taking into account the special shapes of the basis functions $\{\Phi_i\}$ (due to the checkerboard distribution of the coefficients) in the proof of Theorem 1, that the bound for the minimum eigenvalue approximately doubles as the ratio is halved.

References

C. Bernardi, Y. Maday, and A. T. Patera. A new non conforming approach to domain decomposition: The mortar element method. In H. Brezis and J.-L. Lions, editors, *Collège de France Seminar*. Pitman, 1994. This paper appeared as a technical report about five years earlier.

C. Bernardi and R. Verfürth. Adaptive finite element methods for elliptic equations with nonsmooth coefficients. *Numerische Mathematik*, 85(4):579–608, 2000.

S. C. Brenner. Two-level additive Schwarz preconditioners for nonconforming finite element methods. *Mathematics of Computation*, 65(215):897–921, 1996.

R. H. W. Hoppe and B. Wohlmuth. Adaptive multilevel iterative techniques for nonconforming finite element discretizations. *East-West J. Numer. Math.*, 3:179–197, 1995.

L. Marcinkowski. The mortar element method with locally nonconforming elements. *BIT*, 39:716–739, 1999.

T. Rahman, X. Xu, and R. Hoppe. An additive Schwarz method for the Crouzeix-Raviart finite element for elliptic problems discontinuous coefficients. *Submitted to Numerische Mathematik*, 2003.

M. Sarkis. Nonstandard coarse spaces and Schwarz methods for elliptic problems with discontinuous coefficients using non-conforming element. *Numerische Mathematik*, 77:383–406, 1997.

B. F. Smith, P. E. Bjørstad, and W. Gropp. *Domain Decomposition: Parallel Multilevel Methods for Elliptic Partial Differential Equations*. Cambridge University Press, 1996.

X. Xu and J. Chen. Multigrid for the mortar element method for the P1 nonconforming element. *Numerische Mathematik*, 88(2):381–398, 2001.

Part VIII

Minisymposium: FETI and Neumann-Neumann Domain Decomposition Methods

A FETI-DP Method for the Mortar Discretization of Elliptic Problems with Discontinuous Coefficients

Maksymilian Dryja[1] and Wlodek Proskurowski[2]

[1] Warsaw University, Mathematics, Informatics and Mechanics
[2] University of Southern California, Mathematics
(`http://math.usc.edu/~proskuro/`)

Summary. Second order elliptic problems with discontinuous coefficients are considered. The problem is discretized by the finite element method on geometrically conforming non-matching triangulations across the interface using the mortar technique. The resulting discrete problem is solved by a FETI-DP method. We prove that the method is convergent and its rate of convergence is almost optimal and independent of the jumps of coefficients. Numerical experiments for the case of four subregions are reported. They confirm the theoretical results.

1 Introduction

In this paper we discuss a second order elliptic problem with discontinuous coefficients defined on a polygonal region $\Omega \subset R^2$ which is a union of many polygons Ω_i. The problem is discretized by the finite element method on geometrically conforming non-matching triangulations across $\Gamma = \cup_i \partial\Omega_i \backslash \partial\Omega$ using the mortar technique, see Bernardi et al. [1994]. The resulting discrete problem is solved by a FETI-DP method, see Farhat et al. [2001], Klawonn et al. [2002], Mandel and Tezaur [2001] for the matching triangulation and Dryja and Widlund [2002], Dryja and Widlund [2003] for the non-matching one. The method is discussed under the assumption of continuity of the solution at vertices of Ω_i. We prove that the method is convergent and its rate of convergence is almost optimal and independent of the jumps of coefficients.

The presented results are a generalization of results obtained in Dryja and Widlund [2002], Dryja and Widlund [2003] for continuous coefficients and many subregions, and in Dryja and Proskurowski [2003] for discontinuous coefficients and two subregions Ω_i. In the first two papers two different preconditioners, a standard one and a generalized one, are analyzed for the mortar discretization which is not standard. The mortar condition there is modified at the vertices of Ω_i using the continuity of the solution at these vertices. In the present paper we consider a standard mortar discretization and a standard preconditioner. Numerical experiments for the case of four subregions are reported. They confirm the theoretical results.

The paper is organized as follows. In Section 2, the differential and discrete problems are formulated. In Section 3, a matrix form of the discrete problem is given. The preconditioner is described and analyzed in Section 4. Numerical experiments are presented in Section 5.

2 Differential and discrete problem

We consider the following differential problems. Find $u^* \in H_0^1(\Omega)$ such that

$$a(u^*, v) = f(v), \quad v \in H_0^1(\Omega), \tag{1}$$

where $a(u,v) = (\rho(x) \nabla u, \nabla u)_{L^2(\Omega)}, \quad f(v) = (f,v)_{L^2(\Omega)}$.

We assume that Ω is a polygonal region and $\overline{\Omega} = \cup_{i=1}^N \overline{\Omega}_i$, Ω_i are disjoint polygonal subregions of diameter H_i, $\rho(x) = \rho_i$ is a positive constant on Ω_i and $f \in L^2(\Omega)$. We solve (1) by the FEM on non–matching triangulation across $\partial\Omega_i$. To describe a discrete problem the mortar technique is used, see Bernardi et al. [1994].

We impose on Ω_i a triangulation with triangular elements and parameter h_i. The resulting triangulation in Ω is non-matching across $\partial\Omega_i$. We assume that the triangulation on each Ω_i is quasiuniform and additionally that the parameters h_i and h_j on a common edge of Ω_i and Ω_j are proportional. Let $X_i(\Omega_i)$ be a finite element space of piecewise linear continuous functions defined on the introduced triangulation. We assume that functions of $X_i(\Omega_i)$ vanish on $\partial\Omega_i \cap \partial\Omega$. Let

$$X^h(\Omega) = X_1(\Omega_1) \times \ldots \times X_N(\Omega_N). \tag{2}$$

Note that $X^h(\Omega) \subset L^2(\Omega)$ but $X^h(\Omega) \not\subset H_0^1(\Omega)$. To formulate a discrete problem for (1) we use the mortar technique for geometrically conforming case. For that the following notation is used. Let Γ_{ij} be a common edge of two substructures Ω_i and Ω_j, $\Gamma_{ij} = \partial\Omega_i \cap \partial\Omega_j$. Let $\Gamma = (\cup_i \partial\Omega_i) \backslash \partial\Omega$. We now select open edges $\gamma_m \subset \Gamma$, called *mortar* such that $\overline{\Gamma} = \cup \overline{\gamma}_m$ and $\gamma_m \cap \gamma_n = 0$ for $m \neq n$. Let Γ_{ij} as an edge of Ω_i be denoted by $\gamma_{m(i)}$ and called *mortar* (master), and let Γ_{ij} as an edge of Ω_j be denoted by $\delta_{m(j)}$ and called *non-mortar* (slave). The criteria for choosing $\gamma_{m(i)}$ as the mortar side is that $\rho_i \geq \rho_j$, the coefficients on Ω_i and Ω_j, respectively.

Let $M(\delta_{m(j)})$ be a subspace of $W_j(\delta_{m(j)})$, the restriction of $X_j(\Omega_j)$ to $\delta_{m(j)}$, $\delta_{m(j)} \subset \partial\Omega_j$. Functions of $M(\delta_{m(j)})$ are constants on elements of the triangulation on $\delta_{m(j)}$ which touch $\partial\delta_{m(j)}$. We say that $u_i \in X_i(\Omega_i)$ and $u_j \in X_j(\Omega_j)$ on $\delta_m \equiv \delta_{m(j)} = \gamma_{m(i)} = \Gamma_{ij}$, an edge common to Ω_i and Ω_j, satisfy the mortar condition if

$$\int_{\delta_m} (u_i - u_j)\psi ds = 0, \quad \psi \in M(\delta_m). \tag{3}$$

We are now in a position to introduce V^h, the space for discretization of (1). Let $V^h(\Omega)$ be a subspace of $X^h(\Omega)$ of functions which satisfy the mortar condition (3) for each $\delta_m \subset \Gamma$ and which are continuous at common vertices of the substructures. The discrete problem for (1) in V^h is defined as follows.

Find $u_h^* \in V^h$ such that

$$a_H(u_h^*, v_h) = f(v_h), \quad v_h \in V^h \tag{4}$$

where $a_H(u,v) = \sum_{i=1}^{N} a_i(u,v), \quad a_i(u,v) = \rho_i(\nabla u, \nabla v)_{L^2(\Omega_i)}$. The problem has a unique solution and the error bound is known, see Bernardi et al. [1994].

3 FETI-DP equation

To derive FETI-DP method we first rewrite the problem (4) as a saddle-point problem using Lagrange multipliers. For $u = \{u_i\}_{i=1}^N \in X^h(\Omega)$ and $\psi = \{\psi_p\}_{p=1}^P \in M(\Gamma) = \prod_m M(\delta_m)$, the mortar condition (3) can be rewritten as

$$b(u,\psi) \equiv \sum_{i=1}^{N} \sum_{\delta_{m(i)} \subset \partial\Omega_i} \int_{\delta_{m(i)}} (u_i - u_j)\psi_k ds = 0, \tag{5}$$

where $\delta_{m(i)} = \gamma_{m(j)} = \Gamma_{ij}, \psi_k \in M(\delta_{m(i)})$. Let $\tilde{X}^h(\Omega)$ denote a subspace of $X^h(\Omega)$ of functions which are continuous at common vertices of substructures.

The problem now consists of finding $(u_h^*, \lambda_h^*) \in \tilde{X}^h(\Omega) \times M(\Gamma)$ such that

$$a(u_h^*, v_h) + b(v_h, \lambda_h^*) = f(v_h), \quad v_h \in \tilde{X}^h(\Omega), \tag{6}$$

$$b(u_h^*, \psi_h) = 0, \quad \psi_h \in M(\Gamma). \tag{7}$$

It can be proved that u_h^*, the solution of (6) - (7) is the solution of (4) and vice versa. Therefore the problem (6) - (7) has a unique solution.

To derive a matrix form of (6) - (7) we first need a matrix formulation of (7). Using the nodal basis functions $\varphi^{(l)}_{\delta_{m(i)}} \in W_i(\delta_{m(i)})$, $\varphi^{(k)}_{\gamma_{m(j)}} \in W_j(\gamma_{m(j)})$ and $\psi^{(p)}_{\delta_{m(i)}} \in M_m(\delta_{m(i)})$ $(\delta_{m(i)} = \gamma_{m(j)} = \Gamma_{ij})$ the equation (7) can be rewritten on $\bar{\delta}_{m(i)}$ as

$$B_{\delta_{m(i)}} u_{i\delta_{m(i)}} - B_{\gamma_{m(j)}} u_{j\gamma_{m(j)}} = 0, \tag{8}$$

where $u_{i\delta_{m(i)}}$ and $u_{j\gamma_{m(j)}}$ are vectors which represent $u_i|_{\delta_{m(i)}} \in W_i(\delta_{m(i)})$ and $u_j|_{\gamma_m(j)} \in W_j\ (\gamma_{m(j)})$, and $(n_{\delta_{(i)}} \equiv n_{\delta_{m(i)}}$ and $n_{\gamma_{(j)}} \equiv n_{\gamma_{m(j)}})$

$$\begin{aligned} B_{\delta_{m(i)}} &= \{(\psi^{(p)}_{\delta_{m(i)}}, \varphi^{(k)}_{\delta_{m(i)}})_{L^2(\delta_{m(i)})}\},\ p = 1,..,n_{\delta(i)},\ k = 0,..,n_{\delta(i)}+1, \\ B_{\gamma_{m(j)}} &= \{(\psi^{(p)}_{\delta_{m(i)}}, \varphi^{(l)}_{\gamma_{m(j)}})_{L^2(\gamma_{m(j)})}\},\ p = 1,..,n_{\delta(i)},\ l = 0,..,n_{\gamma_{(j)}}+1. \end{aligned} \tag{9}$$

Here $n_{\delta(i)}, n_{\delta(i)}+2$ and $n_{\gamma(j)}+2$ are the dimensions of $M_m(\delta_{m(i)})$, $W_i(\delta_{m(i)})$ and $W_j(\gamma_{m(j)})$, respectively. Note that $B_{\delta_{m(i)}}$ and $B_{\gamma_{m(j)}}$ are rectangular matrices. We split the vectors $u_{i\delta_{m(i)}}$ and $u_{j\gamma_{m(j)}}$ into vectors $u^{(r)}_{i\delta_{m(i)}}$, $u^{(c)}_{i\delta_{m(i)}}$ and $u^{(r)}_{j\gamma_{m(j)}}$, $u^{(c)}_{j\gamma_{m(j)}}$, respectively, where $u^{(c)}_{i\delta_{m(i)}}$ and $u^{(c)}_{j\gamma_{m(j)}}$ represent values of functions u_i and u_j at the end points of $\delta_{m(i)}$ and $\gamma_{m(j)}$, and $u^{(r)}_{i\delta_{m(i)}}$ and $u^{(r)}_{j\gamma_{m(j)}}$ represent values of u_i and u_j at the interior nodal points of $\delta_{m(i)}$ and $\gamma_{m(j)}$. Using this notation one can rewrite (8) as

$$(B^{(r)}_{\delta_{m(i)}} u^{(r)}_{i\delta_{m(i)}} + B^{(c)}_{\delta_{m(i)}} u^{(c)}_{i\delta_{m(i)}}) - (B^{(r)}_{\gamma_{m(j)}} u^{(r)}_{j\gamma_{m(j)}} + B^{(c)}_{\gamma_{m(j)}} u^{(c)}_{j\gamma_{m(j)}}) = 0. \quad (10)$$

Note that

$$B^{(r)}_{\delta_{m(i)}} = \{(\psi^{(p)}_{\delta_{m(i)}}, \varphi^{(k)}_{\delta_{m(i)}})_{L^2(\delta_{m(i)})}\}, \quad p,\ k = 1, \ldots, n_{\delta(i)} \quad (11)$$

is a square tridiagonal matrix $n_{\delta(i)} \times n_{\delta(i)}$, symmetric and positive definite and $cond(B^{(r)}_{\delta_{m(i)}}) \sim 1$, while the remaining matrices $B^{(c)}_{\delta_{m(i)}}$, $B^{(c)}_{\gamma_{m(j)}}$, $B^{(r)}_{\gamma_{m(j)}}$ are rectangular with dimensions $n_{\delta(i)} \times 2$, $n_{\delta(i)} \times 2$, $n_{\delta(i)} \times n_{\gamma(j)}$, respectively.

Let $K^{(l)}$ be the stiffness matrix of $a_l(.,.)$. It is represented as

$$K^{(l)} = \begin{pmatrix} K^{(l)}_{ii} & K^{(l)}_{ir} & K^{(l)}_{ic} \\ K^{(l)}_{ri} & K^{(l)}_{rr} & K^{(l)}_{rc} \\ K^{(l)}_{ci} & K^{(l)}_{cr} & K^{(l)}_{cc} \end{pmatrix}, \quad (12)$$

where the rows correspond to the interior unknowns $u^{(i)}_l$ of Ω_l, $u^{(r)}_l$ to its edges, and $u^{(l)}_c$ to its vertices. Let $S^{(l)}$ denote the Schur complement of $K^{(l)}$ with respect to the second and third rows, i.e. to the unknowns $u^{(r)}_l$ and $u^{(c)}_l$. This matrix is represented as

$$S^{(l)} = \begin{pmatrix} S^{(l)}_{rr} & S^{(l)}_{rc} \\ S^{(l)}_{cr} & S^{(l)}_{cc} \end{pmatrix}, \quad (13)$$

where the first row corresponds to the unknowns $u^{(r)}_l$ while the second one to $u^{(c)}_l$. Let

$$S = \text{diag}\ \{S^{(l)}\}^N_{l=1}, \quad S_{rr} = \text{diag}\ \{S^{(l)}_{rr}\}^N_{l=1}, \quad S_{cr} = (S^{(1)}_{cr}, \ldots, S^{(N)}_{cr}), \quad (14)$$

and the solution u^*_h of (6) - (7) be represented as $(u^{(i)}, u^{(r)}, u^{(c)})$ where these global sub-vectors correspond to the local unknowns $u^{(i)}_l$, $u^{(r)}_l$, $u^{(c)}_l$, respectively. We have taken into account that the values of $u^{(c)}_l$ at the common vertices of substructures are equal.

We set $\tilde{\lambda}^* = \{B^{(r)}_{\delta_{m(i)}} \lambda^*_{\delta_{m(i)}}\}$, $\delta_{m(i)} \subset \Gamma$, where $\lambda^* = \{\lambda^*_{\delta_{m(i)}}\}$ is the solution of (6) - (7). The mortar condition is represented by $B = (B_r, B_c)$, where these global diagonal matrices are represented by the local ones $(I^{(r)}_{\delta_{m(i)}}, -(B^{(r)}_{\delta_{m(i)}})^{-1}B^{(r)}_{\gamma_{m(j)}})$ and $((B^{(r)}_{\delta_{m(i)}})^{-1}B^{(c)}_{\delta_{m(i)}}, -(B^{(r)}_{\delta_{m(i)}})^{-1}B^{(c)}_{\gamma_{m(j)}})$, respectively and $I^{(r)}_{\delta_{m(i)}}$ is an identity matrix of $n_{\delta(i)} \times n_{\delta(i)}$. The form of these matrices follows from (10) after multiplying it by $(B^{(r)}_{\delta_{m(i)}})^{-1}$.

To represent (6) - (7) in the matrix form we first eliminate unknowns corresponding to the interior nodal points of Ω_l, then use the assumption that the unknowns corresponding to the common vertices of Ω_l are the same (the continuity at the vertices) and finally setting $\tilde{\lambda}^* = \{B^{(r)}_{\delta_{m(i)}}\lambda^*_{\delta_{m(i)}}\}$ we get

$$S_{rr}u^{(r)} + S_{rc}u^{(c)} + B_r^T\tilde{\lambda}^* = g_r, \tag{15}$$

$$S_{cr}u^{(r)} + \tilde{S}_{cc}u^{(c)} + B_c^T\tilde{\lambda}^* = g_c, \tag{16}$$

$$B_r u^{(r)} + B_c u^{(c)} = 0. \tag{17}$$

Here S_{rr} and S_{cr} ($S_{rc} = S_{cr}^T$) are defined in (14) while $\tilde{S}_{cc}$ is defined by $S^{(l)}_{cc}$, see (13), taking into account that $u_l^{(c)}$ at common vertices of substructures are the same.

Eliminating $u^{(r)}$ and $u^{(c)}$ from (15) - (17) we get

$$F\tilde{\lambda}^* = d, \tag{18}$$

where $F = B\tilde{S}^{-1}B^T$, $d = B\tilde{S}^{-1}g$, $B = (B_r, B_c)$, $g = (g_r, g_c)^T$ and

$$\tilde{S} = \begin{pmatrix} S_{rr} & S_{rc} \\ S_{cr} & \tilde{S}_{cc} \end{pmatrix}. \tag{19}$$

We check straightforwardly that F and d can be represented as follows:

$$F = F_{rr} - F_{rc}F_{cc}^{-1}F_{cr}, \quad F_{rc}^T = F_{cr}, \tag{20}$$

where

$F_{rr} = B_r S_{rr}^{-1} B_r^T, \quad F_{rc} = B_c - B_r S_{rr}^{-1} S_{rc}, \quad F_{cc} = \tilde{S}_{cc} - S_{cr} S_{rr}^{-1} S_{rc},$

$d = d_r - F_{rc}F_{cc}^{-1}d_c, \quad d_r = B_r S_{rr}^{-1} g_r, \quad d_c = g_c - S_{cr}S_{rr}^{-1}g_r.$

In the next section we analyze the preconditioner for F.

4 FETI-DP preconditioner

The preconditioner M for (18) is defined as

$$M^{-1} = B_r S_{rr} B_r^T. \tag{21}$$

An ordering of substructures Ω_l is called Neumann-Dirichlet (N-D) ordering (a check board coloring) if all sides of a fixed Ω_l are mortar while all sides of the neighboring substructures of Ω_l are non-mortar.

Theorem 1. *Let the mortar side be chosen where the coefficient ρ_i is larger. Then for $\lambda \in M(\Gamma)$ the following holds*

$$c_0 \left(1 + \log \frac{H}{h}\right)^{\alpha} \langle M\lambda, \lambda\rangle \leq \langle F\lambda, \lambda\rangle \leq c_1 \left(1 + \log \frac{H}{h}\right)^{2} \langle M\lambda, \lambda\rangle, \qquad (22)$$

where $\alpha = 0$ for N-D ordering of substructures and $\alpha = -2$ in the general case; c_0 and c_1 are positive constants independent of h_i, H_i, and the jumps of ρ_i; $h = \min_i h_i, H = \max_i H_i$.

In the proof of Theorem 1 we will need the following lemmas.

Lemma 1. *For $w \in X_1(\partial\Omega_1) \times \ldots \times X_N(\partial\Omega_N)$ with the same values at the vertices of Ω_i the following holds*

$$||B_r^T B_r z_r||_{S_{rr}}^2 \leq C(1 + \log \frac{H}{h})^2 ||w||_S^2, \qquad (23)$$

where $z_r = w - I_H w$ and $I_H w$ is a linear interpolant of w on edges of $\partial\Omega_i$ with values w at the end points of the edges.

Lemma 2. *For $\lambda \in M(\Gamma)$*

$$C(1 + \log \frac{H}{h})^{\alpha} \langle M\lambda, \lambda\rangle \leq \langle F_{rr}\lambda, \lambda\rangle, \qquad (24)$$

where $\alpha = 0$ for the N-D ordering of substructures Ω_l and $\alpha = -2$ in the general case, C is independent of h, H and the jumps of ρ_i.

Proofs of these estimates are slight modifications of the proofs of statements in Dryja and Widlund [2002]. The only item one needs to take into account is that the coefficients ρ_i are larger on the mortar sides. Therefore the proofs of these lemmas are omitted.

Proof. To prove the RHS of Theorem 1 we proceed as follows. For $-\lambda \in M(\Gamma)$ we compute $w = (w^{(r)}, w^{(c)})$ by solving (15) - (16) with $g_r = 0$ and $g_c = 0$. Note that this problem has a unique solution under the assumption that $u^{(c)}$ is continuous at the cross points. Using this, after some manipulations we obtain

$$\langle F\lambda, \lambda\rangle = \langle B_r w^{(r)} + B_c w^{(c)}, \lambda\rangle = \langle Bw, \lambda\rangle. \qquad (25)$$

Let $I_H w$ be a linear interpolant of w on edges with values w at the end points of each edge. Note that $Bw = B(w - I_H w) = B_r z_r$ since $z_r \equiv w - I_H w = 0$ at the end points of the edges. Using that in (25), we get

$$\langle F\lambda, \lambda\rangle = \langle Bw, \lambda\rangle = \langle B_r z_r, \lambda\rangle. \qquad (26)$$

On the other hand, using that $\tilde{S}w = B^T\lambda$ and $\langle \tilde{S}w, w\rangle = \langle Sw, w\rangle$, see (15) - (17), we have

$$\langle Bw, \lambda \rangle = \frac{\langle Bw, \lambda \rangle^2}{\langle Bw, \lambda \rangle} = \frac{\langle B_r z_r, \lambda \rangle^2}{\langle Sw, w \rangle} \leq \frac{||M^{1/2}\lambda||^2 ||M^{-1/2} B_r z_r||^2}{||w||_S^2}. \tag{27}$$

By Lemma 1

$$||M^{-1/2} B_r z_r||^2 = ||B_r^T B_r z_r||_{S_{rr}}^2 \leq C(1 + \log \tfrac{H}{h})^2 ||w||_S^2. \tag{28}$$

Substituting this into (27) we have

$$\langle Bw, \lambda \rangle \leq C(1 + \log \frac{H}{h})^2 ||M^{1/2}\lambda||^2. \tag{29}$$

Using this in (26) we get the RHS estimate of Theorem 1.

To prove the LHS of Theorem 1 we first note that, $F \leq F_{rr}$, see (20), and then use Lemma 2.

5 Numerical results

The test example for all our experiments is the weak formulation, see (1), of

$$-div(\rho(x)\nabla u) = f(x) \text{ in } \Omega, \tag{30}$$

with the Dirichlet boundary conditions on $\partial\Omega$, where Ω is a union of four disjoint square subregions Ω_i, $i = 1, \ldots, 4$, of a diameter one, and $\rho(x) = \rho_i$ is a positive constant in each Ω_i. The mortar and non-mortar sides are chosen such that $\rho_\gamma \geq \rho_\delta$, see Theorem 1. The region Ω is cut into 4 subregions in a checkerboard coloring way: two equidistant grids (with the ratios 1:1, 2:1, 4:1, etc.) are imposed, one on the black, the other on the white squares. A random right hand side to f of (30) is chosen. Numerical experiments have been carried out with different scaling of the coefficients in the preconditioner. The best results were obtained for the preconditioner with $\rho_\delta = \rho_\gamma = 1$. They are reported in Table 1 and Table 2, and they confirm the theory.

Table 1. Iteration count without (denoted NO) and with the FETI-DP preconditioner (denoted DP) for different grid ratios $\frac{h_\delta}{h_\gamma}$; continuous coefficients

h_δ	$\frac{h_\delta}{h_\gamma} = 1:1$		$\frac{h_\delta}{h_\gamma} = 2:1$		$\frac{h_\delta}{h_\gamma} = 4:1$		$\frac{h_\delta}{h_\gamma} = 8:1$		$\frac{h_\delta}{h_\gamma} = 16:1$	
	NO	DP	NO	DP	NO	DP	NO	DP	NO	DP
1/8	12	3	12	8	11	10	12	10	11	10
1/16	17	3	16	9	15	11	15	11	15	11
1/32	23	4	22	9	20	11	20	12	19	11
1/64	28	4	28	9	27	11	25	12	-	-
1/128	38	4	36	9	33	11	-	-	-	-
1/256	49	4	47	9	-	-	-	-	-	-
1/512	63	5	-	-	-	-	-	-	-	-

Table 2. Iteration count without (denoted NO) and with the FETI-DP preconditioner (denoted DP) for different grid ratios $\frac{h_\delta}{h_\gamma}$ and $\frac{\rho_\gamma}{\rho_\delta} = 1000$

	$\frac{h_\delta}{h_\gamma} = 1:1$		$\frac{h_\delta}{h_\gamma} = 2:1$		$\frac{h_\delta}{h_\gamma} = 4:1$		$\frac{h_\delta}{h_\gamma} = 8:1$		$\frac{h_\delta}{h_\gamma} = 16:1$	
h_δ	NO	DP	NO	DP	NO	DP	NO	DP	NO	DP
1/8	13	5	10	6	9	5	9	5	9	5
1/16	17	6	14	7	12	6	12	5	12	5
1/32	22	6	18	7	16	6	15	5	16	5
1/64	29	7	25	7	21	6	21	5	-	-
1/128	37	7	30	7	27	6	-	-	-	-
1/256	49	7	42	7	-	-	-	-	-	-
1/512	63	7	-	-	-	-	-	-	-	-

References

C. Bernardi, Y. Maday, and A. T. Patera. A new non conforming approach to domain decomposition: The mortar element method. In H. Brezis and J.-L. Lions, editors, *Collège de France Seminar*. Pitman, 1994. This paper appeared as a technical report about five years earlier.

M. Dryja and W. Proskurowski. On preconditioners for mortar discretization of elliptic problems. *Numerical Linear Algebra with Applications*, 10:65–82, 2003.

M. Dryja and O. Widlund. A FETI-DP method for a mortar discretization of elliptic problems. In P. L. and T. A., editors, *Recent developments in domain decomposition methods*, pages 41–52. Springer Verlag, 2002.

M. Dryja and O. Widlund. A generalized FETI-DP method for the mortar discretization of elliptic problems. In I. H. et al., editor, *Fourteenth International Conference in Domain Decomposition methods*, pages 27–38. UNAM, Mexico City, Mexico, 2003.

C. Farhat, M. Lesoinne, P. LeTallec, K. Pierson, and D. Rixen. FETI-DP: A dual-primal unified FETI method - part i: A faster alternative to the two-level FETI method. *Int. J. Numer. Meth. Engrg.*, 50:1523–1544, 2001.

A. Klawonn, O. B. Widlund, and M. Dryja. Dual-Primal FETI methods for three-dimensional elliptic problems with heterogeneous coefficients. *SIAM J.Numer.Anal.*, 40, 159-179 2002.

J. Mandel and R. Tezaur. On the convergence of a dual-primal substructuring method. *Numer. Math.*, 88:543–558, 2001.

A FETI-DP Formulation for Two-dimensional Stokes Problem on Nonmatching Grids *

Hyea Hyun Kim[1] and Chang-Ock Lee[2]

[1] Division of Applied Mathematics, KAIST, Daejeon, 305-701, Korea (mashy@amath.kaist.ac.kr)
[2] Division of Applied Mathematics, KAIST (colee@amath.kaist.ac.kr)

Summary. We consider a FETI-DP formulation of the Stokes problem with mortar methods. To solve the Stokes problem correctly and efficiently, redundant continuity constraints are introduced. Lagrange multipliers corresponding to the redundant constraints are treated as primal variables in the FETI-DP formulation. We propose a preconditioner for the FETI-DP operator and show that the condition number of the preconditioned FETI-DP operator is bounded by $C \max_{i=1,\cdots,N} \left\{(1+\log(H_i/h_i))^2\right\}$, where H_i and h_i are the subdomain size and the mesh size, respectively, and C is a constant independent of H_i and h_i.

1 Introduction

Recently, FETI-DP methods, which were originally developed by Farhat et al. [2001], have been applied to nonconforming discretizations(Dryja and Widlund [2002, 2003], Kim and Lee [2002]). Nonconforming discretizations are important for multiphysics simulations, contact-impact problems, the generation of meshes and partitions aligned with jumps in diffusion coefficients, *hp*-adaptive methods, and special discretizations in the neighborhood of singularities. For the elliptic problems in $2D$, Dryja and Widlund [2002] showed that the Dirichlet preconditioner gives the condition number bound $C(1+\log(H/h))^4$, where H and h denote the subdomain size and the mesh size, respectively. Further, Dryja and Widlund [2003] proposed a different preconditioner which is similar to the one in Klawonn and Widlund [2001], and proved the condition number bound $C(1+\log(H/h))^2$ with a restriction that the mesh sizes on the nonmortar side and the mortar side are comparable. For the same problem, Kim and Lee [2002] formulated a FETI-DP operator in a different way from Dryja and Widlund [2002, 2003] and proposed a Neumann-Dirichlet preconditioner, which gives the condition number bound $C(1+\log(H/h))^2$ without the restriction on mesh sizes between neighboring subdomains. For the elliptic problems with heterogeneous coefficients, they

* This work was partially supported by KOSEF R01-2000-00008.

obtained the same condition number bound which does not depend on the coefficients.

In this paper, we extend the result in Kim and Lee [2002] to the Stokes problem. We use the inf-sup stable $P_1(h) - P_0(2h)$ finite elements in each subdomain. For the optimality of the approximation under nonmatching discretizations, we impose mortar matching conditions on the velocity functions using the standard Lagrange multiplier space introduced in Bernardi et al. [1994].

2 FETI-DP formulation

Let Ω be a bounded polygonal domain in $\mathbb{R}^2$. We assume that Ω is partitioned into nonoverlapping bounded polygonal subdomains $\{\Omega_i\}_{i=1}^N$ and the partition is geometrically conforming. Let $H_D^1(\Omega_i)$ be a space of functions in $H^1(\Omega_i)$ with zero traces on $\partial\Omega_i \cap \partial\Omega$, $L_0^2(\Omega_i)$ be a space of functions in $L^2(\Omega_i)$ with zero average and Π^0 be a space of functions in $L_0^2(\Omega)$ which are constants in each subdomain. Then, we consider the following variational form of Stokes' problem: Find $(\mathbf{u}, p_I, p^0) \in \prod_{i=1}^N \left[H_D^1(\Omega_i)\right]^2 \times \prod_{i=1}^N L_0^2(\Omega_i) \times \Pi^0$ such that

$$
\begin{aligned}
\sum_{i=1}^N (\nabla \mathbf{u}, \nabla \mathbf{v})_{\Omega_i} - \sum_{i=1}^N (p_I + p^0, \nabla \cdot \mathbf{v})_{\Omega_i} = \sum_{i=1}^N (\mathbf{f}, \mathbf{v})_{\Omega_i} \quad \forall\, \mathbf{v} \in \prod_{i=1}^N \left[H_D^1(\Omega_i)\right]^2, \\
-\sum_{i=1}^N (\nabla \cdot \mathbf{u}, q_I)_{\Omega_i} = 0 \quad \forall\, q_I \in \prod_{i=1}^N L_0^2(\Omega_i), \\
-\sum_{i=1}^N (\nabla \cdot \mathbf{u}, q^0)_{\Omega_i} = 0 \quad \forall\, q^0 \in \Pi^0,
\end{aligned}
\tag{1}
$$

and the velocity $\mathbf{u}$ is continuous across the subdomain interfaces $\Gamma = \bigcup_{i,j=1}^N (\partial\Omega_i \cap \partial\Omega_j)$. Here, $(\cdot,\cdot)_{\Omega_i}$ denotes the inner product in $[L^2(\Omega_i)]^n$ for $n = 1, 2$.

We associate Ω_i with quasi-uniform triangulations $\Omega_i^{h_i}$ and $\Omega_i^{2h_i}$. Then we consider the inf-sup stable $P_1(h_i) - P_0(2h_i)$ finite elements and denote them by X_i and Q_i, respectively. In addition, Q_i^0 is defined as a subspace of Q_i with zero average on Ω_i. Let $W_i := X_i$ for all $i = 1, \cdots, N$. To get a FETI-DP formulation, we define the following spaces:

$$
X := \left\{ \mathbf{v} \in \prod_{i=1}^N X_i \;:\; \mathbf{v} \text{ is continuous at subdomain corners} \right\},
$$

$$
W = \left\{ \mathbf{w} \in \prod_{i=1}^N W_i \;:\; \mathbf{w} \text{ is continuous at subdomain corners} \right\},
$$

$$Q := \prod_{i=1}^{N} Q_i^0.$$

In this paper, we will use the same notation for a finite element function and the vector of nodal values of that function. The same applies to the notations W_i, X, W, etc. For $\mathbf{v}_i \in X_i$, we write

$$\mathbf{v}_i^t = \left((\mathbf{v}_I^i)^t \, (\mathbf{v}_r^i)^t \, (\mathbf{v}_c^i)^t \right),$$

where the symbol I, r and c represent the d.o.f.(degrees of freedom) at the interior nodes, nodes on edges and corners, respectively. Since $\mathbf{v} \in X$ is continuous at subdomain corners, there exists a vector $\mathbf{v}_c$ such that $\mathbf{v}_c^i = L_c^i \mathbf{v}_c$ for all $i = 1, \cdots, N$ with a restriction map L_c^i. The vector $\mathbf{v}_c$ has the d.o.f. corresponding to the union of subdomain corners. Let

$$\mathbf{v}_I^t = \left((\mathbf{v}_I^1)^t \cdots (\mathbf{v}_I^N)^t \right), \quad \mathbf{v}_r^t = \left((\mathbf{v}_r^1)^t \cdots (\mathbf{v}_r^N)^t \right).$$

We define the spaces X_I, W_r and W_c which consist of vectors $\mathbf{v}_I$, $\mathbf{v}_r$ and $\mathbf{v}_c$, respectively. For $\mathbf{w} \in W$, we define $\mathbf{w}_r \in W_r$ and $\mathbf{w}_c \in W_c$ similarly to $\mathbf{v}_r$ and $\mathbf{v}_c$.

On each $\Gamma_{ij}(= \partial\Omega_i \cap \partial\Omega_j)$, we determine mortar and nonmortar sides and define

$$m_i := \left\{ j \, : \, \Omega_i^h|_{\Gamma_{ij}} \text{ is the nonmortar side of } \Gamma_{ij} \right\},$$
$$s_i := \left\{ j \, : \, \Omega_i^h|_{\Gamma_{ij}} \text{ is the mortar side of } \Gamma_{ij} \right\}.$$

We consider the standard Lagrange multiplier space M_{ij} and let

$$M := \prod_{i=1}^{N} \prod_{j \in m_i} M_{ij}.$$

Then the following mortar matching conditions are imposed on $v \in X$:

$$\int_{\Gamma_{ij}} (\mathbf{v}_i - \mathbf{v}_j) \cdot \boldsymbol{\lambda}_{ij} \, ds = 0 \quad \forall \boldsymbol{\lambda}_{ij} \in M_{ij}, \forall\, i = 1, \cdots, N, \forall j \in m_i. \tag{2}$$

Now, we rewrite (2) into a matrix form. Let B_i^{ij} be a matrix with entries

$$(B_i^{ij})_{lk} = \pm \int_{\Gamma_{ij}} \psi_l \cdot \phi_k \, ds \quad \forall l = 1, \cdots, L, \forall k = 1, \cdots, K, \tag{3}$$

where $\{\psi_l\}_{l=1}^{L}$ is basis for M_{ij} and $\{\phi_k\}_{k=1}^{K}$ is nodal basis for $W_i|_{\Gamma_{ij}}$. In (3), the +sign is chosen if $\Omega_i|_{\Gamma_{ij}}$ is a nonmortar side, otherwise the −sign is chosen. Let $E_{ij} : M_{ij} \to M$ be an extension operator by zero and $R_{ij}^l : W_l \to W_l|_{\Gamma_{ij}}$ for $l = i, j$ be a restriction operator and $B_i = \sum_{j \in m_i \cup s_i} E_{ij} B_i^{ij} R_{ij}^i$. Then (2) is written into

$$B\mathbf{w} = \mathbf{0}, \tag{4}$$

where $B = \begin{pmatrix} B_1 \cdots B_N \end{pmatrix}$ and $\mathbf{w} = \begin{pmatrix} \mathbf{w}_1^t \cdots \mathbf{w}_N^t \end{pmatrix}^t$ with $\mathbf{w}_i = \mathbf{v}_i|_{\partial \Omega_i}$. Let $B_{i,r}$ and $B_{i,c}$ be matrices that consist of the columns of B_i corresponding to the d.o.f. on edges and corners, respectively. Then (4) is written into

$$B_r \mathbf{w}_r + B_c \mathbf{w}_c = \mathbf{0},$$

where $B_r = \begin{pmatrix} B_{1,r} \cdots B_{N,r} \end{pmatrix}$ and $B_c = \sum_{i=1}^{N} B_{i,c} L_c^i$.

Borrowing the idea of Li [2001], we add the following redundant continuity constraints to the coarse problem

$$\int_{\Gamma_{ij}} (\mathbf{v}_i - \mathbf{v}_j)\, ds = \mathbf{0} \quad \forall i = 1, \cdots, N, \forall j \in m_i. \tag{5}$$

and rewrite (5) as

$$R^t(B_r \mathbf{w}_r + B_c \mathbf{w}_c) = \mathbf{0}, \tag{6}$$

with a suitable matrix R. Let S be the Lagrange multiplier space corresponding to the constraints (6).

Then, the following is induced from the Galerkin approximation to (1): Find $(\mathbf{u}_I, p_I, \mathbf{u}_r, \mathbf{u}_c, p^0, \boldsymbol{\mu}, \boldsymbol{\lambda}) \in X_I \times Q \times W_r \times W_c \times \Pi^0 \times S \times M$ such that

$$\begin{pmatrix} A_{II} & G_{II} & A_{Ir} & A_{Ic} & \mathbf{0} & \mathbf{0} & \mathbf{0} \\ G_{II}^t & \mathbf{0} & G_{rI}^t & G_{cI}^t & \mathbf{0} & \mathbf{0} & \mathbf{0} \\ A_{rI} & G_{rI} & A_{rr} & A_{rc} & G_{r0} & B_r^t R & B_r^t \\ A_{cI} & G_{cI} & A_{cr} & A_{cc} & G_{c0} & B_c^t R & B_c^t \\ \mathbf{0} & \mathbf{0} & G_{r0}^t & G_{c0}^t & \mathbf{0} & \mathbf{0} & \mathbf{0} \\ \mathbf{0} & \mathbf{0} & R^t B_r & R^t B_c & \mathbf{0} & \mathbf{0} & \mathbf{0} \\ \mathbf{0} & \mathbf{0} & B_r & B_c & \mathbf{0} & \mathbf{0} & \mathbf{0} \end{pmatrix} \begin{pmatrix} \mathbf{u}_I \\ p_I \\ \mathbf{u}_r \\ \mathbf{u}_c \\ p^0 \\ \boldsymbol{\mu} \\ \boldsymbol{\lambda} \end{pmatrix} = \begin{pmatrix} \mathbf{f}_I \\ \mathbf{0} \\ \mathbf{f}_r \\ \mathbf{f}_c \\ \mathbf{0} \\ \mathbf{0} \\ \mathbf{0} \end{pmatrix}.$$

Let

$$\mathbf{z}_r^t = \begin{pmatrix} \mathbf{u}_I^t \; p_I^t \; \mathbf{u}_r^t \end{pmatrix}, \quad \mathbf{z}_c^t = \begin{pmatrix} \mathbf{u}_c^t \; (p^0)^t \; \boldsymbol{\mu}^t \end{pmatrix}.$$

In the FETI-DP formulation, we regard $\mathbf{z}_c$ as a primal variable. After eliminating $\mathbf{z}_r$ and $\mathbf{z}_c$, we obtain the following equation for $\boldsymbol{\lambda}$

$$F_{DP} \boldsymbol{\lambda} = \mathbf{d} \tag{7}$$

and call F_{DP} a FETI-DP operator.

3 Preconditioner

We define S_i as the discrete Schur complement operator of the Stokes problem in Ω_i obtained by eliminating interior velocity and pressure unknowns. Let

$$S := \text{diag}(S_1, \cdots, S_N)$$

and it can be seen easily that S is s.p.d. on W. Hence, we define a norm for $\mathbf{w} \in W$ as

$$\|\mathbf{w}\|_W^2 := \sum_{i=1}^{N} < S_i \mathbf{w}_i, \mathbf{w}_i > .$$

Let

$$W^0 = \prod_{i=1}^{N} \prod_{j \in m_i} W_{ij}^0,$$

where W_{ij}^0 consists of functions in $W_i|_{\Gamma_{ij}}$ with zero value at the end points of Γ_{ij}. For a function $\mathbf{w}_{ij} \in W_{ij}^0$ with $j \in m_i$, let $\widetilde{\mathbf{w}}_{ij} \in W_i$ be the zero extension of $\mathbf{w}_{ij}$. Using this, we define the zero extension $\widetilde{\mathbf{w}} \in W$ of $\mathbf{w} \in W^0$ by

$$\widetilde{\mathbf{w}} = (\widetilde{\mathbf{w}}_1, \cdots, \widetilde{\mathbf{w}}_N) \text{ with } \widetilde{\mathbf{w}}_i = \sum_{j \in m_i} \widetilde{\mathbf{w}}_{ij}$$

and a norm on W^0 by

$$\|\mathbf{w}\|_{W^0} := \|\widetilde{\mathbf{w}}\|_W .$$

We introduce the following subspaces with the norms induced from the spaces W and W^0:

$$\begin{aligned} W_R &:= \left\{ \mathbf{w} \in W \ : \ R^t(B_r \mathbf{w}_r + B_c \mathbf{w}_c) = \mathbf{0} \right\}, \\ W_{R,G} &:= \left\{ \mathbf{w} \in W_R \ : \ G_{r0}^t \mathbf{w}_r + G_{c0}^t \mathbf{w}_c = \mathbf{0} \right\}, \\ W_R^0 &:= \left\{ \mathbf{w} \in W^0 \ : \ \widetilde{\mathbf{w}} \in W_R \right\}. \end{aligned}$$

Let us define

$$M_R = \left\{ \boldsymbol{\lambda} \in M \ : \ R^t \boldsymbol{\lambda} = \mathbf{0} \right\}$$

and a dual norm for $\boldsymbol{\lambda} \in M_R$ by

$$\|\boldsymbol{\lambda}\|_{M_R}^2 := \max_{\mathbf{w} \in W_R^0 \setminus \{0\}} \frac{< \boldsymbol{\lambda}, \mathbf{w} >_m^2}{\|\mathbf{w}\|_{W^0}^2},$$

where $< \boldsymbol{\lambda}, \mathbf{w} >_m = \sum_{i=1}^{N} \sum_{j \in m_i} \int_{\Gamma_{ij}} \boldsymbol{\lambda}_{ij} \cdot \mathbf{w}_{ij} \, ds$ is a duality pairing. From this dual norm, we can find an operator $\widehat{F}_{DP}$ which gives

$$< \widehat{F}_{DP} \boldsymbol{\lambda}, \boldsymbol{\lambda} > = \|\boldsymbol{\lambda}\|_{M_R}^2 \tag{8}$$

and propose $\widehat{F}_{DP}^{-1}$ as a preconditioner for the FETI-DP operator in (7). To give a matrix form of $\widehat{F}_{DP}^{-1}$, we define $R_{ij} : W^0 \rightarrow W_{ij}^0$ as a restriction operator and $E_{ij}^i : W_{ij}^0 \rightarrow W_i$ as an extension operator by zero. Let $\widehat{B}_i^{ij}$ be a matrix obtained from B_i^{ij} after deleting the columns corresponding to the end points of Γ_{ij}. Since, we restrict $\boldsymbol{\lambda} \in M_R$ and $\mathbf{w} \in W_R^0$, we need l^2-orthogonal projections

$$P_{W_R^0}^{ij} : W^0|_{\Gamma_{ij}} \rightarrow W_R^0|_{\Gamma_{ij}}, \quad P_{M_R}^{ij} : M|_{\Gamma_{ij}} \rightarrow M_R|_{\Gamma_{ij}}.$$

Let $\widehat{B}_{ij} = P^{ij}_{M_R} \widehat{B}^{ij}_i P^{ij}_{W^0_R}$ and $\widehat{B}_i = \sum_{j \in m_i} E^i_{ij} \widehat{B}^{-1}_{ij} R_{ij}$. Then, we obtain

$$\widehat{F}^{-1}_{DP} = \sum_{i=1}^{N} \widehat{B}^t_i S_i \widehat{B}_i.$$

Thus, the computation of $\widehat{F}^{-1}_{DP}\boldsymbol{\lambda}$ can be done parallely in each subdomain.

4 Condition number estimation

In this section, we only state lemmas that are used to analyze the condition number bound without proofs. In the following, C is a generic constant independent of h_i and H_i.

Lemma 1. *For $\boldsymbol{\lambda} \in M_R$, we have*

$$< F_{DP}\boldsymbol{\lambda}, \boldsymbol{\lambda} >= \max_{\mathbf{w} \in W_{R,G} \setminus \{0\}} \frac{< B\mathbf{w}, \boldsymbol{\lambda} >^2}{\|\mathbf{w}\|^2_W}.$$

Lemma 2. *For $\boldsymbol{\lambda} \in M_R$, we have*

$$\max_{\mathbf{w} \in W_{R,G} \setminus \{0\}} \frac{< B\mathbf{w}, \boldsymbol{\lambda} >^2}{\|\mathbf{w}\|^2_W} \geq \|\boldsymbol{\lambda}\|^2_{M_R}.$$

Let us define a notation $|\cdot|_{S_i} :=< S_i \cdot, \cdot >^{1/2}$. Then the following lemma can be found in Bramble and Pasciak [1990].

Lemma 3. *For $\mathbf{w}_i \in W_i$, we have*

$$C_1 \beta |\mathbf{w}_i|_{S_i} \leq |\mathbf{w}_i|_{1/2, \partial\Omega_i} \leq C_2 |\mathbf{w}_i|_{S_i},$$

where β is the inf-sup constant for the finite elements of subdomain Ω_i and the constants C_1 and C_2 are independent of h_i and H_i.

We also have the following result which is derived from Lemma 5.1 in Mandel and Tezaur [2001].

Lemma 4. *For $\mathbf{w} \in W$, we have*

$$\|\mathbf{w}_i - \mathbf{w}_j\|^2_{H^{1/2}_{00}(\Gamma_{ij})} \leq C \max_{l \in \{i,j\}} \left\{ \left(1 + \log \frac{H_l}{h_l}\right)^2 \right\} \left(|\mathbf{w}_i|^2_{1/2,\partial\Omega_i} + |\mathbf{w}_j|^2_{1/2,\partial\Omega_j} \right).$$

From Lemma 3, Lemma 4 and the continuity of mortar projection in $H^{1/2}_{00}(\Gamma_{ij})$, we have

Lemma 5. *For $\boldsymbol{\lambda} \in M_R$,*

$$\max_{\mathbf{w} \in W_{R,G} \setminus \{0\}} \frac{< B\mathbf{w}, \boldsymbol{\lambda} >^2}{\|\mathbf{w}\|^2_W} \leq C \max_{i=1,\cdots,N} \left\{ \left(1 + \log \frac{H_i}{h_i}\right)^2 \right\} \|\boldsymbol{\lambda}\|^2_{M_R}.$$

From Lemma 1, Lemma 2, Lemma 5 and (8), we obtain the following condition number bound:

Theorem 1.

$$\kappa(\widehat{F}_{DP}^{-1}F_{DP}) \leq C \max_{i=1,\cdots,N} \left\{ \left(1+\log\frac{H_i}{h_i}\right)^2 \right\}.$$

5 Numerical Results

Let $\Omega = [0,1]\times[0,1] \subset \mathbb{R}^2$ and consider Stokes problem with an exact solution

$$\mathbf{u} = \begin{pmatrix} \sin^3(\pi x)\sin^2(\pi y)\cos(\pi y) \\ -\sin^2(\pi x)\sin^3(\pi y)\cos(\pi x) \end{pmatrix} \quad \text{and} \quad p = x^2 - y^2.$$

Let N denote the number of subdomains. We only consider the uniform partition of Ω. The notation $N = 4\times 4$ means that Ω is partitioned into 4×4 square subdomains. Let n denote the number of nodes on subdomain edges including end points, which is associated with $\Omega_i^{h_i}$, a triangulation for velocity functions. We solve the FETI-DP operator with and without preconditioner varying N and n under nonmatching discretizations. Those cases are denoted by PFETI-DP and FETI-DP, respectively. The CG(Conjugate Gradient) iteration is stopped when the relative residual is less than 10^{-6}.

In Tables 1 and 2, the number of CG iterations and the corresponding condition number are shown varying N and n. From Table 1, we observe that PFETI-DP performs well and the condition numbers seem to behave $\log^2$-growth as n increases. In Table 2, as N increases with $n = 5$ or $n = 9$, the CG iteration becomes stable for both cases with and without preconditioner. Hence, we can see that the developed preconditioner gives the condition number bound as confirmed in theory.

Table 1. CG iterations(condition number) when $N = 4\times 4$

n	FETI-DP	PFETI-DP
5	16(8.35)	12(3.75)
9	50(1.15e+2)	15(5.79)
17	86(5.01e+2)	17(7.93)
33	119(1.31e+3)	20(9.88)
65	153(3.29e+3)	22(1.20e+1)

Table 2. CG iterations(condition number) when $n = 5, 9$

N	$n = 5$ FETI-DP	PFETI-DP	$n = 9$ FETI-DP	PFETI-DP
4×4	16(8.35)	12(3.75)	50(1.15e+2)	15(5.79)
8×8	16(9.18)	12(3.68)	53(1.19e+2)	15(6.21)
16×16	16(9.57)	11(3.42)	57(1.34e+2)	16(6.27)
32×32	16(10.88)	12(3.78)	56(1.25e+2)	16(6.24)

References

C. Bernardi, Y. Maday, and A. T. Patera. A new nonconforming approach to domain decomposition: the mortar element method. In *Nonlinear partial differential equations and their applications. Collège de France Seminar, Vol. XI (Paris, 1989–1991)*, volume 299 of *Pitman Res. Notes Math. Ser.*, pages 13–51. Longman Sci. Tech., Harlow, 1994.

J. H. Bramble and J. E. Pasciak. A domain decomposition technique for Stokes problems. *Appl. Numer. Math.*, 6(4):251–261, 1990.

M. Dryja and O. B. Widlund. A FETI-DP method for a mortar discretization of elliptic problems. In *Recent developments in domain decomposition methods (Zürich, 2001)*, volume 23 of *Lect. Notes Comput. Sci. Eng.*, pages 41–52. Springer, Berlin, 2002.

M. Dryja and O. B. Widlund. A generalized FETI-DP method for a mortar discretization of elliptic problems. In *Domain decomposition methods in Science and Engineering (Cocoyoc, Mexico, 2002)*, pages 27–38. UNAM, Mexico City, 2003.

C. Farhat, M. Lesoinne, P. LeTallec, K. Pierson, and D. Rixen. FETI-DP: a dual-primal unified FETI method. I. A faster alternative to the two-level FETI method. *Internat. J. Numer. Methods Engrg.*, 50(7):1523–1544, 2001.

H. H. Kim and C.-O. Lee. A FETI-DP preconditioner for elliptic problems on nonmatching grids. In *KAIST DAM Research Report 02-9.* KAIST, 2002.

A. Klawonn and O. B. Widlund. FETI and Neumann-Neumann iterative substructuring methods: connections and new results. *Comm. Pure Appl. Math.*, 54(1):57–90, 2001.

J. Li. A dual-primal FETI method for incompressible Stokes equations. In *Technical Report 816.* Department of Computer Science, Courant Institute, New York Unversity, 2001.

J. Mandel and R. Tezaur. On the convergence of a dual-primal substructuring method. *Numer. Math.*, 88(3):543–558, 2001.

Some Computational Results for Dual-Primal FETI Methods for Elliptic Problems in 3D

Axel Klawonn[1], Oliver Rheinbach[1], and Olof B. Widlund[2]

[1] Universität Duisburg-Essen, Campus Essen, Fachbereich Mathematik (http://www.uni-essen.de/ingmath/Axel.Klawonn, www.uni-essen.de/ingmath/people/rheinbach.html)
[2] Courant Institute of Mathematical Sciences, New York University (http://cs.nyu.edu/cs/faculty/widlund/)

Summary. Iterative substructuring methods with Lagrange multipliers for elliptic problems are considered. The algorithms belong to the family of dual-primal FETI methods which were introduced for linear elasticity problems in the plane by Farhat et al. [2001] and were later extended to three dimensional elasticity problems by Farhat et al. [2000]. Recently, the family of algorithms for scalar diffusion problems was extended to three dimensions and successfully analyzed by Klawonn et al. [2002a,b]. It was shown that the condition number of these dual-primal FETI algorithms can be bounded polylogarithmically as a function of the dimension of the individual subregion problems and that the bounds are otherwise independent of the number of subdomains, the mesh size, and jumps in the diffusion coefficients. In this article, numerical results for some of these algorithms are presented and their relation to the theoretical bounds is studied. The algorithms have been implemented in PETSc, see Balay et al. [2001], and their parallel scalability is analyzed.

1 Elliptic model problem, finite elements, and geometry

Let $\Omega \subset \mathbb{R}^3$, be a bounded, polyhedral region, let $\partial\Omega_D \subset \partial\Omega$ be a closed set of positive measure, and let $\partial\Omega_N := \partial\Omega \setminus \partial\Omega_D$ be its complement. We impose homogeneous Dirichlet and general Neumann boundary conditions, respectively, on these two subsets and introduce the Sobolev space $H_0^1(\Omega, \partial\Omega_D) := \{v \in H^1(\Omega) : v = 0 \text{ on } \partial\Omega_D\}$.

We decompose Ω into non-overlapping subdomains $\Omega_i, i = 1, \ldots, N$, where each is the union of shape-regular elements with the finite element nodes on the boundaries of neighboring subdomains matching across the interface Γ. The interface Γ is the union of subdomain faces, which are shared by two subregions, edges which are shared by more than two subregions and vertices which form the endpoints of edges. All of them are regarded as open sets.

For simplicity, we will only consider a piecewise trilinear, conforming finite element approximation of the following scalar, second order model problem:

find $u \in H_0^1(\Omega, \partial\Omega_D)$, such that

$$a(u,v) = f(v) \quad \forall v \in H_0^1(\Omega, \partial\Omega_D), \tag{1}$$

where

$$a(u,v) = \sum_{i=1}^{N} \rho_i \int_{\Omega_i} \nabla u \cdot \nabla v dx, \quad f(v) = \sum_{i=1}^{N} \Big(\int_{\Omega_i} f v dx + \int_{\partial\Omega_i \cap \partial\Omega_N} g_N v ds \Big), \tag{2}$$

where g_N is the Neumann boundary data defined on $\partial\Omega_N$. We further assume that the diffusion coefficient ρ_i is a positive constant on each subregion Ω_i.

For the theoretical analysis, we also make a number of further technical assumptions; see Klawonn et al. [2002a,b] for details.

2 The FETI-DP Method

For each subdomain $\Omega_i, i = 1, \ldots, N$, we assemble local stiffness matrices $K^{(i)}$ and local load vectors $f^{(i)}$. We denote by $u^{(i)}$ the local solution vectors of nodal values. The local stiffness matrices $K^{(i)}$ can be partitioned according to vertex and remaining degrees of freedom, denoted by subscript c and r, respectively.

$$K^{(i)} = \begin{bmatrix} K_{rr}^{(i)} & K_{rc}^{(i)} \\ K_{rc}^{(i)T} & K_{cc}^{(i)} \end{bmatrix}, \quad u^{(i)} = \begin{bmatrix} u_r^{(i)} \\ u_c^{(i)} \end{bmatrix}, \quad f^{(i)} = \begin{bmatrix} f_r^{(i)} \\ f_c^{(i)} \end{bmatrix}, \qquad i = 1, \ldots, N.$$

By assembling the stiffness matrix contributions from the vertices, we obtain from the local submatrices $K_{cc}^{(i)}$ the global matrix $\widetilde{K}_{cc}$ and from the local matrices $K_{rc}^{(i)}$ the partially assembled matrices $\widetilde{K}_{rc}^{(i)}$. Here, we choose to assemble at all vertices. It is also possible to take only a sufficient number of them; for details, see Klawonn et al. [2002a]. We introduce the following notation $K_{rr} := diag_{i=1}^{N}(K_{rr}^{(i)})$ and $\widetilde{K}_{rc} := [\widetilde{K}_{rc}^{(1)T} \cdots \widetilde{K}_{rc}^{(N)T}]^T$. The global vectors $\widetilde{u}_c$ and $\widetilde{f}_c$ are defined accordingly. We note that the FETI-DP iterates will be continuous at all vertices throughout the iterations.

To guarantee continuity at the remaining interface nodes, i.e., those which are not vertices, we introduce the jump operator $B_r = [B_r^{(1)}, \ldots, B_r^{(N)}]$. The entries of this matrix are $0, 1, -1$ and it is constructed such that components of any vector u_r, which are associated with the same node on the interface Γ, coincide when $B_r u_r = \sum_{i=1}^{N} B_r^{(i)} u_r^{(i)} = 0$.

We can now reformulate the finite element discretization of (1) as

$$\begin{bmatrix} K_{rr} & \widetilde{K}_{rc} & B_r^T \\ \widetilde{K}_{rc}^T & \widetilde{K}_{cc} & 0 \\ B_r & 0 & 0 \end{bmatrix} \begin{bmatrix} u_r \\ \widetilde{u}_c \\ \lambda \end{bmatrix} = \begin{bmatrix} f_r \\ \widetilde{f}_c \\ 0 \end{bmatrix}. \tag{3}$$

Elimination of the primal variables u_r and $\widetilde{u}_c$ leads to a reduced linear system of the form

$$F_A \lambda = d_A,$$

where the matrix F_A and the right hand side d_A are formally obtained by block Gauss elimination. Let us note that the matrix F_A is never built explicitly but that in every iteration appropriate linear systems are solved; see Farhat et al. [2000] or Klawonn et al. [2002a] for more details.

To obtain better convergence properties for three dimensional problems, a larger coarse problem was suggested by introducing additional optional constraints of the form

$$Q_r u_r = 0. \tag{4}$$

Here, $Q_r := [Q_r^{(1)} \ldots Q_r^{(N)}]$, $Q_r^{(i)} := [O \quad Q_\Delta B_\Delta^{(i)}]$, and Q_Δ is a rectangular matrix which has as many columns as there are remaining degrees of freedom which are on the interface; for the latter set, we will also use the subscript Δ. The number of rows is determined by the number of primal edges and faces. A primal edge is an edge where the average of u is the same across this edge whichever component of the product space is used in its computation. Analogously, we define a primal face. The matrix Q_Δ is constructed such that (4) guarantees that certain linear combinations of the rows of $B_\Delta u_\Delta$ are zero. These linear combinations are related to primal edges and faces. Then, (4) enforces that averages at primal edges and faces have common values across the interface.

Introducing additional optional Lagrange multipliers μ to enforce the extra constraints given in (4), we obtain from (3) the following linear system

$$\begin{bmatrix} K_{rr} & \widetilde{K}_{rc} & Q_r^T & B_r^T \\ \widetilde{K}_{rc}^T & \widetilde{K}_{cc} & 0 & 0 \\ Q_r & 0 & 0 & 0 \\ B_r & 0 & 0 & 0 \end{bmatrix} \begin{bmatrix} u_r \\ \widetilde{u}_c \\ \mu \\ \lambda \end{bmatrix} = \begin{bmatrix} f_r \\ \widetilde{f}_c \\ 0 \\ 0 \end{bmatrix}. \tag{5}$$

Elimination of $u_r, \widetilde{u}_c$, and μ leads again to a reduced linear system of the form

$$F\lambda = d, \tag{6}$$

where the matrix F and the right hand side d are again formally obtained by block Gauss elimination.

Let us now define the Dirichlet preconditioner. We need a scaled jump operator $B_{D,r}$. It is obtained from $B_r = [O \quad B_\Delta]$ by scaling B_Δ subdomain-wise with appropriate diagonal scaling matrices $D^{(i)}$ and setting $B_{D,\Delta} := [D^{(1)} B_\Delta^{(1)} \ldots D^{(N)} B_\Delta^{(N)}]$. The scaling matrices $D^{(i)}$ are defined using the diffusion coefficients ρ_i; for details, see Klawonn et al. [2002a]. Finally, we add a zero column to $B_{D,r}$ for each vertex node. From the local stiffness matrices $K^{(i)}$, we obtain local Schur complements $S^{(i)}$, by eliminating the interior variables, which operate on the degrees of freedom belonging to the interface nodes. Let us define the block diagonal matrix $S := diag_{i=1}^{N}(S^{(i)})$. The Dirichlet preconditioner is then defined as

$$M^{-1} := B_{D,r} S B_{D,r}^T.$$

The FETI-DP algorithms are preconditioned conjugate gradient methods for solving the preconditioned linear system

$$M^{-1} F \lambda = M^{-1} d.$$

Following the notation in Klawonn et al. [2002a,b], we denote the algorithm using just vertex constraints by Algorithm A. For those methods which additionally use optional constraints, we denote the method choosing all edges and faces as primal by Algorithm B, the one using all edges by Algorithm C, and finally the algorithm which uses just faces by Algorithm E. We denote the corresponding matrix F in (6) by F_B, F_C, and F_E.

3 Theoretical Estimates

For Algorithms A, B, C, and E, we have the following estimates; cf. Klawonn et al. [2002a,b].

Theorem 1. *The condition numbers satisfy*

1. $\kappa(M^{-1}F_A) \leq C(H/h)(1+\log(H/h))^2$,
2. $\kappa(M^{-1}F_B) \leq C(1+\log(H/h))^2$,
3. $\kappa(M^{-1}F_C) \leq C(1+\log(H/h))^2$,
4. $\kappa(M^{-1}F_E) \leq C\max((1+\log(H/h))^2, TOL*(1+\log(H/h)))$,

where $C > 0$ is a constant which is independent of H, h, TOL, and the values of the coefficients ρ_i.

We note that the condition number estimate for Algorithm E is only valid if, for all pairs of substructures Ω_i, Ω_k, which have an edge $\mathcal{E}^{ik}$ in common, we have an *acceptable face path.* An acceptable face path is a path from Ω_i to Ω_k, possibly via several other substructures Ω_j, which do not necessarily touch the edge in question, such that the associated coefficients ρ_j, ρ_i, and ρ_k satisfy $TOL * \rho_j \geq \min(\rho_i, \rho_k)$ for some chosen tolerance TOL.

4 Computational results

We have applied the FETI-DP algorithms A, B, C, and E to the model problem (1), where $\Omega := [0,1]^3$ is the unit cube. We decompose the unit cube into $N \times N \times N$ cubic subdomains with sidelength $H := 1/N$. The diffusion

Fig. 1. Model domain decomposed into cubes with discontinuous diffusion coefficients $\rho_i = 1$ and $\rho_i = 10^4$

coefficients ρ_i alternate between 1 and 10^4 and are distributed in a threedimensional checkerboard pattern; cf. Figure 1. On the front, left, and bottom part, homogeneous Dirichlet boundary conditions are applied. On all the remaining parts of the boundary, we imposed homogeneous Neumann boundary conditions. The coefficients are constant on each subdomain and (1) is discretized by conforming trilinear elements with finite element diameter h. All algorithms are implemented in PETSc, see Balay et al. [2001]. We use the preconditioned conjugate gradient method with a zero initial guess. The stopping criterion is the relative reduction of the initial residual by 10^{-7} in the Euclidean norm. In order to analyze the numerical scalability of our algorithms, we have carried out two different types of experiments. In our first set of runs, we kept the subdomain size H/h fixed and increased the number of subdomains and thus the overall problem size; cf. Tables 1,2,3,4. Our second series of experiments is carried out with a fixed number of subdomains and an increasing subdomain size H/h resulting in an increased $1/h$; cf. Tables 5 and 6 and Figure 2. From both set of runs, we see that our computational results support the theoretical condition number estimates. However, for Algorithm E, we cannot decide if the growth of the condition number is polylogarithmic. From the range of H/h used in the experiments, it rather looks linear than polylogarithmic. We note that for this problem, the bound of Theorem 1 is basically meaningless since $TOL = 10^4$. Experiments for an isotropic material, i.e., with no jumps in the coefficients show the same polylogarithmic growth

Table 1. Algorithm A - Constant H/h

Subdomains	Dof/Subdom.	Dof	Iterations	$\lambda_{\min}$	$\lambda_{\max}$
8	1000	6,859	**9**	1.00035	11.5539
27	1000	21,952	**14**	1.00051	28.8335
64	1000	50,653	**19**	1.00361	25.0130
125	1000	97,336	**22**	1.00283	28.8335
216	1000	166,375	**24**	1.00231	25.0127
343	1000	262,144	**26**	1.00188	28.8335
512	1000	389,017	**25**	1.00161	25.0127
729	1000	551,368	**26**	1.00138	28.8335
1000	1000	753,571	**24**	1.00125	25.0127

Table 2. Algorithm B - Constant H/h

Subdomains	Dof/Subdom.	Dof	Iterations	λ_{min}	λ_{max}
8	1000	6,859	**7**	1.00085	1.47091
27	1000	21,952	**8**	1.00049	1.55036
64	1000	50,653	**8**	1.00025	1.47011
125	1000	97,336	**8**	1.00022	1.55036
216	1000	166,375	**8**	1.00013	1.46995
343	1000	262,144	**8**	1.00013	1.55036
512	1000	389,017	**8**	1.00009	1.46989
729	1000	551,368	**8**	1.00010	1.55036
1000	1000	753,571	**7**	1.00014	1.46985

Table 3. Algorithm C - Constant H/h

Subdomains	Dof/Subdom.	Dof	Iterations	λ_{min}	λ_{max}
8	1000	6,859	**8**	1.00030	1.61492
27	1000	21,952	**9**	1.00040	2.06800
64	1000	50,653	**9**	1.00020	1.93210
125	1000	97,336	**10**	1.00012	2.06875
216	1000	166,375	**9**	1.00009	1.93192
343	1000	262,144	**10**	1.00008	2.06875
512	1000	389,017	**9**	1.00006	1.93210
729	1000	551,368	**10**	1.00005	2.06875
1000	1000	753,571	**9**	1.00005	1.93210

Table 4. Algorithm E - Constant H/h

Subdomains	Dof/Subdom.	Dof	Iterations	λ_{min}	λ_{max}
8	1000	6,859	**8**	1.00102	11.4671
27	1000	21,952	**10**	1.00185	16.2107
64	1000	50,653	**14**	1.00129	16.2191
125	1000	97,336	**16**	1.00113	16.2246
216	1000	166,375	**19**	1.00089	16.2281
343	1000	262,144	**19**	1.00079	16.2304
512	1000	389,017	**20**	1.00067	16.2319
729	1000	551,368	**20**	1.00060	16.2329
1000	1000	753,571	**20**	1.00054	16.2335

as Algorithms B and C. This is an interesting point which needs some further analysis. In a third set of experiments, we have tested our algorithms for parallel scalability. We considered a decomposition into 216 subdomains with 13824 degrees of freedom for each subdomain which yields an overall problem size of 2 685 619 degrees of freedom; cf. Table 7.

Table 5. Algorithms A and C - Constant H

Subdomains	H/h	Dof	**Algorithm A** Iter	λ_{min}	λ_{max}	**Algorithm C** Iter	λ_{min}	λ_{max}
216	4	6,859	14	1.00018	4.20279	6	1.00001	1.28960
216	8	79,507	22	1.00147	16.7662	8	1.00029	1.75693
216	12	300,763	27	1.00306	34.0512	10	1.00010	2.08459
216	16	753,571	31	1.00371	53.9590	11	1.00017	2.34317
216	20	1,520,875	32	1.00519	75.7574	11	1.00024	2.55999
216	24	2,685,619	34	1.00651	99.0372	12	1.00029	2.74869
216	28	4,330,747	36	1.00660	123.530	12	1.00035	2.91716
216	32	6,539,203	36	1.00677	149.054	13	1.00034	3.07033

Table 6. Algorithms B and E - Constant H

Subdomains	H/h	Dof	**Algorithm B** Iter	λ_{min}	λ_{max}	**Algorithm E** Iter	λ_{min}	λ_{max}
216	4	6,859	5	1.01252	1.06768	13	1.00006	4.19816
216	8	79,507	7	1.00052	1.31862	19	1.00044	12.1453
216	12	300,763	8	1.00021	1.62065	22	1.00058	20.3391
216	16	753,571	10	1.00021	1.90164	23	1.00054	28.5889
216	20	1,520,875	10	1.00033	2.14742	23	1.00066	36.8711
216	24	2,685,619	11	1.00040	2.36688	25	1.00062	45.1044
216	28	4,330,747	12	1.00040	2.61352	24	1.00081	53.3703
216	32	6,539,203	12	1.00046	2.80160	24	1.00097	61.5779

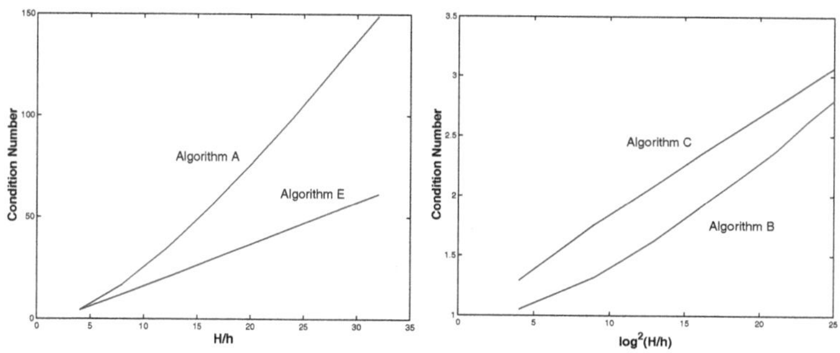

Fig. 2. Condition number growth for varying H/h for Algorithms A and E (left) and Algorithms B and C (right)

Table 7. Parallel Scalability - Algorithms A, B, C and E with 216 subdomains, 13824 dof for each subdomain (2,685,619 dof)

	Algorithm			
Processors	**A**	**B**	**C**	**E**
27	223s	207s	205s	216s
54	113s	106s	106s	110s
108	57.0s	54.2s	53.8s	55.4s
216	29.1s	28.9s	27.2s	29.1s

The experiments in Tables 1,2,3,4 were carried out on two dual Athlon MP 2200+ PCs with 2 GByte memory each. The experiments in Tables 5,6 and 7 were computed on the 350 node Linux cluster Jazz at the Argonne National Laboratory. Each node is a 2.4 GHz Pentium Xeon where half of the nodes has 2 GByte memory and the other half has 1 GByte.

The experiments show that all algorithms have a good parallel scalability for our model problem. For this problem and the number of degrees of freedom considered, the CPU times are not significantly different, although Algorithm C is always slightly faster. To decide which method is the best, more extensive testing with different model problems and geometries is needed. This is currently ongoing research and will be published elsewhere.

Acknowledgement. The authors gratefully acknowledge the use of "Jazz", a 350 node computing cluster operated by the Mathematics and Computer Science Division at Argonne National Laboratory as part of its Laboratory Computing Resource Center.

References

S. Balay, K. Buschelman, W. D. Gropp, D. Kaushik, M. Knepley, L. C. McInnes, B. F. Smith, and H. Zhang. PETSc home page. URL `http://www.mcs.anl.gov/petsc`. 2001.

C. Farhat, M. Lesoinne, P. LeTallec, K. Pierson, and D. Rixen. FETI-DP: A dual-primal unified FETI method - part i: A faster alternative to the two-level FETI method. *Int. J. Numer. Meth. Engrg.*, 50:1523–1544, 2001.

C. Farhat, M. Lesoinne, and K. Pierson. A scalable dual-primal domain decomposition method. *Numer. Lin. Alg. Appl.*, 7:687–714, 2000.

A. Klawonn, O. B. Widlund, and M. Dryja. Dual-Primal FETI methods for three-dimensional elliptic problems with heterogeneous coefficients. *SIAM J.Numer.Anal.*, 40, 159-179 2002a.

A. Klawonn, O. B. Widlund, and M. Dryja. Dual-Primal FETI methods with face constraints. In L. F. Pavarino and A. Toselli, editors, *Recent developments in domain decomposition methods*, pages 27–40. Springer-Verlag, Lecture Notes in Computational Science and Engineering, Volume 23, 2002b.

The FETI Based Domain Decomposition Method for Solving 3D-Multibody Contact Problems with Coulomb Friction*

Radek Kučera[1], Jaroslav Haslinger[2], and Zdeněk Dostál[3]

[1] VŠB-TU Ostrava, Department of Mathematics, radek.kucera@vsb.cz
[2] Charles University, Department of Metal Physics, haslin@met.mff.cuni.cz
[3] VŠB-Technical University Ostrava, Department of Applied Mathematics, zdenek.dostal@vsb.cz

Summary. The contribution deals with the numerical solving of contact problems with Coulomb friction for 3D bodies. A variant of the FETI based domain decomposition method is used. Numerical experiments illustrate the efficiency of our algorithm.

1 Introduction

The FETI method was proposed by Farhat and Roux [1992] for parallel solution of problems described by elliptic partial differential equations. The key idea is elimination of the primal variables so that the original problem is reduced to a small, relatively well conditioned quadratic programming problem in terms of the Lagrange multipliers. Then the iterative solver is used to compute the solution.

Our recent papers (see Dostál et al. [2002] or Haslinger et al. [2002]) apply the FETI procedure to the contact problems with Coulomb friction in 2D. It leads to the sequence of quadratic programming problems with simple inequality bounds so that the fast algorithm based on an active set strategy and an adaptive precision control (see Dostál and Schöberl [2003]) can be used directly. The situation is not so easy in 3D. The reason is that the tangential contact stress has two components in each contact node which are subject to quadratic inequality constraints. Fortunately the structure of this constraints is relatively simple: the vector whose components are the tangential contact stresses belongs to a circle in $\mathbb{R}^2$ with the center at the origin and a given radius. A convenient piecewise linear approximation of the circle can be defined by the intersection of squares rotated of a constant angle α. Doing this approximation at all the contact nodes, we obtain a new quadratic

* Supported by grant GAČR 101/01/0538, MSM ME641 and MSM 272400019.

programming problem with bound and equality constraints that can be efficiently solved by the algorithm based on the augmented Lagrangian (see Dostál et al. [2003]). The implementation details for the problem with one body on the rigid foundation can be found in Haslinger et al. [2003]. Here, we shall extend our method to 3D-multibody problems and show how to reduce the size of the solved quadratic programming problem by means of the mixed finite element method.

2 Formulation of the problems

Let us consider a system of elastic bodies that occupy in the reference configuration bounded domains $\Omega^p \subset \mathbb{R}^3$, $p = 1, 2, \ldots, s$, with sufficiently smooth boundaries Γ^p that are split into three disjoint parts Γ_u^p, Γ_t^p and Γ_c^p so that $\Gamma^p = \overline{\Gamma_u^p} \cup \overline{\Gamma_t^p} \cup \overline{\Gamma_c^p}$. Let us suppose that the zero displacements are prescribed on Γ_u^p and that the surface tractions of density $\mathbf{t}^p \in (L^2(\Gamma_t^p))^3$ act on Γ_t^p. Along Γ_c^p the body Ω^p may get into unilateral contact with some other of the bodies. Finally we suppose that the bodies Ω^p are subject to the volume forces of density $\mathbf{f}^p \in (L^2(\Omega^p))^3$.

To describe non-penetration of the bodies, we shall use linearized non-penetration condition that is defined by a mapping $\chi : \Gamma_c \longrightarrow \Gamma_c$, $\Gamma_c = \bigcup_{p=1}^s \Gamma_c^p$, which assigns to each $\mathbf{x} \in \Gamma_c^p$ some nearby point $\chi(\mathbf{x}) \in \Gamma_c^q$, $p \neq q$. Let $\mathbf{v}^p(\mathbf{x}), \mathbf{v}^q(\chi(\mathbf{x}))$ denote the displacement vectors at $\mathbf{x}, \chi(\mathbf{x})$, respectively. Assuming the small displacements, the *non-penetration condition* reads

$$v_n^p(\mathbf{x}) \equiv (\mathbf{v}^p(\mathbf{x}) - \mathbf{v}^q(\chi(\mathbf{x}))) \cdot \mathbf{n}^p(\mathbf{x}) \leq \delta^p(\mathbf{x}),$$

where $\delta^p(\mathbf{x}) = (\chi(\mathbf{x}) - \mathbf{x}) \cdot \mathbf{n}^p(\mathbf{x})$ is the initial gap and $\mathbf{n}^p(\mathbf{x})$ is the critical direction defined by $\mathbf{n}^p(\mathbf{x}) = (\chi(\mathbf{x}) - \mathbf{x})/\|\chi(\mathbf{x}) - \mathbf{x}\|$ or, if $\chi(\mathbf{x}) = \mathbf{x}$, by the outer unit normal vector to Γ_c^p.

We start with the weak formulation of an auxiliary problem, called the contact problem with *given* friction. To this end we introduce the space of *virtual displacements*

$$V = \{\mathbf{v} = (\mathbf{v}^1, \ldots, \mathbf{v}^s) \in \prod_{p=1}^{s} (H^1(\Omega^p))^3 : \mathbf{v}^p = 0 \text{ on } \Gamma_u^p\}$$

and its closed convex subset of *kinematically admissible* displacements

$$\mathcal{K} = \{\mathbf{v} \in V : v_n^p(\mathbf{x}) \leq \delta^p(\mathbf{x}) \text{ for } \mathbf{x} \in \Gamma_c^p\},$$

where $\mathbf{n}^p \in (L^\infty(\Gamma_c^p))^3$ and $\delta^p \in L^\infty(\Gamma_c^p)$. Let us assume that the normal contact stress $T_\nu \in L^\infty(\Gamma_c)$, $T_\nu \leq 0$, is known *apriori* so that one can evaluate the slip bound g on Γ_c by $g = \mathcal{F}(-T_\nu)$, where $\mathcal{F} = \mathcal{F}^p > 0$ is a coefficient of friction on Γ_c^p. Denote $g^p = g|_{\Gamma_c^p}$.

The *primal* formulation of the contact problem with given friction reads

$(\mathcal{P})$ $$\min \mathcal{J}(\mathbf{v}) \quad \text{subject to} \quad \mathbf{v} \in \mathcal{K},$$

where

$$\mathcal{J}(\mathbf{v}) = \tfrac{1}{2}a(\mathbf{v},\mathbf{v}) - b(\mathbf{v}) + j(\mathbf{v})$$

is the total potential energy functional with the bilinear form a representing the inner energy of the bodies and with the linear form b representing the work of the applied forces $\mathbf{t}^p$ and $\mathbf{f}^p$, respectively. The sublinear functional j represents the work of friction forces

$$j(\mathbf{v}) = \sum_{p=1}^{s} \int_{\Gamma_c^p} g^p \|\mathbf{v}_t^p\| \, ds,$$

where $\mathbf{v}_t^p$ is the projection of the displacement $\mathbf{v}^p$ on the plane tangential to the unit outer normal vector to Γ_c^p denoted by $\boldsymbol{\nu}^p \in (L^\infty(\Gamma_c^p))^3$. Let us introduce unit tangential vectors $\mathbf{t}_1^p, \mathbf{t}_2^p \in (L^\infty(\Gamma_c^p))^3$ such that the triplet $\mathcal{B} = \{\boldsymbol{\nu}^p(\mathbf{x}), \mathbf{t}_1^p(\mathbf{x}), \mathbf{t}_2^p(\mathbf{x})\}$ is an orthonormal basis in $\mathbb{R}^3$ for almost all $\mathbf{x} \in \Gamma_c^p$ and denote $v_{t_1}^p = \mathbf{v}^p \cdot \mathbf{t}_1^p$, $v_{t_2}^p = \mathbf{v}^p \cdot \mathbf{t}_2^p$. Then $\mathbf{v}_t^p = (0, v_{t_1}^p, v_{t_2}^p)$ with respect to the basis $\mathcal{B}$ so that the norm appearing in j reduces to the Euclidean norm in $\mathbb{R}^2$. More details about the formulation of contact problems can be found in Hlaváček et al. [1988].

The Lagrangian $\mathcal{L} : V \times \Lambda_t \times \Lambda_n \longrightarrow \mathbb{R}$ of the problem $(\mathcal{P})$ is defined by

$$\mathcal{L}(\mathbf{v}, \boldsymbol{\mu}_t, \boldsymbol{\mu}_n) = \frac{1}{2}a(\mathbf{v},\mathbf{v}) - b(\mathbf{v}) + \sum_{p=1}^{s} \int_{\Gamma_c^p} \boldsymbol{\mu}_t^p \cdot \mathbf{v}_t^p \, ds + \sum_{p=1}^{s} \langle \mu_n^p, v_n^p - \delta^p \rangle_{\Gamma_c^p},$$

where

$$\Lambda_t = \{\boldsymbol{\mu}_t = (\boldsymbol{\mu}_t^1, \ldots, \boldsymbol{\mu}_t^s) \in \prod_{p=1}^{s} (L^\infty(\Gamma_c^p))^2 : \|\boldsymbol{\mu}_t^p\| \le g^p, \boldsymbol{\mu}_t^p = (\mu_{t_1}^p, \mu_{t_2}^p)\},$$

$$\Lambda_n = \{\boldsymbol{\mu}_n = (\mu_n^1, \ldots, \mu_n^s) \in \prod_{p=1}^{s} H^{-1/2}(\Gamma_c^p) : \mu_n^p \ge 0\}$$

and $\langle \cdot, \cdot \rangle_{\Gamma_c^p}$ denotes the duality pairing between $H^{-1/2}(\Gamma_c^p)$ and $H^{1/2}(\Gamma_c^p)$.

The *Lagrange multipliers* $\boldsymbol{\mu}_t$, $\boldsymbol{\mu}_n$ are considered as functionals on the contact parts of the boundaries. While the first one accounts for the non-penetration condition, the second one removes the non-differentiability of the sublinear functional as

$$j(\mathbf{v}) = \sup_{\boldsymbol{\mu}_t \in \Lambda_t} \sum_{p=1}^{s} \int_{\Gamma_c^p} \boldsymbol{\mu}_t^p \cdot \mathbf{v}_t^p \, ds, \quad \mathbf{v} \in V.$$

Thus the problem $(\mathcal{P})$ can be replaced by the saddle-point problem as

$$\min_{\mathbf{v} \in \mathcal{K}} \mathcal{J}(\mathbf{v}) = \min_{\mathbf{v} \in V} \sup_{(\boldsymbol{\mu}_t, \boldsymbol{\mu}_n) \in \Lambda_t \times \Lambda_n} \mathcal{L}(\mathbf{v}, \boldsymbol{\mu}_t, \boldsymbol{\mu}_n).$$

By the *mixed* formulation of the problem $(\mathcal{P})$, we mean a problem of finding a saddle-point of the Lagrangian $\mathcal{L}$:

$$(\mathcal{M}) \begin{cases} \text{Find } (\mathbf{u}, \boldsymbol{\lambda}_t, \boldsymbol{\lambda}_n) \in V \times \Lambda_t \times \Lambda_n \text{ such that} \\ a(\mathbf{u}, \mathbf{v}) = b(\mathbf{v}) - \displaystyle\sum_{p=1}^{s} \int_{\Gamma_c^p} \boldsymbol{\lambda}_t^p \cdot \mathbf{v}_t^p \, ds - \sum_{p=1}^{s} \langle \lambda_n^p, v_n^p \rangle_{\Gamma_c^p}, \quad \forall \mathbf{v} \in V \\ \displaystyle\sum_{p=1}^{s} \int_{\Gamma_c^p} (\boldsymbol{\lambda}_t^p - \boldsymbol{\mu}_t^p) \cdot \mathbf{u}_t^p \, ds + \sum_{p=1}^{s} \langle \lambda_n^p - \mu_n^p, u_n^p - \delta^p \rangle_{\Gamma_c^p} \le 0, \\ \hfill \forall (\boldsymbol{\mu}_t, \boldsymbol{\mu}_n) \in \Lambda_t \times \Lambda_n. \end{cases}$$

It is well-known that there is a unique saddle-point $(\mathbf{u}, \boldsymbol{\lambda}_t, \boldsymbol{\lambda}_n)$ and its first component $\mathbf{u}$ solves the problem $(\mathcal{P})$.

Let us point out that the solution $\mathbf{u} \equiv \mathbf{u}(g)$ of $(\mathcal{P})$ depends on a particular choice of $g \in L^\infty(\Gamma_c)$, $g \ge 0$. We can define a mapping Φ which associates with every g the product $F(-T_\nu(\mathbf{u}(g)))$, where $T_\nu(\mathbf{u}(g)) \le 0$ is the normal contact stress related to $\mathbf{u}(g)$. The classical Coulomb's law of friction corresponds to the fixed point of Φ which is defined by $g = F(-T_\nu(\mathbf{u}(g)))$. To find it, we can use the *method of successive approximations* which starts from a given $g^{(0)}$ and generates the iterations $g^{(l)}$ by

$$\text{(MSA)} \qquad g^{(l+1)} = \Phi(g^{(l)}), \; l = 1, 2, \ldots.$$

This iterative process converges provided Φ is *contractive*, that is guaranteed for sufficiently small F (see Haslinger [1983]).

3 Discretizations

We shall discretize the contact problem with given friction by means of one of the following two approximations.

Approximation I is based on the finite element method applied to the primal formulation $(\mathcal{P})$. We divide the bodies Ω^p into tetrahedron finite elements $\mathcal{T}$ with the maximum diameter h and assume that the partitions are regular and consistent with the decompositions of $\partial\Omega^p$ into Γ_u^p, Γ_t^p and Γ_c^p. Moreover, we restrict ourselves to the geometrical conforming situation where the intersection between the boundaries of any two different bodies $\partial\Omega^p \cap \partial\Omega^q$, $p \ne q$, is either empty, a vertex, an entire edge, or an entire face. On the partitions, we introduce the finite element subspace of V by

$$V_h = V_h^1 \times \cdots \times V_h^s$$

with

$$V_h^p = \{\mathbf{v}^p \in (C(\Omega^p))^3 : \mathbf{v}^p|_{\mathcal{T}} \in (P_1(\mathcal{T}))^3 \text{ for all } \mathcal{T} \subset \Omega^p\},$$

where $P_m(\mathcal{T})$ denotes the set of all polynomials on $\mathcal{T}$ of degree $\leq m$. Replacing V by V_h, we can rewrite the approximative primal formulation $(\mathcal{P})$ into an algebraic form. Then we can proceed to the dual formulation introducing the algebraic mixed formulation (analogously to the continuous setting) and eliminating the primal variables (displacements).

Approximation II is based on the mixed finite element method applied to the mixed formulation $(\mathcal{M})$. The space V is approximated by the same V_h as in *Approximation I*. In addition, to approximate the sets Λ_t and Λ_n, we introduce regular partitions of Γ_c^p formed by rectangles $\mathcal{R}$ with the maximum diameter H. Let us point out that this partitions are independent on the partitions of Ω^p used for the approximation of V. Let us define

$$\Lambda_H^p = \{\lambda^p \in L^2(\Gamma_c^p) : \lambda^p|_{\mathcal{R}} \in P_0(\mathcal{R}) \text{ for all } \mathcal{R} \subset \Gamma_c^p\}.$$

Repleacing $L^\infty(\Gamma_c^p)$ and $H^{-1/2}(\Gamma_c^p)$ by Λ_H^p in the definitions of Λ_t and Λ_n, we obtain their approximations $\Lambda_{t,H}$ and $\Lambda_{n,H}$, respectively. The approximative mixed formulation $(\mathcal{M})$ can be reduced again to the dual formulation eliminating the primal variables. If the partitions of Γ_c^p are defined by restrictions of the partitions of the bodies Ω^p then we obtain a variant of the so called mortar method, see Krause and Wohlmuth [2002].

The dual formulations arising from both *Approximation I* and *Approximation II* are represented by the quadratic programming problems of the same type:

$$(\mathcal{D}) \qquad \min \Theta(\boldsymbol{\lambda}) \quad \text{s.t.} \quad \boldsymbol{\lambda} \in \boldsymbol{\Lambda} \quad \text{and} \quad \mathbf{R}^\top(\mathbf{f} - \mathbf{B}^\top\boldsymbol{\lambda}) = \mathbf{0},$$

with

$$\Theta(\boldsymbol{\lambda}) = \frac{1}{2}\boldsymbol{\lambda}^\top \mathbf{B}\mathbf{K}^\dagger\mathbf{B}^\top\boldsymbol{\lambda} - \boldsymbol{\lambda}^\top(\mathbf{B}\mathbf{K}^\dagger\mathbf{f} - \mathbf{c}),$$

$$\boldsymbol{\Lambda} = \{\boldsymbol{\lambda} = (\boldsymbol{\lambda}_{t_1}^\top, \boldsymbol{\lambda}_{t_2}^\top, \boldsymbol{\lambda}_n^\top)^\top : \ \|((\boldsymbol{\lambda}_{t_1})_k, (\boldsymbol{\lambda}_{t_2})_k)\| \leq g_k, \ \boldsymbol{\lambda}_n \geq \mathbf{0}\},$$

$$\mathbf{B} = \begin{bmatrix} \mathbf{N} \\ \mathbf{T}_1 \\ \mathbf{T}_2 \end{bmatrix}, \quad \mathbf{c} = \begin{bmatrix} \mathbf{d} \\ \mathbf{0} \\ \mathbf{0} \end{bmatrix}.$$

Here, $\mathbf{K}^\dagger$ denotes a generalized inverse to the symmetric positive semidefinite stiffness matrix $\mathbf{K} = \operatorname{diag}(\mathbf{K}_1, \ldots, \mathbf{K}_s)$, $\mathbf{R}$ is the full rank matrix whose columns span the kernel of $\mathbf{K}$, the full rank matrices $\mathbf{N}, \mathbf{T}_1, \mathbf{T}_2$ describe projections of displacements at the nodes lying on Γ_c to the normal and tangential directions, respectively, $\mathbf{f}$ represents the nodal forces, $\mathbf{d}$ is the vector of distances between the bodies and g_k are the values of the slip bound at the contact nodes. The difference between *Approximation I* and *Approximation II* consists in the different contents of $\mathbf{B}$ and $\mathbf{c}$.

The minimized functional in $(\mathcal{D})$ is strictly convex and quadratic but the feasible set contains the non-linear constraints $\|((\boldsymbol{\lambda}_{t_1})_k, (\boldsymbol{\lambda}_{t_2})_k)\| \leq g_k$. This constraints can be treated by the method described in Haslinger et al. [2003] so that the efficient algorithm based on the augmented Lagrangian (see Dostál et al. [2003]) can be applied.

The advantage of *Approximation II* is that the number of the dual variables is lower compared with *Approximation I*, i.e. the size of the problem $(\mathcal{D})$ is considerably reduced. This happens if $H/h > 1$, i.e. if the partitions of Γ_c^p used for the definitions of Λ_H^p are coarser than the partitions of the bodies Ω^p restricted to Γ_c^p. The coarser partitions are related to the satisfaction of the *Ladyzhenskaya–Babuška–Brezzi* condition that guarantees the existence and the uniquenees of the solution. For our particular choice of the spaces, this condition is satisfied if the ratio H/h is sufficiently large (see Haslinger and Hlaváček [1982]). On the other hand, the ratio H/h should not be too large in order to avoid violation of the non-penetration condition that is satisfied in the weak sense only.

The method of successive approximation (MSA) can be implemented so that the problem $(\mathcal{D})$ is solved to evaluate the mapping Φ. We shall use a more efficient version of this method, in which two outer loops (i.e. the iterative steps of (MSA) and the outer loop of the algorithm for solving $(\mathcal{D})$) can be connected in one loop. The resulting algorithm can be viewed as the method of successive approximation with an *inexact* solving of the auxiliary problems with given friction.

4 Numerical experiments and conclusions

Let us consider two bricks Ω^1 and Ω^2 as in Figure 1 made of an elastic isotropic, homogeneous material characterized by Young modulus $E = 21.2 \times 10^{10}$ and Poisson's ratio $\sigma = 0.277$ (steel). The brick Ω^1 is unilaterally supported by the rigid foundation Ω^0. The applied surface tractions are in Figure 1, the volume forces vanish. Both contact interfaces $\Gamma_c^1 = \Omega^0 \cap \Omega^1$ and $\Gamma_c^2 = \Omega^1 \cap \Omega^2$ are partitioned by two meshes as in Figure 2. The mesh defined by restriction of the partitions of the bodies Ω^1 and Ω^2 is triangular (dotted) while the mesh used for approximation of the Lagrange multipliers is rectangular (solid). Let us point that the meshes on the interfaces do match for the sake of simple implementation of the model problem. Our method can be applied directly to the problems with nonmathing meshes.

Table 1 compares behaviour of our algorithm for *Approximation I* and *Approxiamtion II* with $H/h = 2, 4$. All computations are carried out with 12150 primal variables while the number of the dual variables n_d is different. From the results, we conclude that the performance of the algorithm is not too sensitive to the value of the coefficient of fricrion and the efficiency of our algorithm is comparable to solving of the linear problems. *Approximation II* reduces the size of the dual problem n_d with relatively minor effect on the

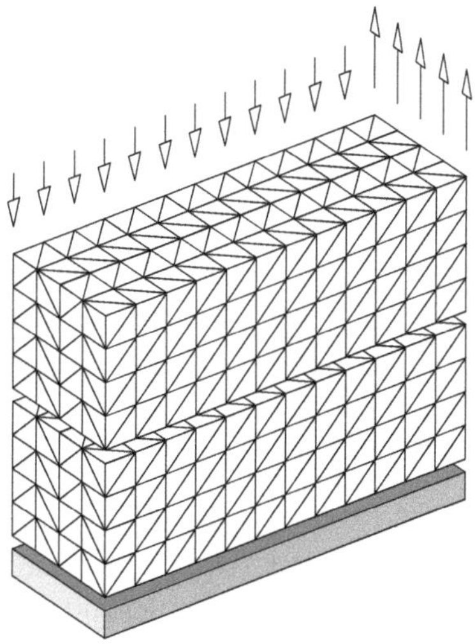

Fig. 1. The model problem with two loaded bodies

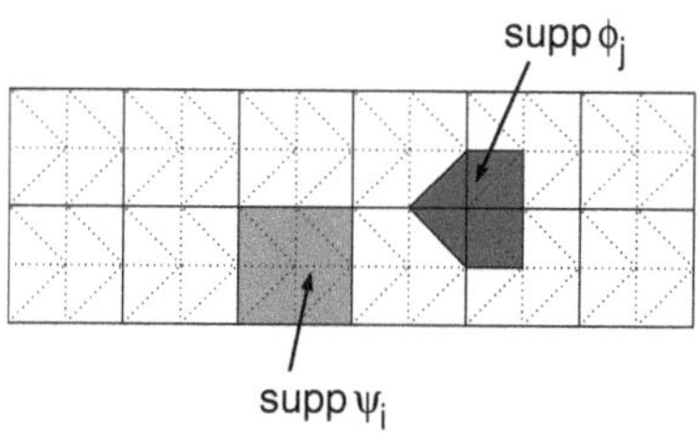

Fig. 2. The meshes on Γ_c^2 for $H/h = 2$ and the supports of FEM basis functions

Table 1. *Iter* denotes the number of outer iterations; *Cg* is the total number of the conjugate gradient steps; *err* is the relative error

	Approx. I		*Approx. II*					
	$n_d = 2592$		$H/h = 2,\ n_d = 576$			$H/h = 4,\ n_d = 144$		
F	*Iter*	*Cg*	*Iter*	*Cg*	*err*	*Iter*	*Cg*	*err*
0.001	24	373	32	299	0.0100	23	147	0.0178
0.01	22	332	26	291	0.0102	16	114	0.0175
0.1	19	331	20	315	0.0120	19	137	0.0147
1	21	931	21	711	0.0078	24	242	0.0104
10	16	229	26	213	0.0206	16	117	0.0413

solutions compared by $err = \|\mathbf{u}_{II} - \mathbf{u}_I\|/\|\mathbf{u}_I\|$, where $\mathbf{u}_I$ and $\mathbf{u}_{II}$ are results of *Approximation I* and *Approximation II*, respectively.

Using auxiliary decomposition, results on natural coarse space projections (see Mandel and Tezaur [1996], Klawonn and Widlund [2001]) and quadratic programming (see Dostál and Schöberl [2003], Dostál [2003]), it is possible to show that our algorithm for the problem with given friction is scalable.

References

Z. Dostál. Inexact semi-monotonic augmented Lagrangians with optimal feasibility convergence for quadratic programming with simple bounds and equality constraints. *SIAM J. Num. Anal.*, 2003. submitted.

Z. Dostál, A. Friedlander, and S. A. Santos. Augmented Lagrangians with adaptive precision control for quadratic programming with simple bounds and equality constraints. *SIAM J. Opt.*, 13:1120–1140, 2003.

Z. Dostál, J. Haslinger, and R. Kučera. Implementation of fixed point method for duality based solution of contact problems with friction. *J. Comput. Appl. Math.*, 140:245–256, 2002.

Z. Dostál and J. Schöberl. Minimizing quadratic functions over non-negative cone with the rate of convergence and finite termination. *Comput. Opt. Appl.*, 2003. submitted.

C. Farhat and F. X. Roux. An unconventional domain decomposition method for an efficient parallel solution of large-scale finite element systems. *SIAM J. Sc. Stat. Comput.*, 13:379–396, 1992.

J. Haslinger. Approximation of the Signorini problem with friction, obeying Coulomb law. *Math. Methods Appl. Sci*, 5:422–437, 1983.

J. Haslinger, Z. Dostál, and R. Kučera. On splitting type algorithm for the numerical realization of contact problems with Coulomb friction. *Comput. Methods Appl. Mech. Eng.*, 191:2261–2281, 2002.

J. Haslinger and I. Hlaváček. Approximation of the Signorini problem with friction by a mixed finite element method. *J. Math. Anal. Appl.*, 86:99–122, 1982.

J. Haslinger, R. Kučera, and Z. Dostál. An algorithm for the numerical realization of 3D contact problems with Coulomb friction. *J. Comput. Appl. Math.*, 2003. accepted.

I. Hlaváček, J. Haslinger, J. Nečas, and J. Lovíšek. *Solution of Variational Inequalities in Mechanics.* Springer, Berlin, 1988.

A. Klawonn and O. B. Widlund. FETI and Neumann-Neumann iterative substructuring methods: connections and new results. *Communic. Pure Appl. Math.*, 54:57–90, 2001.

R. Krause and B. Wohlmuth. A Dirichlet–Neumann type algorithm for contact problems with friction. *CVS*, 5:139–148, 2002.

J. Mandel and R. Tezaur. Convergence of substructuring method with Lagrange multipliers. *Numerische Mathematik*, 73:473–487, 1996.

Choosing Nonmortars: Does it Influence the Performance of FETI-DP Algorithms?

Dan Stefanica

City University of New York, Baruch College, Department of Mathematics

Summary. We investigate whether different choices of nonmortar sides for the geometrically conforming partitions inherent to FETI–DP influence the convergence of the algorithms for four different preconditioners. We conclude experimentally that they do not, although better condition number estimates exist for a Neumann-Dirichlet choice of nonmortars.

1 Introduction

The dual–primal FETI (FETI–DP) method is an iterative substructuring method using Lagrange multipliers. It was introduced in Farhat et al. [1999] for two dimensional problems as a FETI–type algorithm which does not require solving singular problems on individual subdomains, and was extended to three dimensional problems in Farhat et al. [2000] and Klawonn et al. [2002]. The scalability and optimal convergence properties of FETI–DP were established in Mandel and Tezaur [2001] and Klawonn et al. [2002].

Mortar finite elements were first introduced by Bernardi et al. [1994] and are actively used in practice for their advantages over the conforming finite elements, e.g., flexible mesh generation and straightforward local refinement. Extending FETI and FETI–DP algorithms to mortar discretizations is a natural idea and such work can be traced back to Lacour and Maday [1997]. Most of this work was computational, investigating the convergence properties of various FETI preconditioners for mortar algorithms; see, e.g., Stefanica [2001, 2002]. Condition number estimates were established for two FETI–DP preconditioners for mortar methods by Dryja and Widlund [2002a,b]. Recently, similar bounds were obtained by Dryja and Proskurowski [2004] for problems with discontinuous coefficients.

The FETI–DP algorithms discussed here are based on geometrically conforming partitions of the computational domain. Across every subdomain side there is exactly one edge belonging to a different subdomain. For mortar methods, either one of these sides is chosen to be a nonmortar, with the other one being a mortar side. In Dryja and Widlund [2002a], the condition number estimate for a FETI–DP preconditioner depends on whether the choice of nonmortars is random or is made according to a Neumann-Dirichlet ordering of the subdomains, i.e., with the sides of any subdomain being either all mortars or all nonmortars.

In this paper, we investigate experimentally this result and conclude that, in practice, the special choice of nonmortars does not influence the numerical convergence of the FETI–DP algorithm. We also compare the numerical performance of three other possible preconditioners, a generalized one suggested in Dryja and Widlund [2002b] and two other similar to the Dirichlet and generalized preconditioners for FETI mortar algorithms, and conclude that the generalized preconditioners have the best convergence properties.

The notations in this paper are related to those of Dryja and Widlund [2002b] and of Farhat et al. [1999]. For details omitted here due to space constraints, we refer the reader to the same two papers.

2 Mortar Finite Element Spaces of First Type

To construct the mortar finite element space W, the computational domain Ω is partitioned into nonoverlapping rectangular subdomains Ω_i, $i = 1 : N$. For the FETI–DP algorithms considered here, the partition must be geometrically conforming, i.e, the intersection between the closures of any two subdomains is either empty, or consists of a vertex or an entire edge, and the mortars must be of the first type, i.e., continuity is required at the corner nodes.

Across the interface Γ, i.e., the set of points that belong to the boundaries of at least two subregions, we do not require pointwise continuity. Since the partition is geometrically conforming, the edges of the subdomains are pairwise opposite. From each pair, one edge, denoted by $\delta_{m(i)}$ and assumed to belong to the subdomain Ω_i, is chosen to be a nonmortar side, while the other edge, denoted by $\gamma_{m(j)}$ and belonging to Ω_j, is a mortar side.

The restriction of a mortar function $v \in W$ to any subdomain is a P_1 or a Q_1 finite element function. We assume that each subdomain Ω_i has a diameter of order H and that its triangulation has a mesh size of order h. Let v_i and v_j be the restrictions of v to an arbitrary nonmortar side $\delta_{m(i)}$ and to its opposite mortar side $\gamma_{m(j)}$, respectively. Then v_i and v_j have the same values at the left and right end points of $\delta_{m(i)}$ and $\gamma_{m(j)}$, respectively, and the following mortar conditions have to be satisfied:

$$\int_{\delta_{m(i)}} (v_i - v_j)\ \psi\ ds\ =\ 0, \quad \forall\, \psi \in M(\delta_{m(i)}), \tag{1}$$

where $M(\delta_{m(i)})$ is a space of test functions having the same dimension as the number of interior nodes of $\delta_{m(i)}$, i.e., piecewise linear functions on $\delta_{m(i)}$ which are constant in the first and last mesh interval.

For algorithmic purposes, we derive a matrix formulation for the mortar conditions (1). Let $v_{i,r}$ be the vector of interior nodal values of v_i on $\delta_{m(i)}$, and let $v_{i,c}$ be the vector of corner nodal values of v_i on $\delta_{m(i)}$. We define $v_{j,r}$ and $v_{j,c}$ similarly for v_j on $\gamma_{m(j)}$. The matrix formulation of (1) is

$$B_{\delta_{m(i)},r} v_{i,r} \; + \; B_{\delta_{m(i)},c} v_{i,c} \; - \; B_{\gamma_{m(j)},r} v_{j,r} \; - \; B_{\gamma_{m(j)},c} v_{j,c} \; = \; 0, \qquad (2)$$

where the matrix $B_{\delta_{m(i)},r}$ is banded for classical mortars and equal to the identity for the biorthogonal mortars of Wohlmuth [2000].

3 FETI–DP Algorithms for Mortars

As model problem we choose the Poisson problem with mixed boundary conditions on Ω. Given $f \in L^2(\Omega)$, find $u \in H^1(\Omega)$ such that

$$-\Delta u = f \text{ on } \Omega, \quad \text{with } u = 0 \text{ on } \partial\Omega_D \text{ and } \partial u/\partial n = 0 \text{ on } \partial\Omega_N, \qquad (3)$$

where $\partial\Omega_N$ and $\partial\Omega_D$ are the parts of $\partial\Omega = \partial\Omega_N \cup \partial\Omega_D$ where Neumann and Dirichlet boundary conditions are imposed, respectively,

To discretize (3), let W_i be the restriction to Ω_i of the mortar finite element space W. The primal variables space is $\widehat{W}$, the subspace of $\Pi_{i=1}^n W_i$ of functions continuous at each corner node. Lagrange multipliers are used to enforce the mortar conditions (1). The dual variables space is $\Pi_m M(\delta_m)$, where the product is considered over all the nonmortar sides. We partition the nodal values of $v \in \widehat{W}$ into the corner nodal values v_c and the remainder nodal values v_r. Note that v_r can be further split into interior nodal values v_{int} and remainder boundary nodal values v_{b_r}. The continuity conditions at the subdomain corners are enforced by using a global vector of degrees of freedom v_c^g and a global to local map L_c with one nonzero entry per line equal to 1, and by requiring that $v_c = L_c v_c^g$. Therefore, $v = [v_{int}; v_{b_r}; v_{b_c}] = [v_r; v_{b_c}] = [v_r; L_c v_c^g]$.

Let K be the stiffness matrix of the discrete problem and let K_{rr}, K_{rc}, and K_{cc} be its blocks corresponding to a decomposition of v into v_r and v_c. We use a Lagrange multiplier matrix B to enforce the mortar conditions (1). The matrix B has one horizontal block, $B_{\delta_{m(i)}}$, for each nonmortar side $\delta_{m(i)}$, built from the columns of $B_{\delta_{m(i)},r}$, $B_{\delta_{m(i)},c}$, $B_{\gamma_{m(j)},r}$, and $B_{\gamma_{m(j)},c}$, with all the other entries zero; cf. (2). We can also write $B = [B_r \; B_c]$ using vertical blocks corresponding to the remainder and corner nodes.

The saddle point formulation of the model problem is

$$\begin{bmatrix} I & 0 & 0 \\ 0 & L_c^T & 0 \\ 0 & 0 & I \end{bmatrix} \begin{bmatrix} K_{rr} & K_{rc} & B_r^T \\ K_{rc}^T & K_{cc} & B_c^T \\ B_r & B_c & 0 \end{bmatrix} \begin{bmatrix} I & 0 & 0 \\ 0 & L_c & 0 \\ 0 & 0 & I \end{bmatrix} \begin{bmatrix} u_r \\ u_c^g \\ \lambda \end{bmatrix} = \begin{bmatrix} f_r \\ L_c^T f_c \\ 0 \end{bmatrix}. \qquad (4)$$

After eliminating the primal variables u_r and u_c^g we obtain the dual problem

$$F_{I_{rr}} \; + \; \widetilde{F}_{I_{rc}} (K_{cc}^*)^{-1} \widetilde{F}_{I_{rc}}^T \lambda \; = \; d, \qquad (5)$$

where $F_{I_{rr}} = B_r K_{rr}^{-1} B_r^T$, $\widetilde{F}_{I_{rc}} = B_r K_{rr}^{-1} K_{rc} - B_c L_c$, and $K_{cc}^* = L_c^T (K_{cc} - K_{cr} K_{rr}^{-1} K_{rc}) L_c$.

As pointed out by Kim and Lee [2002], the FETI–DP algorithms of Dryja and Widlund are not applied on the mortar space W, but on a space very close to W. We call this the DW setting throughout the rest of the paper. Using the notations from section 2, the condition (3.4) from Dryja and Widlund [2002b] (the same as (6) from Dryja and Widlund [2002a]) used to build the block of B corresponding to the nonmortar $\delta_{m(i)}$ can be expressed as

$$B_{\delta_{m(i)},r} v_{i,r} \; + \; B_{\gamma_{m(j)},r} v_{j,r} \; = \; 0,$$

which is different from the mortar condition (2). Thus, the discrete problem solved in Dryja and Widlund [2002b,a] can be written as

$$\begin{bmatrix} I & 0 & 0 \\ 0 & L_c^T & 0 \\ 0 & 0 & I \end{bmatrix} \begin{bmatrix} K_{rr} & K_{rc} & B_r^T \\ K_{rc}^T & K_{cc} & 0 \\ B_r & 0 & 0 \end{bmatrix} \begin{bmatrix} I & 0 & 0 \\ 0 & L_c & 0 \\ 0 & 0 & I \end{bmatrix} \begin{bmatrix} u_r \\ u_c^g \\ \lambda \end{bmatrix} = \begin{bmatrix} f_r \\ L_c^T f_c \\ 0 \end{bmatrix} \tag{6}$$

and corresponds to (4) for $B_c = 0$. The dual problem for the DW setting is

$$B_r \widetilde{S}^{-1} B_r^T \lambda \; = \; \widetilde{d}, \tag{7}$$

where $B_r \widetilde{S}^{-1} B_r^T = F_{I_{rr}} + F_{I_{rc}} (K_{cc}^*)^{-1} F_{I_{rc}}^T$, with $F_{I_{rc}} = B_r K_{rr}^{-1} K_{rc}$ obtained from $\widetilde{F}_{I_{rc}}$ by setting $B_c = 0$. We note that our matrix B_r is the same as the matrix B from Dryja and Widlund [2002b,a].

Let $S_{rr} = K_{b_r b_r} - K_{b_r \text{int}} K_{\text{int,int}}^{-1} K_{b_r \text{int}}^T$ be a Schur complement–type matrix. The Dirichlet preconditioner M_D^{-1} and the generalized preconditioner $\widetilde{M}_{\text{gen}}^{-1}$ were introduced in Dryja and Widlund [2002b] for the DW setting dual problem (7):

$$M_D^{-1} = B_r S_{rr} B_r^T; \tag{8}$$

$$\widetilde{M}_{\text{gen}}^{-1} = \text{diag}(B_r \widetilde{B}_r^T)^{-1} \widetilde{B}_r S_{rr} \widetilde{B}_r^T \, \text{diag}(\widetilde{B}_r B_r^T)^{-1}. \tag{9}$$

The generalized matrix $\widetilde{B}_r$ is obtained by scaling the $B_{\gamma_{m(j)},r}$ in the block corresponding to the nonmortar side $\delta_{m(i)}$ by $h_{\delta_{m(i)}} / h_{\gamma_{m(j)}}$; see (3.13) from Dryja and Widlund [2002b] for more details.

In Dryja and Widlund [2002a], it was shown that, for a random choice of the nonmortar sides, $\kappa(M_D^{-1}) \leq C(1 + \log H/h)^4$, while $\kappa(M_D^{-1}) \leq C(1 + \log H/h)^2$ if the nonmortar sides are chosen according to a Neumann–Dirichlet ordering, i.e., with all the sides of any subdomain being either all mortars or all nonmortars. In Dryja and Widlund [2002b], it was shown that $\kappa(\widetilde{M}_{\text{gen}}^{-1}) \leq C(1 + \log H/h)^2$. All constants denoted by C are independent of H and h.

For the dual problem (5), based on the numerical performance of FETI algorithms for mortars, see Stefanica [2001], we suggest the following Dirichlet and generalized preconditioners:

$$F_D^{-1} = B_r S_{rr} B_r^T; \tag{10}$$

$$F_{\text{gen}}^{-1} = \text{diag}(B_r B_r^T)^{-1} B_r S_{rr} B_r^T \, \text{diag}(B_r B_r^T)^{-1}. \tag{11}$$

4 Numerical Results

We tested the numerical performance of the preconditioners F_{gen}^{-1} (11) and F_D^{-1} (10) for the mortar dual problem (5), and the preconditioners $\widetilde{M}_{\text{gen}}^{-1}$ (9) and M_D^{-1} (8) for the approximate dual problem (7).

Our interests were four-fold:

- check the convergence and scalability properties of the resulting algorithms;
- compare the performance of the algorithms for mortars to that of the algorithms for the DW setting;
- investigate whether a Neumann–Dirichlet choice of nonmortars improves convergence, in particular for the Dirichlet preconditioner M_D^{-1};
- decide which of the four preconditioners performs best.

The model problem was the Poisson equation on the unit square Ω with mixed boundary conditions. We partitioned Ω into $N = 16$, 36, 64, and 121 congruent squares, and Q_1 elements were used in each square. For each partition, the number of nodes on each edge, H/h, was taken to be, on average, 5, 10, 20, and 40, respectively, for different sets of experiments. The meshes did not match for any neighboring subdomains. The preconditioned conjugate gradient iteration was stopped when the residual norm decreased by a factor of 10^{-6}. The experiments were carried out in MATLAB.

We report iteration counts, condition number estimates, and flop counts for two different sets of experiments: for randomly chosen nonmortars, in Table 1, and for a Neumann–Dirichlet choice of nonmortars, in Table 2.

Table 1. Convergence results, randomly chosen nonmortars

		Generalized			Dirichlet			DW Generalized			DW Dirichlet		
N	H/h	Iter	Cond	Mflops	Iter	Cond	Mflops	Iter	Cond	Mflops	Iter	Cond	Mflops
16	5	7	2.6	3.0e-1	24	32	1.0e+0	8	2.3	3.3e-1	26	32	1.1e+0
16	10	9	3.4	1.7e+0	22	41	4.3e+0	9	3.1	1.7e+0	28	40	5.4e+0
16	20	10	4.5	1.2e+1	23	52	2.8e+1	10	4.0	1.2e+1	30	51	3.6e+1
16	40	11	5.7	9.5e+1	25	65	2.2e+2	10	4.9	8.6e+1	32	62	2.8e+1
36	5	8	2.5	9.9e-1	25	31	3.1e+0	9	2.6	1.0e+0	30	33	3.5e+0
36	10	10	3.5	5.1e+0	26	40	1.3e+1	11	3.4	5.4e+0	32	41	1.6e+1
36	20	12	4.5	3.5e+1	29	51	8.6e+1	12	4.4	3.4e+1	35	52	1.0e+2
36	40	13	5.7	2.7e+2	30	63	6.3e+2	13	5.5	2.6e+2	38	64	7.8e+2
64	5	10	2.8	2.8e+0	28	36	7.9e+0	10	2.8	2.5e+0	32	37	8.0e+0
64	10	12	3.7	1.2e+1	29	47	3.0e+1	12	3.7	1.2e+1	37	48	3.6e+1
64	20	13	4.8	7.1e+1	32	60	1.8e+2	13	4.8	6.9e+1	41	61	2.2e+2
64	40	15	6.1	5.6e+2	34	76	1.3e+3	15	6.0	5.5e+2	45	76	1.7e+3
121	5	9	2.7	6.4e+0	29	36	2.1e+1	11	3.4	6.7e+0	37	41	2.3e+1
121	10	12	3.7	2.8e+1	30	45	7.2e+1	13	4.5	2.8e+1	41	52	8.9e+1
121	20	13	4.8	1.5e+2	33	61	3.8e+2	14	5.6	1.5e+2	46	68	5.0e+2
121	40	15	6.2	1.1e+3	36	77	2.6e+3	16	6.9	1.1e+3	51	84	3.7e+3

Table 2. Convergence results, Neumann–Dirichlet choice for nonmortars

		Generalized			Dirichlet			DW Generalized			DW Dirichlet		
N	H/h	Iter	Cond	Mflops	Iter	Cond	Mflops	Iter	Cond	Mflops	Iter	Cond	Mflops
16	5	7	2.6	3.0e-1	20	21	8.7e-1	7	2.3	2.9e-1	22	21	4.6e+0
16	10	8	3.5	1.6e+0	19	26	3.7e+0	8	3.1	1.5e+0	24	25	4.6e+0
16	20	9	4.5	1.1e+1	19	33	2.3e+1	9	4.0	1.1e+1	24	32	2.9e+1
16	40	10	5.7	8.6e+1	19	42	1.7e+2	9	5.0	7.7e+1	26	40	2.3e+1
36	5	8	2.5	9.9e-1	25	33	3.1e+0	9	2.6	1.0e+0	28	33	3.2e+0
36	10	10	3.4	5.1e+0	26	42	1.3e+1	11	3.5	5.4e+0	31	43	1.5e+1
36	20	12	4.5	3.5e+1	29	54	8.6e+1	12	4.4	3.4e+1	34	54	9.8e+1
36	40	13	5.7	2.7e+2	31	67	6.5e+2	13	5.8	2.6e+2	37	68	7.6e+2
64	5	9	2.7	2.5e+0	29	38	8.2e+0	10	2.9	2.5e+0	33	39	8.0e+0
64	10	12	3.7	1.2e+1	29	49	3.0e+1	12	3.8	1.2e+1	36	49	3.5e+1
64	20	13	4.8	7.1e+1	30	63	1.7e+2	13	4.8	6.9e+1	42	63	2.3e+2
64	40	15	6.2	5.6e+2	31	79	1.2e+3	15	6.1	5.5e+2	46	80	1.7e+3
121	5	9	2.7	6.4e+0	30	40	2.2e+1	10	3.2	6.1e+0	35	38	2.2e+1
121	10	12	3.7	2.8e+1	31	52	7.4e+1	12	4.2	2.6e+1	39	48	8.4e+1
121	20	13	4.8	1.4e+2	33	66	3.8e+2	14	5.3	1.5e+2	44	61	4.8e+2
121	40	15	6.2	1.1e+3	35	83	2.5e+3	16	6.6	1.1e+3	49	77	3.6e+3

The convergence patterns reported in Table 1 and Table 2, showed that all preconditioners yielded scalable algorithms. When the number of nodes on each subdomain edge, H/h, was fixed and the number of subdomains, N, was increased, the iteration count showed only a slight growth. When H/h was increased, while the partition was kept unchanged, the small increase in the number of iterations was satisfactory. The condition number estimates exhibited a similar dependence, or lack thereof, on N and H/h. Note that the Dirichlet preconditioner for the DW setting, M_D^{-1}, albeit scalable, required the largest number of iterations, about three times as many as F_{gen}^{-1}, and had condition numbers about one order of magnitude larger than for F_{gen}^{-1}.

The generalized preconditioners F_{gen}^{-1} and $\widetilde{M}_{\text{gen}}^{-1}$ had almost the same iteration counts and flop counts and were cheapest to implement. This was due in part to the fact that, for FETI–DP, the matrices $\text{diag}(B_r B_r^T)$ and $\text{diag}(B_r \widetilde{B}_r^T)$ were block diagonal. The Dirichlet preconditioner for mortars, F_D^{-1}, was noticeably more efficient than its DW counterpart, M_D^{-1}.

By comparing the convergence results from Table 1 to those from Table 2 for each preconditioner, we concluded that the answer to the question from the title of the paper is that choosing nonmortars does not influence the performance of the FETI–DP algorithms. A relatively small improvement in terms of iteration counts was achieved consistently for M_D^{-1}, the preconditioner for which a tighter condition number estimate was proved for the Neumann–Dirichlet choice of nonmortars in Dryja and Widlund [2002a].

The generalized mortar preconditioners F_{gen}^{-1} and $\widetilde{M}_{\text{gen}}^{-1}$ performed very similarly and were clearly better in terms of iteration and flop counts and condition number estimates than the Dirichlet preconditioners F_D^{-1} and M_D^{-1}.

We conclude by investigating the robustness of the FETI–DP algorithms for mortar discretizations in a more complicated setting, e.g., for an elliptic problem with jump coefficients, $-div(\rho(x)\nabla u) = f$. The domain Ω was partitioned into four equal squares and $\rho(x)$ was chosen to be constant in each of these squares. The ratio of the constants in neighboring squares was 1000. The mortar discretizations considered were similar to those used previously.

The results reported in Table 3 confirm the scalability of the FETI–DP algorithms with respect to the number of subdomains and to the number of nodes on each edge, for the generalized and Dirichlet preconditioners, modified as suggested in Klawonn et al. [2002] to account for the jump coefficients, for randomly chosen mortars. As expected, the generalized preconditioner performs better in terms of both condition numbers and computational costs.

Due to space constraints, we do not present the numerical results for the DW–type preconditioners, or for a Neumann–Dirichlet choice of nonmortars, but we note that exactly the same type of convergence behavior as for the Poisson problem was observed for the elliptic problem with jump coefficients.

Table 3. Convergence results, coefficients with jumps

		Generalized			Dirichlet		
N	H/h	Iter	Cond	Mflops	Iter	Cond	Mflops
16	5	11	9.2	4.7e-1	35	54	1.6e+0
16	10	13	10.8	2.6e+0	37	68	7.4e+0
16	20	14	12.1	1.6e+1	38	80	4.8e+1
16	40	15	13.3	1.3e+2	41	92	3.7e+2
36	5	12	9.6	1.5e+0	36	56	4.5e+0
36	10	14	11.3	7.2e+0	39	71	2.0e+1
36	20	15	12.9	4.5e+1	42	78	1.3e+2
36	40	16	13.6	3.3e+2	46	91	1.0e+3
64	5	14	9.9	4.0e+0	40	61	1.2e+1
64	10	15	12.1	1.6e+1	43	74	4.4e+1
64	20	17	13.4	9.4e+1	47	89	2.7e+2
64	40	19	13.9	7.2e+2	50	98	2.0e+3
128	5	14	10.3	1.0e+1	42	63	3.0e+1
128	10	15	12.0	3.6e+1	45	76	1.1e+2
128	20	17	13.5	2.0e+2	48	91	5.6e+2
128	40	19	13.9	1.5e+3	52	100	3.7e+3

Acknowledgement. This work was supported in part by the National Science Foundation under Grant NSF–DMS–0103588 and by the Research Foundation of the City University of New York under Award PSC-CUNY 65463–00 34.

References

C. Bernardi, Y. Maday, and A. Patera. A new non conforming approach to domain decomposition: The mortar element method. In H. Brezis and J.-L. Lions, editors, *Collège de France Seminar*. 1994.

M. Dryja and W. Proskurowski. A FETI–DP method for the mortar discretization of elliptic problems with discontinuous coefficients. In R. K. et al., editor, *Proceedings of the 15th International Conference on Domain Decomposition Methods*, Berlin, 2004. Springer-Verlag.

M. Dryja and O. B. Widlund. A FETI–DP method for a mortar discretization of elliptic problems. In L. Pavarino and A. Toselli, editors, *Proceedings of the ETH Workshop on DDM*. Springer-Verlag, 2002a.

M. Dryja and O. B. Widlund. A generalized FETI–DP method for the mortar discretization of elliptic problems. In I. H. et al., editor, *Fourteenth International Conference on Domain Decomposition Methods*, pages 27–38. ddm.org, 2002b. Mexico, 2002.

C. Farhat, M. Lesoinne, P. Le Tallec, K. Pierson, and D. Rixen. FETI-DP: A dual-primal unified FETI method – part I: A faster alternative to the two-level FETI method. Technical Report U–CAS–99–15, University of Colorado at Boulder, 1999.

C. Farhat, M. Lesoinne, and K. Pierson. A scalable dual-primal domain decomposition method. *Numer. Lin. Alg. Appl.*, 7:687–714, 2000.

H.-H. Kim and C.-O. Lee. A FETI–DP preconditioner for elliptic problems on nonmatching grids. Technical Report 02-9, KAIST, 2002.

A. Klawonn, O. B. Widlund, and M. Dryja. Dual-Primal FETI methods for three-dimensional elliptic problems with heterogeneous coefficients. *SIAM J. Numer. Anal.*, 40:159–179, 2002.

C. Lacour and Y. Maday. Two different approaches for matching nonconforming grids: the mortar element method and the FETI method. *BIT*, 37: 720–738, 1997.

J. Mandel and R. Tezaur. On the convergence of a Dual-Primal substructuring method. *Numer. Math.*, 88:543–558, 2001.

D. Stefanica. A numerical study of FETI algorithms for mortar finite element methods. *SIAM J. Sci. Comp.*, 23(4):1135–1160, 2001.

D. Stefanica. FETI and FETI-DP methods for spectral and mortar spectral elements: A performance comparison. *J. Sci. Comp.*, 17(1-4):629–637, 2002.

B. Wohlmuth. A mortar finite element method using dual spaces for the Lagrange multiplier. *SIAM J. Numer. Anal.*, 38:989–1012, 2000.

Part IX

Minisymposium: Heterogeneous Domain Decomposition with Applications in Multiphysics

Domain Decomposition Methods in Electrothermomechanical Coupling Problems

Ronald H.W. Hoppe[1,2], Yuri Iliash[2], Siegfried Ramminger[3], and Gerhard Wachutka[4]

[1] University of Houston, Department of Mathematics (http://www.math.uh.edu/~rohop/)
[2] University of Augsburg, Institute for Mathematics (http://wwwhoppe.math.uni-augsburg.de)
[3] Siemens AG, Corporate Technology (http://www.mchp.siemens.de)
[4] Munich University of Technology, Physics of Electrotechnology (http://www.tep.e-technik.tu-muenchen.de)

Summary. In this contribution, we are concerned with electrothermomechanical coupling problems as they arise in the modeling and simulation of high power electronic devices. In particular, we are faced with a hierarchy of coupled physical effects in so far as electrical energy is converted to Joule heat causing heat stresses that have an impact on the mechanical behavior of the devices and may lead to mechanical damage. Moreover, there are structural coupling effects due to the sandwich-like construction of the devices featuring multiple layers of specific materials with different thermal and mechanical properties. The latter motivates the application of domain decomposition techniques on nonmatching grids based on individual finite element discretizations of the substructures. We will address in detail the modeling aspects of the hierarchy of coupling phenomena as well as the discretization-related couplings in the numerical simulation of the operating behavior of the devices.

1 Introduction

We consider the application of heterogeneous domain decomposition methodologies in the simulation of electrothermomechanical coupling problems. Such multiphysics coupling problems occur in many applications such as Micro-Electro-Mechanical-Systems (MEMS) and in high power electronics. In the latter case, a characteristic feature of the operational behavior of the devices and systems is that electric energy is converted to Joule heat causing heat stresses which in turn lead to deformations of the underlying mechanical structure and even to damage, if no appropriate cooling is provided.

Basically, the modeling is done in the macroscopic regime by using a continuum mechanical approach. On the other hand, failure mechanisms such as crack initiation and propagation strongly depend on microstructural details which additionally are taken into account by means of an empirical crack model.

As algorithmic tools in the numerical simulation of the appropriately discretized coupled system of PDEs, we use adaptive multilevel methods and domain decompositions on nonmatching grids. As we shall see, the decomposition of the computational domain is in a natural way given by the geometrical structure of the devices featuring subdomains of strongly different aspect ratios and consisting of materials with largely different thermomechanical properties.

The paper is organized as follows: In Section 2, as an example for a device whose operational behavior is based on electrothermomechanical coupling, we consider an Integrated High Voltage Module. Section 3 provides the mathematical modeling of the coupling problem, whereas Section 4 addresses the algorithmic tools used in the numerical simulation. Finally, in Section 5 we present simulation results illustrating the distribution of the temperature and equivalence stresses as well as the initiation of cracks in critical parts of the module.

2 Integrated High Voltage Modules

In high power electronics, Integrated High Voltage (IHV) Modules are used as converters for high power electromotors. They consist of specific semiconductor devices, as for instance, Insulated Gate Bipolar Transistors (IGBTs) and power diodes serving as switches for the electric currents (see the topmost blocks in Figure 1 referred to as Ω_1 in the sequel).

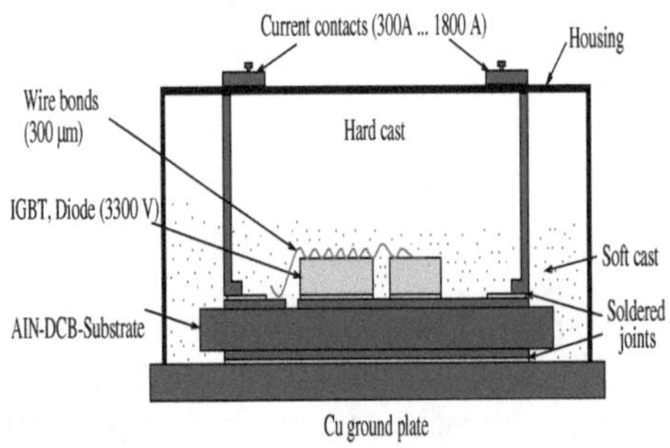

Fig. 1. Schematic representation of an Integrated-High-Voltage Module

Due to high currents up to several kiloamperes, electric energy is converted to Joule heat which leads to a considerable self-heating of the device. In order to facilitate an appropriate distribution of the heat, these blocks are fixed on several layers of different materials (copper and aluminum-nitride) attached to each other by thin soldered joints. The union of these blocks will be denoted by Ω_2. Finally, the copper ground plate is mounted on a cooling device. With regard to failure, the critical parts of the device are the wire bonds connecting the current contacts with the semiconductor devices and the soldered joints.

3 The mathematical model

The operational behavior of the IHV Module involves processes on two different time scales: There is a fast time scale which is the operation of the semiconductor devices having switching times of less than 100 nanoseconds, and there is a slow time scale with regard to the temporal evolution of the temperature in the module which occurs in the range of minutes.
As a model simplification, these two processes are decoupled in the sense that the semiconductor device equations, considered in Ω_1, are treated first to compute the generated Joule heat as an input for the heat equation considered in Ω_2.
We use the classical drift-diffusion model consisting of a potential equation for the electric potential ψ that is coupled with the continuity equations for the carrier concentrations n and p where $\mathbf{J}_n$ and $\mathbf{J}_p$ denote the densities of the electrons and holes, respectively.

$$-\nabla \cdot \varepsilon \nabla \psi + N_{dop}(n,p) + q(n-p) = 0 , \tag{1}$$

$$\frac{\partial n}{\partial t} = + q^{-1} \nabla \mathbf{J}_n + G(\mathbf{J}_n, \mathbf{J}_p, \mathbf{E}) - R(n,p) , \tag{2}$$

$$\frac{\partial p}{\partial t} = - q^{-1} \nabla \mathbf{J}_p + G(\mathbf{J}_n, \mathbf{J}_p, \mathbf{E}) - R(n,p) , \tag{3}$$

$$\mathbf{J}_n = - q \, \mu_n \, n \, \nabla \psi \quad , \quad \mathbf{J}_p = - q \, \mu_p \, p \, \nabla \psi . \tag{4}$$

Here, q stands for the elementary charge, whereas μ_n and μ_p refer to the mobilities of the electrons and holes. Moreover, $\mathbf{E} = -\nabla \psi$ is the electric field whereas N_{dop} , G and R refer to the doping profile, the generation and the recombination. The dominant heat source is Joule heat $H_J = |\mathbf{J}_n|^2/(q\mu_n n) + |\mathbf{J}_p|^2/(q\mu_p p)$, while other sources based on the Seebeck and Nernst effect can be neglected.
The temporal and spatial distribution of the temperature T is described by the heat equation considered in the domain Ω_2 occupied by the aluminum-nitride and copper blocks as well as the joints.

$$\rho \, c \, \frac{\partial T}{\partial t} = \nabla \cdot (\kappa \, \nabla T) \quad \text{in } Q_2 := \Omega_2 \times (t_0, t_1) \quad , \tag{5}$$

$$\mathbf{n} \cdot \kappa \, \nabla T = H_J(t) \quad \text{on } \Gamma_0 \times (t_0, t_1) \quad , \tag{6}$$

$$\mathbf{n} \cdot \kappa \, \nabla T = h \, (T^* - T) \quad \text{on } \Gamma_1 \times (t_0, t_1) \quad , \tag{7}$$

$$\mathbf{n} \cdot \kappa \, \nabla T = 0 \quad \text{on } \Gamma_2 \times (t_0, t_1) \quad , \tag{8}$$

$$T(\cdot, t_0) = T_0(\cdot) \quad \text{in } \Omega_2 \quad . \tag{9}$$

Here, ρ , c, and κ stand for the density, heat capacity, and heat conductivity, respectively. The Joule heat through the part Γ_0 of the upper boundary of the computational domain Ω_2 attached to the semiconductor devices serves as the source term. We further assume a heat exchange at the lower boundary Γ_1 between the copper ground plate and the cooling device where h stands for the heat transition coefficient and T^* denotes the ambient temperature. On the other boundaries of Ω_2 we assume perfect insulation. We note that the thermal properties of the materials such as the heat capacity and the heat conductivity are quite different which means that we experience jumping coefficients across subdomain boundaries. As mentioned before, the self-heating of the devices leads to heat stresses causing mechanical deformations. For all parts of the module except the wire bonds and the joints, we assume the equations of linear elasticity:

$$\operatorname{div} \boldsymbol{\sigma}(\mathbf{u}) = \frac{\alpha\,(1-\nu)\,E}{(1+\nu)\,(1-2\nu)}\,\nabla(T-T_0) \quad \text{in } \Omega_2 \quad , \tag{10}$$

$$\mathbf{u} = 0 \quad \text{on } \Gamma_1 \quad , \quad \mathbf{n}\cdot\boldsymbol{\sigma}(\mathbf{u}) = 0 \quad \text{on } \Gamma_0 \cup \Gamma_2 \quad . \tag{11}$$

Here, $\mathbf{u}$ and $\boldsymbol{\sigma}(\mathbf{u})$ stand for the displacement vector and the stress tensor. Moreover, α denotes the thermal expansion coefficient, and E and ν refer to Young's modulus and Poisson's ratio which are also strongly different for the various materials. The wire bonds and the solders are possibly subject to plastic deformation. Here, we assume stationary plasticity with the von Mises yield criterion where the set K of admissible stresses is given in terms of the Frobenius norm $\|\cdot\|_F$ of the deviatoric stress tensor and the von Mises yield stress σ_Y by $K := \{\sigma \mid \|\mathrm{dev}(\sigma)\|_F \le \sqrt{2/3}\,\sigma_Y\}$. The computational domains are those occupied by the wires and the joints, respectively.
Cracks typically occur in the bonding zone where the wires are attached to the chips and in the soldered joints ([Ramminger, Seliger, and Wachutka, 2000]). There exist empirical crack models that are based on macroscopic data combined with microstructural data due to the nucleation and growth of pores. For instance, the modified Gurson-Model ([Tvergaard, 1989]) consists of a flow rule that reduces to the von Mises yield rule in case of vanishing voids in the microstructure of the material:

$$\frac{\sigma_E^2}{\sigma_Y^2} - 1 + 2\,q\,f\,\cosh((2\sigma_Y)^{-1}\sigma_{ii}) - (q\,f)^2 = 0 \quad . \tag{12}$$

Here, σ_E refers to the von Mises equivalence stress and σ_Y stands for the yield stress whereas σ_{ii} is the trace of the Cauchy stress tensor. Moreover, f denotes the pore volume fraction and q is a material parameter.

In case of plastic deformation, micropores nucleate and grow at places of defects in the crystallographic structure. The pore evolution consists of two parts, namely the nucleation of pores and the growth of already existing pores. For the latter, the growth rate is assumed to be proportional to the hydrostatic part of the stress tensor whereas the nucleation part is controlled by the plastic deformation. Altogether, this leads to an evolution equation for the pore volume fraction f:

$$\frac{\partial f}{\partial t} = (1-f)\,\frac{\partial \varepsilon_{ii}^p}{\partial t} + d_N \exp(-\frac{1}{2}\,\frac{(\varepsilon_M^p - \varepsilon_N)^2}{s_N})\,\frac{\partial \varepsilon_N}{\partial t}\ ,\ f(t_0) \ = \ f_0\ . \quad (13)$$

Here, $\frac{\partial \varepsilon_{ii}^p}{\partial t}$ is the trace of the plastic equivalent rate tensor, ε_N stands for the mean nucleation equivalent plastic strain and ε_M^p denotes the equivalent plastic strain of the matrix material. Moreover, s_N refers to the standard deviation and d_N is a material parameter depending on the volume fraction of void nucleating particles.

4 Algorithmic tools for numerical simulation

The discretization of the drift diffusion model (1)-(4) is done by conforming P1 elements for the potential equation and mixed hybrid finite elements involving the lowest order Raviart-Thomas elements $RT_0(K), K \in \mathcal{T}_h$, for the continuity equations with respect to an adaptively generated hierarchy of triangulations $\mathcal{T}_h$ of Ω_1. Denoting by $P_k(D), k \in \mathbb{N}_0$, the set of polynomials of degree k on D and by $\mathcal{F}_h^{int}$ the set of interior faces of $\mathcal{T}_h$, we set

$$\begin{aligned} RT_0^{-1}(\Omega_1;\mathcal{T}_h) &:= \prod_{K\in\mathcal{T}_h} RT_0(K) \quad , \\ W_0(\Omega_1;\mathcal{T}_h) &:= \{v_h : \Omega_1 \to \mathbb{R} \mid v_h|_K \in P_0(K)\ ,\ K \in \mathcal{T}_h\} \quad , \\ M_0(\Omega_1;\mathcal{F}_h^{int}) &:= \{\mu_h : \cup_{F\in\mathcal{F}_h^{int}} F \to \mathbb{R} \mid \mu_h|_F \in P_0(F)\ ,\ F \in \mathcal{F}_h^{int}\}\ . \end{aligned}$$

The discretized continuity equations are solved by a Gummel type iteration where each iteration step requires the solution of the following problem (cf., e.g., [Brezzi, Marini, and Pietra, 1989]):
Find $(\mathbf{j}_h, u_h, \lambda_h) \in RT_0^{-1}(\Omega_1;\mathcal{T}_h) \times W_0(\Omega_1;\mathcal{T}_h) \times M_0(\Omega_1;\mathcal{F}_h^{int})$ such that for all $\mathbf{q}_h \in RT_0^{-1}(\Omega_1;\mathcal{T}_h)\ ,\ v_h \in W_0(\Omega_1;\mathcal{T}_h)$, and $\mu_h \in M_0(\Omega_1;\mathcal{F}_h^{int})$ there holds

$$\begin{aligned} &\sum_{K\in\mathcal{T}_h} \Big(\int_K a^{-1}\mathbf{j}_h \cdot \mathbf{q}_h dx + \int_K u_h\ \mathrm{div}\mathbf{q}_h dx - \sum_{F\in\mathcal{F}_h(K)} \int_F \lambda_h [\mathbf{n}_F \cdot \mathbf{q}_h]_J d\sigma \Big) = 0, \\ &\sum_{K\in\mathcal{T}_h} \Big(\int_K \mathrm{div}\mathbf{j}_h\ v_h dx - \int_K b u_h v_h\ dx \Big) \ = \ - \int_\Omega f v_h\ dx, \\ &\sum_{K\in\mathcal{T}_h} \sum_{F\in\mathcal{F}_h(K)} \int_F \mu_h\ [\mathbf{n}_F \cdot \mathbf{j}_h]_J d\sigma \ = \ 0\ . \end{aligned}$$

Here, $[\mathbf{n}_F \cdot \mathbf{j}_h]_J$ denotes the jump of the normal component of $\mathbf{j}_h$ across innerelement faces $F \in \mathcal{F}_h(K)$. Static condensation of $\mathbf{j}_h$ and u_h in the resulting algebraic saddle point problem leads to a Schur complement system which can be shown to be equivalent to a nonconforming Petrov-Galerkin approach ([Brezzi, Marini, and Pietra, 1989]). Denoting by $CR_1(\Omega_1;\mathcal{T}_h)$ the lowest order nonconforming Crouzeix-Raviart space and by $B(\Omega_1;\mathcal{T}_h)$ the space of quartic bubble functions associated with each $K \in \mathcal{T}_h$, the problem is to find $u_{NC} \in CR_1(\Omega_1;\mathcal{T}_h) \oplus B(\Omega_1;\mathcal{T}_h)$ such that for all $v_h \in CR_1(\Omega_1;\mathcal{T}_h)$

$$\sum_{K\in\mathcal{T}_h} \int_K \Big[P_{a^{-1}}(a\, \nabla u_{NC}) \cdot \nabla v_h \; + \; b\; P u_{NC}\; P v_h \Big]\, dx \;=\; (Pf, v_h)_{0;\Omega_1}\ .$$

where $P : L^2(\Omega_1) \to W_0(\Omega_1;\mathcal{T}_h)$ and $P_{a^{-1}} : L^2(\Omega_1)^2 \to RT_0^{-1}(\Omega_1;\mathcal{T}_h)$ are the orthogonal L^2- resp. weighted L^2-projection. Taking advantage of this equivalence, a multilevel preconditioned iterative solver can be used, where the multilevel preconditioner is the associated conforming one, put into effect by transforming the nonconforming Crouzeix-Raviart space onto its conforming counterpart (for details as well as for the realized adaptive grid refinement based on a residual-type a posteriori error estimator we refer to [Hoppe and Wohlmuth, 1997]).

As far as the discretization of the thermomechanical coupling problem is concerned, we discretize in time by embedded Singly Diagonally Implicit Runge Kutta (SDIRK) methods. For discretization in space, we use domain decomposition methods on nonmatching grids. We consider a nonoverlapping, geometrically conforming decomposition $\Omega_2 = \cup_{i=1}^n \Omega_{2,i}\ ,\ \Omega_{2,i} \cap \Omega_{2,j} = \emptyset, 1 \leq i \neq j \leq n$, of the computational domain given by the sandwich like structure of the module (cf. Figure 1) and refer to $S = \cup_{i=1}^n (\partial\Omega_{2,i} \setminus \partial\Omega_2)$ as the skeleton of the decomposition. As we can see from the schematic representation of the IHV Module, we are faced with subdomains of different aspect ratios. Moreover, we know that the thermal and mechanical properties of the materials in the individual subdomains are quite different resulting in strongly discontinuous coefficient of the heat and mechanical equations across subdomain boundaries. Therefore, we use individual triangulations $\mathcal{T}_i$ of the subdomains $\Omega_{2,i}$ that do not necessarily match on the interfaces between adjacent subdomains and take care of the resulting nonconformity by mortar element methods based on discretizations of the subdomain problems by continuous, piecewise linear finite elements denoting by $S_{1,\Gamma_D}(\Omega_{2,i};\mathcal{T}_i)$ the associated finite element spaces. For $\Gamma_{ij} \subset S$, we refer to γ_{ij}^m and γ_{ij}^{nm} as the mortar and nonmortar inheriting its triangulations from $\mathcal{T}_i$ and $\mathcal{T}_j$, respectively. We construct the multiplier space $M_h(\gamma_{ij}^{nm})$ in the meanwhile standard way under special consideration of cross points. Setting $V_h = \prod_{i=1}^n S_{1,\Gamma_D}(\Omega_{2,i};\mathcal{T}_i)$ and $M_h = \prod_{\Gamma_{ij}\subset S} M_h(\gamma_{ij}^{nm})$, the mortar finite element approach reads as follows: Find $(u_h, \lambda_h) \in V_h \times M_h$ such that

$$a_h(u_h, v_h) \ + \ b_h(v_h, \lambda_h) = \ell(v_h) \ , \ v_h \in V_h \quad , \tag{14}$$

$$b_h(u_h, \mu_h) \qquad\qquad = \ 0 \ , \ \mu_h \in M_h \ , \tag{15}$$

where $a(\cdot,\cdot) : V_h \times V_h \rightarrow R$ is the bilinear form associated with the FE discretized subdomain problems and

$$b_h(v_h, \mu_h) \ = \ - \sum_{\Gamma_{ij} \subset S} \int_{\Gamma_{ij}} \mu_h \ [v_h]_J \ d\sigma$$

is the bilinear form that realizes the weak continuity constraints across the interfaces. The resulting algebraic saddle point problem is solved by multilevel preconditioned Lanczos iterations with a block diagonal preconditioner, where the first diagonal block consists of subdomain preconditioners, that can be chosen as, for instance, BPX-preconditioners for the discretized subdomain problems, and the second diagonal block is an interface preconditioner being spectrally equivalent to the Schur complement resulting from static condensation (for details we refer to [Hoppe, Iliash, Kuznetsov, Vassilevski, and Wohlmuth, 1998]).
The stationary plasticity problems for the joints and the wire bonds have been solved by the standard return-mapping algorithm.

5 Simulation results

Based on the mathematical models and the numerical methods described in the previous sections, we have performed simulations of the operational behavior of the IHV Module. Figure 2 displays the distribution of the temperature and the von Mises equivalence stresses in a cross section of the upper soldered joints. Temperature peaks of more than 100^0C and the largest equivalence stresses occur in the center of the joints located below the IGBTs and power diodes. The simulation results are in good agreement with experimentally observed data.

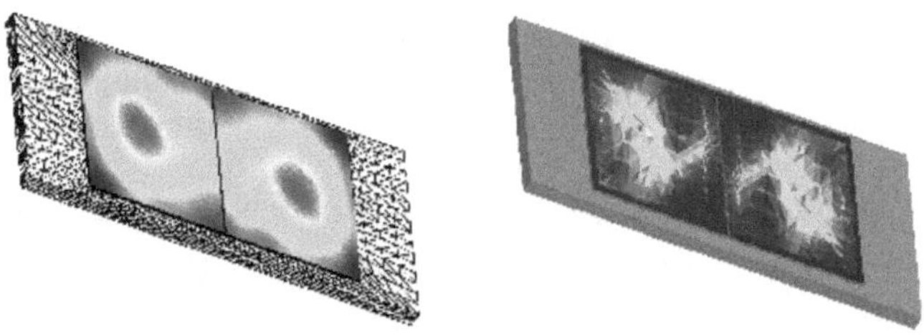

Fig. 2. Temperature distribution (left) and distribution of the equivalence stresses (right) in the upper soldered joint

We further consider the initiation and propagation of cracks in the wire bonding zone. Figure 3 shows a light microscopy of a crack opening in the wire bonding zone (left) as well as the plastic strain behavior at the beginning of the bonding zone (right) along the interface direction (solid lines) and perpendicular to it (dotted lines). Close to the crack tip, the wire is under tension in x-direction (upper curves) and under compression in y-direction (lower curves).

Acknowledgement. The work of the authors has been supported by the Federal Ministry for Education and Research (BMBF) under Grant No. 03HOM3A1.

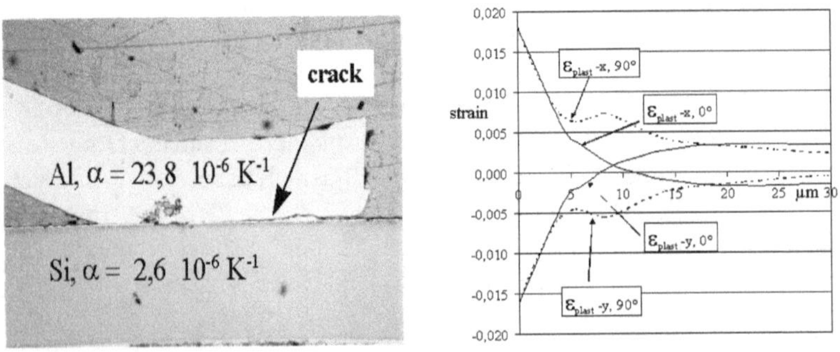

Fig. 3. Light microscopy of a crack (left) and the computed plastic strain at the beginning of the bonding zone (right)

References

F. Brezzi, D. Marini, and P. Pietra. Numerical simulation of semiconductor devices. *Comp. Math. Appl. Mech. Engrg.*, 75:493–514, 1989.

R. Hoppe, Y. Iliash, Y. Kuznetsov, Y. Vassilevski, and B. Wohlmuth. Analysis and parallel implementation of adaptive mortar element methods. *East-West J. Numer. Math.*, 6:223–248, 1998.

R. Hoppe and B. Wohlmuth. Adaptive multilevel techniques for mixed finite element discretizations of elliptic boundary value problems. *SIAM J. Numer. Anal.*, 34:1658–1681, 1997.

S. Ramminger, N. Seliger, and G. Wachutka. Reliability model for al wire bonds subjected to heel crack failures. *Microelectronics Reliability*, 40:1521–1525, 2000.

V. Tvergaard. Material failure by void growth to coalescence. *Advances in Applied Mechanics*, 27:83–151, 1989.

A Multiphysics Strategy for Free Surface Flows

Edie Miglio, Simona Perotto, and Fausto Saleri

MOX, Modeling and Scientific Computing, Department of Mathematics,
Politecnico of Milano, via Bonardi 9, I-20133 Milano, Italy
(http://mox.polimi.it/)

Summary. This work is the first step towards a multiphysics strategy for free-surface flows simulation. In particular, we present a strategy to couple one and two-dimensional hydrostatic free surface flow models. We aim to reduce the computational cost required by a full 2D model. After introducing the two models along with suitable a priori error estimates, we discuss the choice of convenient matching conditions stemming from the results obtained in Formaggia et al. [2001]. The numerical results in the last section confirm the soundness of our analysis.

1 Introduction

Final aim of our research is an efficient and accurate numerical simulation of the motion of water in a complex system of channels such as, for instance, the well-known Venice lagoon. A hydrodynamic configuration of this type involves a wide spectrum of space and time scales related to the presence of different physical phenomena. It is well known that in hydrodynamics there exists a hierarchy of models derived from the Navier-Stokes equations for an incompressible free-surface fluid. Essentially we can distinguish among 1D, 2D and 3D models of hydrostatic and non-hydrostatic type. In descending order of complexity, for the 3D case we can consider either the free surface Navier-Stokes or the hydrostatic 3D shallow water equations; concerning the 2D situation the Boussinesq, Serre or Saint-Venant equations can be adopted; finally the 1D counterpart of these latter models can be used (see, e.g., Miglio et al. [1999], Vreugdenhil [1998], Whitham [1974]). In particular, in this paper we consider only shallow water models, suitable for configurations where the vertical scales are much smaller than the corresponding horizontal ones.
Ideally one should use a full 3D model to capture all the physical features of the problem at hand. However, this approach is characterized by a huge computational effort. Thus, the basic idea is to reduce the computational cost by solving the more expensive model only in some parts of the domain. In this work we deal with the coupling of the 2D and 1D shallow water models.

This choice turns out to be reasonable for instance in the presence of a river bifurcation such as the one shown in Fig. 1 (right). We extend the analysis provided in Formaggia et al. [2001], where the 3D Navier-Stokes equations are coupled with a convenient 1D model for the description of blood flow in a compliant vessel, to the case of free surface flows. Even if, in our case, the dimension of the coupled models is different, we resort to a similar analysis to derive the suitable coupling conditions.

The outline of the paper is as follows. Sect. 2 deals with the 2D model. In Sect. 3, we provide the 1D model and a corresponding stability analysis. In Sect. 4, a set of interface conditions for sub-critical flows is proposed in order to couple the two models. Finally, numerical results are presented in Sect. 5.

2 The 2D Model

We consider the description of the motion of a free-surface viscous incompressible fluid when the vertical scales are much smaller compared with the corresponding horizontal ones. This allows us to consider the *shallow water* theory whose leading hypothesis is the *hydrostatic* approximation of the pressure, i.e., the pressure of the fluid is assumed to depend on the total water depth only. Many of the 1D and 2D hydrodynamic models, successfully used in practical applications, depart from this assumption.

The 2D model is represented by the Saint-Venant or shallow water equations whose conservative form reads as follows

$$\begin{cases} \dfrac{\partial (h\,\mathbf{U})}{\partial t} + \nabla \cdot (h\,\mathbf{U} \otimes \mathbf{U}) + g\,h\,\nabla h = \mathbf{0} & \text{with } \mathbf{x} \in \Omega \text{ and } t > 0, \\ \dfrac{\partial h}{\partial t} + \nabla \cdot (h\,\mathbf{U}) = 0 & \text{with } \mathbf{x} \in \Omega \text{ and } t > 0, \end{cases} \tag{1}$$

where $\mathbf{x} = (x, y)^T$, $\mathbf{U} = (u, v)^T$ is the average velocity, h denotes the total water depth and $\Omega \subset \mathbb{R}^2$ is a bounded open set. Of course, system (1) has to be provided with suitable initial and boundary conditions (see, e.g., Agoshkov et al. [1993]). We assume to be in the presence of a flat bottom and the effect of the friction is neglected. Moreover, we are interested in sub-critical flow regimes. The theory on hyperbolic systems can be applied to compute the eigenvalues and eigenfunctions of (1). With this aim, by considering a region of smooth flow, we can obtain the quasi-linear form of (1)

$$\frac{\partial \mathbf{W}}{\partial t} + A(\mathbf{W})\,\frac{\partial \mathbf{W}}{\partial x} + B(\mathbf{W})\,\frac{\partial \mathbf{W}}{\partial y} = \mathbf{0},$$

where $\mathbf{W} = (u, v, h)^T$,

$$A = \begin{bmatrix} u & 0 & g \\ 0 & u & 0 \\ h & 0 & u \end{bmatrix} \quad \text{and} \quad B = \begin{bmatrix} v & 0 & 0 \\ 0 & v & g \\ 0 & h & v \end{bmatrix}.$$

It is well known that the eigenvalues of system (1) are

$$\mu_1 = -(\cos(\phi)\, u + \sin(\phi)\, v), \quad \mu_{2,3} = -(\cos(\phi)\, u + \sin(\phi)\, v) \pm \sqrt{g\, h}\,,$$

ϕ being the direction of the characteristic lines, while the associated eigenfunctions are given by

$$w_1 = \begin{bmatrix} \sin(\phi) \\ -\cos(\phi) \\ 0 \end{bmatrix}, \qquad w_{2,3} = \begin{bmatrix} \pm\sqrt{g/h}\,\cos(\phi) \\ \pm\sqrt{g/h}\,\sin(\phi) \\ 1 \end{bmatrix}.$$

Concerning the stability analysis, a priori results are available in the literature for the Saint-Venant equations in the conservative form and provided with suitable boundary conditions (see, for instance, Agoshkov et al. [1993]).

3 The 1D Model

In the case Ω is an open channel, the 2D Saint-Venant equations (1) can be replaced by a 1D shallow water model, by assuming that the velocity is uniform over any cross section, that the channel is sufficiently straight and its slope sufficiently mild and uniform throughout the region. Moreover, the streamwise bottom slope and the lateral inflow are assumed equal to zero and the bottom friction is neglected as in the 2D model.

We focus on the case of one-dimensional channels with a rectangular cross-section. This choice turns out not to be so restrictive in realistic situations. Indeed, even if the cross-section is irregular, a sophisticated channel schematization can be employed by resorting to rectangular sections (see Schulz and Steinebach [2002]). In such a case the 1D model reduces to the system

$$\begin{cases} \dfrac{\partial A}{\partial t} + \dfrac{\partial Q}{\partial x} = 0 & \text{with } x \in (a,b) \text{ and } t > 0, \\ \dfrac{\partial Q}{\partial t} + \dfrac{\partial}{\partial x}\left(\dfrac{Q^2}{A}\right) + gA\dfrac{\partial h}{\partial x} = 0 & \text{with } x \in (a,b) \text{ and } t > 0, \end{cases} \tag{2}$$

where A is the area of the wet cross-section, Q is the discharge and h denotes the total water depth. System (2) has to be supplied with proper boundary conditions. Without reducing the generality of our analysis, we assume that the algebraic relation

$$h = \psi(A) + h_0, \quad \text{with} \quad \frac{\partial \psi}{\partial A} > 0 \quad \text{and} \quad \psi(A_0) = 0, \tag{3}$$

holds between the total water depth and the area. Here h_0 stands for the constant undisturbed water depth, A_0 is the area of the corresponding wet section while $\psi(A) = (A - A_0)/L$, L being the width of the section.

Assumption (3) allows us to rewrite (2) as

$$\begin{cases} \dfrac{\partial A}{\partial t} + \dfrac{\partial Q}{\partial x} = 0 & \text{with } x \in (a,b) \text{ and } t > 0, \\ \dfrac{\partial Q}{\partial t} + 2\dfrac{Q}{A}\dfrac{\partial Q}{\partial x} + \left(g\,A\,\dfrac{\partial h}{\partial A} - \dfrac{Q^2}{A^2} \right) \dfrac{\partial A}{\partial x} = 0 & \text{with } x \in (a,b) \text{ and } t > 0. \end{cases} \tag{4}$$

We start from the quasi-linear form (4) to study the mathematical properties of the solutions of the 1D model. Under the assumption (3), it can be proved that system (2) is hyperbolic since it has two real eigenvalues $\lambda_{1,2} = u \pm c$, where $u = Q/A$ while

$$c(A) = \sqrt{g\,A\,\frac{\partial \psi(A)}{\partial A}} = \sqrt{gh}$$

is the celerity of the system. In such a case it is also possible to compute the characteristic variables given by

$$W_{1,2} = u \pm \int_{A_0}^{A} \frac{c(\tau)}{\tau}\, d\tau = u \pm \int_{A_0}^{A} \sqrt{\frac{g\tau}{L}}\,\frac{1}{\tau}\, d\tau = u \pm 2\sqrt{\frac{g}{L}}\left[\sqrt{A} - \sqrt{A_0}\right].$$

3.1 Stability Analysis

In this section we provide an *a priori estimate* for system (2).
We assume that, for any time $t > 0$, the area A remains positive and that the eigenvalues λ_1 and λ_2 are of opposite sign ($\lambda_1 > 0$, $\lambda_2 < 0$), that is we consider a sub-critical and unidirectional flux. This is the most interesting situation in view of the coupling with the 2D model. We endow system (2) with the following general initial and boundary conditions:

$$\begin{aligned} &A(x,0) = A^*(x), \quad Q(x,0) = Q^*(x) \quad \text{with} \quad a < x < b, \\ &W_1 = g_1(t) \quad \text{at} \quad x = a, \quad W_2 = g_2(t) \quad \text{at} \quad x = b, \quad \text{with} \quad t > 0. \end{aligned} \tag{5}$$

Let us introduce the *energy* associated with model (2), defined, for any $t > 0$, as

$$E(t) = \frac{1}{2g}\int_a^b A(x,t)\,u^2(x,t)\,dx + \int_a^b \Psi(A(x,t))\,dx,$$

with $\Psi(A) = \int_{A_0}^{A} \psi(\tau)\,d\tau$. Thanks to (3), we can guarantee that $\Psi(A)$ and the energy $E(t)$ are positive functions, for any $t > 0$ and for any Q and A strictly positive.

Thus, the following conservation property can be proved.

Lemma 1. *Let us assume that relation* (3) *holds. Then for any* $T > 0$, *we have*

$$E(T) + \int_0^T Q\Big((h - h_0) + \frac{1}{2\,g}\,u^2\Big)\Big|_a^b\, dt = E(0), \tag{6}$$

$E(0)$ *depending only on the initial values* A^* *and* Q^*.

We refer to Miglio et al. [2003] for the proof of this result as well as for inequality (7). Result (6) can be used to derive an energy estimate for the 1D problem (2).

Proposition 1. *Let us assume that the boundary data g_1 and g_2 satisfy the following restrictions*

$$g_1(t) > -2\sqrt{\frac{gA_0}{L}} \quad \textit{and} \quad g_2(t) < 2\sqrt{\frac{gA_0}{L}}\,.$$

Then, there exists a positive function $F = F\Big(g_1, g_2, \dfrac{A_0}{L}\Big)$ such that

$$E(T) \leq E(0) + \int_0^T F\left(g_1(t), g_2(t), \frac{A_0}{L}\right)\, dt\,, \tag{7}$$

i.e., the 1D model problem (2) *provided with conditions* (5) *is stable.*

Remark 1. If homogeneous boundary conditions are chosen in (5), estimate (7) simplifies to $E(T) \leq E(0)$, provided that $2\sqrt{A_0}/3 < \sqrt{A} < 2\sqrt{A_0}$.

We remark that no energy estimate is available for a general cross-section.

4 Coupling of the Two Models

After having proved the well-posedness of both the 1D and the 2D problems, we can analyze the coupling of the two models.

Let us consider the coupling sketched on the left of Fig. 1. We denote with a the *matching point* of the two models. The cross-section at $x = a$ is assumed to be a rectangle and its outward normal is along the x-direction. At the right-hand side of a, i.e., in ω, we solve the 1D model (2) which provides the physical quantities $A_{1\mathrm{D}}$, $Q_{1\mathrm{D}}$ and $h_{1\mathrm{D}}$ (and, as a consequence, $u_{1\mathrm{D}} = Q_{1\mathrm{D}}/A_{1\mathrm{D}}$). At the left-hand side of a, that is in Ω, the 2D Saint-Venant equations (1) are solved and the associated physical quantities are $A_{2\mathrm{D}}$, $Q_{2\mathrm{D}}$ and $h_{2\mathrm{D}}$ to be defined shortly.

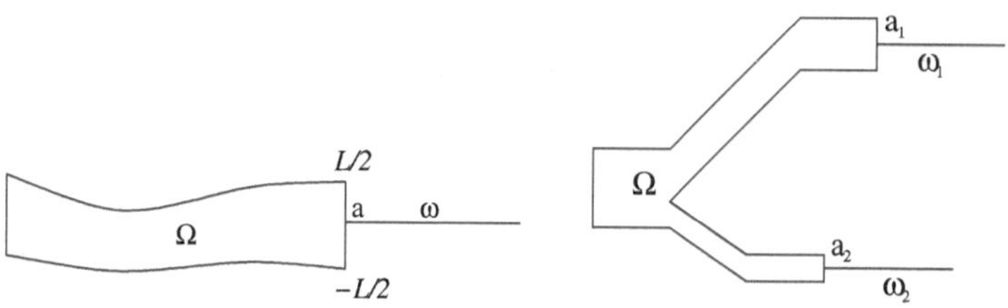

Fig. 1. Coupling of 2D with 1D models

As we are linking quantities of different dimension, we can, for instance, reduce the 2D ones to one-dimensional information by averaging the two-dimensional terms. With this aim, let us introduce

$$\overline{u}_{2D} = \frac{1}{L}\int_{-\frac{L}{2}}^{\frac{L}{2}} u(a,y)\,dy\,,\ \overline{h}_{2D} = \frac{1}{L}\int_{-\frac{L}{2}}^{\frac{L}{2}} h(a,y)\,dy = \frac{A_{2D}}{L}\,,\ \overline{Q}_{2D} = A_{2D}\,\overline{u}_{2D}\,,$$

i.e., the mean velocity, the mean total water depth and the mean discharge, where $A_{2D} = \int_{-\frac{L}{2}}^{\frac{L}{2}} h(a,y)\,dy$. Due to the unidirectional flow assumption, from the 2D to the 1D model, we have a subcritical outflow for the 2D system (with *two outgoing characteristics*) and a subcritical inflow for the one-dimensional problem (with *an incoming characteristic*). It seems reasonable from a physical view-point to demand the continuity of the following quantities at the interface $x = a$:

- C1. cross-section area: $A_{2D} = A_{1D}$, with $A_{1D} = \overline{h}_{2D}L$;
- C2. discharge: $\overline{Q}_{2D} = Q_{1D}$;
- C3. entering characteristic: $2\sqrt{\overline{h}_{2D}\,g} + \overline{u}_{2D} = 2\sqrt{h_{1D}\,g} + \dfrac{Q_{1D}}{A_{1D}}$.

Notice that all the mean variables are considered in an average form on the 2D problem. On the other hand, concerning the choice of the matching conditions, we remark that C1. and C3. would suffice as C2. is automatically guaranteed when C1. and C3. are satisfied.

4.1 The Sub-Domain Iteration Algorithm

To develop a *splitting procedure* to solve the coupled 1D-2D problem, we enforce at the matching point $x = a$ only those conditions which guarantee the well-posedness of each subproblem in Ω and ω. With this aim, we exploit the results of the stability analysis above. In particular,

- C1. is used for imposing the total depth at the *outflow* of the *2D model*;
- C3. is used at the *inflow* of the *1D model*.

Then each subproblem is completed with other boundary conditions:

- *condition a*: at the *inflow* of the *2D model* we assign the *total water depth* $h(t)$ as a function of time;
- *condition b*: at the *outflow* of the *1D model* a *non-reflecting boundary condition* is employed.

Moreover, we recall that on the rigid walls of the channel, no slip boundary conditions are assigned.
Thus the main steps of the algorithm are: given the solution of the coupled problem at time t^n, for $k = 1, 2, \ldots$

1. solve the 2D problem with C1. plus *condition a* in order to obtain h_k^{n+1}, U_k^{n+1};

2. compute $2\sqrt{h_k^{n+1} g} + \overline{U}_k^{n+1}$, that is an approximation for the left-hand side of C3.;
3. solve the 1D problem with C3. plus *condition b.*

We iterate until the coupling conditions are satisfied within a fixed tolerance. In practice, it can be verified that, after 2 or 3 iterations, the difference between the 1D and averaged 2D values is very small.

5 Numerical Assessment

To test the effectiveness of the proposed algorithm we consider the case of a river bifurcation as sketched on the right of Fig. 1. We want to solve the 2D model only in Ω, i.e., near the bifurcation while the one-dimensional problem is solved in ω_1 and ω_2. The numerical solution of the 2D model is obtained by using the 2D counterpart of the approach proposed in Miglio et al. [1999]. As for the 1D model a finite volume method is employed. As initial condition

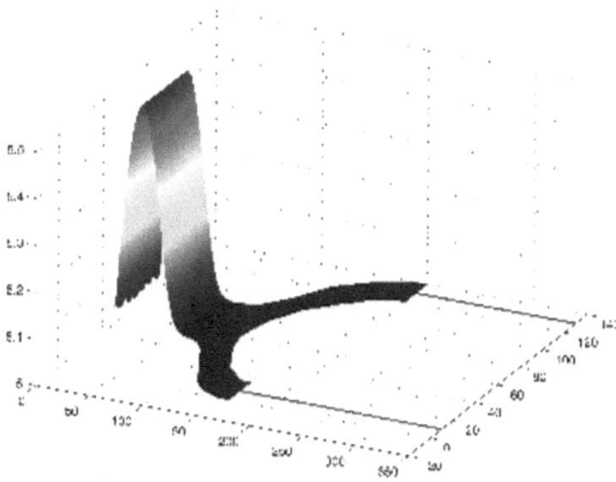

Fig. 2. Initial elevation profile

for the elevation we choose the profile shown in Fig. 2, while the time step is chosen equal to $\Delta t = 0.1s$ and for the space discretization of both the 1D and 2D models a mesh size $h = 0.1m$ is used. In Fig. 3 we show two snapshots of the approximate elevations provided by the full 2D shallow water model (on the left) and by the coupled 2D-1D one (on the right), respectively, corresponding to two different times ($t = 250s$ and $t = 300s$). These results confirm the soundness of the algorithm proposed in Sect. 4.1. The wave travels from the 2D to the 1D model without any significant distortion: no wave amplitude reduction and no phase difference is evident.

Acknowledgement. This work has been supported by the project MIUR 2001 "Numerical Methods in Fluid Dynamics and Electromagnetism". We thank Prof. Luca Formaggia for useful comments and suggestions.

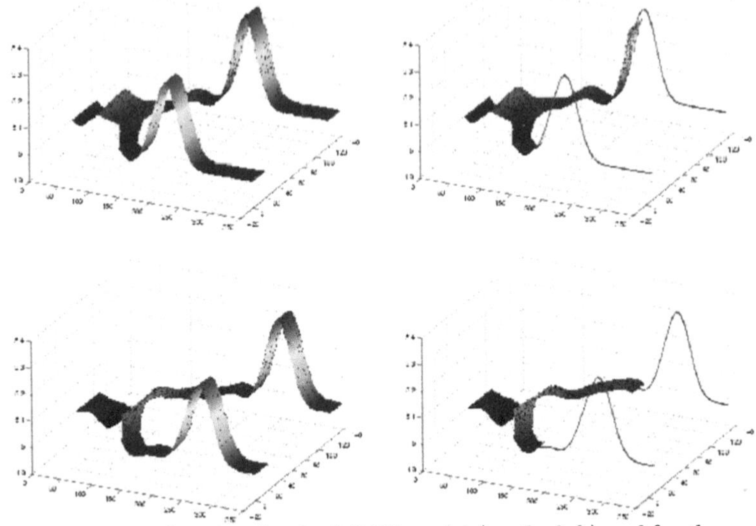

Fig. 3. Approximate elevation for the full 2D model (on the left) and for the coupled 2D-1D model (on the right) corresponding to $t = 250s$ (top) and to $t = 300s$ (bottom)

References

V. Agoshkov, D. Ambrosi, V. Pennati, A. Quarteroni, and F. Saleri. Mathematical and numerical modelling of shallow water flow. *Comput. Mech.*, 11:280–299, 1993.

L. Formaggia, J. Gerbeau, F. Nobile, and A. Quarteroni. On the coupling of 3D and 1D Navier-Stokes equations for flow problems in compliant vessels. *Comput. Methods Appl. Mech. Engrg.*, 191:561–582, 2001.

E. Miglio, S. Perotto, and F. Saleri. On the coupling of 2D and 1D shallow water models for the simulation of free surface flows. *in preparation*, 2003.

E. Miglio, A. Quarteroni, and F. Saleri. Finite element approximation of quasi-3D shallow water equations. *Comput. Methods Appl. Mech. Engrg.*, 174(3-4):355–369, 1999.

M. Schulz and G. Steinebach. Two-dimensional modelling of the river Rhine. *J. Comput. Appl. Math.*, 145(1):11–20, 2002.

C. Vreugdenhil. *Numerical methods for shallow-water flows.* Kluwer Academic Press, Dordrecht, 1998.

G. Whitham. *Linear and nonlinear waves.* Wiley, New York, 1974.

Part X

Minisymposium: Robust Decomposition Methods for Parameter Dependent Problems

Weighted Norm-Equivalences for Preconditioning

Karl Scherer

University of Bonn, Institut fuer Angewandte Mathematik

Summary. The theory of *multilevel methods* for solving Ritz-Galerkin equations arising from discretization of elliptic boundary value problems is by now well developed. There exists a variety of survey talks and books in this area (see e.g.Xu [1992],Yserentant [1993],Oswald [1994]). Among them the *additive methods* are based on a suitable decomposition of the underlying projection operator (thus including also domain decomposition methods). In particular there is a close connection with classical concepts in approximation theory via so- called Jackson and Bernstein inequalities. These provide norm equivalences with the bilinear form underlying the Ritz- Galerkin procedure and thus preconditioners for the arising stiffness matrix.

The size of the constants in this equivalence is crucial for the stability of the resulting iteration methods. In this note we establish *robust norm equivalences* with constants which are *independent* of the mesh size and depend only *weakly* on the ellipticity of the problem, including the case of strongly varying coefficients. Extensions to the case of coefficients with discontinuities are possible, see Scherer [2003/4]. In the case of piecewise constant coefficients on the initial coarse grid there exist already estimates of the condition numbers of BPX-type preconditioners independent of the coefficients (see Yserentant [1990], Bramble and J.Xu [1991]) however they depend still on the mesh size (of the finest level).

1 Introduction

Given coefficients $a_{i,k} \in L_\infty(\Omega), \Omega \subset \mathcal{R}^2$ consider the *bilinear form*

$$a(u,v) := \int_\Omega \sum_{i,k=1}^{2} (a_{i,k}(D_i u)(D_k v)) \qquad \text{for } u,v \in H^1(\Omega) = W_2^1(\Omega). \qquad (1)$$

Here $W_p^r(\Omega)$ denotes the usual *Sobolev space* with norm $(1 \le p < \infty)$

$$\|u\|_{r,p;\Omega} := \|u\|_{p,\Omega} + |u|_{r,p;\Omega}, \qquad |u|_{r,p;\Omega} := \sum_{|\alpha|=r} \|D^\alpha u\|_{p;\Omega}.$$

If $a(u,v)$ is coercive (or L strongly elliptic) the *Lax-Milgram-Theorem* states that the equation $Lu := \sum_{i,k} \partial_i(a_{i,k}\partial_k u) = f$ has a unique generalized solution u satisfying weakly Dirichlet boundary conditions, i.e. $u \in H_0^1(\Omega)$.

Let $\psi_1, \cdots, \psi_N$ be a basis of a finite-dimensional subspace $\mathcal{V}$ of $H_0^1(\Omega)$. The *Ritz-Galerkin-equations* compute an approximate solution $u_N \in \mathcal{V}$ by

$$a(u_N, \psi_k) = (f, \psi_k), \qquad u_N := \sum_{i=1}^{N} \alpha_i \psi_i, \qquad 1 \le k \le N,.$$

These equations are solved *iteratively* for $\nu = 0, 1, 2, \cdots$:

$$u_N^{(\nu+1)} = u_N^{(\nu)} - \omega\, \mathcal{C} r^{(\nu)}, \qquad r^{(\nu)} := \mathcal{A} u^{(\nu)} - \mathsf{b}, \qquad \mathsf{b} := \{(f, \psi_k)\}.$$

Here ω is a relaxation factor and the matrix $\mathcal{C}$ acts as a preconditioner for the *stiffness matrix* $\mathcal{A} := \Big(a(\psi_i, \psi_k)\Big)_{i,k}$. The speed of convergence of this iteration scheme is governed by the condition number $\kappa(\mathcal{CA})$.

In the theory of additive multi-level- methods preconditioners for $\mathcal{A}$ have been constructed for which $\kappa(\mathcal{CA}) = \kappa\ (\mathcal{C}^{1/2}\mathcal{A}\mathcal{C}^{1/2}) = \mathcal{O}(1)$, independent of the mesh-size of the underlying FE-space. Thereby the matrix $\mathcal{C}$ is derived via a norm equivalence with $a(u,u)$ and the size of $\kappa(\mathcal{CA})$ depends on the equivalence constants.

2 Norm equivalences and Approximation processes

Given a hierarchical sequence of subspaces

$$\mathcal{V}_0 \subset \mathcal{V}_1 \subset \cdots \subset \mathcal{V}_J := \mathcal{V} \subset X := L_2(\Omega), \tag{2}$$

assume that there exist of bounded projections $P_j : \mathcal{V} \longrightarrow \mathcal{V}_j$ satisfying

$$\beta_0\ a(u,u) \le \|P_0\|_X^2 + \sum_{j=1}^{J} d_j\ \|P_j u - P_{j-1} u\|_X^2 \le \beta_1\ a(u,u) \tag{3}$$

with suitable coefficients $\{d_j\}$ and constants β_0, β_1 independent of d_j, $u \in \mathcal{V}$ or J. Define via $(u, Bu) := \|P_0\|_X^2 + \sum_{j=1}^{J} d_j\ \|P_j u - P_{j-1}u\|_X^2$ a positive definite operator B for $u \in \mathcal{V}$ and let $\mathcal{C}$ above be the matrix representing the inverse B^{-1}. It is well known that then $\kappa(\mathcal{CA}) \le \beta_1/\beta_0$, showing that $\mathcal{C}$ is a suitable preconditioner. More generally one can chose the matrix $\mathcal{C}$ as the discrete analogue of an operator C which is spectrally equivalent to B^{-1}. The derivation of the norm equivalence (3) proceeds in a meanwhile standard manner (cf. Dahmen and Kunoth [1992], Bornemann and Yserentant [1993], Oswald [1994]):

1.Step: use the equivalence of $a(u,v)$ with a Sobolev-norm, i.e.

$$A_1\ a(u,u) \leq \|u\|_{1,2,\Omega}^2 \leq A_2\ a(u,u), \qquad \forall u \in H_0^1(\Omega), \tag{4}$$

with positive constants A_1, A_2 (needed also in the Lax-Milgram theorem).

2.Step: describe the Sobolev-norm via the *K-functional* of J.Peetre

$$K(t,f;X,Z) := \inf_{g\in Z}(\|f-g\|_X + t|g|_Z), \qquad t>0, \quad f\in X.$$

for normed linear spaces X, Z with $Z \subset X$ and seminorm $|\cdot|_Z$ such that Z is complete under norm $\|\cdot\|_X + |\cdot|_Z$. In the K-method of interpolation theory (see Bennett-R.Sharpley [1988], chapter 5) one defines for any integer r and $0 < \theta < r$:

$$\|f\|_{(L_2(\Omega),W_p^r(\Omega))_{\theta/r,p,q}} := \left(\sum_{n=0}^{\infty}[2^{n\theta}K(2^{-n\theta},f;L_2(\Omega),W_p^r(\Omega))]^q\right)^{1/q}. \tag{5}$$

There holds the equivalence with Besov seminorms (see Johnen and Scherer)

$$\|f\|_{(L_2(\Omega),W_p^r(\Omega))_{\theta/r,p,q}} \approx \|f\|_{\theta,p,q,\Omega} := \left(\sum_{n=0}^{\infty}[2^{n\theta}\omega_r(2^{-n},f)_p]^q\right)^{1/q}$$

where $\approx$ denotes equivalence up to constants not depending on f. Further the equivalence of special Besov- norms with (fractional) Sobolev norms is known (see Triebel [1992],p.9):

$$\|f\|_{\theta,2,2,\Omega} \approx \|f\|_{\theta,2;\Omega} \qquad \text{for } \ \theta > 0. \tag{6}$$

3.Step: describe the *interpolation norms* created by the K-functional via *approximation processes* $\mathbf{V}$, i.e. sequences of linear bounded operators $\{V_j\}$ defined on a Banach space X satisfying $\lim_{n\to\infty} V_n f = f$ for all $f \in X$. Then define *approximation norms* describing certain rates of approximation by

$$\|f\|_{\theta,q;V} := \left\{\sum_{n=0}^{\infty}[2^{n\theta}\|V_n f - f\|_X]^q\right\}^{1/q}, \qquad \theta \geq 0,\ 1 \leq q \leq \infty$$

and introduce Jackson- and Bernstein- inequalities:

Definition 1. *An approximation process satisfies a* Jackson-inequality *with respect to the pair X, Z and* order $\alpha > 0$ *if there exists a constant C_V*

$$\|V_n f - f\|_X \leq C_V\ 2^{-\alpha n}\ |f|_Z, \qquad \forall f \in Z. \tag{7}$$

and a corresponding Bernstein-inequality *if there exists D_V such that*

$$V_n f \in Z, \qquad |V_n f|_Z \leq D_V\ 2^{\alpha n}\ \|f\|_X, \qquad \forall f \in X. \tag{8}$$

Under these assumptions it has been shown (see Butzer and Scherer [1968], Butzer and Scherer [1972]))

Theorem 1. *For the operator sequences V_n defining an approximation process and satisfying Jackson- and Bernstein-inequalities of order α for a pair X, Z there holds for all $\theta > 0$*

$$\Big\{\sum_{n=0}^{\infty}[2^{n\theta}\|V_n f - V_{n-1}f\|_X]^q\Big\}^{1/q} \approx \|f\|_{\theta,q,V} \approx \Big\{\sum_{n=0}^{\infty}[2^{n\theta}K(2^{-n\alpha},f;X,Z)]^q\Big\}^{1/q}$$

with equivalence constants only depending on α, θ, C_V, D_V and $\sup \|V_n\| < \infty$.

The upper bound in the second equivalence follows from the Jackson-type inequality. For the lower one uses the decomposition $f = \sum_{k=n+1}^{\infty} V_k f - V_{k-1} f$,

$$K(2^{-n\alpha}, V_k f - V_{k-1} f; X, Z) \ \leq \ \min\ (1, 2^{(k-n)\alpha})\|V_k f - V_{k-1} f\|_X \tag{9}$$

and Hardy's inequalities to estimate the arising double sum (cf. below).

We consider now the case of uniformly bounded linear projections $V_j = P_j : \mathcal{V} \longrightarrow \mathcal{V}_j$ in (3). For later use we assume $\mathcal{V}_j \subset Z = H_0^2(\Omega), \alpha = q = 2$ and $\theta = 1$. Then we obtain in combination with (4), (5) and (6)

Corollary 1. *Given the elliptic bilinear form $a(u,u)$ in (1) suppose that the above projections satisfy Jackson- and Bernstein-inequalities of order 2 for the pair $L_2(\Omega), H_0^2(\Omega)$. Then there holds for $u \in \mathcal{V}$ the equivalence ($P_{-1}u := 0$)*

$$\sum_{j=1}^{J} 4^J\ \|P_j u - P_{j-1} u\|_{2,\Omega}^2 \approx \sum_{n=0}^{\infty}[2^n\ K(2^{-2n}, f; L_2(\Omega), H_0^2(\Omega))]^2 \approx a(u,u),$$

and the equivalence constants do not depend on the level J and u.

We apply this to the case of FE- spaces consisting of piecewise polynomial functions of degree k in (2) with respect to the sequence of triangulations

$$\mathcal{T}_0 \subset \mathcal{T}_1 \subset \cdots \subset \mathcal{T}_J := \mathcal{T}. \tag{10}$$

The coarse initial triangulation $\mathcal{T}_0$ is adaptively refined by dividing each triangle either into 4 congruent triangles or halving it such that each triangle in $\mathcal{T}_k$ is geometrically similar to a triangle of $\mathcal{T}_0$.

Jackson- inequalities for projections into such spaces with respect to the pair $X = L_2(\Omega), Z = H_0^{k+1}(\Omega)$ of order $k+1$ are well known (cf. Ciarlet [1978]) whereas corresponding Bernstein-inequalities are only possible of maximal order k. However the modified inequality (9) can be proved (in case of maximal smoothness of u) with order $\alpha = k + 1/2$ which suffices for a proof of Theorem 1. Such an inequality follows from a corresponding one for the L_p modulus of continuity $\omega_k(t,f)_p$ (see Oswald [1994]) and the equivalence (see Johnen and Scherer):

$$K(t^k, f; L_p(\Omega), W_p^k(\Omega)) \ \approx\ \omega_k(t,f)_p.$$

3 Weighted norm equivalences

The disadvantage of above approach is that the "equivalence constants" in (3) depend on the ellipticity constants A_1, A_2 in (4). In the following we want to study *robust* norm equivalences, i.e. how they depend on these constants. (In the following constants C will only depend on the initial coarse triangulation $\mathcal{T}_0$). Thereby we restrict us to subspaces $\mathcal{V}_j$ consisting of piecewise linear functions. The basic idea is to introduce for the triangulations (10)

Assumption A: There exist weights $\underline{\omega}_i, \overline{\omega}_i$ such that the bilinear form (1) satisfies on the triangles Z_i of $\mathcal{T}_J$ the ellipticity condition

$$\underline{\omega}_i \sum_{\nu=1}^{2} \xi_\nu^2 \leq \sum_{\nu,\mu=1}^{2} a_{\nu,\mu}(x)\xi_\nu \xi_\mu \leq \overline{\omega}_i \sum_{\nu=1}^{2} \xi_\nu^2, \qquad \text{for all} \quad x \in Z_i. \tag{11}$$

Then one wants to establish Jackson- and Bernstein- inequalities for suitable projections P_j in (3) with respect to a "weighted norm" arising from this assumption. This will be described shortly in the following (for more details see Scherer [2003/4]). Ideally one should take for P_j the Ritz projection Q_j^a defined on $\mathcal{V}$ by

$$a(Q_j^a u, v) = a(u, v), \qquad u \in \mathcal{V}, \quad v \in \mathcal{V}_j \tag{12}$$

since then $v_j := Q_j^a u - Q_{j-1}^a u$ with $Q_{-1}^a u := 0$ satisfies $a(u,u) := ||u||_a^2 = \sum_{j=0}^{J} ||v_j||_a^2$. Then the idea is to replace Q_j^a by projections Q_j^ω with respect to a *weighted norm* (from now on we omit subscript and superscript on ω):

$$(Q_j^\omega u, v)_\omega = (u, v)_\omega := \sum_{Z_i \in \mathcal{T}_J} \omega_i \int_{Z_i} u \cdot v \, dx, \qquad v \in \mathcal{V}_j.$$

Essential for our analysis are also the *average weights* $\omega_T := \frac{1}{\mu(T)} \sum_{Z_i \subset T} \mu(Z_i)\omega_i$ with corresponding *weighted norms*

$$||v||_{j,\omega}^2 := \sum_{T \in \mathcal{T}_j} \omega_T \int_T |v|^2$$

At first two *Bernstein-type inequalities of order 1/2* are proved. To this end we assume a continous weight $\omega(x)$ in Assumption A and work with average weights $\omega_i^* := \frac{1}{\mu(Z_i)} \int_{Z_i} \omega$ as well as corresponding ones ω_T^* for $T \in \mathcal{T}_j$.

Lemma 1. *Define the semi-norm*

$$||u||_{1/2,\omega,l} := \Big(\sum_{T \in \mathcal{T}_l} \omega_T^* \int_{\partial T} |u|^2 \Big)^{1/2}$$

for $u \in \mathcal{V}_J \subset H_0^1(\Omega)$. Then there holds for $v_j := Q_j^a u - Q_{j-1}^a u$ and any $w \in \mathcal{V}_j$

$$\|Q^a_j u - Q^a_{j-1} u\|_a \le E^j_\omega \; \|u - w\|_a \; + \; C \; 2^{j/2} \; \|u\|_{1/2,\omega,j}, \tag{13}$$

where $E^j_\omega := \max_{T \in \mathcal{T}_j} \max_{x,y \in T} |[\omega(x) - \omega(y)]/\omega(y)|$ *is a modulus of continuity of* $\omega(x)$.

Lemma 2. *There holds for any* $u \in \mathcal{V}_k$ *and* $k \ge j$

$$\|u\|_{1/2,\omega,j} \le C \; 2^{k/2} \; \|u\|^*_{k,j}, \qquad \|u_k\|^*_{k,j} := \Big(\sum_{U \in \mathcal{T}_j} \omega^*_U \int_{S_k(U)} |u_k|^2 \Big)^{1/2}.$$

Here $S_k(U)$ *denotes the strip along* ∂U *consisting of triangles* $T \in \mathcal{T}_k$.

We apply the first lemma to each term $\|v_j\|^2_a$ with $j \ge j_0$ for some $j_0 \ge 1$ to be chosen later, take $u - Q^\omega_{j-1} u$ instead of u, $w = Q^a_{j-1} u - Q^\omega_{j-1} u$ and obtain

$$\sum_{j=j_0}^J \|v_j\|^2_a \Big(1 \; - \; 2 \sum_{j=j_0}^J \big(E^j_\omega\big)^2\Big) \le 2C \sum_{j=j_0}^J 2^j \Big(\sum_{k=j}^J \|u_k\|_{1/2,\omega,j} \Big)^2. \tag{14}$$

Next we use the decomposition of $u - Q^\omega_{j-1} u = \sum_{k=j}^J u_k$, with $u_k := Q^\omega_k u - Q^\omega_{k-1} u$ and apply the second Bernstein-type inequality. After inserting this into the right hand side of (14) we obtain a double sum which is estimated with a refined version of Hardy's inequality giving

$$\sum_{j=0}^J 2^j \Big(\sum_{k=j}^J 2^{k/2} \; \|u_k\|^*_{k,j} \Big)^2 \Big[1 \; - \; 2E^j_\omega\Big] \le 4 \sum_{j=j_0}^J \; 4^j \; \|u_j\|^2_{j,\omega} \tag{15}$$

In addition, by an other application of Lemma 1, one can obtain a bound for the remaining sum $\sum_{j=0}^{j_0-1} \|v_j\|^2_a = \|Q^a_{j_0}\|^2_a$. The final estimate is then

$$a(u,u) \; \le \; C \left\{ \frac{1 + 3(E^{j_0}_\omega)^2}{[1 - 2E^{j_0}_\omega]\Big[1 - 2\sum_{j=j_0}^J \big(E^j_\omega\big)^2\Big]} \sum_{j=j_0}^J 4^j \|u_j\|^2_{j,\omega} \right\} + \|Q^\omega_{j_0-1} u\|^2_a.$$

Theorem 2. *If there exists* $j_0 \ge 1$ *such that* $\sum_{j=j_0}^J \big(E^j_\omega\big)^2 \le 1/4$ *there holds*

$$a(u,u) \le C \sum_{j=j_0}^J \; 4^j \; \|Q^\omega_j u - Q^\omega_{j-1} u\|^2_{j,\omega} + a(Q^\omega_{j_0-1} u, Q^\omega_{j_0-1} u).$$

Remarks: The assumption of the theorem can be fulfilled for continuous ω. In case $\omega \in C^1$ or $\omega \in C^\alpha$ a more quantitative description can be given, e.g. for $\omega(x) := \exp\{q(x)\}$ we have $E^j_\omega \le c2^{-j} \|\nabla q\|_\infty \; \exp\{c2^{-j}\|\nabla q\|_\infty\}$. The continuity of ω also justifies the choice $\underline{\omega}_i = \overline{\omega}_i = \omega^*_i$ in (11) since then $a(v,v)$ is norm-equivalent to $\tilde{a}(v,v) := \sum_{Z_i \in \mathcal{T}_J} \omega^*_i \int_{Z_i} \|\nabla v\|^2$. Finally we remark that the above argument can be extended also to the case of a weight function $\omega(x)$ which is continous on Ω up to a (smooth) curve. If this curve coincides with the edges of the initial coarse grid the above argument can be applied to each of the two subregions separately with corresponding moduli of continuity. But also the more general case can be treated (see Scherer [2003/4]).

We turn now to the problem of a lower bound for $a(u,u)$ (more details can be found in Scherer [2003/4]). At first observe that by Hardy's inequality

$$\sum_{j=1}^{J}\left[2^j\|u_j\|_\omega\right]^2 \le \sum_{j=0}^{J}\left[2^j\|Q_j^a u - u\|_\omega\right]^2 \le 4\sum_{j=1}^{J}\left(2^j\|Q_j^a u - Q_{j-1}^a u\|_\omega\right)^2.$$

In order to pass from $\|\cdot\|_\omega$ -norm to $\|\cdot\|_{j,\omega}$ - norm use

Theorem 3. *Suppose that there exists a constant $\gamma \in (0,1]$ such that for any $T \in \mathcal{T}_j$ the weights ω_i satisfy*

$$\omega_E \;/\; \omega_T \le \mu(T) \;/\; 2\mu(E), \qquad \forall\ E \subset T \text{ with } \mu(E) \le \gamma\ \mu(T)\ . \tag{16}$$

Then there holds for $v \in \mathcal{V}_j$

$$[1+\ C\ \gamma^{-1}]^{-1}\ \|v\|_{j,\omega;T} \le\ \|v\|_{\omega;T} \le\ [1+\ \sqrt{6}\ C]\ \|v\|_{j,\omega;T}. \tag{17}$$

Application of this theorem gives

$$\sum_{j=1}^{J}\left[2^j\|Q_j^\omega u - Q_{j-1}^\omega u\|_\omega\right]^2 \le\ C\ \sum_{j=1}^{J}\left(2^j\|Q_j^a u - Q_{j-1}^a u\|_{j,\omega}\right)^2. \tag{18}$$

The crucial step is the following local estimate

Theorem 4. *Let be U be the support of a nodal function in $\mathcal{V}_{j-1}$, $\psi_l^{(j-1)}$ say. Then there holds ($S, S' \in \mathcal{T}_j$)*

$$\|Q_j^a u - Q_{j-1}^a u\|_{j,\omega,U} \le C\left(\max_{S',S\subset U}\sqrt{\frac{\omega_S}{\omega_{S'}}}\right)\ 2^{-j}\ \|\nabla(Q_j^a u - Q_{j-1}^a u)\|_{j,\omega,U}.$$

Sketch of the proof: Using the duality technique of Aubin-Nitsche one has

$$\|v_j\|_{j,\omega,U} = \sup_{g\in L_\omega(U)}\frac{|(g,\omega\cdot v_j)_U|}{\|g\|_{j,\omega,U}} = \sup_{g\in L_\omega(U)}\frac{|(-\Delta\varphi_g,\omega\cdot v_j)_U|}{\|g\|_{j,\omega,U}}.$$

where $-\Delta\varphi_g = \tilde{g}$ on $\tilde{U} \ge U$, $\varphi_g|_{\partial\tilde{U}} = 0$ and $\|\tilde{g}\|_{\tilde{U}} \le C\|g\|_U$ with some absolute constant $\tilde{C}$. Partial integration on each $S \subset U$ gives

$$|(-\Delta\varphi_g,\omega\cdot v_j)_U| \le |\sum_{S\subset U}\omega_S\int_S(\nabla\varphi_g,\nabla v_j)| + |\sum_{S\subset U}\omega_S\int_{\partial S}v_j(\nabla(\varphi_g - v),n_{\partial S})|$$

The bound for the first term uses $\|\nabla\varphi_g\|_{\tilde{U}} \le C\sqrt{\mu(\tilde{U})}\|g\|_U$. In the second supremum one chooses $v = v^* \in \mathcal{V}_{j-1}$ with supp $v^* \subset U$ as interpolant of φ_g. Then (cf. Ciarlet [1978])

$$\|\nabla(\varphi_g - v^*)\|_{\infty,T} \leq C \text{ diam } T \sum_{|\alpha|=3} \|D^\alpha \varphi_g\|_T, \qquad T \in \mathcal{T}_{j-1},$$

and by the special choice $g = g^* := v_j$ the estimate of the theorem follows.

Summing with respect to U and inserting the result into (18) yields

Theorem 5. *Under assumption A and for uniformly refined triangulations there holds for $a(u,u)$ the lower bound*

$$\sum_{j=1}^{J} \left[2^j \|Q_j^\omega u - Q_{j-1}^\omega u\|_{j,\omega}\right]^2 \leq C \left(\max_{1 \leq j \leq J} \max_{U \in \mathcal{T}_{j-1}} \max_{S', S \subset U} \sqrt{\frac{\omega_S}{\omega_{S'}}} \right) a(u,u)$$

with C depending only on the shape of the triangles of the initial triangulation.

4 Application to preconditioning

Theorems 2 and 5 can be combined via Theorem 3 to the following

Theorem 6. *Under assumption A assume further that $\omega(x)$ satisfies the assumptions of Theorems 2 and 3. Then for uniformly refined triangulations there holds for $a(u,u)$ and j_0 depending on $\omega(x)$*

$$C_1 \ a(u,u) \ \leq \ a(u_{j_0}, u_{j_0}) + \sum_{j=1}^{J} \left[2^j \|Q_j^\omega u - Q_{j-1}^\omega u\|_\omega\right]^2 \leq C_2 \ C_\omega \ a(u,u)$$

where $C_\omega := \max_{1 \leq j \leq J} \max_{U \in \mathcal{T}_{j-1}} \max_{S', S \subset U} \sqrt{\omega_S}/\sqrt{\omega_{S'}}$ and C_1, C_2 independent of J and $\omega(x)$.

It seems impossible to dispense with any condition starting from Assumption A. The most restrictive conditions appear in Theorem 2 where the choice of j_0 depends on the decrease of the moduli of continuity E_ω^j in j. This has been discussed in the remarks following it. A discussion of the further condition (16) in Theorem 3 is given in Scherer [2003/4]. The weakest one is probably that of Theorem 5 which requires the boundedness of the constant there. The case of non-uniformly refined meshes can be reduced to that one of uniformly refined meshes by the technique in Bornemann and Yserentant [1993].

For preconditioning we can proceed as in the BPX approach (cf.Bramble et al. [1990]) taking

$$B^{-1} = (Q_{j_0}^\omega)^{-1} + \sum_{j=j_0+1}^{J} 4^{-j} \ (Q_j^\omega \ - \ Q_{j-1}^\omega) \approx (Q_{j_0}^\omega)^{-1} \ + \sum_{j=1}^{J} 4^{-j} \ Q_j^\omega.$$

According to Yserentant [1990] this operator can be replaced by the cheaper one

$$\mathsf{C}\, r := (Q^{\omega}_{j_0})^{-1}\, r + \sum_{j=1}^{J} 4^{-j}\, M_j r, \qquad M_j v := \sum_{i \in \mathcal{N}_j} \frac{(v, \psi_i^{(j)})_\omega}{(1, \psi_i^{(j)})_\omega} \psi_i^{(j)}$$

provided the 'quasi-interpolant' $M_j v$ is spectrally equivalent to Q^{ω}_j uniformly in j. This property holds if the $\psi_l^{(j)}$ form a *Riesz-basis* with respect to the weighted norm, i.e.

$$\| \sum_{l \in \mathcal{N}_j} \alpha_l \psi_l^{(j)} \|^2_\omega \approx \sum_{l \in \mathcal{N}_j} |\alpha_l|^2 (1, \psi_l^{(j)})_\omega$$

The proof follows from the norm equivalence $\| \cdot \|_\omega \approx \| \cdot \|_{\omega,j}$ stated in (17). Then the $\mathcal{C}$ can be taken as a discretized version of the operator C above.

References

C. Bennett-R.Sharpley. *Interpolation of Operators.* Acad. Press, 1988.

F. Bornemann and H. Yserentant. A basic norm equivalence for the theory of multilevel methods. *Numer. Math.*, 64:455–476, 1993.

J. Bramble and J.Xu. Some estimates for a weighted l^2 projection. *Math. of Comp.*, 56:463–476, 1991.

J. Bramble, J. Pasciak, and J.Xu. Parallel multilevel preconditioners. *Math. of Comp.*, 55:1–22, 1990.

P. Butzer and K. Scherer. *Approximationsprozesse und Interpolationsmethoden*, volume 826/826a. BI–Hochschulskripten, Mannheim, 1968.

P. Butzer and K. Scherer. Jackson and Bernstein type inequalities for families of commutative operators in banach spaces. *J. Approx. Theory*, 5:308–342, 1972.

P. Ciarlet. *The finite element method for elliptic problems.* North-Holland, Amsterdam, 1978.

W. Dahmen and A. Kunoth. Multilevel preconditioning. *Numer. Math.*, 63: 315–344, 1992.

H. Johnen and K. Scherer. On the equivalence of the k-functional and moduli of continuity and some applications. *Lect. Notes in Math.*, 571:119–140.

P. Oswald. *Multilevel Finite Approximation: Theory and Application.* Teubner Skripten zur Numerik, 1994.

K. Scherer. Weighted Jackson and Bernstein inequalities. 2003/4.

H. Triebel. *Theory of Function Spaces.* Birkhäuser, 1992.

J. Xu. Iterative methods by space decomposition and subspace correction. *SIAM Rev.*, 34:581–613, 1992.

H. Yserentant. Two preconditioners based on the multi-level splitting of finite element spaces. *Numer. Math.*, 58:163–184, 1990.

H. Yserentant. Old and new convergence proofs for multigrid methods. *Acta Numerica*, pages 285–326, 1993.

Preconditioning for Heterogeneous Problems

Sergey V. Nepomnyaschikh[1] and Eun-Jae Park[2]

[1] Institute of Computational Mathematics and Mathematical Geophysics, SD Russian Academy of Sciences, Novosibirsk, 630090, Russia. svnep@yonsei.ac.kr
[2] Department of Mathematics, Yonsei University, Seoul 120-749, Korea. ejpark@yonsei.ac.kr

Summary. The main focus of this paper is to suggest a domain decomposition method for mixed finite element approximations of elliptic problems with anisotropic coefficients in domains. The theorems on traces of functions from Sobolev spaces play an important role in studying boundary value problems of partial differential equations. These theorems are commonly used for a priori estimates of the stability with respect to boundary conditions, and also play very important role in constructing and studying effective domain decomposition methods. The trace theorem for anisotropic rectangles with anisotropic grids is the main tool in this paper to construct domain decomposition preconditioners.

1 Introduction

In order to present the basic idea of the algorithm, let us consider the following simple model problem. Let a domain Ω be the union of two non-overlapping subdomains which are rectangles, i.e., $\overline{\Omega} = \cup_{i=1}^{2} \overline{\Omega}_i$, where

$$\Omega_i = \{(x,y) | i-1 < x < i,\ 0 < y < 1\},\ i = 1,2.$$

In Ω we consider the following problem. Find p such that

$$\begin{aligned} -\operatorname{div}(a\nabla p) &= f \qquad \text{in } \Omega, \\ p &= 0 \qquad \text{on } \partial\Omega, \end{aligned} \tag{1}$$

where the matrix a is given as follows: $a = \begin{pmatrix} a^x & 0 \\ 0 & a^y \end{pmatrix}$ with $a^x = a_i^x$ and $a^y = a_i^y$ being positive constants in each Ω_i, $i = 1,2$. Denote the interface by $\gamma = \partial\Omega_1 \cup \partial\Omega_2 \setminus \partial\Omega$.

For the problem (1) we introduce a flux variable, $u = -a\nabla p$, which is of interest in many applications. Writing $\alpha = a^{-1}(x)$, the inverse matrix of a, the problem (1) is equivalent to seeking (u, p) such that

$$\begin{aligned} \alpha u + \nabla p &= 0 \quad \text{in } \Omega, \\ \operatorname{div} u &= f \quad \text{in } \Omega, \\ p &= 0 \quad \text{on } \partial\Omega. \end{aligned} \tag{2}$$

Multiplying by test functions and integrating by parts we obtain the following weak formulation of problem (2)

$$\begin{aligned} (\alpha u, v) - (\operatorname{div} v, p) &= 0, \quad \forall v \in H(\operatorname{div};\Omega), \\ -(\operatorname{div} u, q) &= -(f, q), \quad \forall q \in L^2(\Omega), \end{aligned} \tag{3}$$

where $H(\operatorname{div};\Omega) = \{v \in L^2(\Omega)^2 \ : \ \operatorname{div} v \in L^2(\Omega)\}$.

The variational formulation (3) fits the abstract framework that is generally used for mixed methods. It is well known that under the LBB condition the abstract framework is well-posed.

We consider the rectangular Raviart-Thomas mixed finite element spaces $V_h \subset H(\operatorname{div};\Omega)$ and $Q_h \subset L^2(\Omega)$ associated with the triangulation $\mathcal{T}_h$ of Ω. The lowest order rectangular Raviart-Thomas elements are defined as follows:

$$RT(T) = Q_{1,0}(T) \times Q_{0,1}(T), \quad Q(T) = Q_{0,0}(T) \quad \text{for rectangle } T.$$

Define

$$V_h = \{v \in H(\operatorname{div};\Omega) \ : \ v|_T \in RT(T), \quad \forall T \in \mathcal{T}_h\}$$

and

$$Q_h = \left\{q \in L^2(\Omega) \ : \ q|_T \in Q(T), \quad \forall T \in \mathcal{T}_h\right\}.$$

For simplicity, we will consider the uniform rectangular decomposition $\mathcal{T}_h$, where mesh steps $h_x = \frac{1}{m}$, $h_y = \frac{1}{n}$ for some positive integers m, n so that $\mathcal{T}_h = \mathcal{T}_{1h} \cup \mathcal{T}_{2h}$. For $i = 1, 2$

$$\begin{aligned} V_{hi} &= \{v \in H(\operatorname{div};\Omega_i) \ : \ v|_T \in RT(T), \quad \forall T \in \mathcal{T}_{ih}\}, \\ Q_{hi} &= \left\{q \in L^2(\Omega_i) \ : \ q|_T \in Q(T), \quad \forall T \in \mathcal{T}_{ih}\right\}. \end{aligned} \tag{4}$$

The standard mixed finite element approximation $(u_h, p_h) \in V_h \times Q_h$ is defined by

$$\begin{aligned} (\alpha u_h, v) - (\operatorname{div} v, p_h) &= 0, \quad \forall v \in V_h, \\ -(\operatorname{div} u_h, q) &= -(f, q), \quad \forall q \in Q_h. \end{aligned} \tag{5}$$

Note that the normal component of the members in V_h is continuous across the interior boundaries in γ. We relax this constraint on V_h by introducing Lagrange multipliers; see Arnold and Brezzi [1985]. Let E_h be the set of edges which belongs to γ. The Lagrange multipliers space Λ_h to enforce the required continuity on γ is defined by

$$\Lambda_h = \left\{\mu \in L^2\Big(\bigcup_{e \in E_h} e\Big) : \mu|_e \in V_h \cdot \nu|_e \text{ for each } e \in E_h\right\}.$$

Then, the hybridized form of domain decomposition method for mixed finite elements is to find $(u_{ih}, p_{ih}, \lambda_h) \in V_{ih} \times W_{ih} \times \Lambda_h$ such that

$$\begin{aligned}\sum_{i=1}^{2}(\alpha u_{ih}, v_i) - \sum_{i=1}^{2}(\operatorname{div} v_i, p_{ih}) + \sum_{i=1}^{2} < \lambda_h, v_i \cdot \nu_i >= 0, \ \forall v_i \in V_{ih} \\ -\sum_{i=1}^{2}(\operatorname{div} u_{ih}, q_i) = -(f, q_i), \ \forall q_i \in W_{ih} \\ \sum_{i=1}^{2} < \mu, u_{ih} \cdot \nu_i >= 0, \ \forall \mu \in \Lambda_h.\end{aligned} \tag{6}$$

With the standard ordering of the unknowns, the matrix equation for (6) is given by

$$\begin{bmatrix} A_x & 0 & B_x^T & C^T \\ 0 & A_y & B_y^T & 0 \\ B_x & B_y & 0 & 0 \\ C & 0 & 0 & 0 \end{bmatrix} \begin{bmatrix} U_x \\ U_y \\ P \\ \Lambda \end{bmatrix} = \begin{bmatrix} 0 \\ 0 \\ F \\ 0 \end{bmatrix}. \tag{7}$$

2 General Approach to Preconditioning Saddle Point Problems

Let V and Q be Hilbert spaces. Let an operator $A : V \to V$ be linear, symmetric, positive definite, bounded and let a linear operator B map V into Q. Denote by B^T the transpose operator for B. Let us consider the following saddle point problem: find $(u, p) \in V \times Q$ such that

$$\mathcal{A}\chi := \begin{bmatrix} A & B^T \\ B & 0 \end{bmatrix} \begin{bmatrix} u \\ p \end{bmatrix} = \begin{bmatrix} g \\ f \end{bmatrix}, \quad g \in V, f \in Q.$$

For the operator $\mathcal{A}$ we search for a preconditioner R in the block-diagonal form

$$R = \begin{bmatrix} A & 0 \\ 0 & \Sigma \end{bmatrix},$$

where Σ maps Q to Q.

Consider the spectral problem

$$\mathcal{A}\chi = \lambda R \chi$$

with $\Sigma = BA^{-1}B^T$. Then the eigenvalues of the problem belongs to the set

$$\{\frac{1-\sqrt{5}}{2}, 1, \frac{1+\sqrt{5}}{2}\}.$$

According to Rusten and Winther [1992] and Kuznetsov and Wheeler [1995], we have

Lemma 1. *Let $\hat{R}$ be a symmetric operator. If there are positive constants c_1 and c_2 such that*

$$c_1(\mathcal{R}\chi, \chi) \le (\hat{\mathcal{R}}\chi, \chi) \le c_2(\mathcal{R}\chi, \chi), \quad \forall \chi \in V \times Q,$$

then the eigenvalues λ of $\hat{R}^{-1}\mathcal{A}$ belong to two segments

$$\lambda \in [-d_1, -d_2] \cup [d_3, d_4],$$

where

$$d_1 = \frac{1}{c_1}(\frac{\sqrt{5}-1}{2}),\ d_2 = \frac{1}{c_2}(\frac{\sqrt{5}-1}{2}),$$

$$d_3 = \frac{1}{c_2}, \qquad d_4 = \frac{1}{c_1}(\frac{\sqrt{5}+1}{2}).$$

To solve the problem $\mathcal{A}\chi = b$, we use the Lanczos method with a preconditioner $\hat{R}$ satisfying the lemma. Denote

$$\theta = \frac{\max\{d_1, d_4\}}{\min\{d_2, d_3\}}, \qquad q = \frac{\theta - 1}{\theta + 1}.$$

Then, from general theory of iterative methods, if χ^0 is an initial vector and χ^n is an approximation after n iterations by the Lanczos method, the following estimate holds:

$$\|\chi^n - \chi\|_{\hat{R}} \le 2q^n \|\chi^0 - \chi\|_{\hat{R}},$$

where

$$\|\chi\|_{\hat{R}} = (\hat{R}\chi, \chi)^{\frac{1}{2}}.$$

It means that the construction of an effective preconditioner for $\mathcal{A}$ has been reduced to the construction of an effective preconditioner for the Schur complement $BA^{-1}B^T$.

Remark 1. If the cost of the multiplication of A^{-1} by a vector is small (for example if A is a diagonal matrix), then instead of solving the system with $\mathcal{A}$ by the Lanczos method, we can solve the system with the Schur complement $BA^{-1}B^T$ by a preconditioned conjugate gradient method.

3 Preconditioning for the Schur complement

Let us denote by $\hat{p}$, $\hat{q} \in R^{(n \cdot m+n+n \cdot m)}$ vectors in block form:

$$\hat{p} = [p_1\ \lambda\ p_2]^T, \quad \hat{q} = [q_1\ \mu\ q_2]^T$$

where

$$p_1 = [p_{1,1}\ p_{1,2}\ \cdots\ p_{1,n}\ p_{2,1}\ p_{2,2}\ \cdots\ p_{m,1}\ p_{m,2}\ \cdots\ p_{m,n}]^T,$$

$$\lambda = [\lambda_1\ \lambda_2\ \cdots\ \lambda_n]^T,$$

$$p_2 = [p_{m+1,1}\ p_{m+2,2}\ \cdots\ p_{m+1,n}\ p_{m+2,1}\ p_{m+2,2}\ \cdots p_{2m,1}\ \cdots\ p_{2m,n}]^T,$$

and $\hat{q}$ is similarly denoted.

Elimination of the flux variables in (7) reduces to the Schur complement which we denote by the matrix S according to the ordering of unknowns $[p_1\ \lambda\ p_2]^T$. Let

$$\tilde{a}_i^x = a_i^x \frac{h_y}{h_x}, \quad \tilde{a}_i^y = a_i^y \frac{h_x}{h_y}, \quad i = 1, 2.$$

Then we define the $(n \cdot m + n + n \cdot m) \times (n \cdot m + n + n \cdot m)$ matrix $\tilde{S}$ so that

$$\tilde{S}\hat{p} = \hat{q},$$

where

$$\begin{aligned}
-\tilde{a}_1^x p_{i-1,j} - \tilde{a}_1^x p_{i+1,j} - \tilde{a}_1^y p_{i,j-1} - \tilde{a}_1^y p_{i,j+1} + 2(\tilde{a}_1^x + \tilde{a}_1^y) p_{i,j} &= q_{i,j}, \\
i = 1, 2, \cdots, m, \quad j = 1, 2, \cdots, n&, \\
-\tilde{a}_2^x p_{i-1,j} - \tilde{a}_2^x p_{i+1,j} - \tilde{a}_2^y p_{i,j-1} - \tilde{a}_2^y p_{i,j+1} + 2(\tilde{a}_2^x + \tilde{a}_2^y) p_{i,j} &= q_{i,j}, \\
i = m+1, m+2, \cdots, 2m, \quad j = 1, 2, \cdots, n&, \\
-\tilde{a}_1^x p_{m,j} - \tilde{a}_2^x p_{m+1,j} + (\tilde{a}_1^x + \tilde{a}_2^x)\lambda_j &= \mu_j, \\
j = 1, 2, \cdots, n&.
\end{aligned}$$

Here

$$p_{0,j} = 0, \quad p_{2m+1,j} = 0, \quad j = 1, 2, \cdots, n,$$
$$p_{i,0} = 0, \quad p_{i,n+1} = 0, \quad i = 1, 2, \cdots, 2m.$$

The following is an analogue of Cowsar et al. [1995] and Kwak et al. [2003] for anisotropic case:

Lemma 2. *There exist constant c_1, c_2, independent of a^x, a^y, h_x, h_y, such that for any $\hat{p}$*

$$c_1(S\hat{p}, \hat{p}) \le (\tilde{S}\hat{p}, \hat{p}) \le c_2(S\hat{p}, \hat{p}).$$

With the block representation of $\hat{p}$, we can consider a block form of $\tilde{S}\hat{p}$

$$\begin{aligned}
\tilde{S}\hat{p} &= \begin{bmatrix} B_1 & B_{10} & 0 \\ B_{01} & (B_0^{(1)} + B_0^{(2)}) & B_{02} \\ 0 & B_{20} & B_2 \end{bmatrix} \begin{bmatrix} p_1 \\ \lambda \\ p_2 \end{bmatrix} \\
&= \left(\begin{bmatrix} B_1 & B_{10} & 0 \\ B_{01} & B_0^{(1)} & 0 \\ 0 & 0 & 0 \end{bmatrix} + \begin{bmatrix} 0 & 0 & 0 \\ 0 & B_0^{(2)} & B_{02} \\ 0 & B_{20} & B_2 \end{bmatrix} \right) \begin{bmatrix} p_1 \\ \lambda \\ p_2 \end{bmatrix} \\
&= (\tilde{S}_1 + \tilde{S}_2)\hat{p}.
\end{aligned}$$

According to the Additive Schwarz Method; see Matsokin and Nepomnyaschikh [1989], we can define a preconditioner $\tilde{\tilde{S}}$ for $\tilde{S}$ as

$$\tilde{\tilde{S}}^{-1} = \begin{bmatrix} \tilde{B}_1^{-1} & 0 & 0 \\ 0 & 0 & 0 \\ 0 & 0 & \tilde{B}_2^{-1} \end{bmatrix} + \begin{bmatrix} 0 & t_1 & 0 \\ t_1^T & \Sigma^{-1} & t_2^T \\ 0 & t_2 & 0 \end{bmatrix}.$$

Here $\tilde{B}_1, \tilde{B}_2$ are spectrally equivalent to B_1, B_2 and Σ is spectrally equivalent to the Schur complements for $\tilde{S}_1 + \tilde{S}_2$:

$$(B_0^{(1)} - B_{01}B_1^{-1}B_{10}) + (B_0^{(2)} - B_{02}B_2^{-1}B_{20}) = \Sigma_1 + \Sigma_2$$

and t_1, t_2 extension operators of functions from γ to Ω_1 and Ω_2 respectively such that

$$c_1(\Sigma_1\lambda, \lambda) \le (\tilde{S}_1 t_1\lambda, t_1\lambda) \le c_2(\Sigma_1\lambda, \lambda),$$
$$c_1(\Sigma_2\lambda, \lambda) \le (\tilde{S}_2 t_2\lambda, t_2\lambda) \le c_2(\Sigma_2\lambda, \lambda),$$

for any λ.

For optimal convergence of the corresponding iterative method, all constants of spectral equivalence should be independent of a^x, a^y, h_x, h_y.

Now we consider only one subdomain Ω_1. We omit subindex for the subdomain and denote by $\hat{p}$

$$\hat{p} = [p_1 \ \lambda]^T$$

with block vectors p_1, λ defined as before and denote by A_0 the $n \times n$ matrix

$$A_0 = \begin{bmatrix} 2 & -1 & & & \\ -1 & 2 & -1 & & \\ & & & & \\ & & -1 & 2 & -1 \\ & & & -1 & 2 \end{bmatrix}$$

and by I the $n \times n$ identity matrix. Consider the following $(n{\cdot}m{+}n) \times (n{\cdot}m{+}n)$ matrix $\tilde{S}$

$$\tilde{S} = \tilde{a}_x \begin{bmatrix} (\sigma A_0 + 2I) & -I & & \\ -I & (\sigma A_0 + 2I) & -I & \\ & & & \\ & -I & (\sigma A_0 + 2I) & -I \\ & & -I & I \end{bmatrix} := \tilde{a}_x \begin{bmatrix} B_{11} & B_{12} \\ B_{21} & B_{22} \end{bmatrix}$$

where

$$\sigma = \frac{\tilde{a}^y}{\tilde{a}^x}, \quad \tilde{a}^x = a^x \frac{h_y}{h_x}, \quad \tilde{a}^y = a^y \frac{h_x}{h_y}, \quad B_{22} = I.$$

Using the eigenvectors and eigenvalues of A_0

$$A_0 q_i = \lambda_i q_i, \ i = 1, \cdots, n,$$

$$q_i = \begin{bmatrix} q_i(1) \\ q_i(2) \\ \cdot \\ \cdot \\ q_i(n) \end{bmatrix}, \quad q_i(j) = \sqrt{\frac{2}{n+1}} \sin \frac{i\pi j}{n+1}, \ j = 1, 2, \cdots, n,$$

$$\lambda_i = 4 \sin^2 \frac{\pi i}{2(n+1)},$$

we have

$$A_0 = Q \Lambda Q^T \tag{8}$$

$$Q = Q^T = [q_1, \cdots, q_n], \quad \Lambda = \mathrm{diag}\{\lambda_1, \cdots, \lambda_n\}.$$

To compute the Schur complement for $\tilde{S}$

$$\Sigma = \tilde{a}^x (B_{22} - B_{21} B_{11}^{-1} B_{12})$$

we can use (8) and find a diagonal matrix D such that

$$\Sigma = Q D Q^T.$$

Using the technique from Matsokin and Nepomnyaschikh [1989], we have

Lemma 3. *The diagonal matrix D has the following elements.*

$$D = diag\{\mu_1(\Sigma), \mu_2(\Sigma), \cdots, \mu_n(\Sigma)\}$$

$$\mu_i(\Sigma) = \tilde{a}^x \left(1 - \frac{U_{m-1}(\beta_i)}{U_m(\beta_i)}\right),$$

where $\beta_i = \frac{1}{2}\sigma\lambda_i + 1$ and U_j is the Chebyshev polynomial of the second kind of degree j so that

$$U_j(x) = \frac{1}{2\sqrt{x^2-1}}\left((x + \sqrt{x^2-1})^{j+1} - (x + \sqrt{x^2-1})^{-(j+1)}\right).$$

Using the lemma for both subdomains Ω_1 and Ω_2, we can define

$$\Sigma_1 = Q D_1 Q^T \quad \text{for subdomain} \ \ \Omega_1$$

and

$$\Sigma_2 = Q D_2 Q^T \quad \text{for subdomain} \ \ \Omega_2.$$

Then, put

$$\Sigma = Q(D_1 + D_2) Q^T$$

and so

$$\Sigma^{-1} = Q(D_1 + D_2)^{-1} Q^T.$$

Hence, finally we have the following theorem:

Theorem 1. *Let preconditioner $\tilde{\tilde{S}}$ be defined as*

$$\tilde{\tilde{S}}^{-1} = \begin{bmatrix} \tilde{B}_1^{-1} & 0 & 0 \\ 0 & 0 & 0 \\ 0 & 0 & \tilde{B}_2^{-1} \end{bmatrix} + \begin{bmatrix} 0 & t_1 & 0 \\ t_1^T & \Sigma^{-1} & t_2^T \\ 0 & t_2 & 0, \end{bmatrix}$$

with Σ defined as above. Then, there exist constant c_1, c_2, independent of a^x, a^y, h_x, h_y, such that for any $\hat{p}$

$$c_1(S\hat{p}, \hat{p}) \leq (\tilde{\tilde{S}}\hat{p}, \hat{p}) \leq c_2(S\hat{p}, \hat{p}).$$

To summarize, we have presented an optimal algorithm for a model anisotropic problem such that a condition number of the preconditioned problem is independent of parameters coefficients, grid sizes and the arithmetical cost of implementation of this algorithm is proportional to the number of degrees of freedom.

Acknowledgement. This work was supported by the Brain Korea 21 Project through Department of Mathematics, Yonsei University. EJP was supported in part by Com²MaC-KOSEF.

References

D. N. Arnold and F. Brezzi. Mixed and nonconforming finite element methods: implementation, postprocessing and error estimates. *RAIRO Math. Model. Numer. Anal.*, 19:7–32, 1985.

L. C. Cowsar, J. Mandel, and M. F. Wheeler. Balancing domain decomposition for mixed finite elements. *Math. Comp.*, 64(211):989–1015, July 1995.

Y. A. Kuznetsov and M. F. Wheeler. Optimal order substructuring preconditioners for mixed finite element methods on nonmatching grids. *East-West J. Numer. Math.*, 3(2):127–143, 1995.

D. Kwak, S. Nepomnyaschikh, and H. Pyo. Domain decomposition for model heterogeneous anisotropic problem. *Numer. Linear Algebra*, 10:129–157, 2003.

A. M. Matsokin and S. V. Nepomnyaschikh. On using the bordering method for solving systems of mesh equations. *Sov. J. Numer. Anal. Math. Modeling*, 4:487–492, 1989.

T. Rusten and R. Winther. A preconditioned iterative method for saddle point problems. *SIAM J. Matrix Anal.*, 13:887–904, 1992.

Part XI

Minisymposium: Recent Advances for the Parareal in Time Algorithm

On the Convergence and the Stability of the Parareal Algorithm to Solve Partial Differential Equations

Guillaume Bal

Columbia University, APAM (`http://www.columbia.edu/~gb2030/`)

Summary. After stating an abstract convergence result for the parareal algorithm used in the parallelization in time of general partial differential equations, we analyze the stability and convergence properties of the algorithm for equations with constant coefficients. We show that suitably damping coarse schemes ensure unconditional stability of the parareal algorithm and analyze how the regularity of the initial condition influences convergence in the absence of sufficient damping.

1 Introduction

The parareal algorithm pioneered in Lions et al. [2000] and slightly modified in Bal and Maday [2002] allows us to speed up the numerical simulation of solutions to time dependent equations provided that we have enough processors. We refer to Baffico et al. [2002], Bal [2003], Farhat and Chandesris [2003], Maday and Turinici [2002] for additional detailed presentations of the method and applications; see also Section 2 below. Natural questions then arise related to the stability and convergence of the method. Here are some elements of answers to these questions. In Section 2 an abstract result in a general setting shows convergence of the algorithm provided that regularity conditions are satisfied. In the simplified setting of linear partial differential equations with constant coefficients, more refined estimates are provided for the convergence and stability of the parareal algorithm in Section 3. A typical result we can show is as follows: whereas the parareal algorithm is unconditionally stable for most discretizations of parabolic equations, it is not for hyperbolic equations.

2 An abstract convergence result

Let us consider a possibly non-linear partial differential equation of the form

$$\frac{du}{dt} = A(t,u), \qquad t > 0 \\ u(0) = u_0, \tag{1}$$

where $u(t)$ takes values in a Banach space B and $A(t,u)$ is a possibly time-dependent partial differential operator. Let us assume that we want to simulate this equation on an interval of time $(0,T)$ and define the discretization $0 = T^0 < T^1 < \cdots < T^N = T$. We assume that the equation (1) admits a solution operator $g(T^n, v)$, which to $v \in B$ associates $u(T^{n+1}) \in B$ solution of

$$\frac{du}{dt} = A(t,u), \qquad T^n < t < T^{n+1} \\ u(T^n) = v. \tag{2}$$

Let us now assume that we have at our disposal an approximate solution operator $g_\Delta(T^n, v)$. We then define the approximate sequence

$$u_1^{n+1} = g_\Delta(T^n, u_1^n), \qquad n \in I = \{0 \le n \le N-1\}, \tag{3}$$

and $u_1^0 = u_0$. Let ΔT be the maximal lag between successive time steps T^n and assume that $T^{n+1} - T^n \ge \eta_0 \Delta T$ for some positive constant η_0 for all $n \in I$. The convergence of u^N to $u(T^n)$ as ΔT converges to 0 is obtained under the following hypotheses. Let us assume that g_Δ is Lipschitz in a Banach space B_0 and is an approximation of order m of g in the sense that

$$\sup_{n\in I} \|g_\Delta(T^n,u) - g_\Delta(T^n,v)\|_{B_0} \le (1 + C\Delta T)\|u-v\|_{B_0}, \tag{4}$$

$$\sup_{n\in I} \|\delta g(T^n,u)\|_{B_0} \le C(\Delta T)^{m+1}\|u\|_{B_1}, \tag{5}$$

where $\delta g(T^n,u) = g(T^n,u) - g_\Delta(T^n,u)$, C is a constant independent of ΔT, $\|\cdot\|_B$ denotes a norm on the Banach space B and B_1 is another Banach space (usually a strict subset of B_0).

If in addition, (1) is stable in B_1, in the sense that $u(t) \in B_1$ uniformly in t provided that $u_0 \in B_1$, then we have the classical result

$$\|u(T^N) - u_1^N\|_{B_0} \le C(\Delta T)^m \|u_0\|_{B_1}, \tag{6}$$

based on the above regularity assumptions and the decomposition

$$u(T^{n+1}) - u_1^{n+1} = \delta g(T^n, u(T^n)) + g_\Delta(T^n, u(T^n)) - g_\Delta(T^n, u_1^n).$$

We now consider the *parareal algorithm*, which allows us to speed up the calculation of $u(t)$ provided that we have access to a sufficiently large number of processors. The parareal algorithm is given by

$$u_{k+1}^{n+1} = g_\Delta(T^n, u_{k+1}^n) + \delta g(T^n, u_k^n), \qquad n \in I, \quad k \ge 1 \tag{7}$$

with initial condition $u_{k+1}^0 = u_0$. The idea of the algorithm is to add to the prediction term $g_\Delta(T^n, u_{k+1}^n)$ a correction involving the previous iteration

u_k^n and a "fine" calculation that can be done in parallel on every time step (T^n, T^{n+1}) of the coarse discretization since all the terms u_k^n are known when the calculation of u_{k+1}^n starts. This requires to have N processors available. In practice, we cannot simulate $\delta g(T^n, u)$ exactly, but rather an approximation of the form $g_\delta(T^n, u) - g_\Delta(T^n, u)$, where $g_\delta(T^n, u)$ is a sufficiently accurate approximation of $g(T^n, u)$ so that in all the results that follow we can safely replace $g(T^n, u)$ by $g_\delta(T^n, u)$. See the bibliographical references for additional details.

We now show that the above algorithm ideally replaces the accuracy of order m of the non-parallel algorithm by an accuracy of order km to solve (1). Such an accuracy cannot be obtained in general solely under the above assumptions for g and g_Δ. It rather requires much stronger regularity constraints. Let us define a scale of Banach spaces B_0, B_1, ..., B_k, where in practice $B_k \subset B_{k-1} \subset \cdots B_1 \subset B_0$. We have then the following result:

(H1) The equation (1) is stable in all spaces B_j for $0 \le j \le k$, in the sense that $\|u(t)\|_{B_j} \le C\|u_0\|_{B_j}$ where C is independent of u_0 and $t \in (0, T)$.

(H2) The operator g_Δ is Lipschitz in the sense that

$$\sup_{n \in I} \|g_\Delta(T^n, u) - g_\Delta(T^n, v)\|_{B_j} \le (1 + C\Delta T)\|u - v\|_{B_j}, \quad 0 \le j \le k-1,$$

where C is independent of ΔT, u, and v.

(H3) The operator δg is an approximation of order m in the sense that

$$\sum_{n \in I} \|\delta g(T^n, u)\|_{B_j} \le C(\Delta T)^{m+1}\|u\|_{B_{j+1}}, \quad 0 \le j \le k-1. \tag{8}$$

Theorem 1. *Under hypotheses (H1)-(H3), the order of accuracy of the parallel algorithm (7) is mk. More precisely, for* $u_0 \in \bigcap_{0 \le j \le k} B_j$, *we have*

$$\|u(T^N) - u_k^N\|_{B_0} \le C(\Delta T)^{mk}\|u_0\|_{B_k},$$

where C is a constant independent of ΔT and u_0.

Proof. The result is obtained by induction. We know it to hold when $k = 1$ thanks to (H1)-(H3) and assume that it holds for k given. We then apply the result with the sequence of Banach spaces $B_1, \ldots, B_{k+1}$, so that

$$\|u(T^N) - u_k^N\|_{B_1} \le C(\Delta T)^{mk}\|u_0\|_{B_{k+1}}.$$

By definition (7), we have

$$u(T^{n+1}) - u_{k+1}^{n+1} = g_\Delta(T^n, u(T^n)) - g_\Delta(T^n, u_{k+1}^n) + \delta g(T^n, u(T^n)) - \delta g(T^n, u_k^n).$$

Using (H1)-(H3), this implies that

$$\begin{aligned}\|u(T^{n+1}) - u_{k+1}^{n+1}\|_{B_0} &\le (1 + C\Delta T)\|u(T^n) - u_{k+1}^n\|_{B_0} + \\ &\qquad C(\Delta T)^{m+1}\|u(T^n) - u_k^n\|_{B_1} \\ &\le (1 + C\Delta T)\|u(T^n) - u_{k+1}^n\|_{B_0} + \\ &\qquad C(\Delta T)^{m(k+1)+1}\|u_0\|_{B_{k+1}}.\end{aligned}$$

Since $u(T^0) = u_{k+1}^0 = u_0$, it is then a routine calculation to obtain (8). □

3 Stability and convergence for linear operators

The above abstract result shows that the parareal algorithm allows us to replace a scheme of order m by a scheme of order km *provided that* regularity constraints are met. In practice we need to choose B_{k+1} as a strict subset of B_k in (H3); see below. These constraints may not be optimal as they do not account for possible dissipative effects of the coarse scheme.

To address this issue, we consider a pseudo-differential operator $P(D)$ with symbol $P(\xi)$ such that $\Re(P(\xi)) \geq 0$ (otherwise consider $P+\alpha$ with α sufficiently large and solve for $v = e^{-\alpha t}u$), and define $A(t,u) = P(D)u$. To simplify notation, we assume that $u(t)$ is a function on $\mathbb{R}$. In the Fourier domain, the evolution of $\hat{u}(t,\xi) = \int_{\mathbb{R}} e^{-ix\xi}u(t,x)dx$ is thus given by

$$\begin{aligned} &\frac{\partial \hat{u}}{\partial t}(t,\xi) + P(\xi)\hat{u}(t,\xi) = 0 \qquad \xi \in \mathbb{R},\ t>0 \\ &\hat{u}(0,\xi) = \hat{u}_0(\xi), \qquad \xi \in \mathbb{R}. \end{aligned} \tag{9}$$

The evolution operator is independent of time T^n ($= n\Delta T$ from now on to simplify) and is in the frequency domain

$$g(T^n, \hat{u}) = e^{-P(\xi)\Delta T}\hat{u}. \tag{10}$$

We define $\delta(\xi) = P(\xi)\Delta T$ and using the same notation g define the propagator

$$g(\delta(\xi)) = e^{-\delta(\xi)}. \tag{11}$$

We now want to define approximate solutions to the above equation. Let us assume that the symbol $P(\xi)$ is approximated by $P_H(\xi)$ and that the time propagator $g(\delta)$ is approximated by $g_\Delta(\delta_H)$, where $\delta_H(\xi) = P_H(\xi)\Delta T$. For instance $g_\Delta(\delta) = (1+\delta)^{-1}$ for implicit first-order Euler. We then define the parareal scheme as

$$\hat{u}_{k+1}^{n+1}(\xi) = g_\Delta(\delta_H(\xi))\hat{u}_{k+1}^n(\xi) + (g(\delta(\xi)) - g_\Delta(\delta_H(\xi)))\hat{u}_k^n(\xi) \tag{12}$$

for $n \in I$ and $k \geq 0$. The boundary conditions are $\hat{u}_{k+1}^0(\xi) = \hat{u}_0(\xi)$ and $\hat{u}_0^n(\xi) \equiv 0$. In the above equations, ξ is a parameter so stability and error of convergence can be analyzed for each frequency separately. We verify that the exact solution $\hat{u}^n(\xi)$ also satisfies (12) with different initial conditions so that the error term $\varepsilon_k^n(\xi) = \hat{u}^n(\xi) - \hat{u}_k^n(\xi)$ satisfies the following equation

$$\varepsilon_{k+1}^{n+1}(\xi) = g_\Delta(\delta_H(\xi))\varepsilon_{k+1}^n(\xi) + (g(\delta(\xi)) - g_\Delta(\delta_H(\xi)))\varepsilon_k^n(\xi) \tag{13}$$

with boundary conditions $\varepsilon_{k+1}^0(\xi) = 0$ and $\varepsilon_0^n(\xi) = \hat{u}^n(\xi)$. We verify that $\varepsilon_1^n(\xi) = (g^n(\delta(\xi)) - g^n(\delta_H(\xi))\hat{u}_0(\xi)$. Upon defining

$$\theta_k^n = \frac{1}{g_\Delta^{n-k+1}(\delta_H)}\frac{\varepsilon_k^n}{(g(\delta) - g_\Delta(\delta_H))^{k-1}}, \qquad \theta_1^n = \frac{(g^n(\delta) - g_\Delta^n(\delta_H))}{g_\Delta^n(\delta_H)}\hat{u}_0(\xi),$$

we find that $\theta_{k+1}^{n+1} = \theta_{k+1}^{n} + \theta_k^n$ and $\theta_{k+1}^0 = 0$. The constraint $\Re(P(\xi)) \geq 0$ implies that $g^n(\delta)$ is uniformly bounded and we assume that $g_\Delta^n(\delta_H)$ is uniformly bounded for $n \in I$ (this is nothing but stability of the coarse scheme). We also assume that $|(g(\delta)/g_\Delta(\delta_H))(\xi)|$ is uniformly bounded. This allows us to obtain the following bound

$$|\theta_1^n(\xi)| \leq C(n|g(\delta) - g_\Delta(\delta_H)| \wedge 1)|\hat{u}_0(\xi)|. \tag{14}$$

Here $a \wedge b = \min(a,b)$. This implies then the following bound on the error

$$|\varepsilon_{k+1}^{n+1}|(\xi)| \leq C\Big(n|g(\delta) - g_\Delta(\delta_H)| \wedge 1\Big)|g_\Delta(\delta_H)|^{n-k}|g(\delta) - g_\Delta(\delta_H)|^k \binom{n}{k}, \tag{15}$$

for $n \in I$ and $k \geq 1$, where C is proportional to $|\hat{u}_0(\xi)|$ only. This estimate gives us optimal bounds to prove convergence and stability of the parareal scheme. For $k+1=1$ in the above formula we recover that the coarse scheme is stable. We can also obtain the maximal order of convergence of the scheme after k parareal iterations. Assuming that $\delta = \delta_H$ and that g_Δ is of order m so that $|(g - g_\Delta)(\delta)| \leq C(\xi)(\Delta T)^{m+1}$, we deduce that $|\varepsilon_k^N|$ is of order $N(\Delta T)^{m+1}[(\Delta T)^{m+1}]^{k-1}N^{k-1} = (\Delta T)^{km}$. We recover that the parareal algorithm replaces an algorithm of order m by an algorithm of order km.

A central difficulty with the iterative scheme is that for $k \geq 2$, the error term $\varepsilon_k^n(\xi)$, hence the solution $\hat{u}_k^n$, may blow up for large frequencies. We now consider this stability analysis.

The above parallel algorithm requires to solve a fine scale problem $k-1$ times to obtain an accuracy of order km on the coarse time grid (i.e. at the times T^n, $n \in N$). The best use of the available processors is thus obtained for $k=2$; see Bal [2003]. In any case the algorithm is useful when the value of k is small. We therefore assume from now on that $1 \sim k \ll n$. This implies that $\binom{n}{k} \sim n^k$.

Let us assume here that $\delta_H = \delta$, i.e., the spatial discretization is the same for the coarse and fine schemes. Stability at all frequencies is thus ensured provided that

$$R_{k+1,n}(\delta) = \Big(|g_\Delta|^{n-k}|g - g_\Delta|^k\Big)(\delta)n^k, \tag{16}$$

remains bounded for all values of $\delta(\xi) = P(\xi)\Delta T$, $\xi \in \mathbb{R}$. The first term $|g_\Delta|^{n-k}$ is clearly bounded since the coarse algorithm is stable. The second term $|g - g_\Delta|^k$ is also bounded. However it may not be small for values of δ of order 1. The relation (16) indicates how the parareal algorithm blows up. Unless high frequencies (δ of order $O(1)$ or higher) are damped by the coarse scheme $|g_\Delta|$, an instability of size n^k will appear at the iteration step $k+1$.

Consider a real-valued $P(\xi) > 0$ for $|\xi| > 0$ and the centered scheme

$$g_\Delta(\delta) = \frac{1 - \delta/2}{1 + \delta/2} = e^{-\delta} + O(\delta^3). \tag{17}$$

As $\delta \to \infty$, $|g_\Delta(\delta)| \to 1$ and $g(\delta) \to 0$. We thus observe that high frequencies will grow like n^{k-1} in (16) and the parareal scheme is unstable as soon as

$k \geq 2$ although the scheme is unconditionally stable for $k = 1$. However we can still apply the general theory. For instance for $P(\xi) \leq |\xi|^M$ and a coarse scheme of order m we can verify that all the hypotheses of Theorem 1 are verified provided that $B_k = H^{(m+1)Mk}(\mathbb{R})$. The instability of the parareal scheme can thus be overcome by assuming sufficient regularity of the initial conditions.

The growth n^{k-1} may be compensated when the coarse scheme is dissipative. A result that covers many classical examples is the following:

Theorem 2. *Let us assume that the coarse scheme is an approximation of order m of the exact propagator and that it is* dissipative *in the sense that there exist three constants $C, \gamma > 0$ and $1 \leq \beta \leq m+1$ such that for all $\xi \in \mathbb{R}$,*

$$|g(\delta(\xi)) - g_\Delta(\delta(\xi))| \leq C(\delta(\xi)^{m+1} \wedge 1) \tag{18}$$

$$|g_\Delta(\delta(\xi))| \leq (1 + C\Delta T)e^{-\gamma(|\delta(\xi)|^\beta \wedge 1)}. \tag{19}$$

Then the parallel algorithm is stable in the sense that $R_{k,n}(\delta)$ is bounded uniformly in $k = O(1)$, $n \in N$, and $\delta = \delta(\xi)$ for $\xi \in \mathbb{R}$.

Proof. Consider the case $|\delta| \geq 1$ first. We observe that $|R_{k+1,n}(\delta)|$ is bounded by $e^{-\gamma n} n^k$ which is clearly bounded independent of $k = O(1)$, n, and $|\delta| \geq 1$.

For $|\delta| \leq 1$ we obtain that $|R_{k+1,n}(\delta)|$ is bounded by $|\delta|^{(m+1)k} n^k e^{-n\gamma|\delta|^\beta}$. Upon differentiating the above majorizing function with respect to $|\delta|$ we obtain that the maximum is reached for $|\delta_0|^\beta = (k(m+1))/(\gamma\beta n)$, so that a bound for $|R_{k+1,n}(\delta)|$ is given by

$$|R_{k+1,n}(\delta)| \leq Ce^{-k(m+1)/\beta}\Big(n^{1-(m+1)/\beta}\Big)^k.$$

The latter power of n does not grow as $n \to \infty$ provided that $\beta \leq m+1$. □

The above result shows that sufficient exponential damping of the large frequencies is sufficient to ensure stability. Notice that the centered scheme defined in (17) does not verify the hypotheses of the theorem since large values of δ are not damped at all by the coarse scheme. For real valued non-negative symbols $P(\xi)$, we can use Theorem 2 to deduce that the θ scheme

$$g_\Delta(\delta) = \frac{1 - (1-\theta)\delta}{1 + \theta\delta}, \tag{20}$$

makes the parareal algorithm stable as soon as $\theta > 1/2$. Indeed we have then $|\delta(\xi)| = \delta(\xi)$. Since $g'_\Delta(0) = -1$, we verify that $\beta = 1$ and γ sufficiently small (all the more that $\theta \to 1/2+$) does the job. This covers then all parabolic equations (such as the Laplancian $P(\xi) = \xi^2$) and many spatial discretizations (such as the centered finite difference scheme $P(\xi) = 2h^{-2}[1 - \cos(h\xi)]$).

The result also applies to more general equations with complex-valued symbol. Consider the same θ scheme given in (20). We define $\delta = \delta_r + i\delta_i$. The assumption on $P(\xi)$ implies that $\delta_r \geq 0$. We now find that

$$|g_\Delta(\delta)| = \sqrt{\frac{(1+(1-\theta)^2|\delta|^2 - 2(1-\theta)\delta_r)}{(1+\theta^2|\delta|^2 + 2\theta\delta_r)}} \leq \sqrt{\frac{(1+(1-\theta)^2|\delta|^2)}{1+\theta^2|\delta|^2}}. \quad (21)$$

Asymptotically as $\delta \to 0$ we find for the θ scheme that $|g_\Delta(\delta)| \leq 1 - \frac{\theta^2-(1-\theta)^2}{2}|\delta|^2 + O(|\delta|^3)$. We see that we have to choose here $\beta = 2$ to verify the assumptions of Theorem 2 (instead of $\beta = 1$ when δ is real-valued). Since $m = 1$, $\beta = 2$ is the only value allowed. It is then easy to find a value of γ such that the hypotheses of Theorem 2 are satisfied. The theorem thus addresses the case of the transport equation $P(\xi) = ia\xi$ and of the advection diffusion equation $P(\xi) = 1 + ia\xi + b\xi^2$, $b > 0$.

In the latter case we can actually do better. Indeed consider $P(\xi) = \alpha_0 + \sum_{k=1}^{M-1}\alpha_k\xi^k + |\xi|^M$ with $\alpha_k \in \mathbb{C}$ arbitrary and $\alpha_0 > 0$ such that $\Re(P(\xi)) > 0$. Then there exists a constant $\rho > 0$ such that $|\delta_i| \leq \rho\delta_r$, whence $|\delta| \leq \sqrt{1+\rho^2}\delta_r$. We then deduce from (21) that, on $(0, \delta_0)$ for $\delta_0 = 2((1-\theta)\sqrt{1+\rho^2})^{-1}$, we have $|g_\Delta(\delta)| \leq (1 + \frac{\theta}{\sqrt{1+\rho^2}}|\delta|)^{-1}$, and still from (21), that $|g_\Delta(\delta)| < 1 - \varepsilon$ for $\varepsilon > 0$ on (δ_0, ∞) as $\theta > 1/2$. So we can choose $\beta = 1$ (this is important in Theorem 3 below) and γ small enough in Theorem 2.

Let us now turn to convergence. We have seen that the error term at final time T is bounded by $|R_{k,N}(\xi)| \leq C|\delta|^{(m+1)k}N^k e^{-N\gamma|\delta|^\beta}$ for $|\delta| < 1$. Let us assume that $\beta = 1$. We then deduce from the above analysis that $|R_{k,N}(\xi)| \leq CN^{-km}$. So all frequencies are uniformly bounded by CN^{-km} and the algorithm has an accuracy of order $(\Delta T)^{km}$ in $\mathcal{L}(H^\alpha(\mathbb{R}))$ for all $\alpha \in \mathbb{R}$. The case $\beta > 1$ is much less favorable. Let us assume that $|P(\xi)| \leq |\xi|^M$ for $M > 0$. The bound for R is then $|R_{k,N}(\xi)| \leq C(\Delta T)^{km}[|\xi|^{M(m+1)k}e^{-\gamma|\xi|^{M\beta}(\Delta T)^{\beta-1}}]$. This implies that there is no damping for all frequencies of order up to $|\xi| \sim (\Delta T)^{(1-\beta)/(M\beta)} \gg 1$. So the L^2 norm (for instance) of the error term is bounded by

$$\int_{|\xi|\leq(\Delta T)^{(1-\beta)/(M\beta)}} |\hat{u}_0(\xi)|^2|\xi|^{2M(m+1)k}d\xi \leq C\|u_0\|^2_{M(m+1)k},$$

where $\|\cdot\|_\alpha$ is the norm in the Hilbert space $H^\alpha(\mathbb{R})$. We have thus proved the following result:

Theorem 3. *Under the assumptions of Theorem 2 we have the following convergence result for all $\alpha \in \mathbb{R}$. When $\beta = 1$, we have*

$$\|u^N - u_k^N\|_\alpha \leq C(\Delta T)^{km}\|u_0\|_\alpha.$$

When $\beta > 1$, we have for $0 \leq \tau \leq 1$,

$$\|u^N - u_k^N\|_\alpha \leq C(\Delta T)^{km\tau}\|u_0\|_{\alpha+\tau M(m+1)k}.$$

The latter estimate follows from stability by interpolation. At $\tau = 1$, this is nothing but Theorem 1. However the result for $\tau < 1$ requires stability.

For parabolic equations ($P(\xi) > 0$ real-valued), we thus obtain that the θ scheme for $\theta > 1/2$ has a very strong convergence property as the error is of order $(\Delta T)^{km}$ in the space where u_0 is defined. This generalizes the results obtained in Bal and Maday [2002] for $\theta = 1$. The same result holds for symbols of the form $P(\xi) = |\xi|^M + \textit{lower order terms}$ with $\Re(P(\xi)) > 0$ since we can then choose $\beta = 1$. However for the transport equation $P(\xi) = ia\xi$ (or symbols with purely imaginary leading term) we see that convergence of implicit Euler ($m = M = 1$, $\beta = 2$) is of order $(\Delta T)^{k\tau}\|u_0\|_{2k\tau}$ in $L^2(\mathbb{R})$.

Let us conclude by a remark. In the above analysis we have assumed that $\delta = \delta_H$, i.e. the spatial discretization is the same for the coarse and the fine steps. This need not be so. For lack of space, we postpone the general analysis to future work and only mention the result where $P(\xi) = \xi^2$, $P_H(\xi) = 2(1 - \cos(H\xi))/H^2$, and the implicit Euler scheme $g_\Delta(\delta_H) = (1 + \delta_H)^{-1}$. We can then show that $\|u^N - u_2^N\|_\alpha \leq C(\Delta T)^2\|u_0\|_{\alpha+4}$ so that the optimal accuracy $(\Delta T)^2$ is attained for $k = 2$ for a coarse spatial discretization $H = (\Delta T)^{1/2}$. The loss of "4" derivatives comes from the fact that the coarse scheme damps frequencies up to H^{-1} only. This is an intermediate result between Theorems 1 and 3. It should be compared with the case $\delta = \delta_H$ where the spatial discretization need be chosen as $h = \Delta T$. So $H \gg h$ for the same final accuracy (but it requires more regularity of the initial condition).

Acknowledgement. I had very interesting discussions on the parareal scheme with, among others, Charbel Farhat, Paul Fischer, Yvon Maday, and Gunnar Staff during this very interesting and very well organized DD15. Support from NSF grant DMS-0239097 and an Alfred P. Sloan fellowship is also greatly acknowledged.

References

L. Baffico, S. Bernard, T. Maday, G. Turinici, and G. Zérah. Parallel-in-time molecular-dynamics simulations. *Phys. Rev. E*, 66:057701, 2002.

G. Bal. Parallelization in time of (stochastic) ordinary differential equations. *Preprint;* `www.columbia.edu/~gb2030/PAPERS/ParTimeSODE.ps`, 2003.

G. Bal and Y. Maday. A "parareal" time discretization for non-linear PDE's with application to the pricing of an american put. *Recent developments in domain decomposition methods (Zürich, 2001), Lect. Notes Comput. Sci. Eng., Springer, Berlin*, 23:189–202, 2002.

C. Farhat and M. Chandesris. Time-decomposed parallel time-integrators: theory and feasibility studies for fluid, structure, and fluid-structure applications. *Int. J. Numer. Meth. Engng.*, 58(9):1397–1434, 2003.

J.-L. Lions, Y. Maday, and G. Turinici. Résolution d'EDP par un schéma en temps "pararéel". *C.R.A.S. Sér. I Math.*, 332(7):661–668, 2000.

Y. Maday and G. Turinici. A parareal in time procedure for the control of partial differential equations. *C.R.A.S. Sér. I Math.*, 335:387–391, 2002.

A Parareal in Time Semi-implicit Approximation of the Navier-Stokes Equations

Paul F. Fischer[1], Frédéric Hecht[2], and Yvon Maday[2]

[1] Argonne National Laboratory, Mathematics and Computer Science Division
[2] Université Pierre et Marie Curie, Laboratoire Jacques Louis Lions, Boite courrier 187 75252 Paris cedex 05, France.

Summary. The "parareal in time" algorithm introduced in Lions et al. [2001] enables parallel computation using a decomposition of the interval of time integration. In this paper, we adapt this algorithm to solve the challenging Navier-Stokes problem. The coarse solver, based on a larger timestep, may also involve a coarser discretization in space. This helps to preserve stability and provides for more significant savings.

1 Introduction

The "parareal in time" algorithm was introduced in Lions et al. [2001] to allow parallel computations based on a decomposition of the interval of time integration. This algorithm, which can be interpreted as a predictor-corrector scheme (see Bal and Maday [2002] and Baffico et al. [2002]), involves a prediction step based on a coarse approximation and propagation of the phenomenon and a correction step computed in parallel and based on a fine approximation. Significant speedups are observed (see, in particular, Bal [2003] on this aspect). A combination of the parareal in time algorithm with more conventional domain decomposition approaches was presented in Maday and Turinici [2003] and exploits both space and time concurrency.

Many applications of the method have already been performed, but this paper is the first that targets the challenging Navier-Stokes problem. The coarse solver is based on a large timestep but also on a coarse discretization in space, which further reduces serial overhead.

2 The Basic Algorithm on a Simple Equation

Consider the following time dependent problem:

$$\frac{\partial y}{\partial t} + \mathcal{A}y = 0, \quad y(T_0) = y_0,$$

where, for the sake of simplicity, $\mathcal{A}$ *does not depend on time.* We introduce the propagator $\mathcal{E}$ such that $y(\tau) = \mathcal{E}_\tau(y_0)$. Let $T_n = n\Delta T$, $n = 0, ..., N$ be instants at which we wish to consider snapshots of the solution. Then $y(T_n) = \mathcal{E}_{T_n}(y^0) = \mathcal{E}_{\Delta T}(y(T_{n-1}))$, from the semigroup property of $\mathcal{E}$.

In most cases $\mathcal{E}$ is not realizable and can only be approximated; for instance, we can introduce a fine and precise propagator $\mathcal{F}$ defined by an Euler scheme, either implicit or explicit. Similar to the continuous solution, we have the approximations $y(T_n) \simeq \lambda_n = \mathcal{F}_{T_n}(y_0) = \mathcal{F}_{\Delta T}(\lambda_{n-1})$. Clearly, the approximation process is sequential.

The parareal algorithm assumes we are given another propagator denoted as $\mathcal{G}$. It is cheaper (and consequently less accurate) than $\mathcal{F}$. One can think of $\mathcal{F}$ as based on an Euler scheme with a very small timestep δt and $\mathcal{G}$ as based on an Euler scheme with the larger timestep ΔT. We present and implement here another possibility, as proposed in Lions et al. [2001], in which $\mathcal{G}$ is based on a coarse approximation in space as well.

The iterative process $\lambda_n^{k+1} = \mathcal{G}_{\Delta T}(\lambda_{n-1}^{k+1}) + \mathcal{F}_{\Delta T}(\lambda_{n-1}^k) - \mathcal{G}_{\Delta T}(\lambda_{n-1}^k)$ provides a converging sequence toward λ_n. Our interest in this predictor-corrector scheme lies in the fact that after iteration k and before iteration $k+1$ starts, we can compute in parallel the corrections $\mathcal{F}_{\Delta T}(\lambda_{n-1}^k) - \mathcal{G}_{\Delta T}(\lambda_{n-1}^k)$ for all n; thus the only sequential part of the algorithm is the evaluation of the coarse operator.

3 The Parareal in Time Algorithm for Navier-Stokes

We apply the parareal scheme to the incompressible Navier-Stokes equations,

$$\frac{\partial \mathbf{u}}{\partial t} + \mathbf{u} \cdot \nabla \mathbf{u} = -\nabla p + \frac{1}{Re} \nabla^2 \mathbf{u} \text{ in } \Omega, \quad \nabla \cdot \mathbf{u} = 0 \text{ in } \Omega, \tag{1}$$

with prescribed boundary and initial conditions for the velocity, $\mathbf{u}$. Here, p is the pressure, Re the Reynolds number and Ω is a regular domain of $\mathbb{R}^d$.

The temporal discretization is based on the high-order operator-splitting methods developed in Maday et al. [1990] that generalize the characteristics method of Pironneau [1982]. The left-hand side of (1) is recast as a material derivative, which is discretized by using a stable rth-order backward difference formula (BDFr):

$$\frac{\partial \mathbf{u}}{\partial t} + \mathbf{u} \cdot \nabla \mathbf{u} = \frac{Du}{Dt} \approx \frac{1}{\Delta t} \left(\beta_0 \mathbf{u}_{n+1} - \sum_{i=1}^{r} \beta_i \overline{\mathbf{u}}_{n+1-i} \right). \tag{2}$$

For BDF1, $(\beta_0, \beta_1) = (1, 1)$, and for BDF2, $(\beta_0, \beta_1, \beta_2) = (\frac{3}{2}, \frac{4}{2}, \frac{-1}{2})$. The values $\overline{\mathbf{u}}_{n+1-i}$ represent the values of $\mathbf{u}$ at the foot of the characteristic associated with each gridpoint and are computed by solving the linear convective subproblem, $(\overline{\mathbf{u}}_j)_t + \mathbf{u} \cdot \nabla \overline{\mathbf{u}}_j = 0$, $t \in (t_{n+1-i}, t_{n+1}]$, with initial condition $\overline{\mathbf{u}}_j(t_{n+1-i}) = \mathbf{u}_{n+1-i}$ for $j = n+1-i$, $i = 1, \ldots, r$. This leads to a linear symmetric Stokes problem to be solved implicitly at each step and permits timestep sizes, Δt, corresponding to convective CFL numbers greater than unity, thus significantly reducing the number of Stokes solves.

3.1 Finite Element Approximation in Space

The finite element method is based on an compatible choice of spaces for the velocity and the pressure: the Taylor-Hood method. The time discretization is a characteristics method of order 1 for the convection and implicit for the Stokes operator. The algebraic system resulting from the discretization is solved through a Cahouet-Chabart algorithm. The problem solved corresponds to a 2-D flow past a cylinder of diameter 2 centered at the origin. The upstream boundary is located at a distance 5 from the center of the cylinder and the downstream boundary at a distance 10.

A special consideration when using the parareal scheme in conjunction with a *semi-implicit* timestepping scheme is that the step size be small enough to ensure stability. This is a particular concern for the coarse solver, $\mathcal{G}$, where one wants to choose a large timestep in order to minimize the serial overhead. Fortunately, the characteristics scheme allows this, provided that the subintegration timestep satisfies the governing stability criterion.

3.2 Spectral Approximation in Space

The spectral element method (SEM) for the Stokes problem is also based on a compatible choice of spaces: it is the $\mathbb{P}_M \times \mathbb{P}_{M-2}$ discretization introduced in Maday and Patera [1989]. The discretization spaces are $X^M := \{\mathbf{v} \in H_0^1(\Omega)^d, \mathbf{v}_{|\Omega^e} \in \mathbb{P}_M(\Omega^e)^d, e = 1, ..., E\}$ for the velocity and $Y^M := \{q \in L^2(\Omega), q_{|\Omega^e} \in \mathbb{P}_{M-2}(\Omega^e), e = 1, ..., E\}$, for the pressure. Here, $\mathbb{P}_M(\Omega^e), e = 1, ..., E$ is the space of polynomials of partial degree $\leq M$ on each of the E nonoverlapping elements, Ω^e, whose union composes Ω. At present, we restrict our attention to cases where Ω is a rectangular domain comprising a tensor-product array of $E = E_x \times E_y$ elements allowing the use of the fast diagonalization method. Details of the SEM formulation and implementation can be found in Fischer [1997].

To implement the parareal scheme, we require a solver of the form $\underline{\mathbf{u}}_{n+1} = \mathcal{F}_{\Delta T}(\underline{\mathbf{u}}_n)$. That is, given an initial value $\underline{\mathbf{u}}_n$, the solver propagates the solution over a time interval $(T_n, T_{n+1}]$ to produce a result $\underline{\mathbf{u}}_{n+1}$. We thus need to "restart" the computation for each application of $\mathcal{F}$ and $\mathcal{G}$: we use Richardson extrapolation and combine two steps of size $\Delta t/2$ with one of size Δt to yield an $O(\Delta t^3)$ local error at the start of each $\mathcal{F}$ ($\Delta t = \delta t$) or $\mathcal{G}$ ($\Delta t = \Delta T$) substep.

4 Further Reduction of the Cost of the Coarse Operator

In nondimensional time units, the parareal single- and P-processor solution times and parallel speedup are, respectively,

$$\tau_1 = \frac{T}{\delta t}, \quad \tau_P = K\left(\frac{1}{P}\frac{T}{\delta t} + \alpha\frac{T}{\Delta T}\right), \quad S_P := \frac{\tau_1}{P\tau_P} = \frac{P}{K\left(1 + \alpha P \frac{\delta t}{\Delta T}\right)}, \tag{3}$$

The P-processor estimates in (3) neglect communication overhead and simply reflect the extra work resulting from additional iterations (K) and the serial coarse propagation ($\alpha T/\Delta T$). The factor α reflects the relative per-step cost of $\mathcal{F}_{\Delta T}$ and $\mathcal{G}_{\Delta T}$. To achieve reasonable scalability we need K and $\alpha P \frac{\delta t}{\Delta T}$ to be order unity.

Here, we propose to use propagators based not only on different timesteps, but also on different spectral degrees.

4.1 The Finite Element Context

The reduction is obtained by using a fine grid defined by dividing each coarse triangle into four triangles. The resulting number of vertices is equal to 1021 in the coarse mesh (H) and 3994 in the fine mesh (h). The coarse operator $\mathcal{G}_{\Delta T}$ is based on ΔT and the coarse grid H. The proposed parareal in time scheme is then

$$\mathbf{U}_{n+1}^{k+1} = \Pi_H^h \mathcal{G}_{\Delta T}(\Pi_h^H \mathbf{U}_n^{k+1}) + \mathcal{F}_{\Delta T}(\mathbf{U}_n^k) - \Pi_H^h \mathcal{G}_{\Delta T}(\Pi_h^H \mathbf{U}_n^k), \tag{4}$$

The operator that allows one to go from the coarse mesh to the fine one (denoted as Π_H^h) and reciprocally (i.e., Π_h^H) can be either the interpolation operator I_h (resp. I_H) or the L^2 projection on discretely divergence free functions Π_h (resp. Π_H).

4.2 The Spectral Context

We assume that we have a coarse operator $\mathcal{G}_{\Delta T}$ based on ΔT and a spectral degree $\tilde{M}$ together with a fine solver $\mathcal{F}_{\Delta T}$ based on δt and a spectral degree $M > \tilde{M}$. The proposed parareal in time scheme is then

$$\mathbf{U}_{n+1}^{k+1} = \Pi_{\tilde{M}}^M \mathcal{G}_{\Delta T}(\Pi_M^{\tilde{M}} \mathbf{U}_n^{k+1}) + \mathcal{F}_{\Delta T}(\mathbf{U}_n^k) - \Pi_{\tilde{M}}^M \mathcal{G}_{\Delta T}(\Pi_M^{\tilde{M}} \mathbf{U}_n^k), \tag{5}$$

where $\Pi_{\tilde{M}}^M$ is the L^2 prolongation operator from $X^{\tilde{M}}$ onto X^M and $\Pi_M^{\tilde{M}}$ is the L^2 projection operator from X^M onto $X^{\tilde{M}}$.

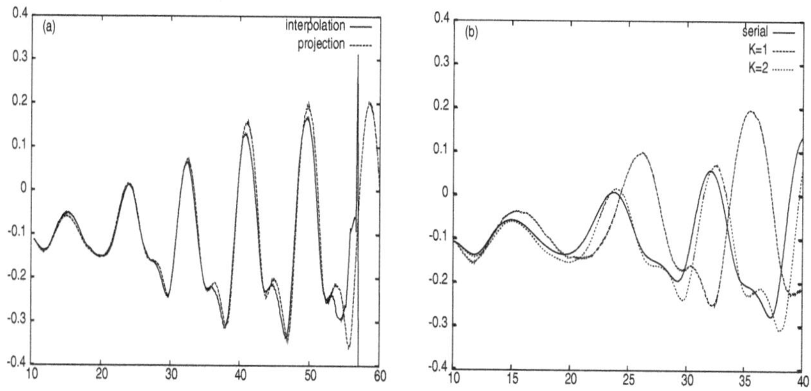

Fig. 1. Time history of the vertical component of the velocity at point (1.5, 0) for the parareal algorithm with coarse operator based on coarse spatial and temporal discretization. (a) Comparison of interpolation and L^2 projection operators. (b) Comparison $K = 1$, $K = 2$, and standard serial algorithms

5 Results

5.1 The Finite Element Context

We have run the cylinder simulation at $Re = 200$, starting with an initial condition made of a flow computed by running the coarse simulation over a time equal to 10. We have used a fine timestep equal to 0.02 and a coarse timestep equal to 0.2. The simulation has been run over successive time intervals of size 2 corresponding to $P = 10$. Over these intervals, we have run the parareal scheme with small numbers of iterations because of the small size of the intervals.

We have first compared the two types of operators to go from one mesh onto the other one. The simple interpolation operator appears to be unstable: the simulation blows up, for example, with the use of the strategy three coarse sweeps alternated with two fine sweeps, corresponding to K=2. On the contrary, the use of an L^2-type projection operator on the discrete divergence-free functions is stable throughout these simulations. These results are illustrated in Fig. 1a, where we plot the time history of the vertical component of the velocity at a point situated on the axis of the flow at a distance 1.5 downstream of the center of the cylinder. To check the accuracy of the method, we have computed the solution corresponding to the fine timestep as a reference and compared the parareal scheme with 2 coarse + 1 fine (K=1) and 3 coarse + 2 fine (K=2) to the reference solution. The results are plotted in Fig. 1b, and the solution for K=2 is quite good. Note that the plot representing the history at this point is much discriminating because of the complexity of the

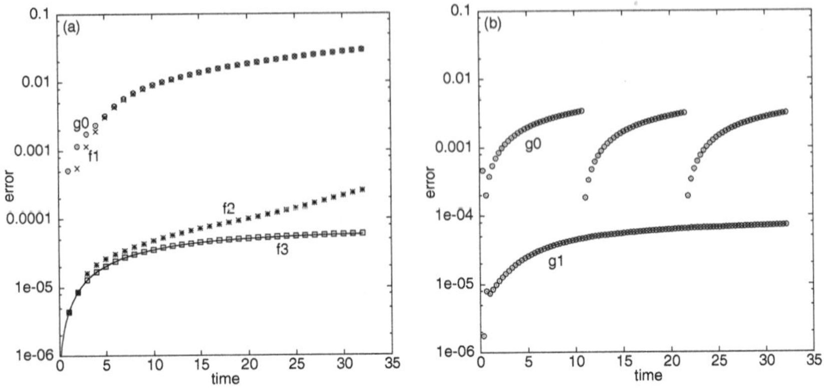

Fig. 2. Relative error in the v-component of the solution, versus time, for the SEM solution of the Orr-Sommerfeld test problem with $E = 15$, $M = 15$, and $\delta t = .005$. (a) single partition with $K = 3$, (b) tripartition with $K = 1$

flow. For longer times, we have checked that the period of the flow and the time to establish the periodic flow are very well captured by the parareal scheme for this situation.

These are only preliminary runs, from which we conclude that the projection operator performs better than the interpolation one. We will seek to optimize the timestep choice in order to be able to get an accurate solution with a lower number of iterations. Let us note that these simulations have been done *effectively* in parallel by using the code Freefem.

5.2 The Spectral Context

We have applied the parareal/SEM algorithm to the Orr-Sommerfeld problem studied in Fischer [1997]. The computational domain is $\Omega = [0, 2\pi] \times [-1, 1]$, with periodic boundary conditions in x and homogeneous Dirichlet conditions on $y = \pm 1$. The growth of a small-amplitude (10^{-5}) Tollmien-Schlichting wave, superimposed on plane Poiseuille channel flow at $Re = 7500$, is monitored and compared with linear theory over the interval $t \in [0, 32]$, which corresponds to ≈ 1.25 periods of the traveling wave solution.

Figure 2a shows the relative L^∞ (maximum pointwise) error in the y-component of velocity versus time for an E=15 element discretization using BDF2 with M=15, $\Delta t_{\mathcal{F}} = \delta t = .005$ for $\mathcal{F}_{\Delta T}$, and $\tilde{M} = 15$, $\Delta t_{\mathcal{G}} = .333 = \Delta T/3$ for $\mathcal{G}_{\Delta T}$. The solid line shows the discretization error for the standard serial algorithm (BDF2,$\Delta t = .005$). The point plots show the error in the solution for the first coarse sweep (g0) and for the first three fine sweeps (f1, f2, and f3). The number of coarse and fine substeps per iteration is 32, corresponding to a $P = 32$ processor simulation. For this problem, the scheme

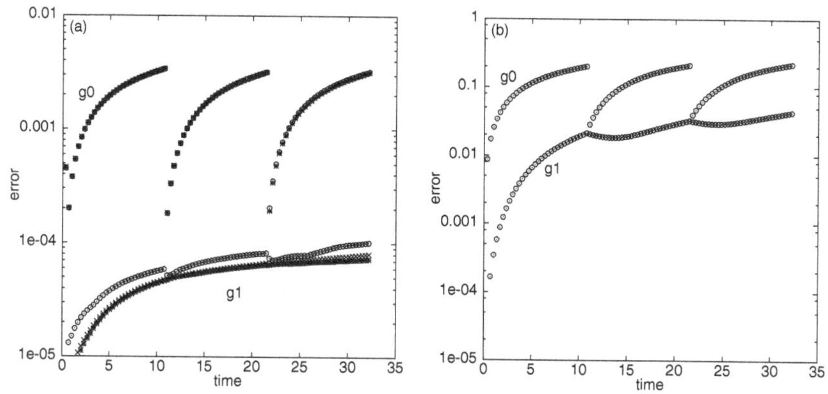

Fig. 3. Error histories for configuration as in Fig. 2b: (a) using varying fine/coarse approximation orders $(M,\tilde{M})$; $\circ = (13{,}13)$, $\times = (15{,}13)$, $* = (15{,}15)$; and (b) $(M,\tilde{M}) = (15{,}15)$ without Richardson extrapolation

converges in $K = 3$ iterations. Each coarse-grid step, $\mathcal{G}_{\Delta T}(\underline{\mathbf{u}}^n)$, is computed by using three steps of size $\Delta t_{\mathcal{G}} = 1/3$, plus two additional steps for the Richardson extrapolation, corresponding to $\alpha = 5$. Based on (3), the estimated speedup is $S_{32} \approx 6$.

To reduce ΔT and, hence, K, we also consider applying the parareal scheme to *subintervals* of $[0,T]$, where the initial condition on each interval is taken to be the $K = 1$ solution from the preceding interval. The scheme requires an initial $\mathcal{G}_{\Delta T}$ sweep ($k = 0$), followed by a single $\mathcal{F}_{\Delta T}$ and $\mathcal{G}_{\Delta T}$ correction ($k = 1 =: K$). The errors after the $\mathcal{G}_{\Delta T}$ sweeps are shown in Fig. 2b. Here, three subintervals are used, with $\delta t = .005$, and $\Delta T = \Delta t_G = 67\delta t$. Given that there are two coarse sweeps per interval, each with a cost of three Navier-Stokes solves (because of the Richardson extrapolation), we have $\alpha = 6$, corresponding to a speedup of $S_{32} = 8.3$.

We next consider the coarse approximation in space for $\mathcal{G}_{\Delta T}$, given by (5). Here, one must be careful that the temporal errors do not dominate the spatial errors. Otherwise, perceived benefits from reducing $\tilde{M}$ could equally well be gained through reductions in both M and $\tilde{M}$. We verify that this is not the case by plotting in Fig. 3a the errors for the tripartition algorithm of Fig. 2b using discretization pairs $(M,\tilde{M})$=(15,15), (15,13), and (13,13). The error for the $(15,13)$ pairing is almost the same as for the $(15,15)$ case. The coarse-grid solve cost, however, is significantly reduced. For the SEM in two dimensions, this cost scales as $\tilde{M}^3$, so we may expect $\alpha \approx (13/15)^3 6$, which implies $S_{32} = 11.2$ for the three-step Richardson scheme.

The Richardson iteration was chosen for programming convenience. Other approaches with lower cost that also have an $O(\Delta t^3)$ local truncation error could be used for the initial coarse step. For example, one could employ a semi-implicit scheme combining Crank-Nicolson and second-order Runge-Kutta that would require only a single set of system solves, thus effectively reducing α threefold. The corresponding speedup for $\alpha \approx 2(13/15)^3$ would be $S_{32} = 19.7$. Note that simply dropping the Richardson extrapolation in favor of BDF1 has disastrous consequences, as illustrated by the error behavior in Fig. 3b.

Acknowledgement. This work was supported in part by the Mathematical, Information, and Computational Sciences Division subprogram of the Office of Advanced Scientific Computing Research, Office of Science, U.S. Department of Energy, under Contract W-31-109-Eng-38.

References

L. Baffico, S. Bernard, Y. Maday, G. Turinici, and G. Zérah. Parallel in time molecular dynamics simulations. *Phys. Rev. E.*, 66, 2002.

G. Bal. Parallelization in time of (stochastic) ordinary differential equations. Math. Meth. Anal. Num. (submitted), 2003.

G. Bal and Y. Maday. A "parareal" time discretization for non-linear PDE's with application to the pricing of an American put. In *Recent developments in domain decomposition methods (Zürich, 2001)*, volume 23 of *Lect. Notes Comput. Sci. Eng.*, pages 189–202. Springer, Berlin, 2002.

P. Fischer. An overlapping schwarz method for spectral element solution of the incompressible navier-stokes equations. *J. Comp. Phys.*, 133:84–101, 1997.

J.-L. Lions, Y. Maday, and G. Turinici. A parareal in time discretization of pde's. *C.R. Acad. Sci. Paris, Serie I*, 332:661–668, 2001. URL `http://www.elsevier.nl/gej-ng/10/37/18/47/27/35/article.pdf`.

Y. Maday and A. T. Patera. Spectral element methods for the Navier-Stokes equations. In A. Noor and J. Oden, editors, *State of the Art Surveys in Computational Mechanics*, New York, 1989. ASME.

Y. Maday, A. T. Patera, and E. M. Rønquist. An operator-integration-factor splitting method for time-dependent problems: application to incompressible fluid flow. *J. Sci. Comput.*, 5(4):263–292, 1990.

Y. Maday and G. Turinici. The parareal in time iterative solver : a further direction to parallel implementation. In *Fifteenth International Conference on Domain Decomposition Methods*, Berlin, 2003. Springer, Lecture Notes in Computational Science and Engineering.

O. Pironneau. On the transport-diffusion algorithm and its applications to the Navier-Stokes equations. *J. Num. Math.*, 38(3):309–332, Mar. 1982.

The Parareal in Time Iterative Solver: a Further Direction to Parallel Implementation

Yvon Maday[1] and Gabriel Turinici[2,3]

[1] Université Pierre et Marie Curie, Laboratoire Jacques Louis Lions, Boite courrier 187, 75252 Paris cedex 05, France.
[2] INRIA Rocquencourt, Domaine de Voluceau, Rocquencourt B.P. 105, 78153 Le Chesnay cedex, France
[3] CERMICS-ENPC, Champs sur Marne, 77455 Marne la Vallée Cedex 2, France

Summary. This paper is the basic one of the series resulting from the minisymposium entitled "Recent Advances for the Parareal in Time Algorithm" that was held at DD15. The parareal in time algorithm is presented in its current version (predictor-corrector) and the combination of this new algorithm with other more classical iterative solvers for parallelization which makes it possible to really consider the time direction as fertile ground to reduce the time integration costs.

1 Introduction

In the seminal paper Lions et al. [2001] the generalization of the concept of domain decomposition for time solution was first proposed. Even though the time direction seems intrinsically sequential, the combination of a coarse and a fine solution procedure have proven to converge and allow for more rapid solution if parallel architectures are available. This has led to the name "parareal in time" that has been proposed for this method. Since then, this scheme has received some attention and a presentation under the format of a predictor-corrector algorithm has been made in Bal and Maday [2002] and also in Baffico et al. [2002]. It is this last presentation that we shall use in what follows. Before this let us mention that a matricial form of the scheme was also presented in Maday and Turinici [2002] were the parareal methodology appears as a preconditioner.

Let us consider the partial differential equation (P.D.E.)

$$\frac{\partial u}{\partial t} + Au = 0, \quad \text{over the time interval } [T_0, T] \tag{1}$$

where A is some functional operator, linear or not, from a Hilbert space V into its dual space V'. This P.D.E. is complemented with initial conditions $u(t = T_0) = u_0$ and appropriate boundary conditions that are implicitly incorporated in the formulation and the space V.

It is well known that, when it exists, the solution of this P.D.E. can be written as

$$u(t) = \mathcal{E}_{t-T_0}(u_0; T_0) \tag{2}$$

and that we have the following semigroup property for any T^*, $T_0 \leq T^* \leq t$

$$u(t) = \mathcal{E}_{t-T^*}(\mathcal{E}_{T^*-T_0}(u_0; T_0); T^*). \tag{3}$$

that formalizes the sequential nature of the this Cauchy problem.

Associated with this formal operator, the numerical solution of this problem leads to an approximate operator $\mathcal{F}$ based on a discretization scheme with time steps δt and some order m. In addition to the time discretization, a discretization in space (with discretization parameter δx) can also be used that leads to an error of the order $\delta t^m + \delta x^\nu$ at any final time T.

Let us assume that a time range $\Delta T >> \delta t$ is being given and that we are interested in the collection of snapshots $\{u(T_n)\}_{0\leq n\leq N}$ where $T_n = T_0 + n\Delta T$ and $T_N = T$. The proper approximation of these values are given by $\{\lambda_n = \mathcal{F}_{n\Delta T}(u_0; T_0)\}_{0\leq n\leq N}$ (hence $\lambda_0 = u_0$) as in (3) we note that

$$\lambda_n = \mathcal{F}_{\Delta T}(\lambda_{n-1}; T_{n-1}). \tag{4}$$

The parareal algorithm makes it possible to define iteratively a sequence λ_n^k that converges toward λ_n as k goes to infinity. It involves a coarse solver $\mathcal{G}$, less accurate than $\mathcal{F}$, but much cheaper. It can be based for example on the time step ΔT (or any coarser time step than δt) together, as was proposed already in Lions et al. [2001], with a coarser discretization in space ΔX (see also Fischer et al. [2003]) or even, a simpler physical model, as was implemented in Maday and Turinici [2003]. The assumptions that are made are that

- $\|D(\mathcal{F}_{\Delta T} - \mathcal{G}_{\Delta T})\| \leq c\varepsilon\Delta T$ where ε depends on both ΔT, δt, ΔX and δx, and goes to zero when ΔT, and ΔX go to zero. The symbol D denotes the first derivative with respect to the first variable.
- $\|D\mathcal{G}_{\Delta T}\| \leq c$

In the parareal algorithm, starting from $\lambda_n^0 = \mathcal{G}_{n\Delta T}(u_0; T_0)$ the sequence λ_n^k, $k \geq 1$ is determined by

$$\lambda_n^k = \mathcal{G}_{\Delta T}(\lambda_{n-1}^k; T_{n-1}) + \mathcal{F}_{\Delta T}(\lambda_{n-1}^{k-1}; T_{n-1}) - \mathcal{G}_{\Delta T}(\lambda_{n-1}^{k-1}; T_{n-1}). \tag{5}$$

and we can prove the following error

$$\|\lambda_n - \lambda_n^k\| \leq C \sum_{m=k}^{n} \binom{n}{m} \|D(\mathcal{F}_{\Delta T} - \mathcal{G}_{\Delta T})\|^m \|D\mathcal{G}_{\Delta T}\|^{n-m}. \tag{6}$$

from which the convergence in k follows since our hypothesis leads to

$$\max_{0\leq n\leq N} \|\lambda_n - \lambda_n^k\| \leq C(T)\varepsilon^k. \tag{7}$$

In order to prove this result we remark that from (5) one obtains:

$$\lambda_n^k - \lambda_n = \left(\mathcal{G}_{\Delta T}(\lambda_{n-1}^k; T_{n-1}) - \mathcal{G}_{\Delta T}(\lambda_{n-1}; T_{n-1})\right) \\ + \ (\mathcal{F}_{\Delta T} - \mathcal{G}_{\Delta T})\,(\lambda_{n-1}^{k-1}; T_{n-1}) - (\mathcal{F}_{\Delta T} - \mathcal{G}_{\Delta T})\,(\lambda_{n-1}; T_{n-1}) \tag{8}$$

Suppose now that $\mathcal{G}_{\Delta T}(\cdot;\cdot)$ and $(\mathcal{F}_{\Delta T} - \mathcal{G}_{\Delta T})\,(\cdot;\cdot)$ are differentiable with respect to the (first) variable uniformly over all its values and over all values of the time parameter. Denoting by $e_n^k = \|\lambda_n^k - \lambda_n\|$ one can write

$$\mathcal{G}_{\Delta T}(\lambda_{n-1}^k; T_{n-1}) - \mathcal{G}_{\Delta T}(\lambda_{n-1}; T_{n-1}) \\ = D\mathcal{G}_{\Delta T}(\lambda_{n-1}; T_{n-1})(\lambda_{n-1}^k - \lambda_{n-1}) + o(e_{n-1}^k)$$

and obtain the estimate

$$\|\mathcal{G}_{\Delta T}(\lambda_{n-1}^k; T_{n-1}) - \mathcal{G}_{\Delta T}(\lambda_{n-1}; T_{n-1})\| \leq 3/2\|D\mathcal{G}_{\Delta T}(\cdot;\cdot)\|e_{n-1}^k$$

for any $e_{n-1}^k \leq \mu_{3/2}^g$. Using the same technique for $(\mathcal{F}_{\Delta T} - \mathcal{G}_{\Delta T})\,(\lambda_{n-1}^{k-1}; T_{n-1})$ one obtains:

$$e_n^k \leq 3/2\|D\mathcal{G}_{\Delta T}(\cdot;\cdot)\|e_{n-1}^k + 3/2\|D\,(\mathcal{F}_{\Delta T} - \mathcal{G}_{\Delta T})\,(\cdot;\cdot)\|e_{n-1}^{k-1} \tag{9}$$

provided that e_{n-1}^k and e_{n-1}^{k-1} are smaller than some constants $\mu_{3/2}^g$ and $\mu_{3/2}^{f-g}$ respectively. Provided that the initial guess λ_n^0, $n = 1, ..., N$ is chosen sufficiently close to the solution λ_n from (4), one can prove by induction the result in (6). Then, we notice that

$$\sum_{m=k}^{n} \binom{n}{m} \|D(\mathcal{F}_{\Delta T} - \mathcal{G}_{\Delta T})\|^m \|D\mathcal{G}_{\Delta T}\|^{n-m} \leq \epsilon^k \sum_{m=k}^{n} \binom{n}{m} (c\Delta T)^m \|D\mathcal{G}_{\Delta T}\|^{n-m}$$

$$\leq \epsilon^k (\Delta T)^k \sum_{m=0}^{n} \binom{n}{m} c^m \|D\mathcal{G}_{\Delta T}\|^{n-m} = \epsilon^k (\Delta T)^k (c + \|D\mathcal{G}_{\Delta T}\|)^n,$$

and thus we obtain the result in (7).

We refer also to Farhat and Chandesris [2003], Staff and Rønquist [2003] and to Bal [2003a] for other issues about stability and approximation of the parareal in time scheme.

2 Combination with domain decomposition – the overlapping case

2.1 The iterative procedure

Let Ω be a domain decomposed into P subdomains that, in this section, we shall assume to be overlapping to make things easier. More precisely let $\overline{\Omega} = \cup_{p=1}^{P} \overline{\Omega}^p$ with $\Omega^p \cap \Omega^q = \emptyset$ whenever $p \neq q$, in addition, we assume that there exists $\omega^{p,q}$ – called here "bandages" – associated with any pair

(p,q) such that $\overline{\Omega}^p \cap \overline{\Omega}^q \neq \emptyset$ so that $\Omega = \{\cup \Omega^p\} \cup \{\cup \omega^{p,q}\}$. Note that those bandages may (and most often have to) overlap.

In what follows, we shall propose a space-time parallel iterative method for solving the following type of problem

$$\begin{aligned} \tfrac{\partial u}{\partial t} - \Delta u &= f, \quad \text{in } \Omega \times [0,T] \\ u(0,x) &= u_0(x), \quad \text{in } \Omega, \\ u(t,x) &= g(t,x), \quad \text{over } \partial\Omega \times [0,T], \end{aligned} \tag{10}$$

To make it easy, for the definition of the algorithm, we assume no discretization is used in space neither for the coarse nor for the fine propagator, similarly we assume that the fine propagator does not involve any discretization in time. We are going to define an iterative procedure that involves the fine and accurate solution (here actually exact) only over each block $\Omega^p \times [T_n, T_{n+1}]$ or $\omega^{p,q} \times [T_n, T_{n+1}]$. The solution at iteration k will be denoted as $u^k_{p,n}$ over $\Omega^p \times [T_n, T_{n+1}]$ and $v^k_{p,q,n}$ over $\omega^{p,q} \times [T_n, T_{n+1}]$. By construction, the function u^k_n built by "concatenation" of the various $\left(u^k_{p,n}\right)_p$ is an element of $H^1(\Omega)$ for almost each time (continuity enforced at the interfaces). We will also have the snapshots λ^k_n available at each time T_n.

The coarse propagator

Once the solution is known at iteration k, the definition of the solution at iteration $k+1$ involves a coarse operator that can be defined as follows:

$$\frac{\mathcal{G}_{\Delta T}(\lambda^k_n) - \lambda^k_n}{\Delta T} - \Delta(\mathcal{G}_{\Delta T}(\lambda^k_n)) = f(T_{n+1}). \tag{11}$$

The fine propagator

The fine propagator actually involves not only the knowledge of λ^k_n but also of u^k_n. It proceeds as follows

Step one. We first propagate the solution over $\omega^{p,q} \times [T_n, T_{n+1}]$ by solving

$$\begin{aligned} \frac{\partial v^{k+1}_{p,q,n}}{\partial t} - \Delta v^{k+1}_{p,q,n} &= f, \quad \text{in } \omega^{p,q} \times [T_n, T_{n+1}] \\ v^{k+1}_{p,q,n}(T_n, x) &= \lambda^k_n(x), \quad \text{in } \omega^{p,q}, \\ v^{k+1}_{p,q,n}(t,x) &= u^k_{p,n}(t,x) + \lambda^k_n(x) - u^k_{p,n}(T_n, x), \\ &\quad \text{over } \partial\omega^{p,q} \cap \partial\Omega^p \times [T_n, T_{n+1}], \end{aligned} \tag{12}$$

Note that the correction: $\lambda^k_n - u^k_{p,n}(T_n,.)$, allows us to have the boundary conditions compatible with the initial condition for each local problem.

Step two. We now define from the various $v^{k+1}_{p,q,n}$ a current global boundary value, named v^{k+1}_n over $(\cup_p \partial\Omega^p) \setminus \partial\Omega$. In the case where the subdomains $\omega^{p,q}$ do not overlap, then v^{k+1}_n is, over each $(\cup_p \partial\Omega^p) \cap \omega^{p,q}$, equal to the unique possible value that is $v^{k+1}_{p,q,n}$. In case of overlapping $\omega^{p,q}$'s, there is a conflict between the $v^{k+1}_{p,q,n}$ that is solved by choosing a continuous convex combination of the different $v^{k+1}_{p,q,n}$'s.

Step three. We now propagate the solution over $\Omega^p \times [T_n, T_{n+1}]$ by solving

$$\begin{aligned}
\frac{\partial u_{p,n}^{k+1}}{\partial t} - \Delta u_{p,n}^{k+1} &= f, \quad \text{in } \Omega^p \times [T_n, T_{n+1}] \\
u_{p,n}^{k+1}(T_n, x) &= \lambda_n^k(x), \quad \text{in } \Omega^p, \\
u_{p,n}^{k+1}(t, x) &= g(x), \quad \text{over } \partial\Omega^p \cap \partial\Omega \times [T_n, T_{n+1}], \\
u_{p,n}^{k+1}(t, x) &= v_n^{k+1}(t, x), \quad \text{over } \partial\Omega^p \cap \partial\omega^{p,q} \times [T_n, T_{n+1}],
\end{aligned} \tag{13}$$

This allows us to define a new global solution u_n^{k+1} over each $\Omega \times [T_n, T_{n+1}]$ since, as we already said, the $u_{p,n}^{k+1}$ do match at the interfaces.

The $k+1$ iteration

The definition of each λ_n^{k+1}, $1 \leq n \leq N$ then proceeds similarly as for (5)

$$\lambda_n^{k+1} = \mathcal{G}_{\Delta T}(\lambda_n^k) + u_n^{k+1}(T_{n+1}) - \mathcal{G}_{\Delta T}(\lambda_n^{k-1}). \tag{14}$$

2.2 Numerical results

The first set of computations has been done on a rectangular domain $]0,4[\times]0,1[$, decomposed into 2 equal rectangles $\Omega^1 =]0,2[\times]0,1[$ and $\Omega^2 =]2,4[\times]0,1[$ plus a rectangular "bandage" $\omega^{1,2}$ of various width ($]1,3[\times]0,1[$ or $]1.5,2.5[\times]0,1[$). The P.D.E. that we have solved is

$$\frac{\partial u}{\partial t} - \nu \Delta u = f, \tag{15}$$

with $\nu = 1$ and $f = 50\sin(2\pi(x+t))\cos(2\pi(y+t))$ over a time range $T - T_0 = 1$. We have used a P_1-finite element discretization in space and an implicit Euler scheme of first order in time. The fine propagator is based on a time step δt that is 50 times smaller than the large time step. In the experiments reported below in Table 1, the size of the large time step $\Delta T = 1/N$ varies. A priori N is related to the number of parallel processors we have. Here this figure should be $2N$ as there are two subdomains that can be run at the same time. Table 1 summarizes the error between λ_n^k and the finite element solution with a very fine discretization in time. Note that in all the situations the error after 5 (resp. 4) iterations remains constant and is (resp. is of the order of) the error resulting from $\delta t = 1/(50N)$. Note that if we double N, achieving thus an error that is, at convergence, twice smaller, the number of iterations remains the same. This indicates the perfect scalability of our global (parareal + Schwarz) scheme.

Note that to be completely legal in the former statement, we assume that the cost of the coarse solvers should be considered as negligible with respect to the cost of the fine solver. To do so a coarse discretization in space should be added, we are currently working in that direction.

Table 1

	width of $\omega^{1,2}$= 2				width of $\omega^{1,2}$ = 1			
k=	N=15	N=30	N=60	N=120	N=15	N=30	N=60	N=120
1	0.95	0.49	0.28	0.17	0.50	0.38	0.31	0.27
2	0.076	0.040	0.031	0.020	0.10	0.068	0.065	0.042
3	0.045	0.022	0.016	0.009	0.056	0.022	0.016	0.009
4	0.045	0.024	0.014	0.005	0.041	0.022	0.016	0.005
5	0.045	0.022	0.011	0.006	0.041	0.020	0.010	0.005
6	0.045	0.022	0.011	0.006	0.041	0.020	0.010	0.005

Another indication on this scalability is that, if we maintain the accuracy, by having the product $n \times N$ constant, then the number of iterations required for convergence remains also constant. Hence provided that you have twice the number of processors, then N can be multiplied by a factor of 2 and the cost of each iteration is divided by 2. Since the number of iterations remains constant, this means that the global time to wait is divided by 2.

We have also performed the same Schwarz method over $]0,4[\times]0,4[$ divided into 4 squares of size 2 (the Ω^p's) and 2 rectangular "bandage" $\omega^{p,q}$: $]1.5,2.5[\times]0,4[$ and $]0,4[\times]1.5,2.5[$. The results are reported in Table 2.

Table 2

k=	N=15	N=30	N=60	N=120
1	0.28	0.11	0.077	0.046
2	0.082	0.032	0.020	0.010
3	0.034	0.014	0.007	0.004
4	0.021	0.009	0.007	0.004
5	0.017	0.009	0.007	0.004
6	0.017	0.009	0.007	0.004

The same conclusion holds for this set of experiments. It is even better since the saturated convergence is achieved for smaller values of k when N (thus here both the accuracy and the number of processors) increases.

We refer also to Farhat and Chandesris [2003] and especially to Bal [2003b] for other issues about scalability of this algorithm.

3 Combination with domain decomposition – the non-overlapping case

We have generalized this approach to a non overlapping situation in the case were we only assume $\overline{\Omega} = \cup_{p=1}^{P} \overline{\Omega^p}$ with $\Omega^p \cap \Omega^q = \emptyset$ whenever $p \neq q$. We have chosen here the Neumann-Neumann strategy as in Bourgat et al. [1989] The approach results also in the fine solution of problems set over P subdomains times a time span of ΔT. The approach here differs from the overlapping case in the sense that the fine propagator involves both a Dirichlet and a Neumann solver:

The fine propagator

Let us assume that $\{\lambda_n^k\}_n$ are given together with the values $\{\beta_n^k\}_n$ corresponding to a predictor for the Dirichlet value of the solution over $\cup\partial\Omega^p \setminus \partial\Omega$. We first propagate the solution over each $\Omega^p \times]T_n, T_{n+1}[$ from λ_n^k with the boundary conditions β_n^k. In order to correct these boundary conditions, as in the Neumann-Neumann algorithm, we transform the jump in the normal derivatives of the solutions that have been computed at interface by a harmonic lifting that provides a corrector for the boundary condition. A relaxation parameter is adjusted (in our case through an optimal gradient approach) in order to minimize the final jump in the solution.

The coarse propagator

The coarse propagator is similar to that of the previous section. We can remark at this level that both in this case or in the overlapping case, there is room for reducing the cost of this global propagation, either by coarsening the spacial mesh size or by using a (sole) domain decomposition approach.

The numerical results

What we report here are only preliminary results that have to be extended to more complex cases. We should also replace the gradient method by a faster (at least a conjugate gradient or a GMRES) methods. We have considered the same problem as for the overlapping strategy, on the rectangular domain $]0, 4[\times]0, 1[$. We assume it is only decomposed into Ω^1 and Ω^2 (without any bandage). We have run the procedure and shown that, by keeping the fine time step constant, thus decreasing the number of fine time step within ΔT as we increase N, the number of iteration for convergence again remains constant. This gives evidences of the scalability of the method. We have to remark that the convergence rate of the iterative procedure is appreciably lower for this non-overlapping strategy than for the overlapping one. We are convinced that by replacing the crude gradient method that we have implemented by a better approach, the method will perform as nicely as in the overlapping case. This is the subject of a forthcoming paper to improve this strategy and extend it to other classes of classical iterative-domain decomposition based- methods in space (as FETI, Dirichlet Neuman, substructuring..).

Acknowledgement. The numerical results have been performed on the Freefem++ platform and we have benefitted from the advices of Frédéric Hecht and Olivier Pironneau.

References

Leonardo Baffico, Stephane Bernard, Yvon Maday, Gabriel Turinici, and Gilles Zérah. Parallel in time molecular dynamics simulations. *Phys. Rev. E.*, 66, 2002.

Guillaume Bal. On the convergence and the stability of the parareal algorithm to solve partial differential equations. In *Fifteen International Conference on Domain Decomposition Methods*, Berlin, 2003a. Springer, Lecture Notes in Computational Science and Engineering (LNCSE).

Guillaume Bal. Parallelization in time of (stochastic) ordinary differential equations. Math. Meth. Anal. Num. (submitted), 2003b.

Guillaume Bal and Yvon Maday. A "parareal" time discretization for non-linear PDE's with application to the pricing of an American put. In *Recent developments in domain decomposition methods (Zürich, 2001)*, volume 23 of *Lect. Notes Comput. Sci. Eng.*, pages 189–202. Springer, Berlin, 2002.

Jean-François Bourgat, Roland Glowinski, Patrick Le Tallec, and Marina Vidrascu. Variational formulation and algorithm for trace operator in domain decomposition calculations. In Tony Chan, Roland Glowinski, Jacques Périaux, and Olof Widlund, editors, *Domain Decomposition Methods*, pages 3–16, Philadelphia, PA, 1989. SIAM.

Charbel Farhat and M. Chandesris. Time-decomposed parallel time-integrators: theory and feasibility studies for fluid, structure, and fluid-structure applications. *Int. J. Numer. Meth. Engng.*, 58(9):1397–1434, 2003.

Paul Fischer, Frédéric Hecht, and Yvon Maday. A parareal in time semi implicit approximation of the Navier Stokes equations. In *Fifteen International Conference on Domain Decomposition Methods*, Berlin, 2003. Springer, Lecture Notes in Computational Science and Engineering (LNCSE).

Jacques-Louis Lions, Yvon Maday, and Gabriel Turinici. A parareal in time discretization of PDE's. *C.R. Acad. Sci. Paris, Serie I*, 332:661–668, 2001.

Yvon Maday and Gabriel Turinici. A parareal in time procedure for the control of partial differential equations. *C. R. Math. Acad. Sci. Paris*, 335 (4):387–392, 2002. ISSN 1631-073X.

Yvon Maday and Gabriel Turinici. Parallel in time algorithms for quantum control: the parareal time discretization scheme. *Int. J. Quant. Chem.*, 93 (3):223–228, 2003.

Gunnar Andreas Staff and Einar M. Rønquist. Stability of the parareal algorithm. In *Fifteen International Conference on Domain Decomposition Methods*, Berlin, 2003. Springer, Lecture Notes in Computational Science and Engineering (LNCSE).

Stability of the Parareal Algorithm

Gunnar Andreas Staff and Einar M. Rønquist

Norwegian University of Science and Technology
Department of Mathematical Sciences

Summary. We discuss the stability of the Parareal algorithm for an autonomous set of differential equations. The stability function for the algorithm is derived, and stability conditions for the case of real eigenvalues are given. The general case of complex eigenvalues has been investigated by computing the stability regions numerically.

1 Introduction

This paper represents one of the contributions at a minisymposium on the Parareal algorithm at this domain decomposition conference. The minisymposium was organized by Professor Yvon Maday, who is also one of the originators of the Parareal algorithm. The main objective is to be able to integrate a set of differential equations using domain decomposition techniques in time. We refer to the review article by Yvon Maday in these proceedings for a more detailed introduction to the ideas and motivation behind this algorithm.

In Section 2, we briefly review the Parareal algorithm and introduce the necessary notation. Our main focus is the stability analysis of this algorithm. In Section 3.1, we briefly review the standard stability analysis of ordinary differential equations, and in Section 3.2, we derive the stability function for the Parareal algorithm. In the remaining part of Section 3, we derive the stability conditions in the case of real and complex eigenvalues.

2 Algorithm

The Parareal algorithm was first presented in Lions et al. [2001]. An improved version of the algorithm was presented in Bal and Maday [2002]. Further improvements and understanding, as well as new applications of the algorithm, were presented in Baffico et al. [2002] and Maday and Turinici [2002]; our point of departure is the version of the Parareal algorithm presented in these papers.

We consider a set of ordinary differential equations that we would like to integrate from an initial time $t_0 = 0$ to a final time T. The time interval is first decomposed as

$$t_0 = T_0 < T_1 < \cdots < T_n = n\Delta T < T_{n+1} < T_N = T.$$

The Parareal algorithm is then given as the predictor-corrector scheme

$$\lambda_n^k = \mathcal{F}_{\Delta T}(\lambda_{n-1}^{k-1}) + \mathcal{G}_{\Delta T}(\lambda_{n-1}^{k}) - \mathcal{G}_{\Delta T}(\lambda_{n-1}^{k-1}), \tag{1}$$

where subscript n refers to the time subdomain number, superscript k refers to the (global) iteration number, and λ_n^k represents an approximation to the solution at time level n at iteration number k. The fine propagator $\mathcal{F}_{\Delta T}$ represents a fine time discretization of the differential equations, with the property that

$$\lambda_n = \mathcal{F}_{\Delta T}(\lambda_{n-1}) \ , n = 1, ..., N,$$

while the coarse propagator $\mathcal{G}_{\Delta T}$ represents an approximation to $\mathcal{F}_{\Delta T}$.

Notice that $\mathcal{F}_{\Delta T}$ operates on initial conditions λ_{n-1}^{k-1}, which are known. This implies that $\mathcal{F}_{\Delta T}(\lambda_{n-1}^{k-1})$ can be implemented in parallel. The coarse propagator $\mathcal{G}_{\Delta T}$, on the other hand, operates on initial conditions λ_{n-1}^{k} from the current iteration, and is therefore strictly serial.

3 Stability analysis

In Farhat and Chandesris [2003], an investigation of the stability for an autonomous problem is presented. We will here use more of the tools provided by the ODE theory, and extend the stability analysis a bit further.

The departure of our stability analysis is the predictor-corrector scheme (1). A stability analysis is performed on the autonomous differential equation

$$y' = \mu y, \quad y(0) = y_0, \quad \mu < 0 \ . \tag{2}$$

The exact solution to this problem is $y(t) = e^{\mu t} y_0$. Since $\mu < 0$, this is a decaying function for increasing t. The numerical solution of (2) is an approximation to the exact solution. It is well known that a convergent numerical scheme can be arbitrarily accurate by choosing sufficiently small time-steps. A numerical scheme which results in an non-increasing approximation for the chosen time-step is called stable. For a more precise definition of stability, the reader is referred to Hairer et al. [2000].

3.1 Stability analysis for ordinary ODE schemes

To better understand the derivation of the stability properties of the Parareal algorithm, we start by deriving the stability properties for two well known numerical schemes, namely the explicit and implicit Euler methods. Applied to our differential equation, the two schemes can be written as

$$\begin{array}{ll} y_n = y_{n-1} + \Delta T \mu\, y_{n-1} = (1 + \Delta T \mu)^n y_0 = R(z)^n\, y_0 & \text{explicit Euler} \\ y_n = y_{n-1} + \Delta T \mu\, y_n \;= (1 - \Delta T \mu)^{-n}\, y_0 = R(z)^n y_0 & \text{implicit Euler} \end{array}$$

where ΔT is the time-step, $z = \Delta T \mu$ and $R(z)$ is called the *stability function* of the chosen scheme. Obviously, $|R(z)| \leq 1$ will prevent the numerical schemes from blowing up for increasing n.

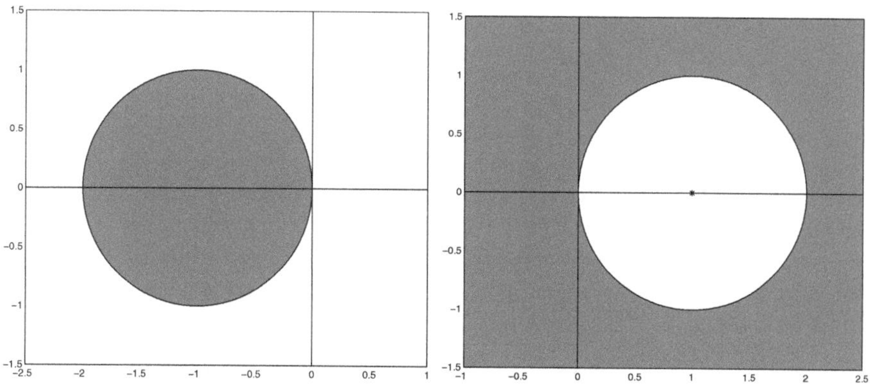

Fig. 1. Stability domain for explicit (left) and implicit(right) Euler. The dark region is the stability domain, i.e., those values of z in the complex plane where $|R(z)| \leq 1$

From Figure 1 we see that explicit Euler suffers from time-step restrictions, while implicit Euler is stable for all possible choices of the time-step ΔT ($\mu < 0$). In the context of the Parareal algorithm, the coarse propagator $\mathcal{G}_{\Delta T}$ is forced to take large time-steps, which clearly indicates that implicit Euler is a better choice then explicit Euler for the coarse propagator.

Consider now a linear system of M autonomous differential equations

$$y' = Ay, \quad y(0) = y_0. \tag{3}$$

Assuming that a spectral factorization is possible, we may write the system matrix $A \in I\!R^{M \times M}$ as

$$A = VDV^{-1}$$

where D is a diagonal matrix containing the eigenvalues $\{\mu_1, \ldots, \mu_M\}$ of A, and V is a matrix containing the corresponding eigenvectors of A.

The exact solution of (3) may then be written as

$$y(t) = e^{tA} y_0 = V e^{tD} V^{-1} y_0 \,,$$

while the approximation of the (3) using implicit Euler can be expressed as

$$y_n = V\,(I - \Delta T D)^{-n}\, V^{-1}\, y_0\,.$$

Obviously, a method is stable for systems of ODE's if $|R(z_i)| \leq 1,\ i = 1,\ldots,M$, where $z_i = \Delta T \mu_i$ and μ_i is the ith eigenvalue of A.

3.2 Stability analysis for the Parareal algorithm

In the following analysis, we assume that we may use different integration schemes for the fine and the coarse propagator. Within each coarse time step ΔT, we will use several fine time steps δt with the fine propagator.

Our first aim is to write the predictor-corrector scheme (1) on the form

$$\lambda_n^k = V\, H(n,k,r(D\delta t), R(D\Delta T)\, V^{-1}\, \lambda_0,$$

where n is the subdomain number (in time), k is the iteration number, H is the "stability function" for the Parareal scheme, r is the stability function for the fine propagator $\mathcal{F}_{\Delta T}$, R is the stability function for the coarse propagator $\mathcal{G}_{\Delta T}$ and D is the diagonal matrix containing all the eigenvalues for the system matrix. To do this we first apply the predictor-corrector scheme (1) to the model problem (2); this gives

$$\lambda_n^k = \bar{r}(\mu\delta t)\lambda_{n-1}^{k-1} + R(\mu\Delta T)\lambda_{n-1}^{k} - R(\mu\Delta T)\lambda_{n-1}^{k-1}, \tag{4}$$

where $\bar{r}(\mu\delta t) = r(\mu\delta t)^s$ is the stability function for the fine operator after $s = \frac{\Delta T}{\delta t}$ fine time-steps δt, and $R(\mu\Delta T)$ is the stability function for the coarse operator $\mathcal{G}_{\Delta T}$. For simplicity we will write $\bar{r} = \bar{r}(\mu\delta t)$ and $R = R(\mu\Delta T)$.

We rearrange (4) and write

$$\lambda_n^k = R\lambda_{n-1}^{k} + (\bar{r} - R)\lambda_{n-1}^{k-1} = R\lambda_{n-1}^{k} + S\lambda_{n-1}^{k-1}. \tag{5}$$

Obviously, the recursion is solved like this:

$$\begin{array}{ccccc}
\lambda_n^k & & & & \\
\downarrow\searrow & & & & \\
\lambda_{n-1}^k & \lambda_{n-1}^{k-1} & & & \\
\downarrow\searrow & \downarrow\searrow & & & \\
\lambda_{n-2}^k & \lambda_{n-2}^{k-1} & \lambda_{n-2}^{k-2} & & \\
\downarrow\searrow & \downarrow\searrow & \downarrow\searrow & & \\
\lambda_{n-3}^k & \lambda_{n-3}^{k-1} & \lambda_{n-3}^{k-2} & \lambda_{n-3}^{k-3} & \\
\downarrow\searrow & \downarrow\searrow & \downarrow\searrow & \downarrow\quad\searrow & \\
\lambda_{n-4}^k & \lambda_{n-4}^{k-1} & \lambda_{n-4}^{k-2} & \lambda_{n-4}^{k-3} & \lambda_{n-4}^{k-4}
\end{array}
\qquad
\begin{array}{cc}
\lambda_n^k & \\
\downarrow R \searrow S & \\
\lambda_{n-1}^k & \lambda_{n-1}^{k-1}
\end{array}$$

We recognize the Pascal tree, and we may write (5) as

$$\lambda_n^k = \left(\sum_{i=0}^{k} \binom{n}{i} (\bar{r} - R)^i R^{n-1}\right) \lambda_0,$$

where we identify the "stability function" H as

$$H(n,k,r,R) = \sum_{i=0}^{k} \binom{n}{i} (\bar{r} - R)^i R^{n-1}.$$

The extension to solve the system (3) is straightforward,

$$\lambda_n^k = V\, H(n,k,r,R)\, V^{-1}\, \lambda_0.$$

Stability is achieved if

$$\sup_{1 \le n \le N} \sup_{1 \le k \le N} |H(n,k,r,R)| \le 1 \qquad \forall \mu_i \ , \ i = 1, ..., M. \tag{6}$$

3.3 Special case: μ_i real

In the case of real eigenvalues, the stability condition (6) can be expressed as

$$\begin{aligned} |H| = \left| \sum_{i=0}^{k} \binom{n}{i} (\bar{r} - R)^i R^{n-i} \right| &\le \sum_{i=0}^{k} \binom{n}{i} |(\bar{r} - R)|^i |R|^{n-i} \\ &\le \sum_{i=0}^{n} \binom{n}{i} |(\bar{r} - R)|^i |R|^{n-i} \\ &= (|\bar{r} - R| + |R|)^n \le 1 \qquad \forall \mu_i \ , \ i = 1, ..., M, \end{aligned}$$

where $|\bar{r} - R| + |R|$ is either $|(\bar{r} - R) + R|$ or $|(\bar{r} - R) - R|$. [1] The condition $|(\bar{r} - R) + R| = |\bar{r}| \le 1$ is the stability condition for the fine operator, and this should be true independent of the use of the Parareal algorithm.

The condition $|(\bar{r} - R) - R| = |2R - \bar{r}| \le 1$ can be rewritten as

$$\frac{\bar{r} - 1}{2} \le R \le \frac{\bar{r} + 1}{2}. \tag{7}$$

Theorem 1. *Assume we want to solve the autonomous differential equation*

$$y' = \mu y\,, \quad y(0) = y_0\,, \quad 0 > \mu \in \mathbb{R}\,,$$

and that $-1 \le r, R \le 1$ where $r = r(\mu \delta t)$ is the stability function for the fine propagator $\mathcal{F}_{\Delta T}$ using time-step δt and $R = R(\mu \Delta T)$ is the stability function for the coarse propagator $\mathcal{G}_{\Delta T}$ using time-step ΔT. Then the Parareal algorithm is stable for all possible values of number of subdomains N and all number of iterations $k \le N$ as long as

$$\frac{\bar{r} - 1}{2} \le R \le \frac{\bar{r} + 1}{2}$$

where $\bar{r} = r(\mu \delta t)^s$ and $s = \frac{\Delta T}{\delta t}$.

[1] This is due to Harald Hanche Olsen, Dept. of Mathematical Sciences, NTNU

It is not obvious from (7) which solvers will fulfil this stability condition. However, Theorem 2 gives some insight by considering a special case.

Theorem 2. *Assume we want to solve the autonomous differential equation*

$$y' = \mu y \,, \quad y(0) = y_0 \,, \quad 0 > \mu \in \mathbb{R} \,,$$

using the Parareal algorithm. Assume also that the system is stiff, meaning that $z = \mu \Delta T \ll -1$, *and that the fine propagator is close to exact. Then the "stability function" can be written as*

$$H(n,k,R) = (-1)^k \binom{n-1}{k} R^n \,,$$

and stability is guaranteed if the following property is fulfilled:

$$R_\infty = \lim_{z \to -\infty} |R(z)| \leq \frac{1}{2} \,. \tag{8}$$

The proofs of Theorem 1 and 2 are not included due to space limitation, but will be included in a future article.

We have tested the condition (8) by solving the one-dimensional unsteady diffusion equation using a spectral Galerkin method in space and a Crank-Nicolson scheme for the fine propagator $\mathcal{F}_{\Delta T}$. We then tested the following schemes for the coarse propagator $\mathcal{G}_{\Delta T}$: the implicit Euler method ($R_\infty = 0$), the Crank-Nicolson scheme ($R_\infty = 1$), and the θ-scheme where we can vary the degree of "implicitness," and hence R_∞; see Hairer et al. [2000] and Hairer and Wanner [2002]. The numerical results demonstrated that rapid convergence of the Parareal scheme is obtained for implicit Euler, while Crank-Nicolson first gives convergence, and then starts to diverge when k increases. However, as k approaches N, the results again start to converge; this is expected since the Parareal algorithm gives precisely the fine solution after N iterations. By varying the degree of "implicitness" in the θ-scheme, we observe that our results are consistent with the "stability condition" (8).

3.4 General case: μ_i complex

Notice that Theorem 1 is true for ODE's and systems of ODE's where the eigenvalues of the system matrix have pure real eigenvalues. For complex eigenvalues, (6) needs to be fulfilled. This is done numerically in Figure 2 and 3 for the two-stage third order Implicit Runge-Kutta-Radau scheme (Radau3); see Hairer and Wanner [2002]. This scheme is chosen because it represents the typical asymptotic behaviour of a scheme which fulfils Theorem 2. The difference in behavior between the various possible schemes lies in the size of the instability regions in the real direction, and in the number of instability regions in the imaginary direction. For example, implicit Euler will have smaller instability regions compared to Radau3.

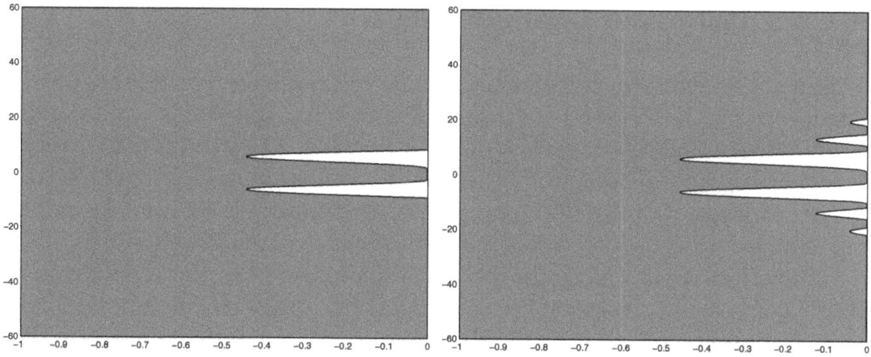

Fig. 2. Stability plots using Radau3 for both $\mathcal{G}_{\Delta T}$ and $\mathcal{F}_{\Delta T}$. The x-axis is $\mathrm{Re}(\mu\Delta T)$ and the y-axis is $\mathrm{Im}(\mu\Delta T)$. The dark regions represent the regions in the complex plane where (6) is satisfied. Here, $N = 10$, and $s = 10$ (left) and $s = 1000$ (right)

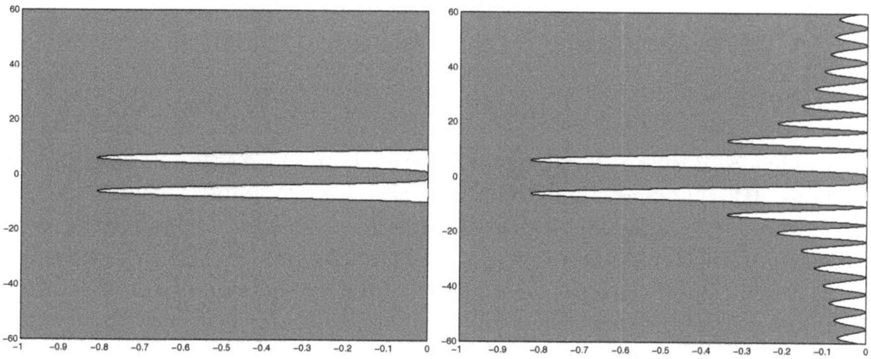

Fig. 3. Stability plots using Radau3 for both $\mathcal{G}_{\Delta T}$ and $\mathcal{F}_{\Delta T}$. The x-axis is $\mathrm{Re}(\mu\Delta T)$ and the y-axis is $\mathrm{Im}(\mu\Delta T)$. The dark regions represent the regions in the complex plane where (6) is satisfied. Here, $N = 1000$, and $s = 10$ (left) and $s = 1000$ (right)

From Figure 2 and 3 we notice that the Parareal algorithm is unstable for pure imaginary eigenvalues, as well as for some complex eigenvalues where the imaginary part is much larger then the real part (notice the difference in scalings along the real and the imaginary axes). No multistage scheme has yet been found that makes the presented formulation of the Parareal algorithm stable for all possible eigenvalues. This means that the numerical solution of some hyperbolic problems, and convection-diffusion problems with highly dominant convection (e.g Navier-Stokes with high Reynolds numbers), are probably unstable using the Parareal algorithm. This is also consistent with the results reported in Farhat and Chandesris [2003].

4 Conclusion and final comments

For an autonomous set of differential equations, we have derived the stability conditions for the Parareal algorithm. The stability conditions corresponding to the case of real eigenvalues are explicitly given, while the general case has been investigated by computing the stability regions numerically. These latter results indicate that the Parareal algorithm is unstable for pure imaginary eigenvalues, which is also consistent with previously reported results.

Numerical results have also been obtained using the Parareal algorithm in the context of solving partial differential equations such as the nonlinear, viscous Burger's equation, and where the coarse propagator incorporates a coarse discretization in space as well as in time. However, a discussion of these results will be reported elsewhere due to space limitation.

Acknowledgement. We thank Professor Yvon Maday for bringing the Parareal algorithm to the attention of the authors, and for many valuable discussions.

References

L. Baffico, S. Bernard, Y. Maday, G. Turinici, and G. Zérah. Parallel in time molecular dynamics simulations. *Physical Review. E*, 66, 2002.

G. Bal and Y. Maday. A "parareal" time discretization for non-linear pde's with application to the pricing of an american put. In L. F. Pavarino and A. Toselli, editors, *Recent Developments in domain Decomposition Methods*, volume 23 of *Lecture Notes in Computational Science and Engineering*, pages 189–202. Springer, 2002.

C. Farhat and M. Chandesris. Time-decomposed parallel time-integrators: theory and feasibility studies for fluid, structure, and fluid-structure applications. to appear in *International Journal for Numerical Methods in Engineering*, 2003.

E. Hairer, S. N. rsett, and G. Wanner. *Solving Ordinary Equations I*, volume 8 of *Springer Series in Computational Mathematics.* Springer, 2. edition, 2000.

E. Hairer and G. Wanner. *Solving Ordinary Equations II*, volume 14 of *Springer Series in Computational Mathematics.* Springer, 2. edition, 2002.

J.-L. Lions, Y. Maday, and G. Turinici. Résolution d'edp par un schéma en temps pararéel. *C.R.Acad Sci. Paris Sér. I Math*, 332:661–668, 2001.

Y. Maday and G. Turinici. A parareal in time procedure for the control of partial differential equations. *C.R.Acad Sci. Paris Sér. I Math*, 335:387–392, 2002.

Part XII

Minisymposium: Space Decomposition and Subspace Correction Methods for Linear and Nonlinear Problems

Multilevel Homotopic Adaptive Finite Element Methods for Convection Dominated Problems*

Long Chen, Pengtao Sun, and Jinchao Xu

The Pennsylvania State University, Department of Mathematics.

Summary. A multilevel homotopic adaptive finite element method is presented in this paper for convection dominated problems. By the homotopic method with respect to the diffusion parameter, the grids are iteratively adapted to better approximate the solution. Some new theoretic results and practical techniques for the grid adaptation are presented. Numerical experiments show that a standard finite element scheme based on this properly adapted grid works in a robust and efficient manner.

1 Introduction

In this paper, we shall present a class of adaptive finite element methods (FEMs) for the convection-dominated problems. One simple model is the following convection-diffusion problem:

$$-\epsilon \Delta u + b(x) \cdot \nabla u = f(x), \tag{1}$$

which is posed on a bounded domain $\Omega \subset R^2$ with a proper boundary condition.

We are interested in adaptive finite element methods for the convection-dominated case, namely ϵ is sufficiently smaller than $b(x)$. It is well-known that one major difficult for this type of problem is that a standard finite element discretization scheme usually fail and specialized methods, such as upwinding scheme and streamline diffusion methods, need to be adapted. One conclusion from the study in this paper is that a standard finite element scheme still works reasonably well if the grid is properly adapted so that sharp boundary or internal layers presented in the solution will be fully resolved. To obtain such a properly adapted grid, we are going to use a homotopic

* This work was supported in part by NSF DMS-0074299, NSF DMS-0209497,NSF DMS-0215392 and the Center for Computational Mathematics and Application at Penn State.

method with respect to the diffusion parameter ϵ. Namely, we first start our computation for large ϵ, say $\epsilon = 1$ and use adaptive grid technique for elliptic problems to obtain a good initial grid. We then start to decrease the value of ϵ and use the current grid as an initial grid to obtain a new adaptive grid. We continue in this way until the desired value of ϵ is reached. As a general approach, the homotopic method of this type is commonly used in many different application areas and there have been existing works for grid adaptation (c.f. Habashi et al. [1997]). What is of importance here is how such a continuation procedure is carried out in a robust and efficient manner. In this paper, we will first present an interpolation error estimate and then develop local mesh improvement techniques such as refinement, coarsening and smoothing and global moving mesh strategy which aims to minimize the interpolation error.

2 Theoretical foundation

In this section, we include an interpolation error estimate from Chen et al. [2003]. (Similar error estimates in special cases can also be found in, e.g, D'Azevedo and Simpson [1989], Habashi et al. [1997], Huang and Sun [2003]). This estimate can be viewed as the theoretical foundation of this paper, namely our algorithms are aimed at minimizing (or at least reducing) this interpolation error by iteratively modifying our grids.

The estimate.

Let Ω be a bounded domain in R^n. Given a function $u \in C^2(\bar{\Omega})$, we call a symmetric positive definite matrix $H \in R^{n\times n}$ to be a majorizing Hessian of u if

$$|\xi^t(\nabla^2 u)(x)\xi| \le c_0 \xi^t H(x)\xi, \quad \xi \in R^n, x \in \Omega$$

for some positive constant c_0.

We then use the majorizing Hessian to define a new metric

$$H_p = (\det H)^{-\frac{1}{2p+n}} H, \, p \ge 1.$$

There are two conditions for a triangulation $\mathcal{T}_N$, where N is the number of simplexes, to be a nearly optimal mesh in the sense of minimizing the interpolation error in L^p norm. The first assumption asks the mesh to capture the high oscillation of the Hessian metric, namely H does not change very much on each element.

(A1) There exist two positive constants α_0 and α_1 such that

$$\alpha_0 \xi^t H_\tau \xi \le \xi^t H(x)\xi \le \alpha_1 \xi^t H_\tau \xi, \quad \xi \in R^n,$$

where H_τ is the average of H over τ.

The second condition asks that $\mathcal{T}_N$ is quasi-uniform under the new metric induced by H_p.

(A2) There exists two positive constants β_0 and β_1 such that

$$\frac{\sum_i \tilde{d}_{\tau,i}^2}{|\tilde{\tau}|^{2/n}} \leq \beta_0, \ \forall \tau \in \mathcal{T}_N \text{ and } \frac{\max_{\tau\in\mathcal{T}} |\tilde{\tau}|}{\min_{\tau\in\mathcal{T}} |\tilde{\tau}|} \leq \beta_1, \tag{2}$$

where $|\tilde{\tau}|$ is the volume of τ and $\tilde{d}_{\tau,i}$ is the length of the i-th edge of τ under the new metric H_p, respectively.

The first inequality in (2) means that each τ is isotropic i.e. shape-regular under the metric H_p. The second inequality (2) means that all elements τ are of comparable size (under the new metric), which is a global condition and known as the equidistribution principal Huang [2001]. It means that the mesh will concentrate at the region where $\det H_p(\mathbf{x})$ is large. We proved in Chen et al. [2003] that a triangulation which satisfies both local and global conditions yields a good approximation.

Theorem 1. *Let $u \in \mathcal{C}^2(\bar{\Omega})$, $\mathcal{T}_N$ satisfy assumptions (A1) and (A2) and u_I is the linear finite element interpolation of u based on the triangulation $\mathcal{T}_N$, the following error estimate holds:*

$$\|u - u_I\|_{L^p(\Omega)} \leq CN^{-2/n} \| \sqrt[n]{\det(H)} \|_{L^{\frac{pn}{2p+n}}(\Omega)}$$

for some constant $C = C(n, p, c_0, \alpha_0, \alpha_1, \beta_0, \beta_1)$. This error estimate is optimal in the sense that for a strictly convex (or concave) function, the above inequality holds in a reversed direction.

It is well known that for diffusion dominated problems, there will be some sharp boundary layers or internal layers with width in ϵ scale. We would like to emphasis that the error bound in Theorem 1 is independent of ϵ. Some numerical results about the ϵ independence of the interpolation error can be found at Chen et al. [2003].

As mentioned before, Theorem 1 will be the basis of grid adaptation algorithms. Roughly speaking, for a given function u, we will adapt our grids in such a way the assumption (A1) and (A2) will be better and better satisfied. One important remark we need to make is that the validity of Theorem 1 allows for a few exceptions of the assumption (A2) for $p < \infty$, see Chen et al. [2003] for details. This is particularly important since in practice it is very difficult to guarantee that (A2) is satisfied everywhere. We note that the theory and algorithms in D'Azevedo and Simpson [1989] and Habashi et al. [1997] are only for $p = \infty$ which requires that (A2) be satisfied on each element in the triangulation.

Postprocessing: recovery of Hessian.

In this section, we will discuss how the Hessian matrix of the solution can be obtained when linear finite element approximation is used for the discretization of partial differential equations. Since taking piecewise second derivatives to piecewise linear functions will given no approximation to Hessian matrix, special Postprocessing techniques need to be used to obtain reasonable Hessian matrix approximation from linear finite elements.

One most popular technique is a patch recovery technique proposed by Zienkiewicz and J.Z.Zhu [1992a,b] which is based on the least squares fitting locally. Results from their application demonstrate that it is robust and efficiency. The theoretical reason for ZZ method to work is largely understood to be related to the superconvergence phenomenon for second order elliptic boundary value problems discretized on a finite element grid that has certain local symmetry, see Wahlbin [1995], Chen and Huang [1995], Babuska and Strouboulis [2001]. These classic superconvergence results can be used to justify the effectiveness of Zienkiewicz-Zhu method, see, for example, Zhang [1999], Li and Zhang [1999] for nearly structured grids. A significant improvement of this type of analysis was given recently by Bank and Xu [2003a,b]. In Bank and Xu [2003a] they gave superconvergence estimates for piecewise linear finite element approximation on quasi-uniform triangular meshes where most pairs of triangles sharing a common edge form approximate parallelograms. They also analyze a postprocessing gradient recovery scheme, showing that $Q_h \nabla u_h$ is a superconvergent approximation to ∇u. Here Q_h is the global L^2 projection. This result leads to a theoretical justification of ZZ method for such type of grids, see Xu and Zhang [2004]. Recently, Carstensen and Bartels [2002] also gave theoretical and numerical support for the robust reliability of all averaging techniques on unstructured grids.

The gradient recovery algorithm used in the numerical examples of this paper is based on a new approach due to Bank and Xu [2003b] where they use the smoothing iteration of the multigrid method to develop a postprocessing gradient recovery scheme. This scheme proves to be very efficient for recovering Hessian matrix. All the above methods can be extended to anisotropic grids with some proper modifications, but a theoretical justification of such extensions is still lacking. Nevertheless, numerical experiments have given satisfactory results.

3 Mesh adaptation

In this section, we will discuss techniques which aim at improving the mesh quality. Here we define the mesh quality for a triangulation $\mathcal{T}$ by the interpolation error:

$$Q(\mathcal{T}, u, p) = \|u - u_{I,\mathcal{T}}\|_{L^p(\Omega)}, 1 \le p \le \infty.$$

Local mesh optimizations.

There are mainly three types mesh improvements: (1) refinement or coarsening Bank et al. [1983], Rivara [1984], Kornhuber and Roitzsch [1990], (2) edge swapping Lawson [1977], and (3) mesh smoothing Bank and Smith [1997],

Jones et al. [1995]. We will derive those techniques by minimizing the interpolation error in L^p norm, which can be achieved by equidistributing edge lengths under the new metric.

We compute edge lengths under the new metric H_p and mark edges whose lengths are greater than $r_1 d$, where $r_1 \geq 1$ is a parameter and d is a fixed edge length. We connect marked edges element-wise according to different situations; See Fig. 1.

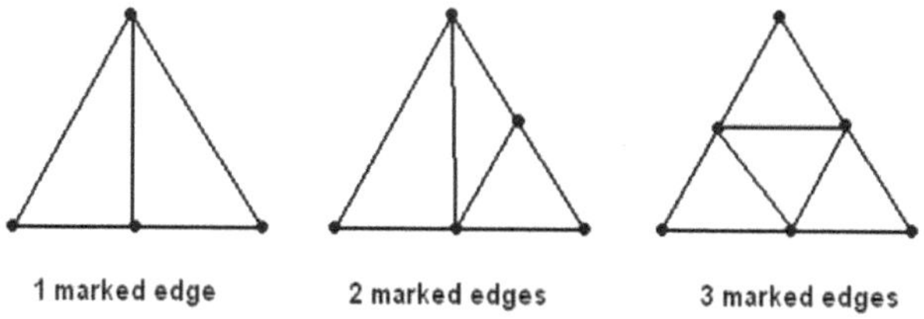

Fig. 1. Edge-based refinement

The coarsening operates like an inverse procedure of refinement. It marks the one whose length is less than $r_2 d$ for another parameter $r_2 \leq 1$. We then shrink this edge to a point and connect to the vertices of the patch of the edge.

Now we consider the edge swapping for four points $\{\mathbf{v}_i\}_{i=1}^4$ which form two adjacent triangles and a convex quadrilateral. Let $\mathcal{T}_1 = \triangle_{123} \cup \triangle_{134}$ and $\mathcal{T}_2 = \triangle_{124} \cup \triangle_{234}$, where $\triangle_{ijk}$ stands for the triangle made up by $\mathbf{v}_i, \mathbf{v}_j$, and $\mathbf{v}_k$. We choose triangulation $\mathcal{T}_1$ if and only if $Q(\mathcal{T}_1, u, p) \leq Q(\mathcal{T}_2, u, p)$, for some $1 \leq p \leq \infty$. In Chen and Xu [2004a], we show this criteria is equivalent to the empty circle criteria when $u(\mathbf{x}) = \|\mathbf{x}\|^2$. Thus it is an appropriate generalization of the edge swapping used in the isotropic case to the anisotropic case.

Local mesh smoothing adjusts the location of a vertex in its patch Ω_i, which consists of all simplexes containing vertex $\mathbf{x}_i$, without changing the connectivity. Moving vertex to the new location will provably or heuristically improve the mesh quality. Several sweeps through the whole mesh can be performed to improve the overall mesh quality. By minimizing the interpolation error in Ω_i, we move $\mathbf{x}_i$ to $\mathbf{x}^*$ such that

$$\nabla u(\mathbf{x}^*) = -\frac{1}{|\Omega_i|} \sum_{\tau_j \in \Omega_i} \left(\nabla |\tau_j| \sum_{\mathbf{x}_k \in \tau_j, \mathbf{x}_k \neq \mathbf{x}_i} u(\mathbf{x}_k)\right). \tag{3}$$

The derivation of this formula can be found at Chen and Xu [2004a]. In the application to numerical solution, we use $Q_h \nabla u_h$ and u_h in (3) to perform the calculation.

Global moving mesh strategy.

Another global approach to improve the mesh to better approximate a solution has been carried out in the study of the so-called moving mesh method Huang [2001], Huang and Sun [2003], Huang and Russell [1999]. Let Ω^c be the computational domain with a quasi-uniform (under the standard Euclidean metric) triangulation T_N^c. The mesh on Ω can be viewed as the image of a transformation $x = x(\xi) : \Omega^c \to \Omega$. Then to ask the transformed mesh to be quasi-uniform respect to the metric $G(x)$ is more or less equivalent to ask $x = x(\xi)$ to be the global minimizer of the minimizing problem:

$$\min_x \int_{\Omega^c} \left(\sum_i (\nabla x_i)^t G(x) \nabla x_i \right)^q d\xi, \ q > n/2.$$

The minimizers of the above functionals is expected to satisfy both equidistribution and isotropy conditions simultaneously Chen et al. [2003]. We note that the $q = 1$ case corresponds to the harmonic mapping but we ask $q > n/2$ here. When $n \geq 3$, these minimization problems (which is more or less p-Laplacian with $p > n$) is significantly different from the harmonic mapping which has been most commonly used in the literature for moving mesh method Liseikin [1999], Dvinsky [1991]. If we choose $G = [\det(H)]^{-1/(2p+n)} H$ in the functional, we can get a nearly optimal mesh which minimizes the interpolation error $\|u - u_I\|_{L^p(\Omega)}$ by solving above functional.

4 Numerical examples

Our multilevel homotopic adaptive grid method for convection dominated problem $-\epsilon \Delta u + b(x) \cdot \nabla u = f(x)$ can be roughly described as follows.

Given $\rho = \epsilon_0 \gg \epsilon$ and $h = h_0$, generate an initial mesh $\mathcal{T}_h$.

1. Discretize the PDE on mesh $\mathcal{T}_h$ and solve it to get the solution u_h.
2. Global or local move $\mathcal{T}_h$ using u_h and its recovered derivatives.
3. If $\rho = \epsilon$, locally improve the grid using the estimated new metric. Otherwise go to (4).
4. Global refine $\mathcal{T}_h$, and set $\rho \leftarrow \gamma\rho$ $(\gamma < 1)$, $h \leftarrow h/2$. Go to (1).

Let's considering the following convection dominated model problem:

$$\begin{cases} -\epsilon \Delta u + u_x = 1 \ x \in \Omega \\ u = 0 \ x \in \partial\bar{\Omega} \end{cases} \tag{4}$$

on the unit square domain $\Omega = (0, 1)^2$ with $\epsilon = 0.001$. By applying above algorithm we solve this convection-diffusion problem and try to catch out the singular sections of solution by means of moving mesh and local optimized mesh. The following pictures in Fig. 2 describe that how the adaptive mesh

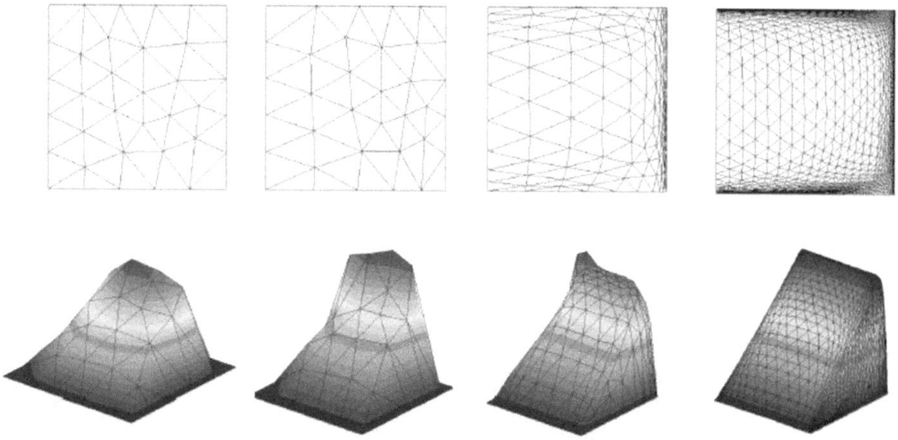

Fig. 2. Continuation adaptive meshes and corresponding solutions

and numerical solution change in this multilevel homotopic adaptive process step by step.

In this process of multilevel homotopic adaptive mesh, we start from a very coarse initial grid with respect to $\epsilon = 0.1$. On the following each level, we decrease ϵ once by dividing by 2 firstly, then resolve the original P.D.E (4) on this level's mesh with the standard finite element method, and get more significant improvements of numerical solution u_h around the upper, lower and right boundary layer. After that, we calculate the modified Hessian matrix H_p (in this example, we choose $p = 1$ i.e. we measure the error in L^1 norm) of the solution via the recovery technique we mentioned in Section 3. Then according to the value of Hessian matrix on each mesh element, we move grids to upper, lower and right boundary layer by virtue of moving mesh method. On the other hand, when we decrease ϵ, we apply global mesh refinement for the whole domain in order to get more grids to move.

Keep running this process until ϵ equals to its original value 0.001, we then begin to do the local mesh optimizations. Still utilizing the Hessian matrix H_p of each mesh element, we catch and mark those edges whose lengths under the new matrix is relative large, then apply our local refinement technique on these marked edges and thus locally generate a finer mesh to resolve the singularity of the solution.

Eventually the singularity of the solution of (4) is resolved, no oscillation any more. To show the numerical optimal convergence rate of our algorithm, we list the error in L^1 norm and its convergence rate in Table 1.

Since the analytical solution to (4) is not available, we compute a solution on a very fine Shishkin mesh Shishkin [1990] for which the near optimal convergence result is known Roos [2002] and use it as the real solution to compute the error. We apply our algorithm with different initial meshes to obtain a

Table 1. Errors of FEM on adapted grids

N	error	Rate
4712	2.400665E-04	0.99
7043	9.871067E-05	1.04
10102	9.144469E-05	1.01
14329	7.822302E-05	0.99
17929	9.253636E-05	0.95
22256	6.092786E-05	0.97

sequence of near optimal meshes for the linear interpolant. The first column of Table 1 is the number of nodes (unknowns) and the second one is the L^1 norm of the error. In the third column, we list the ratio $\ln error/\ln N$. It is clear that the standard finite element on the nearly optimal meshes obtain an optimal convergence rate.

5 Concluding remarks

In this paper, we have shown an optimal interpolation error estimates in L^p norm and, based on the estimate, we have developed new techniques, including local mesh optimizations and global moving mesh strategy, to improve the mesh to better approximate the solution. Those techniques with the homotopy with respect to diffusion coefficient are successfully applied to convection dominated problem. One main observation in our work is that a properly adapted mesh will enhance the stability of the standard finite element methods which often fails for convection dominated problems on quasi-uniform grids. This phenomenon has been observed in other simpler situations (see books Miller et al. [1996], Roos et al. [1996]). In the current work, the mesh is adapted to optimize the interpolation error. We expect that the discretization error will inherit the optimality of the interpolation error on a nearly optimal mesh for linear interpolant. We have obtained some preliminary results for a 1-d convection dominated model problem Chen and Xu [2004b] which shows that the optimality of the convergence rate is sensitive to the perturbation of meshes in the smooth part.

Another critical question that needs to be addressed is that how to solve these sequences of systems efficiently since in the adaptive procedure described in this paper, we need to solve many systems of algebraic system of equations. Hence how to solve these sequence of systems efficiently is crucial to the entire adaptive procedure. It is in fact the main research interest of the authors to develop efficient methods for such systems. We need to develop techniques how to make use of the intermediate grids and equations together with their discrete solutions. This is a subject of our ongoing research.

References

I. Babuska and T. Strouboulis. *The finite element method and its reliability.* Numerical Mathematics and Scientific Computation. Oxford Science Publications, 2001.

R. E. Bank, A. H. Sherman, and A. Weiser. Refinement algorithms and data structures for regular local mesh refinement. In R. S. et al., editor, *Scientific Computing*, pages 3–17. IMACS/North-Holland Publishing Company, Amsterdam, 1983.

R. E. Bank and R. K. Smith. Mesh smoothing using a posteriori error estimates. *SIAM Journal on Numerical Analysis*, 34:979–997, 1997.

R. E. Bank and J. Xu. Asymptotically exact a posteriori error estimators, Part I: Grids with superconvergence. *SIAM J. on Numerical Analysis*, 41 (6):2294–2312, 2003a.

R. E. Bank and J. Xu. Asymptotically exact a posteriori error estimators, Part II: General unstructured grids. *SIAM J. on Numerical Analysis*, 41 (6):2313–2332, 2003b.

C. Carstensen and S. Bartels. Each averaging technique yields reliable a posteriori error control in FEM on unstructured grids. I. low order conforming, nonconforming, and mixed FEM. *Math. Comp.*, 71(239):945–969, 2002.

C. Chen and Y. Huang. *High accuracy theory of finite element methods.* Hunan, Science Press, Hunan, China (in Chinese), 1995.

L. Chen, P. Sun, and J. Xu. Optimal anisotropic simplicial meshes for minimizing interpolation errors in L^p-norm. *Submitted to Math. Comp.*, 2003.

L. Chen and J. Xu. Optimal Delaunay triangulation. *J. of Comp. Math.*, 22 (2):299–308, 2004a.

L. Chen and J. Xu. Stability and accuracy of adapted finite element methods for singularly perturbed problems. Technical report, 2004b.

E. D'Azevedo and R. Simpson. On optimal interpolation triangle incidences. *SIAM Journal on Scientific and Statistical Computing*, 6:1063–1075, 1989.

A. S. Dvinsky. Adaptive grid generation from harmonic maps on riemannian manifolds. *J. Comput. Phys.*, 95(2):450–476, 1991.

W. Habashi, M. Fortin, J. Dompierre, M. Vallet, D. A. A. Yahia, Y. Bourgault, M. Robichaud, A. Tam, and S. Boivin. Anisotropic mesh optimization for structured and unstructured meshes. In *28th Computational Fluid Dynamics Lecture Series 1997-02.* von Karman Institute for Fluid Dynamics, 1997.

W. Huang. Variational mesh adaptation: isotropy and equidistribution. *J. Comput. Phys.*, 174:903–924, 2001.

W. Huang and R. D. Russell. Moving mesh strategy based on a gradient flow equation for two-dimensional problems. *SIAMJournal on Scientific Computing*, 20(3):998–1015, 1999.

W. Huang and W. Sun. Variational mesh adaptation II: Error estimates and monitor functions. *J. Comput. Phys.*, 2003.

M. Jones, L. Freitag, and P. Plassmann. An efficient parallel algorithm for mesh smoothing. In *4th Int. Meshing Roundtables*, pages 47–58. Sandia Labs, 1995.

R. Kornhuber and R. Roitzsch. On adaptive grid refinement in the presence of internal or boundary layers. *IMPACT Comput. Sci. Engrg.*, 2:40–72, 1990.

C. Lawson. Software for C^1 surface interpolation. In J. Rice, editor, *Mathematical Software III*, pages 161–194. Academic Press, 1977.

B. Li and Z. Zhang. Analysis of a class of superconvergence patch recovery techniques for linear and bilinear finite elements. *Numerical Methods for Partial Differential Equations*, pages 151–167, 1999.

V. Liseikin. *Grid Generation Methods.* Springer Verlag, Berlin, 1999.

J. Miller, E. O'Riordan, and G. Shishkin. *Fitted Numerical Methods For Singular Perturbation Problems.* World Scientific, 1996.

M. Rivara. Mesh refinement processes based on the generalized bisection of simplexes. *SIAM J. Numer. Anal.*, 21:604–613, 1984.

H. Roos. Optimal convergence of basic schemes for elliptic boundary value problems with strong parabolic layers. *J. Math. Anal. Appl.*, 267:194208, 2002.

H. Roos, M. Stynes, and L. Tobiska. *Numerical Methods for Singularly Perturbed Differential Equations*, volume 24 of *Springer Series in Computational Mathematics.* Springer Verlag, 1996.

G. Shishkin. *Grid approximation of singularly perturbed elliptic and parabolic equations (in Russian).* PhD thesis, Second doctorial thesis, Keldysh Institute, Moscow, 1990.

L. Wahlbin. *Superconvergence in Galkerkin finite element methods.* Springer Verlag, Berlin, 1995.

J. Xu and Z. Zhang. Analysis of recovery type a posteriori error estimators for mildly structured grids. *Math. Comp.*, 73 (247):1139–1152, 2004. URL `http://www.ams.org/mcom/2004-73-247/S0025-5718-03-01600-4/home.html`.

Z. Zhang. Ultraconvergence of the patch recovery technique II. *Mathematics Of Computation*, 69(229):141–158, 1999.

O. Zienkiewicz and J.Z.Zhu. The superconvergence patch recovery and a posteriori error estimates. Part 1: The recovery techniques. *Int. J. Number. Methods Engrg.*, 33:1331–1364, 1992a.

O. Zienkiewicz and J.Z.Zhu. The superconvergence patch recovery and a posteriori error estimates. Part 2: Error estimates and adaptivity. *Int. J. Number. Methods Engrg.*, 33:1365–1382, 1992b.

A Convergent Algorithm for Time Parallelization Applied to Reservoir Simulation

Izaskun Garrido, Magne S. Espedal, and Gunnar E. Fladmark[1]

University of Bergen, LIM / CIPR, (`http://www.mi.uib.no/~izaskun/`)

Summary. Parallel methods are not usually applied to the time domain because the sequential nature of time is considered to be a handicap for the development of competitive algorithms. However, this sequential nature can also play to our advantage by ensuring convergence within a given number of iterations. The novel parallel algorithm presented in this paper acts as a predictor corrector improving both speed and accuracy with respect to the sequential solvers. Experiments using our in house fluid flow simulator in porous media, **Athena**, show that our parallel implementation exhibit an optimal speed up relative to the method.

1 Time parallel reservoir simulator: Athena

Sub-stepping is common practice in reservoir simulation, in this technique some unknowns of a given system are computed in time-steps smaller than the normal step size, such that the overall system resolution will be comparable to that obtained with the small time-step. We propose a novel modified algorithm where this sub-stepping is computed in parallel following a technique of type Parareal, as studied in Baffico et al. [2002], Lions et al. [2001], Bal and Maday [2002]. This technique is a predictor corrector, PC, where the corrector runs in parallel. Thus, the time domain is subject to the standard treatment of domain decomposition Briggs et al. [1990], Keyes [2002], Brenner and Sung [2000]; being separated into sub-domains where different numerical solvers and discretizations may be applied. At each predictor corrector iteration the coarse solver acts as a predictor handling the sub-domain interfaces by providing initial boundary conditions for the parallel-fine system of equations. Each parallel solution will be used as a corrector determining the modification of the coarse system for the next iteration. Due to the sequential nature of time, one of these solutions is independent from the rest. This particularity will be exploited to modify the algorithm so as to obtain convergence rendering it suitable to dynamic load balancing schemes Lan et al. [2002]. The motivation for this type of parallelization is not only to improve both computational speed and convergence properties but aims to implement multi-grid with nested parallelism.

The paper is organized as follows: The numerical model implemented in our reservoir simulator `Athena` is presented in Section 2. This model contains a sub-stepping technique, which will be modified to run in parallel according to the method described in Section 3. Numerical examples illustrate in Section 4 the performance of these algorithms. Finally, conclusions of this work are given in Section 5.

2 The numerical model and standard method in `Athena`

Athena is a multicomponent multi-phase flow simulator in porous media. This simulator is based on a mathematical model consisting of three coupled non-linear differential equations which solve for the primary variables: molar masses, temperature and water pressure. The equations are derived from the mass conservation, energy conservation and volume balance method. These systems will be decoupled and discretized using Finite Volume in space and a Backward Euler scheme in time. However, the energy equation has coefficients dominated by the rock temperature, which is almost constant, so that their values at time $t^{[n+1]}$, may be approximated by those at previous time, $t^{[n]}$, leading to an explicit equation of the form

$$\mathcal{J}^{[n]} \Delta \mathbf{T}^{[n]} = -\mathbf{f}^{[n]} , \tag{1}$$

where the temperature increment $\Delta \mathbf{T}^{[n]} = \mathbf{T}^{[n+1]} - \mathbf{T}^{[n]}$. Solving the pressure equation with the *Newton-Raphson* method, we get the expression

$$\mathcal{J}^{[n(k)]} \Delta \mathbf{p}^{[n(k)]} = -\mathbf{f}^{[n(k)]} , \tag{2}$$

where $n(k)$ denotes the kth *Newton-Raphson* iteration at the nth time level

$$\Delta \mathbf{p}^{[n(k)]} = \mathbf{p}^{[n(k+1)]} - \mathbf{p}^{[n(k)]} ,$$

$$\mathcal{J}^{[n(k)]} = \left(\frac{\partial \mathbf{f}}{\partial \mathbf{p}} \right)^{[n(k)]} \simeq \frac{\mathcal{D}^{[n(k)]}}{\Delta t^{[n]}} + \mathcal{A}^{[n]} ,$$

and the right hand side looks like

$$\mathbf{f}^{[n(k)]} = \mathcal{D}^{[n(k)]} \frac{\mathbf{p}^{[n(k)]} - \mathbf{p}^{[n]}}{\Delta t^{[n]}} + \mathcal{A}^{[n]} \mathbf{p}^{[n(k)]} - \mathbf{b}^{[n]} .$$

Note that at each Newton iteration we consider an approximation by updating only the diagonal, $\mathcal{D}$, of the Jacobian matrix $\mathcal{J}$. Finally, the molar mass equations are decoupled considering the cross-derivatives between different components to be negligible. This assumption enables to solve sequentially the following residual equations for the molar masses

$$\left(\frac{\mathcal{I}}{\Delta t^{[n]}} + \mathcal{A}_{\nu}^{[n(k)]}\right)\Delta \mathbf{N}_{\nu}^{[n(k)]} = \mathbf{b}_{\nu}^{[n(k)]}\,, \qquad \nu = 1,\ldots,n_c \tag{3}$$

where for a chemical system consisting of n_c components located on a domain decomposed into n_{cv} cells, $\mathcal{I}$ is the identity matrix and each matrix $\mathcal{A}_\nu$ has n_{cv}^2 entries $(\mathcal{A}_\nu)_{ij}^{[n(k)]} = \sum_l \alpha_{\nu_i,\nu_j}^{l\,[n(k)]}$, $i\,,j \in \{1,\ldots,n_{\text{cv}}\}$, which are derived from an analytical expression.

The set of equations (1), (2) and (3), results in a compact numerical model, since they allow to find all the primary and secondary variables for the new time step. As the molar masses change very fast in relation to either the temperature or pressure, computational stability requires small time-steps and a greater number of Newton iterations for the mass conservation equation. In order to mitigate the time step restriction that the molar mass equation, see (3), imposes on the overall system, it is common practice to use sub-stepping. This technique involves computing with a coarse time-step the implicit solution for the temperature and pressure, whilst the molar masses are calculated for the same overall time step with several smaller sub-steps. For further details about `Athena`, its numerical model and implementation we refer the interested reader to a number of previous publications Fladmark [1997], Øye and Reme [1999], Garrido et al. [2003]. The rest of this paper will be devoted to the development and implementation of a Parareal-type algorithm which will parallelize the sequential sub-stepping technique.

3 General algorithm

Denoting by $\bar{\Omega} = \Gamma \cup \Omega$ an arbitrary time-step where $\Gamma = t^n$ and $\Omega = \Omega^n = (t^n, t^{n+1}]$, equation (3) is commonly solved by sub-stepping over a number, N, of sub-domains $\Omega_i = (t^n+(i-1)\Delta t, t^n+i\Delta t]$, where $\Delta t = (t^{n+1}-t^n)/N$, have either artificial boundaries Γ_i, $i > 1$ or real one $\Gamma_1 = \Gamma$ and are discretized independently of one another. The convergence of the sub-stepping over Ω^n determines adaptively the size of the next time step Ω^{n+1}.

The algorithm to be presented is of the Parareal form as proposed by Maday and Lyons Baffico et al. [2002], Lions et al. [2001], Bal and Maday [2002] and uses a PC where the corrector runs in parallel. In what follows, the PC will be described on equation (3) for a general time-step domain, $\bar{\Omega}$, and a given component, ν. For simplicity of notation we rewrite (3) as a residual equation of the form

$$\begin{aligned} \mathcal{J}\Delta u &= f\,, \quad \bar{\Omega}\,,\\ u &= g\,, \quad \Gamma\,. \end{aligned} \tag{4}$$

Denoting the kth PC iteration with the superscript k, the method begins by predicting a solution of

$$\begin{aligned} \mathcal{J}^k\Delta u^k &= f^k\,, \quad \bar{\Omega}\,, \quad k=1\,,\\ u^k_{|\Gamma} &= g\,, \quad \Gamma\,, \quad k=1\,, \end{aligned} \tag{5}$$

sub-stepping a number G of times, with $G << N$ and $\Omega = \cup_{i=1}^{G} \Omega_i$, a coarsening of the original domain decomposition. Note that system (5) is solved over each Ω_i using the *Newton-Raphson* method so that $\mathcal{J}^k$ has to be updated at every Newton iteration. This sub-stepping gives an approximation to the intermediate values $u^k_{|\Gamma_i}, i = 2, \dots, G$ which together with the initial boundary condition, $u^k_{|\Gamma} = u^k_{|\Gamma_1} = g$ serve as initial guesses for the boundary conditions of each independent system

$$\left.\begin{aligned} \mathcal{J}^k_{|\bar{\Omega}_i} \Delta u^k_i &= f^k_{|\bar{\Omega}_i} \\ u^k_{i|\Gamma_i} &= g + \textstyle\sum_{j=1}^{i-1} \Delta u^k_{|\Omega_j} \end{aligned}\right\} \qquad i = 1, \dots, G\,. \tag{6}$$

This set of systems will be solved in parallel so that the ith processor solves the ith system by sub-stepping on Ω_i a number $F(i)$ of times $\Omega_i = \cup_{j=1}^{F(i)} \Omega_j$, giving an approximation to the values $u^k_{|\Gamma_i}, i = 2, \dots, G$. If these approximations do not differ more than a given tolerance to those obtained previously from system (5) convergence for time step Ω has been achieved. Else, a new PC iteration, for systems (5) and (6) is computed, where the predictor equation (5) is corrected with data from the previous iteration as

$$\mathcal{J}^k_{|\bar{\Omega}_i} \Delta u^k = f^k_{|\bar{\Omega}_i} + \mathcal{J}^{k-1}_{|\bar{\Omega}_i} (u^{k-1}_{|\Gamma_{i+1}} - u^{k-1}_{i|\Gamma_{i+1}})\,. \tag{7}$$

Due to the non-linear nature of the PDE system under study, the correction term $\mathcal{J}^{k-1}_{|\bar{\Omega}_i} (u^{k-1}_{|\Gamma_{i+1}} - u^{k-1}_{i|\Gamma_{i+1}})$ is also non-linear and needs to be updated at every Newton iteration. This correction is equivalent to a Jacobi iteration for a linearized system with matching discretization along the artificial boundaries.

Another modification of this scheme can be obtained by adding in equation (5) all the corrections obtained from previous iterations to obtain schemes of the Parareal form

$$\mathcal{J}^k_{|\bar{\Omega}_i} \Delta u^k = f^k_{|\bar{\Omega}_i} + \sum_{j=1}^{k-1} \mathcal{J}^j_{|\bar{\Omega}_i} (u^j_{|\Gamma_{i+1}} - u^j_{i|\Gamma_{i+1}})\,. \tag{8}$$

Even more, after the first iteration the active domain is redefined to be the reduced

$$\Omega = \bar{\Omega} | \bar{\Omega}_1\,, \tag{9}$$

$u^k_{|\Gamma_1} = g$ and the initial boundary condition satisfies

$$u^k_{|\Gamma_2} = u^1_{1|\Gamma_2}\,. \tag{10}$$

Assuming that after iteration $k = G - 1$ it is satisfied

$$u^k_{|\Gamma_1} = g, \dots, u^k_{|\Gamma_G} = u^{G-1}_{G-1|\Gamma_G} \tag{11}$$

where $\Omega = \bar{\Omega} | \cup_{i=1}^{G-1} \bar{\Omega}_i$; at the Gth iteration both predictor and corrector are defined over the same active domain and share the same initial boundary

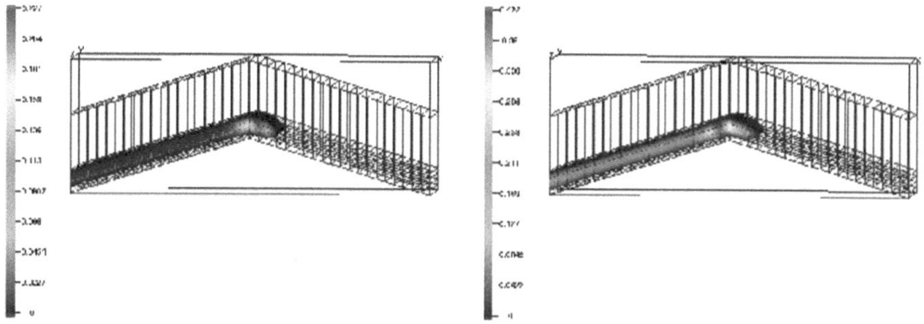

a. Gas saturation b. Oil saturation

Fig. 1. Hydrocarbon migration simulated by **Athena** for $100\,y$ with $0.0047\,y$ as time step

condition, so that the best approximation is given by the solution to the fine-parallel system, $u^G_{G|\Gamma_{G+1}}$. Therefore, this algorithm converges in at most G iterations, where G denotes the amount of systems to be solved in parallel and its accuracy is determined by the corrector approximation.

4 Numerical examples

In this section we will illustrate with two different experiments the scalability and performance of the algorithm implemented for the molar mass equations within the `Athena` fluid flow migration simulator. We have carried out the experiments on a Linux cluster, with PIII processors to explore the behavior of the methods.

Before proceeding with the numerical experiments, the geological domain and boundary conditions shall be described. The three dimensional domain has 50 m depth on the ends, and a size of $1000\,\mathrm{m} \times 100\,\mathrm{m} \times 70\,\mathrm{m}$. There are four different layers in the z direction: shale, sandstone, shale and sandstone again. The lithology for the sandstone has a porosity of $\phi = 0.5$ and a permeability of $K_x = 500\,\mathrm{mD}$, $K_y = 500\,\mathrm{mD}$ and $K_z = 500\,\mathrm{mD}$ while the corresponding values for the shale are $\phi = 0.5$ and $K_x = 5 \cdot 10^{-6}\,\mathrm{mD}$, $K_y = 5 \cdot 10^{-6}\,\mathrm{mD}$ and $K_z = 5 \cdot 10^{-6}\,\mathrm{mD}$. The domain is initially filled with water and the boundary conditions consist of an explicitly given flux of oil and gas with value $5 \cdot 10^{-5}\,\mathrm{mol/m^2s}$ going inwards on the left hand side and an outwards water flux with value $6.5 \cdot 10^{-4}\,\mathrm{mol/m^2s}$ in the right hand side. There are also temperature boundary conditions of 450°K at the top and 460°K at the bottom. The domain is uniformly subdivided in each direction as is shown in Fig. 1, which serve as an illustration of the `Athena` output for a simulated time of 100 years.

We consider the algorithm as described in Section 3 where, due to implementation issues in our particular simulator, the modification term, $S^{[n],k}_{|\bar{\Omega}_i}$, will be an approximation of that given in (7) by using the current matrix

$$\mathcal{J}^k_{|\bar{\Omega}_i} \Delta u^k = f^k_{|\bar{\Omega}_i} + \mathcal{J}^k_{|\bar{\Omega}_i}(u^{k-1}_{|\Gamma_{i+1}} - u^{k-1}_{i|\Gamma_{i+1}}) \,. \tag{12}$$

Computational results: Load balance

In this experiment we are mainly interested in the scalability of the implementation. By varying the number of processors used, we want to explore the speedup compared to the sequential program. When increasing the number of processors the wall clock time is expected to decrease, but there is a limit where adding more processors will not decrease the wall clock time. Besides this MPI implementation has a master-slave structure where the number of sub-domains equals the number of parallel (slave) processors used and the master deals with the non-parallelizable part of the algorithm.

The domain is partitioned considering a fixed underlying grid with a total of $s = \sum s_G = 2^4$ cells. Therefore, letting the number of sub-domains to double, $G = 2^i$, with $i = 0, \ldots, 4$ implies that the number of cells in each sub-domain halves, $s_G = 2^{4-i}$. Besides, since each sub-domain uses only one processor, increasing the number of sub-domains is equivalent to increase the overhead. In in the left of Fig. 2 the run time for the slaves is plotted against

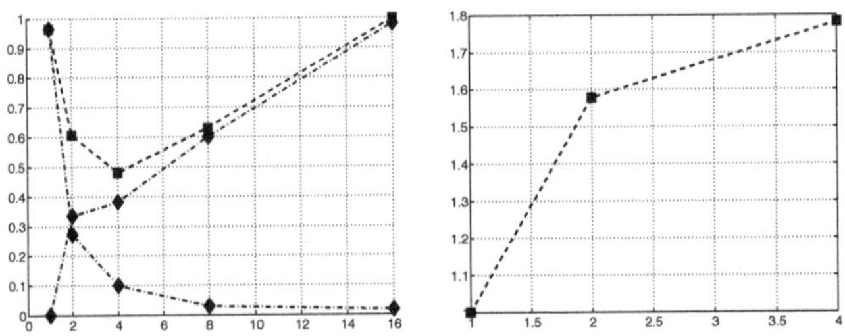

Fig. 2. Space domain with 200 cells. Run time vs. the number of sub-domains. In the left hand side, communication (.- increasing), calculation (.- decreasing) and addition of both times (- -) for the slaves. In the right, the parallel speedup for the addition of slave and master run times

the number of sub-domains, considering the run time values to be the average of all parallel processors. It can be seen that as the number of sub-domains (or processors) doubles, the calculation time at each processor halves (see the monotonic decreasing curve), whilst the time for collective broadcasting increases (see the monotonic increasing curve). The relevant values correspond to the addition of both communication and calculation times; these are plotted by the curve marked with squares which indicates that the method is competitive up to a certain parallelization degree when the communication time overrides the computational time.

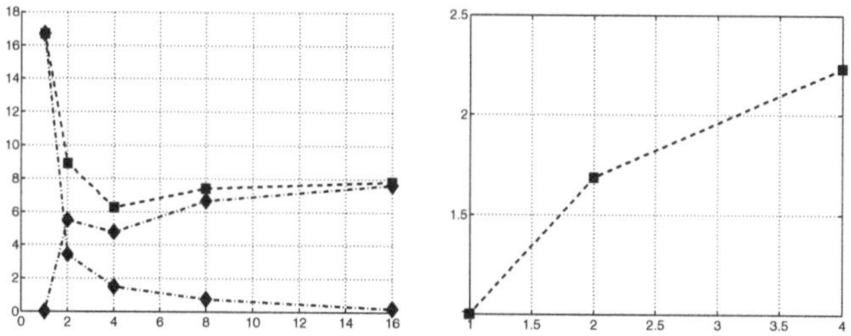

Fig. 3. Space domain with 800 cells. Run time vs. the number of sub-domains. In the left hand side, communication (.- increasing), calculation (.- decreasing) and addition of both times (- -) for the slaves. In the right, the parallel speedup for the addition of slave and master run times

The speedup for the overall method, addition of master (coarse solver) and slave (parallel-fine solver) run times, is displayed in the right of Fig. 2 where the best time is clearly obtained when $G = s_G$, which can easily be proven to coincide with the case when optimal load balance occurs.

Computational Results: Overhead

Given a fixed spatial discretization, the previous section studies the scalability in time as the underlying time-mesh remains constant whilst the ratio coarse to fine cells varies. We aim to demonstrate that the scalability properties hold as the spatial domain increases, therefore decreasing the space overhead. Even when the communication is seen in the left of Fig. 3 to became negligible with respect to the time for calculations, the master-slave structure of the algorithm gives a fixed overhead due to the communication master-slave and it can be seen in the right of Fig. 3 that the total run time is best when the load balance reaches optimality.

5 Conclusions

In this paper a convergent parallel algorithm has been derived in a constructive manner. It acts as a predictor corrector and the numerical experiments indicate that even for highly non-linear problems this parallel formulation improves both speed and accuracy with respect to the standard sequential solvers. Besides, the sequential nature of time allows to ensure convergence within a given number of iterations. The implementation of this algorithm in our fluid flow simulator, `Athena`, has optimal speed-up.

A Galerkin-type of algorithm based on Additive Schwarz Bjørstad et al. [2002], Xu and Zikatanov [2002, 2003], Cai et al. [2002] is under current consideration for space-time parallelization.

Acknowledgement. This work was supported by the Norwegian Research Council.

References

L. Baffico, S. Bernard, Y. Maday, G. Turinici, and G. Zérah. Parallel in time molecular dynamics simulations. *Phys. Rev. E.*, 66, 2002.

G. Bal and Y. Maday. A parareal time discretization for non-linear pde's with application to the pricing of an american put. In *Proceedings of a Workshop on Domain Decomposition*, LNCSE. Springer Verlag Zurich, 2002.

P. E. Bjørstad, M. Dryja, and T. Rahman. Additive schwarz methods for elliptic mortar finite element problems. *Submitted to Numerische Mathematik*, 2002. http://www.ii.uib.no/p̃etter/reports/pbmdtr2002AddSchMort.ps.gz.

S. C. Brenner and L.-Y. Sung. Lower bounds for non-overlapping domain decomposition preconditioners in two dimensions. *Math. Comput.*, 69(232): 1319–1339, 2000.

W. L. Briggs, V. E. Henson, and S. F. McCormick. *A Multigrid Tutorial.* SIAM, 1990.

X.-C. Cai, D. E. Keyes, and L. Marcinkowski. Non-linear additive schwarz preconditioners and application in computational fluid dynamics. *Int. J. Numer. Meth. Fluids*, pages 1463–1470, 2002.

G. E. Fladmark. Secondary oil migration. mathematical and numerical modeling in som simulator. Technical Report R-077857, Norsk Hydro, Bergen, 1997.

I. Garrido, E. Øian, M. Chaib, G. E. Fladmark, and M. S. Espedal. Implicit treatment of compositional flow. *Computational Geosciences*, 2003. To appear.

D. E. Keyes. Domain decomposition in the mainstream of computational science. In *Proceedings of the 14 international conference on Domain Decomposition Methods*, 2002.

Z. Lan, V. E. Taylor, and G. Bryan. A novel dynamic load balancing scheme for parallel systems. *J. Parallel Distrib. Comput.*, 62(12):1763–1781, 2002.

J.-L. Lions, Y. Maday, and G. Turinici. Rèsolution d'edp par un schéma en temps pararéel. *C. R. Acad. Sci. Paris*, 332(1):1–6, 2001.

G. Å. Øye and H. Reme. Parallelization of a compositional simulator with a galerkin coarse/fine method. In P.Amestoy et al., editors, *Euro-Par'99*, volume 1685. Springer-Verlag, Berlin, 1999.

J. Xu and L. Zikatanov. The method of alternating projections and the method of subspace corrections in Hilbert space. *J. of AMS*, 15, 2002. Technical report, PennState, November 2000a.

J. Xu and L. Zikatanov. Some observations on Babuska and Brezzi theories. *Num. Math.*, 94, 2003. Technical report, PennState, September 2000b.

Nonlinear Positive Interpolation Operators for Analysis with Multilevel Grids

Xue-Cheng Tai*

Department of Mathematics, University of Bergen, Johannes Brunsgate 12, 5007, Bergen, Norway. (`http://www.mi.uib.no/~tai/`)

Summary. We introduce some nonlinear positive and negative interpolation operators. The interpolation need to preserve positivity or negativity of a function. In addition, the interpolation must be pointwise below or above the function. Some of the operators also have the pointwise monotone property over refined meshes. It is also desirable that the interpolation have the needed approximation and stability estimates. Those operators could be used in the convergence analysis for domain decomposition and multigrid methods for obstacle problems.

1 Introduction

We are interested in the convergence rate analysis of multigrid and domain decomposition methods for variational inequalities, i.e. we want to solve a convex minimization problems with some convex constraints, c.f. Kornhuber [1994], Tai [2003]. It is well known that both domain decomposition and multigrid methods can be regarded as space decomposition and subspace correction techniques. For a given space decomposition technique, we need two constants to measure the quality of the decomposition. One constant is called the constant for the strengthened Cauchy-Schwarz inequality. The other constant is for the partition lemma, which is also called Lions's lemma. For linear problems, these constants are well established, see Xu [1992]. The concepts of using these constants to analyse the convergence rate for space decomposition techniques was extended to nonlinear problems in Tai [1994b], Tai and Xu [1999], Tai [1994a], Tai and Xu [2002], Tai and Tseng [2002], Tai [2003]. To be more specific, we shall consider the following problem in this work:

$$\min_{v\in K} F(v). \tag{1}$$

For simplicity, we just assume that

$$K = \{v|\ v \in H_0^1(\Omega),\ v \geq 0\}, \quad F(v) = \int_\Omega \frac{1}{2}|\nabla v|^2 - fv. \tag{2}$$

* Financial support from the Norwegian Research Council is gratefully acknowledged.

In order to use domain decomposition or multigrid methods for the above problem, we need to construct finite element or finite difference meshes that are nested and refined (the problem with non-nested mesh is much more complicated and shall not be considered here). For the partition lemma for the above problem, we need to interpolate functions from K to the different meshes or we need to interpolate functions from fine meshes to coarser meshes. The interpolation operators need to satisfy the following properties:

1. (Positivity): It shall preserve the positivity or negativity, i.e. the interpolation of a positive function shall be positive or the interpolation of a negative function shall be negative.
2. (Approximation): The interpolation shall have the needed approximation properties.
3. (Stability): The interpolation shall be stable in the needed norms.
4. (Pointwise above or below): The interpolation of a given function shall be pointwise below or above the function.
5. (Monotonicity with mesh refinement): When interpolating a function to finer or coarser meshes, it is desirable that the interpolation over a finer mesh should be pointwise bigger or smaller than the interpolation over a coarser mesh.

For problem (1)–(2), the standard nodal point linear Lagrangian finite element interpolation operators are not applicable in many context. In Nochetto and Wahlbin [2002] and Chen and Nochetto [2000], some interpolation operators are given which preserve positivity. These operators are linear and satisfy the approximation and stability requirements, but do not have properties that the interpolation is below or above the interpolated functions and also do have pointwise monotonicity with respect to refined meshes. In Nochetto and Wahlbin [2002], it was proved that linear positive interpolation operators may not exist if we require more than first order accuracy at extreme points. In this work, we shall introduce some operators which are not linear, but satisfy all the needed properties.

2 Some nonlinear positive interpolation operators

Let $\mathcal{T}_h$ be a quasi-uniform triangulation of the domain $\Omega \subset R^d, d = 1,2,3$ with a mesh size h and $S_h \subset H_0^1(\Omega)$ be the corresponding piecewise linear finite element space on $\mathcal{T}_h$. In the analysis, we need to use finite element spaces with different mesh sizes. It will be assumed that h is always the smallest mesh size. For an $H > h$, we consider the case that $\mathcal{T}_h$ is a refinement of $\mathcal{T}_H$. In the following, the definition of some nonlinear interpolation operators from S_h to S_H will be given. Denote by $\mathcal{N}_H = \{x_0^i\}_{i=1}^{n_0}$ all the interior nodes for $\mathcal{T}_H$. For a given x_0^i, let ω_i be the union of the mesh elements of $\mathcal{T}_H$ having x_0^i as one of its vertexes, i.e.

$$\omega_i := \cup\{\tau \in \mathcal{T}_H, x_0^i \in \bar{\tau}\}. \tag{3}$$

Let $\{\phi_0^i\}_{i=1}^{n_0}$ be the associated nodal basis functions satisfying $\phi_0^i(x_0^k) = \delta_{ik}$, $\phi_0^i \geq 0$, $\forall i$ and $\sum_i \phi_0^i(x) = 1$. It is clear that ω_i is the support of ϕ_0^i.

In the following, standard notations for Sobolev norms will be used, i.e. $\|\cdot\|_0$ stands for the $L^2(\Omega)$ norm, $\|\cdot\|_1$ and $|\cdot|_1$ are the norms and seminorms for $H^1(\Omega)$, etc.

2.1 A nonlinear positive interpolation operator below the function

Given a nodal point $x_0^i \in \mathcal{N}_H$ and a $v \in S_h$, let

$$I_i v = \min_{\bar{\omega}_i} v(x). \tag{4}$$

The interpolated function is then defined as

$$I_H^{\ominus} v := \sum_{x_0^i \in \mathcal{N}_H} (I_i v) \phi_0^i(x).$$

From the definition, it is easy to see that

$$I_H^{\ominus} v \leq v, \quad \forall v \in S_h, \tag{5}$$
$$I_H^{\ominus} v \geq 0, \quad \forall v \geq 0, v \in S_h. \tag{6}$$

Moreover, the interpolation for a given $v \in S_h$ on a finer mesh is always no less than the corresponding interpolation on a coarser mesh due to the fact that each coarser mesh element contains several finer mesh elements, i.e.

$$I_{h_1}^{\ominus} v \leq I_{h_2}^{\ominus} v, \quad \forall h_1 \geq h_2 \geq h, \quad \forall v \in S_h. \tag{7}$$

In addition, the interpolation operator also has the following approximation properties, c.f. p. 767 of Tai [2003].

Theorem 1. *For any $u, v \in S_h$, it is true that*

$$\|I_H^{\ominus} u - I_H^{\ominus} v - (u-v)\|_0 \leq c_d H |u-v|_1, \tag{8}$$
$$\|I_H^{\ominus} v - v\|_0 \leq c_d H |v|_1, \tag{9}$$
$$|I_H^{\ominus} u - I_H^{\ominus} v|_1 \leq c_d |u-v|_1, \tag{10}$$

where $c_d = C$ if $d=1$; $c_d = C\left(1 + \left|\log \frac{H}{h}\right|^{\frac{1}{2}}\right)$ if $d=2$ and $c_d = C\left(\frac{H}{h}\right)^{\frac{1}{2}}$ if $d=3$. Here and later, the generic constant C is used to denote constants that are independent of the mesh parameters.

2.2 A nonlinear negative interpolation operator above the function

However, if we define

$$I_i v = \max_{\bar{\omega}_i} v(x) \qquad I_H^{\oplus} v := \sum_{x_0^i \in \mathcal{N}_H} (I_i v) \phi_0^i(x). \tag{11}$$

Then it is easy to see that

$$I_H^{\oplus} v \geq v, \quad \forall v \in S_h, \qquad I_H^{\oplus} v \leq 0, \quad \forall v \leq 0, v \in S_h. \tag{12}$$

Moreover, the interpolation for a given $v \in S_h$ on a finer mesh is always no bigger than the corresponding interpolation on a coarser mesh, i.e.

$$I_{h_1}^{\oplus} v \geq I_{h_2}^{\oplus} v, \quad \forall h_1 \geq h_2 \geq h, \quad \forall v \in S_h. \tag{13}$$

From theorem 1, it is easy to see that the following is correct (Tai [2003]).

Theorem 2. *There exists an interpolation operator* $I_H^{\oplus} : S_h \mapsto S_H$ *such that*

$$\begin{aligned}
& I_H^{\oplus} v \geq v, \quad \forall v \in S_h, \\
& I_H^{\oplus} v \leq 0, \forall v \leq 0, v \in S_h, \\
& \|I_H^{\oplus} u - I_H^{\oplus} v - (u-v)\|_0 \leq c_d H |u-v|_1, \\
& \|I_H^{\oplus} v - v\|_0 \leq c_d H |v|_1, \qquad |I_H^{\oplus} u - I_H^{\oplus} v|_1 \leq c_d |u-v|_1, \forall v \in S_h.
\end{aligned}$$

2.3 A nonlinear interpolation operator above or below the function

For some cases, we need an interpolation operator which has the properties of $I_H^{\oplus}$ in some part of the domain Ω and has the properties of $I_H^{\ominus}$ in the rest of Ω. The operator we shall define in the following is a simplified version of the operator used in p.133 of Tai et al. [2002]. For any given $v \in S_h$, we let

$$v^+(x) = max(0, v(x)), \qquad v^-(x) = min(0, v(x)).$$

It is easy to see that $v(x) = v^+(x) + v^-(x)$. The new interpolation operator is then defined as:

$$\mathcal{I}_H v := \sum_{x_0^i \in \mathcal{N}_H} (\min_{\bar{\omega}_i} v^+ + \max_{\bar{\omega}_i} v^-) \phi_0^i(x). \tag{14}$$

We have $\min_{\bar{\omega}_i} v^+ \geq 0$ and $v^-|_{\omega_i} = 0$ if $v \geq 0$ in ω_i. We have $\max_{\bar{\omega}_i} v^- \leq 0$ and $v^+|_{\omega_i} = 0$ if $v \leq 0$ in ω_i. In case that v has both negative and positive values in ω_i, then we have $\min_{\bar{\omega}_i} v^+ = 0$ and $\max_{\bar{\omega}_i} v^- = 0$. For a given v, we let

$$\Omega^+ = \{x | \ v(x) \geq 0\}, \qquad \Omega^0 = \{x | \ v(x) = 0\}, \qquad \Omega^- = \{x | \ v(x) \leq 0\}.$$

It is easy to see that

$$\mathcal{I}_H v \geq 0 \text{ in } \Omega^+, \quad \mathcal{I}_H v \leq 0 \text{ in } \Omega^-, \quad \mathcal{I}_H v = 0 \text{ in } \Omega^0.$$

Moreover, we have that

$$\mathcal{I}_H v \leq v \text{ in } \Omega^+, \quad \mathcal{I}_H v \geq v \text{ in } \Omega^-.$$

If we interpolate a function into a sequence of refined meshes, then the interpolation value is increasing on finer meshes over the region Ω^+ and the interpolation value is decreasing on finer meshes over the region Ω^-. These pointwise monotone properties are visualized in Figure 1. Similarly, the following approximation and stability properties are valid:

$$\begin{aligned}\|\mathcal{I}_H u - \mathcal{I}_H v - (u - v)\|_0 &\le c_d H |u - v|_1, \forall u, v \in S_h, \\ |\mathcal{I}_H u - \mathcal{I}_H v|_1 &\le c_d |u - v|_1, \forall u, v \in S_h.\end{aligned}$$

The proof for the above estimations can be done similarly as in Tai [2003].

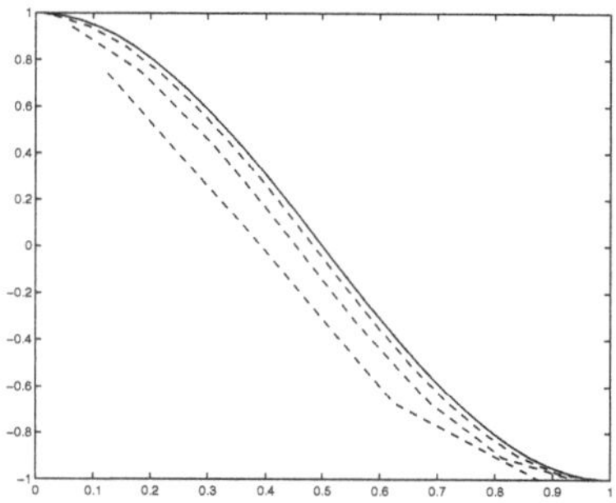

a) Plot of $I_h^\ominus$.

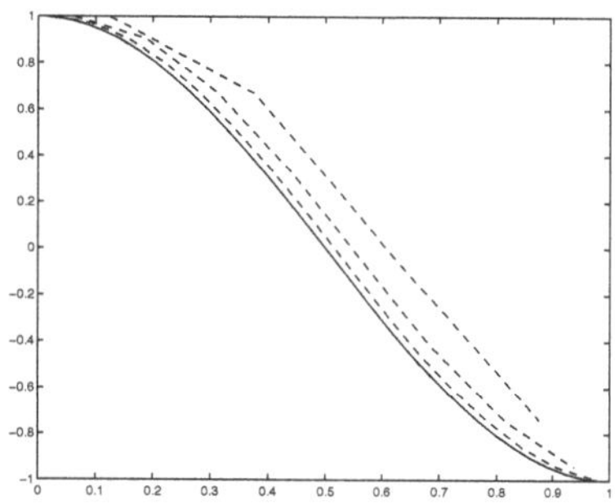

b) Plot of $I_h^\oplus$.

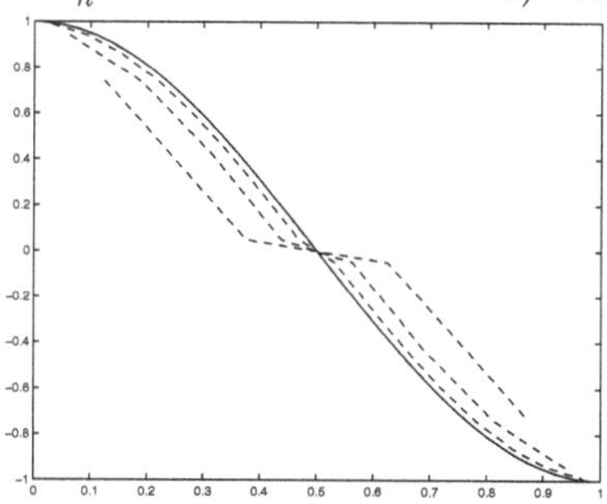

c) Plot of $\mathcal{I}_h$.

Fig. 1. Plots of the interpolation operators over a sequence of refined meshes. If the mesh is refined, the interpolation $I_H^\ominus v$ increases, while $I_H^\oplus v$ decreases. The interpolation $\mathcal{I}_H v$ increases in Ω^+ and decreases in Ω^-. $I_H^\ominus v$ is always below v, while $I_H^\oplus v$ is always above v. $\mathcal{I}_H v$ is below v in Ω^+ and above v in Ω^-

3 Some other nonlinear interpolation operators

The interpolation operators given in Section 2 only preserve constants locally and this can only have first order of convergence. In this section, we will introduce an operator which preserves linear functions locally and thus it can have higher oder approximation accuracy, but we will lose the pointwise monotone property enjoyed by the operators defined in Section 2.

For a given $v \in S_h$, let $v_0^I = I_H v$ to be the standard nodal Lagrangian interpolation of v into S_H. For the coarser mesh S_H, let x_i^0 and ω_i be as defined in Section 2, c.f. (3). We shall construct a new interpolation function v_0 by defining its nodal values as

$$v_0(x_i^0) = v_0^I(x_i^0) - \max_{x\in\omega_i} \left(v_0^I(x) - v(x)\right), \quad \forall x_i^0. \tag{15}$$

For simplicity, we define $\rho_0(x) \in S_H$ to be the coarse mesh function having the nodal values

$$\rho_0(x_i^0) = \max_{x\in\omega_i} \left(v_0^I(x) - v(x)\right), \quad \forall x_i^0.$$

It is easy to see that $v_0 = v_0^I - \rho_0$. Moreover, $\rho_0(x) \geq v_0^I(x) - v(x)$, which implies

$$v_0(x) = v_0^I(x) - \rho_0(x) \leq v_0^I(x) - (v_0^I(x) - v(x)) = v(x).$$

In addition,

$$\|v_0 - v\|_0 \leq \|v_0^I - v\|_0 + \|\rho_0\|_0.$$

As $\rho_0 \in S_H$, it is known that the L^2-norm is equivalent to

$$\|\rho_0\|_0^2 = CH^d \sum_{i=1}^{n_0} |\rho_0(x_i^0)|^2.$$

Using a linear mapping to transform ω_i into a domain of unit size and applying the well-known estimate of Bramble and Xu [1991], we get that

$$\|\rho_0\|_0^2 \leq CH^d \sum_{i=1}^{n_0} \|v_0^I - v\|_{0,\infty,\omega_i}^2 \leq CH^2 c_d^2 |v|_1^2.$$

In the above inequality, we have used the regularity of the meshes, i.e. under the minimum angle condition, the number of elements around a nodal point is always less than a constant. Using the inverse inequality, we know that $\|\rho_0\|_1 \leq CH^{-1}\|\rho_0\|_0$. In case that we want to use the H^2 norm for v, we have

$$\|\rho_0\|_0^2 \leq CH^d \sum_{i=1}^{n_0} \|v_0^I - v\|_{0,\infty,\omega_i}^2 \leq CH^2 |v|_2^2.$$

Denote v_0 by $\mathcal{I}_H^a v$. Combining these estimates with standard estimates for $v - v_0^I$, we have proved the following lemma.

Theorem 3. *Let S_H and S_h be defined as above. There exists an interpolation operator $\mathcal{I}_H^a : S_h \mapsto S_H$ such that*

$$\mathcal{I}_H^a v \leq v, \quad \|\mathcal{I}_H^a v\|_1 \leq c_d \|v\|_1,$$
$$\|\mathcal{I}_H^a v - v\|_0 \leq c_d H |v|_1, \ \|\mathcal{I}_H^a v - v\|_0 \leq H^2 |v|_2,$$

From the inequality

$$|\max_{\bar{\omega}_i} u - \max_{\bar{\omega}_i} v| \leq \|u - v\|_{0,\infty,\omega_i},$$

it is also easy to prove the following estimates using the techniques of Tai [2003]

Theorem 4. *For any u, v from S_h, we have*

$$\|\mathcal{I}_H^a u - \mathcal{I}_H^a v - (u - v)\|_0 \leq c_d H |u - v|_1,$$
$$\|\mathcal{I}_H^a u - \mathcal{I}_H^a v - (u - v)\|_0 \leq c H^2 |u - v|_2,$$
$$\|\mathcal{I}_H^a u - \mathcal{I}_H^a v\|_1 \leq c_d \|u - v\|_1.$$

In addition, the operator $\mathcal{I}_H^a$ have the following property which is not valid for the operators given in Section 2:

$$\mathcal{I}_H^a (v + v_H) = \mathcal{I}_H^a v + v_H, \quad \forall v \in S_h, v_H \in S_H, \tag{16}$$

i.e. the operator $\mathcal{I}_H^a$ is invariant for functions from the coarse mesh space S_H.

The interpolation $\mathcal{I}_H^a v$ is always below the function v, but it may not preserve positivity. It is also easy to define another operators which are always above the function or above the function in part of the domain and below the function in the rest of the domain.

4 Applications to multigrid decomposition

Assume that we have a sequence of shape regular meshes $\mathcal{T}_{h_j}$ that are produced by refining a coarse mesh. The mesh sizes $h_j, j = 1, 2, \cdots, J$ are decreasing and satisfies $c_1 \gamma^{2j} \leq h_j \leq c_2 \gamma^{2j}$ and $0 < \gamma < 1$. Let $\mathcal{M}_j$ be the piecewise linear finite element spaces over the meshes. For a given $v \geq 0$ and $v \in \mathcal{M}_J$ we shall decompose it into $v = \sum_{j=1}^J v_j$ such that $v_j \geq 0 \; \forall j$. In addition, we also need that $\|v_j\|_1 \leq c_d \|v\|_1$. Such a decomposition is needed for the proof of the partition lemma for Tai [2003] and Tai et al. [2002]. Using the operators defined in Section2, we see that the following functions v_j satisfy the needed properties:

$$v_j = I_{h_j}^{\ominus} v - I_{h_{j-1}}^{\ominus} v, \; j = 1, 2, \cdots J - 1, \; v_J = v - I_{h_{J-1}}^{\ominus} v.$$

In order to show that $v_j \geq 0$ we need to use the pointwise monotone properties. In order to show the stability of v_j in H^1 we need the corresponding estimates for $I^{\ominus}$. In fact, the operators defined in Section 2 and Section 3 can be used in different context in the convergence analysis of domain decomposition and multigrid methods for problems like (1).

The interpolation operator $I_H^{\oplus}$ is needed for the analysis given above if we change the constraint set K given in (2) to $K = \{v | \; v \in H_0^1(\Omega), \; v \leq 0\}$. The interpolation operator $\mathcal{I}_H$ is needed if we shall work with two-obstacles, i.e. one obstacle above the solution and one obstacle below the solution.

References

J. H. Bramble and J. Xu. Some estimates for a weighted L^2 projection. *Math. Comp.*, 56:463–476, 1991.

Z. Chen and R. H. Nochetto. Residual type a posteriori error estimates for elliptic obstacle problems. *Numer. Math.*, 84(4):527–548, 2000. ISSN 0029-599X.

R. Kornhuber. Monotone multigrid methods for elliptic variational inequalities I. *Numer. Math*, 69:167–184, 1994.

R. H. Nochetto and L. B. Wahlbin. Positivity preserving finite element approximation. *Math. Comp.*, 71(240):1405–1419 (electronic), 2002. ISSN 0025-5718.

X. C. Tai. Domain decomposition for linear and nonlinear elliptic problems via function or space decomposition. In *Domain decomposition methods in scientific and engineering computing (University Park, PA, 1993)*, volume 180 of *Contemp. Math.*, pages 355–360. Amer. Math. Soc., Providence, RI, 1994a.

X. C. Tai. Parallel function decomposition and space decomposition methods. In *Finite element methods (Jyväskylä, 1993)*, volume 164 of *Lecture Notes in Pure and Appl. Math.*, pages 421–432. Dekker, New York, 1994b.

X.-C. Tai. Rate of convergence for some constraint decomposition methods for nonlinear variational inequalities. *Numer. Math.*, 93(4):755–786, 2003. ISSN 0029-599X.

X.-C. Tai, B.-o. Heimsund, and J. Xu. Rate of convergence for parallel subspace correction methods for nonlinear variational inequalities. In *Thirteenth international domain decomposition conference*, pages 127–138. CIMNE, Barcelona, Spain, 2002. Available online at: http://www.mi.uib.no/7%Etai/.

X.-C. Tai and P. Tseng. Convergence rate analysis of an asynchronous space decomposition method for convex minimization. *Math. Comp.*, 71(239): 1105–1135 (electronic), 2002. ISSN 0025-5718.

X.-C. Tai and J. Xu. Subspace correction methods for convex optimization problems. In *Eleventh International Conference on Domain Decomposition Methods (London, 1998)*, pages 130–139 (electronic). DDM.org, Augsburg, 1999.

X.-C. Tai and J. Xu. Global and uniform convergence of subspace correction methods for some convex optimization problems. *Math. Comp.*, 71(237): 105–124 (electronic), 2002. ISSN 0025-5718.

J. Xu. Iterative methods by space decomposition and subspace correction. *SIAM Review*, 34(4):581–613, December 1992.

Part XIII

Minisymposium: Discretization Techniques and Algorithms for Multibody Contact Problems

On Scalable Algorithms for Numerical Solution of Variational Inequalities Based on FETI and Semi-monotonic Augmented Lagrangians

Zdeněk Dostál[1] and David Horák[1]

VŠB-Technical University Ostrava, Applied Mathematics Tr17.listopadu, CZ-70833,Ostrava,CzechRepublic,zdenek.dostal@vsb.cz,david.horak@vsb.cz

Summary. Theoretical and experimental results concerning a new FETI based algorithm for numerical solution of variational inequalities are reviewed. A discretized model problem is first reduced by the duality theory of convex optimization to the quadratic programming problem with bound and equality constraints. The latter is then optionally modified by means of orthogonal projectors to the natural coarse space introduced by Farhat and Roux in the framework of their FETI method. The resulting problem is then solved by a new variant of the augmented Lagrangian type algorithm with the inner loop for the solution of bound constrained quadratic programming problems. Recent theoretical results are reported that guarantee scalability of the algorithm. The results are confirmed by numerical experiments.

1 Introduction

The FETI method was originally proposed by Farhat and Roux [1992] for parallel solution of linear problems described by elliptic partial differential equations. Its key ingredient is decomposition of the spatial domain into non-overlapping subdomains that are "glued" by Lagrange multipliers, so that after eliminating the primal variables, the original problem is reduced to a small, relatively well conditioned, typically equality constrained quadratic programming problem that is solved iteratively. Observing that the equality constraints may be used to define so called "natural coarse grid", Farhat et al. [1994] modified the basic FETI algorithm so that they were able to prove its numerical scalability.

If the FETI procedure is applied to an elliptic variational inequality, the resulting quadratic programming problem has not only the equality constraints, but also the non-negativity constraints. Even though the latter is a considerable complication as compared with the linear problem, the resulting problem is still easier to solve than the contact problem in displacements as it is smaller, better conditioned and only bound constrained. Promising experimental results by Farhat and Dureisseix [2002], who used a coarse grid initial approximation, supported this claim and even indicated numerical scalability of their method. Scalability was later proved for an algorithm that combined FETI with optimal dual penalty Dostál and Horák [2003b]. A different approach yielding experimental evidence of scalability was based on the augmented Lagrangian method was used by Dostál et al. [2000a,b], Dostál and Horák [2003a]. It should be noted that the effort to develop scalable solvers for variational inequalities was not restricted to FETI. For example, using ideas related to Mandel [1984], Kornhuber [1997], Kornhuber and Krause [2001], Wohlmuth and Krause. [2002] gave an experimental evidence of numerical scalability of the algorithm based on monotone multigrid. Probably the first theoretical results concerning development of scalable algorithms were proved by Schoeberl [1998b,a].

Here we review recent improvements to show scalability for this type of algorithms. We start our exposition by describing a simple model problem and the FETI methodology Dostál et al. [2000a] that turns the variational inequality into the quadratic programming problem with bound and equality constraints. Then we briefly review recent results concerning a new variant of the augmented Lagrangian method. Finally we report the results of numerical experiments that are in agreement with the theory and indicate high and numerical scalability of the algorithm presented.

2 Model problem

Let $\Omega = \Omega^1 \cup \Omega^2$, $\Omega^1 = (0,1) \times (0,1)$ and $\Omega^2 = (1,2) \times (0,1)$ denote open domains with boundaries Γ^1, Γ^2 and their parts Γ_u^i, Γ_f^i, Γ_c^i formed by the sides of Ω^i, $i = 1,2$, so that $\Gamma_u^1 = \{0\} \times (0,1)$, $\Gamma_u^2 = \{2\} \times (0,1), \Gamma_c^i \{1\} \times (0,1)$, and Γ_f^i are formed by the other sides of Ω^i, $i = 1,2$. Let $H^1(\Omega^i), i = 1,2$ denote the Sobolev space of the first order in the space $L^2(\Omega^i)$ of functions on Ω^i whose squares are integrable in the sense of Lebesgue. Let

$$V^i = \{v^i \in H^1(\Omega^i) : v^i = 0 \quad \text{on} \quad \Gamma_u^i\}$$

denote the closed subspaces of $H^1(\Omega^i), i = 1,2$, and let

$$V = V^1 \times V^2 \qquad \text{and} \qquad \mathcal{K} = \{(v^1, v^2) \in V : v^2 - v^1 \geq 0 \quad \text{on} \quad \Gamma_c\}$$

denote the closed subspace and the closed convex subset of $\mathcal{H} = H^1(\Omega^1) \times H^1(\Omega^2)$, respectively. The relations on the boundaries are in terms of traces. On $\mathcal{H}$ we shall define a symmetric bilinear form

$$a(u,v) = \sum_{i=1}^{2} \int_{\Omega^i} \left(\frac{\partial u^i}{\partial x}\frac{\partial v^i}{\partial x} + \frac{\partial u^i}{\partial y}\frac{\partial v^i}{\partial y} \right) d\Omega$$

and a linear form

$$\ell(v) = \sum_{i=1}^{2} \int_{\Omega^i} f^i v^i d\Omega,$$

where $f^i \in L^2(\Omega^i), i = 1, 2$ are the restrictions of

$$f(x,y) = \left\{ \begin{array}{c} -3 \text{ for } (x,y) \in (0,1) \times [0.75,1) \\ 0 \text{ for } (x,y) \in (0,1) \times [0,0.75) \text{ and } (x,y) \in (1,2) \times [0.25,1) \\ -1 \text{ for } (x,y) \in (1,2) \times [0,0.25) \end{array} \right\}.$$

Thus we can define a problem to find

$$\min \quad q(u) = \frac{1}{2}a(u,u) - \ell(u) \quad \text{subject to} \quad u \in \mathcal{K}. \tag{1}$$

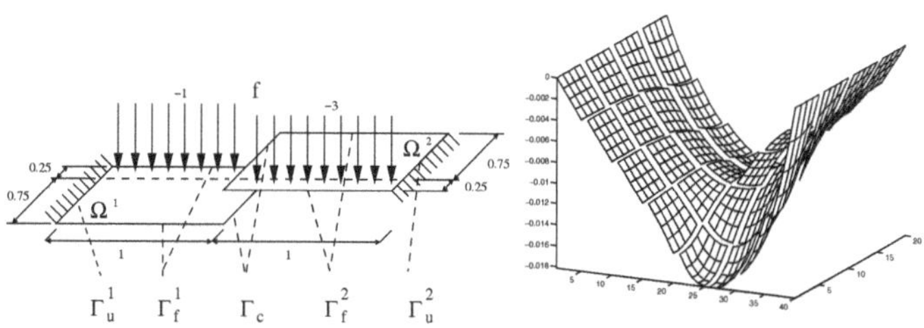

Fig. 1. Model problem and its solution

The solution of the model problem may be interpreted as the displacement of two membranes under the traction f. The membranes are fixed on the outer edges as in Figure 1 and the left edge of the right membrane is not allowed to penetrate below the right edge of the left membrane. Since the Dirichlet conditions are prescribed on parts $\Gamma_u^i, i = 1, 2$ of the boundaries with positive measure, the quadratic form a is coercive which guarantees existence and uniqueness of the solution Hlaváček et al. [1988].

3 Domain decomposition and discretized problem with a natural coarse grid

To enable efficient application of the domain decomposition methods, we can optionally decompose each Ω^i into square subdomains $\Omega^{i1}, \dots, \Omega^{ip}, p = s^2 > 1, i = 1, 2$. The continuity in Ω^1 and Ω^2 of the global solution assembled from the local solutions u^{ij} will be enforced by the "gluing" conditions $u^{ij}(x) = u^{ik}(x)$ that should be satisfied for any x in the interface $\Gamma^{ij,ik}$ of Ω^{ij} and Ω^{ik}. After modifying appropriately the definition of problem (1), introducing regular grids in the subdomains Ω^{ij} that match across the interfaces $\Gamma^{ij,kl}$, indexing contiguously the nodes and entries of corresponding vectors in the subdomains, and using the finite element discretization, we get the discretized version of problem (1) with the auxiliary domain decomposition that reads

$$\min \frac{1}{2} u^\top A u - f^\top u \quad \text{s.t.} \quad B^I u \leq 0 \quad \text{and} \quad B^E u = 0. \tag{2}$$

In (2), A denotes a positive semidefinite stiffness matrix, the full rank matrices B^I and B^E describe the discretized inequality and gluing conditions, respectively, and f represents the discrete analog of the linear term $\ell(u)$. Denoting

$$\lambda = \begin{bmatrix} \lambda^I \\ \lambda^E \end{bmatrix} \quad \text{and} \quad B = \begin{bmatrix} B^I \\ B^E \end{bmatrix},$$

we can write the Lagrangian associated with problem (2) briefly as

$$L(u, \lambda) = \frac{1}{2} u^\top A u - f^\top u + \lambda^\top B u.$$

It is well known that (2) is equivalent to the saddle point problem

$$\text{Find} \quad (\overline{u}, \overline{\lambda}) \quad \text{s.t.} \quad L(\overline{u}, \overline{\lambda}) = \sup_{\lambda_I \geq 0} \inf_u L(u, \lambda). \tag{3}$$

After eliminating the primal variables u from (3), we shall get the minimization problem

$$\min \Theta(\lambda) \quad \text{s.t.} \quad \lambda_I \geq 0 \quad \text{and} \quad R^\top (f - B^\top \lambda) = 0, \tag{4}$$

where

$$\Theta(\lambda) = \frac{1}{2} \lambda^\top B A^\dagger B^\top \lambda - \lambda^\top B A^\dagger f, \tag{5}$$

$A^\dagger$ denotes a generalized inverse that satisfies $AA^\dagger A = A$, and R denotes the full rank matrix whose columns span the kernel of A. We shall choose R so that its entries belong to $\{0, 1\}$ and each column corresponds to some floating auxiliary subdomain Ω^{ij} with the nonzero entries in the positions corresponding to the indices of nodes belonging to Ω^{ij}.

Even though problem (4) is much more suitable for computations than (2), further improvement may be achieved by adapting some simple observations and the results of Farhat et al. [1994], Mandel and Tezaur [1996]. Let us denote

$$F = BA^\dagger B^\top, \quad \widetilde{G} = R^\top B^\top, \quad \widetilde{e} = R^\top f, \quad \widetilde{d} = BA^\dagger f,$$

and let $\widetilde{\lambda}$ solve $\widetilde{G}\widetilde{\lambda} = \widetilde{e}$, so that we can transform the problem (4) to minimization on the subset of the vector space by looking for the solution in the form $\lambda = \mu + \widetilde{\lambda}$. Since

$$\frac{1}{2} \lambda^\top F \lambda - \lambda^\top \widetilde{d} = \frac{1}{2} \mu^\top F \mu - \mu^\top (\widetilde{d} - F\widetilde{\lambda}) + \frac{1}{2} \widetilde{\lambda}^\top F \widetilde{\lambda} - \widetilde{\lambda}^\top \widetilde{d},$$

problem (4) is, after returning to the old notation, equivalent to

$$\min \quad \frac{1}{2}\lambda^\top F\lambda - \lambda^\top d \quad \text{s.t} \quad G\lambda = 0 \quad \text{and} \quad \lambda^I \geq -\widetilde{\lambda}^I \tag{6}$$

where $d = \widetilde{d} - F\widetilde{\lambda}$ and $G = T\widetilde{G}$ denotes a matrix arising from the orthonormalization of the rows of $\widetilde{G}$ by the Schmidt process defined by the regular matrix T.

Our final step is based on observation that the problem (6) is equivalent to

$$\min \quad \frac{1}{2}\lambda^\top PFP\lambda - \lambda^\top Pd \quad \text{s.t} \quad G\lambda = 0 \quad \text{and} \quad \lambda^I \geq -\widetilde{\lambda}^I \tag{7}$$

where

$$Q = G^\top G \qquad \text{and} \qquad P = I - Q$$

denote the orthogonal projectors on the image space of $G^\top$ and on the kernel of G, respectively, so that $P\lambda = \lambda$ for feasible λ.

4 Semi-monotonic augmented Lagrangian method and scalability

In this section we shall describe a recent modification Dostál [2003] of the algorithm introduced earlier by Dostál et al. [2003]. The algorithm treats each type of constraints separately, so that efficient algorithms using projections and adaptive precision control Dostál and Schoeberl [2004] may be used for the bound constrained QP problems.

Let us recall that the augmented Lagrangian for (7) and its gradient are given by

$$L(\lambda, \mu, \rho) = \frac{1}{2}\lambda^\top PFP\lambda - \lambda^\top Pd + \mu^\top G\lambda + \frac{1}{2}\rho||Q\lambda||^2$$

$$g(\lambda, \mu, \rho) = PFP\lambda - Pd + G^\top(\mu + \rho G\lambda).$$

The *projected gradient* $g^P = g^P(\lambda, \mu, \rho)$ of L at λ is given componentwise by

$$g_i^P = g_i \ \text{ for } \ \lambda_i > -\overline{\lambda}_i \ \text{ or } \ i \notin \mathcal{I} \ \text{ and } \ g_i^P = g_i^- \ \text{ for } \ \lambda_i = -\overline{\lambda}_i \ \text{ and } \ i \in \mathcal{I}$$

with $g_i^- = \min(g_i, 0)$, where $\mathcal{I}$ is the set of indices of constrained entries of λ. The Hessian of the augmented Lagrangian $L(\lambda, \mu, \rho)$ is given by $H^\rho = PFP + \rho Q$.

Algorithm 1. Semi-monotonic augmented Lagrangian method (SALM).

Step 0. Set $\eta > 0$, $1 < \beta$, $\rho_0 > 0$, $M > 0$, μ^0 and $k = 0$.
Step 1. Find λ^k so that $\|g^P(\lambda^k, \mu^k, \rho_k)\| \leq \min\{M\|G\lambda^k\|, \eta\}$.
Step 2. If $\|g^P(\lambda^k, \mu^k, \rho_k)\|$ and $\|G\lambda^k\|$ are sufficiently small, then stop.
Step 3. $\mu^{k+1} = \mu^k + \rho_k G\lambda^k$
Step 4. If $L(\lambda^{k+1}, \mu^{k+1}, \rho_{k+1}) < L(\lambda^k, \mu^k, \rho_k) + \frac{\rho_{k+1}}{2}\|G\lambda^{k+1}\|^2$

Step 4a. then $\rho_{k+1} = \beta\rho_k$
Step 4b. else $\rho_{k+1} = \rho_k$
end if.
Step 5. Increase k by one and return to Step 1.

An implementation of Step 1 is carried out by the minimization of the augmented Lagrangian L subject to $\lambda_I \geq -\overline{\lambda}_I$ by the MPRGP algorithm Dostál and Schoeberl [2004]. The MPRGP algorithm with the choice of parameters $\Gamma = 1$ and $\overline{\alpha} \in (0, \|H^\rho\|^{-1}]$ generates the iterations $\{\widetilde{\lambda}^{ki},\ i = 1, 2, \ldots\}$ for the unique solution $\overline{\lambda}^k$ of the auxiliary minimization problem so that the rate of convergence in the energy norm defined by $\|\lambda\|^2_{H^\rho} = \lambda^\top H^\rho \lambda$ may be expressed by means of the least eigenvalue α_1 of $\|H^\rho\|$ in the form

$$\|\widetilde{\lambda}^{ki} - \overline{\lambda}^k\|^2_{H^\rho} \leq \frac{2\eta^i}{\alpha_1}\left(L(\widetilde{\lambda}^{k0}, \mu^k, \rho_k) - L(\overline{\lambda}^k, \mu^k, \rho_k)\right), \qquad \eta = 1 - \frac{\overline{\alpha}\alpha_1}{4}. \quad (8)$$

Algorithm 1 has been proved Dostál [2003] to converge for any set of parameters that satisfy the prescribed relations. It has also been shown that if $\rho_k \geq M^2/\alpha_1$, then

$$L(\lambda^{k+1}, \mu^{k+1}, \rho_{k+1}) \geq L(\lambda^k, \mu^k, \rho_k) + \frac{\rho^{k+1}}{2}\|G\lambda^{k+1}\|^2,$$

so that it is possible to give an upper bound on ρ^k in terms of α_1. The experiments have shown that the penalty parameter should be sufficiently high to enforce fast convergence of the outer loop. Let us recall that a large penalty parameter need not delay too much the convergence of the inner loop as the image spaces of the projectors P and Q are invariants subspaces of H^ρ so that the arguments of Dostál [1999] may be applied. Moreover, it has been proved Dostál [2003] that there is a bound on the number of outer iterations that are necessary to achieve prescribed relative feasibility error $\epsilon\|Pd\|$. The bound may be expressed in terms of the extreme eigenvalues of the Hessian of the augmented Lagrangian. Since it has been established by Mandel and Tezaur [1996] and more recently by Klawonn and Widlund [2001] that the smallest eigenvalue and the spectral condition number of the restriction of PFP to the kernel of G are $O(1)$ for fixed ratio H/h, it is possible to prove the following theorem.

Theorem 1. *Let C, ρ and ϵ denote given positive numbers, and let $\{\lambda^k_{H,h}\}, \{\mu^k\}$ and $\{\rho_k\}$ be generated by Algorithm 1 with Step 1 implemented by the MPRGP algorithm initiated by $\lambda^0_{H,h} = 0, \mu^0 = 0,\ \eta = \|Pd\|$ and $\rho_0 > 0$ for the solution of the problem (7) arising from the regular discretization of (1) with the decomposition and discretization parameters H and h, respectively. Then there is an integer k independent of h and H such that $H/h \leq C$ implies*

$$\|g^P(\lambda^k_{H,h}, \mu^k, \rho_k)\| \leq \epsilon\|Pd\| \quad \text{and} \quad \|G\lambda^k_{H,h}\| \leq \epsilon\|Pd\|. \quad (9)$$

We have implemented Algorithm 1 for solution of (1). Results of computations to the relative precision 1e-4 are in Table 1. The largest problem discretized by more than two million nodal variables required 167 seconds of 32 processors of SGI Origin. The finest discretization of (1) that we have run so far comprised 8464272 nodal variables and its solution required 65 iterations and 1281 seconds of 64 processors with decomposition into 128 subdomains. The results are the same as in Dostál and Horák [2003a,b] as no update of the penalty parameter was observed in either case.

Table 1. Numerical scalability of AL for H/h=128 and ρ=1e+3

dimension	33282	133128	532512	2130048
subdomains	2	8	32	128
iterations	28	59	36	47

5 Comments and conclusion

We have reviewed our recent results related to application of the augmented Lagrangians with the FETI based domain decomposition method to the solution of variational inequalities using recently developed algorithms for the solution of special QP problems. In particular, we have shown that the solution of the discretized problem to a prescribed precision may be found in a number of iterations bounded independently of the discretization parameter. Numerical experiments with the model variational inequality are in agreement with the theory and indicate that the algorithm may be efficient. Let us point out that similar development may be done also on the ground of DP-FETI. We shall describe it elsewhere together with applications to contact problems of elasticity.

Acknowledgement. This research is supported by grants No. 101/02/0072 and 101/01/0538 of the Grant Agency of the Czech Republic and by projects of the Ministry of Education No. AVOZ2076919 and ME641.

References

Z. Dostál. On preconditioning and penalized matrices. *Num. Lin. Alg. Appl.*, 6: 109–114, 1999.

Z. Dostál. Inexact semi-monotonic augmented Lagrangians with optimal feasibility convergence for quadratic programming with simple bounds and equality constraints. *submitted to SIAM J. Num. Anal.*, 2003.

Z. Dostál, A. Friedlander, and S. Santos. Augmented Lagrangians with adaptive precision control for quadratic programming with simple bounds and equality constraints. *SIAM J. Opt.*, 13:1120–1140, 2003.

Z. Dostál, F. Gomes, and S. Santos. Duality based domain decomp. with natural coarse space for variat. ineq. *J. Comput. Appl. Math.*, 126:397–415, 2000a.

Z. Dostál, F. Gomes, and S. Santos. Solution of contact problems by FETI domain decomp. *Comput. Meth. Appl. Mech. Eng.*, 190:1611–1627, 2000b.

Z. Dostál and D. Horák. Scalability and FETI based algorithm for large discretized variational inequalities. *Math. and Comput. in Simul.*, 61:347–357, 2003a.

Z. Dostál and D. Horák. Scalable FETI with optimal dual penalty for a variational inequality. *to appear in Num. Lin. Alg. and Appl.*, 2003b.

Z. Dostál and J. Schoeberl. Minimizing quadratic functions subject to bound constraints with the rate of convergence and finite termination. *to appear in Comput. Optimiz. and Appl.*, 2004.

C. Farhat and D. Dureisseix. A numerically scalable domain dec. meth. for solution of frictionless contact problems. *to appear in Int. J. Num. Meth. Eng.*, 2002.

C. Farhat, J. Mandel, and F. Roux. Optimal convergence properties of the FETI domain decomp. method. *Comp. Meth. Appl. Mech. Eng.*, 115:367–388, 1994.

C. Farhat and F. Roux. An unconventional domain decomposition method for an efficient parallel solution of large-scale finite element systems. *SIAM J. Sc. Stat. Comput.*, 13:379–396, 1992.

I. Hlaváček, J. Haslinger, J. Nečas, and J. Lovíšek. *Solution of Variational Inequalities in Mechanics.* Springer Verlag Berlin, 1988.

A. Klawonn and O. Widlund. FETI and Neumann-Neumann iterative substructuring methods: connections and new results. *Communic. on Pure and Appl. Math.*, LIV:57–90, 2001.

R. Kornhuber. *Adaptive Monotone Multigrid Methods for Nonlinear Variational Problems.* Teubner, Stuttgart, 1997.

R. Kornhuber and R. Krause. Adaptive multigrid methods for Signorini's problem in linear elasticity. *Comput. Visualiz. in Science*, 4:9–20, 2001.

J. Mandel. Étude algébrique d'une méthode multigrille pour quelques problèmes de frontière libre. *Compt. Rendus de l'Acad. des Scien.*, pages 469–472, 1984.

J. Mandel and R. Tezaur. Convergence of a Substructuring Method with Lagrange Multipliers. *Numer. Math.*, 73:473–487, 1996.

J. Schoeberl. Efficient contact solvers based on domain decomposition techniques. *Comput. and Math.*, 42:1217–1228, 1998a.

J. Schoeberl. Solving the Signorini problem on the basis of domain decomposition techniques. *Part. Diff. Eqs. in Physics and Biology*, 60:323–344, 1998b.

B. Wohlmuth and R. Krause. Monotone methods on nonmatching grids for nonlinear contact problems. Technical report, 2002. Research Report No. 2002/02 of the Stuttgart University, Sonderforsungsbereich 404.

Fast Solving of Contact Problems on Complicated Geometries

Rolf Krause[1] and Oliver Sander[2]

[1] Universität Bonn, Institut für Angewandte Mathematik (`http://www.iam.uni-bonn.de/~krause/`)
[2] FU Berlin, FB Mathematik und Informatik (`http://www.math.fu-berlin.de/~sander/`)

Summary. We consider the numerical simulation of multi-body contact problems in linear elasticity. For the discretization of the transmission conditions at the interface between the bodies by means of a transfer operator nonconforming domain decomposition methods (mortar methods) are used. Here, we focus on the difficulties related to the discrete choice of the transfer operator. We explain in detail how the transfer operator can be implemented in the case of three-dimensional nonplanar contact boundaries. For the numerical solution of the arising nonlinear systems of equations monotone multigrid methods are used, which do not require any regularization of the nonpenetration condition at the contact interface.

1 Introduction

The mathematical formulation of quasistationary contact problems in linear elasticity is given as a system of elliptic partial differential equations with suitable boundary conditions. Of particular importance are the boundary conditions at the interface between the bodies coming into contact. They have to be chosen in a way that the bodies do not penetrate each other. For linear elastic materials and small displacements, usually linearized nonpenetration conditions are considered, giving rise to inequality constraints for displacements and normal stresses at the contact interface. For an overview, see, e.g., Kikuchi and Oden [1988]. For a more detailed description we refer to Eck [1996]. By means of this inequality constraints at the interface, the corresponding elliptic boundary value problem becomes nonlinear and nondifferentiable.

The discretization of the boundary conditions requires a discrete transfer operator, mapping the displacements and stresses from one body to the other and vice versa. Here, for the construction of the discrete transfer operator for displacements and stresses, we use nonconforming domain decomposition methods (mortar methods). They have been successfully applied to contact problems and give rise to discretization schemes of optimal order, see, e.g., Ben Belgacem et al. [1999], Hild [2000], Wohlmuth and Krause [2003].

For contact boundaries in three space dimensions, the bodies on the discrete level are represented by polyhedral meshes. At the contact interface, there is no a priori knowledge about the small–scale relationship of the meshes to each other, which is of crucial importance for the construction of the discrete transfer operator for the discretization by mortar methods. We explain how a mapping between the boundary meshes can be constructed and implemented for nonplanar contact boundaries in three space dimensions.

The paper is organized as follows. In Section 2, we give the formulation of a two body contact problem in linear elasticity as a partial differential equation and the discretization by mortar methods. In Section 3, the construction of the discrete transfer operator is depicted and numerical examples are given. For the solution of the arising nonlinear systems of equations, we use monotone multigrid methods for contact problems using mortar methods, see Kornhuber and Krause [2001], Krause [2001] and Wohlmuth and Krause [2003]. In this case, no regularization of the inequality constraints at the interface is required. The nonlinear system of equations can be solved with multigrid complexity.

2 Problem Formulation and Discretization

For simplicity, we restrict ourselves to the case of two deformable bodies in $\mathbb{R}^3$. We identify the two bodies in their reference configurations with two domains $\Omega_{\text{non}}, \Omega_{\text{mor}} \subset \mathbb{R}^3$ with sufficiently smooth boundaries. The naming stems from the fact that Ω_{non} and Ω_{mor} will later be used as nonmortar and mortar side, respectively. Under the influence of boundary conditions and volume forces, the bodies undergo displacements $\mathbf{u} = (\mathbf{u}_s, \mathbf{u}_m) : \Omega_{\text{non}} \times \Omega_{\text{mor}} \to \mathbb{R}^3$.

The boundary of $\Omega := \Omega_{\text{non}} \cup \Omega_{\text{mor}}$ is partitioned into three disjoint subsets Γ_D, Γ_N, and Γ_C. The set Γ_C represents the region where contact might occur. It therefore consists of two parts $\Gamma_C = \Gamma_{\text{non}} \cup \Gamma_{\text{mor}}$ with $\Gamma_{\text{non}} \subset \partial\Omega_{\text{non}}$ and $\Gamma_{\text{mor}} \subset \partial\Omega_{\text{mor}}$. We assume $\text{meas}(\Gamma_D \cap \Omega_{\text{non}}), \text{meas}(\Gamma_D \cap \Omega_{\text{mor}}) \neq \emptyset$.

The materials are supposed to be linear elastic, homogeneous, and isotropic and the stress tensor $\boldsymbol{\sigma}$ is assumed to depend linearly on the strain tensor $\epsilon_{ij} = \frac{1}{2}(u_{i,j} + u_{j,i})$ via Hooke's law $\sigma_{ij} = E_{ijkl}\epsilon_{kl}$. We use the subscript $,i$ to signify the i-th partial derivative. The material constants are Young's modulus $E > 0$ and the Poisson ratio $0 < \nu < 1/2$.

In $\Omega_{\text{non}} \cup \Omega_{\text{mor}}$ the equilibrium conditions from linear elasticity hold and on $\Gamma_D \cup \Gamma_N$ we have Dirichlet and Neumann boundary conditions, respectively, i.e.,

$$
\begin{aligned}
-\sigma_{ij}(\mathbf{u})_{,j} &= f_i && \text{in } \Omega_{\text{non}} \cup \Omega_{\text{mor}}, \\
\mathbf{u} &= 0 && \text{on } \Gamma_D, \\
\sigma_{ij}(\mathbf{u}) \cdot n_j &= p_i && \text{on } \Gamma_N.
\end{aligned}
\tag{1}
$$

Here, $\mathbf{f} = (f_i)$ is the density of volume forces, $\mathbf{p} = (p_i)$ are prescribed surface tractions and $\mathbf{n}$ is the outward surface normal on Γ_N.

In order to model the contact between the two bodies, further conditions have to be prescribed at Γ_C. To this end, we introduce the contact mapping $\Phi : \Gamma_{\text{non}} \to \Gamma_{\text{mor}}$. We assume Φ to be a C^1-diffeomorphism. It allows us to define the initial gap function $g : \Gamma_{\text{non}} \to \mathbb{R}$ with $g(x) = |\Phi(x) - x|$ and the relative normal displacement

$$[\mathbf{u}]_\Phi = \left(\mathbf{u}|_{\Gamma_{\text{mor}}} \circ \Phi - \mathbf{u}|_{\Gamma_{\text{non}}}, \mathbf{n}_{\text{non}}\right) \tag{2}$$

for a given displacement $\mathbf{u} \in \left(H^1_{0;\Gamma_D}(\Omega)\right)^d$. Here, $\mathbf{n}_{\text{non}}$ is the unit normal on the nonmortar contact boundary and $H^1_{0;\Gamma_D}(\Omega)$ is the Sobolev–space that contains only those functions from $H^1(\Omega)$ which satisfy homogeneous Dirichlet boundary conditions on Γ_D. On Γ_C, we then have the linearized contact conditions

$$[\mathbf{u}]_\Phi \le g, \tag{3}$$

see Eck [1996]. Furthermore, the Kuhn–Tucker like conditions

$$\sigma_{\mathbf{n}_{\text{non}}}(\mathbf{u}|_{\Gamma_{\text{non}}}) = \sigma_{\mathbf{n}_{\text{mor}}}(\mathbf{u}|_{\Gamma_{\text{mor}}}) \le 0 \tag{4}$$

$$0 = \left([\mathbf{u}]_\Phi - g\right) \cdot \sigma_n(\mathbf{u}_s) \tag{5}$$

$$\boldsymbol{\sigma}_T(\mathbf{u}|_{\Gamma_{\text{non}}}) = \boldsymbol{\sigma}_T(\mathbf{u}|_{\Gamma_{\text{mor}}}) = 0, \tag{6}$$

are required to hold on Γ_C, where $\sigma_n = n_i \sigma_{ij} n_j$ and $(\boldsymbol{\sigma}_T)_i = \sigma_{ij} n_j - \sigma_n n_i$, $i = 1, \ldots, d$, are the normal and tangential parts of $\boldsymbol{\sigma}$, respectively. Condition (4) ensures that the surface forces at the contact boundary have the character of a pure pressure. Equation (5) states that there can be non-vanishing surface pressure at Γ_C only if there is contact and equation (6) corresponds to frictionless contact.

We now discretize the two–body contact problem by finite elements. On both subdomains, shape regular triangulations are used, which are allowed to be completely unstructured and contain arbitrary element types. For simplicity we assume that Ω_{non} and Ω_{mor} are polyhedral. By h_{non} and h_{mor} we denote the largest diameter of an element occurring in Ω_{non} respective Ω_{mor}. We use piecewise linear functions on simplices and trilinear functions on hexahedra. We set

$$X_{s;h_{\text{non}}} = \left\{ v \;\middle|\; v \in C(\Omega_{\text{non}}), v \text{ is (tri)linear on each } T \in \mathcal{T} \text{ and } v|_{\Gamma_D} = 0 \right\}$$

and $X_{m;h_{\text{mor}}}$ is defined equivalently. We set

$$\mathbf{X}_{s;h_{\text{non}}} = (X_{s;h_{\text{non}}})^3 \qquad \text{and} \qquad \mathbf{X}_{m;h_{\text{mor}}} = (X_{m;h_{\text{mor}}})^3.$$

One of the main difficulties is the discretization of the boundary conditions (3)–(6) at the contact interface for irregular geometries. The contact boundaries have non-matching grids, are nonplanar and do not coincide. The straight forward approach is to enforce the constraints (3)–(6) pointwise. For linear elliptic problems with linear boundary conditions Bernardi et al. [1994]

showed that this yields discretizations which are in general nonoptimal, i.e., the a priori error in the energy norm does not behave as $O(h^s)$ if the solution is H^{1+s}. Optimality can be recovered by enforcing the transmission conditions at the interface in a weak sense. This is done by enforcing the boundary conditions with respect to a space of suitably chosen functionals, the Lagrange multipliers.

We first rewrite (1) as a saddle point problem. By definition, see (2), the jump $[\mathbf{u}]_\Phi$ is contained in the trace space $H^{1/2}(\Gamma_{\text{non}})$. We introduce a space $\mathbf{M}$ of Lagrange multipliers that will serve to enforce nonpenetration. We choose $\mathbf{M} = \mathbf{H}^{-1/2}(\Gamma_{\text{non}})$ and define the positive cone

$$\mathbf{M}^+ = \left\{\boldsymbol{\mu} \in \mathbf{M} \mid \langle \boldsymbol{\mu} \cdot \mathbf{n}, w \rangle_{\Gamma_{\text{non}}} \geq 0,\ w \in W^+\right\}$$

with $W^+ = \left\{w \in H^{1/2}(\Gamma_{\text{non}}) \mid w \geq 0 \text{ a.e.}\right\}$. Then (1) with (3)–(6) can be restated as, see, e.g., Ben Belgacem et al. [1999], Wohlmuth and Krause [2003]: Find a pair $(\mathbf{u}, \boldsymbol{\lambda}) \in (\mathbf{X}, \mathbf{M}^+)$ with

$$\begin{aligned} a(\mathbf{u}, \mathbf{v}) + b(\boldsymbol{\lambda}, \mathbf{v}) &= f(\mathbf{v}) && \text{for all } \mathbf{v} \in \mathbf{H}^1_{0;\Gamma_D}, \\ b(\boldsymbol{\mu}, \mathbf{u}) &\leq \langle \boldsymbol{\mu} \cdot \mathbf{n}, g \rangle_{\Gamma_{\text{non}}} && \text{for all } \boldsymbol{\mu} \in \mathbf{M}^+. \end{aligned} \tag{7}$$

The bilinear form $b(\cdot,\cdot)$ occurring in (7) is defined by

$$b(\boldsymbol{\mu}, \mathbf{v}) = \langle [\mathbf{v}]_\Phi, \boldsymbol{\mu} \cdot \mathbf{n} \rangle_{\Gamma_{\text{non}}}.$$

For the discretization $\mathbf{M}_h$ of $\mathbf{M}$ we use dual Lagrangian multipliers, see, Wohlmuth [2001]. Let ψ_p, ϕ_q, $\theta_{\tilde{q}}$ be basis functions of the discrete multiplier space $\mathbf{M}_h$ and the discrete trace spaces $X|_{\Gamma_{\text{non}}}$ and $X|_{\Gamma_{\text{mor}}}$, respectively. Then, the algebraic representation of (7) involves the discrete transfer operator $S\colon X|_{\Gamma_{\text{mor}}} \longrightarrow X|_{\Gamma_{\text{non}}}$,

$$S\mathbf{v} = D^{-1} M^T \mathbf{v}, \tag{8}$$

where

$$D_{pq} = \text{Id}_{3\times 3} \int_{\Gamma_{\text{non}}} \psi_p \phi_q \, ds \qquad \text{and} \qquad M_{p\tilde{q}} = \text{Id}_{3\times 3} \int_{\Gamma_{\text{non}}} \psi_p \cdot (\theta_{\tilde{q}} \circ \Phi) \, ds.$$

Now, the dual multipliers are characterized by the biorthogonality relation

$$\int_{\Gamma_{\text{non}}} \psi_p \phi_q \, ds = \delta_{pq} \int_{\Gamma_{\text{non}}} \phi \, ds, \qquad p, q \in \mathcal{V}_{\Gamma_{\text{non}}}.$$

Thus, D becomes a block diagonal matrix, and its inverse can easily be computed. This is in contrast to the standard mortar approach, where the finite element trace space is used as space of discrete Lagrangian multipliers. Then, D is a sparse matrix, which is not as easy to invert as a block–diagonal one.

In Wohlmuth and Krause [2003] the monotone multigrid method for contact problems from Kornhuber and Krause [2001] has been generalized to multi body contact problems using dual mortar methods. Thus, the arising nonlinear systems of equations can be solved with high accuracy and with multigrid efficiency.

3 Implementation and Numerical Results

A crucial component of the discretization is the mapping Φ given in (3). Its purpose is to identify the nonmortar and the mortar side of the contact boundary Γ_C. It also appears in the definition (8) of the transfer operator S.

We first describe our data structure and then the concrete construction of Φ. In the following, we assume Γ_{non} and Γ_{mor} to be triangulated surfaces and their mutual distance to be small. Then Φ is a piecewise smooth homeomorphism. We store Γ_{non} as a list of vertices $\mathcal{V}_{\Gamma_{\text{non}}}$ and triangles $\mathcal{T}_{\Gamma_{\text{non}}}$. We additionally define a plane graph for each $T \in \mathcal{T}_{\Gamma_{\text{non}}}$. The graph on T is the image of the edge graph of $\Phi(T) \subset \Gamma_{\text{mor}}$ under Φ^{-1}. Thus, each vertex of Γ_{mor} appears as a graph node on a triangle T of Γ_{non}. This graph node stores its local position on T and its target position as a vertex in Γ_{mor}. That way, Φ can be evaluated for any point $x \in \Gamma_{\text{non}}$ using a point location algorithm and linear interpolation.

For our implementation of the contact mapping Φ we choose Φ^{-1} to be the projection of Γ_{mor} onto Γ_{non} in normal direction of Γ_{mor}. We define a continuous normal vector field $\mathbf{n} : \Gamma_{\text{mor}} \to \mathbb{R}^3$. If $\tilde{v} \in \mathcal{V}_{\Gamma_{\text{mor}}}$, we set $\mathbf{n}(\tilde{v})$ to the average of the triangle normals of the triangles that have $\tilde{v}$ as a vertex. All other values of $\mathbf{n}$ are then defined via linear interpolation.

The actual construction of Φ consists of three steps.

1.: Computing $\Phi^{-1}(\tilde{v})$ for all $\tilde{v} \in \mathcal{V}_{\Gamma_{mor}}$ The vertices of Γ_{mor} appear as nodes in the graph defined on Γ_{non}. Given a vertex $\tilde{v} \in \mathcal{V}_{\Gamma_{\text{mor}}}$, its exact position on Γ_{non} can be found by considering the ray r normal to Γ_{mor} beginning in $\tilde{v}$. If r intersects one or more triangles of Γ_{non}, the intersection closest to $\tilde{v}$ is the one to choose. If there are no intersections we decide that $\tilde{v}$ should not be part of Γ_{mor}. Special care has to be taken if $v = \Phi^{-1}(\tilde{v})$ is on an edge or a vertex of a triangle of Γ_{non}. Then, several graph nodes of different types have to be added to the data structure on Γ_{non} to keep it consistent (see Sander and Krause [2003]). The search for all possible intersections can be sped up by the use of a suitable spatial data structure.

2.: Computing $\Phi(v)$ for all $v \in \mathcal{V}_{\Gamma_{non}}$ In a second step we have to find the images of the vertices of Γ_{non} under Φ. This is an inverse normal projection. For a given $v \in \mathcal{V}_{\Gamma_{\text{non}}}$ we have to find $\Phi(v) \in \Gamma_{\text{mor}}$ such that $v - \Phi(v)$ is normal to Γ_{mor}. Let $\tilde{T}$ be a triangle of Γ_{mor} with the vertices a, b, c. Denote by $\mathbf{n}_a, \mathbf{n}_b, \mathbf{n}_c$ the respective normal vectors. Then checking whether the inverse normal projection has a solution on $\tilde{T}$ amounts to see if the nonlinear system of equations

$$\eta\lambda\mathbf{n}_a + \eta\mu\mathbf{n}_b + \eta(1-\lambda-\mu)\mathbf{n}_c - \mathbf{v} = 0 \tag{9}$$

has a solution with $\lambda, \mu, \eta \geq 0$ and $\lambda + \mu \leq 1$. In the affirmative case, (λ, μ) yields the intersection point in barycentric coordinates on $\tilde{T}$. System (9) may theoretically have more than one solution, but this did not lead to any practical problems. It can be solved efficiently with a standard Newton algorithm.

3.: Adding the edges We enter the edges of Γ_{mor} into the graph on Γ_{non} by running over all edges $\tilde{e} = (\tilde{p}, \tilde{q})$ in Γ_{mor} and entering them one by one. We

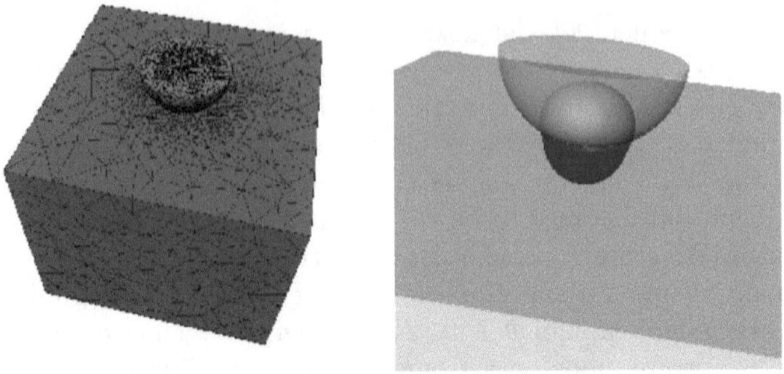

Fig. 1. A Hertzian contact problem

try to 'walk' on Γ_{non} along $\Phi^{-1}(\tilde{e})$ from $p = \Phi^{-1}(\tilde{p})$ to $q = \Phi^{-1}(\tilde{q})$. Since p and q will generally not be on the same triangle of Γ_{non}, we have to find the points where the path from p to q crosses edges of Γ_{non}. For an edge e of Γ_{non} we have to check whether there are points $x \in e$ and $\tilde{x} \in \tilde{e}$ with $x - \tilde{x}$ normal to Γ_{mor}. This can be formulated as a nonlinear system of equations

$$\lambda\tilde{\mathbf{q}} + (1-\lambda)\tilde{\mathbf{p}} + \eta\lambda\mathbf{n}_{\tilde{\mathbf{q}}} + \eta(1-\lambda)\mathbf{n}_{\tilde{\mathbf{p}}} - \mu\mathbf{q} - (1-\mu)\mathbf{p} = 0 \tag{10}$$

which can be solved with a Newton solver. We have found an intersection if (10) has a solution with $0 \leq \lambda, \mu \leq 1$ and $0 \leq \eta$.

Assuming that the Newton solver terminates after a constant number of iterations, the projection algorithm described above requires $O(N_b \log N_b)$ time. Here N_b is the number of unknowns on the contact boundary. Asymptotically, N_b behaves like $N^{2/3}$, where N is the total number of nodes. The construction of Φ therefore takes $O(N^{2/3} \log N^{2/3})$ time. Thus, the overall complexity of the simulation process is still dominated by the nonlinear monotone multigrid method, which requires $O(N)$ time.

Our first numerical example is a Hertzian contact problem. An elastic half–sphere is pressed against an elastic cube, see Figure 1. We model both objects with unstructured tetrahedral grids. Using the boundary parametrization described in Sander and Krause [2003], during the adaptive refinement process the geometry of the sphere is successively approximated. This is done by moving the boundary nodes which are newly created during the refinement process, to their actual position on a corresponding high–resolution half–sphere.

On top of the half–sphere, Dirichlet boundary conditions are applied corresponding to a point load on the upper pole of the corresponding sphere. Homogeneous Dirichlet boundary conditions are applied at the vertical faces of the cube and homogeneous Neumann conditions everywhere else. As material parameters we use $E = 7 \cdot 10^3$ and $\nu = 0.3$ for the sphere and $E = 6.896 \cdot 10^5$ and $\nu = 0.45$ for the cube.

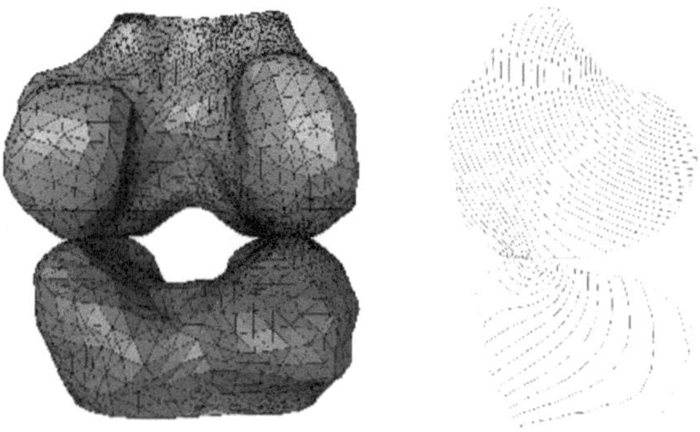

Fig. 2. Femur and tibia meeting in the knee joint

The discrete problem is solved using the monotone multigrid solver described by Kornhuber and Krause [2001], Wohlmuth and Krause [2003]. We perform 3 steps of adaptive mesh refinement using a residual–based error indicator. We compare our nonlinear monotone multigrid method with a standard linear multigrid method. After the nonlinear contact problem has been solved, the computed boundary stresses are taken as boundary data for the linear multigrid method. By means of the linear multigrid method the same solution is computed as by the nonlinear monotone multigrid method. We use 3 pre– and postsmoothing steps on each level $k > 0$. The problems on level 0 are solved by applying one iteration step of an algebraic variant of our nonlinear monotone multigrid method. On subsequent levels $k \geq 1$ the ν-th iterate u_k^ν is accepted if the stopping criterion $\|\mathbf{u}_k^\nu - \mathbf{u}_k^{\nu-1}\| \leq 10^{-12}$ is satisfied. Nested iteration is used. In Table 1, the number of iterations for the nonlinear contact problem and the equivalent linear problem with known boundary stresses are given. For the nonlinear contact problem we observe similar convergence rates as for the corresponding linear problems. The increasing number of iterations might be due to just applying one iteration step of the algebraic multigrid method as basesolver and by decreasing mesh quality caused by recovering the original geometry. In Figure 1, the adaptively refined grid on Level 3 and isosurfaces of the computed displacements are shown.

Table 1. Comparison of nonlinear monotone and linear multigrid method

level	elements	dofs	nonlinear iterations	linear iterations	no. of. contact nodes
0	7.246	4.968	13	16	14
1	18.403	11.577	33	38	39
2	85.567	47.985	66	80	146
3	438517	234.123	100	100	580

Our second example is an application from biomechanics and demonstrates the applicability of our algorithm. The geometry consists of parts of the human proximal femur and tibia meeting in the knee joint. We again use an unstructured tetrahedral grid. The geometry is known in a very high resolution, we can use this to provide a parametrized boundary. The left picture of Figure 2 shows the deformed geometry, in the right picture isolines of the computed displacements are depicted.

Visualization has been done using the visualization environment AMIRA from the Zuse–Institute–Berlin Berlin (ZIB). The monotone multigrid method is implemented in the framework of the finite element package UG, see Bastian et al. [1997], Krause [2001].

References

P. Bastian, K. Birken, K. Johannsen, S. Lang, N. Neuß, H. Rentz–Reichert, and C. Wieners. UG – a flexible software toolbox for solving partial differential equations. *Computing and Visualization in Science*, 1:27–40, 1997.

F. Ben Belgacem, P. Hild, and P. Laborde. Extension of the mortar finite element method to a variational inequality modeling unilateral contact. *Math. Models Methods Appl. Sci.*, 9:287–303, 1999.

C. Bernardi, Y. Maday, and A. T. Patera. A new non conforming approach to domain decomposition: The mortar element method. In H. Brezis and J.-L. Lions, editors, *Collège de France Seminar*. Pitman, 1994.

C. Eck. *Existenz und Regularität der Lösungen für Kontaktprobleme mit Reibung*. PhD thesis, Universität Stuttgart, 1996.

P. Hild. Numerical implementation of two nonconforming finite element methods for unilateral contact. *Comput. Methods Appl. Mech. Eng.*, 184:99–123, 2000.

N. Kikuchi and J. T. Oden. *Contact Problems in Elasticity: A Study of Variational Inequalities and Finite Element Methods*. SIAM, 1988.

R. Kornhuber and R. Krause. Adaptive multigrid methods for Signorini's problem in linear elasticity. *Computing and Visualization in Science*, 4(1): 9–20, 2001.

R. Krause. *Monotone Multigrid Methods for Signorini's Problem with Friction*. PhD thesis, Freie Universität Berlin, Department of Mathematics, July 2001.

O. Sander and R. Krause. Automatic construction of boundary parametrizations for geometric multigrid solvers. *Computing and Visualization in Science*, 2003. Accepted for publication.

B. Wohlmuth. *Discretization Methods and Iterative Solvers Based on Domain Decomposition*. Springer, 2001.

B. Wohlmuth and R. Krause. Monotone multigrid methods on nonmatching grids for nonlinear multibody contact problems. *SIAM Journal on Scientific Computing*, 25(1):324–347, 2003.

Part XIV

Contributed Talks

Generalized Aitken-like Acceleration of the Schwarz Method

Jacques Baranger[1], Marc Garbey[2], and Fabienne Oudin-Dardun[1]

[1] University Lyon 1, Modelisation and Scientific Computing
[2] University of Houston, Computer Science (http://www.cs.uh.edu/~garbey/)

Summary. (**and Introduction**) In this paper, we present a family of domain decomposition based on Aitken like acceleration of the Schwarz method seen as an iterative procedure with linear rate of convergence. This paper is a generalization of the method first introduced in Garbey and Tromeur-Dervout [2001] that was restricted to Cartesian grids. The general idea is to construct an approximation of the eigenvectors of the trace transfer operator associated to dominant eigenvalues and accelerate these components after few Schwarz iterates. We consider here examples with the finite volume approximation on general quadrangle meshes of Faille [1992] and finite element discretization.

1 A General Framework for the Aitken-Schwarz Method

Let us consider formally a linear differential problem

$$L[U] = f \ in \ \Omega, \ U_{|\partial\Omega} = g. \tag{1}$$

We assume that the problem is well posed and has a unique solution U. To simplify the presentation, we restrict ourselves to a domain decomposition of two overlapping subdomains $\Omega_1 \bigcup \Omega_2 = \Omega$, and we consider the additive version of the Schwarz algorithm (Smith et al. [1996]).We assume implicitly in the following notations that the Dirichlet boundary condition in (1) is satisfied by all intermediate subproblems. The Additive Schwarz (AdS) version of the algorithm writes,

$$L[u_1^{n+1}] = f \ in \ \Omega_1, \ u_{1|\Gamma_1}^{n+1} = u_{2|\Gamma_1}^{n}, \tag{2}$$

$$L[u_2^{n+1}] = f \ in \ \Omega_2, \ u_{2|\Gamma_2}^{n+1} = u_{1|\Gamma_2}^{n}. \tag{3}$$

Because L is linear, the following operator T^a is linear:

$$(u_{1|\Gamma_1}^{n} - U_{|\Gamma_1}, u_{2|\Gamma_2}^{n} - U_{|\Gamma_2}) \rightarrow (u_{1|\Gamma_1}^{n+1} - U_{|\Gamma_1}, u_{2|\Gamma_2}^{n+1} - U_{|\Gamma_2}). \tag{4}$$

Let us proceed with the discretized version of the problem (1), with solution U^h. For $i = 1, 2$, let Γ_i^h be the set of mesh nodes corresponding to approximation of U on Γ_i, E_i^h a finite vector space used to approximate the solution restricted to the artificial interface Γ_i^h, and $\{b_i^j, j = 1...N\}$ a set of basis functions for this vector space. We suppose that both vector space E_1^h and E_2^h have the same dimensions and define now the following two linear operators

$$T_l^a : u_{1|\Gamma_1}^n - U_{\Gamma_1} \rightarrow u_{2|\Gamma_2}^{n+1} - U_{\Gamma_2} \tag{5}$$

$$T_r^a : u_{2|\Gamma_2}^n - U_{\Gamma_2} \rightarrow u_{1|\Gamma_1}^{n+1} - U_{\Gamma_1}. \tag{6}$$

Using the discrete representation of the interface Γ_i^h in E_i^h, for $i = 1, 2$, we have

$$(u_{2,j}^{n+1} - U_{j,\Gamma_2})_{j=1,..,N} = P_l \, (u_{1,j}^n - U_{j,\Gamma_1})_{j=1,..,N}, \tag{7}$$

and

$$(u_{1,j}^{n+1} - U_{j,\Gamma_1})_{j=1,..,N} = P_r \, (u_{2,j}^n - U_{j,\Gamma_2})_{j=1,..,N}, \tag{8}$$

with P_l (resp. P_r) square matrix of T_l^a (resp. T_r^a). The matrix of the trace operator T^a has then the characteristic anti diagonal structure

$$P = \begin{pmatrix} 0 & P_r \\ P_l & 0 \end{pmatrix}$$

The *Additive Aitken Schwarz* algorithm is then

- Step AdS0: compute P_l and P_r.
- Step AdS1: from initial artificial interface condition u_1^0 and u_2^0 compute the first Schwarz iterate (2, 3).
- Step AdS2: from u_i^0 and u_i^1 $(i = 1, 2)$ and the linear system (7,8), get the exact interface value U_{j,Γ_1} and U_{j,Γ_2}.
- Step AdS3: starting from the interface condition $U_{|\Gamma_1} = \sum_{j=1..N} U_{j,\Gamma_1} b_1^j$ and $U_{|\Gamma_2} = \sum_{j=1..N} U_{j,\Gamma_2} b_2^j$, apply one last Schwarz iterate (2,3) to get U^h.

Iff $||P_l P_r|| < 1$, the additive Schwarz algorithm converges and the matrix P associated to (7,8) is non singular. This Aitken-Schwarz algorithm is then an exact solver.

Step AdS0 is the critical step of this algorithm; a straightforward and very expansive way to obtain P consist in computing before hand in parallel the solution of $2N$ independent sequences of homogeneous problem; alternatively, one may reconstruct these matrices using $2(N + 1)$ consecutive iterates of the Schwarz method, but existence of the solution and stability of the numerical process is not guaranteed (Garbey and Tromeur-Dervout [2002]).

To find a numerically efficient method to compute P or an approximation of P is the key problem that we will address in the next sections. We are going to simplify the problem and show that our algorithm can be formulated with an approximation of the eigenvectors of the trace transfer operator that has the dominant eigenvalues.

2 Quasi-Diagonal Aitken-Schwarz Procedure

Let us assume that P_l (resp. P_r) can be diagonalized in the basis of eigenvectors V_j corresponding to eigenvalues Λ^l_j (resp. eigenvectors W_j corresponding to eigenvalues Λ^r_j.)

Let us denote by $(\tilde{u}^n_{i,j})_{j=1,\dots,N}$ (resp. $(\hat{u}^n_{i,j})_{j=1,\dots,N}$) $(i = 1, 2)$ the components of $u^n_{i|\Gamma_i}$ in basis $\{V_j, j = 1, ..., N\}$ (resp. $\{W_j, j = 1, ..., N\}$).

Then, we have

$$(\tilde{u}^{n+1}_{2,j} - \tilde{U}_{j|\Gamma_2})_{j=1,..,N} = D^l\,(\tilde{u}^n_{1,j} - \tilde{U}_{j|\Gamma_1})_{j=1,..,N}, \tag{9}$$

with $D^l_{j,j} = \lambda^l_j$, and in the basis of eigenvectors W_j,

$$(\hat{u}^{n+1}_{1,j} - \hat{U}_{j|\Gamma_1})_{j=1,..,N} = D^r\,(\hat{u}^n_{2,j} - \hat{U}_{j|\Gamma_2})_{j=1,..,N}, \tag{10}$$

with $D^r_{j,j} = \lambda^r_j$, $j = 1..N$. In order to compute $U_{j|\Gamma_i}, i = 1..2$, we express both identities (9) and (10) in the same basis. We obtain in vector notations $u = (u_j)_{j=1..N}$, in $b^i_j, i = 1, 2$ basis,

$$U_{\Gamma_1} - \Lambda^r U_{\Gamma_2} = u_1^1 - \Lambda^r\, u_2^0, \tag{11}$$

$$-\Lambda^l\, U_{\Gamma_1} + U_{\Gamma_2} = u_2^1 - \Lambda^l\, u_1^0, \tag{12}$$

where $\Lambda^l = V\, D^l\, V^{-1}$, and $\Lambda^r = W\, D^r\, W^{-1}$.

The *Quasi Diagonal Additive Schwarz algorithm* writes

- Step QD-AdS0: compute approximate main eigenvectors $(\hat{V}_j)_{j=1..q}$ (resp. $(\hat{W}_j)_{j=1..q}$) and corresponding approximate eigenvalues $(\hat{\Lambda}^l_j)_{j=1..q}$ (resp. $(\hat{\Lambda}^r_j)_{j=1..q}$) of P_l (resp. P_r).
- Step QD-AdS1: from initial artificial interface conditions u_1^0 and u_2^0, compute the first Schwarz iterate (2, 3).
- Step QD-AdS2: decompose u_1^0 and u_1^1 into the main components $u^0_{e,1}$ and $u^1_{e,1}$ (projection on $span[\hat{V}_1, ..., \hat{V}_q]$) and the residuals $u^{0/1}_{r,1} = u_1^{0/1} - \sum_{j=1..q} u^{0/1}_{e,j}\, \hat{V}_j$. Decompose $u_2^{0/1}$ in a similar way using the projection on $span[\hat{W}_1, ..., \hat{W}_q]$
- Step QD-AdS3: from the formula (11) restricted to $span[\hat{W}_1, ..., \hat{W}_q]$ and (12) restricted to $span[\hat{V}_1, ..., \hat{V}_q]$ with corresponding approximated eigenvalues, get the (approximated) interface value $\hat{U}_{e,1/2,j}$, $\forall j = 1..q$.
- Step QD-AdS4: recompose the interface conditions from the following approximations $(i = 1, 2)$

$$U_{|\Gamma_1} \approx \hat{U}_{|\Gamma_1} = \sum_{j=1..q} (u^1_{1,e,j} - \hat{\Lambda}^l_j u^0_{1,e,j})/(1 - \hat{\Lambda}^l_j)\, \hat{V}_j + u^1_{r,1},$$

(similarly for $U_{|\Gamma_2}$, with $\hat{\Lambda}^r$ and $span[\hat{W}_1, ..., \hat{W}_q]$), and apply one Schwarz iterate (2,3) to get an approximation of U^h.

The following theorem summarizes the impact of the error on eigenvectors, the error on eigenvalues and the truncation parameter q on the approximation of the artificial interfaces obtained with the quasi-diagonal additive Aitken Schwarz algorithm.

Theorem 1. *Let $\{V_1, ..., V_q\}$ (resp. $\{W_1, ..., W_q\}$) be a set of q independent eigenvectors of the trace transfer operator P_l (resp. P_r). We suppose $||P_{l/r}|| = O(1)$. Let $\{\hat{V}_1, ..., \hat{V}_q\}$ (resp. $\{\hat{W}_1, ..., \hat{W}_q\}$) be a set of q independent vectors such that the matrix ϵ_l (resp. ϵ_r) of column vectors $\epsilon_{l,j} = \hat{V}_j - V_j$ (resp. $\epsilon_{r,j} = \hat{W}_j - W_j$), has norm $||\epsilon_{l/r}|| = o(1)$.*

Let us assume that $\delta_j^{l/r} = |\hat{\lambda}_j^{l/r} - \lambda_j^{l/r}| = o(1), \forall j = 1...q$, and that $dist(u_1^0 - U_{|\Gamma_1}, span[\hat{V}_1, ..., \hat{V}_q]) + dist(u_2^0 - U_{|\Gamma_2}, span[\hat{W}_1, ..., \hat{W}_q]) = \mu$, with $\mu = o(1)$, then

$$||(\hat{U}_{|\Gamma_1} - U_{|\Gamma_1}, \hat{U}_{|\Gamma_2} - U_{|\Gamma_2})|| =$$
$$C^t \, ||(Id - \Lambda^l \, \Lambda^r)^{-1}|| \, O(||\epsilon||) + O(||\beta||) + O(\mu), \qquad (13)$$

with β 2q-vector of components $(\beta_j^l = \frac{\delta_j^l}{|1-\lambda_j^l|}, \beta_j^r = \frac{\delta_j^r}{|1-\lambda_j^r|})$.

Proof. See Garbey [2003]

This theorem suggests to get an approximation of the eigenvectors of the matrices P_l, P_r corresponding to the dominant eigenvectors that is numerically cheap to compute. The Quasi-diagonal Aitken acceleration plays the role of a coarse grid preconditioner and can be iterated until convergence. We introduce in the following two examples of this construction, with respectively finite element discretization and then finite volume approximation.

3 Finite Element on Tensorial Product of Two-D Grid

We consider the homogenous Dirichlet boundary value problem (1) that has a separable second order operator $L = L_1 + L_2$:

$$L_1 = -\partial_x(a_1 \partial_x) + b_1 \partial_x + c_1, \quad L_2 = -\partial_y(a_2 \partial_y) + b_2 \partial_y + c_2. \qquad (14)$$

a_1, b_1, c_1 are functions of x, and a_2, b_2, c_2 are functions of y. Ω is a rectangle with a strip domain decomposition into rectangles. Interfaces of the domain decomposition are therefore parallel to the y direction. The number of subdomains is arbitrary.

Let us consider the semi-discretisation of the operator in y variable, with an irregular mesh in y ($y_i, i = 0, ..., N+1$), L_2^k a discretization of L_2 on the y-mesh, and $u^i(x)$ (resp. $f^i(x)$) an expected approximation of $u(x, y_i)$ (resp. $f(x, y_i)$). The semi-discrete approximation of a subdomain problem analogous to problem (2) or (3) is solved on a rectangle denoted by R=[e,w]x[n,s] in order to simplify the notations:

$$L_1 u^i(x) + L_2^k u^i(x) = f^i(x), \quad x \in]e, w[\tag{15}$$

$$u^i(w) \text{ and } u^i(e) \text{ given}, \quad u^0(x) = u^{N+1}(x) = 0. \tag{16}$$

We introduce the eigenvalue problem:

$$L_2^k \Phi_j = \lambda_j \Phi_j, \ \Phi_j^0 = \Phi_j^{N+1} = 0. \tag{17}$$

We set the hat transform :

$$u^i(x) = \sum_{j=1}^{N} \hat{u}^j(x) \Phi_j^i; i = 1, \cdots, N \tag{18}$$

with a similar expansion for $f^i(x)$. Applying this hat transform to (15-16) gives formally:

$$\sum_{j=1}^{N} \Phi_j^i [(L_1 + \lambda_j)\hat{u}_j(x) - \hat{f}_j(x))] = 0, \quad \sum_{j=1}^{N} \Phi_j^i \hat{u}_j(e/w) \text{ given}$$

Following the notation of Theorem 1, the eigenvectors functions V_j and W_j of the trace transfer operator are identical and equal to the $\Phi_j, j = 1 \cdots N$, functions. More precisely we have the following result for an arbitrary number of P subdomains,

Theorem 2. *Assume problem (17) has N linearly independent real eigenvectors associated to real eigenvalues. Then each semi-discrete approximation of each subdomain problem is constituted of N uncoupled continuous one dimensional linear problems* $j = 1, \cdots, N$*:*

$$[L_1 + \lambda_j]\hat{u}_j(x) = \hat{f}_j(x), \quad \hat{u}_j(e/w) \text{ given} \tag{19}$$

The hat trace transfer operator is affine on $\mathbb{R}^{2N(P-1)}$ *with a block-diagonal matrix of N blocks.*

Proof. See Baranger et al. [2003]

From Theorem 1 one can estimate the number of eigenvectors q that is worth to compute.

Problem (15-16) is closely related to finite difference point of vue. We now show that similar results can be obtained from the variational formulation of the problem.

Let us consider a semi discrete finite element approximation of the variational problem associated to (14). On a y-mesh we have a finite element space with basis function $\varphi_m, m = 1, \cdots, N$. The unknown function is $u^k(x, y) = \sum_{m=1}^{N} u_m(x)\varphi_m(y)$ with $u_m(w/e) = 0$. We obtain then the semi discrete variational problem:

$$\begin{aligned}&\sum_m \int_w^e [a_1 \partial_x u_m \partial_x v + \cdots] dx \int_s^n \varphi_m \varphi_j dy \ldots \\ &+ \sum_m \int_w^e u_m v dx \int_s^n [a_2 \partial_y \varphi_m \partial_y \varphi_j + \cdots] dy = \textstyle\int_R f v \varphi_j dx dy.\end{aligned} \tag{20}$$

Then the semi discrete variational problem (20) is: for all $m = 1, \cdots, N$ find u_m in $H_0^1(w, e)$ such that for all v in $H_0^1(w, e)$ and $i = 1, \cdots, N$

$$\sum_m [\beta_{im} \alpha^1(u_m, v) + \alpha_{im} \beta^1(u_m, v)] = \beta^1(f_i, v) \tag{21}$$

with

$$\alpha^1(u, v) = \int_w^e (a_1 u_x v_x + b_1 u_x v + c_1 uv) dx, \ \alpha^2(u, v) = \int_s^n (a_2 u_y v_y + b_2 u_y v + c_2 uv) dy$$

$$f_i(x) = \int_s^n f \varphi_i dy, \ \beta^1(u, v) = \int_w^e uv dx, \ \beta^2(u, v) = \int_s^n uv dy,$$

$$\gamma_{im} = \gamma^2(\varphi_i, \varphi_m) \ (\gamma = \alpha, \beta).$$

Using the generalized Fourier transform (18) we obtain from equation (21):

$$\sum_m \sum_j [\beta_{im} \alpha^1(\hat{u}_j, v) + \alpha_{im} \beta^1(\hat{u}_j, v) - \beta_{im} \beta^1(\tilde{f}_j, v)] \Phi_{jm} = 0 \tag{22}$$

Choosing the Φ's as the eigenvectors of the spectral problem:

$$\sum_m \alpha_{im} \Phi_{jm} = \lambda_j \sum_m \beta_{im} \Phi_{jm}, j = 1, \cdots, N \tag{23}$$

gives to equation (22) the uncoupled form:

$$\alpha^1(\hat{u}_j, v) + \lambda_j \beta^1(\hat{u}_j, v) = \beta^1(\tilde{f}_j, v). \tag{24}$$

We obtain then a result analogous to Theorem 2. We are going now to show a second variant of this construction for finite volume approximation with general quadrangle meshes.

4 An Example with Finite Volume on General Quadrangle Meshes

We consider an approximation of (1) with the finite volume approximation on general quadrangle meshes of Faille [1992]. For simplicity we restrict ourselves to the Poisson operator with homogeneous Dirichlet boundary conditions. We also consider the multiplicative version of the Schwarz (MuS) algorithm with two subdomains. Those restrictions are not necessary, but they make the understanding of the construction easier. For general quadrangle meshes, an eigenvector basis of the trace transfer operator T^h on the interface $u^n_{h|\Gamma^h} \rightarrow$

$u^{n+1}_{h|\Gamma^h}$, cannot be constructed analytically as in the previous section. We recall that

$$u^n_{h|\Gamma^h} - U_{|\Gamma^h} \rightarrow u^{n+1}_{h|\Gamma^h} - U_{|\Gamma^h}$$

is a linear operator and we denote by P^h its matrix. E^h is the finite vector space used to approximate the solution restricted to the interface Γ^h.

To represent the functions of E^h in a more compact way, we choose the space of approximation $\mathcal{E}^q_1 = span[sin(\tau), sin(2\ \tau), ..., sin(q\ \tau)]$ where $\tau \in (0, \pi)$ is a natural parameterization of the interface Γ. We introduce then the operator $\mathcal{T}_{h/q}$ from $E^{h,0}_1$ to $\mathcal{E}^q_1$ that gives the least square approximation of grid function of Γ^h with q-sine expansion:

$$\sum_{k=1..q} a^k_j sin(k\ \tau).$$

Vice versa the operator $\mathcal{T}_{q/h}$ collocates the sinus expansions of $\mathcal{E}^q_1$ at grid points of Γ^h. The operator $T_{h/q}P^hT_{q/h}$ is linear. Let us denote by P^q its matrix.

If the trace of the sub-domain solution at the artificial interface is regular enough, its approximation in $\mathcal{E}^q$ will have a much lower dimension than an approximation in E^h for the same level of accuracy (Gottlieb and Shu [1997]). Furthermore for the approximation of the Poisson problem or the Helmholtz operator, in deformed rectangle, the *sinus* basis is somehow a natural basis.

We compute therefore directly the matrix P^q with q much smaller than the number of grid points. The column of P^q are obtained by processing q independent Schwarz iterates starting from $sin(k\tau)$ for the artificial boundary condition. We choose the set of basis function $V_j = W_j,\ j = 1...q$ to be the eigenvectors of P^q.

Figure 1 gives the convergence history of our method for a Poisson solver discretized with the Finite Volume method of Faille [1992] in a complex shape domain. The continuous line is the convergence history of MuS with no acceleration. The curve 'o' (resp. '+', '*') is for 2 waves (resp. 4, 8). We see that the convergence improves as the number of modes increases.

The trigonometric representation of the interface may not be the best solution in the general case, and there are many piecewise polynomial spaces of functions that might be more appropriate depending on the space of approximation of the PDE solution. This should be the topic of further investigations.

5 Conclusion

We have shown how to generalized the Aitken-Schwarz method from Cartesian grid with finite differences to other discretization such as finite element on tensorial product of grid or finite volumes on general quadrangle meshes. Let us emphasis that the implementation of our method can reuse the initial coding of the Schwarz method with no change. As a matter of fact, the computation of dominant eigenvectors of the trace transfer operator can be seen as

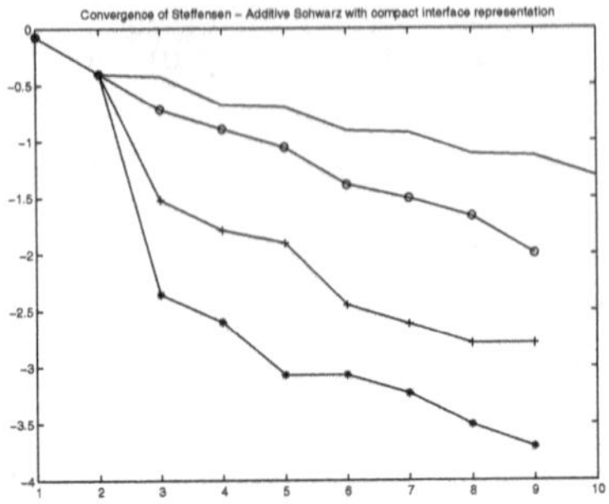

Fig. 1. Central Finite Volume discretisation: compact grid interface, overlap is 5 per cent, grid is 81×81

a pre-processing step. This step may involve few independent parallel executions of one Schwarz iteration with the original code. Further the Aitken-like acceleration procedure itself operates on the trace generated by the Schwarz code and does not require any change in the data structure of the original code. As shown in Garbey and Tromeur-Dervout [2002] this approach gives efficient parallel implementation with slow network. This is the philosophy of our ongoing work on metacomputing of elliptic problems.

Acknowledgement. Research of the second author reported here was supported by Award 0305405 from the National Science Foundation.

References

J. Baranger, M. Garbey, and F. Oudin-Dardun. Acceleration of the Schwarz method : the cartesian grid with irregular space step case. Technical report, CDCSP (Center for the development of parallel scientific computing), 2003.

I. Faille. A control volume method to solve an elliptic equation in two-dimensional irregular meshing. *Comp. Meth. Appl. Mech. Engrg.*, 100:275–290, 1992.

M. Garbey. Acceleration of the Schwarz method for elliptic problems. submitted, 2003.

M. Garbey and D. Tromeur-Dervout. Two level domain decomposition for multi-clusters. In T. Chan, T. Kako, H. Kawarada, and O. Pironneau, editors, *12th Int. Conf. on Domain Decomposition Methods*, pages 325–339. DDM.org, 2001.

M. Garbey and D. Tromeur-Dervout. On some Aitken like acceleration of the Schwarz method. *Int. J. for Numerical Methods in Fluids*, 40 (12): 1493–1513, 2002.

D. Gottlieb and W. Shu, C. On the Gibbs phenomenom and its resolution. *SIAM Review*, 39(4):644–668, 1997.

B. F. Smith, P. E. Bjørstad, and W. Gropp. *Domain Decomposition: Parallel Multilevel Methods for Elliptic Partial Differential Equations.* Cambridge University Press, 1996.

The Fat Boundary Method: Semi-Discrete Scheme and Some Numerical Experiments

Silvia Bertoluzza[1], Mourad Ismail[2], and Bertrand Maury[3]

[1] Istituto di Matematica Applicata e Tecnologie Informatiche del C.N.R. v. Ferrata 1, 27100 Pavia. Italy. (silvia.bertoluzza@imati.cnr.it).
[2] Laboratoire Jacques-Louis Lions, Université Pierre et Marie Curie. Boîte courrier 187, 75252 Paris Cedex 05. France. (ismail@ann.jussieu.fr).
[3] Laboratoire de Mathématiques, Université Paris-Sud. Bâtiment 425, 91405 Orsay. France. (bertrand.maury@math.u-psud.fr).

Summary. The **Fat Boundary Method** (**FBM**) is a fictitious domain like method for solving partial differential equations in a domain with holes $\Omega \setminus \overline{B}$ - where B is a collection of smooth open subsets - that consists in splitting the initial problem into two parts to be coupled via Schwartz type iterations: the solution, with a fictitious domain approach, of a problem set in the whole domain Ω, for which fast solvers can be used, and the solution of a collection of independent problems defined on narrow strips around the connected components of B, that can be performed fully in parallel. In this work, we give some results on a semi-discrete **FBM** in the framework of a finite element discretization, and we present some numerical experiments.

1 The Fat Boundary Method

The Fat Boundary Method (**FBM**) was introduced by Maury [2001] to solve partial differential equations in a domain with holes. For simplicity we present the method in the case of the Poisson problem. Let us denote by $\Omega \subset \mathbb{R}^n$ a Lipschitz bounded domain and $B \subset \Omega$ a collection of smooth subsets (typically balls). The boundaries of Ω and B are respectively denoted by Γ and γ. Our purpose is to solve the problem: Find $u \in H_0^1(\Omega \setminus \overline{B})$, such that

$$-\Delta u = f \text{ in } \Omega \setminus \overline{B}. \tag{1}$$

Solving this problem by **FBM** consists in splitting it into a local resolution in a neighborhood of B, where we can use a fine mesh (in a thin layer around the holes, the dashed subdomain denoted by ω in figure 1), and a global resolution based on a cartesian mesh covering the whole domain Ω. This makes it possible the use of fast solvers and good preconditioners.

The link between the global and the local problem is based on the interpolation of a globally defined field on an artificial boundary which delimits the

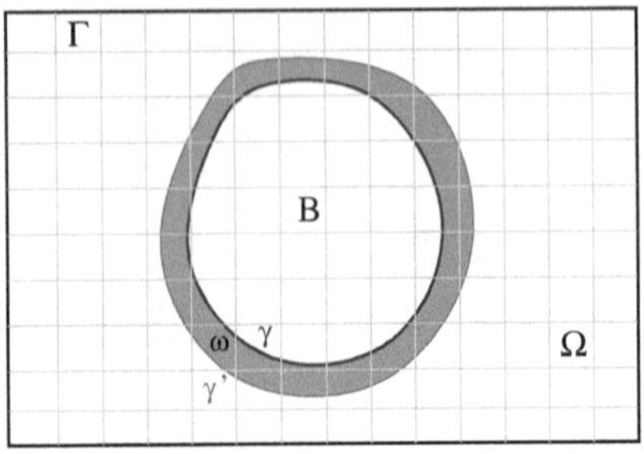

Fig. 1. Domains in the two-dimensional case

local subdomain, and the prescription of the jump of the normal derivative across the boundary of B. More precisely, we introduce a smooth artificial boundary γ' around B, and we denote by ω the (narrow) domain delimited by γ and γ' ($\partial\omega = \gamma \cup \gamma'$). We then introduce the functional space

$$H^1_\gamma(\omega) = \{v \in H^1(\omega), \quad v_{|_\gamma} = 0\}. \tag{2}$$

We can replace problem (1) by two coupled new problems, one of which is set in ω, and the other one in the whole domain Ω: Find $(\widehat{u}, v) \in H^1_0(\Omega) \times H^1_\gamma(\omega)$, such that

$$\begin{cases} a: & \begin{cases} -\Delta v = f & \text{in} \quad \omega, \\ \quad v = \widehat{u} & \text{on} \quad \gamma', \end{cases} \\ b: & -\Delta \widehat{u} = \overline{f} + \dfrac{\partial v}{\partial n}\delta_\gamma \quad \text{in} \quad \Omega, \end{cases} \tag{3}$$

where $\overline{f}$ is the extension of f by 0 in B, and where $\frac{\partial v}{\partial n}\delta_\gamma \in H^{-1}(\Omega)$ stands for the continuous linear form: $w \in H^1_0(\Omega) \mapsto \displaystyle\int_\gamma \frac{\partial v}{\partial n} w$. More precisely, we have this result (Maury [2001])

Theorem 1. *Problems (1) and (3) are equivalent, i.e.*

- *If u is a solution of (1), then the couple $(\overline{u}, u_{|_\omega})$ is a solution of (3).*
- *If $(\widehat{u}, v)$ is a solution of (3), then $\widehat{u}_{|_{\Omega\setminus\overline{B}}}$ is a solution of (1).*

The local problem (3-a) and the global one (3-b) are coupled, and this suggests the use of a fixed point algorithm. Let $\theta \in]0,1[$ be a relaxation parameter. We introduce the following operators: $\mathcal{T}_\theta(\cdot,\cdot; f): H^1_0(\Omega) \times H^1_\gamma(\omega) \longrightarrow H^1_0(\Omega) \times H^1_\gamma(\omega)$ defined by $\mathcal{T}_\theta(\widehat{u}, v; f) = (\widehat{U}, V)$ where $V \in H^1_\gamma(\omega)$ and $\widehat{U} \in H^1_0(\Omega)$ are solutions of

$$-\Delta V = f \text{ in } \omega, \qquad V = \theta v + (1-\theta)\widehat{u} \text{ on } \gamma', \tag{4}$$

$$-\Delta \widehat{U} = \overline{f} + \frac{\partial V}{\partial n}\delta_\gamma \quad \text{in } \Omega. \tag{5}$$

By definition of $\mathcal{T}_\theta$, $(\widehat{u}, v)$ is solution of (3) if and only if $\mathcal{T}_\theta(\widehat{u}, v; f) = (\widehat{u}, v)$. The following convergence result holds. (See Maury [2001] for the proof)

Theorem 2. *There exists $\theta_0 < 1$ such that for all $\theta \in]\theta_0, 1[$ the fixed point procedure*

$$(\widehat{u}^{n+1}, v^{n+1}) = \mathcal{T}_\theta(\widehat{u}^n, v^n; f)$$

converges to the fixed point of the operator $\mathcal{T}_\theta(\cdot, \cdot; f)$.

2 The semi-discrete case

A preliminary step towards the analysis of the discrete **FBM** – where both (4) and (5) are solved numerically – consists in assuming that the local problem (4) is solved exactly, and in focusing then on the discretization of the global problem (5). Letting $\mathbb{U}_h \subset H_0^1(\Omega)$ be a finite dimensional approximation space of finite element type, we propose the following semi-discrete fixed point iteration scheme: let $\mathcal{T}_\theta^h(\cdot, \cdot; f) : \mathbb{U}_h \times H_\gamma^1(\omega) \longrightarrow \mathbb{U}_h \times H_\gamma^1(\omega)$ be defined by $\mathcal{T}_\theta^h(u_h, v; f) = (U_h, V)$ with $V \in H_\gamma^1(\omega)$ and $U_h \in \mathbb{U}_h$ respectively defined by

$$-\Delta V = f \text{ in } \omega, \qquad V = \theta v + (1 - \theta) u_h \text{ on } \gamma', \tag{6}$$

$$\int_\Omega \nabla U_h \cdot \nabla w_h = \int_\Omega \overline{f} w_h + \int_\gamma \frac{\partial V}{\partial n} w_h \qquad \forall w_h \in \mathbb{U}_h. \tag{7}$$

We are interested in studying the existence and uniqueness properties of the solution to the fixed point equation

$$(u_h, v^\star) = \mathcal{T}_\theta^h(u_h, v^\star; f), \tag{8}$$

as well as in giving an estimate on the error $u - u_h$ in Ω.

We briefly sketch here the main steps of the analysis. The first step, in order to analyze the scheme (8) is to introduce an auxiliary fixed point problem. Let us denote by $\boldsymbol{\pi}_h : H_0^1(\Omega) \longrightarrow \mathbb{U}_h$ the Galerkin Projection defined by

$$\int_\Omega \nabla(\boldsymbol{\pi}_h u) \cdot \nabla w_h = \int_\Omega \nabla u \cdot \nabla w_h, \quad \forall w_h \in \mathbb{U}_h. \tag{9}$$

Let, for $\theta \in (0,1)$, $\mathcal{T}_\theta^\star(\cdot, \cdot; f) : H_0^1(\Omega) \times H_\gamma^1(\omega) \longrightarrow H_0^1(\Omega) \times H_\gamma^1(\omega)$ be defined as follows: $\mathcal{T}_\theta^\star(u, v; f) = (U, V)$ with $(U, V) \in H_0^1(\Omega) \times H_\gamma^1(\omega)$ solution to

$$-\Delta V = f \text{ in } \omega, \qquad V = \theta v + (1 - \theta) \boldsymbol{\pi}_h u \text{ on } \gamma', \tag{10}$$

$$-\Delta U = \overline{f} + \frac{\partial V}{\partial n} \delta_\gamma, \quad \text{in } \Omega. \tag{11}$$

Then we consider the problem

$$(u^\star, v^\star) = \mathcal{T}_\theta^\star(u^\star, v^\star; f). \tag{12}$$

The relation between (12) and (8) is the object of the following lemma (Bertoluzza et al.).

Lemma 1. *Let $(u^\star, v^\star)$ be a solution to the auxiliary fixed point problem (12). Then $(\pi_h u^\star, v^\star)$ is a solution to problem (8). Respectively let $(u_h, v^\star)$ be a solution to problem (8) then, letting $u^\star \in H^1_0(\Omega)$ be the unique solution to*

$$-\Delta u^\star = \overline{f} + \frac{\partial v^\star}{\partial n}\delta_\gamma, \quad in\ \Omega, \tag{13}$$

$(u^\star, v^\star)$ is a solution of problem (12).

The key ingredient of the analysis of the auxiliary problem is the following lemma (Bertoluzza et al.), stating that, under suitable assumptions the operator $\mathcal{T}^\star_\theta(\cdot,\cdot;0)$ is a contraction, and whose proof is heavily based on functions which are harmonic in $\Omega \setminus \gamma$.

Lemma 2. *Let $(u, v) \in \mathcal{R}(\mathcal{T}^\star_\theta)$, and let $(U, V) = \mathcal{T}^\star_\theta(u, v; 0)$. Then, if the provided h is sufficiently small, there exists $\theta_0 \in]0,1[$ such that if $\theta > \theta_0$, for some positive $k < 1$, $|U|_{1,\Omega} \leq k|u|_{1,\Omega}$, and $|V|_{1,\omega} \leq \theta|v|_{1,\omega} + C|u|_{1,\Omega}$.*

Existence and uniqueness of the solution of the auxiliary problem (12) (and therefore, thanks to Lemma 1 of the original semidiscrete problem (8)) easily follows. Let us now estimate the error $\widehat{u} - u_h$. Since the mesh is a priori chosen independently of the position of γ, it is clear that, since $\widehat{u}$ has on γ a discontinuity of the normal derivative the best global regularity that we can expect of $\widehat{u}$ is $\widehat{u} \in H^{3/2-\epsilon}(\Omega)$ and therefore the best error estimate that we can expect is $\|\widehat{u} - u_h\|_{1,\Omega} \leq Ch^{1/2-\epsilon}\|u\|_{3/2-\epsilon}$ (h denoting the mesh size of the triangulation which we assume to be regular and quasiuniform). However, assuming that ∂B is sufficiently regular, if $f|_{\Omega\setminus\overline{B}} \in H^{s-2}$ the function $\widehat{u} = u$ is in $H^s(\Omega \setminus \overline{B})$ and then, using the technique introduced by Nitsche and Schatz [1974] in order to estimate local convergence rates, we can hope for a better convergence rate in any open set Ω^* strictly embedded in $\Omega \setminus \overline{B}$. More precisely, assuming that we are using finite elements of order m (either Pm or Qm) the following theorem holds (Bertoluzza et al.)

Theorem 3. *Assume that $f \in H^{s-2}(\Omega \setminus \overline{B})$, with $2 \leq s \leq m+1$, and let $\Omega^* \subset\subset \Omega$. Then, for h sufficiently sufficiently small we have*

$$\|u - u_h\|_{1,\Omega^*} \leq Ch^s\|f\|_{s-2,\Omega\setminus\overline{B}}$$

3 Numerical Experiments and Conclusions

We want to verify that, as stated by Theorem 3, if we use P1 finite elements, **FBM** is of order one in every subdomain $\check{\Omega} \subset\subset \Omega \setminus \overline{B}$. Figure 2 shows the dependence of the errors (in H^1 and L^2 norms) upon the mesh step size h. All tests are carried out using an uniform cartesian grid. We denote global errors the ones computed in the whole domain Ω and local errors the ones computed in the subdomain $\check{\Omega}$ of $\Omega \setminus \overline{B}$. The domain Ω is the box $]-\frac{1}{2},\frac{1}{2}[^3$, the "hole"

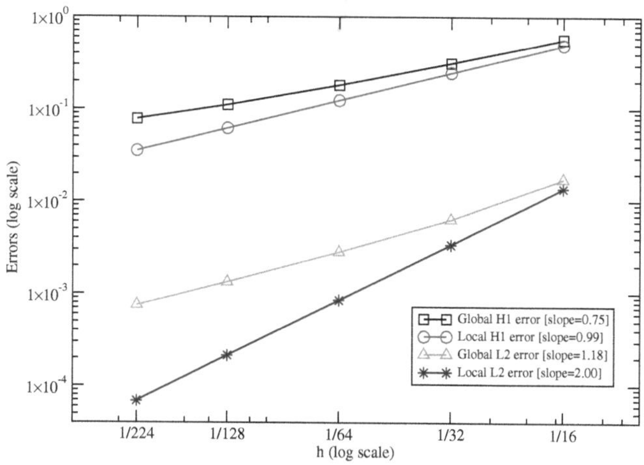

Fig. 2. Errors plots

B is the ball $B(0, R)$, where $R = 0.25$, and the subdomain $\check{\Omega}$ is $\Omega \setminus B(0, 0.3)$. The exact solution u is chosen to be equal to $\sin(2\pi(x^2 + y^2 + z^2 - R^2))$. The analytical solution was selected to be radial in order to eliminate the error due to the local resolution, and thus to be in conformity with the theoretical result.

3.1 Capability to deal with many holes

This numerical experiment illustrates the capability to deal with a domain with "many" holes. We consider the box $]-1, 1[^3$ with 163 disjoint balls disposed in a pseudo-random way. Figure 3-(left) shows the isosurface $u = 0$ of the computed solution, which solves the problem: $-\Delta u = 1$ in $\Omega \setminus \overline{B}$ and $u = 0$ on $\partial(\Omega \setminus \overline{B}) = \Gamma \cup \gamma$. Figure 3-(right) shows the same experiment but with a little larger number of particles: 343 balls disposed in a structured way.

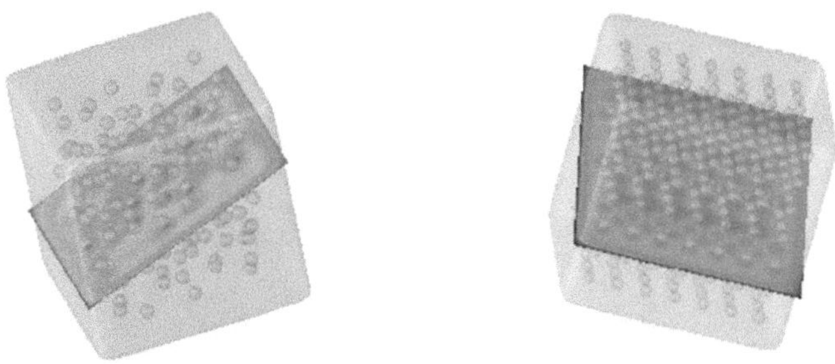

Fig. 3. Isosurface $u = 0$: (left)- 163 balls. (Right)- 343 balls

3.2 Numerical Simulation of convection-diffusion around two moving balls

We consider a parallelepiped Ω in which there are two moving rigid balls $B_1 \cup B_2 = B$. Their trajectories are imposed in advance. On five faces of the box we maintain a temperature equal to 1 and at the sixth face we take homogeneous Neumann boundary conditions. On the surfaces of the two balls we impose (via Dirichlet boundary condition) a null temperature. Heat is convected using a potential field. One expects to have a "trail" of "fresh zones" following the balls in their movements. The problem we solve is the following

$$\begin{cases} \frac{\partial T}{\partial t} - \nu \Delta T + \nabla\phi \cdot \nabla T = 0 \text{ in } \Omega \setminus B, \\ T = 0 \text{ on } \gamma, \\ T = 1 \text{ on } \Gamma \setminus (z = z_{min}), \\ \frac{\partial T}{\partial n} = 0 \text{ on } (z = z_{min}), \end{cases} \tag{14}$$

where ϕ solves

$$\begin{cases} -\Delta\phi = 0 \text{ in } \Omega \setminus B, \\ \frac{\partial \phi}{\partial n} = 0 \text{ on } \gamma, \\ \frac{\partial \phi}{\partial n} = 0 \text{ on } \Gamma \setminus (z = z_{min}, z = z_{max}), \\ \phi = 1 \text{ on } (z = z_{min}), \\ \phi = 1 \text{ on } (z = z_{max}). \end{cases} \tag{15}$$

Figure 4 shows the computed solution at different time iterations.

Fig. 4. Convection-Diffusion around moving balls

3.3 Flow past a sphere

We consider the incompressible Navier-Stokes equations in the parallelepiped $\Omega =]-\frac{3}{2}, \frac{3}{2}[\times]-\frac{3}{2}, \frac{3}{2}[\times]-1, 5[$ containing a spherical obstacle $B((0,0,1), \frac{1}{2})$. The time discretisation is done using the Finite-Element Projection/Lagrange-Galerkin method (see Achdou and Guermond [2000]) which is a projection algorithm combined with the characteristics method (see Pironneau [1982]). At each time step we have to solve, by **FBM**, elliptic problems for the velocity and the pressure. Figure 5 shows the velocity field on the plan $y = 0$ of a flow past a sphere at Reynold's number equal to 100. In order to see the vortices, figure 6 presents a zoom close to the sphere. See Ismail [2003] for more details on numerical simulations of flows past spheres.

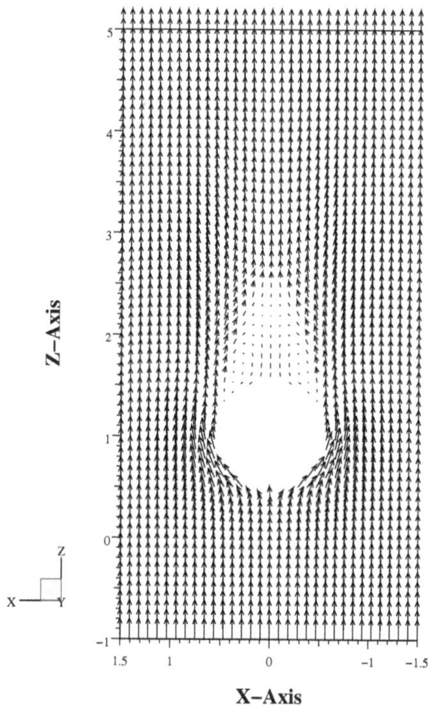

Fig. 5. Flow past a sphere at $Re = 100$

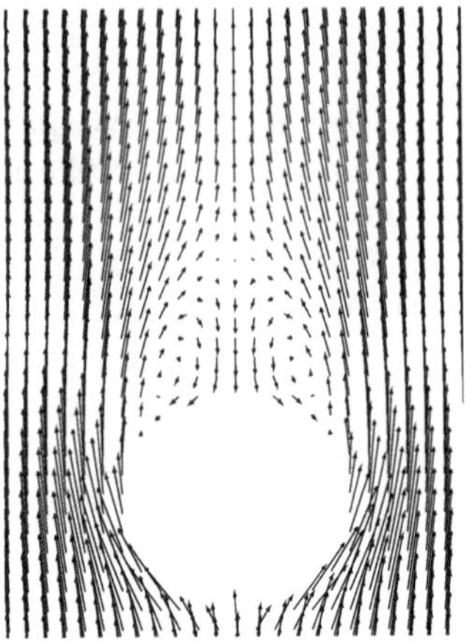

Fig. 6. Zoom

3.4 Conclusions

The numerical results confirm the theoretical estimates and shows the wide applicability of **FBM**. The future work will consist on the theoretical side in taking into account also the error due to the local resolution, thus studying the full discrete scheme. On a practical level we are working on adapting the method to take into account free motion of the bodies in order to be able to simulate fluid-particle flows.

References

Y. Achdou and J.-L. Guermond. Convergence analysis of a finite element Projection/Lagrange-Galerkin method for the incompressible Navier-Stokes equations. *SIAM J. Numer. Anal.*, 37(3):799–826, 2000.

S. Bertoluzza, M. Ismail, and B. Maury. Analysis of the discrete fat boundary method. in preparation.

M. Ismail. *Simulation Numérique d'écoulements fluide-particules par la méthode de la frontière élargie.* PhD thesis (in preparation), to appear in french, Université Pierre et Marie Curie, Paris, France, 2003.

B. Maury. A Fat Boundary Method for the Poisson problem in a domain with holes. *J. of Sci. Comput.*, 16(3):319–339, 2001.

J. A. Nitsche and A. H. Schatz. Interior estimates for Ritz-Galerkin methods. *Math. Comp.*, 28:937–958, 1974.

O. Pironneau. On the transport-diffusion algorithm and its applications to the Navier-Stokes equations. *Numer. Math.*, 38:309–332, 1982.

Modelling of an Underground Waste Disposal Site by Upscaling and Simulation with Domain Decomposition Method

I. Boursier[1], A. Bourgeat[1], and D. Tromeur-Dervout[1]

Modelling and Scientific Computing Laboratory, University Lyon 1, France

Summary. We derive an upscaled but accurate 2D model of the global behavior of an underground radioactive waste disposal. This kind of computation occurs in safety assessment process. Asymptotic development of the solution leads to solve terms of order 1 on more regular and steady-state auxiliary problems. Neumann-Dirichlet domain decomposition methods, with non matching spectral grids, are performed to solve those auxiliary problems. Fourier and Chebychev polynomials approximation of the solution are used depending on boundary conditions implemented on subdomains. Since spectral representation of the solution or its derivatives allows accurate mappings between the interfaces of the different grids, we speed up the convergence of the Neumann-Dirichlet method by the Aitken acceleration which is sensitive to the accuracy of the representation of the iterate solution on the artificial interfaces. In order to enforce regularity for the spectral approximation, some regular extensions and filtering techniques on the artificial interfaces for the right hand side of the problem and the iterate solution are implemented.

1 Field decomposition method applied to an underground waste disposal

The disposal site can be described as a repository array made of a large number of units inside a low permeability layer, called host layer, e.g clay. This clay layer is embedded between layers with higher permeability (Bourgeat et al. [2003]). There is a large number of units, each of them has a small size (10 m) compared with the layer size (100 m). So a direct numerical simulation of the whole field, based on a microscopic model, is not realistic. The ratio between the width l of a single unit and the layer length L can be considered as a small parameter ϵ in the detailed microscopic model. The study of the renormalized model behavior, as ϵ tends to 0, by means of the homogenization method and boundary layers, gives an asymptotic model which could be used as a global repository model for numerical simulations. According to this rescaling, the units have a height of order $\epsilon^{\beta}, \beta > 1$, and are embedded in a layer of thickness ϵ. The leaking of a disposal unit is represented by a hole in the periodic computational domain with a given flux on the boundaries. The radioactive pollutant is transported both by the convection due to the water flowing slowly through the rocks (creeping flow) and by the diffusion due to the dilution into the water. The transport of concentration of an underground pollutant is modelled by the advection-diffusion equation (where the advection velocity is assumed to be given) that follows:

$$\frac{\omega R\partial c}{\partial t} - \nabla \cdot (A\nabla c) + (U\cdot\nabla)c + \lambda\omega Rc = q \text{ in } \Omega \tag{1}$$

where: Ω is the porous medium, c a radioactive component concentration in the water, ω the medium porosity, R the retardation coefficient, U Darcy's velocity, A the diffusive term, λ the radioactive decay, q the source term.

We normalize the geometric dimensions taking into account the ratio ϵ. The process is then described by the following advection-diffusion type equation:

$$\frac{\omega^\epsilon\partial c_\epsilon}{\partial t} - \nabla \cdot (A_\epsilon\nabla c_\epsilon) + (v^\epsilon.\nabla)c_\epsilon + \lambda\omega^\epsilon c_\epsilon = 0 \text{ in } \Omega_\epsilon^T \tag{2}$$

$$c_\epsilon(0,x) = c_0(x), \quad x \in \Omega_\epsilon \tag{3}$$

$$n\cdot\sigma = n\cdot(A^\epsilon\nabla c_\epsilon - v^\epsilon c_\epsilon) = \Phi(t) \quad \text{on } \Gamma_\epsilon^T \tag{4}$$

$$c_\epsilon = 0 \quad \text{on } S_1 \tag{5}$$

$$n\cdot(A^\epsilon\nabla c_\epsilon - v^\epsilon c_\epsilon) = 0 \quad \text{on } S_2 \tag{6}$$

where Ω_ϵ is the adimensionalized domain around the units, $\Omega_\epsilon^T = \Omega_\epsilon\times]0,T[$, $\Gamma_\epsilon^T = \partial B_\epsilon\times]0,T[$ where B_ϵ is the set of the units, $S = \partial\Omega_\epsilon$, S_1 (respect. S_2) represents the bottom (respect. the top) of S, c_ϵ is a radioactive component concentration in the water, ω^ϵ is the adimensionalized medium porosity, v^ϵ is the adimensionalized convection, A^ϵ the adimensionalized diffusion tensor, λ the radioactive decay, Φ the incoming flux of radioactive element.

It was proved in (Bourgeat et al. [2003]) this c^ϵ exists is unique and has a weak limit c. This weak limit c gives the global long time behavior of the process only if the flux Φ is not too large. On the one hand, we expect some fast oscillations of the solution in the vicinity of the containers and therefore we introduce in that region the fast variable $y = \frac{x}{\epsilon}$. On the other hand, c_ϵ is expected to have the same behavior as the weak limit c without any oscillations far from the units area. These behaviors suggest to use matched asymptotic expansions:

The domain is split in two parts:

- $G_\epsilon =]-\delta/2, \delta/2[\times]-\epsilon\log(1/\epsilon), \epsilon\log(1/\epsilon)[$, the inner domain
- $\Omega/\bar{G}_\epsilon$, the outer domain.

δ is defined such that $\partial\Omega\bigcap\partial G_\epsilon = \emptyset$

In G_ϵ, we look for an asymptotic expansion of c_ϵ such as:

$$c_\epsilon \simeq c_\epsilon^0 + \epsilon(\chi_\epsilon^k(\frac{x}{\epsilon})\frac{\partial c_\epsilon^0}{\partial x_k} + w_\epsilon(\frac{x}{\epsilon})\Phi - c_\epsilon^0 \rho_\epsilon^k(\frac{x}{\epsilon})v_k^1) \equiv c_\epsilon^1 \tag{7}$$

where we assume the summation from 1 to 2 over the index k. The function c_ϵ^0 mimics the behavior of the concentration far from the source. χ_ϵ^k represents the correction on the diffusive term in the near field, ρ_ϵ^k, the correction on the convective term and w_ϵ the correction on the source. Their behaviors are described by the way of the following type of auxiliary problem:

$$\begin{cases} -\nabla\cdot(A\nabla u) = f \quad \text{in } G_\epsilon \\ n\cdot(A\nabla u) = g \quad \text{on } \partial M_\epsilon \\ u \text{ is 1-periodic in } y_1 \\ \lim_{y_2\to\infty} A\nabla u = r \end{cases} \tag{8}$$

In order to evaluate the validity of this asymptotic expansion, accurate simulations of these behaviors are needed.

2 Numerical solutions of the auxiliary problems

The behavior of the 1-order terms of the homogenization is represented by a diffusive problem on a domain admitting a hole and periodic conditions in the x-direction. These problems need accurate discretisation, since they represent the oscillations at the beginning of the leak, and will influence the rest of the simulation. Spectral discretization leads to have structured data-blocks on spectral meshes, then the domain is decomposed into three subdomains, Ω_1, Ω_2 and Ω_3 in order to take into account the hole. The physical domain size in x-direction $[0,1]$ is mapped with a linear mapping to the computational domain (Ω_1 and Ω_3) size in x direction $[0,\pi]$. The decomposition of the computational domain is illustrated in figure 1. Ω_2 has Neumann boundary conditions in the x-direction leading to use Chebychev discretisation while Ω_1 and Ω_3 have periodic boundary conditions in x-direction leading to use Fourier discretisation. Thus, meshes between subdomains do not match so spectral mapping techniques are used to represent the solution on both meshes on the artificial interfaces. This accurate representation of the iterate solution on the artificial interfaces generated by the domain decomposition method will allow us to use the Aitken acceleration method developed in (Garbey and Tromeur-Dervout [2002]).

2.1 Computation in subdomains

The diffusion tensor A is assumed to depend only on the vertical direction. The solution in subdomains Ω_1 and Ω_3 is computed on an extended subdomain in order to avoid the 0-mode singularity (because the problems are defined up to a constant). Thus, we compute $\tilde{u} : [0, 2\Pi] \times [-1, 1] \to R$, an odd periodic

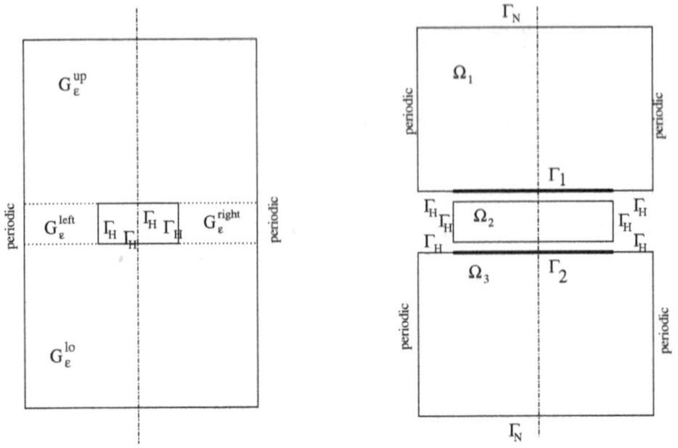

Fig. 1. Decomposition of the computational domain: Ω_1 (resp. Ω_3) corresponds to the upper (resp. lower) part of the domain, Ω_2 corresponds to $G_\epsilon^{right} \cup G_\epsilon^{left}$

function, which is equal to u, solution of problem (8) on $[0, \Pi] \times [-1, 1]$. The extended solution is then, using Chebychev discretisation in y-direction, approximated by: $P_M\tilde{u}(x_i, y_j) = \sum_{0\leq k\leq M}\sum_{0\leq l\leq N}\hat{\hat{u}}_{k,l}T_l(y_j)\sin(kx_i)$, where $x_i = \frac{2i\Pi}{M}$, $i = 0, \ldots, M$, and $y_j = cos(\frac{j\Pi}{N})$, $j = 0, \ldots, N$. So we obtain for subdomain Ω_1:

$$\nabla \cdot (A(y)\nabla\tilde{u}) = \sum_{0\leq k\leq M} (\frac{\partial}{\partial y}(A_2(y)(\frac{\partial}{\partial y})) - k^2 A_1(y))\hat{u}_k(y)\sin(kx) \text{ in } \Omega_1,$$

$$A(y)\nabla\tilde{u} = \sum_{0\leq k\leq M} A_2(y)\frac{\partial}{\partial y}(\hat{u}_k(y))\sin(kx) \text{ on } \Gamma_1,$$

where $\hat{u}_k(y) = \sum_{0\leq j\leq N}\hat{\hat{u}}_{k,j}T_j(y)$. Due to the linearity, the solution $P_M\tilde{u}$ can be decoupled according to the directions. Thus, the subdomain problem is decoupled in M mode-problems of size $(N+1)\times(N+1)$ leading to a well suited situation for the parallelization.

Some difficulties appear due to the boundary condition on $G_1 = \{y = y_N = -1\}$, coming from the decomposition domain:

$$\begin{cases} \frac{\partial u_1(x,y)}{\partial y} = \frac{\partial u_2(x,y)}{\partial y}, & \text{on } \Gamma_1 = [x_1, x_2]\times\{-1\} \\ n\cdot(A\nabla u_1) = g(x,y), \text{ on } G_1 \setminus \Gamma_1 \end{cases}$$

The boundary condition on G_1 is singular. In order to use a discrete Fourier transform, smoothing methods have to be applied. First, a C^2 Hermite interpolation based on two points is computed in the vicinity of x_1 and x_2 like in Garbey and Tromeur-Dervout [1998]. Then the "raised cosinus filter" (Gottlieb and Shu [1996]) is applied on the modes of the traces on the extended boundary condition including G_1,(in the same way as $\tilde{u}$) in order to minimize the Gibbs phenomenon. The third domain Ω_3 is treated in the same way.

On the second subdomain Ω_2, a Chebychev-Chebychev discretisation is used in both directions. Now, the pseudospectral discretisation of the differential operator leads to solve a $((N_1N_2) \times (N_1N_2))$ full but time independent linear system with a PLU factorization. Parallel ADI techniques are also under implementation to save time.

2.2 Methodology on Neumann-Dirichlet Domain Decomposition and Aitken Acceleration method

The Neumann-Dirichlet domain decomposition method leads to solve for the auxiliary problems:

$$\nabla \cdot (A \nabla u_1^{n+1/2}) = f \quad \text{in } \Omega_1 \tag{9}$$

$$\frac{\partial}{\partial n} u_1^{n+1/2} = \frac{\partial}{\partial n} u_2^n \quad \text{on } \Gamma_1 \tag{10}$$

$$\nabla \cdot (A \nabla u_3^{n+1/2}) = f \quad \text{in } \Omega_3 \tag{11}$$

$$\frac{\partial}{\partial n} u_3^{n+1/2} = \frac{\partial}{\partial n} u_2^n \quad \text{on } \Gamma_2 \tag{12}$$

$$\nabla \cdot (A \nabla u_2^{n+1}) = f \quad \text{in } \Omega_2 \tag{13}$$

$$u_2^{n+1} = u_1^{n+1/2} \quad \text{on } \Gamma_1 \tag{14}$$

$$u_2^{n+1} = u_3^{n+1/2} \quad \text{on } \Gamma_2 \tag{15}$$

Using Aitken acceleration method for the Schwarz DDM based on the linear convergence or divergence of the iterative method (Garbey and Tromeur-Dervout [2002]), the convergence of the solution on the artificial interface can be speed up in few iterations. The linear convergence can only be obtained for the iterated solution Fourier modes on the artificial interfaces. The solution u on Γ_1 and Γ_2 on the Chebychev grid being not periodic, $u|_{\Gamma_1}$ is left-extended on $[0, \pi]$ with a fifth-degree Hermite interpolation. Then we seek an odd solution on the extended domain $[0, 2\pi]$.

The distance between the artificial interfaces Γ_1 and Γ_2 is small. Thus these interfaces are coupled for low modes. Let us consider the sequence $\hat{u}_k^n = (\hat{u}_{k,1|\Gamma_1}^n, \hat{u}_{k,3|\Gamma_2}^n)^t$ $k = 0, \cdots, M$. The operator T_k,

$$(\hat{u}_k^n - \hat{U}_k^\infty) \rightarrow (\hat{u}_k^{n+1} - \hat{U}_k^\infty) \tag{16}$$

where U_k^∞ is the k mode of the exact solution, is linear. As long as the artificial interfaces are coupled, the matrix P_k, $k = 0, \cdots, M$ associated to the operator T_k is full.

From (16), we have:

$$\hat{u}_k^{i+2} - \hat{u}_k^{i+1} = P_k(\hat{u}_k^{i+1} - \hat{u}_k^i), \; i = n-1, n; \; k = 0, \cdots, M. \tag{17}$$

We notice that, for each mode, only three iterations are needed to determine the coefficients of the matrix P_k with(17). If the operator $Id - P_k$ is not singular, then the Aitken acceleration will be written as follows for each mode:

$$\hat{U}_k^{\infty} = (Id - P_k)^{-1}(\hat{u}_k^{n+1} - P_k\hat{u}_k^n),\ k = 0, \cdots, M. \tag{18}$$

Finally, a backward Fourier transform on these $\hat{U}_k^{\infty}$ gives the solution in the physical space on the artificial interfaces. The extension of the solution on the artificial interfaces is a C^2 function and it is enough to get a three order rate for the discretisation error of our approximation.
Remark 1: as the smoothing procedure introduces some non linearities, more than one acceleration has to be applied on the Neumann-Dirichlet algorithm.
Remark 2: the linear behavior of the mode error for this Schwarz DDM with smoothing procedures is not obtained directly (because of the Gibbs phenomenon), so that the Aitken speed-up can be applied only after some iterations.

3 Numerical Results

We first checked the accuracy of the method on an analytical solution $u : (x, y) \rightarrow sin(x)sin(y)$ of an elliptical problem defined on the computational domain Ω. Figures 2 and 3 show the Aitken acceleration (at iteration 20) effect on usual Schwarz convergence (the error is the $\|.\|_\infty$ of the difference between two successive iterations for each mode).

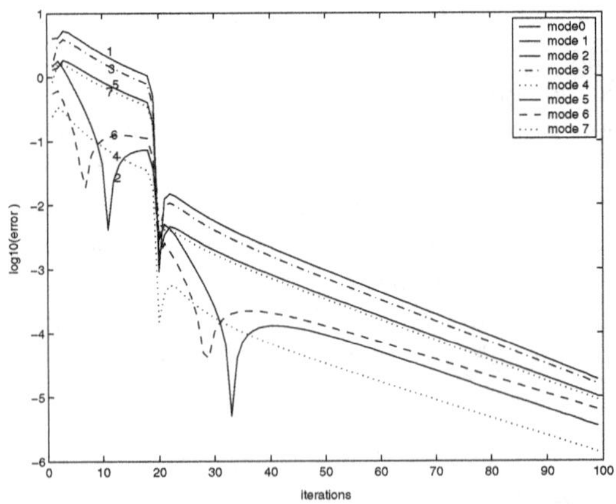

Fig. 2. Aitken acceleration of the mode-error (after 16 iterations) for the analytical problem

The error in the table 1 shows the $\|.\|_\infty$ of the difference between the exact trace and the computed solution on artificial interfaces. The method is of order-2.5 of consistency instead of an expected order-3. Nevertheless, the acceleration of the convergence thanks to the Aitken speed-up is satisfactory (30 iterations instead of 100) for the same accuracy.

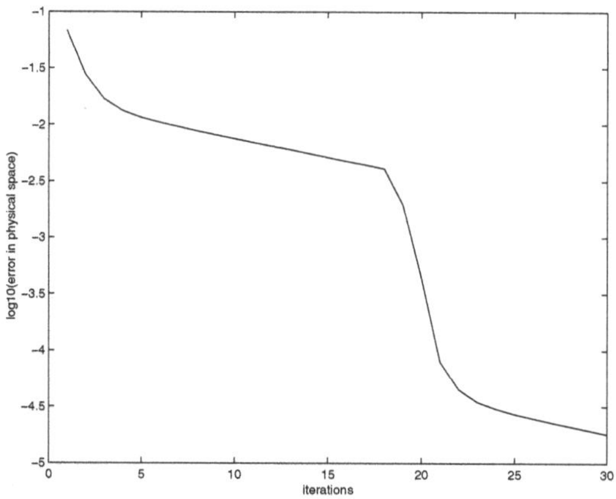

Fig. 3. Aitken acceleration of the error in the physical space (after 16 iterations) for the analytical problem

Table 1. Precision and velocity of the algorithm

Accel./	modes nb.	y-discr.	Precision	iterations nb.	Time(s)
No acceleration	40	32	0.05	100	22,77
No acceleration	80	32	1.e-4	100	27,07
No acceleration	178	32	7.e-5	100	41,64
No acceleration	256	32	4.e-5	100	58,82
Acceleration	178	32	7.e-5	30	13,44

If we now apply our methodology on one of the auxiliary problems, for instance the corrector ρ_ϵ^k on the convective term, Fig. 4 shows the isolines of the computed solution of the problem (19), for k=2, obtained with the present methodology in 12 iterations.

We recall that ρ_ϵ^k follows the equation:

$$\begin{cases} -\nabla \cdot (A \nabla \rho_\epsilon^k) = 0 \quad \text{in } G_\epsilon \\ n \cdot (A \nabla \rho_\epsilon^k + e_k) = 0 \quad \text{on } \partial M_\epsilon \\ \rho_\epsilon^k \text{ is 1-periodic in } y_1 \\ \lim_{y_2 \to \infty} \nabla \rho_\epsilon^k = 0 \end{cases} \tag{19}$$

4 Conclusion

Our aim was to develop a methodology adapted to a physical problem, which cannot be easily simulated directly. The field decomposition splits the solution into regular problems according to the physical situation. In order to speed up

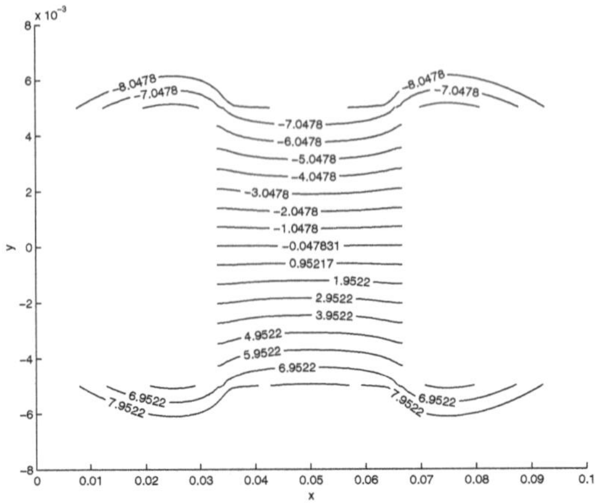

Fig. 4. ρ_k, k=2, isolines in the vicinity of the artificial interfaces for problem (19)

this Schwarz method by Aitken process, spectral methods are helping to obtain an accurate representation of the Neumann-Dirichlet algorithm iterations on a non-matching grid. This methodology has been applied on a problem with an analytical solution and clearly accelerates the speed of convergence of the Schwarz method. The same methodology was also applied successfully on all the auxiliary problems of the 1-order of the asymptotic expansion. The computation of the 0 order term of the model is currently under development (the Aitken-Schwarz method is applied on mixed finite element / spectral element systems).

References

A. Bourgeat, O. Gipouloux, and M. Marusic-Paloka. Modelling of an underground waste disposal site by upscaling. In *Math. Meth. in Appl. Sci.*, pages 152–158. 2003.

M. Garbey and D. Tromeur-Dervout. A new parallel solver for non periodic incompressible navier stockes equation with fourier basis: Application to frontal polymerisation. In *J. Comp. Phys.145*, pages 316–331. 1998.

M. Garbey and D. Tromeur-Dervout. On some aitken like acceleration of the schwarz method. In *Int.J. For Numerical Methods In Fluids 40(12)*, pages 1493–1513. 2002.

D. Gottlieb and C.-W. Shu. On the gibbs phenomenom and its resolution. In *SIAM J. Sci. Comput. 39(4)*, pages 644–668. 1996.

Non-Overlapping DDMs to Solve Flow in Heterogeneous Porous Media

Dan-Gabriel Calugaru and Damien Tromeur-Dervout

University Lyon 1, MCS/CDCSP - ISTIL, 69622 Villeurbanne
({calugaru,dtromeur}@cdcsp.univ-lyon1.fr)

Summary. For flow problems in multi-layered porous media, one can define a natural non-overlapping domain decomposition (DD). The simplest way to obtain DDMs is to distribute interface conditions (pressure and flux continuity) for each pair of adjacent subdomains and to use the Dirichlet-Neumann (D-N) algorithm. A different way is the use of two Robin conditions (RC) also distributed for each subdomain (Robin-Method). The main inconvenience of both methods is that the convergence is not ensured. To obtain efficient methods, we retain from previous works two basic ideas: an acceleration of Aitken type for the D-N algorithm and finding optimized coefficients for the Robin-Method. In the present paper, we analyze these improved algorithms in 1-D and 2-D framework for flow problems in heterogeneous porous media and we present a numerical comparison.

1 Introduction

Flow in heterogeneous porous media is to be solved in many hydrological or engineering applications, as oil recovery (Faille et al. [2001]), earthquake prediction (Calugaru et al. [2002]), radioactive wastes, etc. The reservoir is usually a multi-layered domain composed by some superposed aquifers separated by less permeable layers. In addition, fractured zones can divide the domain in blocks which can slide between each other. The steady one phase flow equation in saturated porous media is derived from the mass conservation law and the linear Darcy's law and can be written (in its simplest form) as:

$$-div\,(k\nabla u) = f \tag{1}$$

where the unknown u is the pressure, k is the permeability and f denotes a possible sink/source term. Obviously, the flow problem is obtained by adding some boundary conditions. For the multi-layered porous media, each layer is assumed homogeneous and the permeability is a piecewise constant function. Therefore, the interfaces between the layers represent the discontinuities of the permeability, but the intrinsic variables (pressure and flux) are continuous in all the domain, notably on interfaces.

2 Non-overlapping DDMs

From the DD point of view, the geological layers induce a natural non-overlapping decomposition. The DD being already set, we must define appropriate algorithms to obtain efficient DDMs. For simplicity of the description, we consider only two layers, i.e. the domain Ω is decomposed in two subdomains Ω_1, Ω_2, with Γ the common interface and k_i the permeability of layer i. The steady flow problem can be written as a transmission problem:

$$-k_1 \Delta u_1 = f \text{ in } \Omega_1 \quad , \quad -k_2 \Delta u_2 = f \text{ in } \Omega_2 \tag{2}$$

$$u_1|_\Gamma = u_2|_\Gamma \quad , \quad k_1 \partial u_1/\partial n_1|_\Gamma = -k_2 \partial u_2/\partial n_2|_\Gamma \tag{3}$$

with appropriate boundary conditions on $\partial\Omega$.

The multiplicative D-N algorithm requires to solve successively ($n \geq 0$):

$$\begin{cases} -k_1 \Delta u_1^{n+1} = f \text{ in } \Omega_1 \\ u_1^{n+1}|_\Gamma = u_2^n|_\Gamma \end{cases} \qquad \begin{cases} -k_2 \Delta u_2^{n+1} = f \text{ in } \Omega_2 \\ -k_2 \partial u_2^{n+1}/\partial n_2|_\Gamma = k_1 \partial u_1^{n+1}/\partial n_1|_\Gamma \end{cases}$$

Since it uses both physical interface conditions (3), this algorithm seems the simplest and the most adapted to the physics of the problem.

An alternative to the D-N algorithm is the Robin-Method in which weighted sums of physical conditions are used:

$$\begin{aligned} \alpha_1\, u_1|_\Gamma + \beta_1\, k_1 \partial u_1/\partial n_1|_\Gamma &= \alpha_1\, u_2|_\Gamma - \beta_1\, k_2 \partial u_2/\partial n_2|_\Gamma \\ \alpha_2\, u_2|_\Gamma + \beta_2\, k_2 \partial u_2/\partial n_2|_\Gamma &= \alpha_2\, u_1|_\Gamma - \beta_2\, k_1 \partial u_1/\partial n_1|_\Gamma \end{aligned} \tag{4}$$

The use of RC for non-overlapping DD has been firstly proposed by P.-L. Lions (Proc. DDM3, 202-223, 1990). It is easy to prove that, if $\alpha_1\beta_2 + \alpha_2\beta_1 \neq 0$, then conditions (3) and (4) are algebraically equivalent. In addition, conditions $\alpha_i\beta_i \geq 0$ have to be verified to obtain well-posed sub-problems. Since conditions (3) can be obtained from (4) by considering $\alpha_1 = \beta_2 = 1$, $\alpha_2 = \beta_1 = 0$, the D-N algorithm could be seen like a particular case of Robin-Method.

Both algorithms present the same inconvenience: the convergence is not ensured. Indeed, as shown in the next sections, the convergence of the Dirichlet-Neumann algorithm depends on the interface conditions distribution between the domains, while the convergence of the second algorithm depends on the choice of the Robin's coefficients α_i, β_i.

3 Improved non-overlapping DDMs

To cure such an inconvenience, several methods have been already proposed in DD literature (with or without overlap) for linear elliptic problems.

A recent method is an acceleration of Aitken type of the iterative solutions obtained by Schwarz algorithm and restricted to the interfaces. This method has been introduced by Garbey and Tromeur-Dervout (Proc. DDM12, 325-339, 2000), and studied theoretically and numerically by Garbey and Tromeur-Dervout [2002] for the additive Schwarz algorithm and 1-D decompositions. Some numerical experiences have also been described for the D-N algorithm. The basic idea of the Aitken technique is to accelerate independently each mode of the sine expansion of the iterative solutions restricted to the interfaces. In these papers, the method is developed at semi-discrete level (the problem is uniformly discretized in the interface direction). Some developments have been proposed to generalize Aitken acceleration for irregular grids (Baranger et al., Proc. DDM13, 287-294, 2001) or for non-matching grids (Baranger et al., Proc. DDM14, 341-348, 2002). The case of 2-D decompositions is treated considering some 1-D decompositions in a recursive manner (Garbey and Tromeur-Dervout, Proc. DDM13, 53-65, 2001) or accelerating the signals obtained by representing the discrete interface solutions in Fourier spaces (Calugaru and Tromeur-Dervout, Proc. Parallel CFD 2003, to appear).

A second method is to find RC which allow a fast convergence. This idea has been introduced by Després et al. [1992] for Helmholtz and Maxwell problems, and by Nataf et al. [1994] for convection-diffusion equations. It has been also used for flow problems in heterogeneous porous media (Faille et al. [2001]). In general, one can suppose $\beta_i = 1$ and then only α_i coefficients are searched. Optimized Robin conditions (ORC) could be also introduced as the best zeroth order approximations of optimal interface conditions (see for instance, Gander et al. [2002] for Helmholtz equation). In this context, it is possible to define other interface conditions (as for example, second order approximations), but which are not investigated in this paper.

A third method (the first in chronological order) was proposed by Funaro, Quarteroni and Zanolli (SIAM J. Num. Anal., 25(6), 1213-1236, 1988) and consists of modifying the multiplicative D-N algorithm by using a relaxation procedure at the end of each iteration. If the relaxation parameter is conveniently chosen, then the convergence is obtained.

In the next sections, for flow problems in multi-layered porous media, we investigate only two multiplicative algorithms: the D-N algorithm accelerated by an Aitken technique (A-D-N) and the Robin-Method. The convergence of these methods is studied in 1-D and 2-D frameworks and a numerical comparison is presented.

4 Convergence in 1-D framework

Let us consider the problem (2)-(3) for $\Omega = (a, b)$, with Dirichlet boundary condition u_D and $\Gamma = \{\lambda\} \subset (a, b)$ the common interface.

The (multiplicative) Robin-Method reads:

$$\begin{cases} -k_1(u_1^{n+1})'' = f \quad \text{in} \quad (a, \lambda) \\ u_1^{n+1}(a) = u_D(a) \\ \alpha_1\, u_1^{n+1}(\lambda) + \beta_1\, k_1(u_1^{n+1})'(\lambda) = \alpha_1\, u_2^n(\lambda) + \beta_1\, k_2(u_2^n)'(\lambda) \end{cases}$$

$$\begin{cases} -k_2(u_2^{n+1})'' = f \quad \text{in} \quad (\lambda, b) \\ u_2^{n+1}(b) = u_D(b) \\ \alpha_2\, u_2^{n+1}(\lambda) - \beta_2\, k_2(u_2^{n+1})'(\lambda) = \alpha_2\, u_1^{n+1}(\lambda) - \beta_2\, k_1(u_1^{n+1})'(\lambda) \end{cases}$$

where $u_2^0(\lambda), (u_2^0)'(\lambda)$ are arbitrarily chosen.

To analyze the convergence of this algorithm, it suffices by linearity to consider homogeneous problem ($f \equiv 0, u_D \equiv 0$) and to analyze convergence to zero. Solving successively the above ODEs and denoting $d_1{=}\lambda{-}a, d_2{=}b{-}\lambda$, we obtain $u_1^{n+2}(\lambda){=}\rho(\alpha_1, \beta_1, \alpha_2, \beta_2)u_1^n(\lambda)$, with the convergence rate:

$$\rho(\alpha_1, \beta_1, \alpha_2, \beta_2) = \frac{\alpha_1 d_2 - \beta_1 k_2}{\alpha_1 d_1 + \beta_1 k_1} \cdot \frac{\alpha_2 d_1 - \beta_2 k_1}{\alpha_2 d_2 + \beta_2 k_2} \tag{5}$$

4.1 Aitken-Dirichlet-Neumann algorithm

For he D-N algorithm the convergence rate becomes: $\rho_{DN} = -d_2/d_1 \cdot k_1/k_2$. Therefore, the convergence of the D-N algorithm is determined only by the ratios of subdomains lengths and of permeabilities. These values being fixed, if the algorithm diverges, it is not possible to adjust any parameter to achieve convergence. The only one possibility is the inter-changing of the interface conditions, but this technique is not easy to handle in practice, where complex basins presenting many porous blocks with extreme contrasts in permeability have to be taken into account. Moreover, in some situations the inter-changing of the interface conditions may lead to an ill-posed problem (only Neumann conditions on all boundaries of a subdomain), which is the typical "danger" of the D-N algorithm.

To transform the D-N algorithm into an attractive algorithm, we use the Aitken acceleration of the traces of the iterative solutions on the common interface. This method is based on the linear behavior of the error at interfaces for the D-N algorithm applied to a linear elliptic operator, as it is here. Indeed, as shown by Garbey and Tromeur-Dervout [2002] in linear cases, the error of the multiplicative D-N algorithm satisfies:

$$u_2^{n+1}(\lambda) - u_2^{\infty}(\lambda) = \delta(u_2^n(\lambda) - u_2^{\infty}(\lambda)) \text{ for all } n \in \mathbb{N} \tag{6}$$

The first step of the Aitken technique is to compute the damping factor δ. For the considered problem, its value is already known (it is exactly ρ_{DN}). Then, we can pass directly to the Aitken acceleration step, which gives the exact value $u_2^{\infty}(\lambda)$ from (6) as follows: $u_2^{\infty}(\lambda){=}(u_2^1(\lambda){-}\delta u_2^0(\lambda))/(1{-}\delta)$, after one D-N iteration. An additional iteration suffices to obtain solution in all domain. In conclusion, for the considered problem, we need only two iterations of D-N algorithm. For other problems, the second step remains non-changed, but only the first step (computation of δ) is modified. This can be done in analytical way, when the operator is still relatively simple. Elsewhere, one can compute the damping factor δ numerically, by performing two iterations of D-N algorithm (and using the obtained iterative solutions on the interface in (6), for $n = 0, 1$). Therefore, in the general linear case, we need three iterations of D-N algorithm to obtain the exact solutions in all domain.

4.2 Optimized Robin-Method

In 1-D framework, the coefficients that minimize the convergence rate are obtained immediately from (5). Indeed, considering $\beta_1=\beta_2=1$, we obtain $\rho_{ORC}=0$ for the following optimal coefficients: $\alpha_{1,opt} = k_2/d_2, \alpha_{2,opt} = k_1/d_1$. Consequently, after only one iteration one obtains exact interface values. As for the A-D-N algorithm, an additional iteration (using exact interface values) allows the complete computation of the solution. Then, for the considered problem, this method requires only two iterations.

If we have a different problem to solve, the convergence rate expression must be analytically deduced and one obtains a relation similar to (5). For usual operators, optimal coefficients can be directly deduced from such an expression in order to obtain a null convergence rate. Then, the exact solution is still obtained after only two iterations. However, a more complicated operator can lead to a more complicated expression of the convergence rate, and it is possible to be not able to deduce analytically the optimal coefficients or/and the optimal rate is not zero. In this case, a numerical optimization procedure can be used or/and the method is not exact, but iterative.

5 Convergence in 2-D framework

Consider now the problem (2)-(3) in 2-D framework. Firstly, we consider an infinite domain $\Omega = \mathbb{R}^2$, with $\Omega_1 = (-\infty, 0) \times \mathbb{R}$, $\Omega_2 = (0, \infty) \times \mathbb{R}$ and suppose that the solution is bounded. Using the Fourier transform in the y direction (with ξ the frequency variable), the Robin-Method yields in Fourier space:

$$\begin{cases} -k_1[(\widehat{u}_1^{n+1})_{xx}(x,\xi) - \xi^2 \widehat{u}_1^{n+1}(x,\xi)] = \widehat{f}(x,\xi), \text{ in } (-\infty, 0) \times \mathbb{R} \\ \alpha_1 \widehat{u}_1^{n+1}(0,\xi) + \beta_1 k_1 (\widehat{u}_1^{n+1})_x(0,\xi) = \alpha_1 \widehat{u}_2^{n}(0,\xi) + \beta_1 k_2 (\widehat{u}_2^{n})_x(0,\xi), \ \xi \in \mathbb{R} \end{cases}$$

$$\begin{cases} -k_2[(\widehat{u}_2^{n+1})_{xx}(x,\xi) - \xi^2 \widehat{u}_2^{n+1}(x,\xi)] = \widehat{f}(x,\xi), \text{ in } (0, \infty) \times \mathbb{R} \\ \alpha_2 \widehat{u}_2^{n+1}(0,\xi) - \beta_2 k_2 (\widehat{u}_2^{n+1})_x(0,\xi) = \alpha_2 \widehat{u}_1^{n+1}(0,\xi) - \beta_2 k_1 (\widehat{u}_1^{n+1})_x(0,\xi), \ \xi \in \mathbb{R} \end{cases}$$

As in 1-D framework, solving successively the above ODEs with boundedness conditions for $\widehat{u}_1^{n+1}$, $\widehat{u}_2^{n+1}$, we obtain the convergence rate:

$$\rho(\alpha_1, \beta_1, \alpha_2, \beta_2, \xi) = \frac{\alpha_1 - \beta_1 k_2 |\xi|}{\alpha_1 + \beta_1 k_1 |\xi|} \cdot \frac{\alpha_2 - \beta_2 k_1 |\xi|}{\alpha_2 + \beta_2 k_2 |\xi|}, \ \forall \xi \in \mathbb{R} \tag{7}$$

5.1 Aitken-Dirichlet-Neumann algorithm

For the D-N algorithm, the convergence rate becomes: $\rho_{DN}(\xi) = -k_1/k_2$, for all $\xi \in \mathbb{R}$, i.e. all frequency components have the same convergence rate. Then, the algorithm converges or diverges according to the distribution of interface conditions in the two subdomains. This convergence rate being also valid in the physical space, the Aitken acceleration can be applied directly to the iterative solutions in each point within the subdomains.

However, the results obtained in analysis of unbounded cases can be not relevant for the bounded case. We can illustrate this situation, considering the domain is bounded only in x direction, as for instance, $\Omega = (0,1) \times \mathbb{R}$, with common interface $\Gamma = \{\lambda\} \times \mathbb{R}$, $0 < \lambda < 1$. On the lateral boundaries, homogeneous Dirichlet conditions are imposed. Using the same Fourier analysis as above, one obtains the convergence rate:

$$\rho_{DN}(\xi) = \frac{k_1}{k_2} \cdot \frac{1 - e^{2|\xi|}}{1 + e^{2|\xi|}} \cdot \frac{1 + e^{-\lambda|\xi|}}{1 - e^{-\lambda|\xi|}}, \quad \forall \xi \in \mathbb{R} \tag{8}$$

In this semi-bounded case, each frequency component has its own own linear damping factor. The Aitken acceleration is no longer possible in the physical space, but can be applied for each frequency component by using the numerical 1-D procedure described in Section 4.1.

For realistic domains (bounded in both directions), we use the Aitken technique as follows: let $\{P_i\}_{i=1,\dots,N}$ the discrete representation of the common interface Γ (we consider a regular discretization in Γ direction). When D-N algorithm is applied, at discrete level we obtain the traces of iterative discrete solutions, denoted $\{u_{2,i}^n\}_{i=1,\dots,N}$ which are transformed in periodic signals, and then represented in mode's Fourier space. Then, the Aitken acceleration is possible for each mode, since each mode is damped linearly. Using accelerated modes, the solution is recomposed in the physical space.

5.2 Optimized Robin-Method

Let β_1=β_2=1 in (7). Applying the technique introduced by Nataf and coauthors in papers cited above, for the considered problem, we obtain optimized Robin coefficients $\alpha_{1,opt}, \alpha_{2,opt}$ by solving the min-max problem:

$$\min_{\alpha_1,\alpha_2 > 0} \left(\max_{\xi \in \mathbb{R}} \left| \frac{\alpha_1 - k_2|\xi|}{\alpha_1 + k_1|\xi|} \cdot \frac{\alpha_2 - k_1|\xi|}{\alpha_2 + k_2|\xi|} \right| \right) \tag{9}$$

Since real computations are performed on bounded domains and discretized operators, the range of ξ can be bounded in an interval (ξ_{min}, ξ_{max}). Even with this simplification, the problem (9) is still difficult to solve analytically. The method given by Faille et al. [2001] divides the previous problem into two auxiliary min-max problems for α_1 and respectively for α_2, which are formulated as for homogeneous media. The problems are similar to optimal parameter search in ADI Peaceman-Racheford method and have the solutions:

$$\alpha_{1,opt} = k_2\sqrt{\xi_{min}\xi_{max}} \quad , \quad \alpha_{2,opt} = k_1\sqrt{\xi_{min}\xi_{max}} \tag{10}$$

Another method retained here is to solve (9) at discrete level, for the heterogeneous case, by considering $\alpha_1{=}\alpha_2{\equiv}\alpha$. For instance, we replaced $L^\infty-$norm by the discrete L^1-norm of frequency components in (ξ_{min}, ξ_{max}) and then, the α_{opt,L^1} is obtained by seeking the minimum for a fine mesh for α.

6 Numerical results

We consider the problem (2)-(3) for $\Omega_1 = (0,\pi)^2$, $\Omega_2 = (\pi, 2\pi) \times (0,\pi)$, $f(x,y) = 2k_1k_2 \sin x \sin y$ and the Dirichlet condition $u = 10$. The exact solution is $u_1(x,y) = 10 + k_2 \sin x \sin y$, $u_2(x,y) = 10 + k_1 \sin x \sin y$. We consider $k_1 = 10$, $k_2 = 1$, and $u_2^0(\pi, y) = (u_2^0)_x(\pi, y) = 0$, for $y \in (0,\pi)$.

Figure 1 shows the evolution of the error with respect to the iterations for the investigated algorithms. One can observe that even if the D-N algorithm diverges rapidly, the Aitken acceleration (A-D-N on the Figure 1) allows a fast convergence of the algorithm.

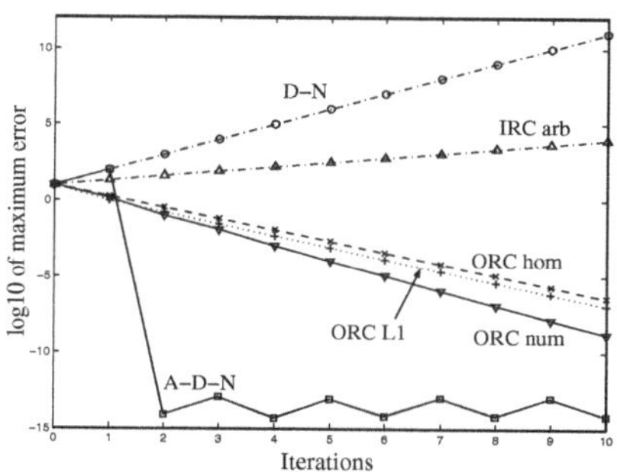

Figure 1.

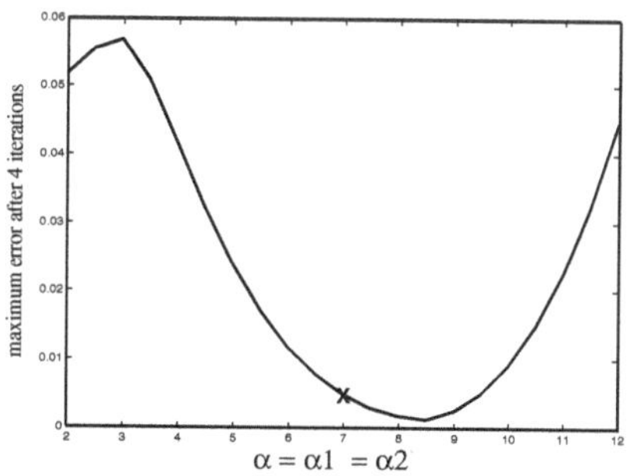

Figure 2.

The Robin-Method can diverge if the RC are chosen arbitrarily, as for instance, using $\alpha_i = k_i$ (curve IRCarb). Now, let us consider optimized RC. Two optimized Robin-Methods have been investigated: the ORChom method which gives $\alpha_{1,opt} = 3$, $\alpha_{2,opt} = 30$ from (10) and the ORCL1 method which gives $\alpha_{opt,L^1} \simeq 7$. For the two methods, the obtained results are relatively close: in 10 iterations the error is reduced by a factor of 10^7.

Both ORC are obtained by solving (9). However, this problem being deduced with a Fourier analysis at continuous level for unbounded domain, the obtained ORC are not necessarily optimal at the discrete level for the bounded domain used in the numerical experiment. To verify how the coefficients obtained with ORCL1 method approach the optimal discrete coefficients, Figure

2 shows the error reduction obtained numerically, after 4 iterations, using various values for $\alpha = \alpha_1 = \alpha_2$. We observe that the coefficient obtained by the optimization procedure (~ 7), is not very close of the discrete optimum, but it can give values which are effective for numerical experiments. Using the discrete optimum (~ 8.5), the obtained error after 10 iterations (curve ORCnum) is better with a factor 10^2 with respect to the ORCL1 method.

7 Conclusions

We studied two non-overlapping methods for flow problem in heterogeneous porous media: A-D-N method and Optimized Robin-Method. Both methods use the Fourier analysis but at different level. In the A-D-N method, the discrete solution is represented in modes's space, accelerated and transformed back in the physical space. For the second method, the Fourier analysis is used only to determine optimized Robin coefficients. Both methods show good convergence properties, especially the Aitken method. However, there are several possibilities to improve the Optimized Robin-Method, as the use of a Krylov acceleration, or the use of second order optimized interface conditions (OIC2). It is also possible to apply Aitken acceleration to the Robin-Method (not necessarily optimized), because it is still linear. The two methods, although studied here for only 2 subdomains have been already used to an arbitrary number of subdomains, in 1-D or 2-D framework (Calugaru and Tromeur-Dervout, Proc. Parallel CFD 2003, to appear, for the A-D-N method and Faille et al. [2001] for the Robin-Method). We are currently investigating a comparison of these extensions in the context of parallelization methods.

References

D.-G. Calugaru, J.-M. Crolet, A. Chambaudet, and F. Jacob. Radon transport as an indicator of seismic activity. An algorithm for inverse problems. *Computational Methods in Water Ressources*, 47:631–638, 2002.

B. Després, P. Joly, and J. E. Roberts. A domain decomposition method for the harmonic Maxwell equations. In *Iterative methods in linear algebra (Brussels, 1991)*, pages 475–484, Amsterdam, 1992. North-Holland.

I. Faille, E. Flauraud, F. Nataf, F. Schneider, and F. Willien. Optimized interface conditions for sedimentary basin modeling. In N. Debit et al., editor, *Proc. of the 13th Int. Conf. on DDM*, pages 461–468, 2001.

M. J. Gander, F. Magoulès, and F. Nataf. Optimized Schwarz methods without overlap for the Helmholtz eq. *SIAM J. Sci. Comput.*, 24(1):38–60, 2002.

M. Garbey and D. Tromeur-Dervout. On some Aitken acceleration of the Schwarz method. *Int. J. for Num. Meth. in Fluids*, 40(12):1493–1513, 2002.

F. Nataf, F. Rogier, and E. de Sturler. Optimal interface conditions for domain decomposition methods. Technical Report 301, CMAP, 1994.

Domain Embedding/Controllability Methods for the Conjugate Gradient Solution of Wave Propagation Problems

H.Q. Chen[1], R. Glowinski[2], J. Periaux[3], and J. Toivanen[4]

[1] University of Nanjing, Institute of Aerodynamics, Nanjing 210016, PR of China
[2] University of Houston, Depart. of Math., Houston, Texas 77204-3008, USA
[3] Dassault-Aviation, Direction de la Prospective, 92552 St Cloud Cedex, France
[4] University of Jyvaskyla, Math. Inf. Tech., Jyvaskyla Finland

Summary. The main goal of this paper is to discuss the numerical simulation of propagation phenomena for time harmonic electromagnetic waves by methods combining controllability and fictitious domain techniques. These methods rely on distributed Lagrangian multipliers, which allow the propagation to be simulated on an obstacle free computational region using regular finite element meshes essentially independent of the geometry of the obstacle and on a controllability formulation which leads to algorithms with good convergence properties to time-periodic solutions. This novel methodology has been validated by the solutions of test cases associated to non trivial geometries, possibly non-convex. The numerical experiments show that the new method performs as well as the method discussed in Bristeau et al. [1998] where obstacle fitted meshes were used.

1 Introduction

Lagrange multiplier based fictitious domain methods have proved to be efficient techniques for the solution of viscous flow problems with moving boundaries (see Glowinski [2003], Chapter 8, Glowinski et al. [2001]). The main goal of this article is to discuss the generalization of this methodology to the simulation of wave propagation phenomena. A motivation for using the fictitious domain approach is that it allows–to some extent–the use of uniform meshes, which is clearly an advantage far from the scatters. In order to capture efficiently time-periodic solutions, the fictitious domain methodology is coupled to exact controllability methods close to those utilized in Bristeau et al. [1998], Glowinski and Lions [1995]. Various formulations of a wave propagation model problem, including a fictitious domain one, will be discussed in Sections 3 and 4. The computation of the gradient of a cost function associated to the control formulation will be briefly addressed in Section 5. The conjugate gradient solution of the control problem will be discussed in Section

6, while the space/time discretization will be discussed in Section 7. Finally, the results of numerical experiments will be presented in Section 8.

2 Formulation of the wave-propagation problem

Let ω be a bounded domain of $\mathbb{R}^d$ $(d = 2, 3)$; we denote by γ the boundary $\partial\omega$ of ω. Consider now $T(> 0)$. We are looking for the $T-$periodic solutions of the following wave equation:

$$\varphi_{tt} - \Delta\varphi = 0 \text{ in } (\mathbb{R}^d \setminus \overline{\omega}) \times (0, T), \ \varphi = g \text{ on } \gamma \times (0, T), \tag{1}$$

completed by additional conditions such as $\lim_{|x|\to+\infty} \varphi(x, t) = 0$. The time-periodicity conditions take then the following form:

$$\varphi(0) = \varphi(T), \ \varphi_t(0) = \varphi_t(T), \tag{2}$$

where $\varphi(t)$ denotes the function $x \to \varphi(x, t)$. From a computational point of view, we imbed ω in a bounded simple-shape domain Ω of boundary Γ_{ext} (see Figure 1) and consider

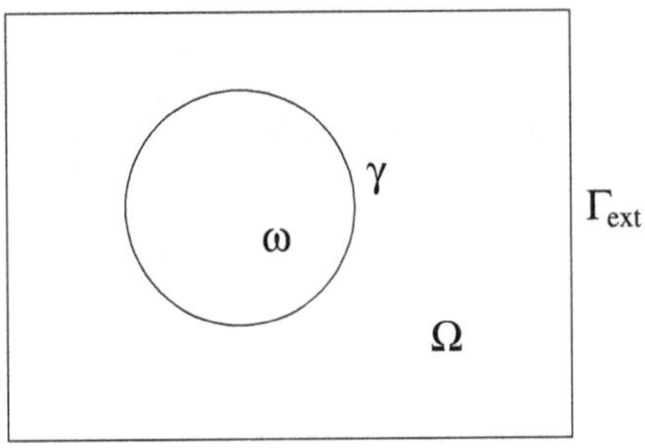

Fig. 1. Imbedding of ω

the following wave problem:

$$\begin{aligned} \varphi_{tt} - \Delta\varphi = 0 \text{ in } (\Omega \setminus \overline{\omega}) \times (0, T), \\ \varphi = g \text{ on } \gamma \times (0, T), \ \frac{\partial\varphi}{\partial n} + \frac{\partial\varphi}{\partial t} = 0 \text{ on } \Gamma_{ext} \times (0, T), \end{aligned} \tag{3}$$

$$\varphi(0) = \varphi(T), \ \varphi_t(0) = \varphi_t(T). \tag{4}$$

3 A fictitious domain formulation of problem (3), (4)

Problem (3), (4) is equivalent to

Find $\{\varphi, \lambda\}$ *verifying:*

$$\int_{\Omega} \varphi_{tt} v dx + \int_{\Omega} \boldsymbol{\nabla}\varphi \cdot \boldsymbol{\nabla} v dx + \int_{\Gamma_{ext}} \frac{\partial \varphi}{\partial t} v d\Gamma + \int_{\omega} \lambda v dx = 0, \ \forall v \in H^1(\Omega),$$
$$\int_{\omega} \mu(\varphi - \tilde{g}) dx = 0, \ \forall \mu \in L^2(\omega), \tag{5}$$

$$\varphi(0) = \varphi(T), \varphi_t(0) = \varphi_t(T), \tag{6}$$

$\tilde{g}(t)$ being an ω-extension of $g(t)$ such that $\tilde{g}(t) \in H^1(\Omega)$.

4 A virtual control/least squares formulation of problem (5), (6)

A virtual control/least squares formulation of problem (5), (6) reads as follows:

$$\begin{gathered} \textit{Find } \mathbf{e} \in E \textit{ such that} \\ J(\mathbf{e}) \leq J(\mathbf{w}), \ \forall \mathbf{w} \ (= \{w_0, w_1\}) \ \in E, \end{gathered} \tag{7}$$

with $E = H^1(\Omega) \times L^2(\Omega)$, and

$$J(\mathbf{w}) = \frac{1}{2}\int_{\Omega} [|\nabla(w_0 - y(T))|^2 + |w_1 - y_t(T)|^2] dx, \tag{8}$$

y being the solution for a.e. t of

$$\int_{\Omega} y_{tt} z dx + \int_{\Omega} \nabla y \cdot \nabla z dx + \int_{\Gamma_{ext}} \frac{\partial y}{\partial t} z d\Gamma + \int_{\omega} \lambda z dx = 0, \ \forall z \in H^1(\Omega),$$
$$\int_{\omega} \mu(y - \tilde{g}) dx = 0, \ \forall \mu \in L^2(\omega), \tag{9}$$

$$y(0) = w_0, \ y_t(0) = w_1. \tag{10}$$

Problem (7), being linear-quadratic, can be solved by a conjugate gradient algorithm operating in E. To implement such an algorithm we need to know $J'(\mathbf{w})$, $\forall \mathbf{w} \in E$. The derivation of $J'(\mathbf{w})$ will be addressed in the following section, while the conjugate gradient solution of problem (7)-(10) will be discussed in Section 6.

5 Derivation of $J'(\mathbf{w})$

It can be shown that if we define p by

$$\int_{\Omega} p_{tt} z dx + \int_{\Omega} \nabla p \cdot \nabla z dx - \int_{\Gamma_{ext}} \frac{\partial p}{\partial t} z d\Gamma + \int_{\omega} \lambda^* z dx = 0,$$
$$\forall z \in H^1(\Omega),$$
$$\int_{\omega} p\mu dx = 0, \ \forall \mu \in L^2(\omega),$$
$$p(T) = y_t(T) - w_1,$$
$$\int_{\Omega} p_t(T) z dx = \int_{\Gamma_{ext}} p(T) z d\Gamma - \int_{\Omega} \boldsymbol{\nabla}(y(T) - w_0) \cdot \boldsymbol{\nabla} z dx, \ \forall z \in H^1(\Omega), \tag{11}$$

then we have the following representation for $J'(\mathbf{w})$:

$$< J'(\mathbf{w}), \mathbf{v} > = \int_\Omega \nabla(w_0 - y(T)) \cdot \nabla v_0 dx - \int_\Omega p_t(0) v_0 dx + \int_{\Gamma_{ext}} p(0) v_0 d\Gamma + \int_\Omega (w_1 - y_t(T)) v_1 dx + \int_\Omega p(0) v_1 dx, \; \forall \mathbf{v} = \{v_0, v_1\} \in E. \tag{12}$$

Relations (11) and (12) are largely "formal"; however it is worth mentioning that the discrete variants of them make sense and lead to algorithms with fast convergence properties.

6 Conjugate gradient solution of problem (7)

As in Section 4, we suppose that $E = H^1(\Omega) \times L^2(\Omega)$. A *conjugate gradient algorithm* for the solution of (7) is given by:

Step 0: Initialization

$$\mathbf{e}^0 = \{e_0^0, e_1^0\} \in E \text{ is given.} \tag{13}$$

Solve the following forward wave problem:

$$\begin{gathered} \int_\Omega y_{tt}^0 z dx + \int_\Omega \boldsymbol{\nabla} y^0 \cdot \boldsymbol{\nabla} z dx + \int_{\Gamma_{ext}} \frac{\partial y^0}{\partial t} z d\Gamma + \int_\omega \lambda^0 z dx = 0, \; \forall z \in H^1(\Omega), \\ \int_\omega \mu (y^0 - \tilde{g}) dx = 0, \; \forall \mu \in L^2(\omega), \\ y^0(0) = e_0^0, \; y_t^0(0) = e_1^0. \end{gathered} \tag{14}$$

Solve next the following backward wave-problem

$$\begin{gathered} \int_\Omega p_{tt}^0 z dx + \int_\Omega \boldsymbol{\nabla} p^0 \cdot \boldsymbol{\nabla} z dx - \int_{\Gamma_{ext}} \frac{\partial p^0}{\partial t} z d\Gamma + \int_\omega \lambda^{*0} z dx = 0, \; \forall z \in H^1(\Omega), \\ \int_\omega p^0 \mu dx = 0, \; \forall \mu \in L^2(\omega), \\ p^0(T) = y_t^0(T) - e_1^0, \\ \int_\Omega p_t^0(T) z dx = \int_{\Gamma_{ext}} p^0(T) z d\Gamma - \int_\Omega \boldsymbol{\nabla}(y^0(T) - e_0^0) \cdot \boldsymbol{\nabla} z dx, \; \forall z \in H^1(\Omega). \end{gathered} \tag{15}$$

Next, define $\mathbf{g}^0 = \{g_0^0, g_1^0\} \in E \; (= H^1(\Omega) \times L^2(\Omega))$ *by*

$$\begin{gathered} \int_\Omega \boldsymbol{\nabla} g_0^0 \cdot \boldsymbol{\nabla} z dx = \int_\Omega \boldsymbol{\nabla}(e_0^0 - y^0(T)) \cdot \boldsymbol{\nabla} z dx - \int_\Omega p_t^0(0) z dx + \int_{\Gamma_{ext}} p^0(0) z d\Gamma, \; \forall z \in H^1(\Omega), \\ g_1^0 = p^0(0) + e_1^0 - y_t^0(T), \end{gathered} \tag{16}$$

and then

$$\mathbf{w}^0 = \mathbf{g}^0. \ \square \tag{17}$$

For $n \geq 0$, suppose that $\mathbf{e}^n$, $\mathbf{g}^n$, $\mathbf{w}^n$ are known; we compute their updates $\mathbf{e}^{n+1}$, $\mathbf{g}^{n+1}$, $\mathbf{w}^{n+1}$ as follows:

Step 1: Descent

Solve

$$\begin{gathered}\int_\Omega \overline{y}^n_{tt} z dx + \int_\Omega \nabla \overline{y}^n \cdot \nabla z dx + \int_{\Gamma_{ext}} \frac{\partial \overline{y}^n}{\partial t} z dx \int_\omega \overline{\lambda}^n z dx = 0, \ \forall z \in H^1(\Omega), \\ \int_\omega \mu \overline{y}^n dx = 0, \ \forall \mu \in L^2(\omega), \\ \overline{y}^n(0) = w^n_0, \ \overline{y}^n_t(0) = w^n_1. \end{gathered} \tag{18}$$

Solve the backward wave problem

$$\begin{gathered}\int_\Omega \overline{p}^n_{tt} z dx + \int_\Omega \nabla \overline{p}^n \cdot \nabla z dx - \int_{\Gamma_{ext}} \frac{\partial \overline{p}^n}{\partial t} z d\Gamma + \int_\omega \overline{\lambda}^{n*} z dx = 0, \ \forall z \in H^1(\Omega), \\ \int_\Omega \overline{p}^n \mu dx = 0, \ \forall \mu \in L^2(\omega), \end{gathered} \tag{19}$$

with $\overline{p}^n(T)$ and $\overline{p}^n_t(T)$ given by

$$\begin{gathered}\overline{p}^n(T) = \overline{y}^n_t - w^n_1, \\ \int_\omega \overline{p}^n_t(T) z dx = \int_\Gamma \overline{p}^n(T) z dx - \int_\Omega \nabla(\overline{y}^n(T) - w^n_0) \cdot \nabla z dx, \ \forall z \in H^1(\Omega),\end{gathered}$$

respectively.

Next define $\overline{\mathbf{g}}^n = \{\overline{g}^n_0, \overline{g}^n_1\} \in H^1(\Omega) \times L^2(\Omega)$ by

$$\begin{gathered}\int_\Omega \nabla \overline{g}^n_0 \cdot \nabla z dx = \int_\Omega \nabla (w^n_0 - \overline{y}^n(T)) \cdot \nabla z dx - \int_\Omega \overline{p}^n_t(0) z dx \\ + \int_{\Gamma_{ext}} \overline{p}^n(0) z d\Gamma, \ \forall z \in H^1(\Omega), \end{gathered} \tag{20}$$

$$\overline{g}^n_1 = \overline{p}^n(0) + w^n_1 - \overline{y}^n_t(T),$$

and then ρ_n by

$$\rho_n = \int_\Omega [|\nabla g^n_0|^2 + |g^n_1|^2] dx \Big/ \int_\Omega (\nabla \overline{g}^n_0 \cdot \nabla w^n_0 + \overline{g}^n_1 w^n_1) dx. \tag{21}$$

We update then $\mathbf{e}^n$ and $\mathbf{g}^n$ by

$$\mathbf{e}^{n+1} = \mathbf{e}^n - \rho_n \mathbf{w}^n, \tag{22}$$

$$\mathbf{g}^{n+1} = \mathbf{g}^n - \rho_n \overline{\mathbf{g}}^n. \tag{23}$$

Step 2: Test of the convergence and construction of the new descent direction

If $\int_{\Omega}(|\nabla g_0^{n+1}|^2+|g_1^{n+1}|^2)dx \Big/ \int_{\Omega}(|\nabla g_0^{n}|^2+|g_1^{n}|^2)dx \leq \varepsilon$, *take* $\mathbf{e}=\mathbf{e}^{n+1}$; *else, compute*

$$\gamma_n = \int_{\Omega}(|\nabla g_0^{n+1}|^2+|g_1^{n+1}|^2)dx \Big/ \int_{\Omega}(|\nabla g_0^{n}|^2+|g_1^{n}|^2)dx \tag{24}$$

and update $\mathbf{w}^n$ *by*

$$\mathbf{w}^{n+1}=\mathbf{g}^{n+1}+\gamma_n\mathbf{w}^n. \ \square \tag{25}$$

Do $n=n+1$ *and go to (18).*

Algorithm (13)-(25) requires the solution of two waves problems at each iteration and also of an elliptic problem such as (20). For more comments see Bristeau et al. [1998], Glowinski and Lions [1995].

7 Finite difference/finite element implementation

Compared to what has been done in Bristeau et al. [1998], Glowinski and Lions [1995] the main difficulty is clearly the numerical implementation of the distributed Lagrange multiplier based techniques used to force Dirichlet boundary conditions. We shall consider the forward wave equations only since the backward ones can be treated by similar methods. Dropping the superscript, the forward wave problems to be solved are all of the following type:

$$\int_{\Omega} y_{tt} z dx + \int_{\Omega} \nabla y \cdot \nabla z dx + \int_{\Gamma_{ext}} \frac{\partial y}{\partial t} z d\Gamma + \int_{\omega} \lambda z dx = 0, \ \forall z \in H^1(\Omega), \tag{26}$$

$$\int_{\Omega} \mu(y-\tilde{g})dx = 0, \ \forall \mu \in L^2(\omega), \tag{27}$$

$$y(0)=e_0, \ y_t(0)=e_1. \tag{28}$$

Approximating spaces $L^2(\Omega)$ and $H^1(\Omega)$ are pretty classical tasks. Let us suppose that Ω is a bounded polygonal domain of $\mathbb{R}^2$; we introduce a triangulation $\mathcal{T}_h$ of Ω and define a space V_h approximating both $H^1(\Omega)$ and $L^2(\Omega)$ by

$$V_h = \{z_h | z_h \in C^0(\overline{\Omega}), z_h|_T \in P_1, \ \forall T \in \mathcal{T}_h\}. \tag{29}$$

Next, in order to implement the fictitious domain methodology, we proceed as follows: we introduce first a set Σ_h of control points belonging to $\overline{\omega}$ and defined as follows:

$$\Sigma_h = \Sigma_h^{\omega} \cup \Sigma_h^{\gamma}, \tag{30}$$

where, in (30), Σ_h^{ω} is the set of the vertices of $\mathcal{T}_h$ belonging to ω and whose distance at γ is more than Ch, C being a positive constant, and where Σ_h^{γ} is a set of points of γ. We suppose that $\Sigma_h = \{p_j\}_{j=1}^{N_h}$, where $N_h = Card(\Sigma_h)$.

Following Glowinski [2003], Glowinski et al. [2001], we shall use as "multiplier" space, Λ_h defined by

$$\Lambda_h = \{\mu_h | \mu_h = \Sigma_{j=1}^{N_h} \mu_j \delta(x - p_j), \ \mu_j \in \mathrm{I\!R}\}. \tag{31}$$

Collecting the above results leads to the following *collocation based* approximation of problem (26)-(28):

$$\int_\Omega \frac{y_h^{n+1} + y_h^{n-1} - 2y_h^n}{\tau^2} z_h dx + \int_\Omega \nabla y_h^n \cdot \nabla z_h dx + \int_{\Gamma_{ext}} \frac{y_h^{n+1} - y_h^{n-1}}{2\tau} z_h d\Gamma$$
$$+ \sum_{j=1}^{N_h} \lambda_j^{n+1} z_h(p_j) = 0, \ \forall z_h \in V_h, \tag{32}$$

$$y_h^{n+1}(p_j) - \tilde{g}_h(p_j, (n+1)\tau) = 0, \ \forall j = 1, ..., N_h, \tag{33}$$

$$y_h^0 = e_{0h}, \ y_h^1 - y_h^{-1} = 2\tau e_{1h}; \tag{34}$$

in (32)-(34), $\tilde{g}_h$, e_{0h}, e_{1h} are approximations–all belonging to V_h–of $\tilde{g}$, e_0 e_1, respectively.

The *finite dimensional linear variational problem* (32), (33) is of the form

$$\begin{aligned} Ax + B^t\lambda &= b, \\ Bx &= c, \end{aligned} \tag{35}$$

where matrix A is symmetric and positive definite. To solve *the saddle point problem* (35), we can use for example the *Uzawa/conjugate gradient algorithms* discussed in, e.g., Glowinski and Lions [1995], Fortin and Glowinski [1982]. Suppose, for simplicity, that functions g and $\tilde{g}$ are time independent. Taking $z_h = (y_h^{n+1} - y_h^{n-1})/2\tau$ in (32) we can easily show that scheme (7.7)-(7.9) is stable if τ verifies a stability condition such as

$$\tau \leq c^{-1}h, \tag{36}$$

where c (which has the dimension of a velocity) is a positive constant *independent* of ω. Related distributed Lagrange multiplier based fictitious domain methods for the solution of wave propagation problems with obstacles are discussed in Bokil [2004].

8 Numerical experiments

In order to validate the methods discussed in the above sections, we will address the solution of three test problems already solved in Bristeau et al. [1998] and Glowinski and Lions [1995] using controllability and obstacle fitted finite element meshes. These problems concern the scattering of planar incident waves by a disk, a convex ogive, and a non-convex reflector (air-intake like).

First Test Problem: For this problem, ω is a disk of radius .25m. This disk is illuminated by an incident planar wave of wavelength .125m (which corresponds to a $2.4 \times GHz$ frequency) coming from the right, the incidence angle with horizontal being zero. The artificial boundary Γ_{ext} is located at a 3 wave length distance from ω. On Figure 2, we have visualized the uniform finite element triangulation used over Ω to define the discrete spaces V_h and Λ_h. It consists of 19,881 vertices and 39,200 triangles; the number of control points used to define Λ_h is 19,881. The value of Δt corresponds to 80 time steps per period, the space discretization $h_x = h_y = 8.928571 \cdot 10^{-3}$ and $\Delta s = 1.6071 \cdot 10^{-2}$ the length between two adjacent point on the disk. The decay of the discrete cost functional is a function of the number of iterations of the discrete analogue of algorithm (13)-(25) is shown on Figure 3. Comparing to the results reported in Glowinski and Lions [1995] shows that the convergence performances are not modified by the addition of the fictitious domain procedure. The computed scattered field has been visualized on Figure 4.

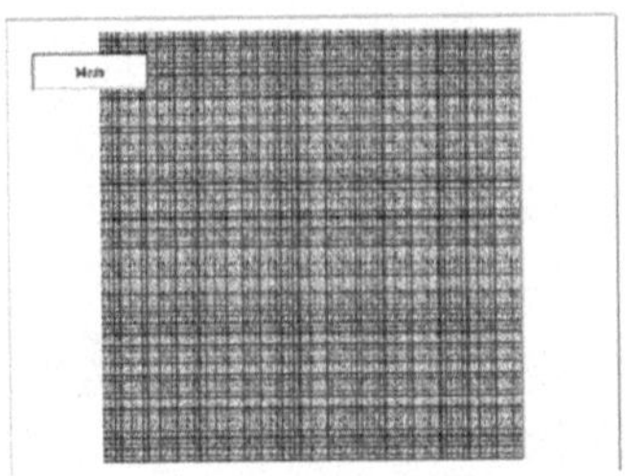

Fig. 2. Uniform finite element triangulation used for the fictitious domain method

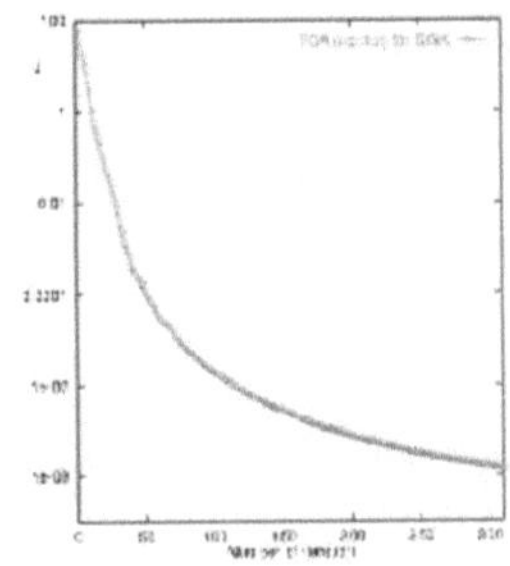

Fig. 3. Disk: convergence of the fictitious domain algorithm

Second Test Problem: For this problem, ω is an ogive-like obstacle of length .6m and thickness .16m, respectively. The wavelength of the incident wave is 0.1m. The artificial boundary is located at a 3 wave-length distance from ω. The finite element triangulation is uniform and has 11,571 vertices and 22,704 triangles. The convergence for a zero degree of incidence monochromatic wave is shown on Figure 5. The imaginary component of the scattered field is shown on Figure 6.

Third Test Problem: Denote by λ the wavelength of the propagation phenomena. For this problem ω is an idealized air intake; it has a semiopen cavity geometry defined by two horizontal plates (length 4λ and thickness equal 0.2λ) and a vertical one (length 1.4λ and thickness $\lambda/5$). We have $f = 1.2GHz$ implying a .25m wavelength. The artificial boundary is located again at a distance of 3λ from ω. The uniform finite element triangulation has 20,202 points and 39,820 triangles. The convergence to the solution for

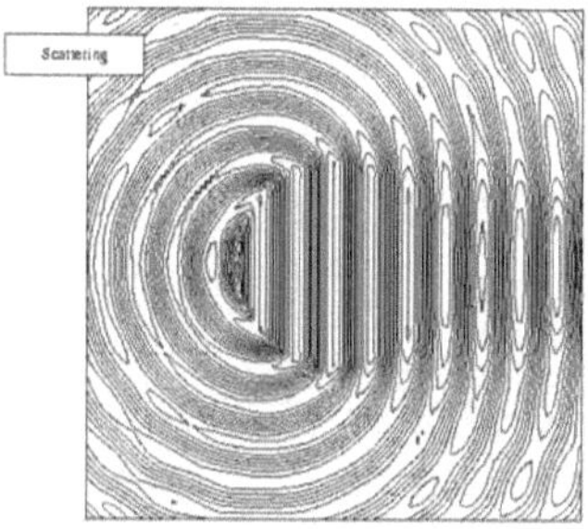

Fig. 4. Disk: visualization of the scattered field

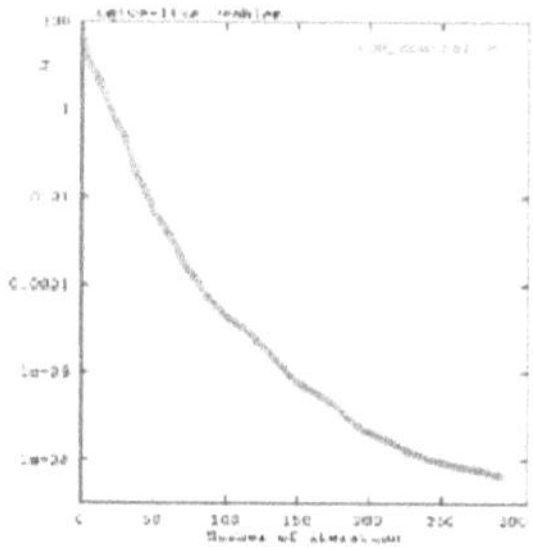

Fig. 5. Ogive: convergence of the fictitious domain algorithm

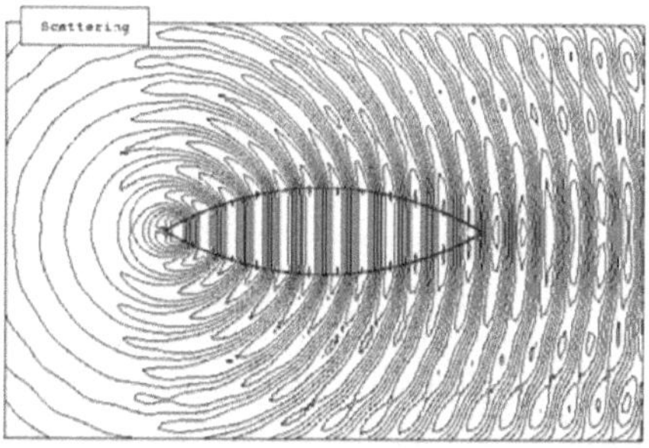

Fig. 6. Ogive: imaginary component of the scattered field

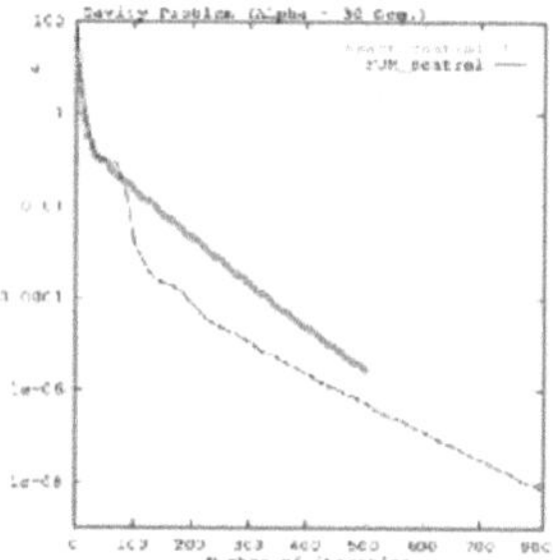

Fig. 7. Idealized air intake: convergence of the fictitious domain algorithm

an illuminating monochromatic wave of incidence $\alpha = 30°$ is shown on Figure 7. We observe from Figure 7 that the convergence of the method combining controllability and fictitious domain is faster than the one of the "pure" controllability method discussed in Glowinski and Lions [1995]. In order to compare the first method with those discussed in Glowinski and Lions [1995], we have visualized on Figures 8 and 9 the scattered field obtained with the methods of Glowinski and Lions [1995] and of this article on Figures 4, 6, and 8. We can observe the extension of the scattered field inside the scatters.

9 Conclusion and Future

The fictitious domain based methods discussed in this article appear to be competitive with the boundary fitted one discussed in Glowinski and Lions [1995]. One of the main advantages of the fictitious domain approach is that it is well-suited to those shape optimization problems with several scatters where we have the shape and position of the obstacles in order to minimize

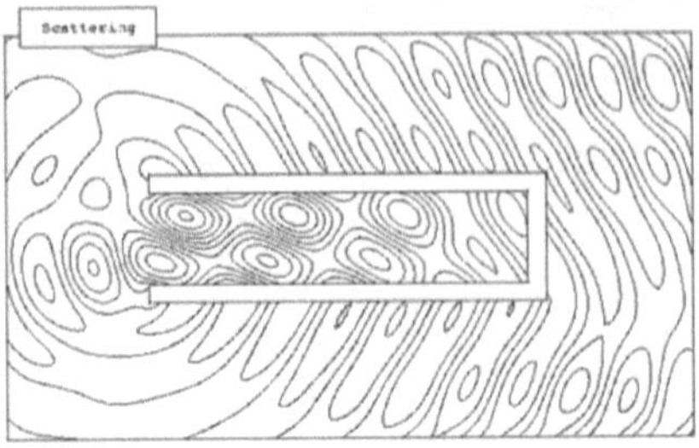

Fig. 8. Idealized air intake: scattered field obtained by the method of reference

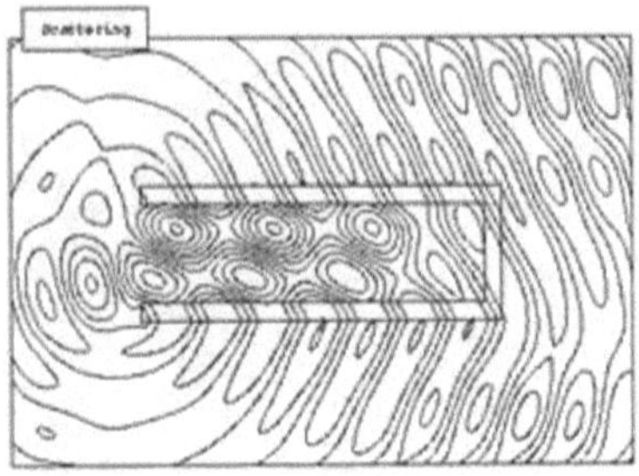

Fig. 9. Idealized air intake: scattered field obtained by the fictitious domain method

for example a Radar Cross section. Only the acoustic wave equation has been considered in this investigation, but we consider generalizing the methods discussed here to Maxwell equations in two and three dimensions.

Acknowledgement. The authors gratefully acknowledge Giovanna Guidoboni for her kind assistance in preparing the paper while visiting University of Houston. We would like also to thank the Association Franco Chinoise pour la Science, la Recherche et la Technologie (AFCSRT) for its support. This work was performed while the first and third author had a PRA SI-O1-O6 from AFCSRT.

References

V. Bokil. *Computational methods for wave propagation problems in unbounded domains.* PhD thesis, University of Houston, Texas, USA, 2004.

M. Bristeau, R. Glowinski, and J.Periaux. Controllability methods for the computation of time periodic solutions; applications to scattering. *Journal of Computational Physics*, 147:265–292, 1998.

M. Fortin and R. Glowinski. *Lagrangiens Augmentés.* Dunod, Paris, 1982.

R. Glowinski. *Finite element methods for incompressible viscous flow*, volume IX of *Handbook of Numerical Analysis.* North-Holland, Amsterdam, 2003.

R. Glowinski and J. L. Lions. Exact and approximate controllability for distributed parameter systems (II). *Acta Numerica*, pages 159–333, 1995.

R. Glowinski, T. Pan, T. Hesla, D. Joseph, and J. Periaux. A fictitious domain approach to the direct numerical simulation of incompressible fluid flow past moving rigid bodies: Application to particulate flow. *J. Comp. Phys.*, pages 363–426, 2001.

An Accelerated Block-Parallel Newton Method via Overlapped Partitioning

Yurong Chen

Lab. of Parallel Computing, Institute of Software, CAS
(http://www.rdcps.ac.cn/~ychen/english.htm)

Summary. This paper presents an overlapped block-parallel Newton method for solving large nonlinear systems. The graph partitioning algorithms are first used to partition the Jacobian into weakly coupled overlapping blocks. Then the simplified Newton iteration is directly performed, with the diagonal blocks and the overlapping solutions assembled in a weighted average way at each iteration. In the algorithmic implementation, an accelerated technique has been proposed to reduce the number of iterations. The conditions under which the algorithm is locally and semi-locally convergent are studied. Numerical results from solving power flow equations are presented to support our study.

1 Introduction

This paper considers the problem of solving the large sparse system of nonlinear equations

$$\mathbf{F}(\mathbf{x}) = \mathbf{0}, \tag{1}$$

where $\mathbf{F}(\mathbf{x}) = (f_1, \ldots, f_N)^T$ is a nonlinear operator from $\mathbb{R}^N$ to $\mathbb{R}^N$. Such systems often arise from scientific and computational engineering problems. It is well-known that Newton methods and its variations (see Ortega and Rheinboldt [1970], etc.) coupled with some direct solution technique such as Gaussian elimination are powerful solvers for these systems when one has a sufficiently good initial guess $\mathbf{x}_0$ and when N is not too large. When the Jacobian is large and sparse, inexact Newton methods (see Dembo et al. [1982], Brown and Saad [1990], Cai and Keyes [2002], etc.) or some kind of nonlinear block-iterative methods (see Zecevic and Siljak [1994],Yang et al. [1997],Chen and Cai [2003], etc.) may be used.

An inexact Newton method (IN) is a generalization of Newton method for solving system (1), in which, each step $\{\mathbf{s}_k\}$ satisfies $\|\mathbf{F}'(\mathbf{x}_k)\mathbf{s}_k + \mathbf{F}(\mathbf{x}_k)\| < \|\mathbf{r}_k\|$, regardless of how $\{\mathbf{s}_k\}$ is determined. In past years, Krylov subspace methods, such as Arnoldi's method (see Saad [1981]), GMRES (see Saad and Schultz [1986]) and so on, have been studied intensively and applied in IN for solving large scale linear systems approximately. This combined method is called inexact Newton-Krylov methods or nonlinear Krylov subspace projection methods. Many works on parallel Newton-Krylov methods have been done by Gropp et al. [2000], Knoll and Keyes [2003], etc.

Parallel nonlinear block-iterative method is another powerful solver for large sparse nonlinear systems, which chiefly consists of block Newton-type and block quasi-Newton methods. The classical nonlinear block-Jacobi algorithm and nonlinear block-Gauss-Seidel algorithm (see Ortega and Rheinboldt [1970]) are two original versions. A block-parallel Newton method via overlapping epsilon decompositions was presented by Zecevic and Siljak [1994]. Yang et al. [1997] described a parallelizable Jacobi-type block Broyden method, and more recently a partially overlapped Broyden method has been proposed by Chen and Cai [2003].

In this paper, we consider a parallelizable block simplified Newton method via overlapped partitioning, which is essentially an additive Schwarz method (block-Jacobi algorithm) with overlapping. In the implementation, an accelerated technique (see Sect. 2) is proposed for each iteration to reduce the number of iterations. Sect. 3 gives the sufficient conditions under which the new method is locally and semi-locally convergent. The numerical results for solving power flow equations are presented in Sect. 4. Finally, we draw conclusions and discuss the future work on this subject in Sect. 5.

2 The Algorithm

In the following discussion, $\mathbf{x}^* \in \mathbb{R}^N$ is an exact solution of system (1), i.e., $\mathbf{F}(\mathbf{x}^*) = \mathbf{0}$. Let $\mathbf{x}^0$ be an initial guess of $\mathbf{x}^*$, and suppose the components of $\mathbf{x}$ and $\mathbf{F}$ are conformally partitioned as follows:

$$\{\mathbf{F}\} = \{\mathbf{F}_1, \ldots, \mathbf{F}_M\}, \ \{\mathbf{x}\} = \{\mathbf{x}_1, \ldots, \mathbf{x}_M\}, \tag{2}$$

where $\mathbf{F}_i = (f_{(i,1)}, \ldots, f_{(i,n_i)})^T : \mathbb{R}^N \to \mathbb{R}^{n_i}$, $\mathbf{x}_i = (x_{(i,1)}, \ldots, x_{(i,n_i)})^T \in \mathbb{R}^{n_i}$ for $i = 1, \ldots, M$. Let $S_i = \{(i,1), \ldots, (i,n_i)\}$, then the partition satisfies that $\bigcup_{i=1}^M S_i = \{1, 2, \ldots, N\}$ and $S_i \cap S_{i+1} \neq \emptyset$ for $i = 1, \ldots, M-1$, which means that the adjacent blocks have partial overlaps. This partition may be obtained by graph-theoretic decomposition algorithms. Several overlapped strategies based on the general graph partitioning scheme included in Chaco, a publicly available graph partitioning software package developed by Hendrickson and Leland [1995], have been chiefly discussed by Chen and Cai [2003].

Let $\mathbf{J}^0$ be the Jacobian matrix of $\mathbf{F}$ at $\mathbf{x}^0$, i.e., $\mathbf{J}^0 = \frac{\partial \mathbf{F}(\mathbf{x})}{\partial \mathbf{x}}|_{\mathbf{x}=\mathbf{x}^0}$, and for $i = 1, \ldots, M$ let $\mathbf{J}_i^0 = \frac{\partial \mathbf{F}_i(\mathbf{x})}{\partial \mathbf{x}_i}|_{\mathbf{x}=\mathbf{x}^0} \in \mathbb{R}^{n_i \times n_i}$ be a nonsingular matrix. An algorithm for the Overlapped Block Simplified Newton method is as follows:

Overlapped Block Simplified Newton (OBSN) algorithm.

1. a. *Partition* $\mathbf{J}^0$ *into* M *blocks* $\mathbf{J}_i^0, i = 1, \ldots, M$.

b. Select weighted average parameter $\alpha = \{\alpha_i\}_{i=1}^{M-1}, 0 \le \alpha_i \le 1$.

2. *For* $k = 0, 1, \ldots$ *until convergence:*
For $i = 1, \ldots, M$, *do:*

a. Solve $\mathbf{J}_i^0 \mathbf{s}_i^k = -\mathbf{r}_i^k$.

b. Assemble the solutions: $\mathbf{x}_i^{k+1} = \mathbf{x}_i^k + \hat{\mathbf{s}}_i^k$, *where for* $j = 1, \ldots, n_i$,

$$\hat{s}_{(i,j)}^k = \begin{cases} \alpha_i s_{(i,j)}^k + (1-\alpha_i) s_{(i+1,j)}^k, & (i,j) \in S_i \cap S_{i+1}; \\ \alpha_{i-1} s_{(i-1,j)}^k + (1-\alpha_{i-1}) s_{(i,j)}^k, & (i,j) \in S_{i-1} \cap S_i; \\ s_{(i,j)}^k, & \text{others}. \end{cases}$$

c. Calculate $\mathbf{r}_i^{k+1} = \mathbf{F}_i(\mathbf{x}^{k+1})$. *If* $\|\mathbf{r}^{k+1}\|$ *is small enough, stop.*

Step 2 of the above algorithm can be essentially replaced by the Newton-type iteration:

$$\mathbf{x}^{k+1} = \mathbf{G}(\mathbf{x}^k) = \mathbf{x}^k - (\mathbf{A}(\mathbf{J}_D^0, \alpha))^{-1} \mathbf{F}(\mathbf{x}^k), \tag{3}$$

where $\mathbf{J}_D^0$ denotes the partially overlapping block diagonal Jacobian and the matrix $(\mathbf{A}(\mathbf{J}_D^0, \alpha))^{-1}$ is determined by $\mathbf{J}_1^{-1}, \ldots, \mathbf{J}_M^{-1}$ and α. To obtain local convergence for OBSN, one only needs to prove the convergence of the iteration (3). However, OBSN is proposed here for solving large sparse nonlinear systems in parallel. The reason is that Step 2 of the algorithm is easily parallelizable despite the use of a direct or an iterative solver.

For most practical problems, increasing the number of blocks will yield a severe increase in the number of iterations for the block-iterative method even the blocks with overlapping. In order to obtain an efficient parallel implementation for OBSN, it is critical to reduce the number of iterations. We therefore propose an accelerated technique based on the zero-nonzero structure of the partitioned Jacobian $\mathbf{J}^0$. Suppose the set

$$\bar{S} = \{\langle i, j \rangle : J_{ij}^0 \neq 0,\ J_{ij}^0 \notin \mathbf{J}_D^0, i, j = 1, \ldots, N\}, \tag{4}$$

to be nonempty, then for all $\langle p, q \rangle \in \bar{S}$, append an updated formula following Step 2.(b) in OBSN as follows:

$$\begin{cases} \mathbf{x}_{(p)}^{k+1} = \mathbf{x}_{(p)}^{k+1} + \gamma \hat{\mathbf{s}}_{(q)}^k, \\ \mathbf{x}_{(q)}^{k+1} = \mathbf{x}_{(q)}^{k+1} + \gamma \hat{\mathbf{s}}_{(p)}^k, \end{cases} \tag{5}$$

where $\gamma \in (0, 1)$ is an accelerated parameter. The algorithm with the updated formula (5) is referred to as AOBSN algorithm.

Note that OBSN is essentially a variation of nonlinear block-multisplitting method presented by Frommer [1989] or additive Schwarz methods (block-Jacobi algorithm) with overlapping. The main difference is that the Jacobian matrix of $\mathbf{F}$ need not be computed at each iteration and there is overlapping only between adjacent blocks in OBSN. Here, the restriction of overlapping just makes the selection of α much easier. The numerical results in Sect. 4 also show that the convergence performance of OBSN is much improved by the formula (5) with γ rather than by α.

3 Local and Semi-local convergence

Let $\|\cdot\|$ be a norm on $\mathbb{R}^N$ and $\bar{\Omega}(\mathbf{x}, r)$ be a close ball of radius r about $\mathbf{x}$. We can immediately obtain the local convergence theorem for OBSN.

Theorem 1. *Let $\mathbf{F} : D \subset \mathbb{R}^N \to \mathbb{R}^N$ be Fréchet differentiable at the zero $\mathbf{x}^* \in D$ of $\mathbf{F}$, and suppose that $\mathbf{J}_i(\mathbf{x}) : \Omega_0 \subset D \to \mathcal{L}(\mathbb{R}^{n_i})$ is defined on some open neighborhood $\Omega_0 \subset D$ of $\mathbf{x}^*$ and is continuous at $\mathbf{x}^*$ with nonsingular $\mathbf{J}_i(\mathbf{x}^*)$ for $i = 1, \ldots, M$. Then there exists a close ball $\Omega = \bar{\Omega}(\mathbf{x}^*, \delta) \subset \Omega_0$, $\delta > 0$, on which for any $\mathbf{x}^0 \in \Omega$ the mapping $\mathbf{G} : \mathbf{x} \in \Omega \to \mathbf{x} - (\mathbf{A}(\mathbf{J}_D^0, \alpha))^{-1}\mathbf{F}(\mathbf{x}) \in \mathbb{R}^N$ is well-defined and $\mathbf{G}$ has at $\mathbf{x}^*$ the Fréchet derivative*

$$\mathbf{G}'(\mathbf{x}^*) = \mathbf{I} - (\mathbf{A}(\mathbf{J}_D^0, \alpha))^{-1}\mathbf{J}(\mathbf{x}^*). \tag{6}$$

If $\rho(\mathbf{G}'(\mathbf{x}^*)) < 1$, then the sequence $\{\mathbf{x}^k\}$ generated by OBSN is well-defined for any $\mathbf{x}^0 \in \Omega$ and it converges to $\mathbf{x}^*$.

Proof. Set $\eta = \max_{i=1,\ldots,M} \|\mathbf{J}_i(\mathbf{x}^*)^{-1}\|$ and for given $\varepsilon > 0$, $2\eta\varepsilon < 1$, choose $\delta > 0$ such that $\Omega = \bar{\Omega}(\mathbf{x}^*, \delta) \subset \Omega_0$ and $\|\mathbf{J}_i(\mathbf{x}) - \mathbf{J}_i(\mathbf{x}^*)\| \leq \varepsilon$ for any $\mathbf{x} \in \Omega$, $i = 1, \ldots, M$. Then $\mathbf{J}_i(\mathbf{x})$ is invertible for all $\mathbf{x} \in \Omega$, and

$$\|(\mathbf{J}_i(\mathbf{x}))^{-1}\| \leq \frac{\eta}{1 - \eta\varepsilon} < 2\eta, \ \mathbf{x} \in \Omega, \ i = 1, \ldots, M, \tag{7}$$

that is, the mapping $\mathbf{G}(\mathbf{x}) = \mathbf{x} - (\mathbf{A}(\mathbf{J}_D^0, \alpha))^{-1}\mathbf{F}(\mathbf{x})$ is well-defined on Ω for any $\mathbf{x}^0 \in \Omega$. In addition, $\rho(\mathbf{G}'(\mathbf{x}^*)) < 1$ which implies the $\mathbf{x}^*$ is an attractor of the iterative formula (3), so the sequence $\mathbf{x}^k$ generated by OBSN is well-defined for any $\mathbf{x}^0 \in \Omega$ and it converges to $\mathbf{x}^*$. □

Furthermore, we can also obtain the semi-local convergence theorem for OBSN from Theorem 12.5.5 in Ortega and Rheinboldt [1970] by the Newton-type iteration (3). The proof is trivial.

Theorem 2. *Let $\mathbf{F} : D \subset \mathbb{R}^N \to \mathbb{R}^N$ be Fréchet differentiable and Lipschitz continuous with Lipschitz constant τ on a close ball $\bar{\Omega}(\mathbf{x}^0, r) \subset D$. Also suppose that there exist κ, η, and μ with $h = \kappa\tau\eta(1-\mu)^2 \leq \frac{1}{2}$ such that*

$$\|(\mathbf{A}(\mathbf{J}_D^0, \alpha))^{-1}\| \leq \kappa, \tag{8}$$

$$\|(\mathbf{A}(\mathbf{J}_D^0, \alpha))^{-1}\mathbf{F}(\mathbf{x}^0)\| \leq \eta, \tag{9}$$

$$\|\mathbf{I} - (\mathbf{A}(\mathbf{J}_D^0, \alpha))^{-1}\mathbf{J}(\mathbf{x}^0)\| \leq \mu < 1, \tag{10}$$

Set

$$r_- = \frac{1-\mu}{\kappa\tau}(1 - \sqrt{1-2h}), \tag{11}$$

$$r_+ = \frac{1-\mu}{\kappa\tau}(1 + \sqrt{1-2h}). \tag{12}$$

If $r \geq r_-$, then the sequence $\{\mathbf{x}^k\}$ generated by OBSN keeps in $\bar{\Omega}(\mathbf{x}^0, r)$ and it converges to the unique root $\mathbf{x}^*$ of $\mathbf{F}$ in $\bar{\Omega}(\mathbf{x}^0, r')$ with $r' = \min\{r, r_+\}$.

4 Numerical Results

In this section, OBSN and AOBSN are applied to the load flow problem in power systems. The relative importance of α, γ on the convergence of (A)OBSN for the IEEE 118-bus system (Problem 1) are studied, and some values are suggested for obtaining good performance with (A)OBSN. Then, the parallel numerical results for the IEEE 662-bus system (Problem 2) on a PC cluster are presented.

For an n-bus system without the slack bus, the load flow problem is described by a system of nonlinear algebraic equations:

$$\mathbf{F}(\mathbf{x}_1, \ldots, \mathbf{x}_n; \mathbf{P}, \mathbf{Q}) = \mathbf{0} \tag{13}$$

where

$$\mathbf{F}_i = (F_{P_i}, F_{Q_i})^T, \ \mathbf{x}_i = (f_i, e_i)^T, \tag{14}$$

$$F_{P_i} = P_i - \mathrm{Re}(E_i \sum_{k=1}^{n} Y_{ik}^* E_k^*), \tag{15}$$

$$F_{Q_i} = Q_i - \mathrm{Im}(E_i \sum_{k=1}^{n} Y_{ik}^* E_k^*). \tag{16}$$

In the above equations, $E_i = e_i + \mathbf{j} f_i$ represents the unknown complex node voltage, $P_i + \mathbf{j}Q_i$ represents the injected power and $Y_{ik} = G_{ik} + \mathbf{j}B_{ik}$ represents the admittance. For PV buses, where the voltage magnitude V_i is fixed, equation (16) is replaced by $F_{Q_i} = V_i^2 - (e_i^2 + f_i^2)$.

In practice, good initial guess can be easily obtained in the load flow computation, especially in the load flow track case. So an approximate start $\mathbf{x}^0$ which is obtained by adding random values ranging from -0.01 to 0.01 to the solutions is considered to evaluate the cases where good initial approximations are provided. In all tests, Gaussian elimination was used to solve the linear subproblems exactly and the nonlinear error tolerance was 10^{-3}.

4.1 Influence of α, γ on Convergence

For the power flow equations (13), using the similarity of the structure of matrix $\mathbf{B} = (B_{ik})_{n \times n}$ and the Jacobian $\mathbf{J}^0$ (see Zecevic and Siljak [1994]), we applied the partially overlapped partitioning to $\mathbf{B}$ to achieve the partition of $\mathbf{J}^0$ which reduces the problem dimension by a factor of 2. The linear-KL partitioning coupled with the boundary-linear strategy (see Chen and Cai [2003]) was chosen as the partition scheme for (A)OBSN in this study. The choice was observed to be better compared to scattered, spectral partitioning and multilevel-KL schemes (see Hendrickson and Leland [1995]) coupled with the boundary-linear strategy. For simplicity, we only considered the case for $\alpha_1 = \cdots = \alpha_{M-1} = \alpha$.

Fig. 1 shows the influence of α on the convergence of OBSN and AOBSN ($\alpha = 0.5$) for Problem 1 in ten differently approximate start cases. From the figure, we can see that the influence of α on the convergence of OBSN is more sensitive to $\mathbf{x}_0$ than that on AOBSN. In addition, the rate of convergence of OBSN is significantly improved by the updated formula (5) with the accelerated parameter $\gamma = 0.5$. Note that the influence of α on the convergence of AOBSN is much less than that of γ, so α can be fixed (for example, $\alpha = 0.5$) if AOBSN is used to solve the power flow equations.

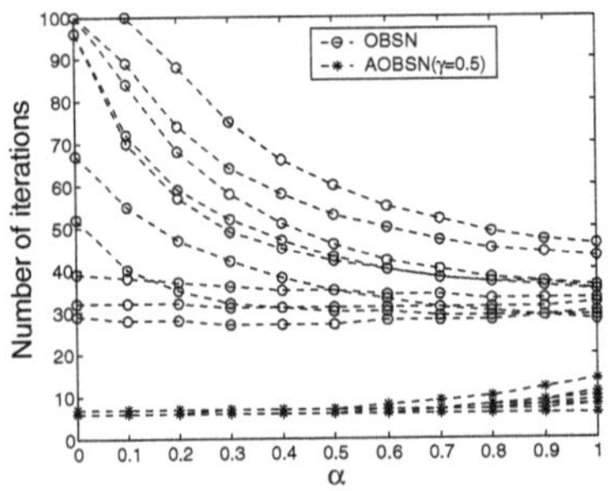

Fig. 1. Influence of α on the convergence of (A)OBSN for Problem 1 (ten approximate starts, $M = 8$)

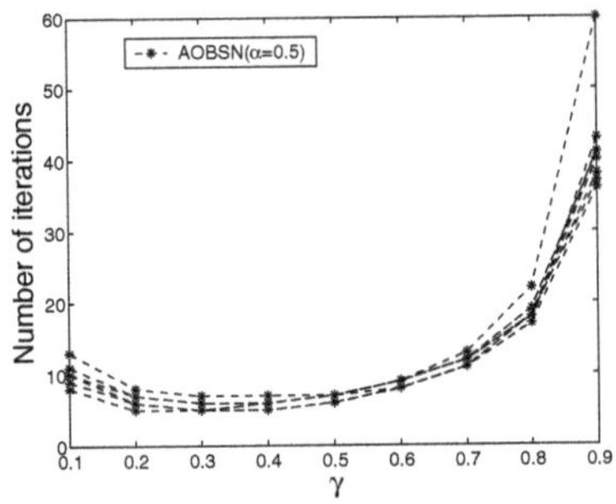

Fig. 2. Influence of γ on the convergence of AOBSN for Problem 1 (ten approximate starts, $M = 8$)

Fig. 2 shows the influence of accelerated parameter γ on the convergence of AOBSN ($\alpha = 0.5$) for Problem 1 in the same ten start cases. The figure also shows that the convergence performance of OBSN is significantly improved by the accelerated technique with $\gamma = 0.1, \ldots, 0.8$. Similar conclusions can be drawn for other ten initial approximations obtained even by a larger disturbed parameter as well.

4.2 Parallel Implementation of AOBSN for Problem 2

Using the above scheme, we partition the matrix B of Problem 2 into 2, 4, 8, 16, 32 and 64 blocks (see Fig. 3 for some cases). Fig. 4 shows the number of iterations of three algorithms mentioned above for Problem 2. We can see that AOBSN has much better convergent performance than the Block Simplified Newton method (BSN) and OBSN, and its number of iterations is less sensitive to the number of blocks. It should be pointed out that AOBSN usually requires more iterations than Newton methods and its simplified version. However, by virtue of the reduction of dimensionality, AOBSN can result in significant computational savings.

In the parallel implementation, we assigned the individual blocks or a group of blocks into per processor in an adequate load balancing way. All

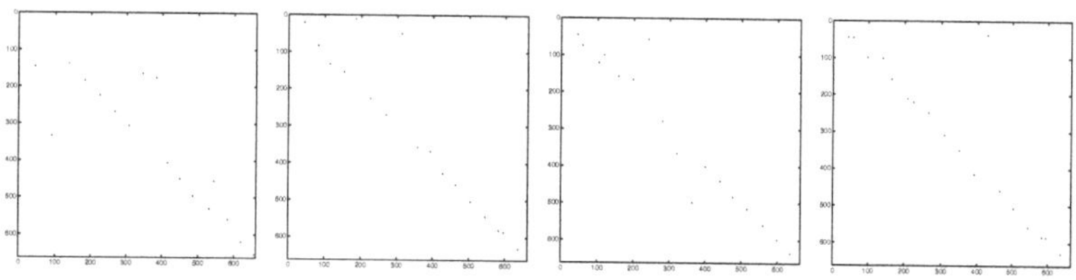

(a) $M = 1$ (b) $M = 8$ (c) $M = 16$ (d) $M = 32$

Fig. 3. Zero-nonzero structure of B and the partitioned B for Problem 2

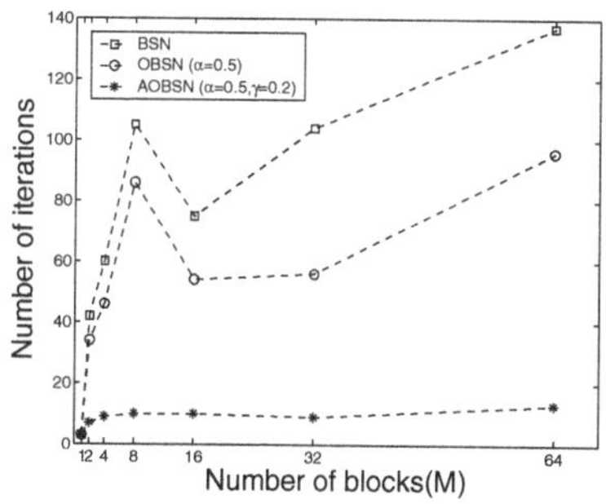

Fig. 4. Comparison of 3 methods for Problem 2 (approximate start)

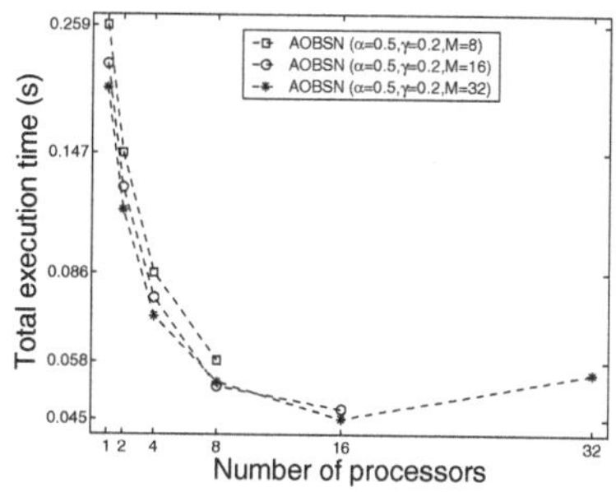

Fig. 5. Parallel computing time of AOBSN for Problem 2

numerical tests were run on an SMP-Cluster (36 nodes, CPU of per node: 4×Intel Xeon PIII700MHz, Myrinet, MPI). The programs were written in C using double precision floating point numbers. Fig. 5 shows the parallel computation time of AOBSN in 8, 16 and 32 block cases. The total execution time reduces with the number of processors and reaches its minimal value when the number of processors is 8 or 16 in 8 or 16 block cases, respectively. The communication time approximately increases with the number of processors and exceeds the computation time with 32 processors in 32 block case, which indicates that Problem 2 is not sufficiently large to be efficiently mapped onto more than 16 processors on the SMP-Cluster.

5 Conclusions and Discussion

This paper has presented an overlapped block-parallel Newton method (OBSN) for solving large nonlinear systems. In the implementation, an accelerated version (AOBSN) is also proposed to reduce the number of iterations. The numerical results of solving power flow equations confirm that AOBSN is indeed effective, particularly for problems where a good initial guess is available.

As mentioned in the previous sections, OBSN and AOBSN are nonlinear methods depending on several parameters, including the partition scheme,

M, α, γ, etc. In this paper, the relative importance of α, γ on the convergence of (A)OBSN were studied for the power flow problem, and some values for obtaining good results with (A)OBSN were suggested. A theoretical study on how these parameters influence the convergence of the algorithms will be carried out in our future work.

Acknowledgement. The author gratefully thanks Dr. B. Hendrickson and Dr. R. Leland of Sandia National Laboratories for allowing him to use the Chaco package for some of the numerical tests performed in this work.

References

P. N. Brown and Y. Saad. Hybrid Krylov methods for nonlinear system of equations. *SIAM J. Sci. Stat. Comput.*, 11(3):450–481, 1990.

X. C. Cai and D. E. Keyes. Nonlinear proconditioned inexact Newton algorithms. *SIAM J. Sci. Comput.*, 24(1):183–200, 2002.

Y. Chen and D. Cai. Inexact overlapped block Broyden methods for solving nonlinear equations. *Appl. Math. Comput.*, 136(2/3):215–228, 2003.

R. Dembo, S. Eisenstat, and T. Steihaug. Inexact Newton methods. *Siam J. Numer. Anal.*, 19(2):400–408, 1982.

A. Frommer. Parallel nonlinear multisplitting methods. *Numer. Math.*, 56: 269–282, 1989.

W. D. Gropp, D. E. Keyes, L. C. McInnes, and M. D. Tidriri. Globalized Newton-Krylov-Schwarz algorithms and software for parallel implicit CFD. *Int. J. High Performance Computing Applications*, 14:102–136, 2000.

B. Hendrickson and R. Leland. The Chaco user's guide, version 2.0. Technical report, Sandia National Laboratories, Albuquerque, NM, July 1995. Tech. Rep. SAND 95-2344.

D. A. Knoll and D. E. Keyes. Jacobian-free Newton-Krylov methods: A survey of approaches and applications. *Int. J. High Performance Computing Applications*, 2003. to appear.

J. M. Ortega and W. C. Rheinboldt. *Iterative Solution of Nonlinear Equations in Several Variables.* Academic Press, 1970.

Y. Saad. Krylov subspace methods for solving unsymmetric linear systems. *Math. Comp.*, 37:105–126, 1981.

Y. Saad and M. Schultz. GMRES: A generalized mininum residual algorithm for solving non-symmetric linear systems. *SIAM J. Sci. Stat. Comput.*, 7: 856–869, 1986.

G. Yang, L. C. Dutto, and M. Fortin. Inexact block Jacobi-Broyden methods for solving nonlinear systems of equations. *SIAM J. Sci. Comput.*, 18(5): 1367–1392, 1997.

A. I. Zecevic and D. D. Siljak. A block-parallel Newton method via overlapping epsilon decompositions. *SIAM J. Matrix Analysis and Applications*, 15(3):824–844, 1994.

Generation of Balanced Subdomain Clusters with Minimum Interface for Distributed Domain Decomposition Applications

Dimos C. Charmpis and Manolis Papadrakakis

Institute of Structural Analysis & Seismic Research, National Technical University of Athens, 9, Iroon Polytechniou Str., Zografou Campus, GR-15780 Athens, Greece

Summary. Balancing and dual Domain Decomposition Methods (DDMs) are used in practice on parallel computing environments with the number of generated subdomains being generally larger than the number of available processors. The present study describes partitioning concepts used to: (a) generate subdomains for such DDMs and (b) organize these subdomains into subdomain clusters, in order to assign each cluster to a processor. The discussion concerns distributed computing environments with dedicated homogeneous processors, as well as with heterogeneous and/or non-dedicated processors. The FETI method is used to obtain numerical results demonstrating the merits of the described partitioning algorithms.

1 Introduction

The practical use of dual DDMs on parallel computing environments with independent numbers of subdomains and processors constitutes today a basic feature of these solution approaches (Lesoinne and Pierson [1998], Farhat et al. [2000], Charmpis and Papadrakakis [2003]). The dominant objective during the mesh partitioning task is to produce subdomains with specific geometric characteristics, in order to improve the conditioning of the arising interface problem. The number of generated subdomains is in general larger than the number of available processors, since the computational performance of dual DDMs is improved both in terms of overall execution time and storage requirements by using mesh partitions with increased numbers of subdomains. This happens because increased numbers of subdomains result in smaller profiles for stiffness and preconditioning matrices. The reduced storage demand of these matrices plays its favorable role in decreasing computational times both during the factorization of the matrices and their application in the iterative solution of the interface problem.

Since dual DDMs are applied in general to partitions with $n_s > n_p$ (n_s is the number of generated subdomains and n_p is the number of utilized processors), an additional computational step has to be performed just after mesh partitioning, in order to appropriately organize the n_s subdomains into n_p subdomain clusters and then assign each cluster to one of the n_p processors. This additional computational step is termed as Subdomain Cluster GENeration (SCGEN) and can be viewed as a second partitioning task.

The issues discussed in the present study are expected to apply to all dual-type and balancing DDMs. Dual-type DDMs are all methods that are either purely based on a dual substructuring formulation (i.e. all FETI versions and variants) or include a dual substructuring kernel in the framework of a DDM of another category. Regarding the balancing DDM, its basic formulation is equivalent to a primal alternative of the FETI method (Fragakis and Papadrakakis [2003]), which indicates that both dual-type and balancing DDMs are similarly affected by the number and shape of the subdomains and the generated subdomain clusters. Therefore, in the sequel of the present work the term DDMs is used to refer collectively to dual-type and balancing DDMs.

The present study overviews the partitioning concepts used to generate subdomains and subdomain clusters for DDMs. The discussion focuses on a recently proposed heuristic approach, which handles the SCGEN task as a graph partitioning optimization problem (Charmpis and Papadrakakis [2003]). The effectiveness of heuristic SCGEN is demonstrated using the FETI method on a distributed computing environment with dedicated homogeneous processors. The applicability of this SCGEN technique is extended also to the case of heterogeneous and/or non-dedicated processors.

2 Mesh partitioning

DDM efficiency is governed by the convergence rate of the iterative algorithm employed for the solution of the arising interface problem. Since convergence is accelerated as the subdomains' aspect ratios improve, DDM performance is sensitive to the geometric characteristics of the subdomains. Therefore, the dominant mesh partitioning criterion that must be optimized to efficiently use DDMs is to produce subdomains with aspect ratios as good as possible. The two-step mesh partitioning strategy described in detail by Farhat et al. [1995] has been followed in the present work to produce subdomains with appropriate geometric characteristics: first the fast and simple Greedy algorithm generates a reasonable mesh partition, then the non-deterministic simulated annealing optimizer post-processes the initial partition to improve mainly the aspect ratios of the subdomains.

3 Subdomain Cluster GENeration (SCGEN)

Two basic requirements are specified, in order to control the SCGEN process and direct it towards the generation of subdomain clusters with suitable properties. We expect the SCGEN task to produce: (a) balanced subdomain clusters, in order to avoid excessive processor idle times and (b) small interface size between subdomain clusters, in an effort to reduce communication overhead during the iterative solution of the DDM interface problem. Thus, two distinct but complementary tasks have to be performed in order to obtain an efficient parallel DDM solution:

- The dominant objective during mesh partitioning is to ensure a favorable numerical DDM behavior. This is achieved by producing subdomains with good aspect ratios, in order to accelerate the convergence of the iterative DDM solution.
- SCGEN focuses on parallel execution aspects of the DDM, since it facilitates the exploitation of the available parallel computing environment by providing the utilized processors with balanced subsets of the DDM problem and small interface size between these subsets.

The following SCGEN approaches can be used to generate equally loaded subdomain clusters:

- The simple *Linear SCGEN* approach simulates the way early FETI runs with $n_s > n_p$ were organized, since it clusters the subdomains according to their position (numbering) in the list of subdomains: the first n_s/n_p subdomains are assigned to cluster 1, the next n_s/n_p to cluster 2, etc.
- Evidence of the importance of the SCGEN task was given with the *Greedy SCGEN* approach (Farhat et al. [2000]), which implements a partitioning algorithm based on the Greedy domain decomposer and exploits to some extent the adjacency information between subdomains.
- The *heuristic SCGEN* approach uses graph partitioning optimization software like METIS, JOSTLE, Chaco, etc., in order to handle in a more explicit and effective way the objectives and constraints of the SCGEN task (Charmpis and Papadrakakis [2003]).

DDM performance is affected by the quality of the generated subdomain clusters. All three aforementioned SCGEN approaches assign roughly n_s/n_p subdomains to each cluster resulting in reasonably balanced processor workloads. However, the Linear and Greedy algorithms often produce rather lengthy interfaces between subdomain clusters, especially when large-size problems with irregular geometry are considered. On the other hand, heuristic partitioning is capable of detecting (near-)optimum SCGEN solutions exhibiting small interface sizes and is therefore the most effective and reliable option for the SCGEN task of DDMs.

4 Graph representation of mesh partition

In order to be able to exploit standard graph partitioning software in the context of heuristic SCGEN, the mesh partitioning output (which defines n_s subdomains with optimal aspect ratios tailored for the DDMs) must be translated into graph data. A graph consists of vertices (associated with computational tasks), which are connected through edges (representing data dependencies between vertices). Vertices and/or edges may be assigned with weights, which implicitly specify the amount of computation associated with each vertex and the communication overhead associated with each edge. Hence, the goal of standard graph partitioning algorithms is to generate subsets of a given graph in a way that the sum of the vertex-weights in each subset is as balanced as possible and the sum of the edge-weights on the interface between the subsets (usually referred to as 'edgecut') is minimized.

In the context of the SCGEN task, the obtained mesh partitioning output is translated into an undirected graph $G = G(V, E)$, in which vertices $v_i \in V$ represent subdomains and edges $e(v_i, v_j) \in E$ are associated with subdomain connectivity (V and E denote the sets of vertices and edges, respectively). Thus, such a graph G consists of n_s vertices and contains two edges for each pair of neighboring subdomains (the adjacency of two subdomains associated with vertices v_i and v_j is described by both edges $e(v_i, v_j)$ and $e(v_j, v_i)$). The vertices and edges of G are associated with weights, which allow the transfer of more complete information from the mesh partitioning output to its graph representation. Hence, each vertex of G is weighted by the total number of degrees of freedom (d.o.f.) in the corresponding subdomain, i.e. each vertex weight implicitly specifies the computing workload of the associated subdomain. Furthermore, each edge is weighted by the number of interface d.o.f. or Lagrange multipliers between the two vertices (i.e. subdomains) the edge interconnects, in order to provide G with information regarding communication overheads during the DDM solution. A characteristic graph representation of a mesh partition is illustrated in Fig. 1.

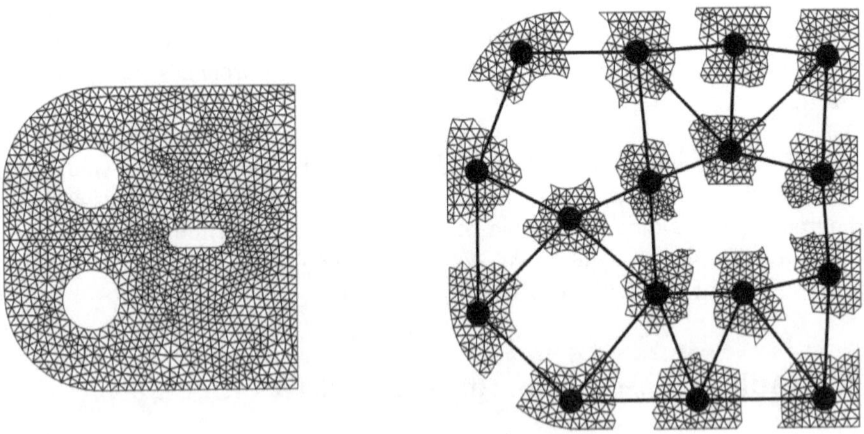

Fig. 1. A finite element mesh (left) and a mesh partition with its graph representation (right)

5 SCGEN using METIS

The construction of the mesh partition's weighted graph reduces the SCGEN task to a standard graph partitioning optimization problem. The objective of the optimization process is to minimize the interface between subdomain clusters, while the requirement of generating balanced subdomain clusters is imposed as a constraint in the partitioning process. This constraint optimization task is performed in the present work using the publicly available graph partitioning software METIS by Karypis and Kumar [1998]. Hence, the constructed weighted graph of a mesh partition is used as input to METIS, which aims in partitioning the n_s vertices of the graph into n_p balanced parts with minimum edgecut. Hence, METIS swaps subdomains between subdomain clusters until a locally optimal SCGEN result is reached. The output obtained contains the requested n_p subdomain clusters, in which the number of d.o.f. is as balanced as possible and the number of interface d.o.f. or Lagrange multipliers between subdomain clusters is small (if not minimum).

When the available parallel computing environment consists of heterogeneous computers, each utilized processor has to be assigned with an amount of workload proportional to its processing speed. This additional requirement has to be taken into account during the SCGEN process, in order to avoid imbalanced DDM computations. A similar situation arises when other users are running a variety of jobs on the available computers resulting in a non-dedicated parallel computing environment, which suffers workloads caused by several processes and therefore utilizes processors that do not have the same processing availability. SCGEN partitions tailored for heterogeneous and/or non-dedicated processors are obtained in the present work with the use of specialized METIS routines (denoted as WMETIS), which can handle prescribed partition weights. The SCGEN solutions yielded by WMETIS heuristics ensure that the computing workload associated with each generated subdomain cluster is proportional to the processing speed or availability of the corresponding processor.

6 Numerical tests

The performance of the described SCGEN approaches is investigated using a six-storey building, which is modeled with a solid mesh of 11485 hexahedral 20-noded elements resulting in 249015 d.o.f. Seven mesh partitions with optimal subdomain aspect ratios are generated for this 3D building test problem. Fig. 2 illustrates the finite element mesh and a characteristic partition of the 3D building. The numerical investigation is conducted on a cluster of 12 ethernet-networked homogeneous PCs (each with a Pentium III 500 MHz processor and 256 MB RAM) using the Linux operating system and the message passing software PVM. A network-distributed one-level FETI implementation described by Charmpis and Papadrakakis [2002] and enhanced by Charmpis

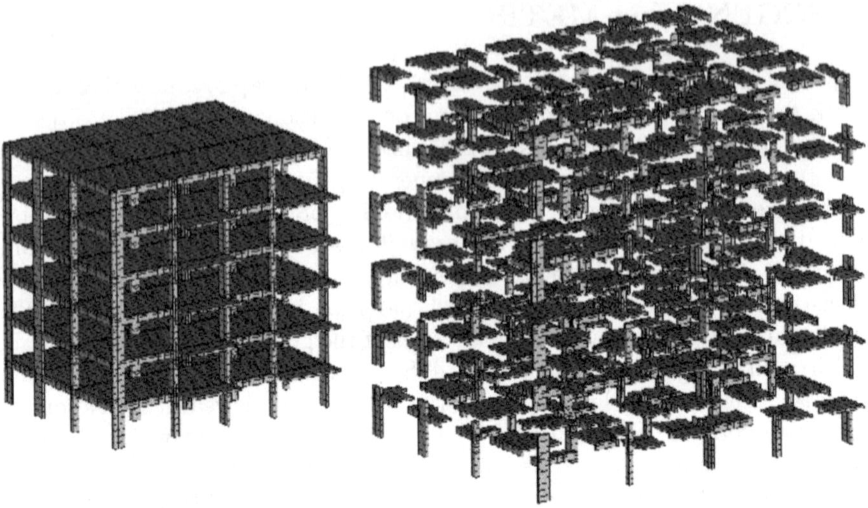

Fig. 2. 3D building test problem: finite element mesh (left) and mesh partition with $n_s = 250$ subdomains (right)

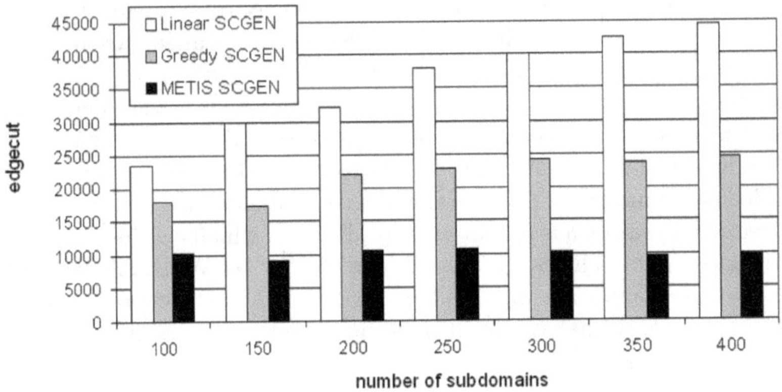

Fig. 3. Edgecuts obtained for several n_s-values ($n_p = 12$)

and Papadrakakis [2003] is executed on the PC cluster using the Dirichlet preconditioner stored in single precision arithmetic and a PCPG convergence tolerance $\epsilon = 10^{-3}$.

Fig. 3 reports edgecut-results for the 3D building test problem obtained by the SCGEN approaches on the dedicated PC cluster. According to these results, the heuristic METIS approach possesses a clear advantage over the Linear and Greedy algorithms giving by far smaller interface sizes between subdomain clusters.

Table 1 presents the time allocation for characteristic FETI runs and demonstrates the effect of low-edgecut SCGEN solutions on FETI perfor-

Table 1. Wall-clock time allocation of FETI runs ($n_s = 250$, $n_p = 12$)

SCGEN algorithm	Mesh partitioning	SCGEN	Formation of FETI matrices	PCPG solution
Linear	4.8s	-	40.9s	204.4s
Greedy	4.8s	0.05s	41.1s	132.9s
METIS	4.8s	0.06s	39.2s	69.0s

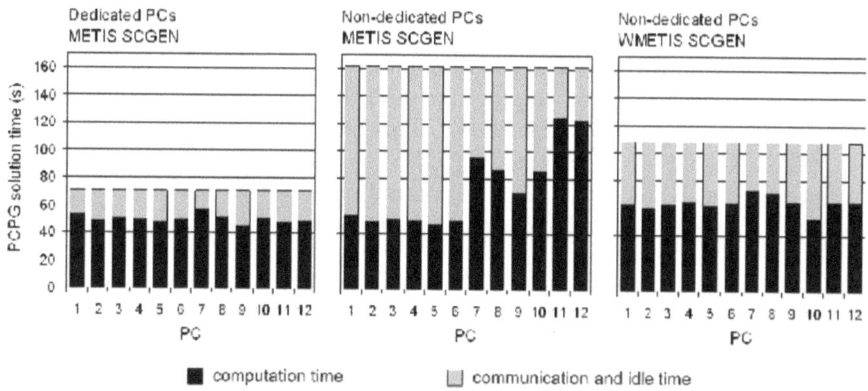

Fig. 4. Time allocation of the PCPG solver executed on dedicated or non-dedicated PCs with METIS or WMETIS SCGEN

mance. METIS markedly outperforms the Linear and Greedy SCGEN algorithms, since the large edgecut-values generated by the last two partitioners lead to excessive communication costs during the solution of the FETI interface problem by the iterative PCPG procedure. Furthermore, Table 1 shows that SCGEN requires a fraction of computing time compared to the overall solution effort even in the case of the METIS heuristic, therefore using parallel graph partitioning software to accelerate the SCGEN task is not necessary.

The left graph of Fig. 4 illustrates the reasonably balanced PCPG computations observed on the 12 dedicated PCs with the use of METIS SCGEN. In order to demonstrate the effectiveness of WMETIS SCGEN, a dummy computational process is started on each of PCs 7-10 and two dummy processes on each of PCs 11 and 12. This results in an artificially non-dedicated parallel computing environment, since the utilized PCs do not exhibit the same processing availability due to the waste of computing power caused by the imposed processor workloads. As shown in the middle graph of Fig. 4, METIS SCGEN cannot effectively handle the reduced processing capability of PCs 7-12 and causes severely imbalanced PCPG computations. This deficiency is alleviated by employing WMETIS SCGEN (see right graph of Fig. 4), which produces subdomain clusters using properly adjusted partition weights and is therefore capable of restoring the balancing of PCPG computations among all processors. It is noted that the communication times of Fig. 4 do not correspond only to message passing through the ethernet network, but include also the times needed to pack the communicated data into send buffers and unpack them from receive buffers.

7 Concluding remarks

The use of graph partitioning optimization algorithms allows the detection of (near-)optimum SCGEN solutions within large-size and irregularly structured search spaces. Such algorithms can effectively handle arbitrarily partitioned unstructured problems, while advantages over other SCGEN alternatives can be provided even for problems meshed and decomposed in a structured way (Charmpis and Papadrakakis [2003]). This consistent performance of heuristic SCGEN leads to distributed DDM runs with minimum communication overheads and reasonably balanced computations on dedicated homogeneous processors, as well as on heterogeneous and/or non-dedicated processors.

References

D. Charmpis and M. Papadrakakis. Enhancing the performance of the FETI method with preconditioning techniques implemented on clusters of networked computers. *Computational Mechanics*, 30(1):12–28, 2002.

D. Charmpis and M. Papadrakakis. Subdomain cluster generation for domain decomposition methods using graph partitioning optimization. *Engineering Computations*, 20(8):932–963, 2003.

C. Farhat, N. Maman, and G. Brown. Mesh partitioning for implicit computations via iterative domain decomposition: impact and optimization of the subdomain aspect ratio. *International Journal for Numerical Methods in Engineering*, 38:989–1000, 1995.

C. Farhat, K. Pierson, and M. Lesoinne. The second generation FETI methods and their application to the parallel solution of large-scale linear and geometrically non-linear structural analysis problems. *Computer Methods in Applied Mechanics and Engineering*, 184:333–374, 2000.

Y. Fragakis and M. Papadrakakis. The mosaic of high performance domain decomposition methods for structural mechanics: formulation, interrelation and numerical efficiency of primal and dual methods. *Computer Methods in Applied Mechanics and Engineering*, 192:3799–3830, 2003.

G. Karypis and V. Kumar. METIS: A software package for partitioning unstructured graphs, partitioning meshes, and computing fill-reducing orderings of sparse matrices – Version 4.0. Technical report, Department of Computer Science, University of Minnesota, USA, 1998.

M. Lesoinne and K. Pierson. An efficient FETI implementation on distributed shared memory machines with independent numbers of subdomains and processors. *Contemporary Mathematics*, 218:318–324, 1998.

Iterative Methods for Stokes/Darcy Coupling

Marco Discacciati

Ecole Polytechnique Fédérale de Lausanne
Institut d'Analyse et Calcul Scientifique
Chair of Modelling and Scientific Computing
marco.discacciati@epfl.ch

Summary. We present iterative subdomain methods based on a domain decomposition approach to solve the coupled Stokes/Darcy problem using finite elements. The dependence of the convergence rate on the grid parameter h and on the physical data is discussed; some difficulties encountered when applying the algorithms are indicated together with possible improvement strategies.

1 Introduction and problem setting

The simulation of incompressible flows in heterogenous media is an interesting topic with many applications: considering the particular case of free fluids which can filtrate through porous media, we recall for example the hydrological environmental applications and mass transfer in biomechanics.

The Stokes/Darcy coupled system provides a linear model to describe such phenomena. We consider a bounded domain $\Omega \subset \mathbb{R}^d$ $(d = 2, 3)$ formed by two non-overlapping subdomains Ω_f and Ω_p separated by a surface $\Gamma = \overline{\Omega}_f \cap \overline{\Omega}_p$. Ω_f is the region occupied by the fluid whose motion is described by the Stokes equations which can be written in adimensional form as:

$$\begin{aligned} -Re_f^{-1}\Delta \mathbf{u}_f + \nabla p_f &= \mathbf{f} \\ \nabla \cdot \mathbf{u}_f &= 0 \end{aligned} \quad \text{in } \Omega_f, \tag{1}$$

where $\mathbf{u}_f$ and p_f are the adimensional velocity and pressure, respectively, while Re_f is the Reynolds number defined as $Re_f = L_f U_f/\nu$, $\nu > 0$ being the fluid kinematic viscosity and L_f, U_f a characteristic length and velocity, respectively.

The filtration through the porous region Ω_p is modeled using Darcy's equations, whose adimensional form reads:

$$\begin{aligned} \mathbf{u}_p &= -\varepsilon Re_p \nabla p_p \\ \nabla \cdot \mathbf{u}_p &= 0 \end{aligned} \quad \text{in } \Omega_p, \tag{2}$$

where $\mathbf{u}_p$ and p_p are the adimensional fluid velocity and pressure, respectively, Re_p is the Reynolds number $Re_p = \delta_p U_p/\nu$, U_p being a characteristic velocity through the porous medium and δ_p a characteristic pore size. Finally, $\varepsilon = \delta_p/L_p$ is the adimensional ratio between the micro and the macro scales in Ω_p.

Across the interface Γ the continuity of normal stresses and fluxes is required; precisely, we impose:

$$\begin{aligned} \mathbf{u}_f \cdot \mathbf{n} &= \mathbf{u}_p \cdot \mathbf{n} \\ -\mathbf{n} \cdot \mathsf{T}(\mathbf{u}_f, p_f) \cdot \mathbf{n} &= p_p \qquad\qquad \text{on } \Gamma \\ -\boldsymbol{\tau}_j \cdot \mathsf{T}(\mathbf{u}_f, p_f) \cdot \mathbf{n} &= \alpha\, \mathbf{u}_f \cdot \boldsymbol{\tau}_j \end{aligned} \tag{3}$$

where $\mathsf{T}(\mathbf{u}_f, p_f)$ is the stress tensor, $\mathbf{n}$ is the unit normal outward vector to $\partial\Omega_f$ and $\boldsymbol{\tau}_j \cdot \mathbf{n} = 0$ $(j = 1, \ldots, d-1)$; α is a dimensionless coefficient depending essentially on ν and the hydraulic conductivity of the porous medium. For an extensive discussion about these coupling conditions we refer to Discacciati et al. [2002], Jäger and Mikelić [1996], Layton et al. [2003], Payne and Straughan [1998].

The mathematical analysis of the coupled problem has been addressed in previous works concerning both the continuous case and the finite element approximation (see Discacciati and Quarteroni [2003], Layton et al. [2003]).

In order to solve the coupled problem, an iterative substructuring method was proposed and analyzed in Discacciati and Quarteroni [2004]. Here, we test it on a model problem in order to investigate its effectiveness and robustness, with particular emphasis on the role that physical and grid parameters play on the convergence properties. We consider the computational domain $\Omega \subset \mathbb{R}^2$, with $\Omega_f = (0,1)\times(1,2)$, $\Omega_p = (0,1)^2$ and interface $\Gamma = (0,1)\times\{1\}$. We rewrite Darcy's equation as $-\nabla \cdot (\varepsilon Re_p \nabla p_p) = 0$ in Ω_p, and consider the following analytic solution: $\mathbf{u}_f = (y^2-2y+1, x^2-x)^T$, $p_f = 2(x+y-1)/Re_f + 1/(3\varepsilon Re_p)$ and $p_p = (x(1-x)(y-1) + y^3/3 - y^2 + y)/(\varepsilon Re_p) + 2x/Re_f$.

Concerning the finite element discretization, $\mathbb{P}_2-\mathbb{P}_1$ Taylor-Hood elements have been used for Stokes equations, while $\mathbb{P}_2$ Lagrangian elements have been adopted for Darcy's problem. All the computational meshes are conforming on Γ.

2 Dirichlet-Neumann (DN) methods

Considering the interface conditions $(3)_1$ and $(3)_2$, we can choose as scalar interface variable λ either $\lambda = \mathbf{u}_f \cdot \mathbf{n}$ on Γ or $\lambda = p_p$ on Γ. These two choices define two different DN-type methods, which can be outlined as follows, respectively:

Algorithm DN$_1$

0. choose $\lambda = \mathbf{u}_f \cdot \mathbf{n}$ on Γ and an initial guess $\lambda^{(0)}$ on Γ ;

For $k = 0, 1, \ldots$ until convergence, Do

1. solve Darcy's equation with b.c. $-\varepsilon Re_p \nabla p_p^{(k+1)} \cdot \mathbf{n} = \lambda^{(k)}$ on Γ ;
2. solve Stokes problem with b.c. $-\mathbf{n} \cdot \mathsf{T}(\mathbf{u}_f^{(k+1)}, p_f^{(k+1)}) \cdot \mathbf{n} = p_p^{(k+1)}$ and $(3)_3$ on Γ;
3. $\lambda^{(k+1)} = \theta \mathbf{u}_f^{(k+1)} \cdot \mathbf{n} + (1-\theta)\lambda^{(k)}$ on Γ , $\theta \in (0,1)$;

End For

Algorithm DN$_2$

0. choose $\lambda = p_p$ on Γ and an initial guess $\lambda^{(0)}$ on Γ ;

   For $k = 0, 1, \ldots$ until convergence, Do
1. solve Stokes problem with b.c. $-\mathbf{n} \cdot \mathsf{T}(\mathbf{u}_f^{(k+1)}, p_f^{(k+1)}) \cdot \mathbf{n} = \lambda^{(k)}$ and $(3)_3$ on Γ;
2. solve Darcy's equation with b.c. $-\varepsilon Re_p \nabla p_p^{(k+1)} \cdot \mathbf{n} = \mathbf{u}_f^{(k+1)} \cdot \mathbf{n}$ on Γ ;
3. $\lambda^{(k+1)} = \theta p_p^{(k+1)} + (1-\theta)\lambda^{(k)}$ on Γ , $\theta \in (0,1)$;

End For

The two DN methods are equivalent to preconditioned Richardson methods to solve the symmetric Steklov-Poincaré equations associated to the coupled problem, and they allow to characterize optimal preconditioners for Krylov type methods (e.g. the Conjugate Gradient) for the corresponding interface problems (see Discacciati and Quarteroni [2004]).

2.1 Numerical results

We consider $Re_f = 1$, $\varepsilon Re_p = 1$ and $tol = 10^{-5}$; in Table 1 we report the number of iterations for both the Richardson and the Preconditioned Conjugate Gradient (PCG) method. These convergence results are satisfactory as they show the optimality of the preconditioners with respect to the grid parameter h.

Number of mesh elements	DN$_1$ $(\theta = 0.7)$	PCG $\lambda = \mathbf{u}_f \cdot \mathbf{n}$	DN$_2$ $(\theta = 0.7)$	PCG $\lambda = p_p$
172	9	4	10	4
688	9	4	10	4
2752	9	4	10	4
11008	9	4	10	4

Table 1. Number of iterations on different grids with $Re_f = 1$ and $\varepsilon Re_p = 1$

However, if the fluid viscosity and the hydraulic conductivity decrease, small relaxation parameters θ must be adopted to guarantee convergence, in accordance with the theoretical estimate of the upper bound θ_{max} given in Discacciati and Quarteroni [2004]. Unfortunately, in some cases θ should be so small that in practice it prevents the numerical scheme from converging. To quote an example, if $Re_f = 10^3$ and $\varepsilon Re_p = 10^{-2}$, then θ should be unreasonably small (smaller than 10^{-4} !) in DN_1 to prevent divergence.

This difficulty should not be ascribed to the non-optimal choice of the relaxation parameter θ. In fact, if we apply the PCG method which embeds the choice of the optimal acceleration parameter (see, e.g., Quarteroni et al. [2000] p. 150), the iterative algorithm converges, but the optimal properties of the preconditioners are lost, since the number of iterations depends on the mesh parameter h, as reported in Table 2.

Table 2. Number of iterations on different grids with $Re_f = 10^3$ and $\varepsilon Re_p = 10^{-2}$

Mesh elements	PCG iterations ($\lambda = \mathbf{u}_f \cdot \mathbf{n}$)
688	82
2752	102
11008	148

On the basis of the numerical results we have obtained we can conclude that DN methods are effective only when the ratio $Re_f/(\varepsilon Re_p)$ is sufficiently small, while dealing with large values causes some difficulties. However, the latter are the very values of interest in real-life applications and, therefore, a robust numerical method is required.

3 Dirichlet-Neumann for a time-dependent problem

We introduce a formal argument to better understand the results obtained in Sect. 2.1 and to set up a more effective numerical scheme. This approach will be treated from a precise mathematical viewpoint in a forthcoming work Discacciati [2004].

The underlying idea is that our difficulties in solving the Stokes/Darcy problem may come from the different structure of equations $(1)_1$ and $(2)_1$, which become even more dissimilar when $Re_f \gg 1$ and $\varepsilon Re_p \ll 1$. In fact, in that case, under the physically reasonable hypothesis that $\Delta \mathbf{u}_f$ and ∇p_p are sufficiently small, $(1)_1$ reduces almost to $C_f \mathbf{I} + \nabla p_f = \mathbf{f}$, while $(2)_1$ becomes $\mathbf{u}_p + C_p \mathbf{I} = \mathbf{0}$, where C_f and C_p denote two positive constants $\ll 1$. We rewrite $(2)_1$ as

$$(\varepsilon Re_p)^{-1} \mathbf{u}_p + \nabla p_p = \mathbf{0} \quad \text{in } \Omega_p \,, \tag{4}$$

and formally comparing (4) to $(1)_1$, we are led to modify the latter by adding a mass term like $(\varepsilon Re_p)^{-1} \mathbf{u}_p$ as follows:

$$\beta(\varepsilon Re_p)^{-1}\mathbf{u}_f - Re_f^{-1}\Delta\mathbf{u}_f + \nabla p_f = \tilde{\mathbf{f}}\,, \quad \beta \in \mathbb{R}^+\,, \tag{5}$$

possibly with a consequent modification of the right hand side (see Remark 1) that we have denoted by $\tilde{\mathbf{f}}$. In this way we obtain a generalized Stokes momentum equation, and note that now (5) has the same behaviour of (4) in the cases of our interest, that is when $Re_f \gg 1$ and $\varepsilon Re_p \ll 1$.

We expect that the mass term $\beta(\varepsilon Re_p)^{-1}\mathbf{u}_f$ would help improving the positivity of the discrete Steklov-Poincaré operator which acts as a preconditioner in the DN_1 method. With this aim, we have carried out some numerical tests using the PCG algorithm with $\lambda = \mathbf{u}_f \cdot \mathbf{n}$ as interface variable to solve the modified problem (2), (5). The convergence results reported in Table 3 show that the numerical scheme has really improved.

Table 3. Number of iterations to solve problem (2), (5) for different values of Re_f, εRe_p and β

Re_f	εRe_p	β	Iterations on the mesh with		
			688 el.	2752 el.	11008 el.
		0.1	17	14	13
10^3	10^{-2}	1	10	9	7
		10	5	5	4
		0.1	19	21	19
10^6	10^{-4}	1	11	10	10
		10	5	5	4

Remark 1. Equation (5) can be regarded as a discretization in time of the time-dependent Stokes momentum equation $\partial_t\mathbf{u}_f - Re_f^{-1}\Delta\mathbf{u}_f + \nabla p_f = \mathbf{f}$ in Ω_f. Precisely, if we consider

$$\beta(\varepsilon Re_p)^{-1}\mathbf{u}_{f,n+1} - Re_f^{-1}\Delta\mathbf{u}_{f,n+1} + \nabla p_{f,n+1} = \tilde{\mathbf{f}}_{n+1} \qquad n \geq 0$$

with $\tilde{\mathbf{f}}_{n+1} = \mathbf{f}(\mathbf{x}, t_{n+1}) + \beta(\varepsilon Re_p)^{-1}\mathbf{u}_{f,n}$, we have a backward Euler discretization in time with $\beta(\varepsilon Re_p)^{-1}$ playing the role of the inverse of a time step.

From the physical viewpoint, since the fluid velocities in Ω_f are much higher than the ones through the porous medium (see Ene and Sanchez-Palencia [1975]), a time-dependent model better represents the phenomena occurring during the filtration process.

3.1 The tDN algorithm

Let $[0, T]$ be a characteristic time interval; using for the sake of simplicity the first-order backward Euler scheme, denoting by $\Delta t > 0$ the time step and

$N = T/\Delta t$, the iterative method that we propose to solve the time-dependent coupled problem reads (the subscript n refers to the nth time level):

Algorithm tDN

For $n = 0, \ldots, N-1$ Do

0. choose an initial guess $\lambda_{n+1}^{(0)}$ for the normal velocity on Γ at the $(n+1)$th time level;

For $k = 0, 1, \ldots$ until convergence, Do

1. solve Darcy's equation with b.c. $-\varepsilon Re_p \nabla p_{p,n+1}^{(k+1)} \cdot \mathbf{n} = \lambda_{n+1}^{(k)}$ on Γ ;
2. solve Stokes problem

$$\left.\begin{aligned} (\Delta t)^{-1}\mathbf{u}_{f,n+1}^{(k+1)} - Re_f^{-1}\Delta \mathbf{u}_{f,n+1}^{(k+1)} + \nabla p_{f,n+1}^{(k+1)} &= (\Delta t)^{-1}\mathbf{u}_{f,n} + \mathbf{f}_{n+1} \\ \nabla \cdot \mathbf{u}_{f,n+1}^{(k+1)} &= 0 \end{aligned}\right\} \quad \text{in } \Omega_f$$

with b.c. $-\mathbf{n} \cdot \mathsf{T}(\mathbf{u}_{f,n+1}^{(k+1)}, p_{f,n+1}^{(k+1)}) \cdot \mathbf{n} = p_{p,n+1}^{(k+1)}$ and $(3)_3$ on Γ ;

3. $\lambda_{n+1}^{(k+1)} = \theta \mathbf{u}_{f,n+1}^{(k+1)} \cdot \mathbf{n} + (1-\theta)\lambda_{n+1}^{(k)}$ on Γ , $\theta \in (0,1)$;

End For

End For

3.2 Numerical tests

We consider the horizontal section of a channel 12 m long and 8 m wide which is partially occupied by a porous medium with discontinuous conductivity, as represented in Fig. 1 (left). A parabolic inflow profile is imposed on the left hand side boundary with maximal velocity equal to 0.1m/s. On the right an outflow condition is imposed. The time interval is $t \in [0, 0.5]$ and the time step $\Delta t = 10^{-3}$ s; for space discretization three different computational meshes have been adopted.

In a first case we have considered $Re_f = 8 \cdot 10^5$ and a discontinuous coefficient $\varepsilon Re_p = 10^{-3}$ in $\Omega_p^{(1)}$, $\varepsilon Re_p = 10^{-7}$ in $\Omega_p^{(2)}$.

In Fig. 1 (right) we have represented the computed solution at time $t = 0.05$ s, while in Fig. 2 a zoom of the velocity field through the porous medium is shown; it can be seen that the velocity is almost null in the less permeable areas of the porous medium. Finally, Table 4 (left) reports the number of iterations obtained for three computational grids at different time levels, showing that the number of iterations is low and independent of h.

The same test has been performed considering different values of the parameters: $Re_f = 8 \cdot 10^2$, $\varepsilon Re_p = 10^{-1}$ in $\Omega_p^{(1)}$ and $\varepsilon Re_p = 10^{-5}$ in the less permeable part of the porous medium $\Omega_p^{(2)}$. The convergence results show that the number of iterations is essentially independent of these parameters, as it can be seen comparing the previous convergence results with those reported in Table 4 (right).

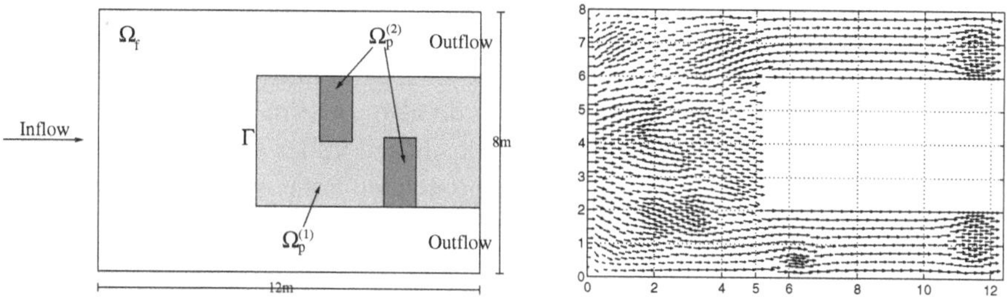

Fig. 1. Computational domain (left) and computed velocity field at $t = 0.05$ s (right)

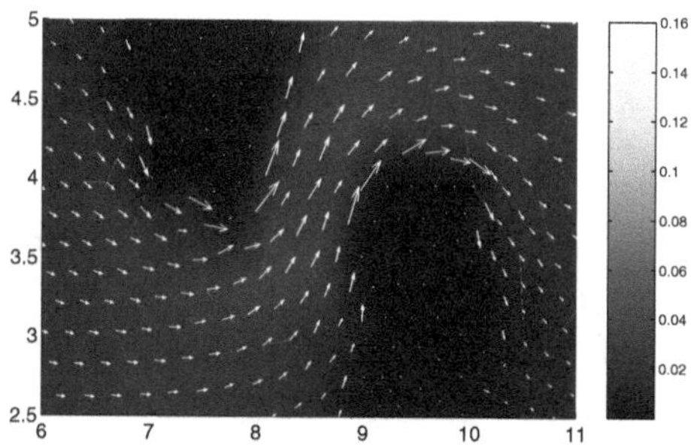

Fig. 2. Zoom of the velocity field through the porous medium

Time level	Iterations on the mesh with 232 el.	928 el.	3712 el.
0.001	21	21	21
0.003	20	19	19
0.006	12	11	11
0.009	10	10	10
0.01	10	10	10

Time level	Iterations on the mesh with 232 el.	928 el.	3712 el.
0.001	22	22	22
0.003	20	20	20
0.006	15	15	15
0.009	15	15	15
0.01	15	15	15

Table 4. Number of iterations on different grids with $Re_f = 8 \cdot 10^5$, $\varepsilon Re_p = 10^{-3}$ and 10^{-7} (left); with $Re_f = 8 \cdot 10^2$, $\varepsilon Re_p = 10^{-1}$ and 10^{-5} (right)

4 Conclusions and perspectives

Numerical results show that considering a time-dependent problem allows to set up a far more efficient DN algorithm for problems with parameters in a range of physical interest. However, as we have shown, the value of Δt generally depends on εRe_p and Re_f, and in some cases we could be forced to consider very small time steps $\Delta t \ll 1$. This could be quite annoying since one might be interested in considering long time scales, for example in modeling the filtration of pollutants in groundwater.

This limitation on Δt drives us to reconsider the steady coupled model. In fact, should we find an algorithm whose behaviour were as much as possible independent of the physical parameters, then not only we would be able to solve the steady problem itself, but we could also use it in the framework of the time-dependent model where Δt would be chosen under the sole requirements of stability and accuracy. A possible approach we are currently considering is a Robin-Robin type method following the ideas presented in Lube et al. [2001] for Oseen equations; its analysis and numerical results will be presented in a future work.

Acknowledgement. I wish to thank Prof. A. Quarteroni, Prof. A. Valli and Dr. N. Neuss for their suggestions and the helpful discussions concerning the subject of this work.

References

M. Discacciati. PhD thesis, Ecole Polytechnique Fédérale de Lausanne, Switzerland, 2004. In preparation.

M. Discacciati, E. Miglio, and A. Quarteroni. Mathematical and numerical models for coupling surface and groundwater flows. *Appl. Numer. Math.*, 43:57–74, 2002.

M. Discacciati and A. Quarteroni. Analysis of a domain decomposition method for the coupling of Stokes and Darcy equations. In F. Brezzi, A. Buffa, S. Corsaro, and A. Murli, editors, *Numerical Mathematics and Advanced Applications, ENUMATH 2001*, pages 3–20. Springer. Milan, 2003.

M. Discacciati and A. Quarteroni. Convergence analysis of a subdomain iterative method for the finite element approximation of the coupling of Stokes and Darcy equations. *Comput. Visual. Sci.*, To be published, 2004.

H. I. Ene and E. Sanchez-Palencia. Equations et phénomenès de surface pour l'écoulement dans un modèle de milieu poreux. *J. Mécanique*, 14(1):73–108, 1975.

W. Jäger and A. Mikelić. On the boundary conditions at the contact interface between a porous medium and a free fluid. *Ann. Scuola Norm. Sup. Pisa Cl. Sci.*, 23:403–465, 1996.

W. L. Layton, F. Schieweck, and I. Yotov. Coupling fluid flow with porous media flow. *SIAM J. Num. Anal.*, 40:2195–2218, 2003.

G. Lube, L. Müller, and F. C. Otto. A nonoverlapping domain decomposition method for stabilized finite element approximations of the Oseen equations. *J. Comput. Appl. Math.*, 132:211–236, 2001.

L. E. Payne and B. Straughan. Analysis of the boundary condition at the interface between a viscous fluid and a porous medium and related modelling questions. *J. Math. Pures Appl.*, 77:317–354, 1998.

A. Quarteroni, R. Sacco, and F. Saleri. *Numerical Mathematics.* Springer-Verlag, New York – Berlin – Heidelberg, 2000.

Preconditioning Techniques for the Bidomain Equations

Rodrigo Weber Dos Santos[1], G. Plank[2], S. Bauer[1], and E.J. Vigmond[3]

[1] Dept. of Biosignals, Physikalisch-Technische Bundesanstalt, Berlin, Germany
email: rodrigo.weber@ptb.de
[2] Inst. für Medizinische Physik und Biophysik, Universität Graz, Austria
[3] Dept. of Electrical and Computer Engineering, University of Calgary, Canada

Summary. In this work we discuss parallel preconditioning techniques for the bidomain equations, a non-linear system of partial differential equations which is widely used for describing electrical activity in cardiac tissue. We focus on the solution of the linear system associated with the elliptic part of the bidomain model, since it dominates computation, with the preconditioned conjugate gradient method. We compare different parallel preconditioning techniques, such as block incomplete LU, additive Schwarz and multigrid. The implementation is based on the PETSc library and we report results for a 16-node HP cluster. The results suggest the multigrid preconditioner is the best option for the bidomain equations.

1 Introduction

The set of bidomain equations (see Keener and Sneyd [1998]) is currently the mathematical model that best reflects the electrical activity of the heart. The non-linear partial differential equations (PDEs) model both the intracellular and extracellular domains of cardiac tissue from an electrostatic point of view. The coupling of the two domains is done via non-linear models describing the current flow through the cell membrane. Such models are based on experimental data that quantify different ionic contributions, as first proposed by Hodgkin and Huxley [1952].

Unfortunately, the bidomain equations are computationally very expensive. Modern membrane models involve more than 20 non-linear equations. One way to avoid the solution of a large non-linear PDE system at every time step is to use an operator splitting approach (Vigmond et al. [2002], Sundnes et al. [2001], Keener and Bogar [1998], Weber dos Santos [2002]). The numerical solution reduces to the solution of a parabolic equation, a non-linear system of ordinary differential equations, and an elliptic system. It is the latter that dominates computation (Vigmond et al. [2002]).

An efficient way of solving the large linear algebraic system that arises from the discretization of the bidomain equations has been a topic of research since 1994 (see Hooke et al. [1994]). Many different approaches have been employed with the preconditioned conjugate gradient method (CG) becoming the standard choice for an iterative method. Diagonal preconditioner was used for the bidomain equations by Eason and Malkin [2000], Skouibine and Krassowska [2000]. Incomplete LU factorization (ILU) was implemented by Pavarino and Franzone [2004], Street and Plonsey [1999], Vigmond et al. [2002]. The Symmetric-Successive-Over-Relaxation (SSOR) preconditioner was investigated by Pennacchio and Simoncini [2002], Weber dos Santos [2002]. In Weber dos Santos [2002] it was shown that incomplete factorization and SSOR achieved comparable results and were at least two times faster than the diagonal preconditioner.

Another way of reducing the computation time is cluster computing. The use of such parallel environment is supported by standard communication libraries such as the Message Passing Interface library [1994] (MPI). The solution of the bidomain equations has been efficiently implemented with MPI on small clusters (16 to 32 processors), see Pavarino and Franzone [2004], Pormann [2000], Yung [2000], Weber dos Santos [2002].

In this work we focus on the solution of the linear algebraic system associated with the elliptic part of the bidomain model. We employ CG and compare different parallel preconditioning techniques, such as block incomplete LU (ILU), the additive Schwarz method (ASM), and multigrid (MG). Both overlapping and non-overlapping domain decomposition techniques are investigated. The implementation is based on the PETSc C library (Balay et al. [2002]), which uses MPI. The results taken from a 16-node HP-Unix cluster indicate that the multigrid preconditioner is at least 13 times faster than the single-level Schwarz-based techniques.

2 Mathematical formulation

We simulated a two-dimensional piece of cardiac tissue in contact with a perfusing bath as previously described by Weber dos Santos et al. [2003], Weber dos Santos and Dickstein [2003]. The model that we present here was successfully used to reproduce an in-vitro experiment which explored the effects of lapine cardiac tissue micro-structure (see Weber dos Santos et al. [2003]). The simulated square region $\Omega = \Omega_t \cup \Omega_o$ with $\bar{\Omega}_t \cap \bar{\Omega}_o = \Gamma_{to}$, where Ω_o accounts for the perfusing bath and Ω_t for the cardiac tissue sample. The geometric information of Ω_t was extracted by image processing techniques and is represented by a mask vector M, $M_{i,j} = 1.0$, for $(i,j) \in \Omega_t$, $M_{i,j} = 0.0$, otherwise. In Figure 1 the tissue regions are represented by the gray color.

The bath was modeled as an isotropic conductor with conductivity σ_o. The electric potential $\phi_o : \Omega_o \times [0, t_f] \rightarrow \Re$ satisfies $\sigma_o \Delta\phi_o = 0$. The cardiac tissue is modeled by the bidomain equations.

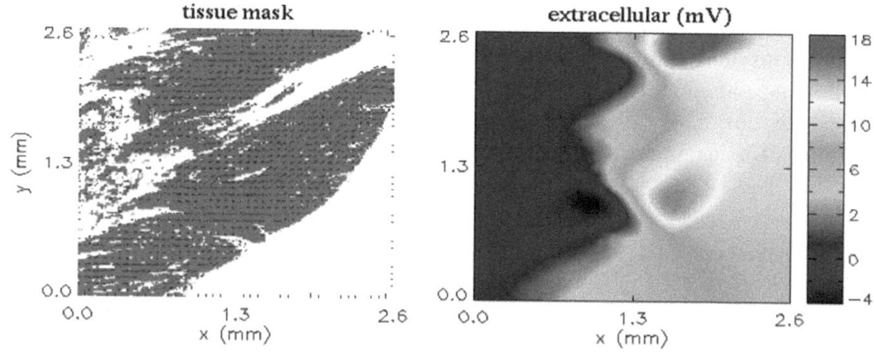

Fig. 1. Tissue geometry and fiber orientation extracted from a histological picture (courtesy of E. Hofer and D. Sanchez-Quintana) (left). Solution of the extracellular potential (ϕ_e and ϕ_o) at time 2.6ms (right)

$$\chi\left(C_m\frac{\partial\phi}{\partial t}+f(\phi,v)\right)=\nabla.(\boldsymbol{\sigma_i}\nabla\phi)+\nabla.(\boldsymbol{\sigma_e}\nabla\phi_e), \tag{1}$$

$$\nabla.((\boldsymbol{\sigma_e}+\boldsymbol{\sigma_i})\nabla\phi_e)=-\nabla.(\boldsymbol{\sigma_i}\nabla\phi), \tag{2}$$

$$\frac{\partial v}{\partial t}=g(\phi,v),\phi=\phi_i-\phi_e, \tag{3}$$

where ϕ_e and ϕ_i: $\Omega_t\times[0,t_f]\rightarrow\Re$ are the extracellular and intracellular potentials; ϕ: $\Omega_t\times[0,t_f]\rightarrow\Re$ is the transmembrane voltage; v: $\Omega_t\times[0,t_f]\rightarrow\Re^m$ represents the ionic current variables; $\boldsymbol{\sigma_i}$ and $\boldsymbol{\sigma_e}$ are conductivity tensors of the intracellular and extracellular spaces; C_m and χ are capacitance per unit area and surface to volume ratio respectively; f: $\Re\times\Re^m\rightarrow\Re$ and g: $\Re\times\Re^m\rightarrow\Re^m$ model ionic currents and specify the cell membrane model. We used the rabbit atrial model ($m=27$) of Lindblad et al. [1996].

An image processing technique (see Weber dos Santos et al. [2003]) was applied to extract the cardiac fiber orientation θ: $\Omega_t\rightarrow\Re$. Conductivity tensors were derived from the curved cardiac fibers based on θ by applying

$$\boldsymbol{\sigma_*}=\begin{pmatrix}\sigma_{*l}\cos^2\theta+\sigma_{*t}\sin^2\theta & (\sigma_{*l}-\sigma_{*t})\cos\theta\sin\theta\\ (\sigma_{*l}-\sigma_{*t})\cos\theta\sin\theta & \sigma_{*t}\cos^2\theta+\sigma_{*l}\sin^2\theta\end{pmatrix},$$

where σ_{*l} and σ_{*t} are longitudinal and transversal fiber conductivities ($*=i,e$). The boundary conditions for the bath to tissue interface were set to (see Krassowska and Neu [1994])

$$\phi_e=\phi_o, \tag{4}$$

$$\boldsymbol{\sigma_e}\nabla\phi_e.\eta=\sigma_o\nabla\phi_o.\eta, \tag{5}$$

$$\boldsymbol{\sigma_i}\nabla\phi_i.\eta=0,\ \text{on}\ \Gamma_{to}\times[0,t_f]. \tag{6}$$

The other boundaries are assumed to be electrically isolated, modeled by imposing homogeneous Neumann-like conditions

$$\boldsymbol{\sigma_i}\nabla\phi_i.\eta = \boldsymbol{\sigma_e}\nabla\phi_e.\eta = \sigma_o\nabla\phi_o.\eta = 0, \text{ on } \partial\Omega \times [0, t_f]. \tag{7}$$

The numerical results presented in the next sections were obtained with the following parameters: $C_m = 1\,\mu\text{F/cm}^2$, $\chi = 2000\,/\text{cm}$, $\sigma_{il} = 3$ mS/cm, $\sigma_{it} = 0.31$ mS/cm, $\sigma_{el} = 2$ mS/cm, $\sigma_{et} = 1.35$ mS/cm and $\sigma_o = 20$ mS/cm. Ω is a square of sides equal to 2.6mm.

3 Operator splitting and boundary conditions

We approached the non-linear system of PDEs with an operator splitting technique (Strang [1968]). The solution reduced to a three-step scheme that involved the solution of a parabolic PDE, an elliptic system, and a non-linear system of ordinary differential equations at each time step. Since the CFL condition of the parabolic PDE severely restricted the time step, we solved this equation via the Crank-Nicolson method. The large non-linear ODE system was solved via a forward-Euler scheme:

$$1.\quad (1 - \frac{\Delta t}{2} A_i)\varphi^{k+1/2} = (1 + \frac{\Delta t}{2} A_i)\varphi^k + \Delta t\, A_i(\varphi_e)^k, \tag{8}$$

$$2.\quad \varphi^{k+1} = \varphi^{k+1/2} - \Delta t\, f(\varphi^{k+1/2}, \nu^k)/(\chi C_m) \tag{9}$$

$$\nu^{k+1} = \nu^k + \Delta t\, g(\varphi^{k+1/2}, \nu^k) \tag{10}$$

$$3.\quad (A_i + A_e)(\varphi_e)^{k+1} = -A_i\varphi^{k+1}, \tag{11}$$

$$A_o{\varphi_o}^{k+1} = 0, \tag{12}$$

where A_i, A_e and A_o are the $\nabla.(\boldsymbol{\sigma_i}\nabla)/(\chi C_m)$, $\nabla.(\boldsymbol{\sigma_e}\nabla)/(\chi C_m)$ and $\sigma_o\Delta$ operators; Δt is the time step; φ^k, φ_e^k, φ_o^k and ν^k are time discretizations of ϕ, ϕ_e, ϕ_o and υ, respectively, for time equal to $k\Delta t$, with $0 \le k \le t_f/\Delta t$. A von Neumann analysis of the linearized system shows that the above scheme is unconditionally stable.

The elliptic equations of the third step are coupled by the boundary conditions (4), (5) and (6). The implementation of these and the other homogeneous Neumann-like conditions deserve further discussion. Without loss of generality, we initially focus on equation (11). By writing the boundary condition in the form $-(\boldsymbol{\sigma_i} + \boldsymbol{\sigma_e})\nabla\varphi_e^{k+1}.\eta = g$, the compatibility condition is

$$\int_{\partial\Omega} \boldsymbol{\sigma_i}\nabla\varphi^{k+1}.\eta = \int_{\Omega} \nabla.(\boldsymbol{\sigma_i}\nabla\varphi^{k+1}) = -\int_{\Omega} \nabla.((\boldsymbol{\sigma_i} + \boldsymbol{\sigma_e})\nabla\varphi_e^{k+1})$$

$$= -\int_{\partial\Omega} (\boldsymbol{\sigma_i} + \boldsymbol{\sigma_e})\nabla\varphi_e^{k+1}.\eta, \Rightarrow \int_{\partial\Omega} (\boldsymbol{\sigma_i}\nabla\varphi^{k+1} - g).\eta = 0. \tag{13}$$

Explicit schemes (see Latimer and Roth [1998], Skouibine and Krassowska [2000], Saleheen and Kwong [1998]) have been implemented with boundary conditions $\boldsymbol{\sigma_i}\nabla\varphi^{k+1}.\eta = -\boldsymbol{\sigma_i}\nabla\varphi_e^k.\eta$, $\boldsymbol{\sigma_e}\nabla\varphi_e^{k+1}.\eta = 0$ on $\partial\Omega$, applied to (8) and (11), respectively. Condition (13) becomes $\int_{\partial\Omega}\boldsymbol{\sigma_e}\nabla(\varphi_e^{k+1} - \varphi_e^k).\eta = 0$

and thus does not always hold. Perhaps such violation was not critical for the explicit schemes, since the CFL condition restricts Δt to very small values. In Sundnes et al. [2001], the following conditions were applied to (8) and (11): $\boldsymbol{\sigma_i}\nabla\varphi^{k+1}.\eta = 0$, $-(\boldsymbol{\sigma_i} + \boldsymbol{\sigma_e})\nabla\varphi_e^{k+1}.\eta = 0$ on $\partial\Omega$, which satisfy (13). These conditions are equivalent to (7) only if $\boldsymbol{\sigma_i} = \alpha\boldsymbol{\sigma_e}$, $\alpha \in \Re$, i.e., when both intra- and extracellular domains have equal anisotropy ratios. Therefore, they were used as approximations of (7), but did not properly implement the electric isolation for the general case of unequal anisotropy ratios, which is the case with cardiac tissue.

We implemented the following boundary conditions

$$\boldsymbol{\sigma_i}\nabla\varphi^{k+1}.\eta = -\boldsymbol{\sigma_i}\nabla\varphi_e^{k}.\eta, \tag{14}$$

$$-(\boldsymbol{\sigma_i} + \boldsymbol{\sigma_e})\nabla\varphi_e^{k+1}.\eta = \boldsymbol{\sigma_i}\nabla\varphi^{k+1}.\eta \,\text{on}\, \partial\Omega, \tag{15}$$

applied to (8) and (11), respectively. (14)-(15) are natural approximations of (7) and satisfy the compatibility condition (13). In addition, since we use the finite element method, conditions (14)-(15) are naturally implemented by the numerical scheme. The same properties apply to our bath-tissue interface conditions and to the homogeneous Neumann condition for ϕ_o:

$$\sigma_o\nabla\varphi_o^{k+1}.\eta = \boldsymbol{\sigma_e}\nabla\varphi_e^{k+1}.\eta, \tag{16}$$

$$\boldsymbol{\sigma_i}\nabla\varphi^{k+1}.\eta = -\boldsymbol{\sigma_i}\nabla\varphi_e^{k+1}.\eta, \,\text{on}\, \Gamma_{to}. \tag{17}$$

$$\sigma_o\nabla\varphi_o^{k+1}.\eta = 0 \,\text{on}\, \partial\Omega. \tag{18}$$

All boundary conditions cancel each other in the variational formulation:

$$\int_{\Omega_t} v\nabla.((\boldsymbol{\sigma_i} + \boldsymbol{\sigma_e})\nabla\varphi_e^{k+1}) + \int_{\Omega_o} v\sigma_o\Delta\varphi_o^{k+1} = -\int_{\Omega_t} v\nabla.(\boldsymbol{\sigma_i}\nabla\varphi^{k+1}),$$

$$\int_{\Omega_t} \nabla v(\boldsymbol{\sigma_i} + \boldsymbol{\sigma_e})\nabla\varphi_e^{k+1} - \int_{\partial\Omega+\Gamma_{to}} v(\boldsymbol{\sigma_i} + \boldsymbol{\sigma_e})\nabla\varphi_e^{k+1}.\eta + \int_{\Omega_o} \nabla v\sigma_o\nabla\varphi_o^{k+1}$$

$$-\int_{\partial\Omega+\Gamma_{to}} v\sigma_o\nabla\varphi_o^{k+1}.\eta = -\int_{\Omega_t} \nabla v\boldsymbol{\sigma_i}\nabla\varphi^{k+1} + \int_{\partial\Omega+\Gamma_{to}} v\boldsymbol{\sigma_i}\nabla\varphi^{k+1}.\eta,$$

where v is a test function. Using (15)-(18), all boundary integrals vanish:

$$\int_{\Omega_t} \nabla v(\boldsymbol{\sigma_i} + \boldsymbol{\sigma_e})\nabla\varphi_e^{k+1} + \int_{\Omega_o} \nabla v\sigma_o\nabla\varphi_o^{k+1} = -\int_{\Omega_t} \nabla v\boldsymbol{\sigma_i}\nabla\varphi^{k+1},$$

which is used to generate the finite element numerical approximation

$$Ax = b, \tag{19}$$

where A is the stiffness matrix, b is the load vector and x is the discretization of ϕ_e and ϕ_o. A uniform mesh of squares and bilinear polynomials were used. Spatial discretization was set to $\Delta x = 3.3\mu$m and the time step to $\Delta t = 10\mu$s.

4 Parallel preconditioners

We used CG to solve the linear system (19). We compare different preconditioners: ILU, additive Schwarz method (ASM) and multigrid (MG). The solution of (19) is implemented in parallel using the PETSc C library v2.1.5 (Balay et al. [2002]), which uses MPI. CG is parallelized via a linear domain decomposition. The spatial domain is decomposed into *proc* domains with equal sizes, where *proc* is the number of processors involved in the simulation.

The non-overlapping parallel version of the ILU preconditioner uses block Jacobi, i.e., an incomplete factorization is performed over the main diagonal block of the local part of the matrix A, thus avoiding extra communication. ILU has as parameter the level of fill-in, $fill$.

The ASM preconditioner implements an overlapping decomposition of the spatial domain Ω. Each processor block overlaps to the neighboring domain block by the amount ovl. An ILU is performed over each processor block. A greater ovl means more communication is necessary between the processors. The way the overlapping regions affect the residual can be controlled by the parameter *method* (*basic*, *restrict*, *interpolate* or *none*, see Balay et al. [2002], Cai and Sarkis [1999]).

The MG preconditioner performs a few iterations of PETSc's native geometric multigrid method. The parameter $levels$ indicates the number of different spatial grids involved in the solution. Based on the finest regular grid G_0, coarser regular grids were successively generated with half of the nodes on each direction (G_l, $l = 0$ to $levels - 1$). For each grid pair, G_l and G_{l+1}, a prolongation rectangular matrix, P_l, was generated using a bilinear interpolation scheme. The restriction operators were set to P_l^T and used to generate coarser tissue masks and conductivity tensors. For every grid level, a matrix A_l was generated by applying the finite element method. The *smoother* used for all but the coarsest level was a number of iterations, *smooth*, of the CG method which was, in turn, also preconditioned by ILU. For the coarsest level, we used a direct LU solver with nested dissection reordering. This was not done in parallel, i.e., it was repeated on every processor, avoiding any communication. In addition to the parameters $levels$, *smooth* and $fill$, we could control the *type* of multigrid (*multiplicative*, *additive*, $full$ and *kaskade*) and the cycle (v or w) (Balay et al. [2002]).

All complete and incomplete factorizations were performed only once, in the first time step of the simulation, so that the cost was amortized over the whole simulation. For the global preconditioned CG algorithm associated with (19) the stop criterion adopted was based on the unpreconditioned and absolute L_2 residual norm, $||Ax_i - b||_2 < tol$, where x_i was the solution at iteration i and *tol* was a tolerance which was set to 10^{-3}. Although this is not the most efficient stop criteria for the CG, it is the fairest one when comparing different preconditioning methods.

5 Results

We performed several comparisons of the different preconditioners on a 16-node HP Unix cluster, each node equipped with McKinley 900 MHz processors and 2 GB of DRAM and connected by a 1Gbit/s Ethernet switch. The electrical activity was initiated on the left side of the 2.6 × 2.6 mm tissue-bath preparation (640,000 nodes) and propagated to the right (see figure 1). The performance measurements reported in this section are relate to the solution of the elliptic system which is responsible for around 70% of the whole simulation time. We simulate 10ms of electrical activity (100 time steps). It is interesting to note that if the parabolic equation (8) was solved with an explicit method such as the forward-Euler method, the CFL condition would severely restrict Δt. An approximation of the CFL condition can be derived by assuming equal anisotropy ratios ($\boldsymbol{\sigma_i} = \alpha\boldsymbol{\sigma_e}$) and straight fibers:

$$\Delta t \leq \frac{\chi C_m \Delta x^2}{2(\sigma_l + \sigma_t)} = 0.07\mu s, \tag{20}$$

where $\sigma_* = \sigma_{i*}\sigma_{e*}/(\sigma_{i*} + \sigma_{e*})$, $(* = l, t)$. Numerically we verified that with $\Delta t = 0.07\mu s$ the forward-Euler scheme already did not converge. Thus, the semi-implicit based scheme allows Δt to be more than 100 times greater than one restricted by an explicit based method.

5.1 Parameter tuning

Several preconditioner parameters were tuned: $fill$ for ILU; *method*, *ovl* and $fill$ for ASM; *levels*, *type*, *cycle*, *smooth* and $fill$ for MG. Table 1 shows for different numbers of processors, the optimal parameter values, those combinations yielding the fastest execution time. The parameter $fill$ was set to 0, 1, 2, 5, 10, and 15; *method* to *basic*, *interpolate*, *restrict* and *none*; *ovl* to 1, 2, 4, 6, 8 and 10; *levels* to values from 2 to 7; *type* to *multiplicative*, *additive*, $full$ and *kaskade*; *cycle* to v and w; and *smooth* to values from 1 to 3. In addition, all parameters were tuned for best execution time on 1, 8 and 16 processors. A total of 3042 simulations were performed during more than two weeks of computation time.

Table 1. Values of parameters leading to the quickest solution time as a function of the number of processors (*proc*)

	ILU	ASM			MG			
proc	*fill*	*method*	*ovl*	*fill*	*levels*	*type*	*smooth*	*fill*
1	15				3	*kaskade*	1	0
8	5	*basic*	4	10	6	*kaskade*	1	0
16	5	*basic*	4	10	6	*kaskade*	1	0

For the ILU preconditioner, as *proc* increased from 1 to 16, the optimal value for $fill$ decreased from 15 to 5, i.e., as the domain was decomposed, it became less effective to increase $fill$ since the preconditioner became more expensive, but did not speed up the convergence. This was improved by ASM which took advantage of higher values of $fill$ by increasing communication, i.e., increasing ovl. The optimal values were $ovl = 4$ and $fill = 10$. For MG, the optimal value of $levels$ depended on *proc*. On a single processor, $levels = 3$ was fastest. In parallel, since the coarsest grid was solved sequentially, the cost of fewer grid $levels$ rivaled the gains of parallelism. Therefore, as *proc* increased, the optimal $levels$ also increased to 6. The other MG parameters indicate that the cheapest MG method was the best option, i.e, the $kaskade$ method with a single iteration of the CG smoother preconditioned by ILU(0). On the other hand, the best ASM method was the expensive $basic$ one which included all off-processor values in the interpolation and restriction processes.

5.2 Performance comparison and parallel speedup

Table 2 shows the execution time and number of CG iterations/time step as well as memory consumption (mem(MB))/processor for all preconditioners with the optimal parameters. The ASM preconditioner achieved better performance results than the non-overlapping ILU. ASM was 1.4 (1.5) times faster than ILU but required 20% (55%) more memory than ILU on 8 (16) *procs*. MG was between 15.5 (*proc* = 1) and 20.6 (*procs*=8) times faster than ILU and it required between 44% less memory (*proc*=1) and 7% more memory (*procs*=8) than ILU. Compared to ASM, MG was 14.9 (13.7) times faster than ASM and required 11% (32%) less memory than ASM on 8 (16) processors. All preconditioners achieved low parallel speedup (execution time with

	1 *proc*			8 *procs*			16 *procs*		
	t(s)	it	mem(MB)	t(s)	it	mem(MB)	t(s)	it	mem(MB)
ILU	96.2	98.7	1157.2	28.8	428.5	96.6	20.1	540.8	50.1
ASM				20.9	205.9	116.2	13.7	228.3	77.7
MG	6.3	7.7	649.6	1.4	11.3	103.4	1.0	13.6	52.5

Table 2. Best results of the preconditioners for different numbers of processors. Execution time per time step in seconds, t(s); CG iterations per time step, it; and memory usage per processor in MBytes, mem(MB)

proc=1/execution time) results with *procs*=16: 4.8 for ILU; 7.0 for ASM (related to ILU with *proc*=1); and 6.3 for MG. The reason was mainly due to the increase of the CG iterations with *proc*. The number of iterations was increased by a factor of 5.6 (2.4) by increasing *proc* from 1 to 16 for ILU (ASM). MG suffered less from this problem with an increase of only 1.7 times. Nevertheless, the speedup was poor. We believe the explanation lies for MG in the sequential direct solver. The cost of this was not reduced by increasing *proc*, and, thus, limited the total parallel speedup. For smaller *proc*, all preconditioners presented reasonable speedup, around 4.5 with *proc*=8.

5.3 Conclusions

In this work, we employed the conjugate gradient algorithm for the solution of the linear system associated with the elliptic part of the bidomain equations and compared different parallel preconditioning techniques, such as ILU, ASM and MG. The results taken from a 16-node HP-Unix cluster indicate that the multigrid preconditioner is at least 13 times faster than the single-level Schwarz based techniques and requires at least 11% less memory.

Acknowledgement. R. Weber dos Santos and S. Bauer acknowledge the support provided by 13N8125 BMFT German Ministry of Research and Technology. G. Plank was supported by the Austrian Science Fund FWF (R21-N04) and by a fund UGP4 of the Austrian Ministry of Education, Science and Culture. E. Vigmond is supported by a grant from the Natural Sciences and Engineering Research Council of Canada.

References

S. Balay, K. Buschelman, W. Gropp, D. Kaushik, M. Knepley, L. McInnes, B. Smith, and H. Zhang. PETSc users manual. Technical Report ANL-95/11 - Revision 2.1.5, Argonne National Laboratory, 2002.

X.-C. Cai and M. Sarkis. A restricted additive Schwarz preconditioner for general sparse linear systems. *SIAM Journal on Scientific Computing*, 21: 239–247, 1999.

J. Eason and R. Malkin. A simulation study evaluating the performance of high-density electrode arrays on myocardial tissue. *IEEE Trans Biomed Eng*, 47(7):893–901, 2000.

A. Hodgkin and A. Huxley. A quantitative description of membrane current and its application to conduction and excitation in nerve. *Journal of Physiology*, 117:500–544, 1952.

N. Hooke, C. Henriquez, P. Lanzkron, and D. Rose. Linear algebraic transformations of the bidomain equations: implications for numerical methods. *Math Biosci*, 120(2):127–45, 1994.

J. Keener and K. Bogar. A numerical method for the solution of the bidomain equations in cardiac tissue. *Chaos*, 8(1):234–241, 1998.

J. Keener and J. Sneyd. *Mathematical physiology.* Springer, 1998.

W. Krassowska and J. Neu. Effective boundary conditions for syncytial tissues. *IEEE Trans. on Biomed. Eng.*, 41:143–150, 1994.

D. Latimer and B. Roth. Electrical stimulation of cardiac tissue by a bipolar electrode in a conductive bath. *IEEE Trans. on Biomed. Eng.*, 45(12): 1449–1458, 1998.

D. Lindblad, C. Murphey, J. Clark, and W. Giles. A model of the action potential and the underlying membrane currents in a rabbit atrial cell. *The American Physiological Society*, (0363-6125):H1666–H1696, 1996.

Message Passing Interface library. MPI, a message-passing interface standard. *Int. J. Supercomp.*, 8:159–416, 1994.

L. Pavarino and P. Franzone. Parallel solution of cardiac reaction-diffusion models. In R. Kornhuber, R. Hoppe, D. Keyes, J. Periaux, O. Pironneau, and J. Xu, editors, *Procedings of the 15th International Conference on Domain Decomposition Methods, Lecture Notes in Computational Science and Engineering.* Springer, 2004.

M. Pennacchio and V. Simoncini. Efficient algebraic solution of reaction-diffusion systems for the cardiac excitation process. *Journal of Computational and Applied Mathematics*, 145(1):49–70, 2002. ISSN 0377-0427.

J. Pormann. Computer simulations of cardiac electrophysiology. In *Proceedings of SC2000*, 2000.

H. Saleheen and Kwong. A new three-dimensional finite-difference bidomain formulation for inhomogeneous anisotropic cardiac tissues. *IEEE Trans. on Biomed. Eng.*, 45(1):15–25, 1998.

K. Skouibine and W. Krassowska. Increasing the computational efficiency of a bidomain model of defibrillation using a time–dependent activating function. *Annals of Biomedical Engineering*, 28:772–780, 2000.

G. Strang. On the construction and comparison of difference scheme. *SIAM Journal on Numerical Analysis*, 5:506–517, 1968.

A. Street and R. Plonsey. Propagation in cardiac tissue adjacent to connective tissue: Two-dimensional modeling studies. *IEEE Transactions on Biomedical Engineering*, 46:19–25, 1999.

J. Sundnes, G. Lines, and A. Tveito. Efficient solution of ordinary differential equations modeling electrical activity in cardiac cells. *Math Biosci*, 172(2): 55–72, 2001.

E. Vigmond, F. Aguel, and N. Trayanova. Computational techniques for solving the bidomain equations in three dimensions. *IEEE Trans Biomed Eng*, 49(11):1260–9, 2002.

R. Weber dos Santos. *Modelling cardiac electrophysiology.* PhD thesis, Federal University of Rio de Janeiro, Mathematics dept., Rio de Janeiro, Brazil, 2002.

R. Weber dos Santos and F. Dickstein. On the influence of a volume conductor on the orientation of currents in a thin cardiac tissue. In I. Magnin, J. Montagnat, P. Clarysse, J. Nenonen, and T. Katila, editors, *Lecture Notes in Computer Science*, pages 111–121. Springer, Berlin, 2003.

R. Weber dos Santos, U. Steinhoff, E. Hofer, D. Sanchez-Quintana, and H. Koch. Modelling the electrical propagation in cardiac tissue using detailed histological data. *Biomedizinische Technik*, 2003.

C. Yung. Application of a stiff, operator-splitting scheme to the computational modeling of electrical propagation of cardiac ventricles. Engineering dept., Johns Hopkins University, Maryland, 2000.

Direct Schur Complement Method by Hierarchical Matrix Techniques

Wolfgang Hackbusch[1], Boris N. Khoromskij[2], and Ronald Kriemann[3]

[1] Max-Planck-Institute for Mathematics in the Sciences (MPI MIS), Leipzig (http://www.mis.mpg.de/scicomp/hackbusch_e.html)
[2] MPI MIS (http://www.mis.mpg.de/scicomp/khoromskij_e.html)
[3] MPI MIS (http://www.mis.mpg.de/scicomp/kriemann_e.html)

Summary. The goal of this paper is the construction of a data-sparse approximation to the Schur complement on the interface corresponding to FEM and BEM approximations of an elliptic equation by domain decomposition. Using the hierarchical ($\mathcal{H}$-matrix) formats we elaborate the *approximate Schur complement inverse* in an explicit form. The required cost $\mathcal{O}(N_\Gamma \log^q N_\Gamma)$ is almost linear in N_Γ – the number of degrees of freedom on the interface. As input, we use the Schur complement matrices corresponding to subdomains and represented in the $\mathcal{H}$-matrix format. In the case of piecewise constant coefficients these matrices can be computed via the BEM representation with the cost $\mathcal{O}(N_\Gamma \log^q N_\Gamma)$, while in the general case the FEM discretisation leads to the complexity $O(N_\Omega \log^q N_\Omega)$.

1 Introduction

In Hackbusch [2003], a direct domain decomposition method was described for rather general elliptic equations based on a traditional FEM. Using $\mathcal{H}$-matrix techniques, almost linear[4] cost in the number N_Ω of degrees of freedom in the computational domain could be achieved. Here we concentrate on the inversion of the Schur complement matrix. We distinguish three approaches to construct and approximate the Schur complement matrix: (a) Methods based on a traditional FEM for rather general variable coefficients (cf. Hackbusch [2003]); (b) Approximation by boundary concentrated FEM for smooth coefficients in subdomains (cf. Khoromskij and Melenk [2003]); (c) BEM based methods for piecewise constant coefficients (cf. Hsiao et al. [2001], Khoromskij and Wittum [2004], Langer and Steinbach [2003]). Below we focus on the cases (a) and (c). In the latter case, which is not covered by Hackbusch [2003], we have the standard advantages of BEM compared to FEM. Furthermore, besides the approximation theory (cf. Theorem 1), we can show (cf. Hackbusch

[4] By "almost linear" we mean $\mathcal{O}(N \log^q N)$ for a fixed q.

et al. [2003, submitted]) the approximability of the Schur complement in the $\mathcal{H}$-matrix format[5]. In both cases we give numerical results.

In a polygonal domain $\Omega \subset \mathbb{R}^2$, we consider the elliptic operator $\mathcal{L} : V \to V'$ for $V = H_0^1(\Omega)$ and $V' = H^{-1}(\Omega)$, with the corresponding V-elliptic bilinear form $a_\Omega : V \times V \to \mathbb{R}$,

$$a_\Omega(u,v) = \int_\Omega (\sum_{i,j=1}^{d} a_{ij}\partial_j u \partial_i v + a_0 uv)dx, \quad a_0 > 0. \tag{1}$$

The corresponding variational equation is: Find $u \in V$ such that

$$a_\Omega(u,v) = \langle f, v\rangle := (f,v)_{L^2(\Omega)} \qquad \text{for all } v \in V, \tag{2}$$

where $f \in H^{-1}(\Omega)$. We suppose the domain Ω to be composed of $M \geq 1$ possibly matching, but non-overlapping polygonal subdomains Ω_i, $\overline{\Omega} = \cup_{i=1}^{M} \overline{\Omega}_i$. We denote the interface (skeleton) of the decomposition of Ω by $\Gamma = \cup \Gamma_i$ ($\Gamma_i := \partial\Omega_i$). Because we focus on the solution of an interface equation, we suppose that the right-hand side f is supported only by the interface, such that

$$\langle f, v\rangle = \sum_{i=1}^{M} \langle \psi_i, v\rangle_{\Gamma_i}, \quad \psi_i \in H^{-1/2}(\Gamma_i). \tag{3}$$

An equation with general f can be reduced to the case (3) by a subtraction of particular solutions in the subdomains which can be performed in parallel.

We may write the bilinear form $a_\Omega(\cdot,\cdot)$ in (1) as a sum of local bilinear forms, $a_\Omega(u,v) = \sum_{i=1}^{M} a_{\Omega_i}(R_i u, R_i v)$, where $R_i : V \to V_i := H^1(\Omega_i)$ is the restriction of functions onto Ω_i and the integration in $a_{\Omega_i} : V_i \times V_i \to \mathbb{R}$ is restricted to Ω_i. Furthermore, a_{Ω_i} is supposed to be V_i-elliptic for $V_i := H_0^1(\Omega_i)$, i.e., there exist $0 < C_1 \leq C_2$ such that (for suitable constants $\mu_i > 0$)

$$C_1\mu_i|u|^2_{H^1(\Omega_i)} \leq a_{\Omega_i}(u,u) \leq C_2\mu_i|u|^2_{H^1(\Omega_i)} \qquad \text{for all } u \in H^1(\Omega_i). \tag{4}$$

We introduce the space $V_\Gamma \subset V$ of piecewise $\mathcal{L}$-harmonic functions by $V_\Gamma := \{v \in V : a_\Omega(v,z) = 0 \ \text{ for all } z \in V_0\}$, with $V_0 := \{v \in V : v(x) = 0 \text{ for all } x \in \Gamma\}$. Note that $V = V_0 + V_\Gamma$ is the orthogonal splitting with respect to scalar product $a_\Omega(\cdot,\cdot)$. The variational equation (2) with f satisfying (3), we next reduce to an *interface equation* (in fact, $u \in V_\Gamma$). To that end, let us introduce the following *trace space* on Γ, $Y_\Gamma := \{u = z_{|\Gamma} : z \in V\}$, $||u||_{Y_\Gamma} = \inf_{z\in V: z_{|\Gamma}=u} ||z||_V$, with the energy norm $||z||_V = \sqrt{a_\Omega(z,z)}$. Next we define the local Poincaré-Steklov operator (Dirichlet-Neumann map) on $\Gamma_i = \partial\Omega_i$, $\mathcal{T}_i : H^{1/2}(\Gamma_i) \to H^{-1/2}(\Gamma_i)$ by $\lambda \in H^{1/2}(\Gamma_i)$, $\mathcal{T}_i(\lambda) := \gamma_{1,i}u$. Here $\gamma_{1,i}u$ is the conormal derivative of u on Γ_i and u solves (2) in Ω_i such

[5] Details will be presented in the forthcoming paper (full version).

that $u|_{\Gamma_i} = \lambda$. Now we reduce (2) to the equivalent *interface problem:* Find $z = u_{|\Gamma} \in Y_\Gamma$ such that

$$b_\Gamma(z,v) := \sum_{i=1}^{M} \langle \mathcal{T}_i z_i, v_i \rangle_{\Gamma_i} = \langle \Psi_\Gamma, v \rangle := \sum_{i=1}^{M} \langle \psi_i, v \rangle_{\Gamma_i}, \quad \forall v \in Y_\Gamma, \tag{5}$$

where $b_\Gamma(\cdot,\cdot) : Y_\Gamma \times Y_\Gamma \to \mathbb{R}$ is a continuous bilinear form, $\Psi_\Gamma \in Y'_\Gamma$ and $z_i = z_{|\Gamma_i}$, $v_i = v_{|\Gamma_i}$.

To apply $\mathcal{H}$-matrix approximations to the discrete version of (5), we represent the inverse operator $\mathcal{L}^{-1}$ using the interface map $\mathcal{B}_\Gamma$ defined by $\langle \mathcal{B}_\Gamma u, v \rangle_\Gamma = b_\Gamma(u,v)$ for all $u, v \in Y_\Gamma$. The following statement describes the structure of the inverse $\mathcal{L}^{-1} : Y'_\Gamma \to V$.

Lemma 1. *The representation* $\mathcal{L}^{-1} = \mathcal{E}^{\mathrm{harm}}_{\Omega\leftarrow\Gamma} \mathcal{B}_\Gamma^{-1}$ *holds, where* $\mathcal{E}^{\mathrm{harm}}_{\Omega\leftarrow\Gamma} : Y_\Gamma \to V_\Gamma$ *is the* $\mathcal{L}$*-harmonic extension from* Y_Γ *to* V_Γ.

Proof. The bilinear form $b_\Gamma(\cdot,\cdot) : Y_\Gamma \times Y_\Gamma \to \mathbb{R}$ is symmetric, continuous and positive definite and thus the same holds for $\mathcal{B}_\Gamma$ and $\mathcal{B}_\Gamma^{-1} : Y'_\Gamma \to Y_\Gamma$. Therefore the operator $\mathcal{L}^{-1} = \mathcal{E}^{\mathrm{harm}}_{\Omega\leftarrow\Gamma} \mathcal{B}_\Gamma^{-1}$ is well-defined. Next, we check that $u = \mathcal{L}^{-1}\Psi_\Gamma$ solves (2). Green's formula yields

$$a_\Omega(u,v) = \sum_{i=1}^{M} a_{\Omega_i}(R_i u, R_i v) = \sum_{i=1}^{M} \langle \mathcal{T}_i u, v_i \rangle_{\Gamma_i} = \sum_{i=1}^{M} \langle \psi_i, v \rangle_{\Gamma_i} \quad \forall v \in V. \tag{6}$$

This also provides $\mathcal{B}_\Gamma^{-1}\Psi_\Gamma = u_{|\Gamma}$ completing the proof.

In the general case, we consider a conventional FEM approximation of (2) by piecewise linear elements on a regular triangulation that aligns with Γ. Let $\mathbf{A}_h \in \mathbb{R}^{I_\Omega \times I_\Omega}$ be the Galerkin-FEM stiffness matrix $\mathbf{A}_h = \begin{pmatrix} \mathbf{A}_{II} & \mathbf{A}_{II_\Gamma} \\ \mathbf{A}_{I_\Gamma I} & \mathbf{A}_{I_\Gamma I_\Gamma} \end{pmatrix}$, corresponding to the chosen FE space $V_h \subset V$. Here I_Γ is the index set corresponding to the interface degrees of freedom and $I = I_\Omega \setminus I_\Gamma$ is the complementary one. Eliminating all interior degrees of freedom corresponding to I, we obtain the so-called *FEM Schur complement* matrix $\mathbf{B}_{\Gamma,h} := \mathbf{A}_{I_\Gamma I_\Gamma} - \mathbf{A}_{I_\Gamma I}\mathbf{A}_{II}^{-1}\mathbf{A}_{II_\Gamma} \in \mathbb{R}^{I_\Gamma \times I_\Gamma}$, where $\mathbf{A}_{II} = \mathit{blockdiag}\{\mathbf{A}_1, ..., \mathbf{A}_M\}$ is the stiffness matrix for $\mathcal{L}$ subject to zero Dirichlet conditions on Γ, hence $\mathbf{A}_{II}^{-1} = \mathit{blockdiag}\{\mathbf{A}_1^{-1}, ..., \mathbf{A}_M^{-1}\}$ can be computed in parallel. In a standard way, each of the "substructure" matrices $\mathbf{A}_i^{-1}$ can be represented by the $\mathcal{H}$-matrix format (cf. Hackbusch [2003]).

Using $\mathbf{B}_{\Gamma,h}$, the original FEM system $\mathbf{A}_h U = F$ is reduced to the interface equation

$$\mathbf{B}_{\Gamma,h} U_\Gamma = F_\Gamma, \qquad U_\Gamma, F_\Gamma \in \mathbb{R}^{I_\Gamma}, \quad \text{where} \quad U_\Gamma = U|_{I_\Gamma}. \tag{7}$$

We construct the approximate direct solver for the Schur complement system (7) focusing on the cases of general and of piecewise constant coefficients. In the latter case, the matrix $\mathbf{B}_{\Gamma,h}$ can be computed by BEM with cost $\mathcal{O}(N_\Gamma \log^q N_\Gamma)$, where $N_\Gamma = card(I_\Gamma)$, while for general coefficients the cost is $O(N_\Omega \log^q N_\Omega)$ (cf. Hackbusch [2003]). Furthermore, $\mathbf{B}_{\Gamma,h}$ is proved to be of almost linear cost in N_Γ concerning operations for storage and for the matrix-by-vector multiplication. Due to the $\mathcal{H}$-matrix arithmetic, our approximate Schur complement inverse matrix $\mathbf{B}_{\Gamma,h}^{-1}$ again needs almost linear complexity $\mathcal{O}(N_\Gamma \log^q N_\Gamma)$.

Notice that our approach can be also viewed as an *approximate direct parallel solver* based on the domain decomposition Schur complement method.

2 FEM- and BEM-Galerkin Approximations

Introduce the FE trace space $Y_N := V_h|_\Gamma \subset Y_\Gamma$ with $N = N_\Gamma = dim Y_N$. Based on the representation in Lemma 1 and using the *$\mathcal{H}$-matrix approximation* to the operators involved, we can construct an approximate direct solver of almost linear complexity in N_Γ that realises the action $\mathcal{B}_\Gamma^{-1}\Psi_\Gamma$. For this purpose we split the *numerical realisation* of $\mathcal{L}^{-1} = \mathcal{E}^{\text{harm}}_{\Omega\leftarrow\Gamma}\mathcal{B}_\Gamma^{-1}$ into three independent steps: (i) Computation of a functional $\Psi_{\Gamma,h} \in Y'_\Gamma$ approximating Ψ_Γ; (ii) An $\mathcal{H}$-matrix approximation to the discrete interface operator $\mathcal{B}_\Gamma^{-1}$; (iii) Implementation of a discrete $\mathcal{L}$-harmonic extension operator $\mathcal{E}^{\text{harm}}_{\Omega\leftarrow\Gamma}$.

In Step (i) we define $\Psi_{\Gamma,h} \in Y'_N$ by $\langle \Psi_{\Gamma,h}, v\rangle_\Gamma := \sum_{i=1}^{M} \langle \psi_{ih}, v\rangle_{\Gamma_i}$ $\forall v \in Y_N$. Given $\Psi_{\Gamma,h} \in Y'_N$, we consider the Schur complement system approximating the interface equation (5). Let us define the local Schur complement operator $\mathcal{T}_{i,N}$ corresponding to the discrete $\mathcal{L}_i$-harmonic extension based on the FEM Galerkin space $V_{ih} := V_h|_{\Omega_i}$, by $\lambda, v \in Y_N|_{\Gamma_i}$: $\langle \mathcal{T}_{i,N}\lambda, v\rangle_{\Gamma_i} = A_{\Omega_i}(\overline{u}_i, \overline{v})$, where $\overline{u}_i \in V_{ih}$, $A_{\Omega_i}(\overline{u}_i, z) = 0$ for all $z \in V_{ih} \cap H_0^1(\Omega_i)$ and with an arbitrary $\overline{v} \in V_{ih}$ such that $\overline{v}|_{\Gamma_i} = v$. With the aid of the local FEM-Galerkin discretisations $\mathcal{T}_{i,N}$ of the Poincaré-Steklov maps $\mathcal{T}_i$, the discrete operator $\mathcal{B}_{\Gamma,N}$ and the corresponding interface equation are given by

$$z \in Y_N : \quad \langle \mathcal{B}_{\Gamma,N} z, v\rangle_\Gamma := \sum_{i=1}^{M} \langle \mathcal{T}_{i,N} z_i, v_i\rangle_{\Gamma_i} = \langle \Psi_{\Gamma,h}, v\rangle_\Gamma \qquad \text{for all } v \in Y_N,$$

where $v_i := v|_{\Gamma_i}$ and z is a desired approximation to the trace $u|_\Gamma$. The corresponding matrix representation to the interface operator $\mathcal{B}_{\Gamma,N}$ reads as

$$\langle \mathbf{B}_{\Gamma,N} U, Z\rangle_{I_\Gamma} = \sum_{i=1}^{M} \langle \mathbf{T}_{i,N} U_i, Z_i\rangle_{I_{\Gamma_i}} := \langle \mathcal{B}_{\Gamma,N} \mathcal{J} U, \mathcal{J} Z\rangle_\Gamma, \quad \mathbf{B}_{\Gamma,N} \in \mathbb{R}^{I_\Gamma \times I_\Gamma}, \tag{8}$$

where $\mathcal{J} : \mathbb{R}^{I_\Gamma} \to Y_N$ is the natural bijection from the coefficient vectors into the FE functions. $\mathbf{T}_{i,N}$ is the local FEM Schur complement matrix and $U_i, Z_i \in \mathbb{R}^{I_{\Gamma_i}}$, $i = 1, ..., M$, are the local vector components defined by $U_i = \mathbf{R}_{\Gamma_i} U$, $Z_i = \mathbf{R}_{\Gamma_i} Z$, where the connectivity matrix $\mathbf{R}_{\Gamma_i} \in \mathbb{R}^{I_{\Gamma_i} \times I_\Gamma}$ provides the restriction of the vector $Z \in \mathbb{R}^{I_\Gamma}$ onto the index set I_{Γ_i}. Let $\mathbf{A}_i$ be the local stiffness matrix corresponding to $a_{\Omega_i}(\cdot,\cdot)$, $\mathbf{A}_i = \begin{pmatrix} \mathbf{A}_{II} & \mathbf{A}_{I\Gamma_i} \\ \mathbf{A}_{\Gamma_i I} & \mathbf{A}_{\Gamma_i\Gamma_i} \end{pmatrix}$, where I and

Γ_i correspond to the interior and boundary index sets in Ω_i, respectively. Then we obtain the FEM Schur complement matrix $\mathbf{T}_{i,N} := \mathbf{A}_{\Gamma_i\Gamma_i} - \mathbf{A}_{\Gamma_i I}\mathbf{A}_{II}^{-1}\mathbf{A}_{I\Gamma_i}$, ($\mathbf{A}_{II}$: stiffness matrix for $\mathcal{L}_i$ subject to zero Dirichlet conditions on Γ_i). Thus, $\mathbf{A}_{II}^{-1}$ can be represented in the $\mathcal{H}$-matrix format (cf. Hackbusch [2003]).

Let us consider the explicit representation of $\mathbf{B}_{\Gamma,N}$ in (8) using the BEM-Galerkin approximation with Lagrange multipliers (cf. Hsiao et al. [2001]). Introduce the classical boundary integral representations involving operators $\mathcal{V}_i$, $\mathcal{D}_i$ and $\mathcal{K}_i$, defined by

$$(\mathcal{V}_i u)(x) = \int_{\Gamma_i} g(x,y)u(y)dy, \qquad (\mathcal{K}_i u)(x) = \int_{\Gamma_i} \frac{\partial}{\partial n_y} g(x,y)u(y)dy,$$

$$(\mathcal{K}_i' u)(x) = \int_{\Gamma_i} \frac{\partial}{\partial n_x} g(x,y)u(y)dy, \quad (\mathcal{D}_i u)(x) = -\frac{\partial}{\partial n_x}\int_{\Gamma_i} \frac{\partial}{\partial n_y} g(x,y)u(y)dy\,,$$

where $g(x,y)$ is the corresponding singularity function (cf. Hackbusch [1995]). In the following, we consider the model case $a_{\Omega_i}(u,v) := \mu_i \int_{\Omega_i} \nabla u \nabla v dx$, $\mu_i > 0$. Introduce the *modified Calderon projection* $\mathcal{C}_{\Gamma_i}$ by

$$C_{\Gamma_i}\begin{pmatrix} u_i \\ \delta_i \end{pmatrix} := \begin{pmatrix} \mu_i \mathcal{D} & \frac{1}{2}I + \mathcal{K}_i' \\ -\frac{1}{2}I - \mathcal{K}_i & \mu_i^{-1}\mathcal{V}_i \end{pmatrix}\begin{pmatrix} u_i \\ \delta_i \end{pmatrix} = \begin{pmatrix} \delta_i \\ 0 \end{pmatrix} \tag{9}$$

(cf. Khoromskij and Wittum [2004] and references therein), applied to the $\mathcal{L}_i$-harmonic function $\overline{u}$ which satisfies $-\Delta\overline{u} = 0$ in Ω_i with $\overline{u}_{|\Gamma_i} = u_i$ and $\delta_i = \mu_i \partial\overline{u}/\partial n$ (see also Costabel [1988], Hackbusch [1995], Wendland [1987]). Note that the Schur complement equation corresponding to (9) reads as

$$\mathcal{T}_i u_i := \mu_i \left(\mathcal{D}_i + \left(\tfrac{1}{2}I + \mathcal{K}_i'\right)\mathcal{V}_i^{-1}\left(\tfrac{1}{2}I + \mathcal{K}_i\right)\right) u_i = \delta_i, \tag{10}$$

providing an explicit symmetric representation to the Poincaré-Steklov map in terms of boundary integral operators.

Let us consider the skew-symmetric interface problem for $M > 1$ (see (11) below). Introducing the trace space $\Sigma_\Gamma := Y_\Gamma \times \Lambda_\Gamma$ with $\Lambda_\Gamma := \prod_{i=1}^{M} H^{-1/2}(\Gamma_i)$ and the weighted norm $\|P\|^2_{\Sigma_\Gamma} = \|u\|^2_{Y_\Gamma} + \sum_{j=1}^{M} \mu_j^{-1}\|\lambda_j\|^2_{H^{-1/2}(\Gamma_j)}$, $P = (u,\lambda) \in \Sigma_\Gamma$, $\lambda = (\lambda_1,\ldots,\lambda_M)$, we define the interface bilinear form $c_\Gamma : \Sigma_\Gamma \times \Sigma_\Gamma \to \mathbb{R}$ by $c_\Gamma(P,Q) := \sum_{i=1}^{M} \langle \mathcal{C}_{\Gamma_i} P_i, Q_i\rangle_{\Gamma_i}$ for all $P = (u,\lambda), Q = (v,\eta) \in \Sigma_\Gamma$, with $\mathcal{C}_{\Gamma_i}$ given by (9). Using the representation $\langle \mathcal{C}_{\Gamma_i} P_i, Q_i\rangle_{\Gamma_i} := \mu_i(\mathcal{D}_i u, v) + ((\frac{1}{2}I + \mathcal{K}_i')\lambda, v) - ((\frac{1}{2}I + \mathcal{K}_i)u, \eta) + \mu_i^{-1}(\mathcal{V}_i\lambda, \eta)$ in each subdomain, the original equation for u (cf. (2)) will be reduced to the skew-symmetric interface equation: *Given $\Psi_\Gamma \in Y_\Gamma'$, find $P \in \Sigma_\Gamma$ such that*

$$c_\Gamma(P,Q) \;=\; \langle \Psi_\Gamma, v\rangle_\Gamma \qquad \text{for all } Q = (v,\eta) \in \Sigma_\Gamma. \tag{11}$$

Let $\Lambda_h := \prod_{i=1}^{M} \Lambda_{ih}$, where Λ_{ih} is the FE space of piecewise linear functions. Introducing the FE Galerkin ansatz space $\Sigma_h := Y_N \times \Lambda_h$, we arrive at the corresponding BEM-Galerkin saddle-point system of equations: *Given* $\Psi_\Gamma \in Y'_\Gamma$, *find* $P_h = (u_h, \lambda_h) \in Y_N \times \Lambda_h$ *such that*

$$c_\Gamma(P_h, Q) = \langle \Psi_\Gamma, v\rangle_\Gamma \qquad \text{for all } Q = (v, \eta) \in Y_N \times \Lambda_h. \tag{12}$$

We further assume $\mathcal{V}_i$, $i = 1, ..., M$, to be positive definite.

Theorem 1. *(i) The bilinear form* $c_\Gamma : \Sigma_\Gamma \times \Sigma_\Gamma \to \mathbb{R}$ *is continuous and* Σ_Γ*-elliptic. (ii) Let* P_h *solve (12), then the optimal error estimate holds:*

$$\|P_h - P\|^2_{\Sigma_\Gamma} \le c \inf_{(w,\mu)\in\Sigma_h} \left[\sum_{i=1}^M \mu_i \|u_i - w_i\|^2_{H^{1/2}(\Gamma_i)} + \sum_{i=1}^M \mu_i^{-1} \|\lambda_i - \mu_i\|^2_{H^{-1/2}(\Gamma_i)}\right].$$

(iii) Let $\mathbf{T}_{i,BEM}$ *be the local BEM Schur complement given by*

$$\mathbf{T}_{i,BEM} := \mu_i \left(\mathbf{D}_{ih} + \left(\tfrac{1}{2}\mathbf{I}_{ih} + \mathbf{K}'_{ih}\right) \mathbf{V}_{ih}^{-1} \left(\tfrac{1}{2}\mathbf{I}_{ih} + \mathbf{K}_{ih}\right)\right), \tag{13}$$

where $\mathbf{D}_{ih}$, $\mathbf{K}_{ih}$, $\mathbf{V}_{ih}$ *are the Galerkin stiffness matrices of the boundary integral operators and* $\mathbf{I}_{ih}$ *is the corresponding mass matrix. Then the BEM Schur complement takes the explicit form* $\mathbf{B}_{\Gamma,N} = \sum_{i=1}^M \mathbf{R}^\top_{\Gamma_i} \mathbf{T}_{i,BEM} \mathbf{R}_{\Gamma_i} \in \mathbb{R}^{I_\Gamma \times I_\Gamma}$ *due to* $\langle \mathbf{B}_{\Gamma,N} Z, V\rangle_{I_\Gamma} = \sum_{i=1}^M \langle \mathbf{T}_{i,BEM} Z_i, V_i\rangle_{I_{\Gamma_i}} = \sum_{i=1}^M \langle \mathbf{R}^\top_{\Gamma_i} \mathbf{T}_{i,BEM} \mathbf{R}_{\Gamma_i} Z, V\rangle_{I_\Gamma}$.

Proof. Statements (i), (ii) are proven in Theorems 2, 3 in Hsiao et al. [2001], while (iii) is the direct consequence of the BEM-Galerkin approximation (12).

3 $\mathcal{H}$-Matrix Approximation to $\mathbf{B}^{-1}_{\Gamma,N}$ and Numerics

Now we discuss the $\mathcal{H}$-matrix approximation to $\mathbf{T}_{i,N}$ and $\mathbf{B}^{-1}_{\Gamma,N}$. In the FEM case let $\mathbf{A}_{II}$ be presented in the hierarchical format. Then we need the formatted multiplication and addition to obtain $\mathbf{T}_{i,N} = \mathbf{A}_{\Gamma_i\Gamma_i} - \mathbf{A}_{\Gamma_i I}\mathbf{A}^{-1}_{II}\mathbf{A}_{I\Gamma_i}$, leading to the cost $\mathcal{O}(N_{\Omega_i} \log^q N_{\Omega_i})$. The matrix $\mathbf{T}_{i,BEM}$ can be computed in $\mathcal{O}(N_{\Gamma_i} \log^q N_{\Gamma_i})$ operations. Note that the $\mathcal{H}$-matrix representations of $\mathbf{T}_{i,N}$ and $\mathbf{T}^{-1}_{i,N}$ can be applied within the so-called BETI iterative method Langer and Steinbach [2003].

Our goal is an algorithm of almost linear complexity in $N_\Gamma := \dim Y_N$ realising the matrix-by-vector multiplication by $\mathbf{B}^{-1}_{\Gamma,N}$. Having all local $\mathcal{H}$-matrices $\mathbf{T}_{i,N}$ available, we first compute the $\mathcal{H}$-matrix representation of $\mathbf{B}_{\Gamma,N}$. To that end, we construct an admissible hierarchical partitioning $P_2(I_\Gamma \times I_\Gamma)$ based on the cluster tree T_{I_Γ} of the skeleton index set I_Γ (cf. Figure 1, left). After some levels the clusters correspond to one-dimensional manifolds. Since a lower spatial dimension leads to better constants in the complexity estimates (cf. Grasedyck and Hackbusch [2003], Hackbusch et al. [2003, submitted]), this property makes the algorithm faster.

To calculate a low-rank approximation of blocks in the hierarchical partitioning $P_2(I_\Gamma \times I_\Gamma)$, we propose an SVD recompression of any block $b \in$

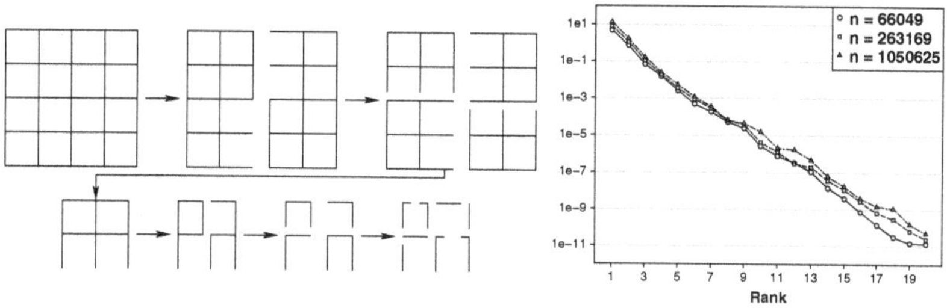

Fig. 1. Clustertree T_{I_Γ} (left); adaptive choice of the local rank (right)

$P_2(I_\Gamma \times I_\Gamma)$ obtained as a sum of fixed number of subblocks extracted as rank-k submatrices in the local Schur complements. This fast algorithm (of almost linear cost) exploits the hierarchical format of the local matrices $\mathbf{T}_{i,N}$ (same for $\mathbf{T}_{i,BEM}$) and will be presented in the next example. The following tables show numerical results for the scaled Laplacian in Ω_i with randomly chosen coefficients $\mu_i \in (0,1]$ (cf. (4)). Presented are the times for computing $\mathbf{T}_{i,N}$, $\mathbf{T}_{i,BEM}$, for the inversion of $B = \mathbf{B}_{\Gamma,N}$ and for its matrix-by-vector multiplication (MV) as well as for the accuracy of this inversion (computed on a SunFire 6800 (900 MHz)). The computing times $\mathbf{T}_{i,BEM}$ for $N_\Omega \approx 4 \cdot 10^6$ and $N_\Omega \approx 16 \cdot 10^6$ are 13.7 s and 36.8 s, respectively[6]. The results correspond to a decomposition of a square into 6×6 subsquares. One can see the almost linear complexity of the inversion algorithm. If we are interested in an efficient preconditioning, the local rank k can be chosen adaptively to achieve the required accuracy ε (cf. Fig. 1 (right) represents ε depending on k).

6×6 domains ($k = 9$)

N_Ω	N_Γ	$t(\mathbf{T}_{i,N})$	$t(\mathbf{T}_{i,BEM})$	$t(\mathbf{B}^{-1}_{\Gamma,N})$	$t(MV)$	$\Vert I - BB^{-1}_{\mathcal{H}}\Vert_2$
16 641	1 245	0.6 s	0.06 s	10.7 s	1.36_{10}-2 s	7.7_{10}-6
66 049	2 525	12.2 s	0.5 s	30.3 s	3.98_{10}-2 s	8.0_{10}-6
263 169	5 085	105.1 s	1.7 s	94.2 s	9.43_{10}-2 s	4.6_{10}-5
1 050 625	10 205	696.2 s	4.9 s	218.1 s	1.85_{10}-1 s	7.1_{10}-5

We present numerical results illustrating an acceleration factor of a *direct multilevel DDM* due to the recursive Schur complement evaluation (see Section 5.2 in Hackbusch [2003]). To reduce the cost of the local Schur complement matrices $\mathbf{T}_{i,N}$ in each subdomain Ω_i, one can apply the same domain decomposition algorithm as in Section 3 to the local inverse $\mathbf{A}_i^{-1}$. This leads to a reduction of the computational time. The following table corresponds to a 4×4 decomposition. We consider $q + 1 \geq 1$ grids $N_i = N_0 4^i$ with the problem size $N_i = N_0 4^i$, $i = 0, 1, ..., q$, and with $N_0 = 16641$, so that $N_3 = 1050625$. On each subdomain of level $\ell = 2, ..., q$ one has the matrix

[6] $t(\mathbf{T}_{i,BEM})$ includes only the dominating cost of two matrix-matrix multiplications and one matrix inversion in the $\mathcal{H}$-matrix format (cf. (13)).

size $N_{\ell-2}$, thus one can recursively apply the algorithm on level $\ell-2$ to compute the local inverse matrix $A_{i,\ell}^{-1}$ on level ℓ. The complexity bound satisfies the recursion $W(A_{i,\ell}^{-1}) = 16W(A_{i,\ell-2}^{-1}) + W(B_{\Gamma,\ell-2}^{-1})$, $W(\cdot)$: cost of the corresponding matrix operation. Based on the table below, the simple calculation $W^{ML}(A_{4,\ell}^{-1}) = 16(16 \times 0.1 + 0.8) + 16.9 \approx 1\,min$ shows an acceleration factor about 33 compared with 2020 *sec* depicted in the last line of our table. Similarly, an extrapolation using the two smaller grids exhibits that our direct solver applied to the problems with $n_\Omega = 4 \cdot 10^6$ and $n_\Omega = 16 \cdot 10^6$ would take about 113 sec. and 1080 sec., respectively, for each subdomain.

4 × 4 domains ($k = 9$)

N_Ω	N_Γ	$t(\mathbf{T}_{i,N})$	$t(\mathbf{B}_{\Gamma,N}^{-1})$	$t(MV)$	$\|I - BB_{\mathcal{H}}^{-1}\|_2$
16 641	753	3.8 s	3.7 s	3.20_{10}-3 s	4.2_{10}-6
66 049	1 521	43.2 s	16.9 s	9.10_{10}-3 s	7.7_{10}-6
263 169	3 057	317.4 s	48.3 s	4.18_{10}-2 s	1.3_{10}-5
1 050 625	6 129	2 020.1 s	118.8 s	8.92_{10}-1 s	2.1_{10}-5

References

M. Costabel. Boundary integral operators on lipschitz domains: elementary results. *SIAM J. Math. Anal.*, 19:613–625, 1988.

L. Grasedyck and W. Hackbusch. Construction and arithmetics of $\mathcal{H}$-matrices. *Computing*, 70:295–334, 2003.

W. Hackbusch. *Integral equations. Theory and numerical treatment*, volume 128 of *ISNM*. Birkhäuser, Basel, 1995.

W. Hackbusch. Direct domain decomposition using the hierarchical matrix technique. In I. Herrera, D. Keyes, O. Widlund, and R. Yates, editors, *DDM14 Conference Proceedings*, pages 39–50, Mexico City, Mexico, 2003. UNAM.

W. Hackbusch, B. Khoromskij, and R. Kriemann. Hierarchical matrices based on a weak admissibility criterion. Technical Report 2, MPI for Math. in the Sciences, Leipzig, 2003, submitted.

G. Hsiao, B. Khoromskij, and W. Wendland. Preconditioning for boundary element methods in domain decomposition. *Engineering Analysis with Boundary Elements*, 25:323–338, 2001.

B. Khoromskij and M. Melenk. Boundary concentrated finite element methods. *SIAM J. Numer. Anal.*, 41:1–36, 2003.

B. Khoromskij and G. Wittum. *Numerical solution of elliptic differential equations by reduction to the interface.* Number 36 in LNCSE. Springer, 2004.

U. Langer and O. Steinbach. Boundary element tearing and interconnecting methods. *Computing*, To appear, 2003.

W. Wendland. Strongly elliptic boundary integral equations. In A. Iserles and M. Powell, editors, *The state of the art in numerical analysis*, pages 511–561, Oxford, 1987. Clarendon Press.

Balancing Neumann-Neumann Methods for Elliptic Optimal Control Problems

Matthias Heinkenschloss and Hoang Nguyen

Rice University, Department of Computational and Applied Mathematics
(`http://www.caam.rice.edu/~heinken/`)

Summary. We present Neumann-Neumann domain decomposition preconditioners for the solution of elliptic linear quadratic optimal control problems. The preconditioner is applied to the optimality system. A Schur complement formulation is derived that reformulates the original optimality system as a system in the state and adjoint variables restricted to the subdomain boundaries. The application of the Schur complement matrix requires the solution of subdomain optimal control problems with Dirichlet boundary conditions on the subdomain interfaces. The application of the inverses of the subdomain Schur complement matrices require the solution of subdomain optimal control problems with Neumann boundary conditions on the subdomain interfaces. Numerical tests show that the dependence of this preconditioner on mesh size and subdomain size is comparable to its counterpart applied to elliptic equations only.

1 Introduction

We are interested in domain decomposition methods for the solution of large-scale linear quadratic problems

$$\text{minimize } \frac{1}{2}\mathbf{y}^T\mathbf{M}\mathbf{y} + \mathbf{c}^T\mathbf{y} + \mathbf{y}^T\mathbf{N}\mathbf{u} + \frac{\alpha}{2}\mathbf{u}^T\mathbf{H}\mathbf{u} + \mathbf{d}^T\mathbf{u}, \tag{1a}$$

$$\text{subject to } \mathbf{A}\mathbf{y} + \mathbf{B}\mathbf{u} + \mathbf{b} = 0, \tag{1b}$$

arising from the finite element discretization of elliptic optimal control problems. In (1) $\mathbf{y} \in \mathbb{R}^m$, $\mathbf{u} \in \mathbb{R}^n$ are called the (discretized) state and the (discretized) control, respectively, and $\mathbf{A}\mathbf{y}+\mathbf{B}\mathbf{u}+\mathbf{b} = 0$ is called the (discretized) state equation. Throughout this paper we assume that

A. $\mathbf{A} \in \mathbb{R}^{m\times m}$ is invertible, $\mathbf{B} \in \mathbb{R}^{m\times n}$, $\mathbf{N} \in \mathbb{R}^{m\times n}$, $\mathbf{M} \in \mathbb{R}^{m\times m}$ is symmetric, $\mathbf{H} \in \mathbb{R}^{n\times n}$ is symmetric and the reduced Hessian $\widehat{\mathbf{H}} = \alpha\mathbf{H} - \mathbf{B}^T\mathbf{A}^{-T}\mathbf{N} - \mathbf{N}^T\mathbf{A}^{-1}\mathbf{B} + \mathbf{B}^T\mathbf{A}^{-T}\mathbf{M}\mathbf{A}^{-1}\mathbf{B}$ is positive definite.

The assumption that $\widehat{\mathbf{H}}$ is positive definite is equivalent to the assumption that the Hessian of (1a) is positive definite on the null-space of the linear constraints (1b). Under the assumption A, the necessary and sufficient optimality conditions for (1) are given by

$$\begin{pmatrix} \mathbf{M} & \mathbf{N} & \mathbf{A}^T \\ \mathbf{N}^T & \alpha\mathbf{H} & \mathbf{B}^T \\ \mathbf{A} & \mathbf{B} & 0 \end{pmatrix} \begin{pmatrix} \mathbf{y} \\ \mathbf{u} \\ \mathbf{p} \end{pmatrix} = \begin{pmatrix} \mathbf{c} \\ \mathbf{d} \\ \mathbf{b} \end{pmatrix}. \tag{2}$$

The system matrix in (2) is also called a KKT (Karush-Kuhn-Tucker) matrix. Large-scale linear quadratic problems of the form (1) arise as subproblems in Newton or sequential quadratic programming (SQP) type optimization algorithms for nonlinear PDE constrained optimization problems. The solution of these subproblems is a very time consuming part in Newton or SQP type optimization algorithms and therefore the development of preconditioners for such problems is of great interest. Domain decomposition methods for steady state optimal control problems were considered in Benamou [1996], Bounaim [1998], Dennis and Lewis [1994], Biros and Ghattas [2000], Lions and Pironneau [1998] and other preconditioners for the system matrix in (2) are discussed, e.g., by Ascher and Haber [2003], Battermann and Sachs [2001], Hoppe et al. [2002], Keller et al. [2000]. Although (2) is a saddle point problem, its structure is quite different from the saddle point problems arising, e.g., from the Stokes problem (see, e.g.,Pavarino and Widlund [2002]) or from mixed finite element discretizations of elliptic PDEs.

We present a Neumann-Neumann (NN) domain decomposition preconditioner for the solution of discretized elliptic linear quadratic optimal control problems. The preconditioner is applied to the optimality system (2). A Schur complement formulation is derived that reformulates (2) as a system in the state and adjoint variables restricted to the subdomain boundaries. The application of the Schur complement matrix requires the solution of subdomain optimal control problems with Dirichlet boundary conditions on the subdomain interfaces. The application of the inverses of the subdomain Schur complement matrices require the solution of subdomain optimal control problems with Neumann boundary conditions on the subdomain interfaces. Our numerical tests in Section 4 show that the dependence of this preconditioner on mesh size and subdomain size is comparable to that of its counterpart applied to elliptic PDEs only. Numerical tests also indicate that, unlike several other KKT preconditioners, the proposed NN preconditioner is rather insensitive to the choice of the penalty parameter α. Unlike several other KKT preconditioners, our preconditioner does not require a preconditioner for the reduced Hessian $\widehat{\mathbf{H}}$, which is often difficult to obtain. Due to page limitations, we only present the algebraic view of the preconditioner. For more details we refer to Heinkenschloss and Nguyen [2004].

2 The Example Problem

We are interested in the solution $y \in H^1(\Omega)$, $u \in L^2(\partial\Omega)$ of the optimal control problem

$$\text{minimize} \quad \frac{1}{2}\int_\Omega (y(x) - \hat{y}(x))^2 dx + \frac{\alpha}{2}\int_{\partial\Omega} u^2(x)dx, \tag{3a}$$

$$\text{subject to} \quad a(y,\psi) + b(u,\psi) = \int_\Omega f(x)\psi(x)dx \quad \forall\psi \in H^1(\Omega), \tag{3b}$$

where $a(y,\psi) = \int_\Omega \nabla y(x)\nabla\psi(x) + y(x)\psi(x)dx$ and $b(u,\psi) = -\int_{\partial\Omega} u(x)\psi(x)dx$. The desired state $\hat{y} \in L^2(\Omega)$ and $f \in L^2(\Omega)$ are given functions, and $\alpha > 0$ is a given parameter. It is shown in Lions [1971] that (3) has a unique solution.

We discretize (3) using conforming finite elements. Let $\{T_l\}$ be a triangulation of Ω. We divide Ω into nonoverlapping subdomains Ω_i, $i = 1,\dots,d$, such that each T_l belongs to exactly one $\overline{\Omega}_i$. We approximate the state y by a function $y_h = \sum_k y_k\psi_k$ which is continuous on Ω and linear on each T_l. Our discretized controls u_h are not chosen to be continuous and piecewise linear on $\partial\Omega$ (see the left plot in Figure 1). A domain decomposition formulation based on such a discretization would introduce 'interface controls' (dotted hat function in the left plot in Figure 1) defined on a 'band' of width $O(h)$ around $\partial\Omega\cap\partial\Omega_i\cap\partial\Omega_j$, $i \neq j$. Since the evaluation of $u \in L^2(\partial\Omega)$ on $\partial\Omega\cap\partial\Omega_i\cap\partial\Omega_j$ does not make sense, we avoid interface controls. We discretize the control u by a function $u_h = \sum_k u_k\mu_k$ which is continuous on each $\partial\Omega_i$, $i = 1,\dots,d$, and linear on each $\partial\Omega\cap\partial\Omega_i\cap T_l$. The discretized control u_h is not assumed to be continuous at $\partial\Omega\cap\partial\Omega_i\cap\partial\Omega_j$, $i \neq j$. In particular, for each point $x_k \in \partial\Omega\cap\partial\Omega_i\cap\partial\Omega_j$, $i \neq j$, there are two discrete controls u_{k_i}, u_{k_j} belonging to subdomains Ω_i and Ω_j, respectively (see the right plot in Figure 1). Hence, our control discretization depends on the partition $\{\Omega_i\}_{i=1}^d$ of the domain Ω.

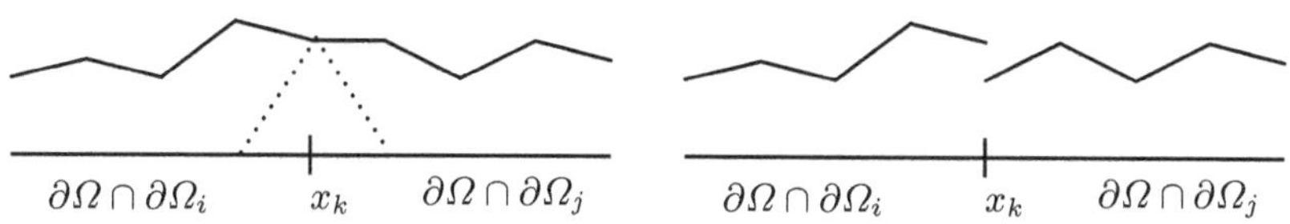

Fig. 1. Sketch of the Control Discretization for the Case $\Omega \subset \mathbb{R}^2$

3 The Domain Decomposition Preconditioners

We define

$$\mathbf{K}^i_{\Gamma\Gamma} = \begin{pmatrix} \mathbf{M}^i_{\Gamma\Gamma} & (\mathbf{A}^i_{\Gamma\Gamma})^T \\ \mathbf{A}^i_{\Gamma\Gamma} & \end{pmatrix}, \quad \mathbf{A}^i = \begin{pmatrix} \mathbf{A}^i_{II} & \mathbf{A}^i_{I\Gamma} \\ \mathbf{A}^i_{\Gamma I} & \mathbf{A}^i_{\Gamma\Gamma} \end{pmatrix}, \quad \mathbf{M}^i = \begin{pmatrix} \mathbf{M}^i_{II} & \mathbf{M}^i_{I\Gamma} \\ \mathbf{M}^i_{\Gamma I} & \mathbf{M}^i_{\Gamma\Gamma} \end{pmatrix},$$

$i = 1, \ldots, d$, and $\mathbf{K}_{\Gamma\Gamma} = \sum_{i=1}^d \mathbf{K}^i_{\Gamma\Gamma}$ $\mathbf{x}_\Gamma = \begin{pmatrix} \mathbf{y}_\Gamma \\ \mathbf{p}_\Gamma \end{pmatrix}$, $\mathbf{g}_\Gamma = \begin{pmatrix} \mathbf{c}_\Gamma \\ \mathbf{b}_\Gamma \end{pmatrix}$. Furthermore, for indices i with $\partial\Omega_i \cap \partial\Omega \neq \emptyset$, we define

$$\mathbf{K}^i_{II} = \begin{pmatrix} \mathbf{M}^i_{II} & \mathbf{N}^i_{II} & (\mathbf{A}^i_{II})^T \\ (\mathbf{N}^i_{II})^T & \mathbf{H}^i_{II} & (\mathbf{B}^i_{II})^T \\ \mathbf{A}^i_{II} & \mathbf{B}^i_{II} & \end{pmatrix}, \mathbf{K}^i_{\Gamma I} = \begin{pmatrix} \mathbf{M}^i_{\Gamma I} & \mathbf{N}^i_{\Gamma I} & (\mathbf{A}^i_{I\Gamma})^T \\ \mathbf{A}^i_{\Gamma I} & \mathbf{B}^i_{\Gamma I} & \end{pmatrix},$$

$$\mathbf{B}^i = \begin{pmatrix} \mathbf{B}^i_{II} \\ \mathbf{B}^i_{\Gamma I} \end{pmatrix}, \quad \mathbf{N}^i = \begin{pmatrix} \mathbf{N}^i_{II} \\ \mathbf{N}^i_{\Gamma I} \end{pmatrix}, \quad \mathbf{x}^i_I = \begin{pmatrix} \mathbf{y}^i_I \\ \mathbf{u}^i_I \\ \mathbf{p}^i_I \end{pmatrix}, \quad \mathbf{g}^i_I = \begin{pmatrix} \mathbf{c}^i_I \\ \mathbf{d}^i_I \\ \mathbf{b}^i_I \end{pmatrix},$$

and for indices i with $\partial\Omega_i \cap \partial\Omega = \emptyset$, we define

$$\mathbf{K}^i_{II} = \begin{pmatrix} \mathbf{M}^i_{II} & (\mathbf{A}^i_{II})^T \\ \mathbf{A}^i_{II} & \end{pmatrix}, \quad \mathbf{K}^i_{\Gamma I} = \begin{pmatrix} \mathbf{M}^i_{\Gamma I} & (\mathbf{A}^i_{I\Gamma})^T \\ \mathbf{A}^i_{\Gamma I} & \end{pmatrix}, \quad \mathbf{x}^i_I = \begin{pmatrix} \mathbf{y}^i_I \\ \mathbf{p}^i_I \end{pmatrix}, \quad \mathbf{g}^i_I = \begin{pmatrix} \mathbf{c}^i_I \\ \mathbf{b}^i_I \end{pmatrix}.$$

Most of this notation is a direct adaption of the notation used for domain decomposition of PDEs (see, e.g., Smith et al. [1996]). For example, $\mathbf{y}^i_I$ is the subvector containing the coefficients y_k of the discretized state belonging to nodes x_k in the interior of Ω_i. Note that in our particular control discretization, all basis functions μ_k for the discretised control u_h have support in only one subdomain boundary $\partial\Omega_i$ (see the right plot in Figure 1). Consequently, there is no $\mathbf{u}_\Gamma$.

After a symmetric permutation, (2) can be written as

$$\begin{pmatrix} \mathbf{K}^1_{II} & & & (\mathbf{K}^1_{\Gamma I})^T \\ & \ddots & & \vdots \\ & & \mathbf{K}^d_{II} & (\mathbf{K}^d_{\Gamma I})^T \\ \mathbf{K}^1_{\Gamma I} & \cdots & \mathbf{K}^d_{\Gamma I} & \mathbf{K}_{\Gamma\Gamma} \end{pmatrix} \begin{pmatrix} \mathbf{x}^1_I \\ \vdots \\ \mathbf{x}^d_I \\ \mathbf{x}_\Gamma \end{pmatrix} = \begin{pmatrix} \mathbf{g}^1_I \\ \vdots \\ \mathbf{g}^d_I \\ \mathbf{g}_\Gamma \end{pmatrix}. \tag{4}$$

Frequently, we use the compact notation

$$\begin{pmatrix} \mathbf{K}_{II} & \mathbf{K}^T_{\Gamma I} \\ \mathbf{K}_{\Gamma I} & \mathbf{K}_{\Gamma\Gamma} \end{pmatrix} \begin{pmatrix} \mathbf{x}_I \\ \mathbf{x}_\Gamma \end{pmatrix} = \begin{pmatrix} \mathbf{g}_I \\ \mathbf{g}_\Gamma \end{pmatrix}, \tag{5}$$

or even $\mathbf{Kx} = \mathbf{g}$ instead of (4). We make the following assumptions.

B. $\mathbf{A}^i_{II} \in \mathbb{R}^{m^I_i \times m^I_i}$ is invertible and $\mathbf{M}^i_{II} \in \mathbb{R}^{m^I_i \times m^I_i}$ is symmetric, $i = 1, \ldots, d$.
For i with $\partial\Omega_i \cap \partial\Omega \neq \emptyset$, $\mathbf{H}^i_{II} \in \mathbb{R}^{n^I_i \times n^I_i}$ is symmetric and $\widehat{\mathbf{H}}^i_{II} = \alpha\mathbf{H}^i_{II} - (\mathbf{B}^i_{II})^T(\mathbf{A}^i_{II})^{-T}\mathbf{N}^i_{II} - (\mathbf{N}^i_{II})^T(\mathbf{A}^i_{II})^{-1}\mathbf{B}^i_{II} + (\mathbf{B}^i_{II})^T(\mathbf{A}^i_{II})^{-T}\mathbf{M}^i_{II}(\mathbf{A}^i_{II})^{-1}\mathbf{B}^i_{II}$ is positive definite.

C. $\mathbf{A}^i \in \mathbb{R}^{m_i \times m_i}$ is invertible and $\mathbf{M}^i \in \mathbb{R}^{m_i \times m_i}$ is symmetric, $i = 1, \ldots, d$.
For i with $\partial\Omega_i \cap \partial\Omega \neq \emptyset$, $\mathbf{H}^i_{II} \in \mathbb{R}^{n^I_i \times n^I_i}$ is symmetric and $\widehat{\mathbf{H}}^i = \alpha\mathbf{H}^i_{II} - (\mathbf{B}^i)^T(\mathbf{A}^i)^{-T}\mathbf{N}^i - (\mathbf{N}^i)^T(\mathbf{A}^i)^{-1}\mathbf{B}^i + (\mathbf{B}^i)^T(\mathbf{A}^i)^{-T}\mathbf{M}^i(\mathbf{A}^i)^{-1}\mathbf{B}^i$ is positive definite.

Assumptions A, B, C are satisfied for our example problem.

Assumption B guarantees that $\mathbf{K}_{II}$ is invertible. Hence, we can form the Schur complement system

$$\mathbf{S}\mathbf{x}_\Gamma = \mathbf{r}, \tag{6}$$

where $\mathbf{S} = \mathbf{K}_{\Gamma\Gamma} - \mathbf{K}_{\Gamma I}\mathbf{K}_{II}^{-1}\mathbf{K}_{\Gamma I}^T$ and $\mathbf{r} = \mathbf{g}_\Gamma - \mathbf{K}_{\Gamma I}\mathbf{K}_{II}^{-1}\mathbf{g}_I$. The Schur complement matrix $\mathbf{S}$ can be written as a sum of subdomain Schur complement matrices. Let $\tilde{\mathbf{R}}_i^y$, $i = 1, \ldots, d$, be the restriction operator which maps from the vector of coefficient unknowns on the artificial boundary, $\mathbf{y}_\Gamma$, to only those associated with the boundary of Ω_i. Let

$$\tilde{\mathbf{R}}_i = \begin{pmatrix} \tilde{\mathbf{R}}_i^y & \\ & \tilde{\mathbf{R}}_i^p \end{pmatrix}, \quad \tilde{\mathbf{R}}_i^p = \tilde{\mathbf{R}}_i^y \tag{7}$$

The Schur complement can be written as $\mathbf{S} = \sum_i \tilde{\mathbf{R}}_i^T \mathbf{S}_i \tilde{\mathbf{R}}_i$, where $\mathbf{S}_i = \mathbf{K}_{\Gamma\Gamma}^i - \mathbf{K}_{\Gamma I}^i (\mathbf{K}_{II}^i)^{-1} (\mathbf{K}_{\Gamma I}^i)^T$. It is shown in Heinkenschloss and Nguyen [2004] that the application $\mathbf{S}_i$ to a vector $\tilde{\mathbf{R}}_i(\mathbf{y}_\Gamma^T, \mathbf{p}_\Gamma^T)^T$ corresponds to solving a subdomain optimal control problem in Ω_i with Dirichlet boundary conditions for the state on $\partial\Omega_i \setminus \partial\Omega$ and then extracting Neumann data of the optimal state and corresponding adjoint on $\partial\Omega_i \setminus \partial\Omega$.

Theorem 1. *If Assumptions* A *and* B *are valid, then the Schur complement matrix* $\mathbf{S}$ *has* $m - \sum_{i=1}^d m_i^I$ *positive and* $m - \sum_{i=1}^d m_i^I$ *negative eigenvalues. If Assumptions* B *and* C *are valid, then the subdomain Schur complement matrix* $\mathbf{S}_i$, $i = 1, \ldots, d$, *has* $m_i - m_i^I$ *positive and* $m_i - m_i^I$ *negative eigenvalues.*

Proof. Recall that $\mathbf{S} = \mathbf{K}_{\Gamma\Gamma} - \mathbf{K}_{\Gamma I}\mathbf{K}_{II}^{-1}\mathbf{K}_{\Gamma I}^T$. It is easy to verify that

$$\begin{pmatrix} \mathbf{K}_{II} & \mathbf{K}_{\Gamma I}^T \\ \mathbf{K}_{\Gamma I} & \mathbf{K}_{\Gamma\Gamma} \end{pmatrix} = \begin{pmatrix} \mathbf{K}_{II} & 0 \\ \mathbf{K}_{\Gamma I} & \mathbf{I} \end{pmatrix} \begin{pmatrix} \mathbf{K}_{II}^{-1} & 0 \\ 0 & \mathbf{S} \end{pmatrix} \begin{pmatrix} \mathbf{K}_{II} & \mathbf{K}_{\Gamma I}^T \\ 0 & \mathbf{I} \end{pmatrix}.$$

The matrix $\mathbf{K}$ is a symmetric permutation of the system matrix in (2) and, hence, both matrices have the same eigenvalues. It is well known that the system matrix in (2) and, hence, $\mathbf{K}$ has $m+n$ positive and m negative eigenvalues (see, e.g., Keller et al. [2000]). Similarly, the matrix $\mathbf{K}_{II}$ has $\sum_{i=1}^d m_i^I + n_i^I$ positive and $\sum_{i=1}^d m_i^I$ negative eigenvalues. By Sylvester's law of inertia, the number of positive [negative] eigenvalues of $\mathbf{K}$ is equal to the number of positive [negative] eigenvalues of $\mathbf{K}_{II}^{-1}$ plus the number of positive [negative] eigenvalues of $\mathbf{S}$. Since $n = \sum_{i=1}^d n_i^I$, this implies that $\mathbf{S}$ has $m - \sum_{i=1}^d m_i^I$ positive and $m - \sum_{i=1}^d m_i^I$ negative eigenvalues.

The second assertion can be proven analogously.

If Assumption C is valid, then $\mathbf{S}_i^{-1}$ exists. It is shown in Heinkenschloss and Nguyen [2004] that the application $\mathbf{S}_i^{-1}$ to a vector $\tilde{\mathbf{R}}_i(\mathbf{v}_\Gamma^T, \mathbf{q}_\Gamma^T)^T$ corresponds to solving a subdomain optimal control problem in Ω_i with Neumann boundary conditions for the state on $\partial\Omega_i \setminus \partial\Omega$ and then extracting Dirichlet data of the optimal state and corresponding adjoint on $\partial\Omega_i \setminus \partial\Omega$.

It is now relatively easy to generalize the Neumann-Dirichlet and Neumann-Neumann preconditioners used in the context of elliptic PDEs to the optimal control context. We focus on Neumann-Neumann (NN) preconditioners.

Let $\mathbf{D}_i^y$ be the diagonal matrix, whose entries are computed as follows. If $x_k \in \partial\Omega_i$, then $(\mathbf{D}_i^y)_{kk}^{-1}$ is the number of subdomains that share node x_k. Note that $\sum_i \mathbf{D}_i^y = \mathbf{I}$. Furthermore, let $\tilde{\mathbf{D}}_i^p = \tilde{\mathbf{D}}_i^y$ and

$$\mathbf{D}_i = \begin{pmatrix} \mathbf{D}_i^y & \\ & \mathbf{D}_i^p \end{pmatrix}.$$

If assumptions B, C are valid, then we can form $\mathbf{S}_i$ and $\mathbf{S}_i^{-1}$. In this case the one-level NN preconditioner is given by

$$\mathbf{P} = \sum_i \mathbf{D}_i \tilde{\mathbf{R}}_i^T \mathbf{S}_i^{-1} \tilde{\mathbf{R}}_i \mathbf{D}_i. \tag{8}$$

It is well known that the performance of one-level NN preconditioners for elliptic PDEs deteriorates fast as the number of subdomains increases. The same is observed for the NN preconditioner (8) in the optimal control context (see Section 4). To avoid this, we include a coarse grid. More precisely, we adapt the balanced NN preconditioner due to Mandel [1993] to the optimal control context. Following the description in [Smith et al., 1996, Sect. 4.3.3], the balanced-NN for the optimal control problem is given by

$$\mathbf{P} = \left(\mathbf{I} - \tilde{\mathbf{R}}_0^T \mathbf{S}_0^{-1} \tilde{\mathbf{R}}_0 \mathbf{S}\right) \left(\sum_{i=1}^{d} \mathbf{D}_i \tilde{\mathbf{R}}_i^T \mathbf{S}_i^{-1} \tilde{\mathbf{R}}_i \mathbf{D}_i\right) \left(\mathbf{I} - \mathbf{S} \tilde{\mathbf{R}}_0^T \mathbf{S}_0^{-1} \tilde{\mathbf{R}}_0\right) + \tilde{\mathbf{R}}_0^T \mathbf{S}_0^{-1} \tilde{\mathbf{R}}_0, \tag{9}$$

where $\mathbf{S}_0 = \tilde{\mathbf{R}}_0 \mathbf{S} \tilde{\mathbf{R}}_0^T$ and $\tilde{\mathbf{R}}_0$ is defined as in (7) with $\tilde{\mathbf{R}}_0^y$ being the restriction operator which returns for each subdomain the weighted sum of the values of all the nodes on the boundary of that subdomain. The weight corresponding to an interface node is one over the number of subdomains the node is contained in.

4 Numerical Results

We consider (3) with $\Omega = (-1,1)^2$, $f(x) = (2\pi^2+1)\sin(\pi x_1)\sin(\pi x_2)$, $\hat{y}(x) = \sin(\pi x_1)\sin(\pi x_2)$. Numerical observations show that the condition number for the system matrix in (2) computed for a fixed discretization is proportional to α^{-1}. Hence, (3) becomes more difficult to solve as $\alpha > 0$ approaches zero.

The domain Ω is partitioned into equal-sized square subdomains in a checkerboard pattern. The side length of each subdomain is denoted by H. Regular meshes consisting of triangular elements of various widths, denoted by h, are generated. The preconditioned system $\mathbf{PS}\mathbf{x}_\Gamma = \mathbf{P}(\mathbf{g}_\Gamma - \mathbf{K}_{\Gamma I}\mathbf{K}_{II}^{-1}\mathbf{g}_I)$ is solved using the symmetric QMR (sQMR) algorithm of Freund and Nachtigal [1995]. The preconditioned sQMR iteration is stopped if the ℓ_2-norm of the residual is less than 10^{-8}. The subdomain problems are solved exactly using a sparse LU decomposition.

Tables 1, 2 show the number of preconditioned sQMR iterations needed to solve for various discretizations h and various subdomain sizes H. As expected, the performance of the NN preconditioner (8) without coarse grid gets worse quickly as the number of subdomain increases while the balanced NN preconditioner (9) remains effective. The number of sQMR iterations for the balanced NN preconditioner

remain nearly constant for a fixed H/h ratio.

The observed performance of the NN preconditioners (8), (9) applied to the optimal control problems is similar to the performance of the NN preconditioners applied to the elliptic PDE (3b) with fixed u. A notable result is that both preconditioners depend only weakly on the regularization parameter α. As α is reduced from 1 to 10^{-8}, the iteration count for the balanced NN preconditioner grows by only a factor of about two.

Table 1. Number of preconditioned sQMR iterations, $\alpha = 1$. Left: NN preconditioner (8). Right: Balanced NN preconditioner (9)

H \ h	1/4	1/8	1/16	1/32	1/64	1/128
1/2	12	15	19	24	26	28
1/4		53	69	94	107	119
1/8			170	226	287	345
1/16				509	679	798
1/32					1578	2233

H \ h	1/4	1/8	1/16	1/32	1/64	1/128
1/2	5	6	8	10	11	12
1/4		5	9	12	14	15
1/8			5	10	13	15
1/16				5	9	13
1/32					4	9

Table 2. Number of preconditioned sQMR iterations, $\alpha = 10^{-8}$. Left: NN preconditioner (8). Right: Balanced NN preconditioner (9)

H \ h	1/4	1/8	1/16	1/32	1/64	1/128
1/2	15	16	16	21	21	23
1/4		58	61	63	74	76
1/8			217	191	202	214
1/16				583	536	584
1/32					1255	1249

H \ h	1/4	1/8	1/16	1/32	1/64	1/128
1/2	9	10	13	15	17	20
1/4		11	17	21	26	30
1/8			13	18	24	30
1/16				12	19	24
1/32					11	16

Acknowledgement. This material was partly supported by Contract No. 74837-001-0349 from the Regents of University of California (Los Alamos National Laboratory) to William Marsh Rice University and by NSF grant ACI-0121360.

References

U. M. Ascher and E. Haber. A multigrid method for distributed parameter estimation problems. *Electron. Trans. Numer. Anal.*, 15:1–17, 2003.

A. Battermann and E. W. Sachs. Block preconditioners for KKT systems in PDE-governed optimal control problems. In K.-H. Hoffmann, R. H. W. Hoppe, and V. Schulz, editors, *Fast solution of discretized optimization problems (Berlin, 2000)*, pages 1–18, Basel, 2001. Birkhäuser.

J.-D. Benamou. A domain decomposition method with coupled transmission conditions for the optimal control of systems governed by elliptic partial differential equations. *SIAM J. Numer. Anal.*, 33:2401–2416, 1996.

G. Biros and O. Ghattas. Parallel Lagrange–Newton–Krylov–Schur methods for PDE–constrained optimization. Part I: The Krylov–Schur solver. Technical report, Carnegie Mellon University, 2000.

A. Bounaim. A Lagrangian approach to a DDM for an optimal control problem. In P. Bjørstad, M. Espedal, and D. Keyes, editors, *DD9 Proceedings*, pages 283–289, Bergen, Norway, 1998. Domain Decomposition Press.

J. E. Dennis and R. M. Lewis. A comparison of nonlinear programming approaches to an elliptic inverse problem and a new domain decomposition approach. Technical Report TR94–33, CAAM Dept., Rice University, 1994.

R. W. Freund and N. M. Nachtigal. Software for simplified Lanczos and QMR algorithms. *Applied Numerical Mathematics*, 19:319–341, 1995.

M. Heinkenschloss and H. Nguyen. Neumann-Neumann domain decomposition preconditioners for linear quadratic elliptic optimal control problems. Technical Report TR04-01, Department of Computational and Applied Mathematics, Rice University, 2004.

R. H. W. Hoppe, S. I. Petrova, and V. Schulz. Primal-dual Newton-type interior-point method for topology optimization. *J. Optim. Theory Appl.*, 114(3):545–571, 2002.

C. Keller, N. I. M. Gould, and A. J. Wathen. Constrained preconditioning for indefinite linear systems. *SIAM J. Matrix Anal. and Appl.*, 21:1300–1317, 2000.

J.-L. Lions. *Optimal Control of Systems Governed by Partial Differential Equations.* Springer Verlag, Berlin, Heidelberg, New York, 1971.

J.-L. Lions and O. Pironneau. Sur le contrôle parallèle des systèmes distribués. *Comptes Rendus de l'Académie des Sciences. Série I. Mathématique*, 327 (12):993–998, 1998.

J. Mandel. Balancing domain decomposition. *Comm. Numer. Meth. Engrg.*, 9:233–241, 1993.

L. F. Pavarino and O. B. Widlund. Balancing Neumann-Neumann methods for incompressible Stokes equations. *Comm. Pure Appl. Math.*, 55(3):302–335, 2002.

B. Smith, P. Bjørstad, and W. Gropp. *Domain Decomposition. Parallel Multilevel Methods for Elliptic Partial Differential Equations.* Cambridge University Press, Cambridge, London, New York, 1996.

Domain Decomposition Preconditioners for Spectral Nédélec Elements in Two and Three Dimensions

Bernhard Hientzsch

Courant Institute of Mathematical Sciences, New York University, 251 Mercer Street, New York, NY 10012, U.S.A., Bernhard.Hientzsch@na-net.ornl.gov, http://www.math.nyu.edu/~hientzsc. This work was supported in part by the U.S. Department of Energy under Contract DE-FC02-01ER25482.

Summary. In this paper, we present several domain decomposition preconditioners for high-order Spectral Nédélec element discretizations for a Maxwell model problem in $H(\text{CURL})$, in particular overlapping Schwarz preconditioners and Balancing Neumann-Neumann preconditioners. For an efficient and fast implementation of these preconditioners, fast matrix-vector products and direct solvers for problems posed on one element or a small array of elements are needed. In previous work, we have presented such algorithms for the two-dimensional case; here, we will present a new fast solver that works both in the two- and three-dimensional case. Next, we define the preconditioners considered in this paper, present numerical results for overlapping methods in three dimensions and Balancing Neumann-Neumann methods in two dimensions. We will also give a condition number estimate for the overlapping Schwarz method.

The model problem is: Find $\mathbf{u} \in H_0(\text{CURL}, \Omega)$ such that for all $\mathbf{v} \in H_0(\text{CURL}, \Omega)$

$$a(\mathbf{u}, \mathbf{v}) := (\alpha \mathbf{u}, \mathbf{v}) + (\beta \text{ CURL } \mathbf{u}, \text{ CURL } \mathbf{v}) = (\mathbf{f}, \mathbf{v}). \tag{1}$$

Here, Ω is a bounded, open, connected polyhedron in $\mathbb{R}^3$ or a polygon in $\mathbb{R}^2$, $H(\text{CURL}, \Omega)$ is the space of vectors in $(L^2(\Omega))^2$ or $(L^2(\Omega))^3$ with CURL in $L^2(\Omega)$ or $(L^2(\Omega))^3$, respectively; $H_0(\text{CURL}, \Omega)$ is its subspace of vectors with vanishing tangential components on $\partial\Omega$; $\mathbf{f} \in (L^2(\Omega))^d$ for $d = 2, 3$, and $(\cdot, \cdot)$ denotes the inner product in $L^2(\Omega)$ of functions or vector fields. For simplicity, we will assume that α and β are piecewise constant.

1 Discretization

We have previously presented the discretization for the two-dimensional case and some fast solvers for it in Hientzsch [2001] and Hientzsch [2003], and we will here concentrate on the three-dimensional case. As in the two-dimensional case, we use a hN-extension of Nédélec elements, parametrized by the values of the vector field on Gauss-Lobatto-Legendre grids inside the elements, with only the appropriate tangential continuity between elements (Nédélec [1980, 1986], Monk [1994], Belgacem and Bernardi [1999], Hientzsch [2001, 2003]). The integrals in the bilinear form and the right hand side are all evaluated by Gauss-Lobatto-Legendre quadrature of arbitrary order. On the reference element, the system reads:

$$\mathbf{Eu} = \tilde{\mathbf{f}} \qquad \text{or} \qquad \begin{pmatrix} E_{11} & E_{12} & E_{13} \\ E_{12}^T & E_{22} & E_{23} \\ E_{13}^T & E_{23}^T & E_{33} \end{pmatrix} \begin{pmatrix} u_1 \\ u_2 \\ u_3 \end{pmatrix} = \begin{pmatrix} \tilde{f}_1 \\ \tilde{f}_2 \\ \tilde{f}_3 \end{pmatrix}$$

with

$$\begin{aligned} E_{11} &= \alpha M_x^{1,1} \otimes M_y^{1,1} \otimes M_z^{1,1} + \beta M_x^{1,1} \otimes K_y^1 \otimes M_z^{1,1} + \beta M_x^{1,1} \otimes M_y^{1,1} \otimes K_z^1 \\ E_{22} &= \alpha M_x^{2,2} \otimes M_y^{2,2} \otimes M_z^{2,2} + \beta K_x^2 \otimes M_y^{2,2} \otimes M_z^{2,2} + \beta M_x^{2,2} \otimes M_y^{2,2} \otimes K_z^2 \\ E_{33} &= \alpha M_x^{3,3} \otimes M_y^{3,3} \otimes M_z^{3,3} + \beta K_x^3 \otimes M_y^{3,3} \otimes M_z^{3,3} + \beta M_x^{3,3} \otimes K_y^3 \otimes M_z^{3,3} \end{aligned}$$

$$\begin{aligned} E_{12} &= -\beta (M_x^{1,2'} D_{m_2^x}) \otimes (D_{m_1^y}^T M_y^{1',2}) \otimes M_z^{1,2} \\ E_{13} &= -\beta (M_x^{1,3'} D_{m_3^x}) \otimes M_y^{1,3} \otimes (D_{m_1^z}^T M_z^{1',3}) \\ E_{23} &= -\beta M_x^{2,3} \otimes (M_y^{2,3'} D_{m_3^y}) \otimes (D_{m_2^z}^T M_z^{2',3}) \end{aligned}$$

$M_{dir}^{i,j}$ is a mass-matrix for direction *dir*, integrating products between components i and j. Primes indicate differentiated components. $K_{dir}^i = D_{m_i^{dir}}^T M_{dir}^{i',i'} D_{m_i^{dir}}$ is the weak 1D Laplacian, D_n is the differentiation matrix of order n, and u_i is of order $m_i^x \times m_i^y \times m_i^z$.

Subassembling such elements in a rectangular array of elements results in a system of the same form, only that the different mass matrices, laplacians, and derivatives are changed into the appropriate matrices for the entire array. In particular, the matrices $M_y^{1,1}$, $M_z^{1,1}$, K_y^1, K_z^1, $M_x^{2,2}$, $M_z^{2,2}$, K_x^2, K_z^2, $M_x^{3,3}$, $M_y^{3,3}$, K_x^3, and K_y^3 are subassembled; all other mass-matrices are block-diagonal, and the differentiation matrices in the cross-terms are mixed, mapping from continuous to discontinuous spaces.

The element-by-element computation of the matrix-vector product with the stiffness matrix can be implemented by dense matrix-matrix multiplications of the factors of the tensor products with the vector field laid out in array form. These multiplications use an optimized BLAS3 kernel and run at close to maximal efficiency on modern computer architectures.

For general geometries, mapped elements are used. The matrix-vector product associated to the discretization can again be implemented by tensor products and entry-by-entry (Hadamard) matrix products, and therefore also has a fast implementation. For some more details in the 2D case, see [Hientzsch, 2002, Section 4].

2 Fast direct solvers in 2D and 3D

We have previously developed a fast direct solver in Hientzsch [2001] and Hientzsch [2003] for the two-dimensional case, where the system in the two components is reduced to a generalized Sylvester equation in one component which is then solved either by a fast diagonalization method or by more stable methods for generalized Sylvester equations. It seems that this solver cannot be extended to three dimensions in the general case.

Instead of a diagonalization technique, we will try to change bases so that a block-diagonal matrix with small blocks is obtained. In two dimensions, the general block is 2×2, in three dimensions, it is 3×3, coupling modes across components. In three dimensions, we look for a basis change matrix

$$\mathbf{V} = \begin{pmatrix} V_x^2 \otimes V_y^1 \otimes V_z^1 & 0 & 0 \\ 0 & V_x^1 \otimes V_y^2 \otimes V_z^1 & 0 \\ 0 & 0 & V_x^1 \otimes V_y^1 \otimes V_z^2 \end{pmatrix}$$

so that in $\mathbf{V}^T\mathbf{E}\mathbf{V}$ all 3×3 blocks in the block tensor product matrix are diagonal (or if they are rectangular, diagonal with an extra block of zeros). Then, in the new basis, the system splits into many 3×3 or smaller systems.

We will only treat the x-direction, the same construction can be repeated for all three directions. Looking at the entries of $\mathbf{E}$, we realize that we can diagonalize all blocks, if we can diagonalize

$$\begin{array}{ll} V_x^{2,T} M_x^{1,1} V_x^2, & V_x^{2,T} M_x^{1,2'} D_{m_2^x} V_x^1, \\ V_x^{1,T} D_{m_2^x}^T M_x^{2',2'} D_{m_2^x} V_x^1, & V_x^{1,T} M_x^{2,2} V_x^1; \end{array}$$

if the second and third component have the same size in x, they are discretized in the same way, and the other discretization parameters are chosen so that mass matrices match. These conditions are not overly restrictive; a large class of generalized Nédélec elements and some newly proposed elements are of that form. In two dimensions, no such degree conditions appear, and the block diagonalization works in the general case.

The question is now if we can find V_x^1 and V_x^2 such that these four matrices are diagonal. If we first consider the terms only in V_x^1, we see that one reasonable choice would be to take the eigenbasis of the following generalized eigenvalue problem:

$$D_{m_2^x}^T M_x^{2',2'} D_{m_2^x} u = \lambda M_x^{2,2} u.$$

Now if $D_{m_2^x} V_x^1$ could be chosen as V_x^2 and had the right size, we would be done. Because the two components do not necessarily have the same size in x, we start with $V_x^2 = I_x^{1,2} D_{m_2^x} V_x^1$, choosing an appropriate $I_x^{1,2}$ as interpolation. For diagonalization to succeed, we need

$$D_{m_2^x}^T M_x^{2',2'} D_{m_2^x} = D_{m_2^x}^T I_x^{1,2,T} M_x^{1,1} I_x^{1,2} D_{m_2^x} = D_{m_2^x}^T I_x^{1,2,T} M_x^{1,2'} D_{m_2^x}$$

which can be satisfied by appropriate choice of discretization parameters and mass matrices, if $m_1^x \geq m_2^x - 1$. (If V_x^1 contains a constant vector, we need to remove this vector before differentiating.) Still, the V_x^2 so constructed is not yet a basis in general, since there may not be enough vectors in it. We construct a full basis for the complement of the range and complement V_x^2 with it.

In special cases, we can give a basis for the complement explicitly, otherwise we start with carefully choosing vectors which we then make orthogonal to $I_x^{1,2} D_{m_2^x} V_x^1$ and each other. The same method works for subassembled problems, and also for essential boundary value problems (also for mixed problems if each face of the box has only one type of boundary condition).

Using the block diagonalization just derived, all factor matrices in the tensor products only have non-zero entries on their diagonals (some factor matrices are rectangular). Therefore, in this basis, the solution of the system decouples into the solution of arrays of 3×3, 2×2 and 1×1 problems. The coefficients in the Gaussian elimination for these symmetric small systems can be precomputed, and the solution reduces to element-wise multiplication and addition.

For instance, a natural boundary value problem on one generalized Nédélec element can be solved in MATLAB like fashion:

```
[fev1,fev2,fev3]=applBasChgT(baschg,fm1,fm2,fm3);
[uev1,uev2,uev3]=nedtwoblslv(blslv,fev1,fev2,fev3);
[u1,u2,u3]=applBasChg(baschg,uev1,uev2,uev3);
```

where applBasChg and applBasChgT apply the basis change and its transpose and the resulting array of 3×3 problems is solve in nedtwoblslv:

```
function [uev1,uev2,uev3]=nedtwoblslv(blslv,fev1,fev2,fev3);
  rhs=fev3+blslv.t34.*fev2+blslv.t35.*fev1;
  uev3=rhs./blslv.t33;
  rhs=fev2+blslv.t24.*fev1-blslv.t23.*uev3;
  uev2=rhs./blslv.t22;
  rhs=fev1-blslv.g12.*uev2-blslv.g13.*uev3;
  uev1=rhs./blslv.g11;
return
```

These element and block solvers run very efficiently; see figure 1.

3 Overlapping Schwarz Methods

To define Schwarz preconditioners (see Smith et al. [1996]), we have to specify subspaces and solvers on them. For the two-dimensional set-up, see Hientzsch [2001] and Hientzsch [2003]. Here we will concentrate on the three-dimensional case. First, a collection of subdomains Ω_i is defined, each subdomain being either one spectral element or a union of several spectral elements. The typical size of a subdomain is denoted H, and each spectral element has a uniform degree N in all components. (The analysis goes through and the methods are implemented for more complicated settings; we chose this case here for simplicity and ease of presentation.) Now, overlapping subregions $\Omega'_{j,\delta} \subset \Omega$ are defined, with an overlap of δ. These subregions can be constructed in several ways, e.g., by extending subdomains by a fixed overlap δ in all directions, or

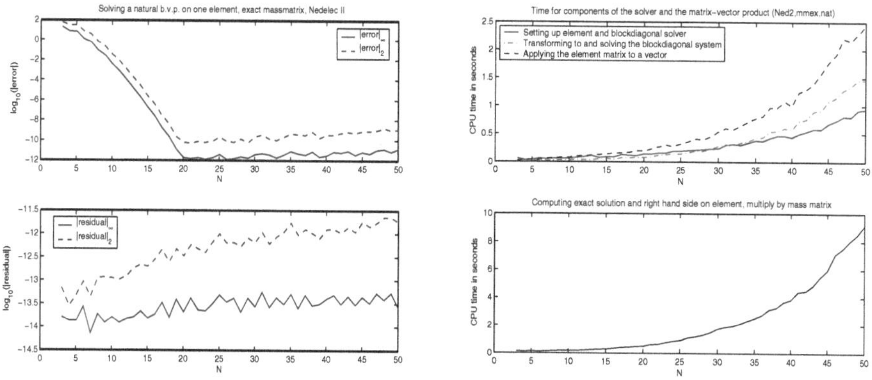

Fig. 1. Fast direct solution of a natural boundary value problem on one Nédélec 2 element of degree N by block tensor block diagonalization in 3D

by finding vertex centered subdomains that overlap by δ. The theory does not require the $\Omega'_{j,\delta}$ to be unions of spectral elements, they can also just contain rectangular subsets of spectral elements. Most of our early computations (and the numerical results that we show in this paper) were performed on $2 \times 2 \times 2$ vertex centered assemblies of subdomains (taken as single spectral elements).

The local spaces V_j are the linear span of the basis functions associated with Gauss-Lobatto-Legendre points in $\Omega'_{j,\delta}$. In general, the support of functions in V_j will be larger than $\Omega'_{j,\delta}$, but if one only considers the Gauss-Lobatto-Legendre grid, they vanish on grid points outside $\Omega'_{j,\delta}$. On the local spaces, we use exact solvers which corresponds to inversion of a submatrix of K. In the $2 \times 2 \times 2$ case, or in all cases where the overlapping regions only contain entire spectral elements, the local solver corresponds to the solution of a standard tangential value problem on the box made from these elements. In any case, the local solver can be implemented using the direct fast solvers introduced in the previous section.

The coarse space V_0 is a low-order Nédélec spectral element space of uniform degree N_0 defined on the coarse (subdomain level) mesh. We use the direct solvers of the last section as exact solvers. In the standard way, the local and the coarse solvers define local projections T_i and T_0 that can be used to implement different overlapping Schwarz methods. In this paper, we only consider the additive operator: a two-level additive Schwarz method T_{as2} defined by

$$T_{as2} = T_0 + \sum_{i \geq 1} T_i$$

We recall that this preconditioner gives optimal results in two dimensions; both iteration numbers and condition numbers are bounded by small constants for an increasing number of subdomains and degree, if there is a generous overlap that does not cut through spectral elements. It is also very robust against changes in α and β over a wide range of magnitudes. For minimal overlap that cuts through elements, iteration and condition numbers increase with increasing degree consistent with a linear growth; the dependence on the relative overlap could be both consistent with linear or quadratic growth (Hientzsch [2003, 2002]).

Table 1 and figure 2 suggest similar behavior in three dimensions.

We refer to Hientzsch [2002] for a proof of a condition number estimate which improves on the one given in Hientzsch [2003], in that it does not depend on the coefficients α and β, to wit for the case of overlapping regions made out of entire spectral elements

$$\kappa(T_{as2}) \leq C\left(1 + \left(\frac{H}{\delta}\right)^2\right)$$

and for the general case, with $\gamma \leq 1$ in two dimensions, and with $\gamma \leq 2$ in three dimensions,

$$\kappa(T_{as2}) \leq CN^{\gamma}\left(1 + \left(\frac{H}{\delta}\right)^2\right)$$

In both cases, C is independent of N, H, δ, α and β.

Table 1. Comparison of different methods for $\alpha = \beta = 1$, $5 \times 5 \times 5$ subdomains, Nédélec 2 elements of degree 10, $2 \times 2 \times 2$ overlapping subdomains, reduction of residual norm by 10^{-6}

# of levels	iter	$\kappa_{est}(K)$	$\|\|\text{error}\|\|_\infty$	t_{CPU} in s
one	25	17.39	4.23e-06	187.26
two ($N_0 = 2$)	22	8.89	3.62e-06	165.07
two ($N_0 = 3$)	21	8.94	8.10e-06	155.62
two ($N_0 = 4$)	21	8.46	9.79e-06	157.18
two ($N_0 = 5$)	20	8.16	1.28e-05	154.51

4 Balancing Neumann-Neumann

Balancing Neumann-Neumann preconditioners are examples of iterative substructuring methods. Here, one iterates on the Schur complement system $Su_S = f_S$ with respect to the shared degrees of freedom u_S on the subdomain interfaces. To apply the matrix-vector product Su_S, one adds up the local contributions from the local Schur complements $S^{(i)}$ from each subdomain. This requires the solution of an essential (Dirichlet) boundary value problem per element. Balancing Neumann-Neumann preconditioners are hybrid methods, with alternating balancing and Neumann-Neumann steps (see, e.g., [Smith et al., 1996, Section 4.3.3]). The Neumann-Neumann step requires the application of the inverse of the local Schur complement which can be implemented

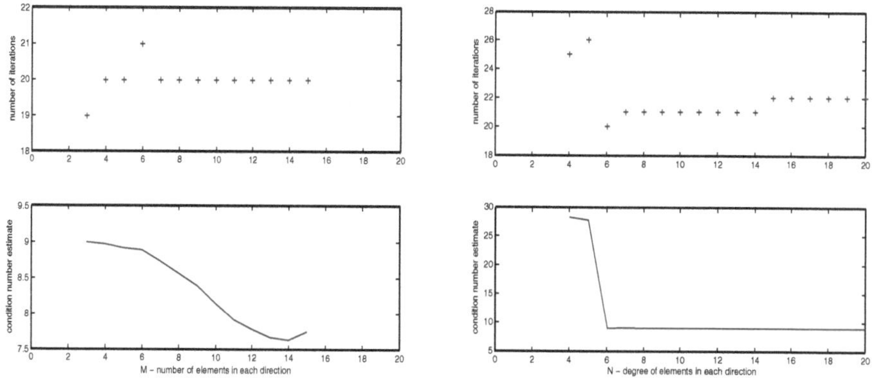

Fig. 2. Two-Level additive overlapping Schwarz method, $\alpha = \beta = 1$, $2 \times 2 \times 2$ overlapping subdomains, reduction of residual norm by 10^{-6}, $M \times M \times M$ subdomains of degree N. On the left, $N = 6$. On the right, $M = 5$

by the solution of a local Neumann problem, and also requires some diagonal scaling. The balancing step constitutes the coarse level correction, and directly inverts the Schur complement restricted to well-chosen coarse basis functions. We use the fast direct solvers from the second section to solve the Dirichlet and Neumann problems; for the coarse grid correction, in general, a general purpose factorization routine has to be used. We use the standard partition-of-unity diagonal scaling. Experiments have been run with different coarse basis functions and variable damping of the coarse grid corrections. We have been able to develop seemingly optimal and efficient damped coarse grid corrections for the two-dimensional case, and made some progress for

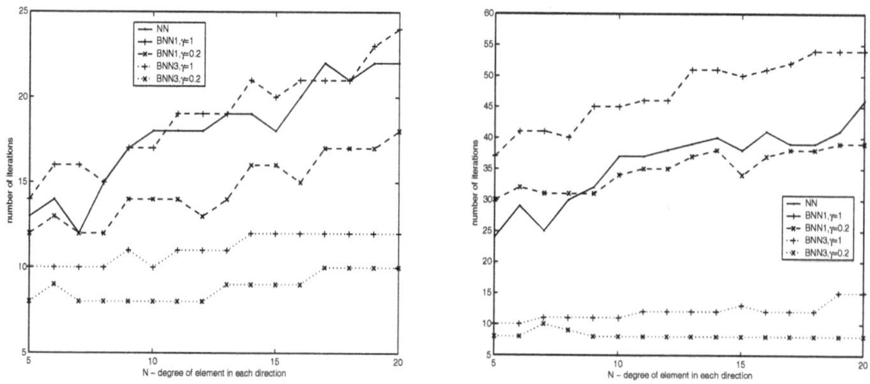

Fig. 3. Balancing Neumann-Neumann methods in two dimensions. Comparison of different coarse bases in the balancing step. $\alpha = \beta = 1$, reduction of residual norm by 10^{-6}, $M \times M \times M$ subdomains of degree N. On the left, $M = 5$. On the right, $M = 10$.

some lower-order cases in the three-dimensional case, but we do not yet have general optimal balancing steps for the three-dimensional case.

Finally, we present in figure 3 some numerical experiments for two cases in two dimensions for several different balancing steps. We intend to present more complete experiments for two and three dimensions and theory for the two-dimensional case in future work.

As coarse grid functions we have chosen the standard partition of unity, either one function for the whole subdomain (BNN1), one for each component (BNN2), or one for each edge/face (BNN3). γ is the damping factor for the coarse grid correction. In figure 3, we see that in both cases, one coarse grid function per subdomain is not enough, even if we allow damping, but that one coarse grid function per edge gives an efficient method.

References

F. B. Belgacem and C. Bernardi. Spectral element discretization of the Maxwell equations. *Math. Comp.*, 68(228):1497–1520, 1999.

B. Hientzsch. *Fast Solvers and Domain Decomposition Preconditioners for Spectral Element Discretizations of Problems in H(curl).* PhD thesis, Courant Institute of Mathematical Sciences, September 2001. Technical Report TR2001-823, Department of Computer Science, Courant Institute.

B. Hientzsch. Overlapping Schwarz preconditioners for spectral Nédélec elements for a model problem in H(curl). Technical Report TR2002-834, Department of Computer Science, Courant Institute of Matical Sciences, November 2002.

B. Hientzsch. Fast solvers and preconditioners for spectral Nédélec element discretizations of a model problem in H(curl). In I. Herrera, D. E. Keyes, O. B. Widlund, and R. Yates, editors, *Domain Decomposition Methods in Science and Engineering*, pages 427–433. National Autonomous University of Mexico (UNAM), Mexico City, Mexico, 2003. Proceedings of the 14th International Conference on Domain Decomposition Methods in Cocoyoc, Mexico, January 6-11, 2002.

P. Monk. On the *p*- and *hp*-extension of Nédélec's curl-conforming elements. *J. Comput. Appl. Math.*, 53(1):117–137, 1994. ISSN 0377-0427.

J.-C. Nédélec. Mixed finite elements in R^3. *Numer. Math.*, 35:315–341, 1980.

J.-C. Nédélec. A new family of mixed finite elements in R^3. *Numer. Math.*, 50:57–81, 1986.

B. F. Smith, P. E. Bjørstad, and W. Gropp. *Domain Decomposition: Parallel Multilevel Methods for Elliptic Partial Differential Equations.* Cambridge University Press, 1996.

Parallel Distributed Object-Oriented Framework for Domain Decomposition

S.P. Kopyssov, I.V. Krasnopyorov, A.K. Novikov, and V.N. Rytchkov[1]

UB RAS, Institute of Applied Mechanics kopyssov@udman.ru

Summary. The aim of this work is to reduce the development costs of new domain decomposition methods and to develop the parallel distributed software adapted to high performance computers. A new approach to development of the domain decomposition software system is suggested; it is based on the object-oriented analysis and middleware CORBA, MPI. In this paper, the main steps of domain decomposition are determined, the object-oriented framework is described, and then it is extended for parallel distributed computing. The given examples demonstrate that the software developed in such a way provides mathematical clarity and rapid implementation of the parallel algorithms.

1 Introduction

The idea of domain decomposition (DD) used not to be applied to parallel algorithms and it gave rise to substructuring (Przemieniecki [1968]), subconstruction, macroelement, superelement, fragment, module-element, reduced element, Schwartz (Sobolev [1936]), capacity matrix and other methods. Usually these methods have been applied to reduce an initial problem in the domain with a complex boundary to the sequence of problems in the subdomains with sufficiently simple boundaries. Nowadays the parallel implementations of DD allow improving the computational performance.

The most of DD software is based on one or another approach to the approximation of a differential problem, mostly on the finite element (FE) method. The complexity of FE models results in the necessity of using suitable programming techniques such as the object-oriented (OO) analysis. At present, there are many publications on the FE OO models (Zimmermann et al. [1992]), but OO analysis is rarely applied to the DD. There are some references to the OO scientific software: Diffpack, PETSc, SPOOLES, Overture. Diffpack (Cai [1998]) is an OO environment aimed at solving partial differential equations (PDE). Overture (Brown et al. [1999]) provides the OO framework for solving PDE on overlapping grids. The fundamental abstractions are divided into functionality groups: data structures, linear and nonlinear solvers, PDE, utilities. Recently, DD and multigrid methods have been included. The MPI-based libraries PETSc (Portable Extensible Toolkit for Scientific computation) and SPOOLES (Sparse Object-Oriented Linear Equations Solver) use the OO style for matrix representation of PDE. In this work the fundamental OO framework consisting of the general FE and DD entities is suggested, and then it is extended by introducing new objects that implement specific algorithms including parallel ones.

For parallel distributed implementation of the DD framework, it is appropriate to apply the existing techniques and middleware. MPI and CORBA are the most commonly used ones. MPI (Message Passing Interface) is used in massively parallel systems. An MPI-based program describes one of the identical processes handling its own portion of the data (SIMD). MPI provides blocking/nonblocking communications between the groups of processes. C++ can be used to implement a parallel distributed object; for that, it is necessary to implement dynamic creation of an object and remote method invocation. The main shortcomings from the point of view of flexibility are the following: the procedure orientation and the primitiveness of the program starting system.

CORBA (Common Object Request Broker Architecture) is used to create OO distributed applications. CORBA provides synchronous/asynchronous remote method invocations and allows creating complex, high performance, cross platform applications. The performance measurements of TAO (CORBA) and MPICH (MPI) middleware were taken on Gigabit Ethernet; they showed the same throughput, with MPICH giving lower latency. The recently published results of comparison of OmniORB and MPICH on Myrinet and SCI communications (Denis et al. [2003]) prove ours. CORBA is an interesting alternative to MPI for flexible and high performance implementation of complex models.

2 Representation of the main steps of DD in the OO framework

The OO analysis was applied for creating an abstract software model, the C++ language was used for programming. Inheritance and polymorphism provided the flexibility of the framework. Data encapsulation brought about creating three subsystems: modeling classes, numerical classes, and analysis classes. The OO model of analysis is shown in Figure 1. Let us determine the main steps of DD and consider them from the point of view of OO programming.

1. Building the finite element model. The `DomainBuilder` class is the base class of design model editors. It provides the methods to create and edit the domain represented by the `Domain` class, which contains a geometry and a FE mesh consisting of nodes (`Node`), different types of elements (`Triangle`, `Tetrahedron`, `Hexahedron` and others), and boundary conditions

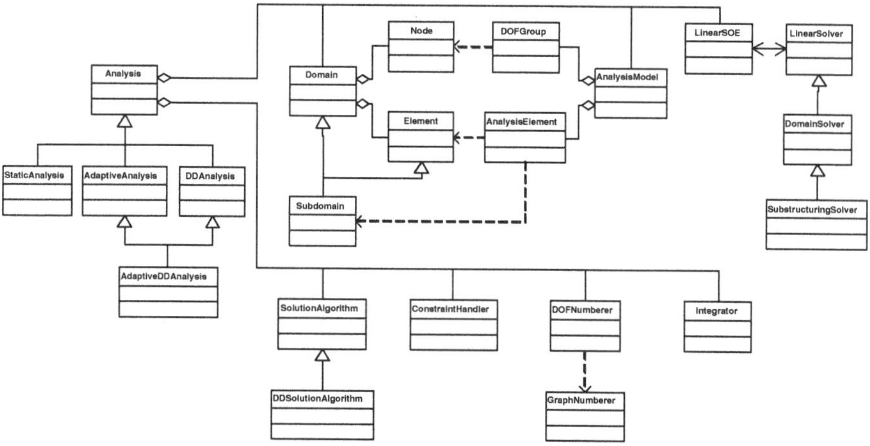

Fig. 1. The object-oriented model of analysis

(`Nodal/ElemanalLoad`, `SP/MPConstraint` and others). The `DomainBuilder` derived classes operate the data from files and CAD systems.

2. Partitioning the domain into subdomains. It is appropriate to represent a FE mesh as a graph of the element connectivity and then to apply any graph partitioning algorithm. The object of the `DomainPartitioner` class gets the element graph `Graph` built by the object of the `PartitionedDomain` class and divides it by any graph partitioning algorithm represented by the object of the `GraphPartitioner` class. The `GraphPartitioner` subclasses are based on the algorithms implemented in METIS and ParMETIS libraries (Karypis and Kumar [1998]). The `Subdomain` class extends the `Domain` interface to make distinction whether the nodes in the subdomain are internal or external. The `PartitionedDomain` and `Subdomain` derived classes are designed to partition the domain into both non-overlapping and overlapping subdomains. For additive Schwartz methods, the best of all would be the synchronous handling of the objects in the intersection; in this case the objects are instantiated once. For multiplicative Schwartz methods, it is the other way round; in each `Subdomain` object the copies of the objects that are included into the intersection are created to be independently calculated and periodically synchronized.

3. Assignment of both local and global equation numbers to nodal degrees-of-freedom. To make such a mapping there is a need to apply any graph numbering algorithm to the mesh node graph. The mapping can have a significant influence on the amount of computation required to solve the system of equations and on the amount of memory required to store it.

The `AnalysisModel` is a container for storing and providing access to the objects of the `DOFGroup` and `AnalysisElement` classes. The `DOFGroup` objects represent the degrees-of-freedom at the nodes and new degrees-of-freedom introduced into the analysis to enforce the constraints. The `AnalysisElement` objects represent the elements and subdomains or they are introduced to add stiffness and/or load to the system of equations in order to enforce the constraints. The `DOFGroups` and `AnalysisElements` remove from the `Node` and `Element` objects the need to worry about the mapping between the degrees-of-freedom and equation numbers. They also have the methods for forming tangent and residual vectors that are used to form the system of equations. Besides, they handle the constraints.

The `DOFNumberer` is responsible for mapping the numbers of equation to the degrees-of-freedom in the `DOFGroup` objects.

4. Assembling the systems of equations using elemental and nodal contributions determined by the integration scheme chosen. Assembling the systems of equations is also based on the FE graphs and on defining the contributions for the different types of element as well. According to the integration scheme, the local systems of equations are formed by the FE contributions. For the DD methods that need assembling the global system of equations, it seems efficient to represent all subdomains as the graph of special-purpose elements (superelements) in order to apply the approach stated above. Different FE contributions and various integration schemes give a wide range of assembling methods for DD. On multiprocessors the contributions from internal elements of the subdomain can be calculated on the processor that handles the subdomain; to determine them from external ones it is necessary to use different approaches to distributed computing.

The `Integrator` is responsible for defining the contributions of the `DOFGroup` and `AnalysisElement` objects to the system of equations and for updating the response quantities in the `DOFGroup` objects with the appropriate values given the solution to the system of equations.

5. Imposing boundary conditions. Applying the constraints may involve transformation of the elemental and nodal contributions or adding new terms and unknowns to the matrix equations.

The `ConstraintHandler` class is responsible for handling the constraints by creating appropriate `DOFGroup` and `AnalysisElement` objects. It also allows to introduce the multiple constraints arising from adaptive refinement.

6. Solving the system of equations. Different DD methods are similar in the presence of the local and possibly, global systems of equations and in the performance of the local and sometimes, global matrix-vector operations according to the solution algorithm.

The `Analysis` class is a container for all of the analysis objects mentioned above. It is responsible for starting the analysis steps specified in the `SolutionAlgorithm` class. The `Analysis` class is associated with either a domain or a subdomain and allows describing either a global solution or one branch of the solution in the subdomain. In the second case, several sets of analysis objects are executed simultaneously.

The `LinearSOE` class stores the matrix, the right hand side and the solution of a linear system of equations. `LinearSOE` derived classes correspond to the systems with different types of matrices (band, profile, etc). The

`LinearSolver` class is responsible for performing the numerical operations on the equations. `LinearSolver` subclasses encapsulate linear algebra libraries LAPACK, PETSc, SuperLU.

7. Update of nodal degrees-of-freedom with the appropriate response quantities.

8. Determining the rated conditions in finite elements.

Nowadays it is generally accepted that the effective solution of applied problems is almost impossible without using an adaptive process when the obtained solution is examined to determine the strategy of further calculations: mesh refinement with the same connectivity (r-version of FEM), local mesh refinement (h-version), increase of the degree of approximation basis functions (p-version) or whatever combinations (h-p, h-r versions). The best choice gives the maximum precision with the minimum computational costs. The OO model for adaptive analysis allows to build the optimal computational model with the given precision and minimum computational costs.

9. A-priori error estimation. The additional data included in the `AdaptiveAnalysis` class are the following: error estimation `ErrorEstimation`, error indicator `ErrorIndicator`, refinement strategy `Refinement`. `ErrorEstimation` subclasses represent a-priori and a-posteriori error estimations based on: residual, interpolation, projection, extrapolation, dual method.

10. Determining the objects to be more precise. The `ErrorIndicator` subclasses provide the selection of the part of the domain to refine: global refinement, strategy of maximum, equidistribution, guaranteed error reduction.

11. Repartitioning the mesh in accordance with the criterion of refinement. Mesh repartitioning gives rise to redistributing the work among the processors, with each processor busy in actual loading as long as possible, in other words, to load balancing. The main difference between the dynamic load balancing and static one is the necessity to redistribute the work among the processors; it brings about considerable computational costs (Kopyssov and Novikov [2001]).

Different improvements of the solution are inherited from the `Refinement` class: relocation of the nodes in 2D/3D area (r-version), local refinement and coarsening for 2D triangle meshes (h-version), increasing the degree of integrated Legendre polynomials for 3D hierarchical hexahedral elements (p-version).

The dynamic load balancing is implemented by including the `Refinement` object to `DomainPartitioner`. Using them on every iteration, one could efficiently redistribute the FE mesh objects, with `Refinement` providing graph weights handling, and `GraphPartitioner` partitioning the graph.

3 Extension for parallel distributed computing

To provide the interprocess communication between objects on multiprocessor computers, it is necessary to implement the remote method invocation. In addition, it is required to implement the migration of the objects representing the distributed data and the asynchronous invocation of the methods of the objects representing parallel functions. Some parallel distributed implementations of objects were examined on MPI and CORBA middleware (Kopyssov et al. [2003]). A new approach is suggested to develop parallel distributed OO software for DD. It is based on CORBA, the AMI (Asynchronous Method Invocation) callback model (Schmidt and Vinoski [1999]) and integration of MPI applications.

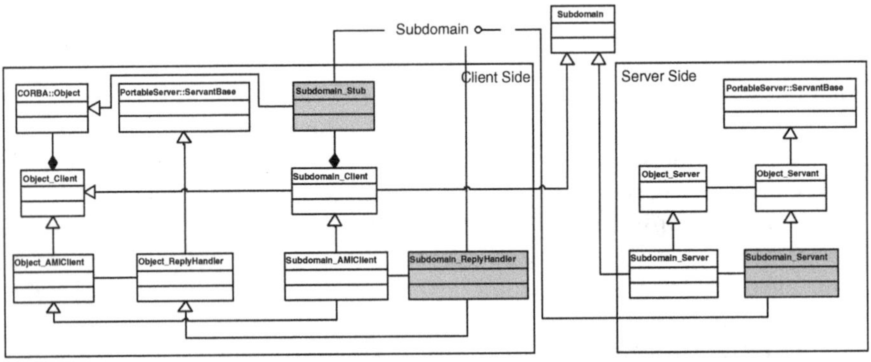

Fig. 2. CORBA implementation of *Subdomain* class

The CORBA implementation of data and function objects consists of:

IDL interface. The objects to be called remotely are specified as IDL interfaces (`interface`), with the migrated objects specified as IDL structures (`struct`). According to the IDL interface CORBA generates the C++ templates (stubs, servants, AMI handlers) designed to develop parallel distributed objects. Figure 2 shows the interface and the client-server (gray filled) templates for the `Subdomain` class.

Client-server library. It is based on the C++ library for the DD and CORBA client-server templates. The client classes are inherited from the DD ones and aggregate the stubs of the CORBA interfaces, with the virtual functions of the DD objects overloaded: the input parameters are converted from the DD types to those of CORBA, the method of the remote CORBA object is invoked, the output parameters are converted backward and returned as the result of the remote invocation. The server classes are inherited from the DD ones and aggregate the CORBA servants, with the pure virtual functions of servants implemented: the input parameters are converted from the CORBA types to those of DD, the method of the DD object is called, the output parameters are converted backward to be sent to the client side. The asynchronous method invocation is implemented with the help of the AMI templates: the client objects send the object reference to the aggregated AMI handlers within asynchronous invocation and return control to the main process. Having addressed to the result objects, the main process is blocked until the remote methods finish and return their results to the AMI handlers. Figure 2 shows the base client-server classes (denoted by `Object_` prefix) and the example of their use for implementing the client-server classes `Subdomain`.

Components. The components are the executable modules that include the CORBA objects for DD. It is available to launch the components with the help of MPI as the set of identical CORBA servers, which provide distributed data and carry out the operations in parallel via MPI communications; the server with a null identifier synchronizes all others.

The parallel distributed OO model for DD is the extension of the original model, of several objects in particular: `Subdomain` (Figure 2), `Node`, `Element`, `Nodal/ElementalLoad`, `SP/MPConstraint` and some others. There are some development principles.

Remote method invocation. The `DomainBuilder` subclass creates the server finite-element objects on the governing computing node. On other nodes the `Subdomain_Server` objects are launched, with an array of `Subdomain_Client` objects, that aggregates the stubs to them, being created on the governing node. Thus, without any modifications of the C++ library, the DD steps could be performed in the distributed address space through the `_Client` and `_Server` objects.

Object migration. In the initial OO model for DD the `DomainPartitioner` handles the pointers to the C++ objects when it is called to distribute the data among the subdomains. The object migration is more complicated; it includes creating the destination remote server object, copying the data of the source object and removing the source one. For that, it is necessary to modify the `DomainPartitioner` class: it has to include virtual functions with an empty body to collect garbage; the `Partition` method has to include the calls to them. The `DomainPartitioner` subclass overloads these functions and thus, it removes all the transferred objects in the end of partitioning.

Asynchronous method invocation. The principal operations to be performed simultaneously in the subdomains are forming the blocks of the global system of equations (if it is need) and solving the local systems of equations. For that, the `Subdomain_AMIClient` object is instantiated; it aggregates the AMI handler `Subdomain_ReplyHandler` for this methods. Invoking the remote method, the `Subdomain_AMIClient` object does not wait for its completion and returns control to the main process. As the main process needs the results, it calls the `Subdomain_AMIClient` object that, in its turn, blocks the main process until the `Subdomain_ReplyHandler` receives the result from the remote `Subdomain_Servant` object.

4 Examples

The parallel distributed OO framework for DD is intended for representation of a wide range of DD methods by using different: types of FE; mesh partitioning algorithms; ordering, storage and solution methods for the system of equations; means of handling of boundary conditions; error estimations; refinement strategies. Let us consider the substructuring method as an example of usage of the framework. It is suited to demonstrate the features of the analysis classes for both a partitioned domain and a subdomain, and the scope for the parallel distributed computing as well.

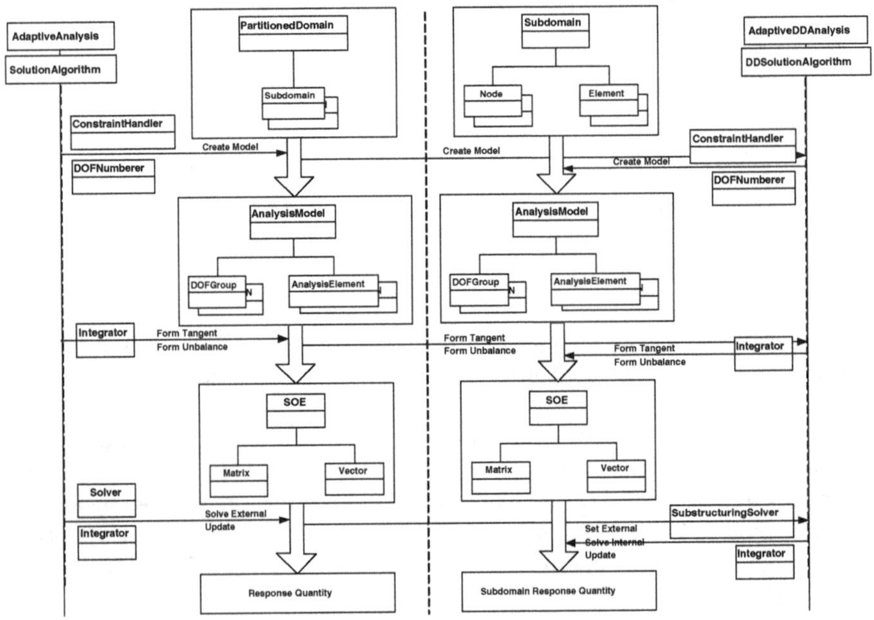

Fig. 3. The object-oriented model of substructuring

After the initial partitioning of the domain, analysis classes are used twice (see Figure 3): to solve the problem in the subdomains and on the interface. In the subdomain the objects aggregated by the `DDAnalysis` cooperate under the `DDSolutionAlgorithm` control in the following way: `ConstraintHandler` and `DOFNumberer` take `Subdomain` as the input data to create the `AnalysisModel`. The Integrator performs static condensation and forms the matrix and the right hand side for the Schur complement system block `LinearSOE` from the `AnalysisModel`.

After that, the interface problem is solved by the `AdaptiveAnalysis` object with its own `SolutionAlgorithm`, `PartitionedDomain`, `AnalysisModel`, `DOFNumberer`, `ConstraintHandler`, `Integrator`, `LinearSOE`, `LinearSolver`

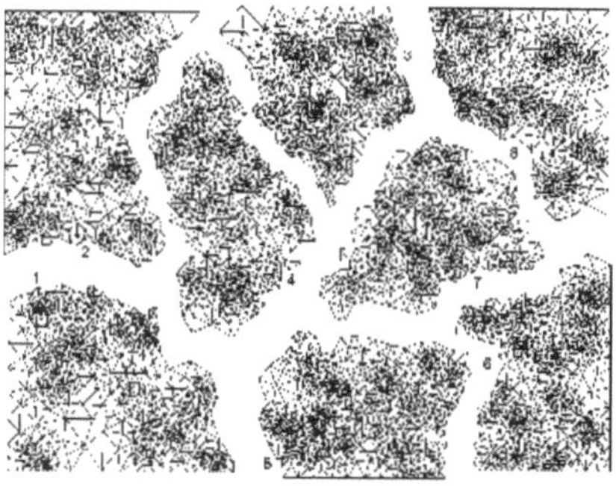

Fig. 4. Initial partition: 8 subdomains, 5959 nodes, 11740 elements

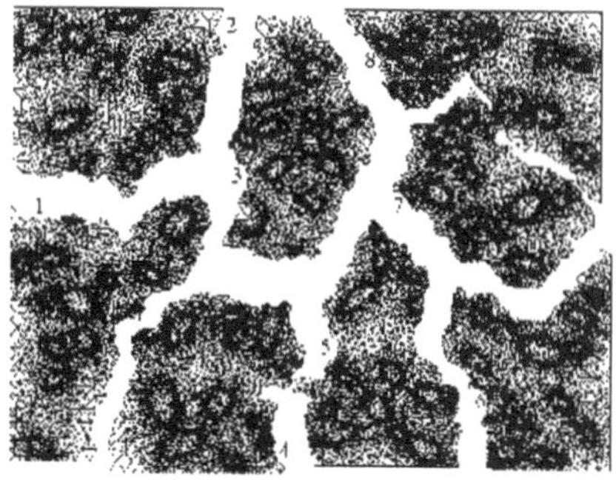

Fig. 5. Partition after 5 adaptive refinements: 42470 nodes, 84621 elements

objects. In that case, the `AnalysisModel` includes the `AnalysisElement` objects corresponding to the subdomains; the `Integrator` forms the system of equations from the interface. The `LinearSolver` object solves the global interface problem.

The `Subdomain` objects get the equations numbers from the response quantities obtained in the previous calculations on the whole `AnalysisModel`. The `SubstructuringSolver` objects solve the internal systems of equations. The `Integrator` objects update the response quantities.

The results of numerical experiments are presented in Figure 4-6. It is 2D/3D strain stress analysis; iterative/direct substructuring method is used; solution is adaptively refined by h-version/p-version of FEM.

5 Conclusions and Further research directions

The analysis allowed us to represent the main steps of DD in the form of objects and their relations. The main features of the OO framework for DD have been described. The framework was extended on CORBA middleware for parallel distributed computing. The given examples demonstrated its expressiveness and flexibility.

Further research directions are as follows:

1. extension of the OO framework for DD by new algorithms
2. inclusion of geometrical data and encapsulation of CAD systems
3. implementation of parallel algorithms of mesh generation; integration of existing mesh generators
4. development of the visual editor for DD

Acknowledgement. The work is supported by Russian Foundation for Basic Research (grants 02-07-90265, 03-07-06119, 03-07-06120, and 03-07-90002) and UB of RAS.

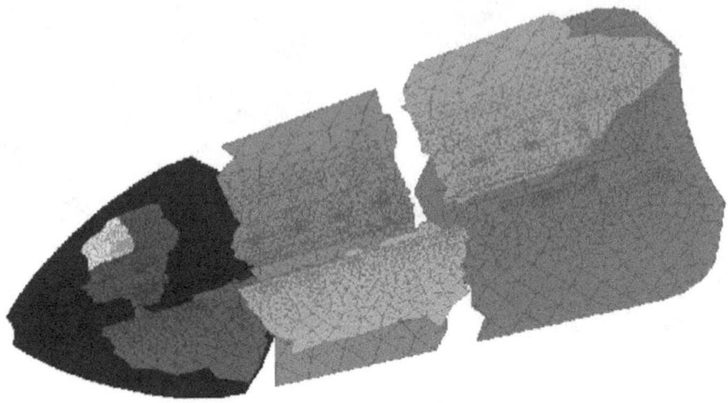

Fig. 6. Domain partition for p-version of FEM based on substructuring

References

D. Brown, W. Henshaw, and D. Quinlan. Overture: An object-oriented framework for solving partial differential equations on overlapping grids. In *SIAM conference on Object Oriented Methods for Scientfic Computing*, UCRL-JC-132017, 1999.

X. Cai. Domain decomposition in high-level parallelization of PDE codes. In *Eleventh International Conference on Domain Decomposition Methods*, pages 388–395, Greenwich, England, 1998.

A. Denis, C. Perez, T. Priol, and A. Ribes. Parallel CORBA objects for programming computational grids. *Distributed Systems Online*, 4(2), 2003.

G. Karypis and V. Kumar. A parallel algorithm for multilevel graph partitioning and sparse matrix ordering. *Journal of Parallel and Distributed Computing*, 48(1):71–95, 1998.

S. Kopyssov, I. Krasnopyorov, and V. Rytchkov. Parallel distributed CORBA-based implementation of object-oriented model for domain decomposition. *Numerical Methods and Programming*, 4(1):194–206, 2003.

S. Kopyssov and A. Novikov. Parallel adaptive mesh refinement with load balancing for finite element method. *Lecture Notes in Computer Science*, 2127:266–276, 2001.

J. Przemieniecki. *Theory of Matrix Structural Analysis.* McGaw-Hill, N.Y., 1968.

D. Schmidt and S. Vinoski. Programming asynchronous method invocations with CORBA messaging. *C++ Report, SIGS*, 11(2), 1999.

S. Sobolev. Schwartz algorithm in elasticity theory. *RAS USSR*, 4(6):235–238, 1936.

T. Zimmermann, Y. Dubois-Pelerin, and P. Bomme. Object-oriented finite element programming: I. governing principles. *Computer Methods in Applied Mechanics and Engeneering*, 98(2):291–303, 1992.

A Domain Decomposition Based Two-Level Newton Scheme for Nonlinear Problems

Deepak V. Kulkarni and Daniel A. Tortorelli*

University of Illinois,
Department of Mechanical and Industrial Engineering
Urbana, Illinois 61801

Summary. We present two non-overlapping domain decomposition based two-level Newton schemes for solving nonlinear problems and demonstrate their effectiveness by analyzing systems with balanced and unbalanced nonlinearities. They both have been implemented in parallel and show good scalability. The implementations accommodate non-symmetric matrices and unstructured meshes.

1 Introduction

One can refer to the paper by Keyes [1992] and the book by Smith et al. [1996] for in-depth reviews of domain decomposition (DD) methods. Of particular interest here are non-overlapping schemes such as iterative substructuring (Bjørstad et al. [2001]) and FETI methods (Farhat et al. [2001]).

When solving non-linear boundary value problems (BVPs) via domain decomposition it is common to use Newton type algorithms and then to apply existing DD approaches to the ensuing linearized problems (Knoll and Keyes [2002]). The NK-Schwarz scheme (Keyes [1995]) as the name suggests, uses a Krylov scheme equipped with a Schwarz preconditioner to solve this linear update equation.

If the non-linear effects are unbalanced, i.e., the nonlinearity has a significant spatial variation, then the Jacobian becomes ill-conditioned and hence the NK-Schwarz scheme is not effective cf. Cai and Keyes [2002]. A scheme that proves effective for problems with unbalanced nonlinearities is the multi level Newton Schwarz (MLN-Schwarz) scheme that was originally introduced to solve multi-physics problems (Bächtold et al. [1995], Aluru and White [1999]). Kim et al. [2003] implemented a serial version of the MLN-Schwarz to solve fluid-structure interaction problems which had the flavor of a multiplicative Schwarz approach. The scheme employs a global consistency equation in

* Both the authors would like to acknowledge the support provided by NSF under grant no. DMR 01-21695

the place of the standard residual equation. In the case of unbalanced nonlinearities, the Jacobian for the global consistency equation appears to be better conditioned (Cai and Keyes [2002]). However, the MLN-Schwarz requires the full solution of the sub-domain residual equations for each global Newton iteration. Hence the scheme is not efficient for problems with balanced nonlinearities.

Another scheme that is used to resolve BVPs with unbalanced nonlinearities is the ASPIN method. Cai et al. [2001] introduced this as a nonlinearly preconditioned version of the NK-Schwarz scheme. In comparison to the MLN-Schwarz, the ASPIN method has been implemented in parallel and accommodates both overlapping and non-overlapping domains. However, like the MLN-Schwarz scheme, the ASPIN method is inefficient for problems with balanced nonlinearities.

In this work we present two non-overlapping DD schemes to solve nonlinear BVPs. The first scheme, which we call the modified Newton Krylov Schur (MNK-Schur) approach, is based on a Newton Krylov Schur (NK-Schur) approach. Since our method uses a two-level Newton scheme, it efficiently solves problems with unbalanced nonlinearities thereby incorporating the advantages of both the NK-Schwarz and ASPIN methods. The second method modifies the MLN-Schwarz method to obtain a non-overlapping DD scheme. We show that this scheme is in fact a special case of our MNK-Schur approach.

2 Two-level Newton Krylov Schur Approach

For our finite element (or similar) computations we partition the domain into n non-overlapping sub-domains and represent the discretized nodal response vector $\mathbf{u}$ and global residual vector $\mathbf{R}(\mathbf{u})$ for the entire domain Ω as:

$$\mathbf{u} = \left\{ \begin{array}{c} \mathbf{u}_1^S \\ \vdots \\ \mathbf{u}_n^S \\ \mathbf{u}^I \end{array} \right\} \qquad \mathbf{R}(\mathbf{u}) = \left\{ \begin{array}{c} \mathbf{R}_1^S(\mathbf{u}_1^S, \mathbf{u}^I) \\ \vdots \\ \mathbf{R}_n^S(\mathbf{u}_n^S, \mathbf{u}^I) \\ \boldsymbol{\mathcal{R}}(\mathbf{u}_1^S, \mathbf{u}_2^S, \ldots, \mathbf{u}_n^S, \mathbf{u}^I) \end{array} \right\} \tag{1}$$

where $\mathbf{u}^I$ corresponds to the interface nodal Degrees Of Freedom (DOF) and $\mathbf{u}_j^S$ corresponds to the internal sub-domain and Neumann boundary nodal DOF of sub-domain j. In the above equation the interface residual $\boldsymbol{\mathcal{R}}$ is assembled as

$$\boldsymbol{\mathcal{R}}(\mathbf{u}_1^S, \mathbf{u}_2^S, \ldots, \mathbf{u}_n^S, \mathbf{u}^I) = \sum_{j=1}^{n} \mathbf{R}_j^I(\mathbf{u}_j^S, \mathbf{u}^I) \tag{2}$$

where $\mathbf{R}_j^I(\mathbf{u}_j^S, \mathbf{u}^I)$ represents the contribution of sub-domain j to the interface nodal residual vector. Note that the residual $\mathbf{R}(\mathbf{u})$ is a reordering of the residual that one would form without decomposition techniques. We apply Newton's method to the nonlinear residual equation $(1)_2$ and obtain the update equation:

$$\begin{bmatrix} \frac{\partial \mathbf{R}_1^S}{\partial \mathbf{u}_1^S} & \cdots & \mathbf{0} & \frac{\partial \mathbf{R}_1^S}{\partial \mathbf{u}^I} \\ \vdots & \ddots & \mathbf{0} & \vdots \\ \mathbf{0} & \mathbf{0} & \frac{\partial \mathbf{R}_n^S}{\partial \mathbf{u}_n^S} & \frac{\partial \mathbf{R}_n^S}{\partial \mathbf{u}^I} \\ \frac{\partial \boldsymbol{\mathcal{R}}^I}{\partial \mathbf{u}_1^S} & \cdots & \frac{\partial \boldsymbol{\mathcal{R}}}{\partial \mathbf{u}_n^S} & \frac{\partial \boldsymbol{\mathcal{R}}}{\partial \mathbf{u}^I} \end{bmatrix} \begin{Bmatrix} \Delta \mathbf{u}_1^S \\ \vdots \\ \Delta \mathbf{u}_n^S \\ \Delta \mathbf{u}^I \end{Bmatrix} = - \begin{Bmatrix} \mathbf{R}_1^S \\ \vdots \\ \mathbf{R}_n^S \\ \boldsymbol{\mathcal{R}} \end{Bmatrix} \tag{3}$$

On applying block Gauss elimination we obtain the Schur's complement representation of the above. We first solve for the interface increment $\Delta \mathbf{u}^I$ from

$$\sum_{j=1}^{n} \left[\frac{\partial \mathbf{R}_j^I}{\partial \mathbf{u}^I} - \frac{\partial \mathbf{R}_j^I}{\partial \mathbf{u}_j^S} \left[\frac{\partial \mathbf{R}_j^S}{\partial \mathbf{u}_j^S} \right]^{-1} \frac{\partial \mathbf{R}_j^S}{\partial \mathbf{u}^I} \right] \Delta \mathbf{u}^I = -\boldsymbol{\mathcal{R}}^I + \sum_{j=1}^{n} \frac{\partial \mathbf{R}_j^I}{\partial \mathbf{u}_j^S} \left[\frac{\partial \mathbf{R}_j^S}{\partial \mathbf{u}_j^S} \right]^{-1} \mathbf{R}_j^S \tag{4}$$

then we solve for the sub-domain increments from

$$\left[\frac{D\mathbf{R}_j^S}{D\mathbf{u}_j^S} \right] \Delta \mathbf{u}_j^S = -\mathbf{R}_j^S - \frac{\partial \mathbf{R}_j^S}{\partial \mathbf{u}^I} \Delta \mathbf{u}^I \tag{5}$$

and finally we update the interface and sub-domain DOF vectors as

$$\mathbf{u}^I = \mathbf{u}^I + \Delta \mathbf{u}^I; \qquad \mathbf{u}_j^S = \mathbf{u}_j^S + \Delta \mathbf{u}_j^S \tag{6}$$

The process repeats until convergence of equation $(1)_2$.

The algorithm as described above is equivalent to a NK-Schur approach. However, as discussed in the previous section this algorithm has poor convergence if the Jacobian in equation (3) is ill-conditioned. To alleviate this problem we augment the algorithm with a lower level sub-domain Newton scheme to obtain our MNK-Schur algorithm. After updating $\mathbf{u}^S$ and $\mathbf{u}^I$ at every Newton iteration (cf. equation (6)) we perform additional sub-domain iterations keeping the interface unknowns fixed via

$$\left[\frac{D\mathbf{R}_j^S}{D\mathbf{u}_j^S} \right] \Delta \mathbf{u}_j^S = -\mathbf{R}_j^S; \qquad \mathbf{u}_j^S = \mathbf{u}_j^S + \Delta \mathbf{u}_j^S \tag{7}$$

We may or may not iterate until sub-domain convergence is obtained, i.e., until $\mathbf{R}_j^S \approx \mathbf{0}$. In either case we revert to the NK-Schur approach and repeat equations (3)-(7) until $\mathbf{R}(\mathbf{u}) \approx \mathbf{0}$ (cf. equation $(1)_2$).

Remark 1: If a particular sub-domain is linear, the tangent matrices $\partial \mathbf{R}_j^S / \partial \mathbf{u}_j^S$, $\partial \mathbf{R}_j^S / \partial \mathbf{u}^I$, $\partial \mathbf{R}^I / \partial \mathbf{u}^I$, $\partial \mathbf{R}^I / \partial \mathbf{u}_j^S$ remain constant in all applications
of equations (4), (5) and (7)

Remark 2: Various criteria can be used to determine if additional sub-domain iterations are required. In our implementation we perform sub-domain iterations if $|\mathbf{R}_i^S| > |\mathbf{R}_j^S|_{avg}$ and $|\mathbf{R}_j^S| > \varepsilon_{sub-domain}$ where $|\mathbf{R}_j^S|_{avg}$ is the average norm of all sub-domain residuals (where the $\mathbf{R}_j^S$ are evaluated after the update of equation (6) is completed) and $\varepsilon_{sub-domain}$ is a tolerance.

Remark 3: The Newton iteration at the sub-domain level of the MLN-Schwarz and ASPIN methods is precisely the augmented sub-domain iteration introduced in our MNK-Schur method (cf. equations (7)). We introduce here, a non-overlapping multi level Newton Schur (MLN-Schur) approach. The global consistency equations of the MLN-Schwarz and ASPIN methods are replaced by the interface residual equation (2) now expressed as

$$\boldsymbol{\mathcal{R}}\left(\mathbf{u}^I; \hat{\mathbf{u}}_1(\mathbf{u}^I), \hat{\mathbf{u}}_2(\mathbf{u}^I), \cdots, \hat{\mathbf{u}}_n(\mathbf{u}^I)\right) = \sum_{j=1}^{n} \mathbf{R}_j^I\left(\hat{\mathbf{u}}_j(\mathbf{u}^I), \mathbf{u}_j^I\right) = \mathbf{0} \tag{8}$$

The update equation for the above problem is

$$\left[\frac{D\boldsymbol{\mathcal{R}}}{D\mathbf{u}^I}\right] \Delta\mathbf{u}^I = -\boldsymbol{\mathcal{R}}\,; \qquad\qquad \mathbf{u}^I = \mathbf{u}^I + \Delta\mathbf{u}^I \tag{9}$$

where, upon applying chain rule to (8) and differentiating the sub-domain residual equation (here expressed as $\mathbf{R}_j^S(\hat{\mathbf{u}}_j^S(\mathbf{u}^I), \mathbf{u}^I) = \mathbf{0}$) we obtain

$$\frac{D\boldsymbol{\mathcal{R}}}{D\mathbf{u}^I} = \sum_{j=1}^{n}\left[\frac{\partial \mathbf{R}_j^I}{\partial \mathbf{u}^I} - \frac{\partial \mathbf{R}_j^I}{\partial \mathbf{u}_j}\left[\frac{\partial \mathbf{R}_j^S}{\partial \mathbf{u}_j^S}\right]^{-1}\frac{\partial \mathbf{R}_j^S}{\partial \mathbf{u}^I}\right] \tag{10}$$

We notice immediately that $D\boldsymbol{\mathcal{R}}/D\mathbf{u}^I$ is the Schur's complement matrix of equation (4). However the right hand side of the above MLN-Schur update equation (9) does not contain the sub-domain residual $\mathbf{R}_j^S$ present in the MNK-Schur update equation (4). This is to be expected in the MLN-Schur scheme because the sub-domain problem is resolved making $\mathbf{R}_j^S \approx \mathbf{0}$. Thus the MLN-Schur scheme is a special case of the MNK-Schur scheme.

3 Implementation

We use a direct solver to resolve the linear sub-domain update equations (5) and (7). It is noted that the use of a direct solver enables us to store the factored sub-domain Jacobian. Obviously, the sub-domain computations are independent of each other and therefore easily parallelized.

We employ an iterative method (e.g. GMRES, Saad and Schultz [1986]) to solve the interface update equation (4) in parallel. The iterative scheme requires multiple evaluations of the matrix-vector product $\left[D\boldsymbol{\mathcal{R}}/D\mathbf{u}^I\right]\mathbf{s}$ until equation (4) converges. Expanding this product we see that

$$\frac{D\boldsymbol{\mathcal{R}}}{D\mathbf{u}^I}\mathbf{s} = \sum_{j=1}^{n}\left[\frac{\partial \mathbf{R}_j^I}{\partial \mathbf{u}^I}\mathbf{s} - \frac{\partial \mathbf{R}^I}{\partial \mathbf{u}_j^S}\overbrace{\left[\frac{\partial \mathbf{R}_j^S}{\partial \mathbf{u}_j^S}\right]^{-1}\left\{\frac{\partial \mathbf{R}_j^S}{\partial \mathbf{u}^I}\mathbf{s}\right\}}^{\mathbf{x}_j}\right] \tag{11}$$

It is emphasized that each matrix-vector product required by the iterative solver involves a back solve with the already factored sub-domain Jacobian matrices to obtain $\mathbf{x}_j$. In fact the interface Jacobian matrix $D\mathcal{R}/D\mathbf{u}^I$ is never assembled, only its effect on the vector $\mathbf{s}$ is evaluated, i.e., at the interface level this is a matrix free method. However, we form a preconditioner $\mathbf{P} = [\widetilde{\frac{D\mathcal{R}}{D\mathbf{u}^I}}]^{-1}$ where

$$\frac{\widetilde{D\mathcal{R}}}{D\mathbf{u}^I} = \sum_{j=1}^{n} \left[\frac{\partial \mathbf{R}_j^I}{\partial \mathbf{u}_j^I} - \left(\frac{\partial \mathbf{R}_j^I}{\partial \mathbf{u}_j^S} \mathbf{D}_j^{-1} \frac{\partial \mathbf{R}_j^S}{\partial \mathbf{u}_j^I} \right) \right] \tag{12}$$

and $\mathbf{D}_j$ is the diagonal of the sub-domain Jacobian $\partial \mathbf{R}_j^S / \partial \mathbf{u}_j^S$.

The proposed scheme is summarized in algorithm (1) where $\varepsilon_{sub-domain}$, ε_{global} and $\varepsilon_{iterative}$ are prescribed tolerances.

Algorithm 1 Modified Newton Krylov Schur Algorithm

Partition the mesh
Initialize $\mathbf{u}^I$, $\mathbf{u}^S$, compute $\mathbf{R}_j^I$, $\mathbf{R}_j^S$, $\partial\mathbf{R}_j^I/\partial\mathbf{u}_j^I$, $\partial\mathbf{R}_j^S/\partial\mathbf{u}_j^I$, $\partial\mathbf{R}_j^I/\partial\mathbf{u}_j^S$, factor $\partial\mathbf{R}_j^s/\partial\mathbf{u}_j^S$ and assemble the interface preconditioner, cf. equation (12)
Repeat Newton iterations
  Repeat iterative solver computation of $\Delta\mathbf{u}^I$
   • Evaluate $(D\mathcal{R}/D\mathbf{u}^I)\,\mathbf{s}$ via equation (11)
   • Update $\mathbf{s}$
  Until $|(D\mathcal{R}/D\mathbf{u}^I)\mathbf{s} + \mathcal{R} - \sum_{j=1}^{n} \left[\partial\mathbf{R}_j^S/\partial\mathbf{u}_j^S\right]^{-1} \mathbf{R}_j^S| < \varepsilon_{iterative}$ (cf. eq. (4))
  • Solve $j = 1, 2, \ldots, n$ local sub-domain update equations (5)
  • Update $\mathbf{u}^I$ and $\mathbf{u}_j^S$ via equation (6)
  Repeat sub-domain Newton iterations
   • Solve Newton update equation (7) and store $\partial\mathbf{R}_j^S/\partial\mathbf{u}_j^S$ in factored form
   • Update sub-domain response $\mathbf{u}_j^S$ via equation (7)
  Until $|\mathbf{R}_j^S| < \varepsilon_{sub-domain}$ or $|\mathbf{R}_j^S| < |\mathbf{R}_j^S|_{avg}$
  • Compute $\mathbf{R}_j^I$, $\mathbf{R}_j^S$, $\partial\mathbf{R}_j^I/\partial\mathbf{u}_j^I$, $\partial\mathbf{R}_j^S/\partial\mathbf{u}_j^I$, $\partial\mathbf{R}_j^I/\partial\mathbf{u}_j^S$, $\left[\mathrm{Diag}(\partial\mathbf{R}_j^s/\partial\mathbf{u}_j^S)\right]^{-1}$
  • Assemble the interface preconditioner, cf. equation (12)
Until $|\mathbf{R}(\mathbf{u})| < \varepsilon_{global}$

---

4 Results

We have developed parallel domain-decomposition codes using MPI (Forum [1994]) to implement the proposed methodologies. We use METIS (Karypis and Kumar) to partition the domain, SuperLU (Demmel et al.) for the sparse solution of the sub-domain problems and PETSc (Balay et al.) for the iterative solution of the interface problem. All computations are performed on a distributed shared memory Origin 2000 machine.

The preconditioner matrix is in general dense and could be computationally expensive to factorize. For problems in which $(dim(\mathbf{u}^I)/max(dim(\mathbf{u}_j^s))) <$

1, i.e., for problems with few sub-domains, we use an LU factorization to obtain $\mathbf{P}$ otherwise we use a Jacobi method[2] to approximate $\mathbf{P}$.

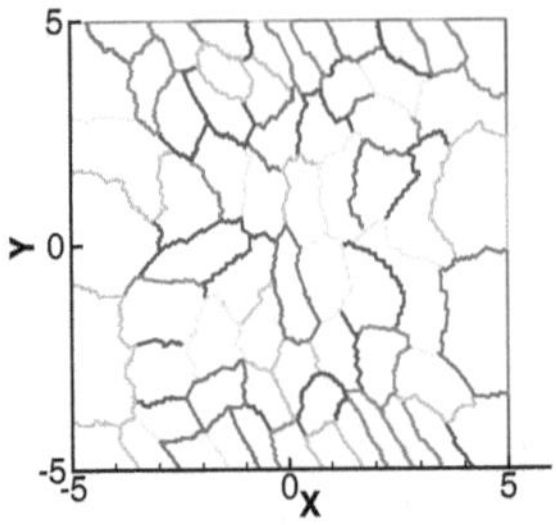

Fig. 1. Domain Partitioning

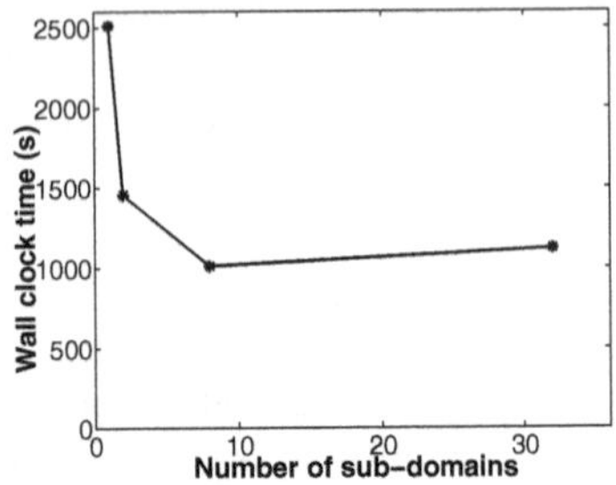

Fig. 2. Single processor efficiency

We consider a steady-state heat conduction problem to demonstrate the proposed algorithms. The nonlinear isotropic heat conduction coefficient κ is defined as: $\kappa(T) = \kappa_0(1 + \gamma\ T)$ where T is the temperature and κ_0 and γ are parameters, the latter of which controls the nonlinearity of the problem. Figure (1) shows the rectangular domain partitioned into 64 sub-domains. We impose zero flux conditions on the north and south boundaries and Dirichlet conditions of $T = 500$ and $T = 0$ on the west and east boundaries respectively. The discretization contains approximately 100,000 elements and 50,000 DOF.

4.1 Single processor efficiency

Figure 2 shows the timing results obtained using the MNK-Schur approach with varying number of sub-domains on a single processor. The problem uses $\kappa_0 = 1$ and $\gamma = 0.01$ and exhibits a balanced nonlinearity. The clock time for the single sub-domain case is obtained using the *dgssv* sparse direct solver of SuperLU (Demmel et al.). As seen from the figure, even on a single processor the DD based MNK-Schur approach performs better than a standard Newton scheme (i.e., the single sub-domain case) equipped with a sparse direct solver.

The timing results obtained for such single processor cases are used as baseline results for evaluating the scalability of the parallel implementations.

4.2 Parallel Scalability

To study the parallel scalability of the MNK-Schur algorithm we analyze the problem described in the previous section with 32 sub-domains and varying number of processors. The MNK-Schur shows near linear scale-up at a 55% to 65% efficiency as shown in figure 3. Note that the ability of our parallel implementation to accommodate multiple sub-domains per processor is

[2] PETSc has several built-in preconditioners that can be chosen at run time in place of the Jacobi method.

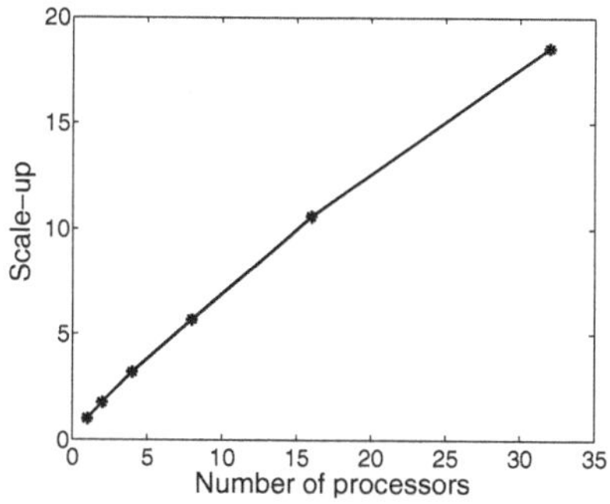

Fig. 3. Scalability of MNK-Schur

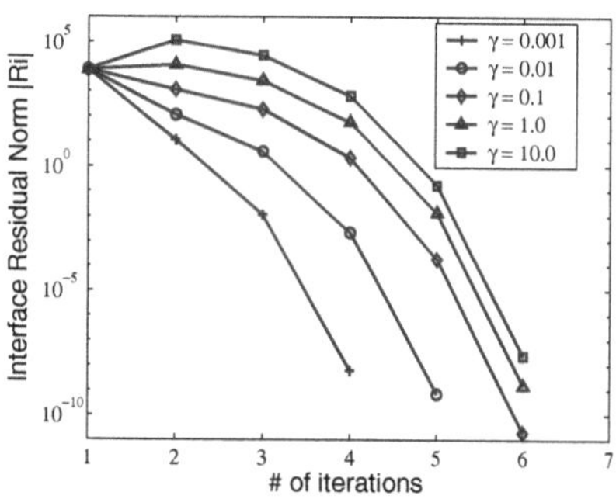

Fig. 4. Nonlinear convergence

demonstrated in these examples as the number of sub-domains is fixed while the number of processors is varied. Figure (4) shows the terminal quadratic convergence of the MNK-Schur scheme for several nonlinear problems.

4.3 Comparison of the two algorithms

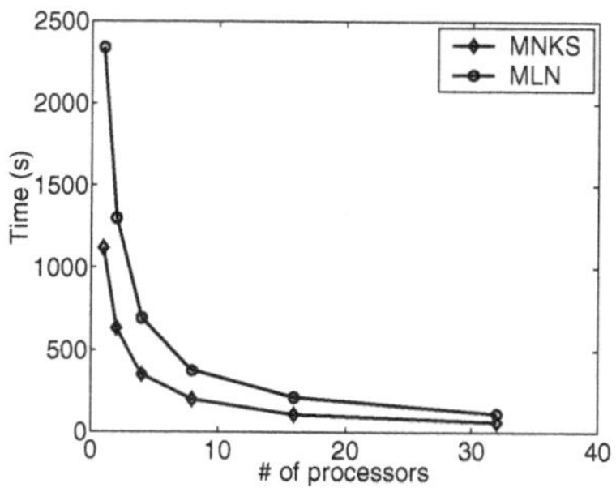

Fig. 5. MLN-Schur vs. MNK-Schur

To compare our MNK-Schur and MLN-Schur algorithms we analyze the 32 sub-domain case of the previous example problem. In figure 5 we plot the wall-clock time of the MLN-Schur and MNK-Schwarz schemes for varying number of processors. We see that the MLN-Schur scheme is less efficient than the MNK-Schur scheme irrespective of the number of processors employed. This difference is attributed to the full resolution of the sub-domain residual equations in the MLN-Schur scheme. Hence, the MLN-Schur requires more computations per interface Newton iteration.

However, for problems with unbalanced nonlinearities, fewer interface Newton iterations may be required when using the MLN-Schur method.

5 Conclusion

We have introduced two non-overlapping DD schemes based on a two-level Newton approach. The MNK-Schur scheme combines the advantages of the MLN-Schur and NK-Schur schemes to provide a general approach that efficiently solves problems with balanced and unbalanced nonlinearities. The DD implementation shows good scalability. By assigning multiple sub-domains to each processor we obtain a scheme that is efficient on a single processor and one that is amenable to load balancing in parallel implementations. The implementations have been designed to accommodate unstructured meshes, nonsymmetric matrices and a variety of iterative solvers and preconditioners.

References

N. Aluru and J. White. A MLN method for mixed-energy domain sim. of MEMS. *Journal of MEMS systems*, 8(3):299–308, 1999.

M. Bächtold, J. Korvink, J. Funk, and H. Baltes. New convergence scheme for self-consistent electromech. analysis of iMEMS. In *IEEE Inter. Electron Devices Meeting*, 1995.

S. Balay, W. Gropp, L. McInnes, and B. Smith. PETSc, v. 2.1.3 code and documentation. URL `http://www-unix.mcs.anl.gov/petsc/`.

P. Bjørstad, J. Koster, and P. Krzyżanowski. DD solvers for large scale industrial finite element problems. In *PARA 2000*, volume 1947 of *LNCS*. Springer, 2001.

X.-C. Cai and D. Keyes. Nonlinearly preconditioned inexact Newton algorithms. *SIAM J. Sci. Comput.*, 24:183–200, 2002.

X.-C. Cai, D. E. Keyes, and D. P. Young. A nonlinear additive Schwarz preconditioned inexact Newton method for shocked duct flow. In *Proc. D.D. Methods-13*, 2001.

J. Demmel, J. Gilbert, and X. Li. SuperLU v. 2.0 code and documentation. URL `http://crd.lbl.gov/~xiaoye/SuperLU/`.

C. Farhat, M. Lesoinne, P. LeTallec, K. Pierson, and D. Rixen. FETI-DP: a dual-primal unified FETI method–part I. *IJNME*, 50:1523–1544, 2001.

M. P. I. Forum. MPI: A message-passing interface standard. *Inter. J. Supercomputing Apps.*, 8(3/4), 1994.

G. Karypis and V. Kumar. METIS: v. 4.0 code and documentation. URL `http://www-users.cs.umn.edu/~karypis/metis`.

D. Keyes. DD methods for PDEs. *SIAM J. Sci. Statistic. Comput.*, 13:967–993, 1992.

D. Keyes. Aerodynamic applications of NKS solvers. In *Proceedings of the 14th Conf. Numer. Methods in Fluid Dynamics*, pages 1–20, Berlin, 1995.

J. Kim, N. Aluru, and D. Tortorelli. Improved MLN solvers for fully-coupled multi-physics problems. *IJNME*, 58, 2003.

D. Knoll and D. Keyes. Jacobian-free Newton-Krylov methods: A survey of approaches and applications. *Journal of Comp. Physics*, 2002.

Y. Saad and M. H. Schultz. GMRES: An algorithm for solving nonsymmetric linear systems. *SIAM J. Sci. Stat. Comput.*, 7:856–869, 1986.

B. Smith, P. Bjørstad, and W. Gropp. *D.D, Parallel multilevel methods for elliptic PDEs.* Cambridge University Press, 1996.

Domain Decomposition for Discontinuous Galerkin Method with Application to Stokes Flow

Piotr Krzyżanowski

Warsaw University, Faculty of Mathematics, Informatics and Mechanics
(http://www.mimuw.edu.pl/piotr.krzyzanowski)

Summary. We report on recent results related to domain decomposition methods based on the Discontinuous Galerkin discretizations of Stokes equations. We analyze the efficiency of a block nonoverlapping Schwarz preconditioner based on the approach by Feng and Karakashian [2001]. We also prove the inf-sup stability of a substructuring method.

1 Introduction

Discontinuous Galerkin (DG) methods have attained a lot of interest in the past years. These nonconforming finite element methods have several advantages over the classical, conforming elements. For example, the finite elements are very easy to construct and allow the use of nonuniform meshes. Moreover, they still guarantee optimal error estimates, requiring only local regularity of the solution. On the other hand, as compared to the conforming methods, the DG methods introduce more degrees of freedom per grid point.

Recently Filippini and Toselli [2002] proved an inf-sup stability result for a Discontinuous Galerkin approximation of Stokes equations on non-matching grids, while at the same time a series of papers by Toselli [2002], Cockburn et al. [2002], Schötzau et al. [2002] developed the stability and approximation theory of DG methods for the Stokes system.

In this paper, we use a DG discretization of the velocity–pressure formulation of the Stokes equations (1), using macroelements as in Filippini and Toselli [2002]. The variational form of the Stokes equations gives rise to a symmetric operator, and our discretization not necessarily retains this property, depending on the choice of the method.

We consider here two domain decomposition methods for the resulting, possibly nonsymmetric, discrete saddle point problem. First we use the approach of block preconditioning, see Krzyżanowski [2001], Klawonn [1998b], Klawonn [1998a]. Using available results related to nonoverlapping Additive Schwarz preconditioning DG discretizations of the second order equations, we prove the convergence rate bounds for the corresponding block preconditioner for the DG Stokes discretization. To the author's knowledge, this is the first such result for domain decomposition preconditioners for DG discretizations of Stokes equations.

Next, we define a substructuring method for the discretization under consideration. We show that the resulting problem has also a saddle point structure. For this problem, we show that the inf-sup constant is independent both of the fine mesh size and of the number of the subdomains. This result is a basis for the analysis of parallel substructuring preconditioners, such as the Neumann–Neumann; this topic, however, is not covered in the present paper.

In the paper, for nonnegative scalars x, y, we shall write $x \lesssim y$ if there exits a positive constant C, independent of x, y and the mesh parameters h, H, such that $x \leq Cy$.

2 DG discretization of the Stokes equation

Let Ω be a bounded open polygon in R^d, $d = 2, 3$. The Stokes equations in Ω read

$$\begin{aligned} -\Delta u + \nabla p &= f, \\ \nabla \cdot u &= 0, \end{aligned} \tag{1}$$

where $u = (u_1, \ldots, u_d)$ denotes fluid velocity and p is the pressure. For simplicity, we assume homogeneous Dirichlet boundary condition on u. The given function $f : \Omega \to R^d$ is the external force.

In what follows, for a domain D, $(\cdot,\cdot)_D$ denotes the usual inner product in $L^2(D)$ (or, depending on the context, $[L^2(D)]^d$), while $\langle\cdot,\cdot\rangle_D$ denotes the inner product in $L^2(\partial D)$ ($[L^2(\partial D)]^d$). We shall omit the subscript, if the integrals are taken over Ω.

2.1 Finite element spaces

Let $\mathcal{T}_H$ be a subdivision of Ω into N disjoint triangles Ω_i, $i = 1, \ldots, N$, such that $\bar{\Omega} = \bigcup_{i=1,\ldots,N} \bar{\Omega}_i$ and $\mathcal{T}_H$ forms an affine, regular triangulation of Ω, with mesh parameter H. Further, let $\mathcal{T}_h$ denote an affine, shape regular triangulation of Ω, $\bar{\Omega} = \bigcup_{\kappa \in \mathcal{T}_h} \bar{\kappa}$, which is derived from $\mathcal{T}_H$ by some refinement procedure. The diameter of a triangle $\kappa \in \mathcal{T}_h$ will be denoted by h_κ and the mesh parameter is $h = \max_{\kappa \in \mathcal{T}_h} h_\kappa$. By $\mathcal{E}_h$ we denote the set of all edges of elements in $\mathcal{T}_h$, while $\mathcal{E}_h^0$ we use to denote the internal edges, that is those not included in $\partial\Omega$; for $e \in \mathcal{E}_h$, we also set $h_e = \mathrm{diam}(e)$). The set of all edges of elements from $\mathcal{T}_H$ will be denoted by Γ.

In order to formulate our domain decomposition method, we shall use the coarse triangulation $\mathcal{T}_H$ of Ω. We shall assume that restricted to each subdomain our finite element spaces consist of continuous functions, and that those functions satisfy the inf-sup condition.

Let us begin with the definition of the local finite element spaces,

$$V_h^{r_i}(\Omega_i) = \{v^i \in [C(\Omega_i)]^d : v_{|\kappa} \in [Pr_i(\kappa)]^d, \forall \kappa \in \mathcal{T}_h, \kappa \subset \Omega_i\},$$

and

$$W_h^{q_i}(\Omega_i) = \{w^i \in C(\Omega_i) : w_{|\kappa} \in Pq_i(\kappa), \forall \kappa \in \mathcal{T}_h, \kappa \subset \Omega_i\}.$$

Then, we set global spaces, in which we shall pose our discrete problem,

$$V_h^H = \{v \in [L^2(\Omega)]^d : v_{|\Omega_i} \in V_h^{r_i}(\Omega_i), \forall \Omega_i \in \mathcal{T}_H\}, \tag{2}$$

and analogously,

$$W_h^H = \{w \in L_0^2(\Omega) : w_{|\Omega_i} \in W_h^{q_i}(\Omega_i), \forall \Omega_i \in \mathcal{T}_H\}. \tag{3}$$

For short, we shall denote by v^i the restriction of $v \in V_h^H$ to Ω_i. We shall make one more assumption, which relates $V_h^{r_i}(\Omega_i)$ to $W_h^{q_i}(\Omega_i)$:

For $i = 1, \ldots, N$, there exist constants λ_i independent of h such that

$$\sup_{v \in V_h^{r_i}(\Omega_i), v \neq 0} \frac{(w, \nabla \cdot v)_{\Omega_i}}{|v|_{1,\Omega_i}} \gtrsim \lambda_i |w|_{0,\Omega_i}, \qquad \forall w \in W_h^{q_i}(\Omega_i), \int_{\Omega_i} w = 0. \tag{4}$$

Since the traces of the functions from V_h^H and W_h^H (and more generally, from $H^1(\mathcal{T}_H)$) are double-valued on the interelement interface $\Gamma_H^0 = \Gamma_H \setminus \partial\Omega$, where $\Gamma_H = \cup_{\Omega_i \in \mathcal{T}_H} \partial\Omega_i$, we shall define, following Arnold et al. [2001/02], their average $\{\cdot\}$ and jump $[\cdot]$ on an edge e shared by two elements $\kappa_1, \kappa_2 \in \mathcal{T}_h$, see [Arnold et al., 2001/02, Section 3.1]. The spaces V_h^H and W_h^H are equipped with the following norms. For $u \in V_h^H$ we set [Arnold, 1982, Lemmas 2.2 and 2.1]

$$|||u|||^2 = \sum_{\kappa \in \mathcal{T}_h} |u|_{1,\kappa}^2 + \sum_{e \in \mathcal{E}_h} \frac{1}{h_e} |[u]|_{0,e}^2.$$

(The corresponding inner product in V_h^H will be denoted by $((\cdot,\cdot))$.) For $p \in W_h^H$ we define its norm as the usual L^2 norm: $|p|^2 = \sum_{\kappa \in \mathcal{T}_h} |p|_{0,\kappa}^2$.

2.2 Discretization

We use the following discontinuous Galerkin finite element approximation to (1):

Problem 1. Find $(u_h, p_h) \in V_h^H \times W_h^H$, such that

$$\begin{aligned} A_h(u_h, v) + B_h(v, p_h) &= (f, v), \\ B_h(u_h, w) &= 0, \end{aligned} \tag{5}$$

for all $(v, w) \in V_h^H \times W_h^H$.

Here, we have some freedom in how to choose the form $A_h(\cdot, \cdot)$ which approximates the Laplacian, see Schötzau et al. [2002] for a discussion. We allow here for two quite popular choices: the symmetric Interior Penalty method as in Douglas and Dupont [1976] (see also Arnold [1982], Arnold et al. [2001/02]) or the nonsymmetric form, which differs from the previous one by a change in the sign of one boundary term, considered, e.g. in Filippini and Toselli [2002]:

$$A_h(u,v)=\sum_{\kappa\in\mathcal{T}_h}(\nabla u,\nabla v)_\kappa\mp\sum_{e\in\mathcal{E}_h}\langle[u],\{\nabla v\}\rangle_e-\sum_{e\in\mathcal{E}_h}\langle\{\nabla u\},[v]\rangle_e+\sum_{e\in\mathcal{E}_h}\langle\mu_e[u],[v]\rangle_e, \tag{6}$$

The penalty scaling μ_e is a properly chosen function, usually of the form $\mu_e=\frac{\delta_e}{h_e}$, $e\in\mathcal{E}_h$, with constant δ_e large enough to preserve the ellipticity of the original problem. The choice of the sign in the definition above results in different DG methods, as described above, and obviously affects the symmetry of this bilinear form.

The approximate divergence form is defined, see e.g. Toselli [2002],

$$B_h(u,p)=-\sum_{\kappa\in\mathcal{T}_h}(p,\nabla\cdot u)_\kappa+\sum_{e\in\mathcal{E}_h^0}\langle\{p\},[u]\rangle_e. \tag{7}$$

Let us introduce a stability result for the discrete problem (5):

Lemma 1. *The pair $V_h^H\times W_h^H$ is inf-sup stable, and the inf-sup constant is independent of both h and H.*

Proof. A similar theorem has been proved for the rectangular elements, in [Filippini and Toselli, 2002, Theorem 4.1] and this lemma validates it for the case of triangular elements. Under assumptions made throughout the paper, there is a quite straightforward way to prove the above Lemma. The proof of course follows the idea of Boland and Nicolaides [1983]. In view of the local inf-sup assumption, the key point of the proof is to specify a globally stable subspace of (V_h^H,W_h^H). We may use for it e.g. the space (V_H^1,W_H^0), consisting of piecewise linear and piecewise constant functions on $\mathcal{T}_H$, respectively, which inf-sup stability can be directly proved (see also Schötzau et al. [2002]). We omit the details due to the lack of space.

3 Nonoverlapping domain decomposition block preconditioner

We follow the general idea of Krzyżanowski [2001] (see also Klawonn [1998b] for the symmetric case analysis). Using the natural formulation of the variational discrete problem (5) in the operator form,

$$\mathcal{M}\begin{pmatrix}u\\p\end{pmatrix}=\begin{pmatrix}A & B^*\\B & 0\end{pmatrix}\begin{pmatrix}u\\p\end{pmatrix}=\begin{pmatrix}F\\G\end{pmatrix},$$

we define a block diagonal preconditioner for $\mathcal{M}$

$$\mathcal{M}_D=\begin{pmatrix}A_0 & 0\\0 & J_0\end{pmatrix},$$

and transform the original system into

Problem 2. Find $(u,p) \in V_h^H \times W_h^H$ such that

$$\mathcal{M}_D^{-1}\mathcal{M}^*\mathcal{M}_D^{-1}\mathcal{M}\begin{pmatrix} u \\ p \end{pmatrix} = \mathcal{M}_D^{-1}\mathcal{M}^*\mathcal{M}_D^{-1}\begin{pmatrix} F \\ G \end{pmatrix}. \tag{8}$$

The operator $\mathcal{P} = \mathcal{M}_D^{-1}\mathcal{M}^*\mathcal{M}_D^{-1}\mathcal{M}$ is positive definite and self-adjoint with respect to the inner product $(\mathcal{M}_D\,\cdot,\cdot)$ induced by $\mathcal{M}_D$, regardless the potential lack of symmetry properties of $\mathcal{M}$, and an iterative method such as the conjugate gradient method can be used to solve this problem efficiently[1].

The building blocks of the preconditioner $\mathcal{M}_D$ will be based on a symmetric nonoverlapping domain decomposition preconditioner for the symmetric DG stiffness matrix. In this way, we will obtain a highly parallelizable, nonoverlapping preconditioner for the whole system. The lack of the overlap is an important feature from the point of view of parallel computer implementation, since this lowers the interprocessor communication cost. Moreover, the symmetry of the preconditioned system will give us a possibility to use cheaper symmetric iterative solvers (on the global level) or direct sparse solvers (on the subdomain level).

For A_0^{-1} we choose the nonoverlapping Additive Schwarz preconditioner for a *symmetric* DG method developed by Feng and Karakashian [2001]. According to [Feng and Karakashian, 2001, Theorem 4.5], we have

$$|||u|||^2 \lesssim ((A_0 u, u)) \lesssim \frac{H}{h}|||u|||^2, \tag{9}$$

We also set, for simplicity, $J_0^{-1} = M^{-1}$, where M is the pressure mass matrix operator. Since M is block diagonal, with each block corresponding to a mass matrix assembled on a given substructure, J_0^{-1} is perfectly parallelizable across the subdomains and can also be relatively cheaply applied using local sparse solvers. Note also that J_0^{-1} could even be further simplified, at the price of reducing its efficiency, e.g. by the mass lumping procedure.

The following theorem estimates the condition number of the preconditioned operator:

Theorem 1. *Under the above assumptions,*

$$\operatorname{cond}(\mathcal{P}) \lesssim \left(\frac{H}{h}\right)^2,$$

regardless of the choice of the sign in $A_h(\cdot,\cdot)$ in Section 2.2.

Proof. We use the technique of Krzyżanowski [2001]. Due to stability results for $A_h(\cdot,\cdot)$, it is sufficient to check only the influence of the quality of the velocity preconditioner. Let us assume that the preconditioner A_0^{-1} satisfies

[1] Another approach, if $\mathcal{M}^* = \mathcal{M}$, could be to solve $\mathcal{M}_D^{-1}\mathcal{M}$ with the conjugate residual method, see Klawonn [1998b]. While the two approaches are comparable in the symmetric case, the symmetrized method can also be applied to DG discretizations which lead to nonsymmetric saddle point problems.

$$a_0|||u|||^2 \le ((A_0 u, u)) \le a_1|||u|||^2,$$

and $a_0 < 1 < a_1$. Observe that

$$\begin{aligned}\left(\mathcal{M}_D \mathcal{P}\begin{pmatrix}u\\p\end{pmatrix}, \begin{pmatrix}u\\p\end{pmatrix}\right)_{V_h^H \times W_h^H} &= ((A_0^{-1}(Au + B^*p), Au + B^*p)) + \\ &\qquad (J_0^{-1}(Bu - Cp), Bu - Cp) \\ &\ge \frac{1}{a_1}|||Au + B^*p|||^2 + |Bu - Cp|^2.\end{aligned} \tag{10}$$

By the stability, we obtain

$$\begin{aligned}\left(\mathcal{M}_D \mathcal{P}\begin{pmatrix}u\\p\end{pmatrix}, \begin{pmatrix}u\\p\end{pmatrix}\right)_{V_h^H \times W_h^H} &\gtrsim \frac{1}{a_1}\left(|||u|||^2 + |p|^2\right) \\ &\gtrsim \frac{a_0}{a_1}\left(\mathcal{M}_D \begin{pmatrix}u\\p\end{pmatrix}, \begin{pmatrix}u\\p\end{pmatrix}\right)_{V_h^H \times W_h^H}.\end{aligned}$$

Similarly, see Krzyżanowski [2001], we prove the upper bound,

$$\left(\mathcal{M}_D \mathcal{P}\begin{pmatrix}u\\p\end{pmatrix}, \begin{pmatrix}u\\p\end{pmatrix}\right)_{V_h^H \times W_h^H} \lesssim \frac{a_1}{a_0}\left(\mathcal{M}_D \begin{pmatrix}u\\p\end{pmatrix}, \begin{pmatrix}u\\p\end{pmatrix}\right)_{V_h^H \times W_h^H}.$$

Now the conclusion follows from (9).

4 Stability of the substructuring method

In the conforming case, it is known that substructuring preconditioners, such as the FETI or the balancing Neumann–Neumann methods, give rise to only polylogarithmic condition number bound. Thus, in the view of the polynomial in $\frac{H}{h}$ condition bound for the nonoverlapping Additive Schwarz proved in Theorem 1, one can hope for a better behavior of the substructuring methods.

In what follows, we shall give a result, see Pavarino and Widlund [2002], which is a basis for further investigation of the substructuring preconditioners for DG discretizations of Stokes equations: we recognize the substructuring as a specific saddle point problem on the interface and prove its inf-sup stability.

Let us first define the substructuring method, restricting ourselves to the symmetric interior penalty DG discretization. Define $V(\Gamma)$ as the space of restrictions of functions from V_h^H to the interface Γ_H. We define the local (discrete) DG Stokes extension operator $S : V(\Gamma) \to V_h^H \times W_h^H$ as $Su_\Gamma = (u, p)$, satisfying $u = u_\Gamma$ on the interface Γ and on each subdomain Ω_i,

$$\begin{aligned}(\nabla u_i, \nabla v_i)_{\Omega_i} - (\operatorname{div} v_i, p_i)_{\Omega_i} &= \langle [u_\Gamma], \{\nabla v_i\}\rangle_{\partial\Omega_i}, \\ (\operatorname{div} u_i, q_i)_{\Omega_i} &= \langle [u_\Gamma], \{q_i\}\rangle_{\partial\Omega_i}.\end{aligned}$$

Note that, in contrast to the conforming FE discretizations, our Stokes extensions are not homogeneous right hand side problems. Defining $V_\Gamma = \{v \in V_h^H : v = Su_\Gamma, \text{ for some } u_\Gamma \in V(\Gamma)\}$, and $W_0 = \{q \in W_h^H : q_{\Omega_i} = \text{const } \forall i\}$ we arrive, following the lines of Pavarino and Widlund [2002], at the following form of the Schur complement of the DG Stokes discretization:

Problem 3. Find $(u_\Gamma, p_0) \in V_\Gamma \times W_0$ such that

$$A_h(u_\Gamma, v_\Gamma) + B_h(v_\Gamma, p_0) = (\tilde{F}, v_\Gamma), \qquad \forall v_\Gamma \in V_\Gamma,$$
$$B_h(u_\Gamma, q_0) = 0, \qquad \forall q_0 \in W_0.$$

This problem looks similar to the one considered in Pavarino and Widlund [2002] and, despite its different origin, has similar stability property, partly because of good stability properties of the DG discretizations.

Theorem 2. *There exists a constant β_Γ, independent of H and h, such that*

$$\sup_{v_\Gamma \in V_\Gamma} \frac{B_h(v_\Gamma, q_0)}{|||v_\Gamma|||} \geq \beta_\Gamma |q_0|, \qquad \forall q_0 \in W_0. \tag{11}$$

Proof. Since, see the proof of Lemma 1, the pair $V_H^1 \times W_H^0$ is inf-sup stable, there exists $\hat{u}$ in $V^1(\mathcal{T}_h)$ such that $B_h(\hat{u}, q_0) = B_h(u, q_0)$ and $|||\hat{u}||| \lesssim |||u|||$. Taking $u_\Gamma = S(\hat{u}_{|\Gamma})$ and using the stability of the Stokes extension we see that $v_\Gamma = u_\Gamma$ satisfies the desired inequality. We skip the details.

5 Concluding remarks

Block preconditioners with high level of parallelism and relatively low intersubdomain communication requirements are relatively easy to derive for DG discretizations of the Stokes equations, and their properties directly reflect those of the building blocks for second order elliptic equations. However, existing DD preconditioners for DG Laplacian discretizations feature only $\frac{H}{h}$ condition bound, which makes substructuring preconditioners potentially more attractive. (Another potentially nice feature of using, e.g. the Neumann–Neumann alike preconditioners would be, in the case of DG discretizations, that the problem of floating subdomains can be totally avoided.) While the question of their performance remains open, we proved that at least the very Schur complement problem is stable independently of the number of subdomains.

Acknowledgement. This work has partially been supported by the KBN research grant 2 P03A 005 24.

References

D. N. Arnold. An interior penalty finite element method with discontinuous elements. *SIAM J. Numer. Anal.*, 19(4):742–760, 1982. ISSN 0036-1429.

D. N. Arnold, F. Brezzi, B. Cockburn, and L. D. Marini. Unified analysis of discontinuous Galerkin methods for elliptic problems. *SIAM J. Numer. Anal.*, 39(5):1749–1779 (electronic), 2001/02. ISSN 1095-7170.

J. M. Boland and R. A. Nicolaides. Stability of finite elements under divergence constraints. *SIAM J. Numer. Anal.*, 20(4):722–731, 1983. ISSN 0036-1429.

B. Cockburn, G. Kanschat, D. Schötzau, and C. Schwab. Local discontinuous Galerkin methods for the Stokes system. *SIAM J. Numer. Anal.*, 40(1): 319–343 (electronic), 2002. ISSN 1095-7170.

J. Douglas, Jr. and T. Dupont. Interior penalty procedures for elliptic and parabolic Galerkin methods. In *Computing methods in applied sciences (Second Internat. Sympos., Versailles, 1975)*, pages 207–216. Lecture Notes in Phys., Vol. 58. Springer, Berlin, 1976.

X. Feng and O. A. Karakashian. Two-level additive Schwarz methods for a discontinuous Galerkin approximation of second order elliptic problems. *SIAM J. Numer. Anal.*, 39(4):1343–1365 (electronic), 2001. ISSN 1095-7170.

L. Filippini and A. Toselli. hp finite element approximations on non-matching grids for the Stokes problem. Technical report, ETH, Zurich, Switzerland, 2002.

A. Klawonn. Block-triangular preconditioners for saddle point problems with a penalty term. *SIAM J. Sci. Comput.*, 19(1):172–184 (electronic), 1998a. ISSN 1095-7197. Special issue on iterative methods (Copper Mountain, CO, 1996).

A. Klawonn. An optimal preconditioner for a class of saddle point problems with a penalty term. *SIAM J. Sci. Comput.*, 19(2):540–552 (electronic), 1998b. ISSN 1095-7197.

P. Krzyżanowski. On block preconditioners for nonsymmetric saddle point problems. *SIAM J. Sci. Comp.*, 23(1):157–169, 2001.

L. F. Pavarino and O. B. Widlund. Balancing Neumann-Neumann methods for incompressible Stokes equations. *Comm. Pure Appl. Math.*, 55(3):302–335, 2002. ISSN 0010-3640.

D. Schötzau, C. Schwab, and A. Toselli. Mixed *hp*-DGFEM for incompressible flows. *SIAM J. Numer. Anal.*, 40(6):2171–2194 (electronic) (2003), 2002. ISSN 1095-7170.

A. Toselli. *hp* discontinuous Galerkin approximations for the Stokes problem. *Math. Models Methods Appl. Sci.*, 12(11):1565–1597, 2002. ISSN 0218-2025.

Hierarchical Matrices for Convection-Dominated Problems

Sabine Le Borne

Tennessee Technological University, Mathematics
(http://math.tntech.edu/~sleborne/)

Summary. Hierarchical matrices provide a technique to efficiently compute and store explicit approximations to the inverses of stiffness matrices computed in the discretization of partial differential equations. In a previous paper, Le Borne [2003], it was shown how standard $\mathcal{H}$-matrices must be modified in order to obtain good approximations in the case of a convection dominant equation with a constant convection direction. This paper deals with a generalization to arbitrary (non-constant) convection directions. We will show how these $\mathcal{H}$-matrix approximations to the inverse can be used as preconditioners in iterative methods.

1 Introduction

Considerable advancements have been achieved in algebraic and geometric multigrid solvers, state-of-the art domain decomposition methods such as FETI (Farhat et al. [2001]), direct and approximate (inverse) factorization solvers (Grote and Huckle [1997], Chow and Saad [1998]) as well as custom strategies for coarsening, partitioning, ordering, pivoting, etc., which improve the effectiveness and robustness of these methods. However, many important challenges remain which in particular include the construction of a robust solver for convection-dominant systems of PDEs.

A completely new and powerful approach for the construction of efficient preconditioners and smoothing iterations has recently been introduced that involve so-called hierarchical matrices, or $\mathcal{H}$-matrices (see, e.g., Hackbusch [1999], Grasedyck and Hackbusch [2002], Le Borne [2003]). The $\mathcal{H}$-matrix technique is a generalization of the panel clustering method and permits the treatment of fully populated matrices while restricting the requirements for storage and arithmetics (approximate matrix-vector multiplication, matrix-matrix multiplication and matrix inversion) to nearly optimal complexity $\mathcal{O}(n \log_2^\alpha n)$ for some (small) constant α. Related methods are the multipole method and the mosaic skeleton method (Tyrtyshnikov [2000]). In this paper we use an $\mathcal{H}$-matrix as a preconditioner in an iterative method to solve a convection-dominant problem. The characteristic feature that distinguishes

$\mathcal{H}$-matrices from other sparse approximate inverse techniques (SPAI) (see, e.g., Grote and Huckle [1997], Chow and Saad [1998], Benzi and Tuma [1998]) is the particular storage format of an $\mathcal{H}$-matrix that will be further explained below. Whereas these sparse (SPAI) methods typically work well if the approximated matrix contains many very small entries, the $\mathcal{H}$-matrix techniques provides convergent approximations if (large) subblocks of the approximated matrix are smooth (but not necessarily have small entries).

The remainder of this paper is organized as follows: After the introduction of the model problem in Section 2.1 we review the construction of an $\mathcal{H}$-matrix in Section 2.2. In Sections 2.3 and 2.4, modifications to the standard $\mathcal{H}$-matrix are developed for the convection-dominant case with constant and non-constant convection directions, resp. In Section 3, we will provide the results of numerical tests where $\mathcal{H}$-matrices have been used in iterative methods.

2 Preliminaries

2.1 The model problem

In this paper we consider the two-dimensional convection-diffusion equation with Dirichlet boundary conditions

$$-\epsilon \Delta u + b \cdot \nabla u = f \text{ in } \Omega = (0,1)^2, \tag{1}$$

$$u = g \text{ on } \partial\Omega \tag{2}$$

for $0 < \epsilon \ll 1$ and an arbitrary convection $b : \mathbb{R}^2 \to \mathbb{R}^2$. An (upwind) finite element discretization leads to a linear system of equations $A_h x_h = f_h$ where the parameter h characterizes the grid width of the underlying mesh. The $\mathcal{H}$-matrix technique is applicable to matrices obtained by a wide range of discretizations since its theory is based upon the approximability of the underlying Green's function by a separable function (and not on a particular discretization technique). Even though the construction of an $\mathcal{H}$-matrix is based on some knowledge on the underlying Green's function, the Green's function need not be known explicitly.

In Le Borne [2003], the case of a constant convection b was analysed. The numerical results showed better results in the case where the convection b aligned with the grid compared to a general, non-aligning convection direction. This can be explained by the numerical diffusion produced by the discretization scheme in the case of a non-aligning convection direction. Therefore, the construction of the $\mathcal{H}$-matrix should not only depend on the continuous problem but also on the amount of numerical diffusion, especially since in the case of an arbitrary, non-constant convection we typically cannot expect the grid to align with the convection.

2.2 $\mathcal{H}$-matrices

We will briefly review the the definition and standard construction of an $\mathcal{H}$-matrix in order to later derive modifications for the convection-dominant case. More details on $\mathcal{H}$-matrix approximations can be found in, e.g. Hackbusch [1999], Hackbusch and Khoromskij [2000b,a], Hackbusch et al. [2003], Grasedyck and Hackbusch [2002], and the references therein. An $\mathcal{H}$-matrix approximation to a given (fully populated) matrix $A \in \mathbb{R}^{I \times I}$ for a finite index set I is obtained by first constructing a certain block partitioning of the matrix index set $I \times I$, and then replacing each subblock $b = b_1 \times b_2 \subset I \times I$ of this partitioning that is larger than a certain threshold by a matrix of low rank $k(b)$. If this rank $k(b)$ is small compared to the number of indices contained in b_1 and b_2, then such a low rank matrix has much lower storage requirements than the approximated full matrix.

Definition 1 (R(k)-matrix representation). *Let $k, n, m \in \mathbb{N}_0$, and let $M \in \mathbb{R}^{n \times m}$ be a matrix of at most rank k. A representation of M in factorised form $M = AB^T$, $A \in \mathbb{R}^{n \times k}, B \in \mathbb{R}^{m \times k}$, with A and B stored as full matrices, is called an $R(k)$-matrix representation of M, or, in short, we call M an $R(k)$-matrix.*

Remark 1. The storage requirement $N_{R,St}(n,m,k)$ and the costs $N_{R\cdot v}(n,m,k)$ for the matrix-vector product with a matrix $M \in \mathbb{R}^{n \times m}$ in R(k)-matrix representation are $N_{R,St}(n,m,k) = k(n+m)$ and $N_{R\cdot v}(n,m,k) = 2k(n+m)-n-k$.

Compared to the respective complexities for full matrices, $\mathcal{O}(nm)$, we have significant savings for the R(k)-matrix if the rank k is small compared to the size of the matrix.

Definition 2 ($\mathcal{H}$-matrix). *Let $n_{\min} \in \mathbb{N}_0$. Let $\mathcal{P}$ be a partition of the block index set $I \times I$. Let $k : \mathcal{P} \to \mathbb{N}_0$ be a mapping that assigns a rank $k(b)$ to each block $b = s \times t \in \mathcal{P}$. The set of $\mathcal{H}$-matrices induced by the partition $\mathcal{P}$ and with minimum block size $n_{\min}$ is defined by*

$$\mathcal{H}(\mathcal{P},k) := \{M \in \mathbb{R}^{I \times I} \, | \forall s \times t \in \mathcal{P} : \operatorname{rank}(M|_{s\times t}) \leq k(s \times t) \text{ or } \min\{\#s, \#t\} \leq n_{\min}\}.$$

A matrix $M \in \mathcal{H}(\mathcal{P},k)$ is said to be given in $\mathcal{H}$-matrix representation if the blocks $M|_{s\times t}$ with rank$(M|_{s\times t}) \leq k(s \times t)$ are stored in $R(k)$-matrix representation and the remaining blocks with $\min\{\#s, \#t\} \leq n_{\min}$ as full matrices.

The accuracy of an $\mathcal{H}$-matrix approximation depends on how well the individual blocks in the partition can be approximated by low rank matrices, which in turn depends on the approximability of the underlying Green's function by separable functions as well as the ordering of the unknowns. To obtain a suitable block partition, we construct a hierarchy of partitionings from which we choose the "coarsest" one that satisfies a certain admissibility condition which shall ensure the approximability by a low rank matrix. The construction of a hierarchy of partitionings of an index set is shown in Figure 1. The hierarchical index set partition of Figure 1 does not state how to divide an index set into two subsets. Typically, the indices are ordered in a certain

Let $I = I_{0,0}$ be a finite index set. If the jth subset on level ℓ, $I_{\ell,j} \subset I$, contains more than one index, we subdivide it into two disjoint successor index sets $I_{\ell+1,j\ell-1}$ and $I_{\ell+1,j\ell}$ of approximately the same size on the next level $\ell+1$ that satisfy $I_{\ell,j} = I_{\ell+1,j\ell-1} \cup I_{\ell+1,j\ell}$.

Fig. 1. Hierarchical index set partitioning

way based upon the geometric information associated with the indices, and then this ordered list of indices is bisected into two sets of approximately the same size. In the case of uniformly elliptic differential operators, it has been shown in Bebendorf and Hackbusch [2003] that a partitioning into subsets with small diameters (with respect to the Euclidean norm) will lead to a convergent $\mathcal{H}$-matrix approximation. Such a partition is obtained if the indices within each index set $I_{\ell,j}$ are ordered as follows:

$$\begin{aligned} &\textit{if} \max_{v,w\in I_{\ell,j}} |x_v - x_w| > \max_{v,w\in I_{\ell,j}} |y_v - y_w| \text{ then} \\ &\qquad n(v) < n(w) \text{ if } x_v < x_w \text{ or } (x_v = x_w \text{ and } y_v < y_w) \\ &\text{else} \quad n(v) < n(w) \text{ if } y_v < y_w \text{ or } (y_v = y_w \text{ and } x_v < x_w). \end{aligned}$$

Here, (x_v, y_v) describes the geometric location associated with an index v, and $n(v) \in \{1, \cdots, \#I_{\ell,j}\}$ assigns the index number. We will refer to this type of bisection as the *standard partition* or *geometric bisection.*

In order to define an admissibility condition, let $B_{\ell,j} := B_{I_{\ell,j}}$ be an axially parallel bounding box that contains the union of the supports of the basis functions corresponding to the indices in $I_{\ell,j}$. Then, the standard *admissibility condition* is given by: $I_{\ell,j} \times I_{\ell,k}$ is admissible if

$$\min\{\operatorname{diam}(B_{\ell,j}), \operatorname{diam}(B_{\ell,k})\} \le \eta \operatorname{dist}(B_{\ell,j}, B_{\ell,k}) \tag{3}$$

for some parameter $\eta > 0$.

Given a hierarchical index set partitioning, a hierarchy of partitionings of the block index set $I \times I$ is obtained in a canonical way as shown in Figure 2.

Let a hierarchical index set partitioning be given. We define a hierarchy of block partitionings by defining $I \times I = I_{0,0} \times I_{0,0}$, and a block $b := I_{\ell,j_1} \times I_{\ell,j_2}$ satisfies exactly one of the following three conditions:

(i) b satisfies an admissibility condition (3),
(ii) $\min\{\#I_{\ell,j_1}, \#I_{\ell,j_2}\} \le n_{\min}$,
(iii) b has (four) successors $I_{\ell+1,k_1} \times I_{\ell+1,k_2}$ where $I_{\ell+1,k_1}$ and $I_{\ell+1,k_2}$ are successors of I_{ℓ,j_1} and I_{ℓ,j_2}, resp..

Fig. 2. Hierarchical block index set partitioning

In terms of the respective matrix blocks, the three cases (i) - (iii) correspond to (i) the approximation of a block that satisfies the admissibility condition (3) by an $R(k)$-matrix, (ii) the representation of small blocks as full matrices, and, (iii) the subdivision of blocks that have successors in the hierarchical block index set partition.

2.3 Modifications for constant convection directions

In Le Borne [2003], modifications to the standard $\mathcal{H}$-matrix have been developed at first for the pure convection case $\epsilon = 0$ and then been generalized for arbitrary $\epsilon > 0$ to an ϵ- and $\mathbf{b}$-dependent partitioning and admissibility condition which produce a gradual transition from the standard partitioning and admissibility condition to their modified counterparts as $\epsilon \to 0$. In order to generate such a gradual transition for a *constant* convection direction $\mathbf{b}$, the (Euclidean) norm that was used for the calculation of the diameter and distance of clusters in the admissibility (3) has been replaced by the norm

$$\|\mathbf{x}\|_{\alpha,\mathbf{b}} := \sqrt{\alpha(\mathbf{b}\cdot\mathbf{x})^2 + (\mathbf{b}^\perp\cdot\mathbf{x})^2} \qquad \text{for} \qquad \mathbf{x} = (x_1, x_2)^T \in \mathbb{R}^2$$

where $\mathbf{b}$ is the convection vector in the convection-diffusion equation, $\mathbf{b}^\perp$ is its orthogonal complement, and $\alpha \in \mathbb{R}^+$ is a parameter that depends on the convection dominance given by ϵ, the mesh width h, and the numerical viscosity induced by the discretization.

In the index partitioning algorithm we will now use bounding boxes that are parallel to the convection $\mathbf{b}$ and its orthogonal complement $\mathbf{b}^\perp$, and the objective is no longer to produce subsets with small diameters but rather to produce subsets *stretched in convection direction.* The modified partition is obtained as follows:

if $\left(\max_{v,w\in B_{\ell,j}} \{\alpha|\mathbf{b}\cdot(v-w)|\} > \max_{v,w\in B_{\ell,j}} |\mathbf{b}^\perp\cdot(v-w)|\right)$ then
  partition cluster $I_{\ell,j}$ along $\mathbf{b}^\perp$ (orthogonal complement of $\mathbf{b}$)
else partition cluster $I_{\ell,j}$ along $\mathbf{b}$ (convection vector);

If we set $\alpha = 1$ and $\mathbf{b} = (1,0)^T$ we obtain the standard partition. Given such a hierarchical index partitioning, we will then construct the hierarchy of block partitioning in the canonical way described in Figure 2.

2.4 Modifications for non-constant convection directions

In the case of a non-constant convection $\mathbf{b}$, we begin our consideration with an example where the convection aligns perfectly with the underlying grid as shown in Figure 3. We will later generalize our strategy to the more realistic case of a convection $\mathbf{b}$ that does not necessarily align with the grid. In the

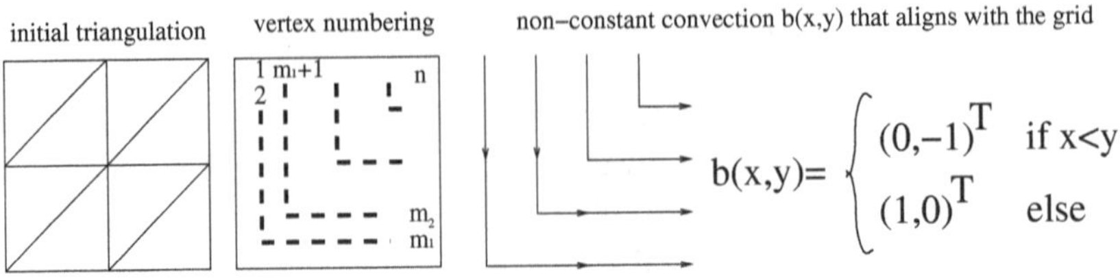

Fig. 3. Non-constant convection that aligns with the grid

given example, we will order the n unknowns with respect to the convection as indicated in Figure 3.

We now construct an $\mathcal{H}$-matrix structure using the index and block index partitioning as described in Figures 1 and 2. Using the weak admissibility condition: $s \times t$ is admissible if $s \neq t$ (i.e., all off-diagonal blocks are admissible), we can represent the exact inverse in the case of $\epsilon = 0$ (pure convection problem) as an $\mathcal{H}$-matrix with local ranks $k(b) = 1$. The storage costs for this $\mathcal{H}$-matrix structure amount to $\mathcal{O}(n \log_2 n)$ as proven in [Hackbusch, 1999, Lemma 3.1]. The fact that we indeed represent the exact inverse results from the particular ordering which guarantees that off-diagonal blocks have at most rank 1.

In the case of a non-zero parameter ϵ or a non-aligning convection direction, the discrete system will contain some artificial diffusion. In the case of a constant convection $\mathbf{b}$, we introduced a parameter α to let the amount of diffusion control the partition and admissibility. In the case of non-constant convection, the general idea to proceed is as follows: For the first p refinement steps, we use a precomputed downwind ordering of the unknowns (along the non-constant convection direction) to partition the index set. Suitable downwind ordering strategies can be found in, e.g, Le Borne [2000]. For any further refinement steps, we use the standard partitioning (trying to obtain subsets with small diameters).

3 $\mathcal{H}$-matrices in iterative methods and numerical results

The $\mathcal{H}$-matrix technique allows to compute a data-sparse approximation $A^{-\mathcal{H}}$ to a (typically fully populated) matrix A^{-1} in nearly optimal complexity. Such an approximation can be used

- in a linear iteration $x_{i+1} = x_i - A^{-\mathcal{H}}(Ax_i - b)$,
- as a preconditioner in a Krylov subspace method (e.g., BiCG-stab, GMRES, etc.), or
- as a smoother in a multigrid iteration,
- for the computation of Schur complements and their inverses, etc.

Here we provide numerical results for the first two applications. The convection-diffusion equation (1) serves as a test problem for various values of ϵ and convection directions $\mathbf{b} = (1,0)^T$ (Table 1) and $\mathbf{b}(x,y) = (0.5-y, x-0.5)^T$

(Table 2). In Tables 1 and 2 we provide the number of iteration steps that are necessary to reduce the Euclidean norm of the residual $\|b - Ax_i\|_2$ to an accuracy of 10^{-8} for $n = 16129$ unknowns (with a maximum number of iterations of 100 and initial iterate $x_0 = 100(1, \cdots, 1)^T$). The time per iteration step is recorded in the second column of Table 1 (computed on a DELL Precision Workstation, 2.4GHz, compare with 0.012s for a classical Gauß-Seidel step).

Table 1. Iteration steps for $\mathbf{b} = (0,1)^T$, modified $\mathcal{H}$-matrix

	time per	basic iteration				bicg-stab			
k(b)/ϵ	step (s)	1	1e-2	1e-4	1e-6	1	1e-2	1e-4	1e-6
1	0.117	100	19	4	2	100	9	2	1
2	0.128	61	7	3	2	11	4	2	1
3	0.143	8	5	3	2	4	3	2	1
4	0.155	6	4	3	2	3	2	2	1
5	0.165	4	3	2	2	2	2	1	1
6	0.183	3	3	2	2	2	2	1	1

As expected, the number of necessary steps decreases considerably as we increase the local rank of the $\mathcal{H}$-matrix. For the numerical tests reported in Table 2 we used the standard $\mathcal{H}$-matrix. For the convection-dominant case $\epsilon = 10^{-6}$, we also provide in parentheses the results for the modified partition. Here, in the first two index partitions the indices have been ordered with respect to their distance to the circle origin $(0.5, 0.5)$. All further partitions were performed in the standard way. We observe slight improvements.

Table 2. Iteration steps for $\mathbf{b}$ =circle, standard $\mathcal{H}$-matrix

	basic iteration				bicg-stab			
k(b)/ϵ	1	1e-2	1e-4	1e-6	1	1e-2	1e-4	1e-6
1	100	100	100	100 (100)	100	100	63	69 (90)
2	57	26	100	100 (100)	11	9	26	40 (29)
3	8	6	39	55 (72)	4	3	11	13 (11)
4	6	4	20	34 (23)	3	2	7	9 (6)
5	4	3	9	14 (10)	2	2	4	5 (4)
6	3	3	8	11 (8)	2	2	4	4 (3)

The $\mathcal{H}$-matrix approximations $A^{-\mathcal{H}}$ have been computed using a block Gauß elimination process and are therefore not necessarily the best possible approximations. When evaluating a preconditioner, the costs for the construction of the preconditioner have to be taken into account. In this case, the construction of the $\mathcal{H}$-matrix $A^{-\mathcal{H}}$ is of nearly optimal complexity $\mathcal{O}(n \log_2^2 n)$, however, with a relatively high constant, see Grasedyck and Hackbusch [2002] (taking 73s ($k = 1$) up to 166s ($k = 6$) for $\epsilon = 1.0$ and 40s ($k = 1$) up to 76s ($k = 6$) for $\epsilon = 1e - 6$). $A^{-\mathcal{H}}$ can, however, be computed more efficiently via a (parallelizable) domain decomposition algorithm, see Hackbusch [2002].

These positive results encourage the further study of $\mathcal{H}$-matrix preconditioners in harder problems involving non-constant, cyclic convection directions in systems of PDEs in three spatial dimensions where there is still a need for efficient iteration methods.

References

M. Bebendorf and W. Hackbusch. Existence of $\mathcal{H}$-matrix Approximants to the inverse FE-Matrix of Elliptic Operators with L^∞-Coefficients. *Numerische Mathematik*, 95:1–28, 2003. available electronically at http://www.mis.mpg.de/preprints/2002/prepr2002_21.html.

M. Benzi and M. Tuma. A sparse approximate inverse preconditioner for nonsymmetric linear systems. *SIAM J. Sci. Comput.*, 19:968–994, 1998.

E. Chow and Y. Saad. Approximate inverse preconditioners via sparse-sparse iterations. *SIAM J. Sci. Comp.*, 19:995–1023, 1998.

C. Farhat, M. Lesoinne, P. L. Tallec, K. Pierson, and D. Rixen. FETI-DP: A dual-primal unified FETI method - part I: A faster alternative to the two-level FETI method. *J. Numer. Methods. Eng.*, 50:1523–1544, 2001.

L. Grasedyck and W. Hackbusch. Construction and arithmetics of $\mathcal{H}$-matrices. *Computing*, 70:295–334, 2002.

M. Grote and T. Huckle. Parallel preconditioning with sparse approximate inverses. *SIAM J. Sci. Comput.*, 18:838–853, 1997.

W. Hackbusch. A sparse matrix arithmetic based on $\mathcal{H}$-matrices. Part I: Introduction to $\mathcal{H}$-matrices. *Computing*, 62:89–108, 1999.

W. Hackbusch. Direct domain decomposition using the hierarchical matrix technique. In *Proceedings of the 14th International Conference on Domain Decomposition Methods in Cocoyoc, Mexico*, pages 39–50. ddm.org, 2002.

W. Hackbusch, L. Grasedyck, and S. Börm. Introduction to hierarchical matrices with applications. *Engineering Analysis with Boundary Elements*, 27: 405–422, 2003.

W. Hackbusch and B. Khoromskij. A sparse $\mathcal{H}$-matrix arithmetic: General complexity estimates. *J. Comp. Appl. Math.*, 125:479–501, 2000a.

W. Hackbusch and B. Khoromskij. A sparse $\mathcal{H}$-matrix arithmetic. Part II: Application to multi-dimensional problems. *Computing*, 64:21–47, 2000b.

S. Le Borne. Ordering techniques for two- and three-dimensional convection-dominated elliptic boundary value problems. *Computing*, 64:123–155, 2000.

S. Le Borne. $\mathcal{H}$-matrices for convection-diffusion problems with constant convection. *Computing*, 70:261–274, 2003.

E. E. Tyrtyshnikov. Incomplete cross approximation in the mosaic-skeleton method. *Computing*, 64:367–380, 2000.

Parallel Performance of Some Two-Level ASPIN Algorithms

Leszek Marcinkowski[1]* and Xiao-Chuan Cai[2]**

[1] Department of Mathematics, Informatics and Mechanics, Warsaw University, Banacha 2, 02-096 Warszawa, Poland
(lmarcin@mimuw.edu.pl, http://www.mimuw.edu.pl/~lmarcin/)

[2] Department of Computer Science, University of Colorado at Boulder, Boulder, CO 80309-0430, USA
(cai@cs.colorado.edu, http://www.cs.colorado.edu/~cai/)

Summary. In this paper we study the parallel performance of some nonlinear additive Schwarz preconditioned inexact Newton methods for solving large sparse system of nonlinear equations arising from the discretization of partial differential equations. The main idea of nonlinear preconditioning is to replace an ill-conditioned nonlinear system by an equivalent nonlinear system that has more balanced nonlinearities. In addition to balance the nonlinearities through nonlinear preconditioning, we also need to make sure that the multilayered iterative solver is scalable with respect to the number of processors. We focus on some two-level nonlinear additive Schwarz preconditioners, and show numerically that these two-level methods can reduce the nonlinearities and at the same time maintain the parallel scalability. Parallel numerical results for some high Reynolds number incompressible Navier-Stokes equations will be presented.

1 Introduction

We study Newton type algorithms for solving a nonlinear system of equations

$$F(u_*) = 0, \tag{1}$$

starting from an initial guess $u^{(0)} \in \Re^n$. Here $F = (F_1, \ldots, F_n)^T$, $F_i = F_i(u_1, \ldots, u_n)$ are given functions which are often the result of the discretization of some nonlinear partial differential equations, such as the incompressible Navier-Stokes equations for fluid flows, using finite element or finite difference methods. For such nonlinear systems, some parallel nonlinear additive

* This work was partially supported by Polish Scientific Grant 2/P03A/005/24.
** The work was partially supported by the Department of Energy, FC02-01ER-25479, and by the National Science Foundation, CCR-0219190, ACI-0072089 and ACI-0305666.

Schwarz preconditioned inexact Newton methods (ASPIN) were recently proposed in Cai and Keyes [2002], and Cai et al. [2002]. ASPIN has been applied successfully to some rather difficult problems such as the transonic full potential flows (Cai et al. [2000]), and the high Reynolds number incompressible Navier-Stokes flows (Hwang and Cai [2003b]) and (Hwang and Cai [2003a]). In this paper we compare the parallel performance of a one-level method (Cai and Keyes [2002]), a two-level method (Cai et al. [2002]), and a slightly modified two-level method to be presented in this paper. In the modified two-level method, the initial guess $u^{(0)}$ is replaced by a fine grid interpolation of the coarse grid solution. It turns out in some situations that the small modification has some major impact on the overall performance of the algorithm. The focus of this paper is on the linear and nonlinear scalability issues of the methods, and our discussions will be based on the numerical results for solving some high Reynolds number incompressible Navier-Stokes equations on distributed memory parallel computers with modest number of processors.

2 Algorithm description

In the rest of the paper we shall refer to the nonlinear algebraic system (1) as the *fine grid system*, or simply the fine system, which has n unknowns and n equations. In order to introduce the two-level algorithm, we assume that there is a "coarse" version of (1) in the following form

$$F^c(u^c_*) = 0, \tag{2}$$

which is a nonlinear algebraic system with n^c unknowns and n^c equations. Usually $n^c << n$. Such a coarse system can be obtained by the discretization of the same differential equations on a coarser grid. The coarse and fine functions $F(u)$ and $F^c(u^c)$ approximate each other in certain sense.

Inexact Newton algorithms (IN) (Eisenstat and Walker [1994]) are commonly used for solving such systems. In this paper, we work in the framework of nonlinearly preconditioned inexact Newton algorithms (PIN) recently introduced in (Cai and Keyes [2002]). In other words, we try to find the solution u_* of equation (1) by solving an equivalent system of nonlinear equations

$$\mathcal{F}(u_*) = 0. \tag{3}$$

(1) and (3) are equivalent in the sense that they have the same solution. Other than having the same solution, the nonlinear functions $F(\ )$ and $\mathcal{F}(\ )$ may have completely different forms. We will define the function $\mathcal{F}$ using the restriction of F on subspaces, and the coarse function F^c in the case of multilevel methods.

2.1 A one-level method

We first introduce the subspaces by an overlapping partition of $S = (1, \ldots, n)$, which is an index set for the system (1); i.e. one integer for each unknown

u_i and F_i. We assume that $S_1, \ldots, S_N$ is a partition of S in the sense that $\bigcup_{i=1}^N S_i = S$, and $S_i \subset S$. Here we allow the subsets to have overlap. Let n_i be the dimension of S_i; then, in general, $\sum_{i=1}^N n_i \geq n$. Using the partition of S, we introduce subspaces of $\Re^n$ and the corresponding restriction and extension matrices. For each S_i we define $V_i \subset \Re^n$ as

$$V_i = \{v | v = (v_1, \ldots, v_n)^T \in \Re^n, v_k = 0, \text{ if } k \notin S_i\}$$

and a $n \times n$ restriction (also extension) matrix I_{S_i} whose kth column is either the kth column of the $n \times n$ identity matrix $I_{n \times n}$ if $k \in S_i$ or zero if $k \notin S_i$. Using the restriction operator, we define the subdomain nonlinear function as $F_{S_i} = I_{S_i} F$. We next define the major component of the algorithm, namely the nonlinearly preconditioned function. For any given $v \in \Re^n$, define $T_i(v) \in V_i$ as the solution of the following subspace nonlinear system

$$F_{S_i}(v - T_i(v)) = 0,$$

for $i = 1, \ldots, N$. Taking the sum of the all T_is, we have a new function

$$\mathcal{F}^{(1)}(u) = \sum_{i=1}^{N} T_i(u), \tag{4}$$

The operators T_i and $\mathcal{F}^{(1)}$ were introduced by Dryja and Hackbusch [1997] in which a version of a nonlinear Richardson method was applied to solve the nonlinear system corresponding to (4).

Algorithm 1 (ASPIN(1)) *Obtain an approximate solution of u_* by solving*

$$\mathcal{F}^{(1)}(u_*) = 0$$

using the inexact Newton method with $u^{(0)}$ as the initial guess (Cai and Keyes [2002]).

It is worth to note that under some assumptions it was proven by Dryja and Hackbusch [1997] and Cai and Keyes [2002] that the local problems have unique solutions, thus T_i are well defined. It is also shown there that the Jacobian of the preconditioned system is well defined.

To apply an inexact Newton method to (4) we have to know how to compute the Jacobian of $\mathcal{F}^{(1)}$. It is shown in (Cai and Keyes [2002]) that one can obtain the Jacobian of $\mathcal{F}^{(1)}$, denoted by $\mathcal{J}^{(1)}$, by the following formula:

$$\mathcal{J}^{(1)}(u) = \sum_{i=1}^{N} J_{S_i}^{-1}(u - T_i(u)) \cdot J(u),$$

where $J(u) = DF(u)$ is the Jacobian of the original function F and $J_{S_i}(u) = I_{S_i} J(u) I_{S_i}$. In practice since $T_i(u)$ converges to zero, we can assume that a good approximation of the Jacobian is given by $\mathcal{J}^{(1)}(u) \approx \sum_{i=1}^N J_{S_i}^{-1}(u) J(u)$, which is, as a matter of fact, the original Jacobian matrix preconditioned by a one-level additive Schwarz method, thus it should be well-conditioned as long as the number of subdomains is not very large.

2.2 Two two-level methods

Similarly, let $S^c = (1, \ldots, n^c)$ be an index set for the coarse system, and we assume that $\{S_1^c, \ldots, S_N^c\}$ is a partition. For simplicity, we partition the fine and the coarse systems into the same number of subsets. Also for simplicity, in our parallel implementation, we allocate the subsystems corresponding to the index sets S_i and S_i^c to the same processor. We introduce the subdomain fine to coarse restriction operator $R_i : S_i \longrightarrow S_i^c$, in the sense that for each vector $v_i \in V_i$, there is a unique vector $v_i^c \in V_i^c$, such that

$$v_i^c = R_i v_i.$$

Assuming the R_is are consistent in the overlapping part of the subdomains, we can define a global fine to coarse restriction operator $R^c : \Re^n \longrightarrow \Re^{n^c}$ as follows: For any $v \in \Re^n$, the k component of $R^c v$ is defined as

$$(R^c v)_k = (R_i v)_k, \text{ if } k \in S_i^c.$$

A global coarse to fine extension operator E^c can be defined as the transpose of R^c. To define the coarse function $T_0 : \Re^n \longrightarrow \Re^n$, we first introduce a projection $T^c : \Re^n \longrightarrow \Re^{n^c}$ as follows: For any given $v \in \Re^n$, $T^c v$ satisfies the coarse nonlinear system

$$F^c(T^c(v)) = R^c F(v). \tag{5}$$

We assume that (5) has a unique solution. Then we define an operator $T_0 : \Re^n \longrightarrow \Re^n$ by

$$T_0(v) = E^c T^c(v). \tag{6}$$

Suppose that T_0 is given as in (6); it is easy to see that $T_0(u_*)$ can be computed without knowing the exact solution u_* itself. In fact, from (5), we have $T_0(u_*) = E^c u_*^c$. Throughout this paper, we assume that the coarse solution u_*^c is given, through a pre-processing step. We can now introduce a new nonlinear function $\Re^n \longrightarrow \Re^n$ by

$$\mathcal{F}^{(2)}(u) = T_0(u) - T_0(u_*) + \sum_{i=1}^{N} T_i(u). \tag{7}$$

Algorithm 2 (ASPIN(2)) *Obtain an approximate solution of u_* by solving*

$$\mathcal{F}^{(2)}(u_*) = 0$$

using the inexact Newton method with $u^{(0)}$ as the initial guess (Cai et al. [2002]).

In this paper, we propose a slight modification of the above algorithm in the selection of the initial guess. The algorithm takes the following form.

Algorithm 3 (ASPIN(2')) *Obtain an approximate solution of u_* by solving*

$$\mathcal{F}^{(2)}(u_*) = 0$$

using the inexact Newton method with $T_0(u_*)$ as the initial guess.

No additional cost is needed to switch from the original initial guess $u^{(0)}$ to $T_0(u_*)$ since the vector $T_0(u_*)$ is needed anyway in the nonlinear function evaluation.

3 Numerical studies

We next present some numerical results on a two-dimensional lid driven cavity flow problem (Hirsch [1990]). Consider the velocity-vorticity formulation of the incompressible Navier-Stokes equations on the unit square $\Omega = (0,1) \times (0,1)$:

$$\begin{cases} -\Delta u - \dfrac{\partial \omega}{\partial y} & = 0 \\ -\Delta v + \dfrac{\partial \omega}{\partial x} & = 0 \\ -\dfrac{1}{Re}\Delta \omega + u\dfrac{\partial \omega}{\partial x} + v\dfrac{\partial \omega}{\partial y} & = 0, \end{cases} \tag{8}$$

where Re is the Reynolds number, $(u,\, v)$ is the velocity and ω is the vorticity. The boundary conditions are: $u = v = 0$ for bottom, left and right, and $u = 1$, $v = 0$ for top. The boundary condition on ω is given by its definition: $\omega(x,y) = -\frac{\partial u}{\partial y} + \frac{\partial v}{\partial x}$. The usual uniform mesh 5-point finite difference approximation is used to discretize the boundary value problem. Upwinding is used for the first derivative terms and central differencing for the second derivative terms. To obtain a nonlinear algebraic system of equations F, we use natural ordering inside each subdomain, and at each mesh point we arrange the unknowns in the order of u, v, and ω. The partitioning of F is through the partitioning of the mesh points in a checkerboard fashion for both the fine and the coarse grids. The coarse to fine interpolation is defined using the coarse grid bilinear finite element basis functions. The implementation is done using PETSc (Balay et al. [2002]), and the results are obtained on an IBM SP supercomputer. Double precision is used throughout the computations. The initial guess $u^{(0)}$ is zero for u, v and ω in ASPIN(1) and ASPIN(2). We stop the global PIN iterations if $\|\mathcal{F}(u^{(k)})\| \le 10^{-10}\|\mathcal{F}(u^{(0)})\|$. The same stopping condition is used for the coarse grid nonlinear systems, which are solved by a Newton-Krylov-Schwarz method based on the same mesh partition. The global Jacobian systems are solved with GMRES restarting at 30. The global linear iteration for solving the global Jacobian system is stopped if the relative tolerance $\|\mathcal{F}(u^{(k)}) - \mathcal{F}'(u^{(k)})p^{(k)}\| \le 10^{-3}\|\mathcal{F}(u^{(k)})\|$ is satisfied. At the kth global nonlinear iteration, nonlinear subsystems

Table 1. Varying Reynolds numbers. Fine mesh size 128 × 128, coarse mesh size 32 × 32, number of processors 16

	Reynolds number	global nonlinear iterations	global linear iterations	average linear iteration per nonlinear step
ASPIN(1)	10^1	3	112	37
	10^2	4	162	40
	10^3	7	216	30
	10^4	6	156	26
ASPIN(2)	10^1	4	38	9
	10^2	6	89	14
	10^3	7	99	14
	10^4	22	9517	432
ASPIN(2')	10^1	3	28	9
	10^2	4	51	12
	10^3	4	48	12
	10^4	3	40	13

$$F_{S_i}\left(u^{(k)} - g_i^{(k)}\right) = 0,$$

have to be solved. We use the standard IN with a cubic line search for such systems with initial guess $g_{i,0}^{(k)} = 0$. The local nonlinear iteration in subdomain S_i is stopped if the following condition is satisfied: $\|F_{S_i}(g_{i,l}^{(k)})\| \leq 10^{-3}\|F_{S_i}(g_{i,0}^{(k)})\|$.

Table 2. Varying the overlapping size. Reynolds number $= 10^3$. Fine mesh size 128 × 128, coarse mesh size 32 × 32, number of processors 16

	overlap	global nonlinear iterations	global linear iterations	average linear iteration per nonlinear step
ASPIN(1)	1	7	216	30
	2	6	141	23
	4	6	112	18
ASPIN(2)	1	8	167	20
	2	8	122	15
	4	7	100	14
ASPIN(2')	1	5	62	12
	2	4	46	11
	4	4	45	11

We first compare the three ASPIN algorithms for different Reynolds numbers. In Tables 1, we report the total number of global nonlinear iterations, the total number of linear iterations, and the average number of linear iterations per nonlinear iteration. For this particular test problem, the nonlinearity is determined mostly by the Reynolds number. As *Re* increases the nonlinear system becomes more difficult to solve with the regular inexact Newton method (Cai and Keyes [2002]). However, as shown in Table 1, the numbers of linear and nonlinear iterations of ASPIN(2') are not very sensitive to the increase of *Re*.

In Table 2, we test the algorithms with different level of overlaps in the Schwarz preconditioner. It is quite interesting to see that ASPIN(2') is not sensitive to this parameter, which is a bit surprising.

To use the two-level algorithms on large number of processors and for large fine meshes, the coarse grid size has to be sufficiently fine. This leads to some difficult coarse grid nonlinear systems to solve. Although the coarse problems are, in general, easier to solve than the fine grid problem but sometimes NKS may not be good enough to converge the coarse nonlinear iterations. In the next set of experiments we use an ASPIN(1) coarse solver. That is instead of solving problem (2) by NKS we solve it using ASPIN(1). The stopping criteria for the coarse solver and the fine solver are the same.

In Table 3 we present some experiments for ASPIN(2') on an 1024×1024 mesh, Reynolds number 10^4, and the coarse mesh is 64×64.

Table 3. ASPIN(2'). Varying the number of processors. Fine mesh size 1024×1024, coarse mesh size 64×64, Reynolds number $= 10^4$

processors #	nonlinear iterations	average linear iter. per nonlin. step	total CPU time (sec)
32	7	34	1377
64	8	32	653
128	8	39	418
256	10	44	374

The results show that both the number of linear and nonlinear iterations are nearly independent of the number of processors, which is the same as the number of subdomains. In terms of the CPU time, the algorithm scales well for up to 128 processors. The CPU time for 256 processors is only slightly smaller than for 128 processors. We suspect that for large number of proces-

Table 4. Performance of the ASPIN(1) based coarse solver. Fine grid 1024×1024, Reynolds number $= 10^4$

processors #	coarse grid	total CPU time (sec)	coarse CPU time (sec)	percentage
32	64×64	1377	20	1.4 %
64	64×64	653	19	2.9%
128	64×64	418	16	3.8%
256	64×64	374	35	9.3%
256	128×128	361	155	42.9%

sors the ASPIN(1) coarse solver becomes less effective. Thus we measure the computational time the coarse solver takes, and the results are summarized in Table 4 in which we also report the percentage of time spent on the coarse solver. It seems that the ASPIN(1) based coarse solver takes much more computing time for large number of processors. Our current approach works fine for modest number of processors, but for larger number of processors a more efficient parallel coarse solver is definitely needed.

References

S. Balay, K. Buschelman, W. D. Gropp, D. Kaushik, M. Knepley, L. C. McInnes, B. F. Smith, and H. Zhang. PETSc users manual. Technical Report ANL-95/11 - Revision 2.1.5, Argonne National Laboratory, 2002.

X.-C. Cai and D. E. Keyes. Nonlinearly preconditioned inexact Newton algorithms. *SIAM J. Sci. Comput.*, 24(1):183–200, 2002.

X.-C. Cai, D. E. Keyes, and L. Marcinkowski. Non-linear additive Schwarz preconditioners and application in computational fluid dynamics. *Internat. J. Numer. Methods Fluids*, 40(12):1463–1470, 2002.

X.-C. Cai, D. E. Keyes, and D. P. Young. A nonlinear additive Schwarz preconditioned inexact Newton method for shocked duct flow. In N. Debit, M. Garbey, R. Hoppe, J. Periaux, D. Keyes, and Y. Kuznetsov, editors, *13th Int. Conference on Domain Decomposition Methods, (Lyon, France, October 9-12 2000)*, pages 343–350. DDM.org, Augsburg, 2000.

M. Dryja and W. Hackbusch. On the nonlinear domain decomposition method. *BIT*, 37(2):296–311, 1997.

S. C. Eisenstat and H. F. Walker. Globally convergent inexact Newton methods. *SIAM J. Optim.*, 4(2):393–422, 1994.

C. Hirsch. *Numerical computation of internal and external Flows: computational methods for inviscid and viscous flows.* John Wiley & Sons, New York, 1990.

F.-N. Hwang and X.-C. Cai. Improving robustness and parallel scalability of Newton's method through nonlinear preconditioning. These proceedings, 2003a.

F.-N. Hwang and X.-C. Cai. A parallel nonlinear additive Schwarz preconditioned inexact Newton algorithm for incompressible Navier-Stokes equations. Preprint, Department of Computer Science, University of Colorado at Boulder, 2003b.

Algebraic Analysis of Schwarz Methods for Singular Systems

Ivo Marek and Daniel B. Szyld

Czech Institute of Technology, School of Civil Engineering

Temple University, Department of Mathematics
(http://www.math.temple.edu/~szyld/)

Summary. During the last few years, an algebraic formulation of Schwarz methods was developed. In this paper this algebraic formulation is used to prove new convergence results for multiplicative Schwarz methods when applied to consistent singular systems of linear equations. Coarse grid corrections are also studied. In particular, these results are applied to the numerical solutions of Markov chains.

1 Introduction

We consider the solution of consistent large sparse linear singular systems of the form

$$Ax = b. \tag{1}$$

We study its solution by means of Schwarz methods. Specifically, we analyze the case where the coefficient matrix $A = I - B$, where I is the identity matrix and B is a nonnegative (column) stochastic matrix, i.e., $B^T e = e$, where $e = (1, 1, \dots, 1)^T \in \mathbb{R}^n$. Thus A is a singular M-matrix; see Section 2 for definitions. In particular we consider the case of $b = 0$, and thus we look for the nonnegative vector v, normalized so that $v^T e = 1$, satisfying $Av = 0$, i.e., such that $Bv = v$. This is the stationary probability distribution of the Markov chain represented by B.

In our analysis we use the algebraic formulation of Schwarz methods developed in Benzi et al. [2001], Frommer and Szyld [1999], and applied, e.g., in Frommer and Szyld [2001], Nabben [2003], Nabben and Szyld [2003].

There is no separate treatment in the literature of Schwarz methods for singular systems in the p.d.e. context. Nevertheless the implementations derived mostly for the non-singular case can be shown to work in the singular case as well, especially when the null space is known. This is the case, for example, when Neumann boundary conditions are present. The convergence theory developed, e.g., in Dryja et al. [1994], Dryja and Widlund [1994], can be applied to these cases with little or no changes.

We believe that this is the first time that singular systems are analyzed using an algebraic approach to Schwarz methods (with overlap), and that Markov chains problems are studied in this context. One of our goals is to present Schwarz iterations as one more possible tool for the numerical solutions of Markov chains. In fact, multiplicative Schwarz iterations reduce to the block Gauss-Seidel method when the overlap is removed. Having the overlap has proved crucial for the fast convergence of these methods in the nonsingular case; see, e.g., Smith et al. [1996], Dryja and Widlund [1994]. In the singular context, having larger overlap may decrease the convergence rate of the iteration. Comparison theorems may be used to prove such decrease in convergence rate; see Marek and Szyld [2000], Marek and Szyld [2002].

We discuss here one approach, namely that multiplicative Schwarz iterations applied directly to the $n \times n$ system (1) converge. Other approaches are discussed in Marek and Szyld [2004], where "coarse-grid" corrections are also considered.

2 Definitions and auxiliary results

In this section we present some notation, definitions, and preliminaries. Concepts on nonnegative matrices not explicitly defined here can be found in the book by Berman and Plemmons [1979].

An $n \times n$ matrix $C = (c_{jk})$ with $c_{jk} \in \mathbb{R}$, is called nonnegative if $c_{jk} \geq 0$, $j,k = 1,\ldots,n$; this is denoted $C \geq O$. When $c_{jk} > 0$, $j,k = 1,\ldots,n$, we say that the matrix is positive and denote it by $C > O$. The same notation is used for nonnegative and positive vectors. By $\sigma(C)$ we denote the spectrum of C and by $\rho(C)$ its spectral radius. By $\mathcal{R}(C)$ and $\mathcal{N}(C)$ we denote the range and null space of C, respectively.

Let $\lambda \in \sigma(C)$ be a pole of the resolvent operator $R(\mu, C) = (\mu I - C)^{-1}$. The multiplicity of λ as a pole of $R(\mu, C)$ is called the index of C with respect to λ and denoted $ind_\lambda C$. Equivalently, $k = ind_\lambda C$ if it is the smallest integer for which $\mathcal{R}((\lambda I - C)^{k+1}) = \mathcal{R}((\lambda I - C)^k)$. This happens if and only if $\mathcal{R}((\lambda I - C)^k) \oplus \mathcal{N}((\lambda I - C)^k) = \mathbb{R}^n$.

Let A be an $n \times n$ matrix. A is an M-matrix if $A = \beta I - B$, B nonnegative and $\rho(B) \leq \beta$. A pair of matrices (M, N) is called a splitting of A if $A = M - N$ and M^{-1} exists. A splitting of a matrix A is called *of nonnegative type* if the matrix $T = M^{-1}N$ is nonnegative (Marek [1970]). If the matrices M^{-1} and N are nonnegative, the splitting is called *regular* (Varga [1962]).

Let T be a square matrix. T is called *convergent* if $\lim\limits_{k\to\infty} T^k$ exists, and *zero-convergent* if $\lim\limits_{k\to\infty} T^k = O$. Standard stationary iterations of the form

$$x^{k+1} = Tx^k + c, \qquad k = 0, 1, \ldots, \tag{2}$$

converge if and only if either $\lim\limits_{k\to\infty} T^k = O$. or, if $\rho(T) = 1$, T is convergent. A square matrix T with unit spectral radius is convergent if the following two

conditions hold:
(i) if $\lambda \in \sigma(T)$ and $\lambda \neq 1$, then $|\lambda| < 1$.
(ii) $ind_1 T = 1$.
When $T \geq O$, (i) can be replaced with T having positive diagonal entries (Alefeld and Schneider [1982]); see Szyld [1994] for equivalent conditions for (ii).

We state a very useful Lemma; see e.g., Bohl and Marek [1995] for a proof. We note that when $\rho(T) = 1$, this lemma can be used to show condition (ii) above. To prove convergence one needs to show in addition that condition (i) also holds, or equivalently, that the diagonal entries are all positive.

Lemma 1. *Let T be a nonnegative square matrix such that $Tv \leq \alpha v$ with $v > 0$. Then $\rho(T) \leq \alpha$. If furthermore $\rho(T) = \alpha$, then $ind_\alpha T = 1$.*

A square nonnegative matrix B is irreducible if for every pair of indices i, j there is a power $k = k(i,j)$ such that the ij entry of B^k is nonzero.

3 Algebraic formulation of Schwarz methods

Given an initial approximation x^0 to the solution of (1), the (one-level) multiplicative Schwarz method can be written as the stationary iteration (2), where

$$T = (I - P_p)(I - P_{p-1}) \cdots (I - P_1) = \prod_{i=p}^{1} (I - P_i) \tag{3}$$

and c is a certain vector. Here

$$P_i = R_i^T A_i^{-1} R_i A, \tag{4}$$

where $A_i = R_i A R_i^T$, R_i is a matrix of dimension $n_i \times n$ with full row rank, $1 \leq i \leq p$; see, e.g., Smith et al. [1996]. In the case of overlap we have $\sum_{i=1}^{p} n_i > n$. Note that each P_i, and hence each $I - P_i$, is a projection operator; i.e., $(I - P_i)^2 = I - P_i$. Each $I - P_i$ is singular and $\rho(I - P_i) = 1$.

We refer the reader to the papers Benzi et al. [2001], Frommer and Szyld [1999] for details of the algebraic formulation of Schwarz methods. What we will say here is that nonsingular matrices M_i, $i = 1, \ldots, p$, are defined so that $A = M_i - N_i$ are regular splittings (and thus of nonnegative type), and furthermore the following equality holds:

$$E_i M_i^{-1} = R_i^T A_i^{-1} R_i, \; i = 1, \ldots, p, \tag{5}$$

where $E_i = R_i^T R_i$. These diagonal matrices E_i have ones on the diagonal in every row where R_i^T has nonzeros. We can thus rewrite (3) as

$$T = (I - E_p M_p^{-1} A)(I - E_{p-1} M_{p-1}^{-1} A) \cdots (I - E_1 M_1^{-1} A). \tag{6}$$

In the context of discretizations of p.d.e.s, the use of Schwarz methods greatly benefit from the use of coarse grid corrections, and they are needed to guarantee a convergence rate independent of the mesh size; see, e.g., Dryja et al. [1994], Dryja and Widlund [1994], Quarteroni and Valli [1999], Smith et al. [1996]. Coarse grid corrections can be additive or multiplicative. Here we restrict our comments to the multiplicative corrections. To that end consider a new projection P_0 of the form (4) onto the "coarse space", i.e., onto a particular subset of states, usually taken in the overlap between the other set of states. Corresponding to these "coarse" states, there correspond a natural matrix R_0 and $A_0 = R_0AR_0^T$, so that $E_0 = R_0^TR_0$ and M_0 is similarly defined; see Benzi et al. [2001]. The multiplicative corrected multiplicative Schwarz iteration operator is then

$$T_{\mu c} = (I - P_0)T_\mu = (I - E_0M_0^{-1}A)T. \tag{7}$$

In Benzi et al. [2001] it was shown that when A is nonsingular, $\rho(T) < 1$, and thus, the method (2) is convergent. The same results hold for $T_{\mu c}$, i.e., with a "coarse grid" correction. In this paper we explore the convergence of (2), using the iterations defined by (6) and (7), when A is singular. Other Schwarz methods for the singular case are studied in Marek and Szyld [2004].

4 Convergence of multiplicative Schwarz

We prove here our main result, namely that when A is irreducible, the multiplicative Schwarz iterations are convergent. As is well known, when $B \geq O$ is irreducible, its Perron eigenvector is strictly positive $v > 0$. If in addition we require that the diagonals of the iteration matrices are positive, we show in the next theorem that the matrix (6) is convergent.

Theorem 1. *Let $A = I - B$, where B is an $n \times n$ column stochastic matrix such that $Bv = v$ with $v > 0$. Let $p \geq 1$ be a positive integer and $A = M_i - N_i$ be splittings of nonnegative type such that the diagonals of $T_i = M_i^{-1}N_i$, $i = 1, \ldots, p$, are positive. Then (6) is a convergent matrix.*

Proof. Let $v > 0$ be such that $Bv = v$, i.e., $Av = 0$. For each splitting $A = M_i - N_i$, we then have th at $M_iv = N_iv$. This implies that $Tv = v$, and by Lemma 1 we have that $\rho(T) = 1$ and that the index is 1. To show that T is convergent, we show that its diagonal is positive. Each factor in (6) is then

$$I - E_i + E_i(I - M_i^{-1}A) = I - E_i + E_iM_i^{-1}N_i,$$

and since $O \leq E_i \leq I$ and $M_i^{-1}N_i \geq O$, each factor is nonnegative. For a row in which E_i is zero, the diagonal entry in this factor has value one. For a row in which E_i has value one, the diagonal entry in this factor is the positive diagonal entry of $M_i^{-1}N_i$. Thus, we have a product of nonnegative matrices, each having positive diagonals, implying that the product T has positive diagonal entries, and therefore it is convergent. □

Corollary 1. *Theorem 1 applies verbatim to the case of "coarse grid" correction, by considering the additional splitting* $A = M_0 - N_0$*, with* $T_0 = M_0^{-1} N_0$ *having positive diagonals, so that* $T_{\mu c}$ *of (7) is convergent.*

Let $\gamma = \max\{|\lambda|, \lambda \in \sigma(T), \lambda \neq 1\}$. The fact that T is convergent implies that $\gamma < 1$; see, e.g., Berman and Plemmons [1979]. Therefore Theorem 1 and Corollary 1 indicate that for multiplicative Schwarz, $\sigma(M^{-1}A) = \sigma(I - T)$ has zero as an isolated eigenvalue with index 1, and the rest of the spectrum is contained in a ball with center 1 and radius γ. Furthermore, the smaller γ is, the smaller this ball around 1 is. This configuration of the spectrum often gives good convergence properties to Krylov subspace methods preconditioned with multiplicative Schwarz.

5 The reducible case

We consider here the general case, where B might not be irreducible. There is a permutation matrix H such that the symmetric permutation of B is lower block-triangular [Gantmacher, 1959, p.341], and in fact it has the form

$$HBH^T = \begin{bmatrix} G_0 & O & \cdots & O \\ G_1 & C_1 & \cdots & O \\ \vdots & \vdots & \ddots & \vdots \\ G_p & O & \cdots & C_p \end{bmatrix}, \tag{8}$$

where $\lim_{k \to \infty} G_0^k = O$ and C_i is an irreducible and stochastic matrix, $i = 1, \ldots, p$. There are efficient algorithms to compute the permutation matrix H, and thus, the form (8). For example, Tarjan's algorithm has almost linear complexity and good software is available for it; see Duff and Reid [1978].

Solving linear systems with the matrix B reduces then to solving systems with each of the diagonal blocks of (8). This can be accomplished using multiplicative Schwarz iterations, which were shown to converge for irreducible stochastic matrices in Section 4.

References

G. Alefeld and N. Schneider. On square roots of M-matrices. *Linear Algebra and its Applications*, 42:119–132, 1982.

M. Benzi, A. Frommer, R. Nabben, and D. B. Szyld. Algebraic theory of multiplicative Schwarz methods. *Numerische Mathematik*, 89:605–639, 2001.

A. Berman and R. J. Plemmons. *Nonnegative Matrices in the Mathematical Sciences*. Academic Press, New York, third edition, 1979. Reprinted by SIAM, Philadelphia, 1994.

E. Bohl and I. Marek. A model of amplification. *Journal of Computational and Applied Mathematics*, 63:27–47, 1995.

M. Dryja, B. F. Smith, and O. B. Widlund. Schwarz analysis of iterative substructuring algorithms for elliptic problems in three dimensions. *SIAM J. Numer. Anal.*, 31(6):1662–1694, December 1994.

M. Dryja and O. B. Widlund. Domain decomposition algorithms with small overlap. *SIAM J. Sci.Comput.*, 15(3):604–620, May 1994.

I. S. Duff and J. K. Reid. An implementation of tarjan's algorithm for the block triangularization of a matrix. *ACM Transactions on Mathematical Software*, 4:337–147, 1978.

A. Frommer and D. B. Szyld. Weighted max norms, splittings, and overlapping additive Schwarz iterations. *Numerische Mathematik*, 83:259–278, 1999.

A. Frommer and D. B. Szyld. An algebraic convergence theory for restricted additive Schwarz methods using weighted max norms. *SIAM Journal on Numerical Analysis*, 39:463–479, 2001.

F. R. Gantmacher. *Application of the Theory of Matrices.* Interscience, New York, 1959.

I. Marek. Frobenius theory of positive operators. comparison theorems and applications. *SIAM Journal on Applied Mathematics*, 19:608–628, 1970.

I. Marek and D. B. Szyld. Comparison of convergence of general stationary iterative methods for singular matrices. *Linear Algebra and its Applications*, 316:67–87, 2000.

I. Marek and D. B. Szyld. Comparison of convergence of general stationary iterative methods for singular matrices. *SIAM Journal on Matrix Analysis and Applications*, 24:68–77, 2002.

I. Marek and D. B. Szyld. Algebraic Schwarz methods for the numerical solution of Markov chains. *Linear Algebra and its Applications*, 2004. To appear.

R. Nabben. Comparisons between additive and multiplicative Schwarz iterations in domain decomposition methods. *Numerische Mathematik*, 95: 145–162, 2003.

R. Nabben and D. B. Szyld. Convergence theory of restricted multiplicative Schwarz methods. *SIAM Journal on Numerical Analysis*, 40:2318–2336, 2003.

A. Quarteroni and A. Valli. *Domain Decomposition Methods for Partial Differential Equations.* Oxford Science Publications, 1999.

B. F. Smith, P. E. Bjørstad, and W. Gropp. *Domain Decomposition: Parallel Multilevel Methods for Elliptic Partial Differential Equations.* Cambridge University Press, 1996.

D. B. Szyld. Equivalence of convergence conditions for iterative methods for singular equations. *Numerical Linear Algebra with Applications*, 1:151–154, 1994.

R. S. Varga. *Matrix Iterative Analysis.* Prentice-Hall, Englewood Cliffs, New Jersey, 1962. Second Edition, Springer, Berlin, 2000.

Schwarz Waveform Relaxation Method for the Viscous Shallow Water Equations

Véronique Martin

LAGA Université Paris XIII
martin@math.univ-paris13.fr

Summary. We are interested in solving time dependent problems using domain decomposition method. In the classical methods, one discretizes first the time dimension and then one solves a sequence of steady problems by a domain decomposition method. In this paper, we study a Schwarz Waveform Relaxation method which treats directly the time dependent problem. We propose algorithms for the viscous Shallow Water equations.

1 Introduction

The principle of domain decomposition methods is to partition the initial domain into several subdomains and then to use a processor per subdomain to solve the equation. The global solution is obtained if the processors exchange informations in an iterative way at the common interfaces. This method is useful to solve problems with a great number of unknowns. And it is more and more used to simulate complex phenomena with different spatial discretizations in each subdomain.

Solving time dependent problems, classical methods discretize the time dimension first and then use domain decomposition methods on the steady problems at each time step. Different strategies rely on the choice of transmission conditions (see Schwarz [1870], Lions [1990], Quarteroni and Valli [1999], Japhet et al. [2001]). In particular, in Japhet et al. [2001] transmission conditions are designed which minimize the convergence rate. This strategy proved to be very useful for many steady problems, for instance convection diffusion, Euler or Helmholtz equations. However the classical strategy to treat evolution equations does not allow to manage different time discretizations for each subdomain.

In some recent works a domain decomposition method for evolution problems quite different from the classical one has been proposed: they apply the iterative algorithm directly to the time dependent problem. This Schwarz Waveform Relaxation (SWR) method, permits to work with different time discretizations in each subdomain and therefore it provides an accurate method

to simulate complex phenomena. This method is a derivation of the Waveform Relaxation method: inspired by the Picard iteration, it has been studied in Lelarasmee et al. [1982] for integrated circuit simulation and its convergence can be accelerated by a multigrid method (see Vandewalle [1993]).
The first SWR algorithm used Dirichlet conditions on the interfaces (see Gander and Zhao [1997], Gander and Stuart [1998]) and more recently more appropriate interface conditions have been written in Gander et al. [1999]. In this paper we apply Schwarz Waveform Relaxation methods to the Shallow Water equations.

The Shallow Water equations are obtained by average of the Navier-Stokes equations when the depth of the water is much smaller than the other dimensions of the basin. If linearized around the velocity field $\mathbf{U} = \mathbf{0}$ this model becomes (see for example Pedlosky [1987])

$$\begin{cases} \partial_t \mathbf{U} - \nu \triangle \mathbf{U} + D\mathbf{U} + c^2 \nabla h = \boldsymbol{\tau_s}/\rho_0, \\ \partial_t h + \mathrm{div}\mathbf{U} \qquad\qquad\qquad\quad = 0. \end{cases} \tag{1}$$

where $\mathbf{U} = (u, v)$ is the velocity field, h the depth of the water, $D = \begin{pmatrix} 0 & -f \\ f & 0 \end{pmatrix}$, c^2 is the speed of internal gravity waves, ν the viscosity of the fluid, $\boldsymbol{\tau_s}$ is the wind stress and f the Coriolis force supposed to be constant for the theory. We introduce the Shallow Water operator $\mathcal{L}_{SW}$ where $\mathbf{W} = (\mathbf{U}, h)$ and we are interested in solving $\mathcal{L}_{SW}\mathbf{W} = \mathbf{F}_W$ in $\Omega \times (0, T)$ with $T < +\infty$, $\mathbf{W}(\cdot, \cdot, 0) = \mathbf{W}_0$ in Ω and with boundary conditions.

In this paper we study Schwarz Waveform Relaxation algorithms to solve the Shallow Water equations. We work on the space $\mathbb{R}^2$ which is split into two half spaces $\Omega_- = (-\infty, L) \times \mathbb{R}$ and $\Omega_+ = (0, +\infty) \times \mathbb{R}$, $L \geq 0$ is the overlap and let $\Gamma_0 = \{y \in \mathbb{R}, x = 0\}$ and $\Gamma_L = \{y \in \mathbb{R}, x = L\}$ denote the interfaces.

In Section 2 we propose an algorithm with Dirichlet interface conditions (which needs an overlap), then we propose in Section 3 an optimized algorithm which can be implemented without overlap. Finally we show numerical results which underline the efficiency of the optimized method (Sect. 4). More details about theorems will be found in Martin [2003].

2 Classical Schwarz Waveform Relaxation Method

Following ideas introduced in Gander and Zhao [1997] for the heat equation, we propose the following algorithm for $L > 0$

$$\begin{cases} \mathcal{L}_{SW}\,\mathbf{W}_-^{k+1} &= \mathbf{F}_W \;\text{ in } \Omega_- \times (0,T), \\ \mathbf{W}_-^{k+1}(\cdot,\cdot,0) &= \mathbf{W}_0 \;\text{ in } \Omega_-, \\ \mathbf{U}_-^{k+1} &= \mathbf{U}_+^k \;\text{ on } \Gamma_L \times (0,T), \end{cases} \quad \begin{cases} \mathcal{L}_{SW}\,\mathbf{W}_+^{k+1} &= \mathbf{F}_W \;\text{ in } \Omega_+ \times (0,T), \\ \mathbf{W}_+^{k+1}(\cdot,\cdot,0) &= \mathbf{W}_0 \;\text{ in } \Omega_+, \\ \mathbf{U}_+^{k+1} &= \mathbf{U}_-^k \;\text{ on } \Gamma_0 \times (0,T), \end{cases} \tag{2}$$

where $\mathbf{F}_W = (F_1, F_2, 0) = (\mathbf{F}, 0)$, $\mathbf{W}_0 = (\mathbf{U}_0, h_0)$ and $k \geq 0$. This algorithm is initialized by $\mathbf{U}_\pm^0$ in $\mathbf{H}^{2,1}(\Omega_\pm \times (0,T))$ such that $\mathbf{U}_\pm^0(\cdot,\cdot,0) = \mathbf{U}_0$ in $\Omega_\pm$.

We recall that we can find in Lions and Magenes [1972] the definition of anisotropic Sobolev spaces and a theorem of extension. If we use moreover a Fourier transform in y, a Laplace transform in t and *a priori* estimates, then we can prove that algorithm (2) is well posed.

Theorem 1. *Let* $\mathbf{F}$ *be in* $\mathbf{L}^2(0,T;\mathbf{L}^2(\Omega))$, $\mathbf{W}_0 = (\mathbf{U}_0, h_0)$ *in* $\mathbf{H}^1(\Omega)\times H^1(\Omega)$. *The algorithm (2) defines two unique sequences* $\mathbf{W}_\pm^k = (\mathbf{U}_\pm^k, h_\pm^k)$ *in* $\mathbf{H}^{2,1}(\Omega_\pm \times (0,T)) \times H^{1,1}(\Omega_\pm \times (0,T))$ *with* $\nabla h_\pm$ *in* $\mathbf{H}^1(0,T;\mathbf{L}^2(\Omega_\pm))$.

We can prove that algorithm (2) converges by computing its convergence rate written in Fourier-Laplace variables.

Theorem 2. *Let* $\mathbf{F}$ *be in* $\mathbf{L}^2(0,T;\mathbf{L}^2(\Omega))$, $\mathbf{W}_0 = (\mathbf{U}_0, h_0)$ *in* $\mathbf{H}^1(\Omega)\times H^1(\Omega)$. *The algorithm (2) converges in* $\mathbf{L}^2(0,T;\mathbf{H}^1(\Omega_\pm)) \times L^2(0,T;L^2(\Omega_\pm))$.

It is well-known that this algorithm is not efficient: the overlap between the two subdomains is necessary and the convergence is slow. In Gander et al. [1999] interface conditions have been introduced which are more appropriate. In the next section we apply this new strategy to the Shallow Water equations.

3 Optimized Schwarz Waveform Relaxation Method

In this section we consider the case without overlap of the subdomains ($L = 0$) and we denote by Γ the common interface. Since physical transmission conditions, (*i.e.* quantities that must be continuous through the interface) are $\mathbf{U}$ and $-\nu\partial_x\mathbf{U} + c^2(h,0)^t$ we propose the algorithm

$$\begin{cases} \mathcal{L}_{SW}\,\mathbf{W}_-^{k+1} = \mathbf{F}_W & \text{in } \Omega_- \times (0,T) \\ \mathbf{W}_-^{k+1}(\cdot,\cdot,0) = \mathbf{W}_0 & \text{in } \Omega_- \\ -\nu\partial_x\mathbf{U}_-^{k+1} + c^2(h_-^{k+1},0)^t - \Lambda^+\mathbf{U}_-^{k+1} = -\nu\partial_x\mathbf{U}_+^k + c^2(h_+^k,0)^t - \Lambda^+\mathbf{U}_+^k & \\ & \text{on } \Gamma \times (0,T) \end{cases}$$

$$\begin{cases} \mathcal{L}_{SW}\,\mathbf{W}_+^{k+1} = \mathbf{F}_W & \text{in } \Omega_+ \times (0,T) \\ \mathbf{W}_+^{k+1}(\cdot,\cdot,0) = \mathbf{W}_0 & \text{in } \Omega_+ \\ \nu\partial_x\mathbf{U}_+^{k+1} - c^2(h_+^{k+1},0)^t - \Lambda^-\mathbf{U}_+^{k+1} = \nu\partial_x\mathbf{U}_-^k - c^2(h_-^k,0)^t - \Lambda^-\mathbf{U}_-^k & \\ & \text{on } \Gamma \times (0,T) \end{cases} \tag{3}$$

with Λ^+ and Λ^- to be defined. The next theorem shows that we can choose the operators $\Lambda^\pm$ in an optimal way.

Theorem 3. *The operators* $\Lambda^\pm$ *can be chosen such that algorithm (3) converges in two iterations. These operators are denoted* $\Lambda^\pm_{exac}$.

These transmission conditions coincide with absorbing boundary conditions (see for example Gander et al. [1999] for time dependent scalar equations). As for many problems the operators $\Lambda^\pm_{exac}$ are not differential and difficult to use, therefore we have to approximate them (see for example Nataf and Rogier [1995]). For low spatial frequencies, small Coriolis force and small viscosity $\Lambda^\pm_{exac}$ are approximated by:

$$\Lambda^{\pm}_{app} = \begin{pmatrix} c + \frac{\nu}{2c}\partial_t & 0 \\ 0 & p \end{pmatrix},$$

with p a constant to be chosen. The following theorem gives a result of well-posedness for the corresponding algorithm. It can be proved by a Fourier-Laplace analysis and by an extension theorem.

Theorem 4. *Let* $\mathbf{F}$ *be in* $\mathbf{H}^{2,1}(\Omega\times(0,T))$, $\mathbf{W}_0 = (\mathbf{U}_0, h_0)$ *in* $\mathbf{H}^3(\Omega)\times H^3(\Omega)$ *and* p *be a strictly positive constant. If algorithm (3) is initialized by* $\mathbf{U}^0_\pm$ *in* $\mathbf{H}^{4,2}(\Omega_\pm \times (0,T))$ *and* $h^0_\pm$ *in* $H^1(0,T;H^3(\Omega_\pm))$ *with some compatibility relations satisfied at* $t = 0$*, then algorithm (3) defines two unique sequences* $(\mathbf{U}^k_\pm, h^k_\pm)$ *in* $\mathbf{H}^{4,2}(\Omega_\pm \times (0,T)) \times H^{3,2}(\Omega_\pm \times (0,T))$ *with* $h^k_\pm$ *in* $H^1(0,T;H^3(\Omega_\pm))$.

By *a priori* estimates we can prove that algorithm (3) converges.

Theorem 5. *Let* $\mathbf{F}$ *be in* $\mathbf{H}^{2,1}(\Omega\times(0,T))$, $\mathbf{W}_0 = (\mathbf{U}_0, h_0)$ *in* $\mathbf{H}^3(\Omega)\times H^3(\Omega)$ *and* p *be a strictly positive constant. If algorithm (3) is initialized by* $\mathbf{U}^0_\pm$ *in* $\mathbf{H}^{4,2}(\Omega_\pm \times (0,T))$ *and* $h^0_\pm$ *in* $H^1(0,T;H^3(\Omega_\pm))$ *with some compatibility relations satisfied at* $t = 0$*, then the sequences* $(\mathbf{U}^{k+1}_\pm, h^{k+1}_\pm)$ *defined by (3) converge in* $\mathbf{L}^2(0,T;\mathbf{H}^1(\Omega_\pm)) \times L^2(0,T;L^2(\Omega_\pm))$.

4 Numerical Results

4.1 Description of the experience

We work on a rectangular basin with closed boundaries, which extends from 0 to 15000 km in the x (east-west) direction and from -1500 km to 1500 km in the y (north-south) direction. The wind stress $\boldsymbol{\tau}_s = (\tau_x, \tau_y)$ is purely zonal ($\tau_y = 0$) and we have $\tau_x = 0.5\tau_0(1 + \tanh((x - x_0)/L))$, with $\tau_0 = 5 \cdot 10^{-2}$ N/m^2 and $x_0 = 3000$ km. The value of the physical parameters are $c = 3$ m/s and $\nu = 500$ m^2/s. For further details about the experience the reader is referred to Jensen and Kopriva [1990].

The Figure 1 shows the evolution in time of the depth of water. At $t = 0$ the ocean is at rest when the wind stress begins to be applied. Towards the equator the upper layer thickness increases. This anomaly travels eastward with a speed $c = 3m/s$ (the speed of Kelvin waves present in the model without viscosity or external stress). After 60 days the wave reaches the eastern wall and the incoming wave is divided into four waves: two coastal Kelvin waves and two Rossby waves (see for example Pedlosky [1987] for more details about these waves).

4.2 Solving by domain decomposition method

We solve now this problem by domain decomposition method with the interface at $x = 7500$ km. The value of the space and time steps is $\Delta x = 25$ km and $\Delta t = 30$ min. The experience lasts 200 days, therefore $200\times 24\times 2 = 9600$ time steps are needed. Schwarz Waveform Relaxation methods work on the whole time interval, but if this one is too large, solving the equation in $(0,T)$

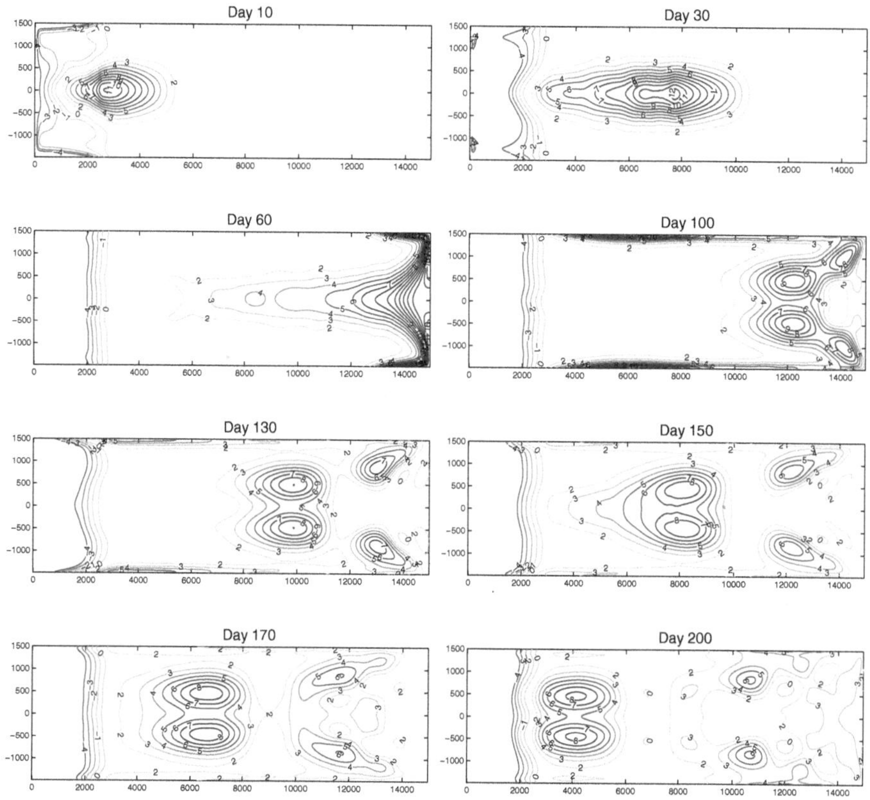

Fig. 1. Width of water at day 10, 30, 60, 100, 130, 150, 170 and 200

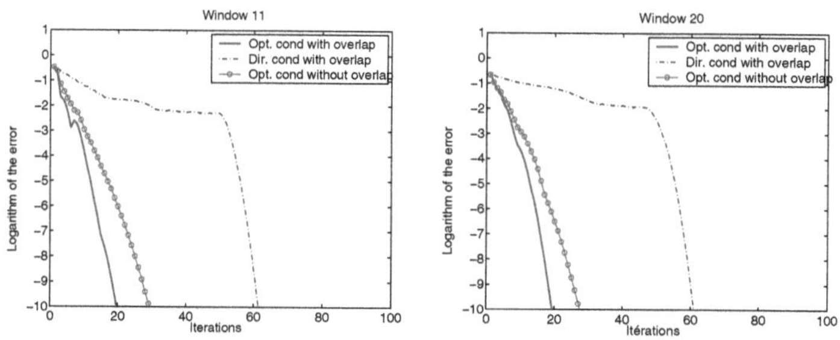

Fig. 2. Evolution of the logarithm of the error $L^2(\Omega)$ at the end of the time windows 11 and 20 versus the iterations

can be too expensive. So, we will split the time interval into several smaller time intervals. We write $(0, T) = \cup_{i=0,N-1}(T_i, T_{i+1})$ with $T_0 = 0$ and $T_N = T$, then we apply our domain decomposition algorithm on each time window; we first solve $\mathcal{L}_{SW}\mathbf{W}^0 = \mathbf{F}$ in $\Omega \times (0, T_1)$ with $\mathbf{W}^0(\cdot, \cdot, 0) = \mathbf{W}_0$ in Ω then for all

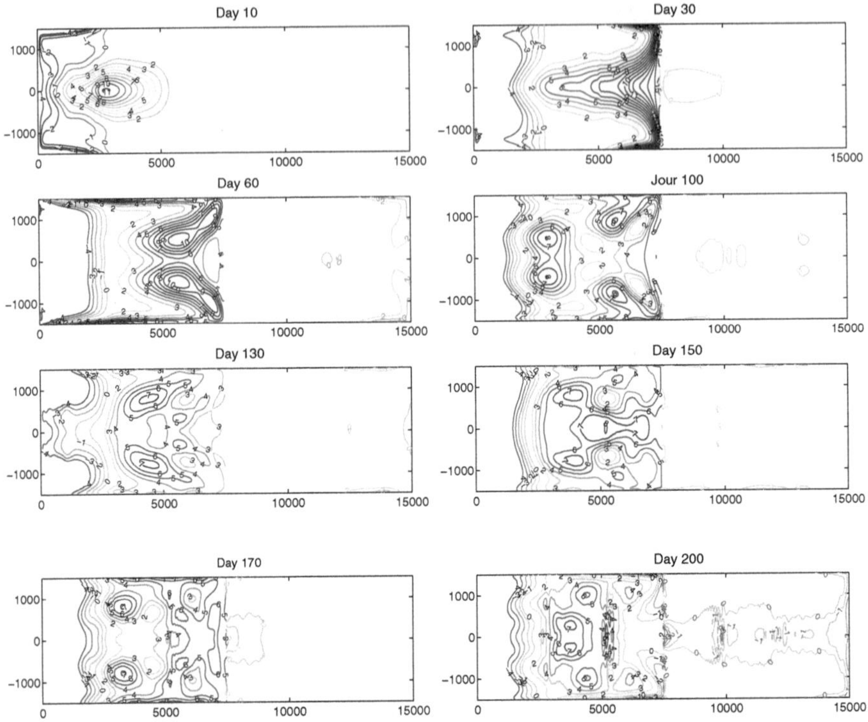

Fig. 3. Solution after two Schwarz iterations and with Dirichlet conditions at day 10, 30, 60, 100, 130, 150, 170 and 200

$i \geq 1$:

$$\begin{cases} \mathcal{L}_{SW}\mathbf{W}^i = \mathbf{F} & \text{in } \Omega \times (T_i, T_{i+1}), \\ \mathbf{W}^i(\cdot,\cdot,T_i) = \mathbf{W}^{i-1}(\cdot,\cdot,T_i) & \text{in } \Omega, \end{cases} \tag{4}$$

Here $T_{i+1} - T_i = 10$ days, *i.e.* we are going to work with 20 windows of 10 days.

When the overlap is $L = \Delta x$, we use the Dirichlet conditions introduced in Section 2 and the optimized conditions of Section 3. When there is no overlap we can only use optimized conditions. The parameter p of algorithm (3) with $\Lambda^{\pm} = \Lambda^{\pm}_{app}$ optimizes the convergence rate of the algorithm (see for example Japhet [1998]). The Figure 2 shows the evolution of the logarithm of the error $L^2(\Omega)$ at the end of the time windows 11 and 20 versus the iterations for each method. We can see how fast is the optimized method compared to the classical Schwarz method. Obviously with an overlap the optimized method is better than without one.

For more realistic simulations where such interface conditions appear, we can not wait for the convergence of the Schwarz algorithm because of the cost of each model, and only a few iterations can be implemented. The Figures 3

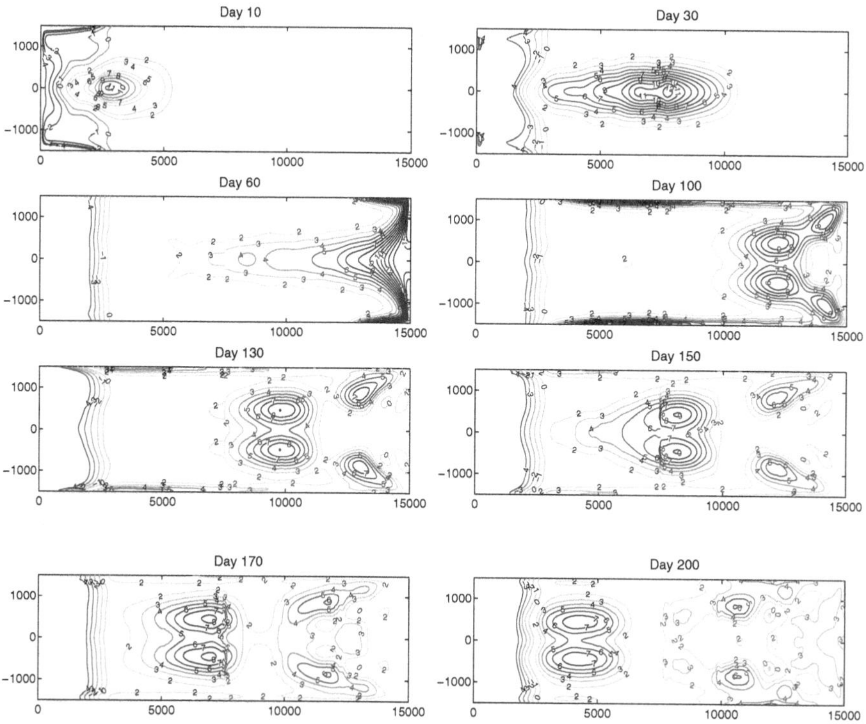

Fig. 4. Solution after two Schwarz iterations and with optimized conditions at day 10, 30, 60, 100, 130, 150, 170 and 200

and 4 show the solution obtained after two Schwarz iterations in each time window with Dirichlet conditions or optimized one. We can see that Dirichlet conditions act like a wall and waves reflect in it, whereas with optimized conditions the solution is admittedly discontinuous at the interface but it is closed to the monodomain solution.

5 Conclusion and perspectives

We have applied a Schwarz Waveform Relaxation method to the viscous Shallow Water equations; we have studied the classical SWR algorithm and a an optimized algorithm. Numerical results have shown that the optimized method is a good one. Perspectives of that work is to improve the interface conditions of the optimized algorithm and apply this method to the Shallow Water equations linearized around any velocity field $\mathbf{U}_0 \neq 0$.

References

M. J. Gander, L. Halpern, and F. Nataf. Optimal convergence for overlapping and non-overlapping Schwarz waveform relaxation. In C.-H. Lai,

P. Bjørstad, M. Cross, and O. Widlund, editors, *Eleventh international Conference of Domain Decomposition Methods*. ddm.org, 1999.

M. J. Gander and A. M. Stuart. Space time continuous analysis of waveform relaxation for the heat equation. *SIAM J.*, 19:2014–2031, 1998.

M. J. Gander and H. Zhao. Overlapping Schwarz waveform relaxation for parabolic problems in higher dimension. In A. Handlovičová, M. Komorníkova, and K. Mikula, editors, *Proceedings of Algoritmy 14*, pages 42–51. Slovak Technical University, September 1997.

C. Japhet. Optimized Krylov-Ventcell method. Application to convection-diffusion problems. In P. E. Bjørstad, M. S. Espedal, and D. E. Keyes, editors, *Proceedings of the 9th international conference on domain decomposition methods*, pages 382–389. ddm.org, 1998.

C. Japhet, F. Nataf, and F. Rogier. The optimized order 2 method. application to convection-diffusion problems. *Future Generation Computer Systems FUTURE*, 18, 2001.

T. G. Jensen and D. A. Kopriva. Comparison of a finite difference and a spectral collocation reduced gravity ocean model. *Modelling Marine Systems*, 2:25–39, 1990.

E. Lelarasmee, A. E. Ruehli, and A. L. Sangiovanni-Vincentelli. The waveform relaxation method for time-domain analysis of large scale integrated circuits. *IEEE Trans. on CAD of IC and Syst.*, 1:131–145, 1982.

J.-L. Lions and E. Magenes. *Nonhomogeneous Boundary Value Problems and Applications*, volume I. Springer, New York, Heidelberg, Berlin, 1972.

P.-L. Lions. On the Schwarz alternating method. III: a variant for nonoverlapping subdomains. In T. F. Chan, R. Glowinski, J. Périaux, and O. Widlund, editors, *Third International Symposium on Domain Decomposition Methods for Partial Differential Equations , held in Houston, Texas, March 20-22, 1989*, Philadelphia, PA, 1990. SIAM.

V. Martin. *Méthode de décomposition de domaine de type relaxation d'ondes pour des équations de l'océanographie*. PhD thesis, 2003.

F. Nataf and F. Rogier. Factorization of the convection-diffusion operator and the Schwarz algorithm. M^3AS, 5(1):67–93, 1995.

J. Pedlosky. *Geophysical Fluid Dynamics*. Springer Verlag, New-York, 1987.

A. Quarteroni and A. Valli. *Domain Decomposition Methods for Partial Differential Equations*. Oxford Science Publications, 1999.

H. A. Schwarz. Über einen Grenzübergang durch alternierendes Verfahren. *Vierteljahrsschrift der Naturforschenden Gesellschaft in Zürich*, 15:272–286, May 1870.

S. Vandewalle. *Parallel multigrid waveform relaxation for parabolic problems*. B. G. Teubner, Stuttgart, 1993.

A Two-Grid Alternate Strip-Based Domain Decomposition Strategy in Two-Dimensions

L. Angela Mihai[1] and Alan W. Craig[2]

[1] University of Durham, Department of Mathematical Sciences (l.a.mihai@durham.ac.uk)
[2] University of Durham, Department of Mathematical Sciences (alan.craig@durham.ac.uk)

Summary. The alternate strip-based iterative substructuring algorithms are preconditioning techniques for the discrete systems which arise from the finite element approximation of symmetric elliptic boundary value problems. The algorithms presented in this paper may be viewed as simple, direct extensions of the two disjoint subdomains case to the multiple domains decomposition with interior cross-points. The separate treatment of vertex points is avoided by dividing the original non-overlapping subdomains into strip-subregions. Both scalability and efficiency are enhanced by alternating the direction of the strips.

1 Introduction

In domain decomposition (DD) work is concentrating both on the improvement of existing algorithms as well as on the development of new ones and it is mainly the treatment of the interface relations between subdomains that distinguishes one method from another (Chan and Mathew [1994], Smith et al. [1996], Xu and Zou [1998], Quarteroni and Valli [1999]). The goal of our work is to construct simple, efficient preconditioners with good parallelism and optimal convergence properties, which draw upon the strengths of both overlapping and non-overlapping DD methodologies. After the model problem is introduced, the rest of this paper is organised as follows: Section 2 is devoted to the description and analysis of the one level strip-based substructuring (SBS_2) algorithm. Each strip is a union of non-overlapping subdomains and the global interface between subdomains is partitioned as a union of edges between strips and edges between subdomains which belong to the same strip. In Section 3 we derive and investigate the two-grid alternate strip-based substructuring ($ASBS_{2g}$) algorithms. The key ingredients are alternate strip-based solvers which generate algorithms in two stages and allow the use of efficient subdomain preconditioners such as a two-grid V or W cycle. We emphasise that a novel feature of our approach is that at each stage the direction of the strips changes and with it, the coupling between vertex points and edges. In Section 4 we illustrate the performance of the new two-grid strategy by some numerical estimates. The techniques introduced here (see also Mihai and Craig [2003]) extend in a straightforward manner to three-dimensional problems (Mihai and Craig [2004]). Once it is understood how and why it works, the alternate strip-based substructuring (*ASBS*) strategy can be regarded more as a principle in DD and extended to more general problems defined on more complex geometries. For an extended discussion on this new strategy, in two and three dimensions, we also refer to Mihai [2004].

The Problem. We consider a second order, symmetric, coercive equation, with homogeneous Dirichlet boundary conditions, defined on a polygonal domain $\Omega \subset \mathbb{R}^2$. With the given domain we associate a uniform square grid Σ^h of mesh-size h. The Galerkin finite element approximation generates the equivalent algebraic problem in the form of the linear system:

$$Au = f, \tag{1}$$

where the matrix A is symmetric and positive definite and f is the load vector. We are interested in the case when A is ill-conditioned and a preconditioned conjugate gradient (PCG) method is employed for solving the original partial differential equation.

The DD Approach. Without loss of generality, we assume Ω to be of unit diameter. A DD without overlapping, of the domain Ω, consists of a number of mutually disjoint open subdomains Ω_i, such that: $\bar{\Omega} = \bigcup_{i=1}^{N} \bar{\Omega}_i$. Let all subdomains Ω_i be of size H ($h < H < 1$) in the sense that there exists constants c and C independent of H and h such that Ω_i contains a ball of diameter cH and it is contained in a ball of diameter CH. Let also the coefficients of the original equation be either constants or piecewise constants. In the latter case the partition into disjoint subdomains is chosen in such a way that the jumps in the coefficients align with subdomain boundaries. We also assume that the triangulation Σ^h is consistent with the original DD in the sense that each $\partial\Omega_i$ can be written as a union of boundaries of elements in Σ^h. Let Γ denote the global interface between all subdomains $\{\Omega_i\}_{i=1}^{N}$. Then the linear system (1) can be written equivalently as:

$$\begin{bmatrix} A_{II} & A_{IE} \\ A_{IE}^T & A_{EE} \end{bmatrix} \begin{bmatrix} u^I \\ u^E \end{bmatrix} = \begin{bmatrix} f^I \\ f^E \end{bmatrix},$$

where the indices I and E are associated with the nodes in $\Omega \setminus \Gamma$ and the nodes in Γ respectively. By eliminating u^I, we obtain:

$$Su^E = f_S, \tag{2}$$

where $S = A_{EE} - A_{IE}^T A_{II}^{-1} A_{IE}$ is the *Schur complement* (SC) matrix and $f_S = f^E - A_{IE} A_{II}^{-1} f^I$. The condition number $\kappa(S)$, although smaller than $\kappa(A)$, deteriorates with respect to the subdomain size H, the finite element mesh-size h, and the coefficients of the model problem (Le Tallec [1994]). Our work aims to solve the SC system (2) by constructing a parallel preconditioner M, via a new DD strategy augmented with *two-grid* iteration. The equation (1) gets solved by the following procedure:

(I) (*Preprocessing*) solve the equation with homogeneous Dirichlet boundary conditions on every subdomain $\Omega_i \subset \Omega$.
(II) (*PCG*) solve the equation on the interface-boundary between all subdomains in Ω.
(III) (*Postprocessing*) update the solution on every subdomain Ω_i, using the boundary conditions given by (II).

Note that the preconditioned matrix $B^{-1}A$ has the same eigenvalues as the matrix $M^{-1}S$, plus the eigenvalue 1.

2 One-Level Strip-Based Substructuring

We begin by associating the non-overlapping subdomains in the initial partition of Ω, into strip-subregions Ω^s, whose vertices are on the boundary $\partial\Omega$ and whose edges align with the edges of the original subdomains. Each strip is a union of non-overlapping subdomains and the global interface between subdomains is partitioned as a union of edges between strips (which include also all the cross points in the initial partition) and edges between subdomains which belong to the same strip (inside strips, the interface edges do not contain their end points), see Fig.1 left.

The One-Level Strip-Based Substructuring (SBS_2) Algorithm. If u^k is a given iteration, we define:

$$u^{k+1} \leftarrow SBS_2(u^k, S, f_S)$$

to be the new approximation for the solution to the SC problem when the following process is applied:

$$u^{k+1} \leftarrow u^k + M^{-1}(f_S - Su^k),$$

where M is a preconditioner, such that the preconditioned system is symmetric and positive definite, hence it can also be used with CG acceleration. The new procedure for solving (2) can be described as follows:

(II_1) solve the one-dimensional equation with homogeneous Dirichlet boundary conditions on every edge between subdomains inside strips, with a preconditioner.
(II_2) solve the one-dimensional equation on every edge between strips, with a preconditioner.
(II_3) update the solution on every edge between subdomains inside strips, with Dirichlet boundary conditions from (II_2).

For a survey of edge preconditioners, see e.g. Keyes and Gropp [1987]. We consider the elements (strip-edges, edges inside strips) of the boundary of a subdomain to be direct projections of the corresponding elements in Γ. Let:

$$\Gamma = \bigcup_k \Gamma_k \cup \bigcup_j \Gamma^j,$$

where Γ_k denotes a generic edge (an edge does not include its end points) inside strips, that is the interface between two adjacent subdomains inside a strip-subregion and Γ^j denotes a generic strip-edge, that is the interface between two adjacent strip-subregions. We denote by $S_h^0(\Gamma_k)$ and $S_h^0(\Gamma^j)$ the subspace of the relevant boundary space consisting of functions whose support is contained in the corresponding edge. Let the following inner product:

$$s(\mathbf{u}^E, \mathbf{v}^E) = (u^E)^T S v^E$$

define the bilinear form associated with the Schur complement matrix under the standard nodal basis functions in $S_h(\Gamma)$. We decompose functions in $S_h^0(\Gamma)$ into $\mathbf{u}^E = \mathbf{u}^e + \mathbf{u}^s$, where

$$\mathbf{u}^e \in V^e = \bigoplus_k S_h^0(\Gamma_k)$$

and it is the solution of the following problem:

$$s(\mathbf{u}^e, \mathbf{v}) = (\mathbf{f}_S, \mathbf{v}), \ \forall \mathbf{v} \in V^e.$$

We solve for $\mathbf{u}_k^e \in S_h^0(\Gamma_k)$, on every edge Γ_k, the following local homogeneous Dirichlet problem:

$$s(\mathbf{u}_k^e, \mathbf{v}) = (\mathbf{f}_S, \mathbf{v}), \ \forall \mathbf{v} \in S_h^0(\Gamma_k).$$

We denote by $\mathbf{u}^s = \mathbf{u}^E - \mathbf{u}^e$ the part of the solution $\mathbf{u}^E$ which lies in the orthogonal complement of V^e in $S_h^0(\Gamma)$:

$$V^s = \{\mathbf{u} \in S_h^0(\Gamma) : s(\mathbf{u}, \mathbf{v}) = 0, \ \forall \mathbf{v} \in V^e\}.$$

Therefore $\mathbf{u}^s$ satisfies:

$$s(\mathbf{u}^s, \mathbf{v}) = (\mathbf{f}_S, \mathbf{v}) - s(\mathbf{u}^e, \mathbf{v}), \ \forall \mathbf{v} \in S_h^0(\Gamma),$$

or equivalently, by the definition of V^e,

$$s(\mathbf{u}^s, \mathbf{v}^s) = (\mathbf{f}_S, \mathbf{v}) - s(\mathbf{u}^e, \mathbf{v}), \ \forall \mathbf{v} \in S_h^0(\Gamma).$$

Note that:

$$s(\mathbf{u}, \mathbf{v}) = s(\mathbf{u}^e, \mathbf{v}^e) + s(\mathbf{u}^s, \mathbf{v}^s)$$

(here $\mathbf{v}^e$ and $\mathbf{v}^s$ are defined similarly as $\mathbf{u}^e$ and $\mathbf{u}^s$ respectively).

In the following lemma, every inequality can be proved by direct integration and with the help of the Cauchy inequality.

Lemma 1. *Let $\Omega \subset \mathbb{R}^2$ be the unit square and let $\Omega^s = (0,1) \times (0,H)$ be a strip-subregion of Ω. For $\mathbf{u} \in H^1(\Omega^s)$, the following inequalities hold:*

(i) if $\mathbf{u}$ is equal to zero along one short side of Ω^s, then:

$$\|\mathbf{u}\|^2_{L^2(\Omega^s)} \leq C|\mathbf{u}|^2_{H^1(\Omega^s)}.$$

(ii) if $\mathbf{u}$ is equal to zero along one long side of Ω^s, then:

$$\|\mathbf{u}\|^2_{L^2(\Omega^s)} \leq CH^2|\mathbf{u}|^2_{H^1(\Omega^s)}.$$

(iii) if Γ^j is a long side of Ω^s, then:

$$\|\mathbf{u}\|^2_{L^2(\Gamma^j)} \leq C\left(\frac{1}{H}\|\mathbf{u}\|^2_{L^2(\Omega^s)} + H|\mathbf{u}|^2_{H^1(\Omega^s)}\right).$$

(iv) if $\mathbf{u} \in H^1(\Omega)$ then:

$$\|\mathbf{u}\|^2_{L^2(\Omega^s)} \leq CH^2\left(\frac{1}{H}\|\mathbf{u}\|^2_{L^2(\Omega)} + |\mathbf{u}|^2_{H^1(\Omega)}\right).$$

Similar inequalities hold if we replace Ω^s by a square of side H, $\Omega^s_i = (iH,(i+1)H) \times (0,H)$, and Ω by Ω^s. C denotes positive constants which are independent of the parameters H and h. The actual value of these constants will not necessarily be the same in any two instances.

Theorem 1. *For the SBS_2 algorithm with exact solvers on the subdomains, the condition number of the preconditioned system grows linearly as $1/H$. For the case of discontinuous coefficients, the bounds are independent of the jumps in the coefficients as long as the jumps align with strip boundaries.*

Corollary 1. *For the case of homogeneous Dirichlet boundary conditions on $\partial\Omega$, if the domain Ω is reduced to at most two strip-subregions of width H, such that each strip is the reunion of $1/H$ non-overlapping subdomains, then, for the SBS_2 algorithm with exact solvers on the subdomains, the condition number of the preconditioned system is bounded independently of the partitioning parameters H and h.*

3 Two-Grid Alternate Strip-Based Substructuring

In this section we extend the SBS_2 algorithm, introduced in the previously, to a *two-stage* algorithm. In order to remove the factor $1/H$ from the order of convergence, at each stage the direction of the strips changes and with it, the coupling between vertex points and edges (e.g. *horizontal strips* at the first stage, *vertical strips* at the second stage). Moreover, at the second stage, the calculations are carried out on a coarser grid. This can reduce considerably the amount of computational work needed to solve the problem to a particular accuracy. Let $\Sigma^{2^p h} \subset \cdots \subset \Sigma^{2h} \subset \Sigma^h$ be a set of nested uniform square grids associated with the original domain Ω, such that $1 \leq p \in \mathbb{N}$ and $2^p h \leq H$. To describe the two-grid algorithms, we introduce the following operators: the projection P is an interpolation from grid $\Sigma^{2^p h}$ to grid Σ^h; the restriction from grid Σ^h to grid $\Sigma^{2^p h}$ is defined as $R = P^*$. Finally, we shall also be using the notation: $S^{(1)}$ and $S^{(2)}$, for the coefficient matrix and $f_S^{(1)}$ and $f_S^{(2)}$, for the load vector of the linear system to be solved at the first and second stage respectively. Figure 1 shows the partition of the unit square $\Omega = (0,1) \times (0,1)$ into $1/H$ disjoint, uniform strips Ω^s, at two different stages.

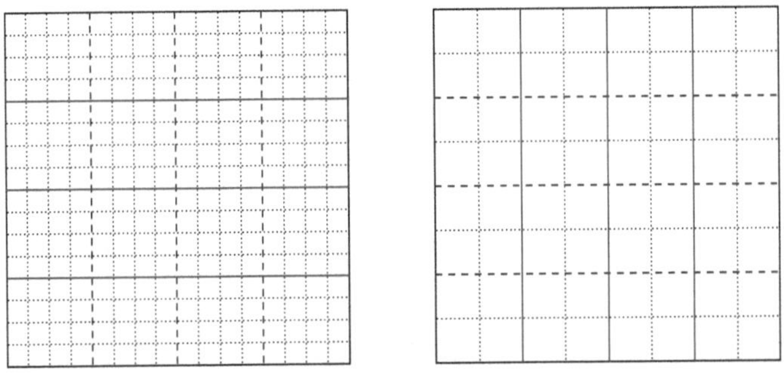

Fig. 1. The horizontal (left) and vertical (right) partition into strips of the domain Ω, with two levels of mesh-refinement

The Two-Grid Alternate Strip-Based Substructuring ($ASBS_{2g}$) Algorithm. Let u^k be a given iteration, then u^{k+1} is the result of the following V-cycle:

- $u^{k+\frac{1}{2}} \leftarrow SBS_2(u^k, S^{(1)}, f_S^{(1)})$
- If $R^{(2)} \leftarrow f_S^{(2)} - S^{(2)} u^k$, $R_c^{(2)} \leftarrow RR^{(2)}$, its restriction to the coarse grid, then $w_c^{(2)} \leftarrow SBS_2(0, S_c^{(2)}, R_c^{(2)})$, and its prolongation to the fine grid is $w^{(2)} \leftarrow P w_c^{(2)}$. We set

$$u^{k+1} \leftarrow u^{k+\frac{1}{2}} + w^{(2)}.$$

This procedure can be regarded as an additive Schwarz process between stages and expressed equivalently as:

$$\begin{aligned} u^{k+\frac{1}{2}} &\leftarrow u^k + M_1^{-1}(f_S^{(1)} - S^{(1)} u^k) \\ w^{(2)} &\leftarrow M_2^{-1}(f_S^{(2)} - S^{(2)} u^k) \\ u^{k+1} &\leftarrow u^{k+\frac{1}{2}} + w^{(2)}, \end{aligned}$$

where M_1 and M_2 are preconditioners. The preconditioned system, which is symmetric and can be used with CG acceleration, can be written as:

$$M_{2g}^{-1}S = M_1^{-1}S + M_2^{-1}S.$$

Note that at the coarse-grid level, if the mesh-size is equal to H, then only equations corresponding to the interface between strip-subregions need to be solved. Therefore, instead of defining the coarse-grid solver on the whole global interface Γ between subdomains, we can consider only one-dimensional coarse-solvers defined on the edges between strip-subregions, then alternate the strips at the fine stage. We note that for problems in three dimensions, the possibility of reducing the size of the coarse solver from three to only two dimensions seems to offer an advantage (in forthcoming Mihai and Craig [2004]).

The performance of the $ASBS_{2g}$ method is illustrated by the following result. Its proof is based on the observation that the preconditioner can be interpreted as a two-level overlapping Schwarz method, where every overlapping subdomain is the union of two adjacent subdomains that share the same edge.

Theorem 2. *For the $ASBS_{2g}$ method, if exact solvers are used for the subproblems, the condition number of the preconditioned system is bounded independently of the partitioning parameters H and h.*

4 Numerical Estimates

Example 1. Consider the model problem

$$-\nabla \cdot \alpha(x)\nabla \mathbf{u}(x) = \mathbf{f}(x), \ in \ \Omega = (0,1) \times (0,1)$$
$$\mathbf{u}(x) = 0, \ on \ \partial\Omega,$$

discretized by piecewise linear finite elements. In the computations, at each stage the unit square Ω is partitioned into $N = 1/H^2$ equal squares; $\alpha(\cdot)$ is 1 (for the Poisson equation) or random constants inside each subdomain. For the interface edges, the coefficients are the average values of all the subdomains adjacent to that interface. The mesh-parameter is h for the fine grid and H for the coarse-grid. The iteration counts are calculated for 10^{-4} reduction in error. All computations were carried out in Matlab.

Discussion. In Table 1, for the SBS_2 algorithm, the condition number of the preconditioned SC system grows like $1/H$ and remains bounded independently of the mesh-size h. In Table 2, for $ASBS_{2g}$, the condition number of the preconditioned SC system is less than 2. The bounds are also independent of the jumps in the coefficients as long as the jumps align with subdomain boundaries.

Table 1. Condition number and iteration counts for the SBS_2 algorithm

N	1/h=32		64		128		256	
4	1.3403	4	1.3408	4	1.3387	4	1.3376	4
16	1.8739	6	1.8717	6	1.8755	6	1.8755	6
64	3.2879	8	3.2801	8	3.2826	8	3.2916	8
256	6.3364	12	6.3376	12	6.3276	12	6.3208	12

Table 2. Condition number and iteration counts for the additive $ASBS_{2g}$ algorithm

N	1/h=32		64		128		256	
4	1.2565	4	1.2582	4	1.2574	4	1.2573	4
16	1.3467	4	1.3362	4	1.3284	4	1.3265	4
64	1.4875	5	1.4715	5	1.4160	5	1.4111	5
256	1.8746	6	1.5948	5	1.5198	5	1.5032	5

References

T. F. Chan and T. P. Mathew. Domain decomposition algorithms. In *Acta Numerica 1994*, pages 61–143. Cambridge University Press, 1994.

D. E. Keyes and W. D. Gropp. A comparison of domain decomposition techniques for elliptic partial differential equations and their parallel implementation. *SIAM J. Sci. Stat. Comput.*, 8(2):s166–s202, 1987.

P. Le Tallec. Domain decomposition methods in computational mechanics. In J. T. Oden, editor, *Computational Mechanics Advances*, volume 1 (2), pages 121–220. North-Holland, 1994.

L. Mihai. *A Class of Alternate Strip-Based Domain Decomposition Methods for Elliptic PDEs.* PhD thesis, in preparation. University of Durham, U.K., 2004.

L. Mihai and A. Craig. Alternate strip-based substructuring algorithms for elliptic PDEs in two-dimensions. *submitted*, 2003.

L. Mihai and A. Craig. Alternate strip-based substructuring algorithms for elliptic PDEs in three-dimensions. *submitted*, 2004.

A. Quarteroni and A. Valli. *Domain Decomposition Methods for Partial Differential Equations.* Oxford Science Publications, 1999.

B. F. Smith, P. E. Bjørstad, and W. Gropp. *Domain Decomposition: Parallel Multilevel Methods for Elliptic Partial Differential Equations.* Cambridge University Press, 1996.

J. Xu and J. Zou. Some nonoverlapping domain decomposition methods. *SIAM Review*, 40:857–914, 1998.

Parallel Solution of Cardiac Reaction-Diffusion Models

Luca F. Pavarino[1] and Piero Colli Franzone[2]

[1] Università di Milano, Dept. of Mathematics (www.mat.unimi.it/~pavarino)
[2] Università di Pavia, Dept. of Mathematics (www-dimat.unipv.it)

Summary. We present and study a parallel iterative solver for reaction-diffusion systems in three dimensions arising in computational electrocardiology, such as the Bidomain and Monodomain models. The models include intramural fiber rotation and anisotropic conductivity coefficients that can be fully orthotropic or axially symmetric around the fiber direction. These cardiac models are coupled with a membrane model for the ionic currents, consisting of a system of ordinary differential equations. The solver employs structured isoparametric Q_1 finite elements in space and a semi-implicit adaptive method in time. Parallelization and portability are based on the PETSc parallel library and large-scale computations with up to $O(10^7)$ unknowns have been run on parallel computers. These simulation of the full Bidomain model (without operator or variable splitting) for a full cardiac cycle are, to our knowledge, among the most complete in the available literature.

1 The cardiac Bidomain and Monodomain models

Cardiac tissue is traditionally modeled as an arrangement of cardiac fibers that rotate counterclockwise from the epicardium to the endocardium, (see Streeter [1979]). Moreover, from LeGrice and et al. [1995], the cardiac tissue has a laminar organization that can be modeled as a set of muscle sheets running radially from epi to endocardium. Therefore, at any point $\mathbf{x}$, it is possible to identify a triplet of orthonormal principal axes $\mathbf{a}_l(\mathbf{x})$, $\mathbf{a}_t(\mathbf{x})$, $\mathbf{a}_n(\mathbf{x})$, with $\mathbf{a}_l(\mathbf{x})$ parallel to the local fiber direction, $\mathbf{a}_t(\mathbf{x})$ and $\mathbf{a}_n(\mathbf{x})$ tangent and orthogonal to the radial laminae respectively and both being transversal to the fiber axis. The macroscopic Bidomain model represents the cardiac tissue as the superposition of two anisotropic continuous media, the intra (i) and extra (e) cellular media, coexisting at every point of the tissue and connected by a distributed continuous cellular membrane; see Keener and Sneyd [1998]. The anisotropic conductivity properties of the tissue are described by the conductivity coefficients in the intra and extracellular media $\sigma_l^{i,e}$, $\sigma_t^{i,e}$, $\sigma_n^{i,e}$, measured along the corresponding directions $\mathbf{a}_l, \mathbf{a}_t, \mathbf{a}_n$, and by the conductivity tensors $D_i(\mathbf{x})$ and $D_e(\mathbf{x})$

$$D_{i,e} = \sigma_l^{i,e}\ \mathbf{a}_l \mathbf{a}_l^T + \sigma_t^{i,e}\ \mathbf{a}_t \mathbf{a}_t^T + \sigma_n^{i,e}\ \mathbf{a}_n \mathbf{a}_n^T.$$

When the media are *axially isotropic*, i.e. when $\sigma_n^{i,e} = \sigma_t^{i,e}$, we have $D_{i,e} = {\sigma_t}^{i,e} I + ({\sigma_l}^{i,e} - {\sigma_t}^{i,e}) \mathbf{a}_l \mathbf{a}_l^T$. The intra and extracellular electric potentials u_i, u_e in an insulated cardiac domain H are described in the Bidomain model by a reaction-diffusion system coupled with a system of ODEs for the ionic gating variables w. Given the applied currents per unit volume $I_{app}^{i,e}$, satisfying the compatibility condition $\int_H I_{app}^i\, \mathrm{dx} = \int_H I_{app}^e \mathrm{dx}$, the initial conditions v_0, w_0, then u_i, u_e, w satisfy the system:

$$\begin{cases} c_m \dfrac{\partial v}{\partial t} - div(D_i \nabla u_i) + I_{ion}(v,w) = I_{app}^i \\ -c_m \dfrac{\partial v}{\partial t} - div(D_e \nabla u_e) - I_{ion}(v,w) = -I_{app}^e \\ \dfrac{\partial w}{\partial t} - R(v,w) = 0, \qquad v(t) = u_i(t) - u_e(t) \\ \mathbf{n}^T D_i \nabla u_i = 0, \quad \mathbf{n}^T D_e \nabla u_e = 0, \\ v(\mathbf{x},0) = v_0(\mathbf{x}), \quad w(\mathbf{x},0) = w_0(\mathbf{x}), \end{cases} \tag{1}$$

where $c_m = \chi * C_m$, $I_{ion} = \chi * i_{ion}$, with χ the ratio of membrane surface area per tissue volume, C_m the membrane capacitance and i_{ion} the ionic current of the membrane per unit area. Existence and regularity results for this degenerate system can be found in Colli Franzone and Savaré [2002]. The system uniquely determines v, while the potentials u_i and u_e are defined only up to a same additive time-dependent constant related to the reference potential, chosen to be the average extracellular potential in the cardiac volume by imposing $\int_H u_e\, \mathrm{dx} = 0$.

If the two media have equal anisotropy ratio, i.e. $D_i = \lambda D_e$ with λ constant, then the Bidomain system reduces to the Monodomain model consisting in a parabolic reaction-diffusion equation for the transmembrane potential v coupled with a system of ODEs for the gating variables:

$$\begin{cases} c_m \dfrac{\partial v}{\partial t} - div(D_m(\mathbf{x}) \nabla v) + I_{ion}(v,w) = I_{app}^m, \\ \dfrac{\partial w}{\partial t} - R(v,w) = 0, \quad w(\mathbf{x},0) = w_0(\mathbf{x}), \\ \mathbf{n}^T D_m \nabla v = 0, \quad v(\mathbf{x},0) = v_0(\mathbf{x}), \end{cases} \tag{2}$$

where $D_m = \sigma_l\ \mathbf{a}_l \mathbf{a}_l^T + \sigma_t\ \mathbf{a}_t \mathbf{a}_t^T + \sigma_n\ \mathbf{a}_n \mathbf{a}_n^T$, with $\sigma_{l,t,n} = \lambda \sigma_{l,t,n}^i / (1+\lambda)$ and $I_{app}^m = (\lambda I_{app}^i + I_{app}^e)/(1+\lambda)$.

The dynamics of S gating variables are described by a so-called membrane model, consisting of ordinary differential equations of the form

$$\frac{\partial w_j}{\partial t} = R_j(v, w_j) = (w_{j\infty}(v) - w_j)/\tau_j(v), \quad j = 1,..,S. \tag{3}$$

In this paper, we consider one of the most used detailed membrane models in the literature, the Luo-Rudy phase I (LR1) model (see Luo and Rudy [1991]), based on six gating variables and one variable for the calcium ionic concentration.

2 Discretization of the models

The Monodomain (2) and Bidomain models (1) are discretized by meshing the cardiac tissue volume H with a structured grid of hexahedral isoparametric Q_1 elements and by introducing the associated finite element space V_h. A semidiscrete problem is obtained by applying a standard Galerkin procedure and choosing a finite element basis $\{\phi_i\}$ for V_h. We denote by $\mathrm{M} = \{\mathrm{m}_{rs} = \int_H \varphi_r \varphi_s \mathrm{dx}\}$ the symmetric mass matrix, by $\mathrm{A}_{m,i,e} = \{\mathrm{a}_{rs}^{m,i,e} = \int_H (\nabla\varphi_r)^T D_{m,i,e} \nabla\varphi_s \mathrm{dx}\}$ the symmetric stiffness matrices and by $\mathrm{I}_{ion}^h, \mathrm{I}_{app}^{(m,i,e),h}$ the finite element interpolants of I_{ion} and $I_{app}^{m,i,e}$, respectively. Integrals are computed with a 3D trapezoidal quadrature rule, so the mass matrix M is lumped to diagonal form; see Quarteroni and Valli [1994] for an introduction to finite element methods. In our implementation, we have actually reordered the unknowns writing for every node the $\mathbf{u}_i$ and $\mathbf{u}_e$ components consecutively, so as to minimize bandwidth of the stiffness matrix.

The time discretization is performed by a semi-implicit method using for the diffusion term the implicit Euler method, while the nonlinear reaction term I_{ion} is treated explicitly. The use of an implicit treatment of the diffusion terms appearing in the Mono or Bidomain models is essential to allow an adaptive change of the time step according to the stiffness of the various phases of the heartbeat. The ODE system for the gating variables is discretized by the semi-implicit Euler method; in this way we decouple the gating variables by solving the gating system first (given the potential $\mathbf{v}^n$ at the previous time-step)

$$(\mathbf{w}^{n+1} - \mathbf{w}^n)/\Delta t = R(\mathbf{v}^n, \mathbf{w}^{n+1})$$

and then solving for $\mathbf{u}_i^{n+1}, \mathbf{u}_e^{n+1}$ in the *Bidomain case*

$$\left(\frac{c_m}{\Delta t}\begin{bmatrix} \mathrm{M} & -\mathrm{M} \\ -\mathrm{M} & \mathrm{M}\end{bmatrix} + \begin{bmatrix} \mathrm{A}_i & 0 \\ 0 & \mathrm{A}_e\end{bmatrix}\right)\begin{pmatrix}\mathbf{u}_i^{n+1} \\ \mathbf{u}_e^{n+1}\end{pmatrix} =$$

$$\frac{c_m}{\Delta t}\begin{pmatrix}\mathrm{M}(\ \mathbf{u}_i^n - \mathbf{u}_e^n) \\ \mathrm{M}[-\mathbf{u}_i^n + \mathbf{u}_e^n]\end{pmatrix} + \begin{pmatrix}\mathrm{M}[-\mathrm{I}_{ion}^h(\mathbf{v}^n, \mathbf{w}^{n+1}) + \mathrm{I}_{app}^{i,h}] \\ \mathrm{M}[\ \ \mathrm{I}_{ion}^h(\mathbf{v}^n, \mathbf{w}^{n+1}) - \mathrm{I}_{app}^{e,h}]\end{pmatrix}, \tag{4}$$

where $\mathbf{v}^n = \mathbf{u}_i^n - \mathbf{u}_e^n$. As in the continuous model, $\mathbf{v}^n$ is uniquely determined, while $\mathbf{u}_i^n$ and $\mathbf{u}_e^n$ are determined only up to the same additive time-dependent constant chosen by imposing the condition $\mathbf{1}^T\mathrm{M}\mathbf{u}_e^n = 0$.

In the *Monodomain case*, we have to solve for $\mathbf{v}^{n+1}$

$$\left(\frac{c_m}{\Delta t}\mathrm{M} + \mathrm{A}_m\right)\mathbf{v}^{n+1} = \frac{c_m}{\Delta t}\mathrm{M}\mathbf{v}^n - \mathrm{M}\,\mathrm{I}_{ion}^h(\mathbf{v}^n, \mathbf{w}^{n+1}) + \mathrm{M}\mathrm{I}_{app}^{m,h}. \tag{5}$$

We employed an adaptive time-stepping strategy based on controlling the transmembrane potential variation $\Delta v = max(\mathbf{v}^{n+1} - \mathbf{v}^n)$ at each time-step, see Luo and Rudy [1991]. If $\Delta v < \Delta v_{min} = 0.05$ then we select $\Delta t = (\Delta v_{max}/\Delta v)\Delta t$ (if smaller than $\Delta t_{max} = 6$ msec), if $\Delta v > \Delta v_{max} = 0.5$ then we select $dt = (\Delta v_{min}/\Delta v)dt$ (if greater than $\Delta t_{min} = 0.005$ msec). In order to guarantee a control on the variation of the gating variables of the LR1 membrane model as well, each gating equation of the system (3) is integrated exactly (see Victorri and et al. [1985]), while the calcium ionic concentration is updated using the explicit Euler method.

The linear system at each time step in the discrete problems is solved iteratively by the preconditioned conjugate gradient (PCG) method, using as initial guess the solution at the previous time step. Parallelization and portability are realized using the PETSc parallel library (Balay et al. [2001]) and a preconditioned conjugate gradient solver at each time step, with block Jacobi preconditioner and ILU(0) on each block, the default one-level preconditioner in the PETSc library. The numerical experiments reported in the next section show that it performs well in the Monodomain case, but not in the Bidomain case. Therefore, more research is needed in order to build better preconditioners, particularly with two or more levels; see Smith et al. [1996].

3 Numerical results

We have conducted several numerical experiments in three dimensions on distributed memory parallel architectures, with both the Monodomain and the Bidomain model coupled with the LR1 membrane model. The parallel machines employed are an IBM SP RS/6000 with 512 processors Power 4 of the Cineca Consortium (www.cineca.it), and an HP SuperDome 64000 with 64 processors PA8700 of the Cilea Consortium (www.cilea.it). We refer to Colli Franzone and Pavarino [2003] for more detailed numerical results. Multigrid preconditioners for the Bidomain system have been studied by Weber dos Santos et al. [2004], while mortar finite element discretizations by Pennacchio [2004]. We studied first the spectrum of the iteration matrices (5) and (4) on a small 15×15×8 mesh in the Monodomain case and 15×15×4 in the Bidomain case (these meshes are chosen in order to have matrices of the same size). The eigenvalues of the stiffness matrices are reported in the left panel of Figure 1, while the eigenvalues of the iteration matrices are reported in the right panel. It is clear that the addition to the stiffness matrix of a term with the mass matrix greatly improves the spectrum of the Monodomain iteration

Table 1. Parameters calibration for numerical tests

ellipsoidal geometry	$a_1 = b_1 = 1.5\ cm,\quad a_2 = b_2 = 2.7\ cm,\quad c_1 = 4.4,\quad c_2 = 5\ cm$ $\phi_{min} = 0,\quad \phi_{max} = 2\pi,\quad \theta_{min} = -3\pi/8,\quad \theta_{max} = \pi/8$
	$\chi = 10^3\ cm^{-1},\quad C_m = 10^{-3}\ mF/cm^2$
Monodomain parameters	$\sigma_l = 1.2 \cdot 10^{-3}\ \Omega^{-1}cm^{-1},\quad \sigma_t = 2.5562 \cdot 10^{-4}\ \Omega^{-1}cm^{-1}$ $G = 1.5\ \Omega^{-1}cm^{-2},\quad v_{th} = 13\ mV,\quad v_p = 100\ mV$ $\eta_1 = 4.4\ \Omega^{-1}cm^{-2},\quad \eta_2 = 0.012,\quad \eta_3 = 1$
Bidomain parameters	$\sigma_l^e = 2 \cdot 10^{-3}\ \Omega^{-1}cm^{-1},\quad \sigma_l^i = 3 \cdot 10^{-3}\ \Omega^{-1}cm^{-1}$ $\sigma_t^e = 1.3514 \cdot 10^{-3}\ \Omega^{-1}cm^{-1},\quad \sigma_t^i = 3.1525 \cdot 10^{-4}\ \Omega^{-1}cm^{-1}$ $\sigma_n^e = \sigma_t^e/\mu_1,\quad \sigma_n^i = \sigma_t^i/\mu_2$ $\mu_1 = \mu_2 = 1$ axial isotropic case, $\quad \mu_1 = 2, \mu_2 = 10$ orthotropic case

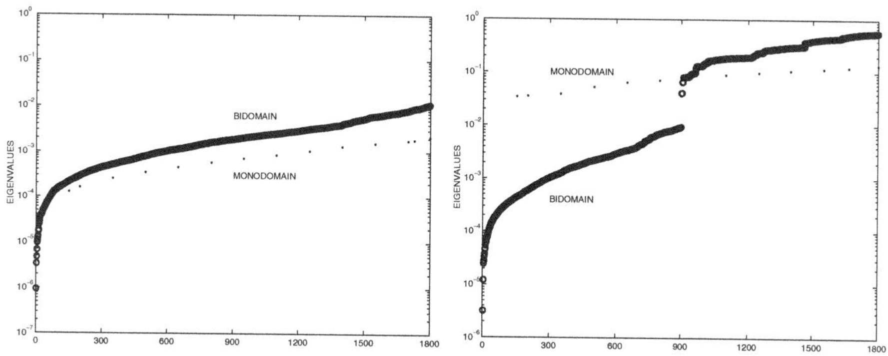

Fig. 1. Nonzero eigenvalue distribution of the stiffness matrices related to elliptic operators with Neumann boundary conditions (left) and of the iteration matrices in (5) and (4) (right) for a small mesh. Monodomain eigenvalues are denoted by dots (·), Bidomain eigenvalues by circles (o)

matrix (5), but not of the Bidomain iteration matrix (4). In fact, the iterative solution of the linear system at each time step turns out to be much harder for the Bidomain model than for the Monodomain model.

3.1 Scaled speedup for Monodomain-LR1 and Bidomain-LR1 solvers

We consider first the Monodomain equation with LR1 ionic model, simulating on the IBM SP4 machine the initial depolarization of some ellipsoidal blocks after one stimulus of 250 mA/cm^3 has been applied for 1 msec on a small area (5 mesh points in each direction) of the epicardium. The blocks are chosen in increasing sizes so as to keep constant the number of mesh points per subdomain (processor). As shown in Figure 2, the domain varies from the smaller block with 8 subdomains to half ventricle with 128 subdomains. We fixed the local mesh in each subdomain to be of 75×75×50 nodes (281.750 unknowns), hence varying the global number of unknowns of the linear system from $2.25 \cdot 10^6$ in the smaller case with 8 subdomains on a global mesh of 150×150×100 nodes to $3.6 \cdot 10^7$ in the larger case with 128 subdomains on a global mesh of 600×600×100 nodes. The model is run for 30 time steps of 0.05 msec each. At each time step, we compute the potential v, the gating and concentration variables $w_1, \cdots, w_7$ and the depolarization time. The results are reported in the upper part of Table 2. The assembling time, average number of PCG iterations per time step and the average time per time step (last three columns) are reasonably small. Up to 64 processors, the algorithm seems practically scalable, and even for 128 processors, the number of PCG iterations grows to just 8.

We then consider the Bidomain system with LR1 ionic model, in the same setting (initial stimulus and domain decomposition) of the previous case. At

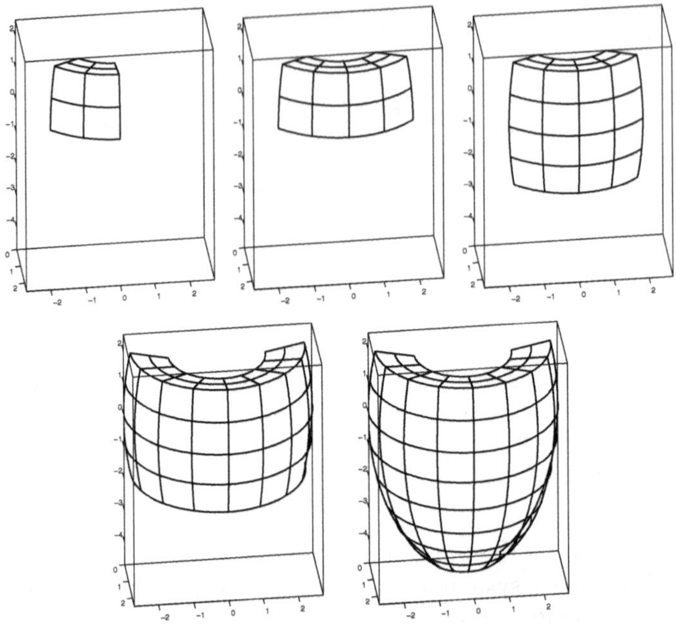

Fig. 2. Scaled speedup test: ellipsoidal domains of increasing sizes decomposed into 8, 16, 32, 64 and 128 subdomains of fixed size

Table 2. Scaled speedup tests for Monodomain - LR1 and Bidomain - LR1 models. Initial depolarization of an ellipsoidal block: 1 stimulus on epicardial surface, 30 time steps of 0.05 msec each, computation of $v, w_1, \cdots, w_7$ and isochrones. t_A = assembly timing, it = average number of PCG iterations at each time step, time = average CPU timing of each time step

Monodomain - LR1					
# proc.	mesh	unknowns (nodes)	t_A	it.	time
8 = 2·2·2	150×150×100	2.250.000	7.7 s	4	2.7 s
16 = 4·2·2	300×150×100	4.500.000	8.5 s	4	3 s
32 = 4·4·2	300×300×100	9.000.000	9.1 s	5	3.6 s
64 = 8·4·2	600×300×100	18.000.000	9.2 s	5	3.6 s
128 = 8·8·2	600×600×100	36.000.000	10.6 s	8	5.1 s

Bidomain - LR1					
# proc.	mesh	unknowns (2× nodes)	t_A	it.	time
8 = 2·2·2	100×100×70	1.400.000	12.9 s	98	40.2 s
16 = 4·2·2	200×100×70	2.800.000	13.3 s	127	55.5 s
32 = 4·4·2	200×200×70	5.600.600	15.7 s	148	72 s
64 = 8·4·2	400×200×70	11.200.000	16.2 s	176	91.9 s
128 = 8·8·2	400×400×70	22.400.000	18.4 s	244	129.7 s

each time step, we now compute the potentials u_i, u_e, the gating and concentration variables and the depolarization time. Due to the larger memory requirements of the Bidomain model, we used a smaller mesh of 50×50×35 nodes in each subdomain (processor), hence varying the global number of unknowns of the linear system from $1.4{\cdot}10^6$ in the smaller case with 8 subdomains on a global mesh of 100×100×70 to $2.24{\cdot}10^7$ unknowns in the larger case with 128 subdomains on a global mesh of 400×400×70 nodes. The results are reported in the lower part of Table 2. While the assembling time remains reasonable (under 20 sec.), the average number of PCG iterations per time step and the average time per time step are now much larger, clearly showing the limits of the one-level preconditioner and the effects of the severe ill-conditioning of the Bidomain iteration matrix.

3.2 Simulation of a full cardiac cycle

We also simulated a complete cardiac cycle (excitation-recovery) in a slab of cardiac tissue of size $2\times2\times0.5\ cm^3$, discretized with a fine mesh 200×200×50. We used 25 processors of the HP SuperDome machine with 64 processors. The fibers rotate intramurally linearly with depth for a total amount of 90°. A stimulus is applied at an epicardial vertex and the excitation of the entire slab requires about 80 msec, while the time interval for simulating the cardiac cycle is on the order of 360 msec. The adaptive time-stepping algorithm automatically adapts, in an efficient way, the time step size in the three main different phases of the heart beat, see Figure 3 (left). While the Monodomain solver is quite efficient, the Bidomain solver is not, since the number of PCG iterations at each time step increases considerably, reaching a maximum of about 250 iterations in the depolarization phase, see Figure 3 (right). The simulation took about 6.4 days for the the Bidomain model and about 5 hours for the Monodomain model. We compared the two computer platforms mentioned above by simulating the Monodomain model on a slab

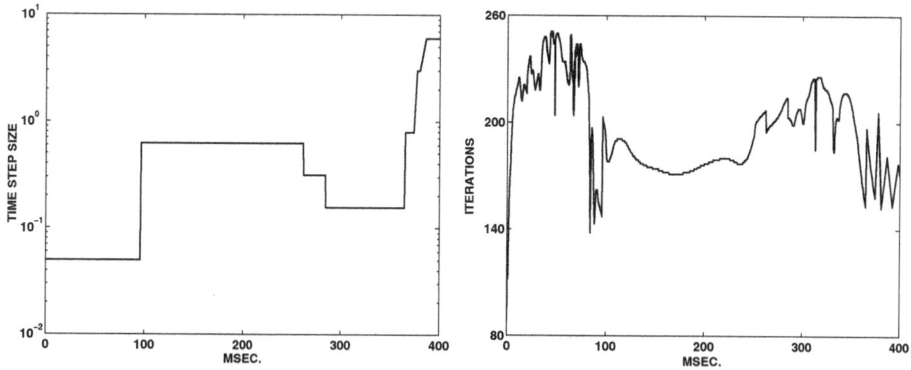

Fig. 3. Full cardiac cycle with Bidomain model and LR1 gating. Time-step size in msec. on a semilogarithmic scale (left), PCG iterations at each time step (right)

with dimensions $4 \times 4 \times 0.5\ cm^3$ and mesh $400 \times 400 \times 50$: the HP SuperDome machine with 32 processors took about 20 hours and the IBM SP4 machine with 64 processors took about 2.5 hours. Therefore, a considerable CPU time reduction in the Bidomain case is to be expected by using the SP4 machine.

References

S. Balay, W. D. Gropp, L. C. McInnes, and B. F. Smith. PETSc users manual. Technical Report ANL-95/11 - Revision 2.1.1, Argonne National Laboratory, 2001.

P. Colli Franzone and L. F. Pavarino. A parallel solver for reaction-diffusion systems in computational electrocardiology. Technical report, IMATI CNR Tech. Rep. 9-PV, 2003.

P. Colli Franzone and G. Savaré. Degenerate evolution systems modeling the cardiac electric field at micro and macroscopic level. In A. Lorenzi and B. Ruf, editors, *Evolution equations, Semigroups and Functional Analysis*, pages 49–78. Birkhauser, 2002.

J. Keener and J. Sneyd. *Mathematical Physiology.* Springer, 1998.

I. LeGrice and et al. Laminar structure of the heart: ventricular myocyte arrangement and connective tissue architecture in the dog. *Am. J. Physiol. (Heart Circ. Physiol)*, 269 (38):H571–H582, 1995.

C. Luo and Y. Rudy. A model of the ventricular cardiac action potential: depolarization, repolarization, and their interaction. *Circ. Res.*, 68 (6): 1501–1526, 1991.

M. Pennacchio. The mortar finite element method for the cardiac bidomain model of extracellular potential. *J. Sci. Comp.*, 20 (2), 2004. To appear.

A. Quarteroni and A. Valli. *Numerical Approximation of Partial Differential Equations.* Springer-Verlag, Berlin, 1994.

B. F. Smith, P. E. Bjørstad, and W. Gropp. *Domain Decomposition: Parallel Multilevel Methods for Elliptic Partial Differential Equations.* Cambridge University Press, 1996.

D. Streeter. Gross morphology and fiber geometry in the heart. In R. Berne, editor, *Handbook of Physiology, vol. 1 The Heart*, pages 61–112. Williams & Wilkins, Baltimore, 1979.

A. Victorri and et al. Numerical integration in the reconstruction of cardiac action potential using the Hodgkin-Huxley type models. *Comp. Biomed. Res.*, 18:10–23, 1985.

R. Weber dos Santos, G. Plank, B. S., and V. E.J. Preconditioning techniques for the bidomain equations. In R. K. et al., editor, *These Proceedings. LNCSE.* Springer, 2004.

Predictor-Corrector Methods for Solving Continuous Casting Problem

J. Pieskä[1], E. Laitinen[1], and A. Lapin[2]

[1] University of Oulu, Department of Mathematical Sciences, Infotech of Oulu, Oulu, Finland, erkki.laitinen@oulu.fi, jali.pieska@oulu.fi
[2] Kazan State University, Department of Computational Mathematics and Cybernetics Kazan, Russia, alapin@ksu.ru

Summary. In this paper we present new numerical approach to solve the continuous casting problem. The main tool is to use IPEC method and DDM similar to Lapin and Pieska [2002] with multilevel domain decomposition. On the subdomains we use multidecomposition of the subdomains. The IPEC is used both in the whole calculation domain and inside the subdomains. The calculation algorithm is presented and numerically tested. Several conclusions are made and discussed.

1 Introduction

Theory of the so-called regional-additive schemes (splitting schemes with domain decomposition) for linear diffusion and convection-diffusion have been studied in Samarskii and Vabischevich [1996] and Vabischevich [1994]. The stability have been proved and error estimates have been deduced. For non-linear problems like our their technique is not available.

Several new finite-difference schemes for a nonlinear convection-diffusion problem are constructed and numerically studied in Lapin and Pieska [2002]. These schemes are constructed on the basis of non-overlapping domain decomposition and predictor-corrector approach.

The paper of Lapin and Pieska [2002] was motivated by Dawson et al. [1991], Rivera et al. [2003], Rivera et al. [2001], where TL[3], EP[4] and EPIC[5] methods have been studied and tested. The EPIC method was proved to be stable and scalable when solving on big number of processors. In the paper of Lapin and Pieska [2002] the scheme from Rivera et al. [2003], Rivera et al. [2001] was modified in such a way, that its implementation leads to IPEC[6] method.

[3] time lagging
[4] explicit predictor
[5] explicit predictor-implicit corrector
[6] implicit predictor-explicit corrector

The main idea of these kind of algorithms is first to solve the problem in artificial boundaries (predictor step). After the solution at the boundaries is known then it can be used as Dirichlet type boundary condition and the noncoupled subdomain problems can be solved parallel. The last step of these methods is to correct the solution at the artificial boundaries (corrector step).

The advantages of predictor-corrector methods (IPEC or EPIC) is that we reduce the amount of information send between processors. We need to send only once the subsolutions from slave processors to master processor. When we use Schwarz alternating methods with overlapping subdomains, the number of sending and receiving is much more bigger. The numerical experiences in Lapin and Pieska [2002] show that the speedup of IPEC method is linear. However, the stability and the rate of convergence for IPEC method is not known but asymptotically the rate of approximation is of the order $O(\tau + h^2)$.

The idea of multidecomposition method MDD is to use DDM with IPEC inside the subdomains. The subdomain is divided to smaller subdomains and then IPEC method is used to solve these smaller subproblems sequently. The main reason and motivation for this kind of method is to economise the number of processors. Nowadays the PC computers have multiprocessor cards but the number of processors in it are very small and limited. Our proposed algorithm gives a good and effective way to decrease calculation times in the case of only few processors.

2 Problem statement

The continuous casting problem can be mathematically formulated as follows. Let the rectangular domain $\Omega \subset \mathbb{R}^2, \Omega = (0, l_1) \times (0, l_2)$ be occupied by a thermodynamically homogeneous and isotropic steel. We denote by $\bar{H}(x,t)$ the enthalpy and by $T(x,t)$ the temperature for $(x,t) \in \bar{\Omega} \times [0, t_f]$. We have constitutive law $\bar{H} = \bar{H}(T) = \rho \int_0^T c(\Theta) d\Theta + \rho L(1 - f_s(T))$, where ρ is the density, $c(T)$ is the specific heat, L is the latent heat and $f_s(T)$ is the solid fraction at temperature T. The graph $\bar{H}(T)$ is an increasing function $\mathbb{R} \to \mathbb{R}$, involving near vertical segment, which corresponds to a phase transition state, namely, for $T \in [T_S, T_L]$. In our numerical example in Section 6 we choose for the enthalpy function phase change interval $T_L - T_S = 0.02$. This does not effect cruisial for convergence properties of our method. Further by $k(T)$ we denote the thermal conductivity coefficient.

Using Kirchoff's transformation $u = K(T) = \int_0^T k(\xi) d\xi$ a continuous casting process can be described by a boundary-value problem, formally written in the following pointwise form: find $u(x,t)$ and $H(x,t)$ such that

$$(\mathrm{P}) \begin{cases} \frac{\partial H(u)}{\partial t} + v \frac{\partial H(u)}{\partial x_2} - \Delta u = 0, \text{ for } x \in \Omega, t > 0, \\ u = z(x,t) \text{ for } x \in \Gamma_D, t > 0, \\ \frac{\partial u}{\partial n} = g, \text{ for } x \in \Gamma_N, t > 0, H = H_0(x) \text{ for } x \in \bar{\Omega}, t = 0, \end{cases}$$

where $v = const > 0$ is a casting speed in x_2-direction, $\Gamma_D \cup \Gamma_N = \partial\Omega$ is the boundary of the domain, below $\Gamma_D = \{x \in \partial\Omega : x_2 = 0 \vee x_2 = l_2\}$. The existence and uniqueness of a weak solution for problem (P) are proved in Rodrigues and Yi [1990].

3 Mesh approximation of continuous casting problem

We approximate problem (P) by an implicit finite difference scheme in time and finite element method in space. Let T_h be a partitioning of Ω in the rectangular elements δ of dimensions $h_1 \times h_2$ and $V_h = \{u_h(x) \in H^1(\Omega) : u_h(x) \in Q_1$ for all $\delta \in T_h\}$, where Q_1 is the space of bilinear functions. By $\Pi_h v(x)$ we denote the V_h-interpolant of a continuous function $v(x)$, i.e. $\Pi_h v(x) \in V_h$ and coincides with $v(x)$ in the mesh nodes (vertices of all $\delta \in T_h$). We also use an interpolation operator P_h, which is defined as follows: for any continuous function $v(x)$ the function $P_h v(x)$ is piecewise linear in x_1, piecewise constant in x_2 and on $\delta = [x_1, x_1 + h_1] \times [x_2, x_2 + h_2]$ it coincides with $v(x)$ at $(x_1, x_2 + h_2)$ and $(x_1 + h_1, x_2 + h_2)$.

Let further $V_h^0 = \{u_h(x) \in V_h : u_h(x) = 0 \text{ for all } x \in \Gamma_D\}$, $V_h^z = \{u_h(x) \in V_h : u_h(x) = z_h \text{ for all } x \in \Gamma_D\}$. Here z_h is the bilinear interpolation of z on the boundary Γ_D. For any continuous function $v(x)$ we define the quadrature formulas: $S_\delta(v) = \int_\delta \Pi_h v dx$, $S_{\partial\delta}(v) = \int_{\partial\delta} \Pi_h v dx$, $E_\delta(v) = \int_\delta P_h v dx$, $S_\Omega(v) = \sum_{\delta \in T_h} S_\delta v$, $S_{\Gamma_N}(v) = \sum_{\partial\delta \in T_h \cap \bar{\Gamma}_N} S_{\partial\delta}(v)$, $E_\Omega(v) = \sum_{\delta \in T_h} E_\delta(v)$. Let also $\omega_\tau = \{t_k = k\tau, 0 \le k \le M, M\tau = t_f\}$ be an uniform mesh in time on the segment $[0, t_f]$ and $\partial_{\bar{t}} H = \frac{1}{\tau}(H(x,t) - H(x,t-\tau))$. When constructing the characteristic mesh scheme we approximate the term $(\frac{\partial}{\partial t} + v\frac{\partial}{\partial x_2})H$ by using the characteristics of the first order differential operator Chen [1991]. Namely, if (x_1, x_2, t) is the mesh point on the time level t we choose $\tilde{x}_2 = x_2 - \int_{t-\tau}^{t} v(\xi)d\xi$ and approximate: $(\frac{\partial}{\partial t} + v\frac{\partial}{\partial x_2})H \approx \frac{1}{\tau}(H(x_1, x_2, t) - \tilde{H}(x, t-\tau))$, where we denote $\tilde{H}(x, t-\tau) = H(x_1, \tilde{x}_2, t-\tau)$. Near the boundary it can happen that $\tilde{x}_2 < 0$. In that case we put $\tilde{H}(x, t-\tau) = H(x_1, 0, t-\tau)$. In what follows we use the notation $d_{\bar{t}} H = \frac{1}{\tau}(H(x,t) - \tilde{H}(x, t-\tau))$ for the difference quotient in each mesh point on time level t.

Now, the characteristic finite difference scheme for problem (P) is: for all $t \in \omega_\tau$, $t > 0$, find $u_h \in V_h^z$ and $H_h \in V_h$ such that

$$S_\Omega(d_{\bar{t}} H_h \eta_h) + S_\Omega(\nabla u_h \nabla \eta_h) = S_{\Gamma_N}(g\eta_h) \text{ for all } \eta_h \in V_h^0 \quad (1)$$

Let $N_0 = \text{card } V_h^0$ and $u \in \mathbb{R}^{N_0}$ be the vector of nodal values for $u_h \in V_h^0$. We use the writing $u_h \Leftrightarrow u$ for this bijection. We define $N_0 \times N_0$ matrices by the following relations: for all $u, \eta \in \mathbb{R}^{N_0}$, $u \Leftrightarrow u_h \in V_h^0$ and $\eta \Leftrightarrow \eta_h \in V_h^0$, $(\tilde{A}u, \eta) = S_\Omega(\nabla u_h \nabla \eta_h)$, $(Mu, \eta) = S_\Omega(u_h \eta_h)$, $A_0 = M^{-1}\tilde{A}$. Let now $\tilde{z}_h(x) \in V_h$ be the function, which is equal to z_h on $\bar{\Gamma}_D$ and 0 for all nodes in $\Omega \cup \Gamma_N$. Then a right hand side vector f is defined by the equality $(f, \eta) = S_{\Gamma_N}(g\eta_h) - S_\Omega(\nabla \tilde{z}_h, \nabla \eta_h)$ $\forall \eta \in \mathbb{R}^{N_0}$, $\eta \Leftrightarrow \eta_h \in V_h^0$, and we set $F = M^{-1}f$. In these notations the algebraic form for characteristic mesh scheme (1) becomes

$$d_{\tilde{t}}H + A_0 u = F, \tag{2}$$

It is easy to see, that A_0 is the standard five-point finite difference approximation of Laplace operator, $A_0 u = -u_{x_1\bar{x}_1} - u_{x_2\bar{x}_2}$ for the internal mesh points with the notations $u_{x_1} = h^{-1}(u(x_1 + h_1, x_2) - u(x_1, x_2))$, $u_{\bar{x}_1} = h^{-1}(u(x_1, x_2) - u(x_1 - h_1, x_2))$, and similarly for u_{x_2} and $u_{\bar{x}_2}$. Furthermore, let $\bar{\omega}$ be the set of all grid points, $\gamma_D = \bar{\Gamma}_D \cap \bar{\omega}$, $\gamma_N = \Gamma_N \cap \bar{\omega}$, $\omega = \Omega \cap \bar{\omega}, \gamma_N^- = \{x \in \gamma_N : x_1 = 0\}$, $\gamma_N^+ = \{x \in \gamma_N : x_1 = l_1\}$.

4 Domain decomposition by straight lines

In this section we present the IPEC algorithm of Lapin and Pieska [2002]. We restrict our discussion to the case of decomposition by unidirect straight lines. More variations and possibilities of decomposition is discussed and tested in Lapin and Pieska [2002].

Let the domain Ω be decomposed into two subdomains Ω_1 and Ω_2 by a straight line S_y in x_2-direction, which is also a grid line. We denote by δ_{S_y} the characteristic function of this line, i.e., the mesh function $\delta_{S_y}(x) = 1$ for $x \in S_y \cap \bar{\omega}$, while $\delta_{S_y}(x) = 0$ for other mesh points. Also, let $\bar{\omega}_k$, $k = 1, 2$ be the corresponding to the subdomains $\bar{\Omega}_k$ sets of grid points, S_y being the common part of their boundaries.

Let $A_2 u = -\delta_{S_y} u_{x_1\bar{x}_1}$ and $A_1 = A_0 - A_2$,

$$A_1 u = \begin{cases} -(1 - \delta_{S_y})u_{x_1\bar{x}_1} - u_{x_2\bar{x}_2} \text{ for } x \in \omega, \\ -2h_1^{-1}u_{x_1} - u_{x_2\bar{x}_2} \text{ for } x \in \gamma_N^-, \\ 2h_1^{-1}u_{\bar{x}_1} - u_{x_2\bar{x}_2} \text{ for } x \in \gamma_N^+. \end{cases}$$

Now, instead of characteristic scheme (2) we consider the following scheme on the time level $t_{n+1} = (n + 1)\tau$:

$$\frac{1}{\tau}(H^{n+\frac{1}{2}} - \tilde{H}^n) + A_1 u^{n+\frac{1}{2}} + A_2 u^n = F, \tag{3}$$

$$\frac{\delta_{S_y}}{\tau}(H^{n+1} - \tilde{H}^n) + \frac{1 - \delta_{S_y}}{\tau}(H^{n+1} - H^{n+\frac{1}{2}}) + \delta_{S_y} A_1 u^{n+\frac{1}{2}} + A_2 u^{n+1} = \delta_{S_y} F, \tag{4}$$

Let us discuss the implementation of scheme (3),(4). In the points of S_y equation (3) has the form:

$$\frac{H^{n+\frac{1}{2}} - \tilde{H}^n}{\tau} - u_{x_2\bar{x}_2}^{n+\frac{1}{2}} - u_{x_1\bar{x}_1}^n = F, \tag{5}$$

i.e. in the points of S_y we have one-dimensional problem (5), that we solve first. After that the equation (3) is splitted in two non-coupled characteristic schemes in the subdomains:

$$\begin{cases} \dfrac{H^{n+\frac{1}{2}} - \tilde{H}^n}{\tau} - u_{x_1\bar{x}_1}^{n+\frac{1}{2}} - u_{x_2\bar{x}_2}^{n+\frac{1}{2}} = F, \text{ for } x \in \omega_1 \cup \omega_2, \\ \dfrac{H^{n+\frac{1}{2}} - \tilde{H}^n}{\tau} - \dfrac{2}{h_1} u_{x_1}^{n+\frac{1}{2}} - u_{x_2\bar{x}_2}^{n+\frac{1}{2}} = F, \text{ for } x \in \gamma_N^-, \\ \dfrac{H^{n+\frac{1}{2}} - \tilde{H}^n}{\tau} + \dfrac{2}{h_1} u_{\bar{x}_1}^{n+\frac{1}{2}} - u_{x_2\bar{x}_2}^{n+\frac{1}{2}} = F, \text{ for } x \in \gamma_N^+, \end{cases} \tag{6}$$

and these equations are accomplished by Dirichlet boundary conditions, given on γ_D and calculated from (5) on S_y. Finally we solve the system of the equations, corresponding to $x \in S_y$: $\frac{H^{n+1} - \tilde{H}^n}{\tau} - u_{x_2\bar{x}_2}^{n+\frac{1}{2}} - u_{x_1\bar{x}_1}^{n+1} = F$. As $u^{n+1}(x) = u^{n+\frac{1}{2}}(x)$ for $x \notin S_y$, this system becomes

$$\begin{cases} \dfrac{H^{n+1} - \tilde{H}^n}{\tau} + 2\dfrac{u^{n+1}(x_1, x_2)}{h_1^2} - u_{x_2\bar{x}_2}^{n+\frac{1}{2}} \\ -\dfrac{u^{n+\frac{1}{2}}(x_1 - h_1, x_2) + u^{n+\frac{1}{2}}(x_1 + h_1, x_2)}{h_1^2} = F, \ x \in S_y. \end{cases} \tag{7}$$

Thus, the algorithm for the implementation of (3),(4) consists of 3 steps:

1) Predictor step: solving one-dimensional problem (5);
2) Main step: concurrent solving subproblems (6);
3) Corrector step: solving the system of scalar equations (7).

5 Multidecomposition method

The general idea of the multidecomposition is to divide the subdomain to smaller subdomains i.e. use two-level decomposition of the calculation domain. The division of the subdomains is presented in the figure 1. We use the

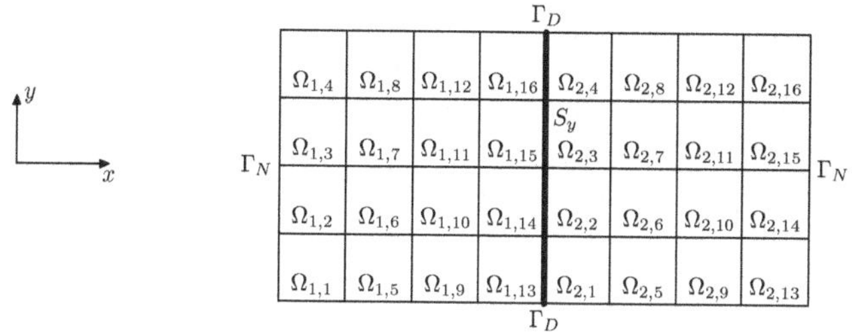

Fig. 1. Used nonoverlapping domain decomposition and multidecomposition of the subdomains

notation $\Omega_i = \cup_{j_i=1}^{p_i} \Omega_{i,j_i}$. The use of high number of subdomains inside the subdomain may increase the error dramatically. To overcome this feature we introduce so called smoothing steps to our method. The calculation algorithm for characteristic mesh scheme (5)-(7) is presented below.

Algorithm 1.
1. Time step n perform on the main processor the predictor step (5) on S_y.
2. Send the values of $u^{n+\frac{1}{2}}$ and $H^{n+\frac{1}{2}}$ on S_y to the slave processors.
3. Concurrently on the slave processors perform the predictor step (5) on the artificial boundaries of the subdomains Ω_{i,j_i}, $i = 1,2$, $j_1 = 1,...,p_1$, $j_2 = 1,...,p_2$.
4. Concurrently on the slave processors perform sequentially the main step (6) on the subdomains Ω_{i,j_i}.
5. Concurrently on the slave processors perform the corrector step (7) on the artificial boundaries of the subdomains Ω_{i,j_i}, $i = 1,2$, $j_1 = 1,...,p_1$, $j_2 = 1,...,p_2$.
6. On the slave processors perform the smoothing step i.e. few iterations of the MSOR-method over the whole subdomain Ω_i.
7. Send the subsolutions u^{n+1} and H^{n+1} to the main processor.
8. On the main processor perform the corrector step (7) on S_y.
9. On the main processor perform few iterations of the MSOR-method in the neighborhood of S_y.
10. Put $n = n+1$, if the final time t_f is reached **STOP**, else **GOTO 1.**

Remark 1. On the step **3.** we do not do the predictor step (5) on S_y.

Remark 2. On the steps **3.-6.** we do the calculations concurrently. Each processor perform the steps asynchronously.

6 Numerical verification

Let $\Omega =]0,1[\times]0,1[$ with the boundary Γ divided in two parts $\Gamma_D = \{x \in \partial\Omega : x_2 = 0 \vee x_2 = 1\}$ and $\Gamma_N = \Gamma \setminus \Gamma_D$. Moreover, let $t_f = 1$ and $u_{SL} = 1$. The phase change interval is $[u_{SL} - \varepsilon, u_{SL} + \varepsilon]$, $\varepsilon = 0.01$, and the velocity is $v(t) = \frac{1}{5}$. Our numerical example is

$$\begin{aligned} \frac{\partial H}{\partial t} - \Delta K + v(t)\frac{\partial H}{\partial x_2} &= f(x;t) && \text{on } \Omega, \\ u(x_1,x_2;t) &= (x_1 - \tfrac{1}{2})^2 - \tfrac{1}{2}e^{-4t} + \tfrac{5}{4} && \text{on } \Gamma_D, \\ \frac{\partial u}{\partial n} &= 1 && \text{on } \Gamma_N, \\ u(x_1,x_2;0) &= (x_1 - \tfrac{1}{2})^2 + (x_2 - \tfrac{1}{2})^2 + \tfrac{1}{2} && \text{on } \Omega, \end{aligned}$$

where Kirchoff's temperature $K(u)$ and enthalpy $H(u)$ are according to their definition

$$K(u) = \begin{cases} u & \text{if } u < u_{SL} - \varepsilon, \\ \frac{3}{2}u - \frac{1-\varepsilon}{2} & \text{if } u \in [u_{SL} - \varepsilon, u_{SL} + \varepsilon], \\ 2u - 1 & \text{if } u > u_{SL} + \varepsilon, \end{cases}$$

and

$$H(u) = \begin{cases} 2u & \text{if } u < u_{SL} - \varepsilon, \\ \left(\frac{1+8\varepsilon}{2\varepsilon}\right)(u-1) + \frac{5+4\varepsilon}{2} & \text{if } u \in [u_{SL} - \varepsilon, u_{SL} + \varepsilon], \\ 6u - 3 & \text{if } u > u_{SL} + \varepsilon. \end{cases}$$

The known right hand side $f(x;t)$ is chosen such that our problem have the exact solution $u(x_1, x_2; t) = (x_1 - \frac{1}{2})^2 + (x_2 - \frac{1}{2})^2 - \frac{1}{2}e^{-4t} + 1$.

The stopping criterion of the calculations was the L_2-norm of residual $\|r\|_{L_2(\Omega)} < 10^{-4}$. We solve our problem by using different methods, Additive Schwarz alternating method (ASAM), Implicit Predictor-Explicit Corrector method (IPEC), Multidomain decomposition method (MDD) and sequential modified SOR method (SEQ). The results are presented in the table 1.

Table 1. Calculation times for ASAM, IPEC, MDD and sequential MSOR methods when the number of processors and calculation grid is changed. Number of inside subdomains, 4 × 4

Grid	SEQ 1 proc.	ASAM 2 proc.	ASAM 4 proc.	ASAM 8 proc.	IPEC 2 proc.	IPEC 4 proc.	IPEC 8 proc.	MDD 2 proc.	MDD 4 proc.
65 × 65 × 128	8.67	7.01	4.76	3.91	3.76 s	2.33 s	1.64 s	5.4 s	4.4
129 × 129 × 256	112.9	77.6	59.3	33.9	47.3 s	25.1 s	14.1 s	25.9 s	14.6
257 × 257 × 512	1425	889	494	281	600 s	285 s	164 s	342 s	179

Table 2. Efficiencies for ASAM, IPEC, MDD and methods when the number of processors and calculation grid is changed. Number of inside subdomains, 4 × 4

Grid	ASAM 2 proc.	ASAM 4 proc.	ASAM 8 proc.	IPEC 2 proc.	IPEC 4 proc.	IPEC 8 proc.	MDD 2 proc.	MDD 4 proc.
65 × 65 × 128	0.62	0.46	0.28	1.15	0.93	0.66	0.8	0.49
129 × 129 × 256	0.73	0.48	0.42	1.19	1.12	1	2.18	1.93
257 × 257 × 512	0.8	0.72	0.63	1.19	1.25	1.09	2.08	1.99

7 Conclusions

The numerical examples show that the multidecomposition method (MDD) is very effective numerical method when solving continuous casting problem. The idea to divide the subdomains to smaller subdomains seems to be very good and profitable. The algebraic dimension of the subproblems inside the subdomains are very small and thus they are very quick to solve.

The introduced smoothing step allows us to use quite big number of subdomains. The accuracy of the different methods, MDD, ASAM and IPEC are the same. However, the smoothing step is economical to perform and calculation times for MDD are roughly half of the calculation times of the IPEC method.

The tables 1 and 2 show very clearly the advantages of the multidecomposition method over other methods. It is extremely quick and accuracy is the same than other methods. Implementation of MDD is straightforward and it do not need huge amount of processors to solve big and complicated problems.

References

Z. Chen. Numerical solutions of a two-phase continuous casting problem. In P. Neittaanmaki, editor, *Numerical Methods for Free Boundary Problem*, pages 103–121, Basel, 1991. International Series of Numerical Mathematics, Birkhuser.

C. N. Dawson, Q. Du, and T. F. Dupont. A finite difference domain decomposition algorithm for numerical solution of the heat equation. *Math. Comput.*, 57:63–71, 1991.

A. V. Lapin and J. Pieska. On the parallel domain decomposition algorithms for time-dependent problems. *Lobachevskii Journal of Mathematics*, 10: 27–44, 2002.

W. Rivera, J. Zhu, and D. Huddleston. An efficient parallel algorithm for solving unsteady nonlinear equations. In T. M. Pinkston, editor, *Proceedings of the International Conference on Parallel Processing Workshops*, pages 79–84. IEEE Computer Society, Los Alamitos, California, 2001.

W. Rivera, J. Zhu, and D. Huddleston. An efficient parallel algorithm with application to computational fluid dynamics. *Computers and Mathematics with Applications*, 45:165–188, 2003.

J. F. Rodrigues and F. Yi. On a two-phase continuous casting stefan problem with nonlinear flux. *Euro. of Applied Mathematics*, 1:259–278, 1990.

A. Samarskii and P. Vabischevich. Factorized regional-additive schemes for convection-diffusion problems (in russian). Technical Report V. 346, Russian Academic of Sciences, 1996.

P. Vabischevich. Parallel domain decomposition algorithms for time-dependent problems of mathematical physics. *Advances in Numerical Methods and Applications*, pages 293–299, 1994.

Editorial Policy

§1. Volumes in the following three categories will be published in LNCSE:

i) Research monographs
ii) Lecture and seminar notes
iii) Conference proceedings

Those considering a book which might be suitable for the series are strongly advised to contact the publisher or the series editors at an early stage.

§2. Categories i) and ii). These categories will be emphasized by Lecture Notes in Computational Science and Engineering. **Submissions by interdisciplinary teams of authors are encouraged.** The goal is to report new developments – quickly, informally, and in a way that will make them accessible to non-specialists. In the evaluation of submissions timeliness of the work is an important criterion. Texts should be well-rounded, well-written and reasonably self-contained. In most cases the work will contain results of others as well as those of the author(s). In each case the author(s) should provide sufficient motivation, examples, and applications. In this respect, Ph.D. theses will usually be deemed unsuitable for the Lecture Notes series. Proposals for volumes in these categories should be submitted either to one of the series editors or to Springer-Verlag, Heidelberg, and will be refereed. A provisional judgment on the acceptability of a project can be based on partial information about the work: a detailed outline describing the contents of each chapter, the estimated length, a bibliography, and one or two sample chapters – or a first draft. A final decision whether to accept will rest on an evaluation of the completed work which should include

- at least 100 pages of text;
- a table of contents;
- an informative introduction perhaps with some historical remarks which should be accessible to readers unfamiliar with the topic treated;
- a subject index.

§3. Category iii). Conference proceedings will be considered for publication provided that they are both of exceptional interest and devoted to a single topic. One (or more) expert participants will act as the scientific editor(s) of the volume. They select the papers which are suitable for inclusion and have them individually refereed as for a journal. Papers not closely related to the central topic are to be excluded. Organizers should contact Lecture Notes in Computational Science and Engineering at the planning stage.

In exceptional cases some other multi-author-volumes may be considered in this category.

§4. Format. Only works in English are considered. They should be submitted in camera-ready form according to Springer-Verlag's specifications.
Electronic material can be included if appropriate. Please contact the publisher.
Technical instructions and/or TeX macros are available via
http://www.springeronline.com/sgw/cda/frontpage/0,10735,5-111-2-71391-0,00.html
The macros can also be sent on request.

General Remarks

Lecture Notes are printed by photo-offset from the master-copy delivered in camera-ready form by the authors. For this purpose Springer-Verlag provides technical instructions for the preparation of manuscripts. See also *Editorial Policy*.

Careful preparation of manuscripts will help keep production time short and ensure a satisfactory appearance of the finished book.

The following terms and conditions hold:

Categories i), ii), and iii):
Authors receive 50 free copies of their book. No royalty is paid. Commitment to publish is made by letter of intent rather than by signing a formal contract. Springer-Verlag secures the copyright for each volume.

For conference proceedings, editors receive a total of 50 free copies of their volume for distribution to the contributing authors.

All categories:
Authors are entitled to purchase further copies of their book and other Springer mathematics books for their personal use, at a discount of 33,3 % directly from Springer-Verlag.

Addresses:

Timothy J. Barth
NASA Ames Research Center
NAS Division
Moffett Field, CA 94035, USA
e-mail: barth@nas.nasa.gov

Michael Griebel
Institut für Angewandte Mathematik
der Universität Bonn
Wegelerstr. 6
53115 Bonn, Germany
e-mail: griebel@iam.uni-bonn.de

David E. Keyes
Department of Applied Physics
and Applied Mathematics
Columbia University
200 S. W. Mudd Building
500 W. 120th Street
New York, NY 10027, USA
e-mail: david.keyes@columbia.edu

Risto M. Nieminen
Laboratory of Physics
Helsinki University of Technology
02150 Espoo, Finland
e-mail: rni@fyslab.hut.fi

Dirk Roose
Department of Computer Science
Katholieke Universiteit Leuven
Celestijnenlaan 200A
3001 Leuven-Heverlee, Belgium
e-mail: dirk.roose@cs.kuleuven.ac.be

Tamar Schlick
Department of Chemistry
Courant Institute of Mathematical
Sciences
New York University
and Howard Hughes Medical Institute
251 Mercer Street
New York, NY 10012, USA
e-mail: schlick@nyu.edu

Springer-Verlag, Mathematics Editorial IV
Tiergartenstrasse 17
D-69121 Heidelberg, Germany
Tel.: *49 (6221) 487-8185
Fax: *49 (6221) 487-8355
e-mail: Martin.Peters@springer-sbm.com

Lecture Notes in Computational Science and Engineering

Vol. 1 D. Funaro, *Spectral Elements for Transport-Dominated Equations.* 1997. X, 211 pp. Softcover. ISBN 3-540-62649-2

Vol. 2 H. P. Langtangen, *Computational Partial Differential Equations.* Numerical Methods and Diffpack Programming. 1999. XXIII, 682 pp. Hardcover. ISBN 3-540-65274-4

Vol. 3 W. Hackbusch, G. Wittum (eds.), *Multigrid Methods V.* Proceedings of the Fifth European Multigrid Conference held in Stuttgart, Germany, October 1-4, 1996. 1998. VIII, 334 pp. Softcover. ISBN 3-540-63133-X

Vol. 4 P. Deuflhard, J. Hermans, B. Leimkuhler, A. E. Mark, S. Reich, R. D. Skeel (eds.), *Computational Molecular Dynamics: Challenges, Methods, Ideas.* Proceedings of the 2nd International Symposium on Algorithms for Macromolecular Modelling, Berlin, May 21-24, 1997. 1998. XI, 489 pp. Softcover. ISBN 3-540-63242-5

Vol. 5 D. Kröner, M. Ohlberger, C. Rohde (eds.), *An Introduction to Recent Developments in Theory and Numerics for Conservation Laws.* Proceedings of the International School on Theory and Numerics for Conservation Laws, Freiburg / Littenweiler, October 20-24, 1997. 1998. VII, 285 pp. Softcover. ISBN 3-540-65081-4

Vol. 6 S. Turek, *Efficient Solvers for Incompressible Flow Problems.* An Algorithmic and Computational Approach. 1999. XVII, 352 pp, with CD-ROM. Hardcover. ISBN 3-540-65433-X

Vol. 7 R. von Schwerin, *Multi Body System SIMulation.* Numerical Methods, Algorithms, and Software. 1999. XX, 338 pp. Softcover. ISBN 3-540-65662-6

Vol. 8 H.-J. Bungartz, F. Durst, C. Zenger (eds.), *High Performance Scientific and Engineering Computing.* Proceedings of the International FORTWIHR Conference on HPSEC, Munich, March 16-18, 1998. 1999. X, 471 pp. Softcover. 3-540-65730-4

Vol. 9 T. J. Barth, H. Deconinck (eds.), *High-Order Methods for Computational Physics.* 1999. VII, 582 pp. Hardcover. 3-540-65893-9

Vol. 10 H. P. Langtangen, A. M. Bruaset, E. Quak (eds.), *Advances in Software Tools for Scientific Computing.* 2000. X, 357 pp. Softcover. 3-540-66557-9

Vol. 11 B. Cockburn, G. E. Karniadakis, C.-W. Shu (eds.), *Discontinuous Galerkin Methods.* Theory, Computation and Applications. 2000. XI, 470 pp. Hardcover. 3-540-66787-3

Vol. 12 U. van Rienen, *Numerical Methods in Computational Electrodynamics.* Linear Systems in Practical Applications. 2000. XIII, 375 pp. Softcover. 3-540-67629-5

Vol. 13 B. Engquist, L. Johnsson, M. Hammill, F. Short (eds.), *Simulation and Visualization on the Grid.* Parallelldatorcentrum Seventh Annual Conference, Stockholm, December 1999, Proceedings. 2000. XIII, 301 pp. Softcover. 3-540-67264-8

Vol. 14 E. Dick, K. Riemslagh, J. Vierendeels (eds.), *Multigrid Methods VI.* Proceedings of the Sixth European Multigrid Conference Held in Gent, Belgium, September 27-30, 1999. 2000. IX, 293 pp. Softcover. 3-540-67157-9

Vol. 15 A. Frommer, T. Lippert, B. Medeke, K. Schilling (eds.), *Numerical Challenges in Lattice Quantum Chromodynamics.* Joint Interdisciplinary Workshop of John von Neumann Institute for Computing, Jülich and Institute of Applied Computer Science, Wuppertal University, August 1999. 2000. VIII, 184 pp. Softcover. 3-540-67732-1

Vol. 16 J. Lang, *Adaptive Multilevel Solution of Nonlinear Parabolic PDE Systems.* Theory, Algorithm, and Applications. 2001. XII, 157 pp. Softcover. 3-540-67900-6

Vol. 17 B. I. Wohlmuth, *Discretization Methods and Iterative Solvers Based on Domain Decomposition.* 2001. X, 197 pp. Softcover. 3-540-41083-X

Vol. 18 U. van Rienen, M. Günther, D. Hecht (eds.), *Scientific Computing in Electrical Engineering.* Proceedings of the 3rd International Workshop, August 20-23, 2000, Warnemünde, Germany. 2001. XII, 428 pp. Softcover. 3-540-42173-4

Vol. 19 I. Babuška, P. G. Ciarlet, T. Miyoshi (eds.), *Mathematical Modeling and Numerical Simulation in Continuum Mechanics.* Proceedings of the International Symposium on Mathematical Modeling and Numerical Simulation in Continuum Mechanics, September 29 - October 3, 2000, Yamaguchi, Japan. 2002. VIII, 301 pp. Softcover. 3-540-42399-0

Vol. 20 T. J. Barth, T. Chan, R. Haimes (eds.), *Multiscale and Multiresolution Methods.* Theory and Applications. 2002. X, 389 pp. Softcover. 3-540-42420-2

Vol. 21 M. Breuer, F. Durst, C. Zenger (eds.), *High Performance Scientific and Engineering Computing.* Proceedings of the 3rd International FORTWIHR Conference on HPSEC, Erlangen, March 12-14, 2001. 2002. XIII, 408 pp. Softcover. 3-540-42946-8

Vol. 22 K. Urban, *Wavelets in Numerical Simulation.* Problem Adapted Construction and Applications. 2002. XV, 181 pp. Softcover. 3-540-43055-5

Vol. 23 L. F. Pavarino, A. Toselli (eds.), *Recent Developments in Domain Decomposition Methods.* 2002. XII, 243 pp. Softcover. 3-540-43413-5

Vol. 24 T. Schlick, H. H. Gan (eds.), *Computational Methods for Macromolecules: Challenges and Applications.* Proceedings of the 3rd International Workshop on Algorithms for Macromolecular Modeling, New York, October 12-14, 2000. 2002. IX, 504 pp. Softcover. 3-540-43756-8

Vol. 25 T. J. Barth, H. Deconinck (eds.), *Error Estimation and Adaptive Discretization Methods in Computational Fluid Dynamics.* 2003. VII, 344 pp. Hardcover. 3-540-43758-4

Vol. 26 M. Griebel, M. A. Schweitzer (eds.), *Meshfree Methods for Partial Differential Equations.* 2003. IX, 466 pp. Softcover. 3-540-43891-2

Vol. 27 S. Müller, *Adaptive Multiscale Schemes for Conservation Laws.* 2003. XIV, 181 pp. Softcover. 3-540-44325-8

Vol. 28 C. Carstensen, S. Funken, W. Hackbusch, R. H. W. Hoppe, P. Monk (eds.), *Computational Electromagnetics.* Proceedings of the GAMM Workshop on "Computational Electromagnetics", Kiel, Germany, January 26-28, 2001. 2003. X, 209 pp. Softcover. 3-540-44392-4

Vol. 29 M. A. Schweitzer, *A Parallel Multilevel Partition of Unity Method for Elliptic Partial Differential Equations.* 2003. V, 194 pp. Softcover. 3-540-00351-7

Vol. 30 T. Biegler, O. Ghattas, M. Heinkenschloss, B. van Bloemen Waanders (eds.), *Large-Scale PDE-Constrained Optimization.* 2003. VI, 349 pp. Softcover. 3-540-05045-0

Vol. 31 M. Ainsworth, P. Davies, D. Duncan, P. Martin, B. Rynne (eds.) *Topics in Computational Wave Propagation.* Direct and Inverse Problems. 2003. VIII, 399 pp. Softcover. 3-540-00744-X

Vol. 32 H. Emmerich, B. Nestler, M. Schreckenberg (eds.) *Interface and Transport Dynamics.* Computational Modelling. 2003. XV, 432 pp. Hardcover. 3-540-40367-1

Vol. 33 H. P. Langtangen, A. Tveito (eds.) *Advanced Topics in Computational Partial Differential Equations.* Numerical Methods and Diffpack Programming. 2003. XIX, 658 pp. Softcover. 3-540-01438-1

Vol. 34 V. John, *Large Eddy Simulation of Turbulent Incompressible Flows.* Analytical and Numerical Results for a Class of LES Models. 2004. XII, 261 pp. Softcover. 3-540-40643-3

Vol. 35 E. Bänsch, *Challenges in Scientific Computing - CISC 2002.* Proceedings of the Conference *Challenges in Scientific Computing*, Berlin, October 2-5, 2002. 2003. VIII, 287 pp. Hardcover. 3-540-40887-8

Vol. 36 B. N. Khoromskij, G. Wittum, *Numerical Solution of Elliptic Differential Equations by Reduction to the Interface.* 2004. XI, 293 pp. Softcover. 3-540-20406-7

Vol. 37 A. Iske, *Multiresolution Methods in Scattered Data Modelling.* 2004. XII, 182 pp. Softcover. 3-540-20479-2

Vol. 38 S.-I. Niculescu, K. Gu, *Advances in Time-Delay Systems.* 2004. XIV, 446 pp. Softcover. 3-540-20890-9

Vol. 39 S. Attinger, P. Koumoutsakos, *Multiscale Modelling and Simulation.* 2004. VIII, 277 pp. Softcover. 3-540-21180-2

Vol. 40 R. Kornhuber, R. Hoppe, J. Périaux, O. Pironneau, O. Wildlund, J. Xu, *Domain Decomposition Methods in Science and Engineering.* 2005. XVIII, 690 pp. Softcover. 3-540-22523-4

For further information on these books please have a look at our mathematics catalogue at the following URL: `www.springeronline.com/series/3527`

Texts in Computational Science and Engineering

Vol. 1 H. P. Langtangen, *Computational Partial Differential Equations.* Numerical Methods and Diffpack Programming. 2nd Edition 2003. XXVI, 855 pp. Hardcover. ISBN 3-540-43416-X

Vol. 2 A. Quarteroni, F. Saleri, *Scientific Computing with MATLAB.* 2003. IX, 257 pp. Hardcover. ISBN 3-540-44363-0

Vol. 3 H. P. Langtangen, *Python Scripting for Computational Science.* 2004. XXII, 724 pp. Hardcover. ISBN 3-540-43508-5

For further information on these books please have a look at our mathematics catalogue at the following URL: `www.springeronline.com/series/5151`

Printing and Binding: Strauss GmbH, Mörlenbach

FSC
www.fsc.org
MIX
Papier | Fördert gute Waldnutzung
FSC® C083411

Kolibri 360 GmbH
Ferdinand-Jühlke-Straße 7
99095 Erfurt, Deutschland
produktsicherheit@kolibri360.de